Pro Android 4

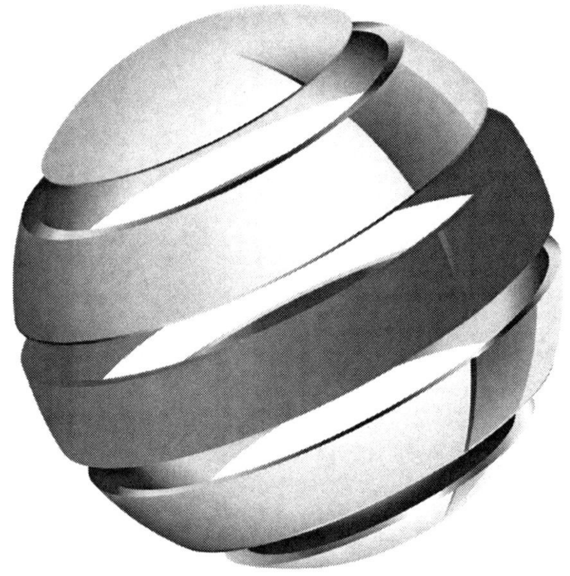

Satya Komatineni
Dave MacLean

Apress®

Pro Android 4

ISBN-13 (pbk): 978-1-4302-3930-7

ISBN-13 (electronic): 978-1-4302-3931-4

President and Publisher: Paul Manning
Lead Editor: Matthew Moodie
Technical Reviewers: Eric Franchomme, Dylan Phillips, Michael Nguyen, Karim Varela
Editorial Board: Steve Anglin, Ewan Buckingham, Gary Cornell, Louise Corrigan, Morgan Ertel, Jonathan Gennick, Jonathan Hassell, Robert Hutchinson, Michelle Lowman, James Markham, Matthew Moodie, Jeff Olson, Jeffrey Pepper, Douglas Pundick, Ben Renow-Clarke, Dominic Shakeshaft, Gwenan Spearing, Matt Wade, Tom Welsh
Coordinating Editor: Corbin Collins
Copy Editor: Tiffany Taylor
Compositor: MacPS, LLC
Indexer: BIM Indexing & Proofreading Services
Artist: SPi Global
Cover Designer: Anna Ishchenko

Distributed to the book trade worldwide by Springer Science+Business Media New York, 233 Spring Street, 6th Floor, New York, NY 10013. Phone 1-800-SPRINGER, fax (201) 348-4505, e-mail orders-ny@springer-sbm.com, or visit www.springeronline.com.

For information on translations, please e-mail rights@apress.com, or visit www.apress.com.

Apress and friends of ED books may be purchased in bulk for academic, corporate, or promotional use. eBook versions and licenses are also available for most titles. For more information, reference our Special Bulk Sales–eBook Licensing web page at www.apress.com/bulk-sales.

Any source code or other supplementary materials referenced by the author in this text is available to readers at www.apress.com. For detailed information about how to locate your book's source code, go to www.apress.com/source-code/

To my Dad, whose license let me take my own path.
—Satya Komatineni

To my wife Rosie, who inspires me to pursue my dreams. To my son Mike, who keeps life interesting every day.
And to my friends the Bighams (Dale, Heather, Eric, and Lizzie), whose amazing generosity has helped make this possible.
—Dave MacLean

Contents at a Glance

Contents

About the Authors

Satya Komatineni (satyakomatineni.com) has over 20 years of programming experience working with small and large corporations. Satya has published more than 30 articles on web development using Java, .NET, and database technologies. He is a frequent speaker at industry conferences on innovative technologies and a regular contributor to the weblogs on Java.net. He is the author of AspireWeb (satyakomatineni.com/aspire), a simplified all-in-one open source tool for Java web development, and the creator of Aspire Knowledge Central (www.knowledgefolders.com), an open source personal web operating system with a focus on individual productivity and publishing. Satya is also a contributing member to a number of Small Business Innovation Research (SBIR) programs. He received a bachelor's degree in electrical engineering from Andhra University, Visakhapatnam, and a master's degree in electrical engineering from the Indian Institute of Technology, New Delhi.

Dave MacLean is a software engineer and architect currently living and working in Orlando, Florida. Since 1980, he has programmed in many languages, developing solutions ranging from robot automation systems to data warehousing, from web self-service applications to EDI transaction processors. Dave has worked for Blue Cross Blue Shield of Florida, Sun Microsystems, IBM, Trimble Navigation, General Motors, and several small companies. He graduated from the University of Waterloo in Canada with a degree in systems design engineering. Visit his blog at http://davemac327.blogspot.com, or contact him at davemac327@gmail.com.

About the Technical Reviewers

Eric Franchomme had his first computer, a Thomson TO7, at the age of 13, and got fascinated by being able to move an object on the monitor just by writing code. After graduating from a French graduate engineering school in the field of information technologies, he moved to the heart of Silicon Valley for a 16-month internship. From there, he never stopped learning and coding, starting with assembly programming on DSP and moving up through systems (Symbian, BREW, Android) and layers, adding C, C++, Java, JavaScript, and PHP to his language set on top of French, English, Spanish, German, and Portuguese. Now working for PacketVideo, he built the first version of the Verizon V CAST video-on-demand product on BREW, and PacketVideo's first Android application to beam videos and music from a cell phone to a TV (Twonky). When he isn't speaking a different language, Eric plays Taiko (the art of Japanese drumming), travels around the world, and sips his espresso with almond cookies. He can be reached at `eric_franchomme@yahoo.com`.

Michael Nguyen is a software developer from Silicon Valley. He has over 10 years of experience developing software for small office to enterprise applications, using a variety of different languages and tools. Most recently, he has been involved in developing mobile applications for iOS, Android, and other platforms to benefit health care professionals. He currently lives in San Jose with his wife Julie and two adorable little girls. He spends his "copious" amounts of spare time with his family. He is involved in the development of smaller projects in the mobile space, one of which is a popular trading-card game variant called Wagic.

Dylan Phillips is a software engineer and architect who has been working in the mobile space for the last 10 years. With a broad range of experience ranging from J2ME to .NET Compact Framework to Android, he is incredibly excited about the opportunity presented by the broad consumer adoption of an array of Android devices. He can be reached at `mykoan@hotmail.com`, `@mykoan` on Twitter, or at lunch, in various Pho Houses around the country.

Karim Varela is an international platform engineer with Wilshire Media Group and has over eight years of mobile application development experience. Karim is currently an engineer on the Android Muve Music project. Previously he was the Android support lead within AT&T Mobility's technical developer support team, focusing on Android, Java ME, and U-verse Enabled. At the dawn of the mobile era, Karim held various roles in the mobile games industry ranging from quality assurance and build engineering to porting and networking engineering. He also serves as a board member for Java Verified and the Unified Testing Initiative, industry-unifying organizations dedicated to improving the quality and usability of mobile applications, as well as the App Developers Conference Advisory Board. He holds a bachelor's degree in computer science from the University of California and is working toward his MBA from the University of Florida.

Acknowledgments

Writing this book took effort not only on the part of the authors, but also from some of the very talented staff at Apress, as well as the technical reviewers. Therefore, we would like to thank Steve Anglin, Matthew Moodie, Corbin Collins, Douglas Pundick, Brigid Duffy, and Tiffany Taylor.

We would also like to extend our appreciation to the technical reviewers—Eric Franchomme, Michael Nguyen, Dylan Phillips, and Karim Varela—for their keen eye for detail and for making us wiser.

When searching for answers on the Android Developers Forum, we were often helped by Dianne Hackborn, Romain Guy, Nick Pelly, Brad Fitzpatrick, and other members of the Android Team at all hours of the day and weekends, and to them we would like to say, "Thank you." They truly are the hardest working team in mobile. The Android community is very much alive and well and was also very helpful in answering questions and offering advice. We hope this book in some way is able to give back to the community. Finally, the authors are deeply grateful to their families for accommodating prolonged irresponsibility.

Preface

Have you ever wanted to be a Rodin, sitting with a chisel and eroding a block of rock to mold it to your vision? Mainstream programmers have kept away from the severely constrained mobile devices for fear of being unable to chisel out a workable application. Well, those times are past.

Android OS places programmatically unconstrained mobile devices (with an incredible reach) at your doorstep. In this book, we want to positively confirm your suspicion that Android is a great OS to program with. If you are a Java programmer, you have a wonderful opportunity to profit from the Android OS, an exciting, capable, and general-purpose computing platform. Despite being a mobile OS, the Android OS has introduced new paradigms in framework design.

This is our fourth edition on the subject of Android, and it's our best edition yet. *Pro Android 4* is an extensive programming guide for the Android 4.0 SDK (the first SDK of Android to cover both phones and tablets). In this edition, we've refined, rewritten, and enhanced a number of things from *Pro Android 3* to create a thoroughly updated guide for both beginners and professionals. *Pro Android 4* is a result of our four years of research. We cover more than 100 topics in 31 chapters.

In this edition, we have beefed up our discussion of Android internals by covering threads, processes, long-running services, broadcast receivers, alarm managers, device configuration changes, and asynchronous tasks. We dedicate over 150 pages of material to fragments, fragment dialogs, ActionBar, and drag and drop. We have significantly enhanced the services and sensor chapters. The animation chapter has been revised to include property-based animations. The chapter on the contacts API is also largely rewritten to take into account the personal profile that paves the way for the Social API.

Concepts, code, and tutorials are the essence of this book. Every chapter reflects this philosophy. The self-contained tutorials in each chapter are annotated with expert advice. All projects in the book are available for download for easy importing into Eclipse.

Finally, in this book we went beyond basics, asked tough questions on every topic, and documented the results (see the table of contents for the extensive list of what we cover). We are also actively updating the supplemental web site (`www.androidbook.com`) with current and future research material on the Android SDK. As you walk through the book, if you have any questions, we are only an e-mail away for a quick response.

Chapter 1

Introducing the Android Computing Platform

Computing is more accessible than ever before. Handheld devices have transformed into computing platforms. Be it a phone or a tablet, the mobile device is now so capable of general-purpose computing that it's becoming the *real* personal computer (PC). Every traditional PC manufacturer is producing devices of various form factors based on the Android OS. The battles between operating systems, computing platforms, programming languages, and development frameworks are being shifted and reapplied to mobile devices.

We are also seeing a surge in mobile programming as more and more IT applications start to offer mobile counterparts. In this book, we'll show you how to take advantage of your Java skills to write programs for devices that run on Google's Android platform (`http://developer.android.com/index.html`), an open source platform for mobile and tablet development.

> **NOTE:** We are excited about Android because it is an advanced Java-based platform that introduces a number of new paradigms in framework design (even with the limitations of a mobile platform).

In this chapter, we'll provide an overview of Android and its SDK, give a brief overview of key packages, introduce what we are going to cover in each chapter, show you how to take advantage of Android source code, and highlight the benefits of programming for the Android platform.

A New Platform for a New Personal Computer

The Android platform embraces the idea of general-purpose computing for handheld devices. It is a comprehensive platform that features a Linux-based operating system stack for managing devices, memory, and processes. Android's Java libraries cover telephony, video, speech, graphics, connectivity, UI programming, and a number of other aspects of the device.

> **NOTE:** Although built for mobile- and tablet-based devices, the Android platform exhibits the characteristics of a full-featured desktop framework. Google makes this framework available to Java programmers through a Software Development Kit (SDK) called the Android SDK. When you are working with the Android SDK, you rarely feel that you are writing to a mobile device because you have access to most of the class libraries that you use on a desktop or a server—including a relational database.

The Android SDK supports most of the Java Platform, Standard Edition (Java SE), except for the Abstract Window Toolkit (AWT) and Swing. In place of AWT and Swing, Android SDK has its own *extensive modern UI framework*. Because you're programming your applications in Java, you could expect that you need a Java Virtual Machine (JVM) that is responsible for interpreting the runtime Java byte code. A JVM typically provides the necessary optimization to help Java reach performance levels comparable to compiled languages such as C and C++. Android offers its own optimized JVM to run the compiled Java class files in order to counter the handheld device limitations such as memory, processor speed, and power. This virtual machine is called the Dalvik VM, which we'll explore in a later section, "Delving into the Dalvik VM."

> **NOTE:** The familiarity and simplicity of the Java programming language, coupled with Android's extensive class library, makes Android a compelling platform to write programs for.

Figure 1–1 provides an overview of the Android software stack. (We'll provide further details in the section "Understanding the Android Software Stack.")

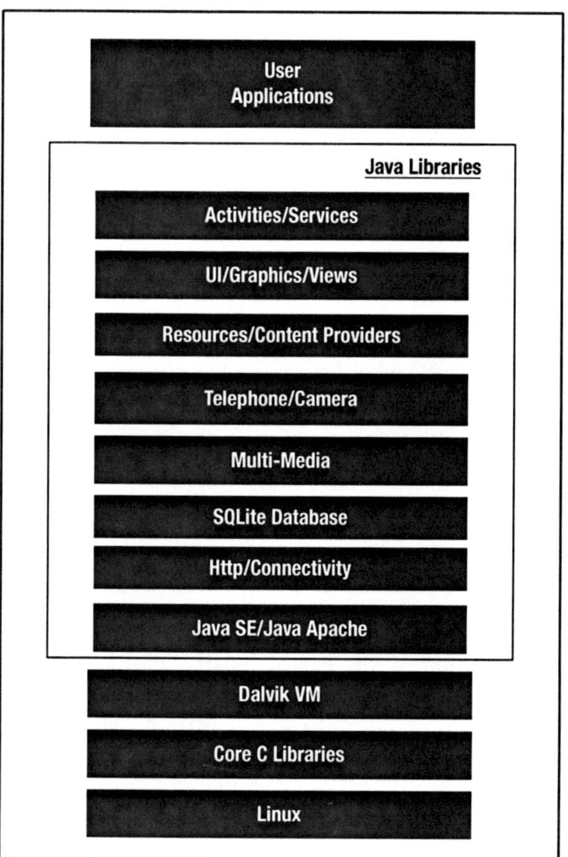

Figure 1–1. *High-level view of the Android software stack*

Early History of Android

Mobile phones use a variety of operating systems, such as Symbian OS, Microsoft's Windows Phone OS, Mobile Linux, iPhone OS (based on Mac OS X), Moblin (from Intel), and many other proprietary OSs. So far, no single OS has become the de facto standard. The available APIs and environments for developing mobile applications are too restrictive and seem to fall behind when compared to desktop frameworks. In contrast, the Android platform promised openness, affordability, open source code, and, more important, a high-end, all-in-one-place, consistent development framework.

Google acquired the startup company Android Inc. in 2005 to start the development of the Android platform (see Figure 1–2). The key players at Android Inc. included Andy Rubin, Rich Miner, Nick Sears, and Chris White.

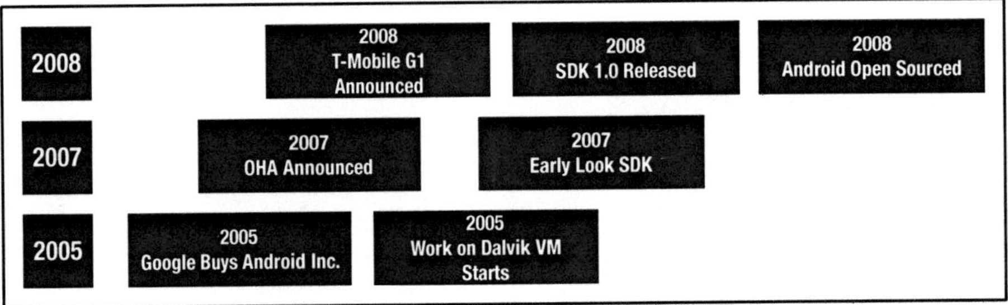

Figure 1–2. *Android early timeline*

The Android SDK was first issued as an "early look" release in November 2007. In September 2008, T-Mobile announced the availability of T-Mobile G1, the first smartphone based on the Android platform. Since then we have seen the SDKs 2.0, 3.0, and now 4.0, roughly one every year. The devices that run Android started out as a trickle but now are a torrent.

One of Androids key architectural goals is to allow applications to interact with one another and reuse components from one another. This reuse applies not only to services, but also to data and the user interface (UI).

Android has attracted an early following and sustained the developer momentum because of its fully developed features to exploit the cloud-computing model offered by web resources and to enhance that experience with local data stores on the handset itself. Android's support for a relational database on the handset also played a part in early adoption.

In releases 1.0 and 1.1 (2008) Android did not support soft keyboards, requiring the devices to carry physical keys. Android fixed this issue by releasing the 1.5 SDK in April 2009, along with a number of other features, such as advanced media-recording capabilities, widgets, and live folders.

In September 2009 came release 1.6 of the Android OS and, within a month, Android 2.0 followed, facilitating a flood of Android devices in time for the 2009 Christmas season. This release introduced advanced search capabilities and text to speech.

In Android 2.3, the significant features include remote wiping of secure data by administrators, the ability to use camera and video in low-light conditions, Wi-Fi hotspot, significant performance improvements, improved Bluetooth functionality, installation of applications on the SD card optionally, OpenGL ES 2.0 support, improvements in backup, improvements in search usability, Near Field Communications support for credit card processing, much improved motion and sensor support (similar to Wii), video chat, and improved Market.

Android 3.0 is focused on tablet-based devices and much more powerful dual core processors such as NVIDIA Tegra 2. The main features of this release include support to use a larger screen. A significantly new concept called fragments has been introduced. Fragments permeate the 3.0 experience. More desktop-like capabilities, such as the action bar and drag-and-drop, have been introduced. Home-screen widgets have been significantly enhanced, and more UI controls are now available. In the 3D space, OpenGL has been enhanced with Renderscript to further supplement ES 2.0. It is an exciting introduction for tablets.

However, the 3.0 experience is limited to tablets. At the time of the 3.0 release, the 2.x branch of Android continued to serve phones while 3.x branches served the tablets. Starting with 4.0, Android has merged these branches and forged a single SDK. For phone users, the primary UI difference is that the tablet experience is brought to phones as well.

The key aspects of the 4.0 user experience are as follows:

- A new type face called Roboto to provide crispness on high-density screens.
- A better way to organize apps into folders on home pages.
- Ability to drag apps and folders into the favorites tray that is always present at the bottom of the device.
- Optimization of notifications based on device type. For small devices, they show up on the top, and for larger devices they show up in the bottom system bar.
- Resizable, scrollable widgets.
- A variety of ways to unlock screens.
- Spell checker.
- Improved voice input with a "speak continuously" option.
- More controls to work with network data usage.
- Enhanced Contacts application with a personal profile much like social networks.
- Enhancements to the calendar application.
- Better camera app: continuous focus, zero shutter lag, face detection, tap to focus, and a photo editor.
- Live effects on pictures and videos for silly effects.
- A quick way to take and share screen shots.
- Browser performance that is twice as fast.
- Improved e-mail.
- A new concept called Android beaming for NFC-based sharing.
- Support for Wi-Fi Direct to promote P2P services.
- Bluetooth health device profile.

Key aspects of developer support for 4.0 include

- Revamped animation based on changing properties of objects, including views
- Fixed number of list-based widget behaviors from 3.0
- Much more mature action bar with integrated search
- Support for a number of mobile standards: Advanced Audio Distribution Profile (A2DP: the ability to use external speakers), Real-time Transport Protocol RTP: to stream audio/video over IP), Media Transfer Protocol (MTP), Picture Transfer Protocol (PTP: for hooking up to computers to download photos and media), and Bluetooth Headset Profile (HSP)
- Full device encryption
- Digital Rights Management (DRM)
- Encrypted storage and passwords
- Social API involving personal profiles
- Enhanced Calendar API
- Voice Mail API

Delving Into the Dalvik VM

As part of Android, Google has spent a lot of time thinking about optimizing designs for low-powered handheld devices. Handheld devices lag behind their desktop counterparts in memory and speed by eight to ten years. They also have limited power for computation. The performance requirements on handsets are severe as a result, requiring handset designers to optimize everything. If you look at the list of packages in Android, you'll see that they are fully featured and extensive.

These issues led Google to revisit the standard JVM implementation in many respects. The key figure in Google's implementation of this JVM is Dan Bornstein, who wrote the Dalvik VM—Dalvik is the name of a town in Iceland. Dalvik VM takes the generated Java class files and combines them into one or more Dalvik Executable (`.dex`) files. The goal of the Dalvik VM is to find every possible way to optimize the JVM for space, performance, and battery life.

The final executable code in Android, as a result of the Dalvik VM, is based not on Java byte code but on `.dex` files instead. This means you cannot directly execute Java byte code; you have to start with Java class files and then convert them to linkable `.dex` files.

Understanding the Android Software Stack

So far we've covered Android's history and its optimization features, including the Dalvik VM, and we've hinted at the Java programming stack available. In this section, we will cover the development aspect of Android. Figure 1–3 shows the Android software stack from a developer's perspective.

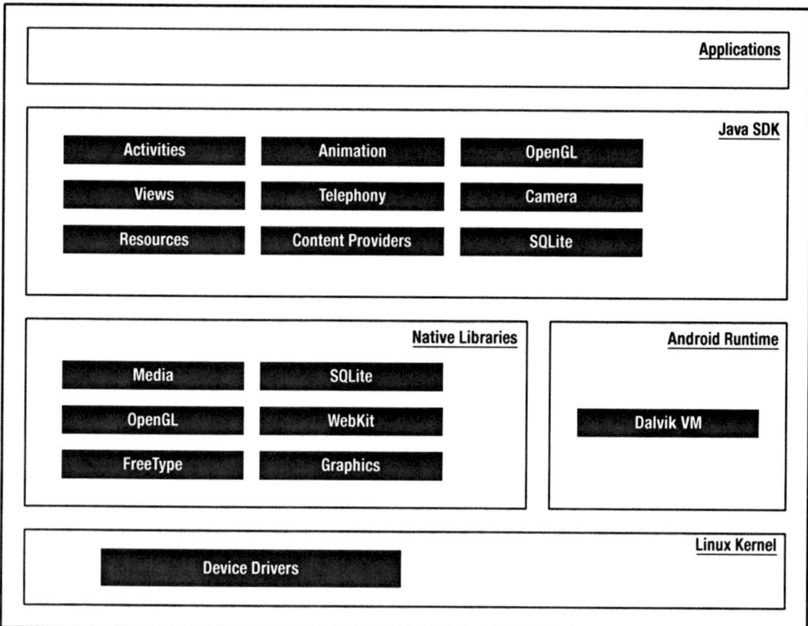

Figure 1–3. *Detailed Android SDK software stack*

At the core of the Android platform is a Linux kernel responsible for device drivers, resource access, power management, and other OS duties. The supplied device drivers include Display, Camera, Keypad, Wi-Fi, Flash Memory, Audio, and inter-process communication (IPC). Although the core is Linux, the majority—if not all—of the applications on an Android device such as a Motorola Droid are developed in Java and run through the Dalvik VM.

Sitting at the next level, on top of the kernel, are a number of C/C++ libraries such as OpenGL, WebKit, FreeType, Secure Sockets Layer (SSL), the C runtime library (`libc`), SQLite, and Media. The system C library based on Berkeley Software Distribution (BSD) is tuned (to roughly half its original size) for embedded Linux-based devices. The media libraries are based on PacketVideo's (`www.packetvideo.com/`) OpenCORE. These libraries are responsible for recording and playback of audio and video formats. A library called Surface Manager controls access to the display system and supports 2D and 3D.

> **NOTE:** These core libraries are subject to change because they are all internal implementation details of Android and not directly exposed to the published Android API. We have indicated these core libraries just to inform you of the nature of the underbelly of Android. Refer to the Android developer site for updates and future insight.

The WebKit library is responsible for browser support; it is the same library that supports Google Chrome and Apple's Safari. The FreeType library is responsible for font support. SQLite (www.sqlite.org/) is a relational database that is available on the device itself. SQLite is also an independent open source effort for relational databases and not directly tied to Android. You can acquire and use tools meant for SQLite for Android databases as well.

Most of the application framework accesses these core libraries through the Dalvik VM, the gateway to the Android platform. As we indicated in the previous sections, Dalvik is optimized to run multiple instances of VMs. As Java applications access these core libraries, each application gets its own VM instance.

The Android Java API's main libraries include telephony, resources, locations, UI, content providers (data), and package managers (installation, security, and so on). Programmers develop end-user applications on top of this Java API. Some examples of end-user applications on the device include Home, Contacts, Phone, and Browser.

Android also supports a custom Google 2D graphics library called Skia, which is written in C and C++. Skia also forms the core of the Google Chrome browser. The 3D APIs in Android, however, are based on an implementation of OpenGL ES from the Khronos group (www.khronos.org). OpenGL ES contains subsets of OpenGL that are targeted toward embedded systems.

From a media perspective, the Android platform supports the most common formats for audio, video, and images. From a wireless perspective, Android has APIs to support Bluetooth, EDGE, 3G, Wi-Fi, and Global System for Mobile Communication (GSM) telephony, depending on the hardware.

Developing an End-User Application with the Android SDK

In this section, we'll introduce you to the high-level Android Java APIs that you'll use to develop end-user applications on Android. We will briefly talk about the Android emulator, Android foundational components, UI programming, services, media, telephony, animation, and more.

Android Emulator

The Android SDK ships with an Eclipse plug-in called Android Development Tools (ADT). You will use this Integrated Development Environment (IDE) tool for developing, debugging, and testing your Java applications. (We'll cover ADT in depth in Chapter 2.) You can also use the Android SDK without using ADT; you'd use command-line tools instead. Both approaches support an emulator that you can use to run, debug, and test your applications. You will not even need the real device for 90% of your application development. The full-featured Android emulator mimics most of the device features. The emulator limitations include USB connections, camera and video capture, headphones, battery simulation, Bluetooth, Wi-Fi, NFC, and OpenGL ES 2.0.

The Android emulator accomplishes its work through an open source "processor emulator" technology called QEMU, developed by Fabrice Bellard (`http://wiki.qemu.org/Index.html`). This is the same technology that allows emulation of one operating system on top of another, regardless of the processor. QEMU allows emulation at the CPU level.

With the Android emulator, the processor is based on Advanced RISC Machine (ARM). ARM is a 32-bit microprocessor architecture based on Reduced Instruction Set Computing (RISC), in which design simplicity and speed is achieved through a reduced number of instructions in an instruction set. The emulator runs the Android version of Linux on this simulated processor.

ARM is widely used in handhelds and other embedded electronics where lower power consumption is important. Much of the mobile market uses processors based on this architecture.

You can find more details about the emulator in the Android SDK documentation at `http://developer.android.com/guide/developing/tools/emulator.html`.

The Android UI

Android uses a UI framework that resembles other desktop-based, full-featured UI frameworks. In fact, it's more modern and more asynchronous in nature. The Android UI is essentially a fourth-generation UI framework, if you consider the traditional C-based Microsoft Windows API the first generation and the C++-based Microsoft Foundation Classes (MFC) the second generation. The Java-based Swing UI framework would be the third generation, introducing design flexibility far beyond that offered by MFC. The Android UI, JavaFX, Microsoft Silverlight, and Mozilla XML User Interface Language (XUL) fall under this new type of fourth-generation UI framework, in which the UI is declarative and independently themed.

> **NOTE:** In Android, you program using a modern user interface paradigm, even though the device you're programming for may be a handheld.

Programming in the Android UI involves declaring the interface in XML files. You then load these XML view definitions as windows in your UI application. This is very much like HTML-based web pages. Much as in HTML, you find (get hold of) the individual controls through their IDs and manipulate them with Java code.

Even menus in your Android application are loaded from XML files. Screens or windows in Android are often referred to as *activities*, which comprise multiple views that a user needs in order to accomplish a logical unit of action. *Views* are Android's basic UI building blocks, and you can further combine them to form composite views called *view groups*.

Views internally use the familiar concepts of canvases, painting, and user interaction. An activity hosting these composite views, which include views and view groups, is the logical replaceable UI component in Android.

Android 3.0 introduced a new UI concept called *fragments* to allow developers to chunk views and functionality for display on tablets. Tablets provide enough screen space for multipane activities, and fragments provide the abstraction for the panes.

One of the Android framework's key concepts is the life cycle management of activity windows. Protocols are put in place so that Android can manage state as users hide, restore, stop, and close activity windows. You will get a feel for these basic ideas in Chapter 2, along with an introduction to setting up the Android development environment.

The Android Foundational Components

The Android UI framework, along with many other parts of Android, relies on a new concept called an *intent*. An intent is an intra- and interprocess mechanism to invoke components in Android.

A component in Android is a piece of code that has a well defined life cycle. An activity representing a window in an Android application is a component. A service that runs in an Android process and serves other clients is a component. A receiver that wakes up in response to an event is another example of a component in Android.

While serving this primary need of invoking components, an intent exhibits the characteristics parallel to those of windowing messages, actions, publish-and-subscribe models, and interprocess communications. Here is an example of using the Intent class to invoke or start a web browser:

```
public static void invokeWebBrowser(Activity activity)
{
    Intent intent = new Intent(Intent.ACTION_VIEW);
    intent.setData(Uri.parse("http://www.google.com"));
    activity.startActivity(intent);
}
```

In this example, through an intent, we are asking Android to start a suitable window to display the content of a web site. Depending on the list of browsers that are installed on the device, Android will choose a suitable one to display the site. You will learn more about intents in Chapter 5.

Android has extensive support for *resources*, which include such things as strings and bitmaps, as well as some not-so-familiar items like XML-based view (layout like HTML) definitions. The Android framework makes use of resources in a novel way so their usage is easy, intuitive, and convenient. Here is an example where resource IDs are automatically generated for resources defined in XML files:

```
public final class R {
    //All string resources will have constants auto generated here
    public static final class string {
        public static final int hello=0x7f070000;
```

```
    }
    //All image files will have unique ids generated here
    public static final class drawable {
        public static final int myanimation=0x7f020001;
        public static final int numbers19=0x7f02000e;
    }
    //View ids are auto generated based on their names
    public static final class id {
        public static final int textViewId1=0x7f080003;
    }
    //The following are two files (like html) that define layout
    //auto generated from the filenames in respective sub directories.
    public static final class layout {
        public static final int frame_animations_layout=0x7f030001;
        public static final int main=0x7f030002;
    }
}
```

Each auto-generated ID in this class corresponds to either an element in an XML file or a whole file itself. Wherever you would like to use those XML definitions, you will use these generated IDs instead. This indirection helps a great deal when it comes to specializing resources based on locale, device size, and so on. (Chapter 3 covers the `R.java` file and resources in more detail.)

Another new concept in Android is the *content provider*. A content provider is an abstraction of a data source that makes it look like an emitter and consumer of RESTful services. The underlying SQLite database makes this facility of content providers a powerful tool for application developers. We will cover content providers in Chapter 4. In Chapters 3, 4, and 5, we'll discuss how intents, resources, and content providers promote openness in the Android platform.

Advanced UI Concepts

XML page-layout definitions (similar to HTML web pages) play a critical role in describing the Android UI. Let's look at an example of how an Android layout XML file does this for a simple layout containing a text view:

```
<?xml version="1.0" encoding="utf-8"?>
<!-- place it in /res/layout/sample_page1.xml -->
<!-- will auto generate an id called: R.layout.sample_page1 -->
<LinearLayout ..some basic attributes..>
<TextView android:id="@+id/textViewId"
    android:layout_width="fill_parent"
    android:layout_height="wrap_content"
    android:text="@string/hello"
    />
</LinearLayout>
```

You will use an ID generated for this XML file to load this layout into an activity window. (We'll cover this process in Chapter 6.) Android also provides extensive support for menus (more on that in Chapter 7), from standard menus to context menus. You'll find it convenient to work with menus in Android, because they are also loaded as XML files and because resource IDs for those menus are auto-generated. Here's how you would declare menus in an XML file:

```
<menu xmlns:android="http://schemas.android.com/apk/res/android">
    <!-- This group uses the default category. -->
    <group android:id="@+id/menuGroup_Main">
        <item android:id="@+id/menu_clear"
            android:orderInCategory="10"
            android:title="clear" />
        <item android:id="@+id/menu_show_browser"
            android:orderInCategory="5"
            android:title="show browser" />
    </group>
</menu>
```

Android supports dialogs, and all dialogs in Android are asynchronous. These asynchronous dialogs present a special challenge to developers accustomed to the synchronous modal dialogs in some windowing frameworks. We'll address menus in Chapter 7 and dialogs in Chapter 9.

Android offers extensive support for animation. There are three fundamental ways to accomplish animation. You can do frame-by-frame animation. Or you can provide tweening animation by changing view transformation matrices (position, scale, rotation, and alpha). Or you can also do tweening animation by changing properties of objects. The property-based animation is introduced in 3.0 and is the most flexible and recommended way to accomplish animation. All of these animations are covered in Chapter 21.

Moreover, Android allows you to define these animations in an XML resource file. Check out this example, in which a series of numbered images is played in frame-by-frame animation:

```
<animation-list xmlns:android="http://schemas.android.com/apk/res/android"
        android:oneshot="false">
    <item android:drawable="@drawable/numbers11" android:duration="50" />
  ......
    <item android:drawable="@drawable/numbers19" android:duration="50" />
</animation-list>
```

Android also supports 3D graphics through its implementation of the OpenGL ES 1.0 and 2.0 standards. OpenGL ES, like OpenGL, is a C-based flat API. The Android SDK, because it's a Java-based programming API, needs to use Java binding to access the OpenGL ES. Java ME has already defined this binding through Java Specification Request (JSR) 239 for OpenGL ES, and Android uses the same Java binding for OpenGL ES in its implementation. If you are not familiar with OpenGL programming, the learning curve is steep. Due to space limitations, we are not able to include the OpenGL coverage in the fourth edition of the book. However we have over 100 pages of coverage in our third edition.

Android has a number of new concepts that revolve around *information at your fingertips* using the home screen. The first of these is *live folders*. Using live folders, you can publish a collection of items as a folder on the homepage. The contents of this collection change as the underlying data changes. This changing data could be either on the device or from the Internet. Due to space limitations, we are not able to cover live folders in the fourth edition of the book. However, the third edition includes extensive coverage.

The second homepage-based idea is the *home screen widget*. Home screen widgets are used to paint information on the homepage using a UI widget. This information can change at regular intervals. An example could be the number of e-mail messages in your e-mail store. We describe home screen widgets in Chapter 25. The home screen widgets are enhanced in 3.0 to include list views that can get updated when their underlying data changes. These enhancements are covered in Chapter 26.

Integrated Android Search is the third homepage-based idea. Using integrated search, you can search for content both on the device and also across the Internet. Android search goes beyond search and allows you to fire off commands through the search control. Due to space limitations, we can't include coverage of the search API in the fourth edition of the book. However, it is covered in the third edition.

Android also supports touchscreen and gestures based on finger movements on the device. Android allows you to record any random motion on the screen as a named gesture. This gesture can then be used by applications to indicate specific actions. We cover touchscreens and gestures in Chapter 27.

Sensors are now becoming a significant part of the mobile experience. We cover sensors in Chapter 29.

Another necessary innovation required for a mobile device is the dynamic nature of its configurations. For instance, it is very easy to change the viewing mode of a handheld between portrait and landscape. Or you may dock your handheld to become a laptop. Android 3.0 has introduced a concept called fragments to deal with these variations effectively. Chapter 8 is dedicated to fragments, and Chapter 12 talks about how to deal with configuration changes.

We also cover the 3.0 feature (which is much enhanced in 4.0) of action bars in Chapter 10. Action bars bring Android up to par with a desktop menu bar paradigm.

Drag-and-drop is introduced for tablets in 3.0. This feature is now available to phones as well. We cover drag-and-drop in Chapter 28.

Handheld devices are fully aware of a cloud-based environment. To make server-side HTTP calls, it is important to understand the threading model to avoid Application Not Responding messages. We cover the mechanisms available for asynchronous processing in Chapter 18.

Outside of the Android SDK, there are a number of independent innovations taking place to make development exciting and easy. Some examples are XML/VM, PhoneGap, and Titanium. Titanium allows you to use HTML technologies to program the WebKit-based Android browser. We covered Titanium in the second edition of this book. However, due to time and space limitations, we are not covering Titanium in this edition.

Android Service Components

Security is a fundamental part of the Android platform. In Android, security spans all phases of the application life cycle—from design-time policy considerations to runtime boundary checks. We cover security and permissions in Chapter 14.

In Chapter 15, we'll show you how to build and consume services in Android, specifically HTTP services. This chapter will also cover interprocess communication (communication between applications on the same device).

Location-based service is another of the more exciting components of the Android SDK. This portion of the SDK provides application developers with APIs to display and manipulate maps, as well as obtain real-time device-location information. We'll cover these ideas in detail in Chapter 22.

Android Media and Telephony Components

Android has APIs that cover audio, video, and telephony components. Chapter 23 will address the telephony API. We'll cover the audio and video APIs extensively in Chapter 24.

Starting with Android 2.0, Android includes the Pico Text-to-Speech engine. Due to space limitations, we are not able to include Text-to-Speech coverage in the fourth edition of the book. The third edition does cover the Text-to-Speech API.

Last but not least, Android ties all these concepts into an application by creating a single XML file that defines what an application package is. This file is called the application's *manifest file* (AndroidManifest.xml). Here is an example:

```
<?xml version="1.0" encoding="utf-8"?>
<manifest xmlns:android="http://schemas.android.com/apk/res/android"
      package="com.ai.android.HelloWorld"
      android:versionCode="1"
      android:versionName="1.0.0">
    <application android:icon="@drawable/icon" android:label="@string/app_name">
        <activity android:name=".HelloWorld"
                  android:label="@string/app_name">
            <intent-filter>
                <action android:name="android.intent.action.MAIN" />
                <category android:name="android.intent.category.LAUNCHER" />
            </intent-filter>
        </activity>
    </application>
</manifest>
```

The Android manifest file is where activities are defined, where services and content providers are registered, and where permissions are declared. Details about the manifest file will emerge throughout the book as we develop each idea.

Android Java Packages

One way to get a quick snapshot of the Android platform is to look at the structure of Java packages. Because Android deviates from the standard JDK distribution, it is important to know what is supported and what is not. Here's a brief description of the important packages that are included in the Android SDK:

- *android.app:* Implements the Application model for Android. Primary classes include `Application`, representing the start and stop semantics, as well as a number of activity-related classes, fragments, controls, dialogs, alerts, and notifications. We work with most of these classes through out this book.
- *android.app.admin:* Provides the ability to control the device by folks such as enterprise administrators.
- *android.accounts:* Provides classes to manage accounts such as Google, Facebook, and so on. The primary classes are `AccountManager` and `Account`. We cover this API briefly in Chapter 30 when we discuss the Contacts API.
- *android.animation:* Hosts all the new property animation classes. These clases are extensively covered in Chapter 21.
- *android.app.backup:* Provides hooks for applications to back up and restore their data when folks switch their devices.
- *android.appwidget:* Provides functionality for home screen widgets. This package is covered extensively in Chapter 25 and Chapter 26 when we talk about home screen widgets, including list-based widgets.
- *android.bluetooth:* Provides a number of classes to work with Bluetooth functionality. The main classes include `BluetoothAdapter`, `BluetoothDevice`, `BluetoothSocket`, `BluetoothServerSocket`, and `BluetoothClass`. You can use `BluetoothAdapter` to control the locally installed Bluetooth adapter. For example, you can enable it, disable it, and start the discovery process. `BluetoothDevice` represents the remote Bluetooth device that you are connecting with. The two Bluetooth sockets are used to establish communication between the devices. A Bluetooth class represents the type of Bluetooth device you are connecting to.
- *android.content:* Implements the concepts of content providers. Content providers abstract out data access from data stores. This package also implements the central ideas around intents and Android Uniform Resource Identifiers (URIs). These classes are covered in Chapter 4.

- *android.content.pm:* Implements package manager–related classes. A package manager knows about permissions, installed packages, installed providers, installed services, installed components such as activities, and installed applications.
- *android.content.res:* Provides access to resource files, both structured and unstructured. The primary classes are `AssetManager` (for unstructured resources) and `Resources`. Some of the classes from this package are covered in Chapter 3.
- *android.database:* Implements the idea of an abstract database. The primary interface is the `Cursor` interface. Some of the classes from this package are covered in Chapter 4.
- *android.database.sqlite:* Implements the concepts from the `android.database` package using SQLite as the physical database. Primary classes are `SQLiteCursor`, `SQLiteDatabase`, `SQLiteQuery`, `SQLiteQueryBuilder`, and `SQLiteStatement`. However, most of your interaction is going to be with classes from the abstract `android.database` package.
- *android.drm:* Classes related to Digital Rights Management.
- *android.gesture:* Houses all the classes and interfaces necessary to work with user-defined gestures. Primary classes are `Gesture`, `GestureLibrary`, `GestureOverlayView`, `GestureStore`, `GestureStroke`, and `GesturePoint`. A `Gesture` is a collection of `GestureStrokes` and `GesturePoints`. Gestures are collected in a `GestureLibrary`. Gesture libraries are stored in a `GestureStore`. Gestures are named so that they can be identified as actions. Some of the classes from this package are covered in Chapter 27.
- *android.graphics:* Contains the classes `Bitmap`, `Canvas`, `Camera`, `Color`, `Matrix`, `Movie`, `Paint`, `Path`, `Rasterizer`, `Shader`, `SweepGradient`, and `TypeFace`.
- *android.graphics.drawable:* Implements drawing protocols and background images, and allows animation of drawable objects.
- *android.graphics.drawable.shapes:* Implements shapes including `ArcShape`, `OvalShape`, `PathShape`, `RectShape`, and `RoundRectShape`.
- *android.hardware:* Implements the physical Camera-related classes. The Camera represents the hardware camera, whereas `android.graphics.Camera` represents a graphical concept that's not related to a physical camera at all.
- *android.hardware.usb:* Lets you talk to USB devices from Android.

- *android.location:* Contains the classes `Address`, `GeoCoder`, `Location`, `LocationManager`, and `LocationProvider`. The `Address` class represents the simplified Extensible Address Language (XAL). `GeoCoder` allows you to get a latitude/longitude coordinate given an address, and vice versa. `Location` represents the latitude/longitude. Some of the classes from this package are covered in Chapter 22.
- *android.media:* Contains the classes `MediaPlayer`, `MediaRecorder`, `Ringtone`, `AudioManager`, and `FaceDetector`. `MediaPlayer`, which supports streaming, is used to play audio and video. `MediaRecorder` is used to record audio and video. The `Ringtone` class is used to play short sound snippets that could serve as ringtones and notifications. `AudioManager` is responsible for volume controls. You can use `FaceDetector` to detect people's faces in a bitmap. Some of the classes from this package are covered in Chapter 24.
- *android.media.audiofx:* Provides audio effects.
- *android.media.effect:* Provides video effects.
- *android.mtp:* Provides the ability to interact with cameras and music devices.
- *android.net:* Implements the basic socket-level network APIs. Primary classes include `Uri`, `ConnectivityManager`, `LocalSocket`, and `LocalServerSocket`. It is also worth noting here that Android supports HTTPS at the browser level and also at the network level. Android also supports JavaScript in its browser.
- *android.net.rtp*: Supports streaming protocols.
- *android.net.sip*: Provides support for VOIP.
- *android.net.wifi*: Manages Wi-Fi connectivity. Primary classes include `WifiManager` and `WifiConfiguration`. `WifiManager` is responsible for listing the configured networks and the currently active Wi-Fi network.
- *android.net.wifi.p2p:* Supports P2P networks with Wi-Fi Direct.
- *android.nfc:* Lets you interact with devices in close proximity to enable touchless commerce such as credit card processing at sales counters.
- *android.opengl:* Contains utility classes surrounding OpenGL ES 1.0 and 2.0 operations. The primary classes of OpenGL ES are implemented in a different set of packages borrowed from JSR 239. These packages are `javax.microedition.khronos.opengles`, `javax.microedition.khronos.egl`, and `javax.microedition.khronos.nio`. These packages are thin wrappers around the Khronos implementation of OpenGL ES in C and C++.

- *android.os:* Represents the OS services accessible through the Java programming language. Some important classes include `BatteryManager`, `Binder`, `FileObserver`, `Handler`, `Looper`, and `PowerManager`. `Binder` is a class that allows interprocess communication. `FileObserver` keeps tabs on changes to files. You use `Handler` classes to run tasks on the message thread and `Looper` to run a message thread.

- *android.preference:* Allows applications to have users manage their preferences for that application in a uniform way. The primary classes are `PreferenceActivity`, `PreferenceScreen`, and various preference-derived classes such as `CheckBoxPreference` and `SharedPreferences`. Some of the classes from this package are covered in Chapter 13 and Chapter 25.

- *android.provider:* Comprises a set of prebuilt content providers adhering to the `android.content.ContentProvider` interface. The content providers include `Contacts`, `MediaStore`, `Browser`, and `Settings`. This set of interfaces and classes stores the metadata for the underlying data structures. We cover many of the classes from the Contacts provider package in Chapter 30.

- *android.sax:* Contains an efficient set of Simple API for XML (SAX) parsing utility classes. Primary classes include `Element`, `RootElement`, and a number of `ElementListener` interfaces.

- *android.speech.*:* Provides support for converting text to speech. The primary class is `TextToSpeech`. You will be able to take text and ask an instance of this class to queue the text to be spoken. You have access to a number of callbacks to monitor when the speech has finished, for example. Android uses the Pico Text-to-Speech (TTS) engine from SVOX.

- *android.telephony:* Contains the classes `CellLocation`, `PhoneNumberUtils`, and `TelephonyManager`. `TelephonyManager` lets you determine cell location, phone number, network operator name, network type, phone type, and Subscriber Identity Module (SIM) serial number. Some of the classes from this package are covered in Chapter 23.

- *android.telephony.gsm:* Allows you to gather cell location based on cell towers and also hosts classes responsible for SMS messaging. This package is called GSM because Global System for Mobile Communication is the technology that originally defined the SMS data-messaging standard.

- *android.telephony.cdma:* Provides support for CDMA telephony.

- *android.test*, *android.test.mock*, *android.test.suitebuilder:* Packages to support writing unit tests for Android applications.

- *android.text:* Contains text-processing classes.
- *android.text.method:* Provides classes for entering text input for a variety of controls.
- *android.text.style:* Provides a number of styling mechanisms for a span of text.
- *android.utils:* Contains the classes Log, DebugUtils, TimeUtils, and Xml.
- *android.view:* Contains the classes Menu, View, and ViewGroup, and a series of listeners and callbacks.
- *android.view.animation:* Provides support for tweening animation. The main classes include Animation, a series of interpolators for animation, and a set of specific animator classes that include AlphaAnimation, ScaleAnimation, TranslationAnimation, and RotationAnimation.Some of the classes from this package are covered in Chapter 21.
- *android.view.inputmethod:* Implements the input-method framework architecture.
- *android.webkit:* Contains classes representing the web browser. The primary classes include WebView, CacheManager, and CookieManager.
- *android.widget:* Contains all of the UI controls usually derived from the View class. Primary widgets include Button, Checkbox, Chronometer, AnalogClock, DatePicker, DigitalClock, EditText, ListView, FrameLayout, GridView, ImageButton, MediaController, ProgressBar, RadioButton, RadioGroup, RatingButton, Scroller, ScrollView, Spinner, TabWidget, TextView, TimePicker, VideoView, and ZoomButton.
- *com.google.android.maps:* Contains the classes MapView, MapController, and MapActivity, essentially classes required to work with Google maps.

These are some of the critical Android-specific packages. From this list, you can see the depth of the Android core platform.

NOTE: In all, the Android Java API contains more than 50 packages and more than 1,000 classes, and it keeps growing with each release.

In addition, Android provides a number of packages in the java.* namespace. These include awt.font, beans, io, lang, lang.annotation, lang.ref, lang.reflect, math, net, nio, nio.channels, nio.channels.spi, nio.charset, security, security.acl, security.cert, security.interfaces, security.spec, sql, text, util, util.concurrent, util.concurrent.atomic, util.concurrent.locks, util.jar, util.logging, util.prefs, util.regex, and util.zip.

Android comes with these packages from the javax namespace: crypto, crypto.spec, microedition.khronos.egl, microedition.khronos.opengles, net, net.ssl, security.auth, security.auth.callback, security.auth.login, security.auth.x500, security.cert, sql, xml, and xmlparsers.

In addition to these, it contains a lot of packages from org.apache.http.* as well as org.json, org.w3c.dom, org.xml.sax, org.xml.sax.ext, org.xml.sax.helpers, org.xmlpull.v1, and org.xmlpull.v1.sax2. Together, these numerous packages provide a rich computing platform to write applications for handheld devices.

Taking Advantage of Android Source Code

Android documentation is a bit wanting in places. Android source code can be used to fill the gaps.

The source code for Android and all its projects is managed by the Git source code control system. Git (http://git-scm.com/) is an open source source-control system designed to handle large and small projects with speed and convenience. The Linux kernel and Ruby on Rails projects also rely on Git for version control.

The details of the Android source distribution are published at http://source.android.com. The code was made available as open source around October 2008. One of the Open Handset Alliance's goals was to make Android a free and fully customizable mobile platform.

Browsing Android Sources Online

Prior to Android 4.0, the Android source distribution was made available at http://android.git.kernel.org/. Android now is hosted on its own Git site at https://android.googlesource.com. However, this is not browsable online as of this writing. There are some posts online indicating that online browsing may be available soon.

Another frequently visited site to browse Android sources online is at

www.google.com/codesearch/p?hl=en#uX1GffpyOZk/core/java/android/

However, there are rumors that the Code Search project may be in the process of getting shut down. Even if it is not, this site does not search the Android 4.0 code yet. For example, we were not able to find the new Contact APIs here.

Another useful site is

www.grepcode.com/search/?query=google+android&entity=project

There seems to be a 4.01 branch of Android available here.

We hope both these sites will continue to have the most recent releases so that you can browse the sources online.

Using Git to Download Android Sources

If all else fails, you may have to install Git on your computer and download the sources yourself. If you have a Linux distribution, you can follow the instructions at `http://source.android.com` to get the latest sources.

If you are on a Windows platform, this gets to be a challenge. You will have to install Git first and then use it to get the Android packages you want.

> **NOTE:** Our research notes on using Git to download Android can be found at `http://androidbook.com/item/3919`.

Installing Git

Use the following URL to install the `msysGit` package on Windows:

`http://code.google.com/p/msysgit/downloads/list`

Once you have installed it, you will see a directory called `C:\git` (assuming you have installed it under `c:\`).

Testing the Git Installation

The key directory is `C:\git\bin`. To see if it is working, you can use the following command to clone a public repository:

```
git clone git://git.kernel.org/pub/scm/git/git.git
```

This should clone the repository to your local drive.

Downloading Android Repositories

Run this command to discover how many Android Git repositories there are:

```
git clone https://android.googlesource.com/platform/manifest.git
```

This will bring down a directory called `manifest`. Look for a file called `manifest\default.xml`.

This file will have many of the names for the Android repositories. Here are a couple of lines from that file:

```
<project path="frameworks/base"
         name="platform/frameworks/base" />
<project path="frameworks/compile/libbcc"
           name="platform/frameworks/compile/libbcc" />
```

You can see this full file for 4.0 at http://androidbook.com/item/3920, where we have posted the contents of the file for quick review. Keep in mind that it is not updated with the latest information.

Now you can get the base android.jar source code by using the command

```
git clone https://android.googlesource.com/platform/frameworks/base.git
```

Using the same logic, you can get the contacts provider package by typing

```
git clone https://android.googlesource.com/platform/packages/providers/ContactsProvider
```

The Sample Projects in this Book

In this book, you will find many, many working sample projects. At the end of each chapter is a "References" section that contains a URL to download sample projects for that chapter. All of these sample projects can be accessed from

http://androidbook.com/proandroid4/projects

If you have any issues downloading or compiling these projects, please contact us by e-mail: satya.komatineni@gmail.com or davemac327@gmail.com.

We are continuously updating the androidbook.com supporting site with what we are learning. It is well worth our efforts if we are able to further contribute to your learning.

Summary

In this chapter, we wanted to pique your curiosity about Android. If you are a Java programmer, you have a great opportunity to profit from this exciting, capable, general-purpose computing platform. We welcome you to journey through the rest of the book for a methodical and in-depth understanding of the Android SDK.

Chapter 2

Setting Up Your Development Environment

The last chapter provided an overview of Android's history and hinted at concepts that are covered in the rest of the book. At this point, you're probably eager to get your hands on some code. You start by seeing what you need to begin building applications with the Android software development kit (SDK) and set up your development environment. Next, you step through a "Hello World!" application. Then the chapter explains the Android application life cycle and ends with a discussion about running your applications with Android Virtual Devices (AVDs) and on real devices.

To build applications for Android, you need the Java SE Development Kit (JDK), the Android SDK, and a development environment. Strictly speaking, you can develop your applications using a primitive text editor, but for the purposes of this book, you use the commonly available Eclipse IDE. The Android SDK requires JDK 5 or JDK 6 (the examples use JDK 6) and Eclipse 3.5 or higher (this book uses Eclipse 3.5, also known as Galileo, and 3.6, also known as Helios).

> **NOTE:** At the time of this writing, Java 7 was available but not yet supported by the Android SDK. The latest version of Eclipse (3.7, a.k.a. Indigo) was also available, but Android has historically not been reliable on the latest Eclipse right away. Check the System Requirements here to find the latest: `http://developer.android.com/sdk/requirements.html`.

The Android SDK is compatible with Windows (Windows XP, Windows Vista, and Windows 7), Mac OS X (Intel only), and Linux (Intel only). In terms of hardware, you need an Intel machine, the more powerful the better.

To make your life easier, you want to use Android Development Tools (ADT). ADT is an Eclipse plug-in that supports building Android applications with the Eclipse IDE.

The Android SDK is made up of two main parts: the tools and the packages. When you first install the SDK, all you get are the base tools. These are executables and supporting files to help you develop applications. The packages are the files specific to a particular version of Android (called a *platform*) or a particular add-on to a platform. The platforms include Android 1.5 through 4.0. The add-ons include the Google Maps API, the Market License Validator, and even vendor-supplied ones such as Samsung's Galaxy Tab add-on. After you install the SDK, you then use one of the tools to download and set up the platforms and add-ons. Let's get started!

Setting Up Your Environment

To build Android applications, you need to establish a development environment. In this section, you walk through downloading JDK 6, the Eclipse IDE, the Android SDK (tools and packages), and ADT. You also configure Eclipse to build Android applications. Google provides a page to describe the installation process (`http://developer.android.com/sdk/installing.html`) but leaves out some crucial steps, as you will see.

Downloading JDK 6

The first thing you need is the Java SE Development Kit. The Android SDK requires JDK 5 or higher; we developed the examples using JDK 6. For Windows, download JDK 6 from the Oracle web site (`www.oracle.com/technetwork/java/javase/downloads/index.html`) and install it. You only need the JDK, not the bundles. For Mac OS X, download the JDK from the Apple web site (`http://developer.apple.com/java/download/`), select the appropriate file for your particular version of Mac OS, and install it. You need to register for free as an Apple developer to get the JDK, and once at the Downloads page, you need to click the Java link at right. To install the JDK for Linux, open a Terminal window and try the following:

```
sudo apt-get install sun-java6-jdk
```

This should install the JDK plus any dependencies such as the Java Runtime Environment (JRE). If it doesn't, it probably means you need to add a new software source and then try that command again. The web page `https://help.ubuntu.com/community/Repositories/Ubuntu` explains software sources and how to add the connection to third-party software. The process is different depending on which version of Linux you have. After you've done that, retry the command.

With the introduction of Ubuntu 10.04 (Lucid Lynx), Ubuntu recommends using OpenJDK instead of the Oracle/Sun JDK. To install OpenJDK, try the following:

```
sudo apt-get install openjdk-6-jdk
```

If this is not found, set up the third-party software as outlined previously and run the command again. All packages on which the JDK depends are automatically added for you. It is possible to have both OpenJDK and the Oracle/Sun JDK installed at the same time. To switch active Java between the installed versions of Java on Ubuntu, run this command at a shell prompt

```
sudo update-alternatives --config java
```

and then choose which Java you want as the default.

Now that you have a Java JDK installed, it's time to set the JAVA_HOME environment variable to point to the JDK install folder. To do this on a Windows XP machine, choose Start ➤ My Computer, right-click, select Properties, choose the Advanced tab, and click Environment Variables. Click New to add the variable or Edit to modify it if it already exists. The value of JAVA_HOME is something like C:\Program Files\Java\jdk1.6.0_27.

For Windows Vista and Windows 7, the steps to get to the Environment Variables screen are a little different. Choose Start ➤ Computer, right-click, choose Properties, click the link for Advanced System Settings, and click Environment Variables. After that, follow the same instructions as for Windows XP to change the JAVA_HOME environment variable.

For Mac OS X, you set JAVA_HOME in the .bashrc file in your home directory. Edit or create the .bashrc file, and add a line that looks like this

```
export JAVA_HOME=path_to_JDK_directory
```

where path_to_JDK_directory is probably /Library/Java/Home. For Linux, edit your .bashrc file and add a line like the one for Mac OS X, except that your path to Java is probably something like /usr/lib/jvm/java-6-sun or /usr/lib/jvm/java-6-openjdk.

Downloading Eclipse 3.6

After the JDK is installed, you can download the Eclipse IDE for Java Developers. (You don't need the edition for Java EE; it works, but it's much larger and includes things you don't need for this book.) The examples in this book use Eclipse 3.6 (on a Windows environment). You can download all versions of Eclipse from www.eclipse.org/downloads/.

The Eclipse distribution is a .zip file that can be extracted just about anywhere. The simplest place to extract to on Windows is C:\, which results in a C:\eclipse folder where you find eclipse.exe. For Mac OS X, you can extract to Applications. For Linux, you can extract to your home directory or have your administrator put Eclipse into a common place where you can get to it. The Eclipse executable is in the eclipse folder for all platforms. You may also find and install Eclipse using Linux's Software Center for adding new applications, although this may not provide you with the latest version.

When you first start up Eclipse, it asks you for a location for the workspace. To make things easy, you can choose a simple location such as C:\android or a directory under your home directory. If you share the computer with others, you should put your workspace folder somewhere underneath your home directory.

Downloading the Android SDK

To build applications for Android, you need the Android SDK. As stated before, the SDK comes with the base tools; then you download the package parts that you need and/or want to use. The tools part of the SDK includes an emulator so you don't need a mobile device with the Android OS to develop Android applications. It also has a setup utility to allow you to install the packages that you want to download.

You can download the Android SDK from `http://developer.android.com/sdk`. It ships as a `.zip` file, similar to the way Eclipse is distributed, so you need to unzip it to an appropriate location. For Windows, unzip the file to a convenient location (we used the `C:` drive), after which you should have a folder called something like `C:\android-sdk-windows` that contains the files as shown in Figure 2–1. For Mac OS X and Linux, you can unzip the file to your home directory. Notice that Mac OS X and Linux do not have an SDK Manager executable; the equivalent of the SDK Manager in Mac OS X and Linux is to run the `tools/android` program.

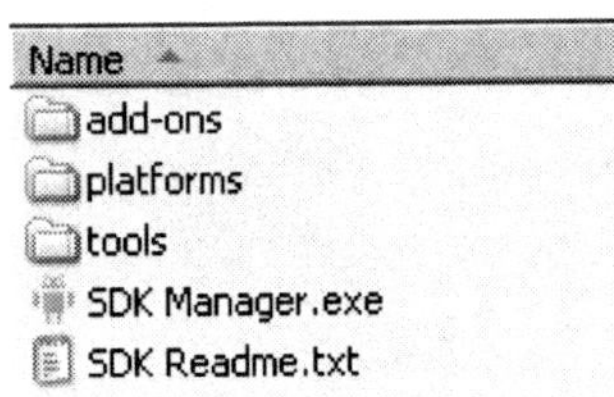

Figure 2–1. *Base contents of the Android SDK*

An alternate approach (for Windows only) is to download an installer EXE instead of the zip file and then run the installer executable. This executable checks for the Java JDK, unpacks the embedded files for you, and runs the SDK Manager program to help you set up the rest of the downloads.

Whether through using the Windows installer or by executing the SDK Manager, you should install some packages next. When you first install the Android SDK, it does not come with any platform versions (that is, versions of Android). Installing platforms is pretty easy. After you've launched the SDK Manager, you see what is installed and what's available to install, as shown in Figure 2–2. You must add Android SDK Tools and Platform-tools in order for your environment to work. Because you use it shortly, add at least the Android 1.6 SDK Platform.

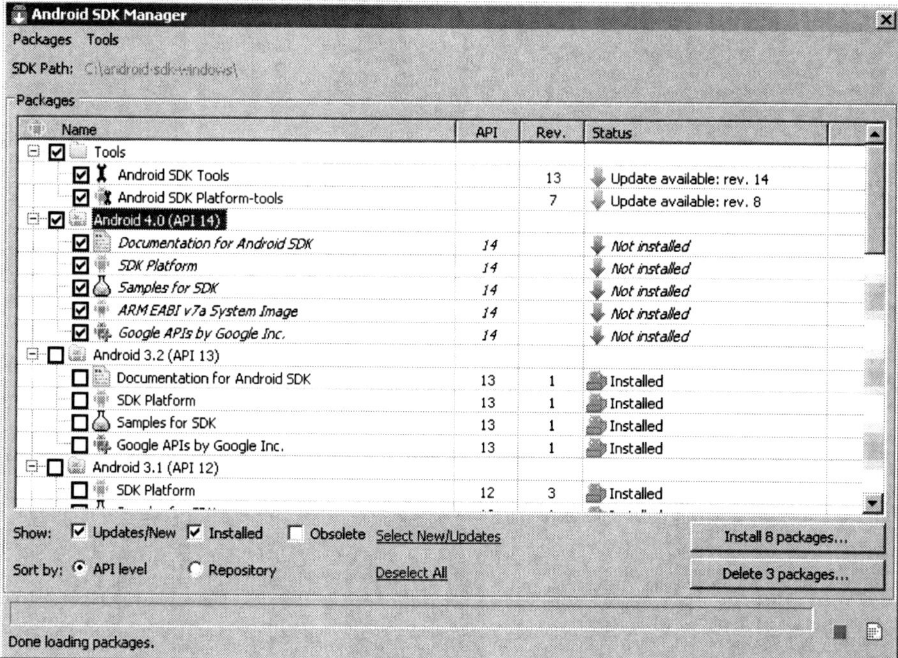

Figure 2–2. *Adding packages to the Android SDK*

Click the Install button. You need to click Accept for each item you're installing (or Accept All) and then click Install. Android then downloads your packages and platforms to make them available to you. The Google APIs are add-ons for developing applications using Google Maps. You can always come back to add more packages later.

Updating Your PATH Environment Variable

The Android SDK comes with a `tools` directory that you want to have in your PATH. You also need in your PATH the `platform-tools` directory you just installed. Let's add them now or, if you're upgrading, make sure they're correct. While you're there, you can also add a JDK bin directory, which will make life easier later.

For Windows, get back to the Environment Variables window. Edit the PATH variable and add a semicolon (;) on the end, followed by the path to the Android SDK tools folder, followed by another semicolon, followed by the path to the Android SDK platform-tools folder, following by another semicolon, and then `%JAVA_HOME%\bin`. Click OK when you're done. For Mac OS X and Linux, edit your `.bashrc` file and add the Android SDK tools directory path to your PATH variable, as well as the Android SDK `platform-tools` directory and the `$JAVA_HOME/bin` directory. Something like the following works for Linux:

```
export PATH=$PATH:$HOME/android-sdk-linux_x86/tools:$HOME/android-sdk-
linux_x86/platform-tools:$JAVA_HOME/bin
```

Just make sure that the PATH component that's pointing to the Android SDK tools directories is correct for your particular setup.

The Tools Window

Later in this book, there are times when you need to execute a command-line utility program. These programs are part of the JDK or part of the Android SDK. By having these directories in your PATH, you don't need to specify the full pathnames in order to execute them, but you need to start up a *tools window* in order to run them (later chapters refer to this tools window). The easiest way to create a tools window in Windows is to choose Start ➤ Run, type in `cmd`, and click OK. For Mac OS X, choose Terminal from your `Applications` folder in Finder or from the Dock if it's there. For Linux, choose Terminal from the Applications ➤ Accessories menu.

You may need to know the IP address of your workstation later. To find this in Windows, launch a tools window and enter the command `ipconfig`. The results contain an entry for IPv4 (or something like that) with your IP address listed next to it. An IP address looks something like this: 192.168.1.25. For Mac OS X and Linux, launch a tools window and use the command `ifconfig`. You find your IP address next to the label `inet addr`.

You may see a network connection called localhost or lo; the IP address for this network connection is 127.0.0.1. This is a special network connection used by the operating system and is not the same as your workstation's IP address. Look for a different number for your workstation's IP address.

Installing Android Development Tools (ADT)

Now you need to install ADT, an Eclipse plug-in that helps you build Android applications. Specifically, ADT integrates with Eclipse to provide facilities for you to create, test, and debug Android applications. You need to use the Install New Software facility in Eclipse to perform the installation. (The instructions for upgrading ADT appear later in this section.) To get started, launch the Eclipse IDE and follow these steps:

1. Select Help ➤ Install New Software.
2. Select the Work With field, type in

   `https://dl-ssl.google.com/android/eclipse/`

   and press Enter. Eclipse contacts the site and populates the list as shown in Figure 2–3.

3. You should see an entry named Developer Tools with four child nodes: Android DDMS, Android Development Tools, Android Hierarchy Viewer, and Android Traceview. Select the parent node Developer Tools, make sure the child nodes are also selected, and click the Next button. The versions you see may be newer than these, and that's okay. You may also see additional tools. These tools are explained further in Chapter 11.
4. Eclipse asks you to verify the tools to install. Click Next.
5. You're asked to review the licenses for ADT as well as for the tools required to install ADT. Review the licenses, click "I accept," and then click the Finish button.

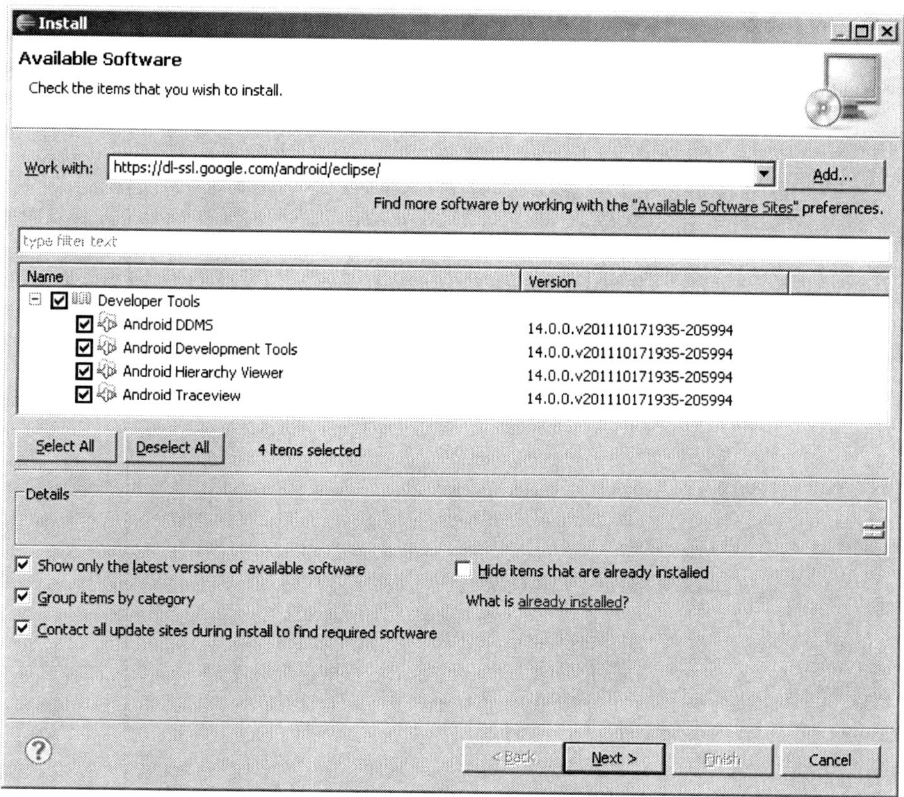

Figure 2–3. *Installing ADT using the Install New Software feature in Eclipse*

Eclipse downloads the Developer Tools and installs them. You need to restart Eclipse for the new plug-in to show up in the IDE.

If you already have an older version of ADT in Eclipse, go to the Eclipse Help menu and choose Check for Updates. You should see the new version of ADT and be able to follow the installation instructions, picking up at step 3.

> **NOTE:** If you're doing an upgrade of ADT, you may not see some of these tools in the list of tools to be upgraded. If you don't see them, then after you've upgraded the rest of the ADT, go to Install New Software and select `https://dl-ssl.google.com/android/eclipse/` from the Works With menu. The middle window should show you other tools that are available to be installed.

The final step to make ADT functional in Eclipse is to point it to the Android SDK. In Eclipse, select Window ➤ Preferences. (On Mac OS X, Preferences is under the Eclipse menu.) In the Preferences dialog box, select the Android node and set the SDK Location field to the path of the Android SDK (see Figure 2–4), and then click the Apply button. Note that you may see a dialog box asking if you want to send usage statistics to Google concerning the Android SDK; that decision is up to you.

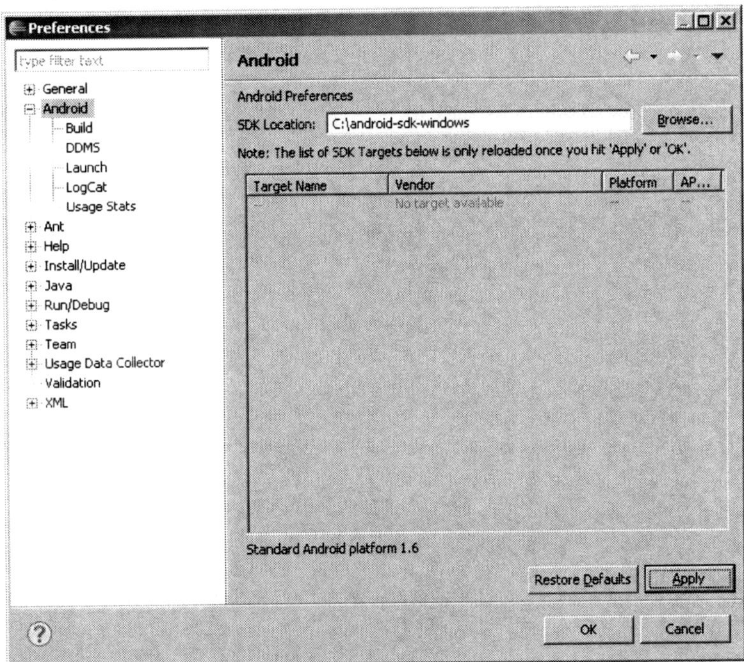

Figure 2–4. *Pointing ADT to the Android SDK*

You may want to make one more Preferences change on the Android ➤ Build page. The Skip Packaging option should be checked if you'd like to make your file saves faster. By default, the ADT readies your application for launch every time it builds it. By checking this option (see Figure 2–5), packaging and indexing occur only when truly needed.

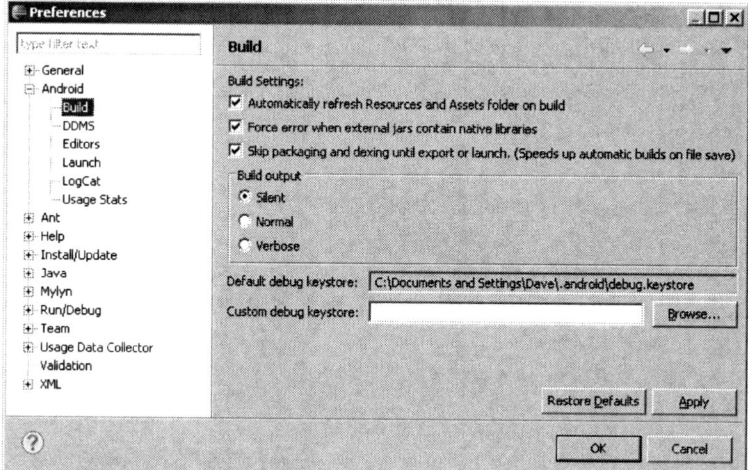

Figure 2–5. *Speeding up the builds*

From Eclipse, you can launch the SDK Manager. To do so, choose Window ➤ Android SDK Manager. You should see the same window as in Figure 2–2.

You are almost ready for your first Android application—but first, you must briefly look at the fundamental concepts of Android applications.

Learning the Fundamental Components

Every application framework has some key components that developers need to understand before they can begin to write applications based on the framework. For example, you need to understand JavaServer Pages (JSP) and servlets in order to write Java 2 Platform, Enterprise Edition (J2EE) applications. Similarly, you need to understand views, activities, fragments, intents, content providers, services, and the `AndroidManifest.xml` file when you build applications for Android. You briefly cover these fundamental concepts here and explore them in more detail throughout the book.

View

Views are user interface (UI) elements that form the basic building blocks of a user interface. A view can be a button, a label, a text field, or many other UI elements. If you're familiar with views in J2EE and Swing, then you understand views in Android. Views are also used as containers for views, which means there's usually a hierarchy of views in the UI. In the end, everything you see is a view.

Activity

An *activity* is a UI concept that usually represents a single screen in your application. It generally contains one or more views, but it doesn't have to. An activity is pretty much like it sounds—something that helps the user do one thing, which could be viewing data, creating data, or editing data. Most Android applications have several activities within them.

Fragment

When a screen is large, it becomes difficult to manage all of its functionality in a single activity. *Fragments* are like sub-activities, and an activity can display one or more fragments on the screen at the same time. When a screen is small, an activity is more likely to contain just one fragment, and that fragment can be the same one used within larger screens.

Intent

An *intent* generically defines an "intention" to do some work. Intents encapsulate several concepts, so the best approach to understanding them is to see examples of their use. You can use intents to perform the following tasks:

- Broadcast a message.
- Start a service.
- Launch an activity.
- Display a web page or a list of contacts.
- Dial a phone number or answer a phone call.

Intents are not always initiated by your application—they're also used by the system to notify your application of specific events (such as the arrival of a text message).

Intents can be explicit or implicit. If you simply say that you want to display a URL, the system decides what component will fulfill the intention. You can also provide specific information about what should handle the intention. Intents loosely couple the action and action handler.

Content Provider

Data sharing among mobile applications on a device is common. Therefore, Android defines a standard mechanism for applications to share data (such as a list of contacts) without exposing the underlying storage, structure, and implementation. Through content providers, you can expose your data and have your applications use data from other applications.

Service

Services in Android resemble services you see in Windows or other platforms—they're background processes that can potentially run for a long time. Android defines two types of services: local services and remote services. Local services are components that are only accessible by the application that is hosting the service. Conversely, remote services are services that are meant to be accessed remotely by other applications running on the device.

An example of a service is a component that is used by an e-mail application to poll for new messages. This kind of service may be a local service if the service is not used by other applications running on the device. If several applications use the service, then it's implemented as a remote service.

AndroidManifest.xml

`AndroidManifest.xml`, which is similar to the `web.xml` file in the J2EE world, defines the contents and behavior of your application. For example, it lists your application's activities and services, along with the permissions and features the application needs to run.

Android Virtual Devices

An Android Virtual Device (AVD) allows developers to test their applications without hooking up an actual Android device (typically a phone or a tablet). AVDs can be created in various configurations to emulate different types of real devices.

Hello World!

Now you're ready to build your first Android application. You start by building a simple "Hello World!" program. Create the skeleton of the application by following these steps:

1. Launch Eclipse, and select File ➤ New ➤ Project. In the New Project dialog box, select Android and then click Next. You see the New Android Project dialog box, as shown in Figure 2–6. (Eclipse may have added Android Project to the New menu, so you can use it if it's there.) There's also a New Android Project button on the toolbar.

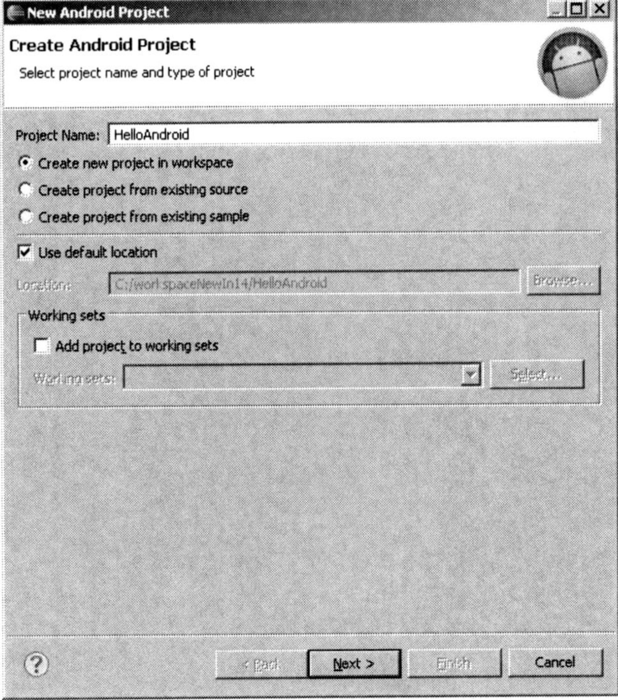

Figure 2–6. *Using the New Project Wizard to create an Android application*

2. As shown in Figure 2–6, enter **HelloAndroid** as the project name. You need to distinguish this project from other projects you create in Eclipse, so choose a name that will make sense to you when you are looking at all the projects in your Eclipse environment. Also note that the default location for the project is derived from the Eclipse workspace location. The New Project Wizard appends the name of the new application to the workspace location. In this case, if your Eclipse workspace is `c:\android`, your new project is at `c:\android\HelloAndroid\`.

3. Leave the Location section alone for now, because you want to create a new project in your workspace in the default location. Click Next.

4. The next window shows you the available Build Targets. Select Android 1.6. This is the version of Android you use as your base for the application. You can run your application on later versions of Android, such as 2.1 and 2.3.3; but Android 1.6 has all the functionality you need, so choose it as your target. In general, it's best to choose the lowest version number you can, because that maximizes the number of devices that can run your application. Click Next to move to the last wizard window.

5. Type **Hello Android** as the application name. This is the name that will appear with the application icon in your application's title bar and in application lists. It should be descriptive but not very long.

6. Use **com.androidbook.hello** as the package name. Your application must have a base package name, and this is it. This package name will be used as an identifier for your application and must be unique across all applications. For this reason, it's best to start the package name with a domain name that you own. If you don't own one, be creative to ensure that your package name won't likely be used by anyone else. However, don't use a package name that starts with com.google, com.android, android, or com.example, because these are restricted by Google and you won't be able to upload your application to Android Market.

7. Type **HelloActivity** as the Create Activity name. You're telling Android that this activity is the one to launch when your application starts up. You may have other activities in your application, but this is the first one the user should see when the application is started.

8. The Min SDK Version value of 4 tells Android that your application requires Android 1.6 or newer. Technically, you can specify a Min SDK Version that is less than the Build Target value. If your application calls for functionality that is not present in the older version of Android, you need to handle that situation gracefully, but this can be done.

9. Click the Finish button, which tells ADT to generate the project skeleton for you. For now, open the `HelloActivity.java` file under the `src` folder and modify the `onCreate()` method as follows:

```
/** Called when the activity is first created. */
    @Override
    public void onCreate(Bundle savedInstanceState) {
        super.onCreate(savedInstanceState);
        /** create a TextView and write Hello World! */
        TextView tv = new TextView(this);
        tv.setText("Hello World!");
        /** set the content view to the TextView */
        setContentView(tv);
    }
```

You probably need to add an `import android.widget.TextView;` statement to the code to get rid of the error reported by Eclipse. Save the `HelloActivity.java` file.

To run the application, you need to create an Eclipse launch configuration, and you need a virtual device on which to run it. You go quickly through these steps and come back later to more details about AVDs. Create the Eclipse launch configuration by following these steps:

1. Select Run ➤ Run Configurations.
2. In the Run Configurations dialog box, double-click Android Application in the left pane. The wizard inserts a new configuration named New Configuration.
3. Rename the configuration **RunHelloWorld**.
4. Click the Browse button, and select the HelloAndroid project.
5. Leave Launch Action set to Launch Default Activity. The dialog should appear as shown in Figure 2–7.

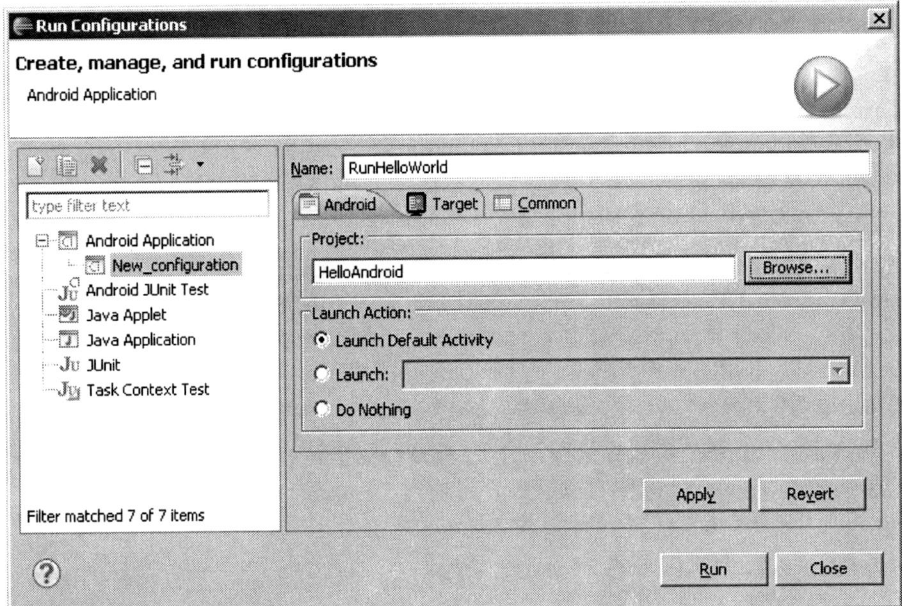

Figure 2–7. *Configuring an Eclipse launch configuration to run the "Hello World!" application*

6. Click Apply and then Run. You're almost there! Eclipse is ready to run your application, but it needs a device on which to run it. As shown in Figure 2–8, you're warned that no compatible targets were found and asked if you'd like to create one. Click Yes.

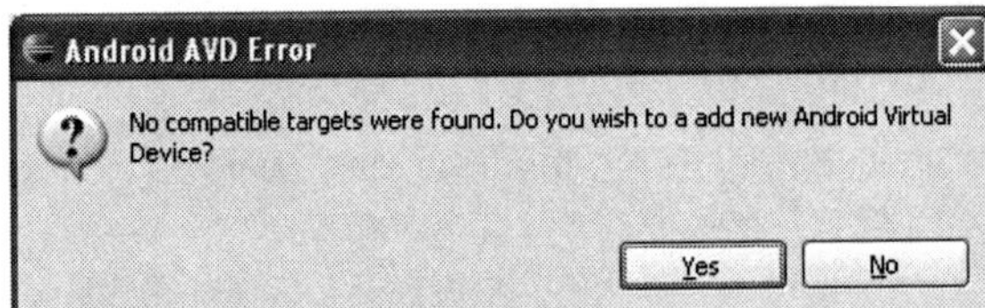

Figure 2–8. *Error message warning about targets and asking for a new AVD*

7. You're presented with a window that shows the existing AVDs (see Figure 2–9). You need to add an AVD suitable for your new application. Click the New button.

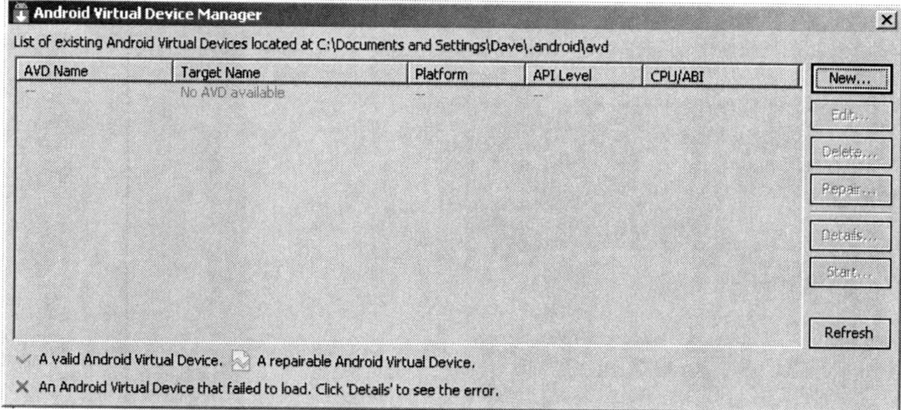

Figure 2–9. *The existing Android Virtual Devices*

8. Fill in the Create AVD form as shown in Figure 2–10. Set Name to Gingerbread, choose Android 2.3.3 - API Level 10 (or some other version) for the Target, set SD Card Size to 10 (for 10MB), enable Snapshots, and go with the default for Skin. Click Create AVD. The Manager may confirm the successful creation of your AVD. Close the AVD Manager window by clicking X in the upper-right corner.

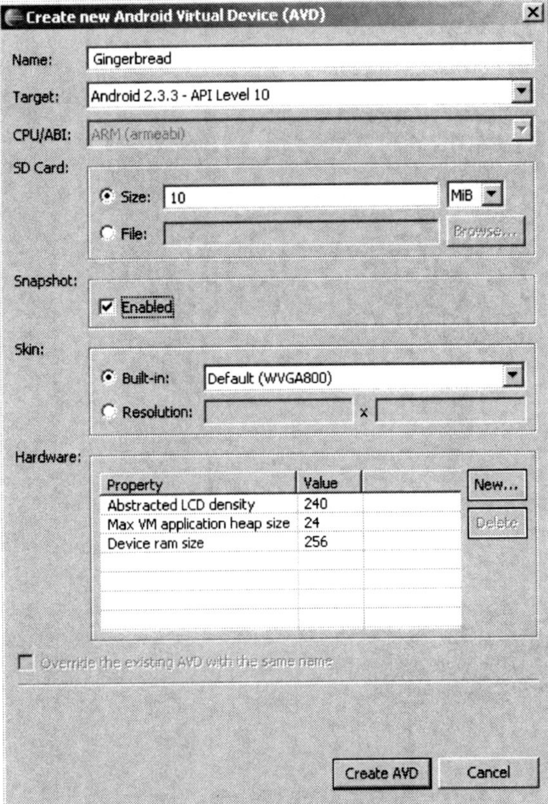

Figure 2–10. *Configuring an Android Virtual Device*

> **NOTE:** You're choosing a newer version of the SDK for your AVD, but your application can also run on an older one. This is okay because AVDs with newer SDKs can run applications that require older SDKs. The opposite, of course, is not true: an application that requires features of a newer SDK won't run on an AVD with an older SDK.

9. Select your new AVD from the bottom list. Note that you may need to click the Refresh button to make any new AVDs to show up in the list. Click the OK button.

10. Eclipse launches the emulator with your very first Android app (see Figure 2–11)!

NOTE: It may take the emulator a while to emulate the device bootup process. Once the bootup process has completed, you typically see a locked screen. Click the Menu button or drag the unlock image to unlock the AVD. After unlocking, you should see HelloAndroidApp running in the emulator, as shown in Figure 2–11. Be aware that the emulator starts other applications in the background during the startup process, so you may see a warning or error message from time to time. If you do, you can generally dismiss it to allow the emulator to go to the next step in the startup process. For example, if you run the emulator and see a message like "application abc is not responding," you can either wait for the application to start or simply ask the emulator to forcefully close the application. Generally, you should wait and let the emulator start up cleanly.

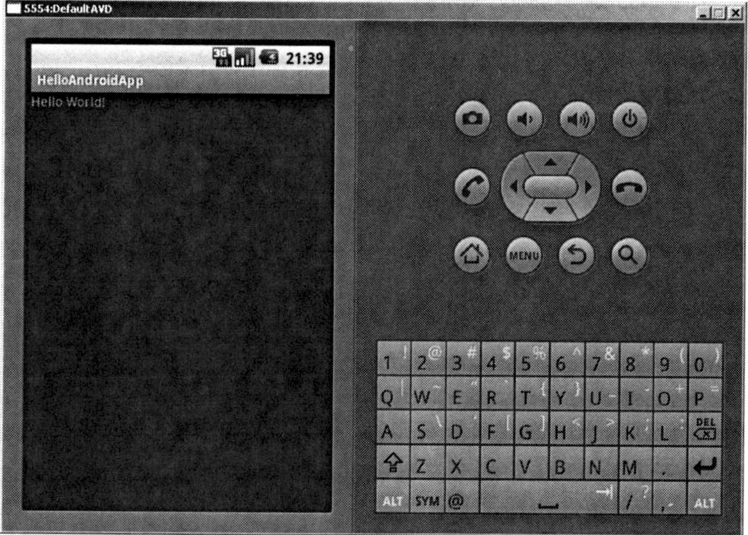

Figure 2–11. *HelloAndroidApp running in the emulator*

Now you know how to create a new Android application and run it in the emulator. Next, you look more closely at AVDs, and also how to deploy to a real device.

Android Virtual Devices

An AVD represents a device configuration. For example, you could have an AVD representing a really old Android device running version 1.5 of the SDK with a 32MB SD card. The idea is that you create AVDs you are going to support and then point the emulator to one of those AVDs when developing and testing your application. Specifying (and changing) which AVD to use is very easy and makes testing with various configurations a snap. Earlier, you saw how to create an AVD using Eclipse. You can make more AVDs in Eclipse by choosing Window ➤ AVD Manager. You can also create AVDs using the command line. Here's how.

To create an AVD, you use a batch file named `android` under the `tools` directory (`c:\android-sdk-windows\tools\`). `android` allows you to create a new AVD and manage existing AVDs. For example, you can view existing AVDs, move AVDs, and so on. You can see the options available for using `android` by running `android -help`. For now, let's just create an AVD.

By default, AVDs are stored under your home directory (all platforms) in a folder called `.android\AVD`. If you created an AVD for the "Hello World!" application you just created, then you find it here. If you want to store or manipulate AVDs somewhere else, you can do that, too. For this example, let's create a folder where the AVD image will be stored, such as `c:\avd\`. The next step is to list your available Android targets using the following command in a tools window:

```
android list target
```

The output of this command is a list of all installed Android versions, and each item in the list has an ID. Using the tools window again, type the following command (using an appropriate path to store the AVD files for your workstation, and using an appropriate value for the `-t` ID argument based on what SDK platform targets you installed):

```
android create avd -n CupcakeMaps -t 2 -c 16M -p c:\avd\CupcakeMaps\
```

The parameters passed to the batch file are listed in Table 2–1.

Table 2–1. *Parameters Passed to the `android.bat` Tool*

Argument/Command	Description
`create avd`	Tells the tool to create an AVD.
`n`	The name of the AVD.
`t`	The target runtime ID. Use the `android list target` command to get the ID for each installed target.
`c`	Size of the SD card in bytes. Use `K` for kilobytes and `M` for megabytes.
`p`	The path to the generated AVD. This is optional.
`A`	Enables snapshots. This is optional. Snapshots are explained later in the "Launching the Emulator" section.

Executing the preceding command generates an AVD; you should see output similar to what's shown in Figure 2–12. Note that when you run the `create avd` command, you are asked if you want to create a custom hardware profile. Answer no to this question for now, but know that answering yes will then prompt you to configure many options for your AVD, such as screen size, presence of a camera, and so on.

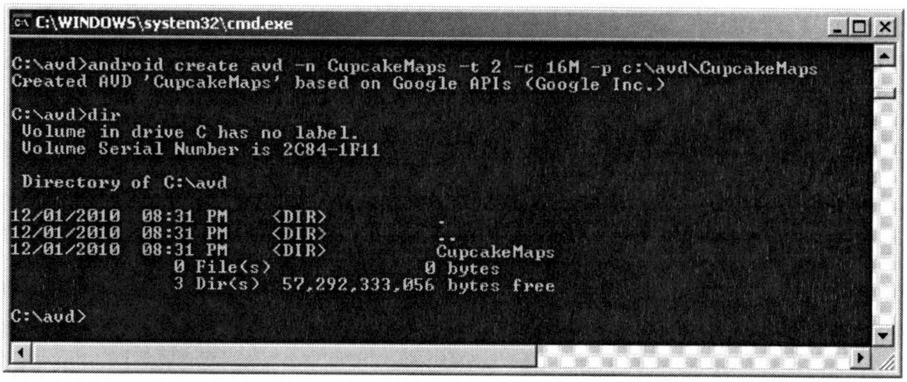

```
C:\WINDOWS\system32\cmd.exe

C:\avd>android create avd -n CupcakeMaps -t 2 -c 16M -p c:\avd\CupcakeMaps
Created AVD 'CupcakeMaps' based on Google APIs (Google Inc.)

C:\avd>dir
 Volume in drive C has no label.
 Volume Serial Number is 2C84-1F11

 Directory of C:\avd

12/01/2010  08:31 PM    <DIR>          .
12/01/2010  08:31 PM    <DIR>          ..
12/01/2010  08:31 PM    <DIR>          CupcakeMaps
               0 File(s)              0 bytes
               3 Dir(s)  57,292,333,056 bytes free

C:\avd>
```

Figure 2–12. *Creating an AVD yields this* `android.bat` *output.*

Even though you specified an alternate location for `CupcakeMaps` using the `android.bat` program, there is a `CupcakeMaps.ini` file under your home directory's `.android/AVD` folder. This is a good thing because if you go back into Eclipse and select Window ➤ AVD Manager, you see all of your AVDs. You can access any of them when running your Android applications in Eclipse.

Take another look at Figure 2–2. Each version of Android has an API level. Android 1.6 has an API level of 4, and Android 2.1 has an API level of 7. These API level numbers do not correspond to the target IDs that the `android create avd` command uses for the `-t` argument. You always have to use the `android list target` command to get the appropriate target ID value for the `android create avd` command.

Running on a Real Device

The best way to test an Android app is to run it on a real device. Any commercial Android device should work when connected to your workstation, but you may need to do a little work to set it up. If you have a Mac, you don't need to do anything except plug it in using the USB cable. Then, on the device itself, choose Settings ➤ Applications ➤ Development and enable USB debugging. On Linux, you probably need to create or modify this file: `/etc/udev/rules.d/51-android.rules`. We put a copy of this file on our web site with the project files; copy it to the proper directory, and modify the username and group values appropriately for your machine. Then, when you plug in an Android device, it will be recognized. Next, enable USB debugging on the device.

For Windows, you have to deal with USB drivers. Google supplies some with the Android packages, which are placed under the `usb_driver` subdirectory of the Android SDK directory. Other device vendors provide drivers for you, so look for them on their web sites. When you have the drivers set up, enable USB debugging on the device, and you're ready.

Now that your device is connected to your workstation, when you try to launch an app, either it launches directly on the device or (if you have an emulator running or other devices attached) a window opens in which you choose which device or emulator to launch into. If not, try editing your Run Configuration to manually select the target.

Exploring the Structure of an Android Application

Although the size and complexity of Android applications can vary greatly, their structures are similar. Figure 2–13 shows the structure of the "Hello World!" app you just built.

Figure 2–13. *The structure of the "Hello World!" application*

Android applications have some artifacts that are required and some that are optional. Table 2–2 summarizes the elements of an Android application.

Table 2–2. *The Artifacts of an Android Application*

Artifact	Description	Required?
AndroidManifest.xml	The Android application descriptor file. This file defines the activities, content providers, services, and intent receivers of the application. You can also use this file to declaratively define permissions required by the application, as well as grant specific permissions to other applications using the services of the application. Moreover, the file can contain instrumentation detail that you can use to test the application or another application.	Yes
src	A folder containing all of the source code of the application.	Yes
assets	An arbitrary collection of folders and files.	No
res	A folder containing the resources of the application. This is the parent folder of drawable, anim, layout, menu, values, xml, and raw.	Yes
drawable	A folder containing the images or image-descriptor files used by the application.	No
animator	A folder containing the XML-descriptor files that describe the animations used by the application. On older Android versions, this is called anim.	No
layout	A folder containing views of the application. You should create your application's views by using XML descriptors rather than coding them.	No
menu	A folder containing XML-descriptor files for menus in the application.	No
values	A folder containing other resources used by the application. Examples of resources found in this folder include strings, arrays, styles, and colors.	No
xml	A folder containing additional XML files used by the application.	No
raw	A folder containing additional data—possibly non-XML data—that is required by the application.	No

As you can see from Table 2–2, an Android application is primarily made up of three pieces: the application descriptor, a collection of various resources, and the application's source code. If you put aside the AndroidManifest.xml file for a moment, you can view an Android app in this simple way: you have some business logic implemented in code, and everything else is a resource. This basic structure resembles the basic structure of a J2EE app, where the resources correlate to JSPs, the business logic correlates to servlets, and the `AndroidManifest.xml` file correlates to the `web.xml` file.

You can also compare J2EE's development model to Android's development model. In J2EE, the philosophy of building views is to build them using markup language. Android has also adopted this approach, although the markup in Android is XML. You benefit from this approach because you don't have to hard-code your application's views; you can modify the look and feel of the application by editing the markup.

It is also worth noting a few constraints regarding resources. First, Android supports only a linear list of files within the predefined folders under `res`. For example, it does not support nested folders under the `layout` folder (or the other folders under `res`). Second, there are some similarities between the `assets` folder and the `raw` folder under `res`. Both folders can contain raw files, but the files in `raw` are considered resources, and the files in `assets` are not. So the files in `raw` are localized, accessible through resource IDs, and so on. But the contents of the `assets` folder are considered general-purpose content to be used without resource constraints and support. Note that because the contents of the `assets` folder are not considered resources, you can put an arbitrary hierarchy of folders and files in this folder. (Chapter 3 talks a lot more about resources.)

> **NOTE:** You may have noticed that XML is used quite heavily with Android. You know that XML is a bloated data format, so this begs the question, does it make sense to rely on XML when you know your target is a device with limited resources? It turns out that the XML you create during development is actually compiled down to binary using the Android Asset Packaging Tool (AAPT). Therefore, when your application is installed on a device, the files on the device are stored as binary. When the file is needed at runtime, the file is read in its binary form and is not transformed back into XML. This gives you the benefits of both worlds—you get to work with XML, and you don't have to worry about taking up valuable resources on the device.

Examining the Application Life Cycle

The life cycle of an Android application is strictly managed by the system, based on the user's needs, available resources, and so on. A user may want to launch a web browser, for example, but the system ultimately decides whether to start the application. Although the system is the ultimate manager, it adheres to some defined and logical guidelines to determine whether an application can be loaded, paused, or stopped. If the user is currently working with an activity, the system gives high priority to that application. Conversely, if an activity is not visible and the system determines that an application must be shut down to free up resources, it shuts down the lower-priority application.

Contrast this with the life cycle of web-based J2EE applications. J2EE apps are loosely managed by the container they run in. For example, a J2EE container can remove an application from memory if it sits idle for a predetermined time period. But the container generally doesn't move applications in and out of memory based on load and/or available resources. A J2EE container usually has sufficient resources to run lots of applications at the same time. With Android, resources are more limited, so Android must have more control and power over applications.

> **NOTE:** Android runs each application in a separate process, each of which hosts its own virtual machine. This provides a protected-memory environment. By isolating applications to an individual process, the system can control which application deserves higher priority. For example, a background process that's doing a CPU-intensive task can't block an incoming phone call.

The concept of application life cycle is logical, but a fundamental aspect of Android applications complicates matters. Specifically, the Android application architecture is component- and integration-oriented. This allows a rich user experience, seamless reuse, and easy application integration, but creates a complex task for the application life-cycle manager.

Let's consider a typical scenario. A user is talking to someone on the phone and needs to open an e-mail message to answer a question. The user goes to the home screen, opens the mail application, opens the e-mail message, clicks a link in the e-mail, and answers the friend's question by reading a stock quote from a web page. This scenario requires four applications: the home application, a talk application, an e-mail application, and a browser application. As the user navigates from one application to the next, the experience is seamless. In the background, however, the system is saving and restoring application state. For instance, when the user clicks the link in the e-mail message, the system saves metadata on the running e-mail message activity before starting the browser-application activity to launch a URL. In fact, the system saves metadata on any activity before starting another so that it can come back to the activity (when the user backtracks, for example). If memory becomes an issue, the system has to shut down a process running an activity and resume it as necessary.

Android is sensitive to the life cycle of an application and its components. Therefore, you need to understand and handle life-cycle events in order to build a stable application. The processes running your Android application and its components go through various life-cycle events, and Android provides callbacks that you can implement to handle state changes. For starters, you should become familiar with the various life-cycle callbacks for an activity (see Listing 2–1).

Listing 2–1. *Life-Cycle Methods of an Activity*

```
protected void onCreate(Bundle savedInstanceState);
protected void onStart();
protected void onRestart();
protected void onResume();
protected void onPause();
protected void onStop();
protected void onDestroy();
```

Listing 2–1 shows the list of life-cycle methods that Android calls during the life of an activity. It's important to understand when each of the methods is called by the system in order to ensure that you implement a stable application. Note that you do not need to react to all of these methods. If you do, however, be sure to call the superclass versions as well. Figure 2–14 shows the transitions between states.

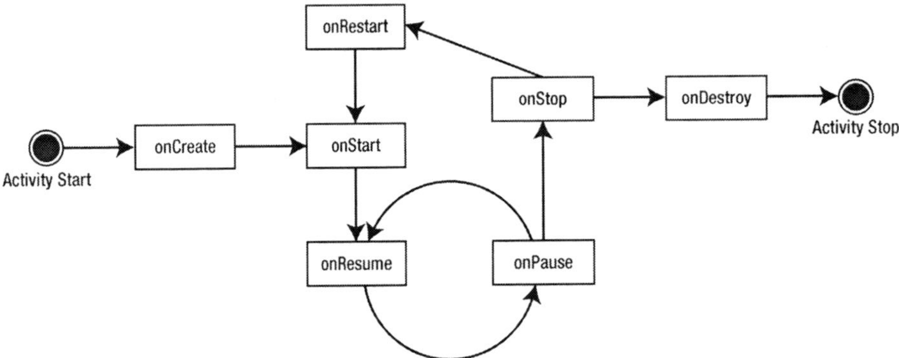

Figure 2–14. *State transitions of an activity*

The system can start and stop your activities based on what else is happening. Android calls the `onCreate()` method when the activity is freshly created. `onCreate()` is always followed by a call to `onStart()`, but `onStart()` is not always preceded by a call to `onCreate()` because `onStart()` can be called if your application was stopped. When `onStart()` is called, your activity is not visible to the user, but it's about to be. `onResume()` is called after `onStart()`, just when the activity is in the foreground and accessible to the user. At this point, the user can interact with your activity.

When the user decides to move to another activity, the system calls your activity's `onPause()` method. From `onPause()`, you can expect either `onResume()` or `onStop()` to be called. `onResume()` is called, for example, if the user brings your activity back to the foreground. `onStop()` is called if your activity becomes invisible to the user. If your activity is brought back to the foreground after a call to `onStop()`, then `onRestart()` is called. If your activity sits on the activity stack but is not visible to the user, and the system decides to kill your activity, `onDestroy()` is called.

The state model described for an activity appears complex, but you are not required to deal with every possible scenario. You mostly handle `onCreate()`, `onResume()`, and `onPause()`. You handle `onCreate()` to create the user interface for your activity. In this method, you bind data to your widgets and wire up any event handlers for your UI components. In `onPause()`, you want to persist critical data to your application's data store: it's the last safe method that is called before the system kills your application. `onStop()` and `onDestroy()` are not guaranteed to be called, so don't rely on these methods for critical logic.

The takeaway from this discussion? The system manages your application, and it can start, stop, or resume an application component at any time. Although the system controls your components, they don't run in complete isolation with respect to your application. In other words, if the system starts an activity in your application, you can count on an application context in your activity.

So far, you've covered the basics of creating a new Android app, running an Android app in the emulator, the basic structure of an Android app, and several of the common features you find in many Android apps. But you haven't seen how to resolve problems that occur in Android apps. The final section of this chapter discusses simple debugging.

Simple Debugging

The Android SDK includes a host of tools that you can use for debugging purposes. These tools are integrated with the Eclipse IDE (see Figure 2–15 for a small sample).

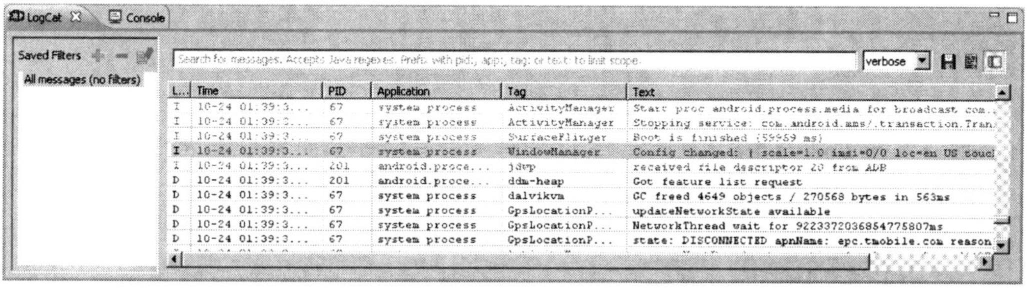

Figure 2–15. *Debugging tools that you can use while building Android applications*

One of the tools that you use throughout Android development is LogCat. This tool displays the log messages you emit using `android.util.Log`, exceptions, `System.out.println`, and so on. Although `System.out.println` works, and the messages appear in the LogCat window, to log messages from your application you should use the `android.util.Log` class. This class defines the familiar informational, warning, and error methods that you can filter in the LogCat window to see just what you want to see. Here is a sample `Log` command:

```
Log.v("string TAG", "This is my verbose message to write to the log");
```

This example shows the static `v()` method of the `Log` class, but there are others for different levels of severity. It's best to use the appropriate call level for the message you want to log, and it generally isn't a good idea to leave a verbose call in an app that you want to deploy to production. Keep in mind that logging uses memory and takes CPU resources.

What's particularly nice about LogCat is that you can view log messages when you're running your application in the emulator, but you can also view log messages when you've connected a real device to your workstation and it's in debug mode. In fact, log messages are stored such that you can even retrieve the most recent messages from a device that was disconnected when the log messages were recorded. When you connect a device to your workstation and you have the LogCat view open, you see the last several hundred messages. Chapter 11 covers more advanced debugging.

Launching the Emulator

Earlier you saw how to launch the emulator from your project in Eclipse. In most cases, you want to launch the emulator first and then deploy and test your applications in a running emulator. To launch an emulator any time, first go to the AVD Manager by running the `android` program from the tools directory of the Android SDK or from the Window menu in Eclipse. Once in the Manager, choose the desired AVD from the list, and click Start.

When you click the Start button, the Launch Options dialog opens (see Figure 2–16). This allows you to scale the size of the emulator's window and change the startup and shutdown options. When you're working with AVDs of small- to medium-screen devices, you can often use the default screen size. But for large and extra-large screen sizes, such as tablets, the default screen size may not fit nicely on the screen of your workstation. If that's the case, you can enable Scale Display to Real Size and enter a value. This label is somewhat misleading, because tablets may have a different screen density than your workstation, and the emulator doesn't perfectly match the actual physical measurement of the emulator window on your screen. For example, on our workstation screen, when emulating a Honeycomb tablet with its 10-inch screen, a "real size" of 10 inches corresponds to a scale of .64 and a screen that is a bit larger than 10 inches on the workstation screen. Pick the value that works for you based on your screen size and screen density.

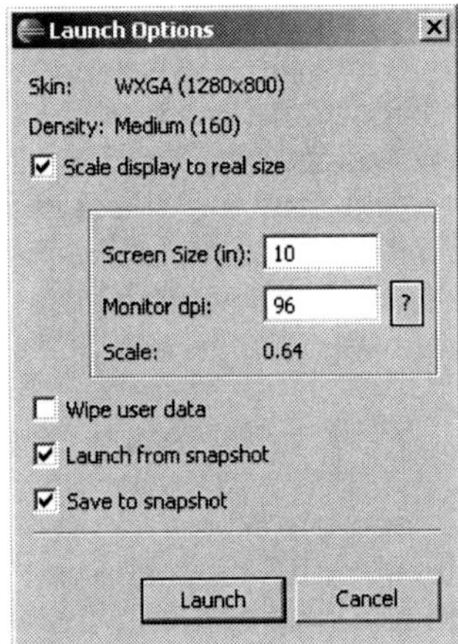

Figure 2–16. *The Launch Options dialog*

You can also work with snapshots in the Launch Options dialog. Saving to a snapshot causes a somewhat longer delay when you exit the emulator. As the name suggests, you are writing out the current state of the emulator to a snapshot image file, which can then be used the next time you launch to avoid going through an entire Android bootup sequence. Launching goes much faster if a snapshot is present, making the delay at save time well worth it—you basically pick up where you left off.

If you want to start completely fresh, you can choose Wipe User Data. You can also deselect Launch from Snapshot to keep the user data and go through the bootup sequence. Or you can create a snapshot that you like and enable *only* the Launch from Snapshot option; this reuses the snapshot over and over so your startup is fast and the shutdown is fast too, because it doesn't create a new snapshot image file every time it exits. The snapshot image file is stored in the same directory as the rest of the AVD image files. If you didn't enable snapshots when you created the AVD, you can always edit the AVD and enable them there.

References

Here are some helpful references to topics you may wish to explore further:

- `http://developer.motorola.com/docstools/`: Motorola site where you can find device add-ons as well as other tools for developing Android for Motorola handsets, including the MOTODEV Studio, which is an alternative to Eclipse.
- `http://developer.htc.com/`: HTC site for Android developers.
- `http://developer.android.com/guide/developing/tools/index.html`: Developer documentation for the Android debugging tools described previously.
- `www.droiddraw.org/`: DroidDraw site. This is a UI designer for Android applications that uses drag-and-drop to build layouts.

Summary

This chapter covered the following topics to get you set up for Android development:

- Downloading and installing the JDK, Eclipse, and the Android SDK
- How to modify your PATH variable and launch a tools window
- Installing the Android Development Tools (ADT), and how to update them if you have an older version
- Fundamental concepts of views, activities, fragments, intents, content providers, services, and the `AndroidManifest.xml` file
- Android Virtual Devices (AVDs), which can be used to test apps when you don't have a device (or the particular device you want to test with)
- Building a "Hello World!" app and deploying it to an emulator
- The basic requirements to initialize any application (project name, Android target, application name, package name, main activity, minimum SDK version)
- Where the run configurations are and how to change them

- The command-line method of creating AVDs
- Connecting a real device to your workstation and running your new apps on it
- The inner structure of an Android app, and the life cycle of an activity
- LogCat, and where to look for the internal messages from apps
- Options available when launching an emulator, such as snapshots and adjusting the screen display size

Interview Questions

Ask yourself the following questions to solidify your understanding of this chapter:

1. Does developing for Android require the JRE or the JDK?
2. Can you do Android development without using Eclipse?
3. What's the difference between the `tools` and `platform-tools` directories under the Android SDK?
4. What is a view in Android?
5. How about an intent?
6. True or false: the build target for an application must be the same as the minimum SDK version. Why or why not?
7. What precautions must you take when choosing the package name for an application?
8. What is an AVD? What do you use it for?
9. What is an AVD snapshot? How do you use it?
10. What source folder would be used to store an MP3 file that is required by an app?
11. Where do application icon files go?
12. What is the first life-cycle callback an activity sees?
13. What is the last life-cycle callback an activity sees?
14. What class is used to write log messages from your application?
15. What are all the methods that can be used to write log messages, and what's different about them?

Chapter 3

Understanding Android Resources

In Chapter 2, we gave you an overview of an Android application and a quick look at some of its underlying concepts. You also learned about the Android SDK, the Eclipse Android Development Tool (ADT), and how to run your applications on emulators identified by Android virtual devices (AVDs).

In this and the next few chapters, we follow that introduction with an in-depth look at Android SDK fundamentals. These fundamentals include resources, content providers, and intents.

Android relies on resources for defining UI components in a declarative manner. This declarative approach is not that dissimilar to the way HTML uses declarative tags to define its UI. In this sense, Android is forward thinking in its approach to UI development. Android further allows these resources to be styled and localized. This chapter covers a variety of resources that are available in Android.

Understanding Resources

Resources play a key role in Android architecture. A resource in Android is a file (like a music file or a file that describes the layout for a window) or a value (like the title of a dialog box) that is bound to an executable application. These files and values are bound to the executable in such a way that you can change them or provide alternatives without recompiling the application.

Familiar examples of resources include strings, colors, bitmaps, and layouts. Instead of hard-coding strings in an application, resources allow you to use their IDs instead. This indirection lets you change the text of the string resource without changing the source code.

There are many, many resource types in Android. Let's start this discussion of resources with a very common resource: a string.

String Resources

Android allows you to define strings in one or more XML resource files. These XML files containing string-resource definitions reside in the /res/values subdirectory. The names of the XML files are arbitrary, although you commonly see the file name as strings.xml. Listing 3–1 shows an example of a string-resource file.

Listing 3–1. *Example* strings.xml

```
<?xml version="1.0" encoding="utf-8"?>
<resources>
    <string name="hello">hello</string>
    <string name="app_name">hello appname</string>
</resources>
```

NOTE: In some releases of Eclipse, the <resources> node needs to be qualified with an xmlns specification. It doesn't seem to matter what the xmlns is pointing to as long as it is there. The following two variations of it work:

```
<resources xmlns="http://schemas.android.com/apk/res/android" >
```

and

```
<resources xmlns="default namespace" >
```

Even the first line of the file indicating that it is an XML file with a certain encoding is optional. Android works fine without that line.

When this file is created or updated, the Eclipse ADT plug-in automatically creates or updates a Java class in your application's root package called R.java with unique IDs for the two string resources specified. Notice the placement of this R.java file in the following example. This is a high-level directory structure for a project like, say, MyProject:

```
\MyProject
   \src
         \com\mycompany\android\my-root-package
         \com\mycompany\android\my-root-package\another-package
   \gen
        \com\mycompany\android\my-root-package\R.java
   \assets
   \res
   \AndroidManifest.xml
...etc
```

NOTE: Regardless of the number of resource files, there is only one R.java file.

For the string-resource file in Listing 3–1, the updated R.java file has the entries in Listing 3–2.

Listing 3–2. *Example of* `R.java`

```
package com.mycompany.android.my-root-package;
public final class R {
   ...other entries depending on your project and application

    public static final class string
   {
      ...other entries depending on your project and application

        public static final int hello=0x7f040000;
        public static final int app_name=0x7f040001;

      ...other entries depending on your project and application
    }
   ...other entries depending on your project and application
}
```

Notice, first, how `R.java` defines a top-level class in the root package: `public static final class R`. Within that outer class of `R`, Android defines an inner class, `static final class string`. `R.java` creates this inner static class as a namespace to hold string resource IDs.

The two `static final ints` defined with variable names `hello` and `app_name` are the resource IDs that represent the corresponding string resources. You can use these resource IDs anywhere in the source code through the following code structure:

```
R.string.hello
```

The generated IDs point to `ints` rather than `strings`. Most methods that take strings also take these resource identifiers as inputs. Android resolves those `ints` to `strings` where necessary.

It is merely a convention that most sample applications define all strings in one `strings.xml` file. Android takes any number of arbitrary files as long as the structure of the XML file looks like Listing 3–1 and the files reside in the `/res/values` subdirectory.

The structure of this file is easy to follow. You have the root node `<resources>` followed by one or more `<string>` child elements. Each `<string>` element or node has a property called `name` that ends up as the `id` attribute in `R.java`.

To see that multiple string resource files are allowed in this subdirectory, you can place another file with the following content in the same subdirectory and call it `strings1.xml` (see Listing 3–3).

Listing 3–3. *Example of an Additional* `strings.xml` *File*

```
<?xml version="1.0" encoding="utf-8"?>
<resources>
    <string name="hello1">hello 1</string>
    <string name="app_name1">hello appname 1</string>
</resources>
```

The Eclipse ADT plug-in validates the uniqueness of these IDs at compile time and places them in `R.java` as two additional constants: `R.string.hello1` and `R.string.app_name1`.

Layout Resources

In Android, the view for a screen is often loaded from an XML file as a resource. This is very similar to an HTML file describing the content and layout of a web page. These XML files are called layout resources. A *layout resource* is a key resource used in Android UI programming. Consider the code segment in Listing 3–4 for a sample Android activity.

Listing 3–4. *Using a Layout File*

```
public class HelloWorldActivity extends Activity
{
    @Override
    public void onCreate(Bundle savedInstanceState)
    {
        super.onCreate(savedInstanceState);
        setContentView(R.layout.main);
        TextView tv = (TextView)this.findViewById(R.id.text1);
        tv.setText("Try this text instead");
    }
    ...
}
```

The line `setContentView(R.layout.main)` points out that there is a static class called `R.layout`, and within that class, there is a constant called `main` (an integer) pointing to a `View` defined by an XML layout resource file. The name of the XML file is `main.xml`, which needs to be placed in the resources' `layout` subdirectory. In other words, this statement expects the programmer to create the file `/res/layout/main.xml` and place the necessary layout definition in that file. The contents of the `main.xml` layout file could look like Listing 3–5.

Listing 3–5. *Example `main.xml` Layout File*

```
<?xml version="1.0" encoding="utf-8"?>
<LinearLayout xmlns:android="http://schemas.android.com/apk/res/android"
    android:orientation="vertical"
    android:layout_width="fill_parent"
    android:layout_height="fill_parent"
    >
<TextView     android:id="@+id/text1"
    android:layout_width="fill_parent"
    android:layout_height="wrap_content"
    android:text="@string/hello"
    />
 <Button     android:id="@+id/b1"
    android:layout_width="fill_parent"
    android:layout_height="wrap_content"
    android:text="@string/hello"
   />
</LinearLayout>
```

The layout file in Listing 3–5 defines a root node called `LinearLayout`, which contains a `TextView` followed by a `Button`. A `LinearLayout` lays out its children vertically or horizontally—vertically, in this example.

You need to define a separate layout file for each screen (or activity). More accurately, each layout needs a dedicated file. If you are painting two screens, you probably need two layout files, such as /res/layout/screen1_layout.xml and /res/layout/screen2_layout.xml.

NOTE: Each file in the /res/layout/ subdirectory generates a unique constant based on the name of the file (extension excluded). With layouts, what matters is the number of files; with string resources, what matters is the number of individual string resources *inside* the files.

For example, if you have two files under /res/layout/ called file1.xml and file2.xml, then in R.java you have the entries shown in Listing 3–6.

Listing 3–6. *Multiple Constants for Multiple Layout Files*

```
public static final class layout {
  .... any other files
    public static final int file1=0x7f030000;
    public static final int file2=0x7f030001;
}
```

The views defined in these layout files, such as a TextView (see Listing 3–5), are accessible in Java code through their resource IDs generated in R.java:

```
TextView tv = (TextView)this.findViewById(R.id.text1);
tv.setText("Try this text instead");
```

In this example, you locate the TextView by using the findViewById method of the Activity class. The constant R.id.text1 corresponds to the ID defined for the TextView. The ID for the TextView in the layout file is as follows:

```
<TextView android:id="@+id/text1"
..
</TextView>
```

The value for the id attribute indicates that a constant called text1 is used to uniquely identify this view among other views hosted by that activity. The plus sign (+) in @+id/text1 means the ID text1 will be created if it doesn't exist already. There is more to this resource ID syntax, as you see next.

Resource Reference Syntax

Regardless of the type of resource (string and layout are the two covered so far), all Android resources are identified (or referenced) by their IDs in Java source code. The syntax you use to allocate an ID to a resource in the XML file is called *resource-reference syntax*. This syntax is not limited to allocating just ids: it is a way to identify any resource such as a string, a layout file, or an image.

How is this general-purpose means of locating a resource or referencing an existing resource tied to IDs? As it turns out, IDs are numbers that are tracked as resources much like strings. Imagine your project holding a bucket of numbers. You can take one of those numbers and allocate it to a control.

Let's first investigate this resource-reference structure a bit further. This resource reference has the following formal structure:

```
@[package:]type/name
```

The type corresponds to one of the resource-type namespaces available in R.java, some of which follow:

- R.drawable
- R.id
- R.layout
- R.string
- R.attr
- R.plural
- R.array

The corresponding types in XML resource-reference syntax are as follows:

- drawable
- id
- layout
- string
- attr
- plurals
- string-array

The name part in the resource reference @[package:]type/name is the name given to the resource (for example, text1 in Listing 3–5); it also gets represented as an int constant in R.java.

If you don't specify any package in the syntax @[package:]type/name, the pair type/name is resolved based on local resources and the application's local R.java package.

If you specify android:type/name, the reference is resolved using the package android and specifically through the android.R.java file. You can use any Java package name in place of the package placeholder to locate the correct R.java file to resolve the reference.

Let's come back now to the way an ID is allocated to a control in Listing 3–5. IDs are considered resources. By this logic, when you write

```
<TextView android:id="@+id/text1" ..../>
```

you are saying, "Tke a resource identified by type id and a value of text1 and allocate it to this instance of TextView." The + indicates that if the id of text1 is not defined as a resource, go ahead and define it with a unique number.

Based on this information, let's analyze a few ID examples. As you go through Listing 3–7, note that the left side of the ID android:id is not part of the syntax. android:id is just how you allocate an ID to a control like TextView.

Listing 3–7. *Exploring Resource Reference Syntax*

```
<TextView  android:id="text">
// Compile error, as id will not take raw text strings.
// More over it is not a valid resource ref syntax.

<TextView  android:id="@text">
// wrong syntax. @text is missing a type name.
// it should have been @id/text or @+id/text or @string/string1.
// However string type here is invalid although it is a valid
// resource reference. This is because the left hand side
// needs an "id" and not a "string"
// you will get an error "No Resource type specified

<TextView  android:id="@id/text">
//Error: No Resource found that matches id "text"
//Unless you have taken care to define "text" as an ID before

<TextView  android:id="@android:id/text">
// Error: Resource is not public
// indicating that there is no such id in android.R.id
// Of course this would be valid if Android R.java were to define
// an id with this name

<TextView  android:id="@+id/text">
//Success: Creates an id called "text" in the local package's R.java
```

In the syntax "@+id/text", the + sign has a special meaning. It tells Android that the ID text may not already exist and, if that's the case, to create a new one and name it text. We are not aware of a place where + is used in resource-reference syntax other than in the context of an ID. This makes sense because you cannot assume a case where a string resource is created with out explicitly specifying what it is. The system cannot create on its own as it does a unique number.

This connection between an ID and the resource-reference syntax is often a source of confusion. To resolve this confusion, remember one thing: an ID is specified as a resource.

Defining Your Own Resource IDs for Later Use

The general pattern for allocating an ID is either to create a new one or to use the one created by the Android package. However, it is possible to create IDs beforehand and use them later in your own packages. Again, this stems from the fact that IDs are resources. If they are resources, they should be allowed to be predefined and made available for later use.

The line `<TextView android:id="@+id/text">` in Listing 3–7 indicates that an ID named `text` is used if it already exists. If the ID doesn't exist, a new one is created. So when might an ID such as `text` already exist in `R.java`, to be reused?

You might be inclined to put a constant like `R.id.text` in `R.java`, but `R.java` is not editable. Even if it were, it gets regenerated every time something is changed, added, or deleted in the `/res/*` subdirectory.

The solution is to use a resource tag called `item` to define an ID without attaching to any particular resource. Listing 3–8 shows an example.

Listing 3–8. *Predefining an ID*

```
<resources>
<item type="id" name="text"/>
</resources>
```

The `type` refers to the type of resource—`id` in this case. Once this ID is in place, the `View` definition in Listing 3–9 will work.

Listing 3–9. *Reusing a Predefined ID*

```
<TextView android:id="@id/text">
..
</TextView>
```

Compiled and Uncompiled Android Resources

Android supports resources primarily through two types of files: XML files and raw files (examples of which include images, audio, and video). You have seen that in some cases, resources are defined as values inside an XML file (strings, for example), and sometimes an XML file as a whole is a resource (a layout resource file to quote).

As a further distinction within the set of XML files, you find two types: one gets compiled into binary format, and the other is copied as-is to the device. The examples you have seen so far—string resource XML files and layout resource XML files—are compiled into binary format before becoming part of the installable package. These XML files have predefined formats where XML nodes can be translated to IDs.

You can also choose some XML files to have their own free format structure; these are not interpreted but have resource IDs generated (`resource type: xml`). However, you do want them compiled to binary formats and also have the comfort of localization. To do this, you can place these XML files in the `/res/xml/` subdirectory to have them compiled into binary format. In this case, you would use Android-supplied XML readers to read the XML nodes.

But if you place files, including XML files, in the `/res/raw/` directory instead, they don't get compiled into binary format. However, because it's a resource, Android generates an ID through `R.java`. The resource type for raw files is `raw`. So, you can access these file identities through `R.raw.`*`some-filename-minus-extension`*. You must use explicit stream-based APIs to read these files. Audio and video files fall into this category.

NOTE: Because the raw directory is part of the /res/* hierarchy, even these raw audio and video files can take advantage of localization and ID generation like all other resources.

As we mentioned in Table 2-1 in the previous chapter, resource files are housed in various subdirectories based on their type. Here are some important subdirectories in the /res folder and the types of resources they host:

- anim: Compiled animation files
- drawable: Bitmaps
- layout: UI and view definitions
- values: Arrays, colors, dimensions, strings, and styles
- xml: Compiled arbitrary XML files
- raw: Noncompiled raw files

The resource compiler in the Android Asset Packaging Tool (AAPT) compiles all the resources except the raw resources and places them into the final .apk file. This file, which contains the Android application's code and resources, correlates to Java's .jar file (*apk* stands for *Android package)*. The .apk file is what gets installed onto the device.

NOTE: Although the XML resource parser allows resource names such as hello-string, you will see a compile-time error in R.java. You can fix this by renaming your resource hello_string (replacing the dash with an underscore).

Enumerating Key Android Resources

Now that you've been through the basics of resources, let's enumerate some of the other key resources that Android supports, their XML representations, and the way they're used in Java code. (You can use this section as a quick reference as you write resource files for each resource.) To begin, take a quick glance at the types of resources and what they are used for in Table 3–1.

Table 3–1. *Types of Resources*

Resource Type	Location	Description
Colors	`/res/values/any-file`	Represents color identifiers pointing to color codes. These resource IDs are exposed in `R.java` as `R.color.*`. The XML node in the file is `/resources/color`.
Strings	`/res/values/any-file`	Represents string resources. String resources allow Java-formatted strings and raw HTML in addition to simple strings. These resource IDs are exposed in `R.java` as `R.string.*`. The XML node in the file is `/resources/string`.
String arrays	`/res/values/any-file`	Represents a resource that is an array of strings. These resource IDs are exposed in `R.java` as `R.array.*`. The XML node in the file is `/resources/string-array`.
Plurals	`/res/values/any-file`	Represents a suitable collection of strings based on the value of a quantity. The quantity is a number. In various languages, the way you write a sentence depends on whether you refer to no objects, one object, few objects, or many objects. The resource IDs are exposed in `R.java` as `R.plural.*`. The XML node in the value file is `/resources/plurals`.
Dimensions	`/res/values/any-file`	Represents dimensions or sizes of various elements or views in Android. Supports pixels, inches, millimeters, density independent pixels, and scale independent pixels. These resource IDs are exposed in `R.java` as `R.dimen.*` . The XML node in the file is `/resources/dimen`.
Images	`/res/drawable/multiple-files`	Represents image resources. Supported images include `.jpg`, `.gif`, `.png`, and so on. Each image is in a separate file and gets its own ID based on the file name. These resource ids are exposed in `R.java` as `R.drawable.*`. The image support also includes an image type called a *stretchable image* that allows portions of an image to stretch while other portions of that image stay static. The stretchable image is also known as a *9-patch file* (`.9.png`).

Resource Type	Location	Description
Color drawables	/res/values/any-file also /res/drawable/multiple-files	Represents rectangles of colors to be used as view backgrounds or general drawables like bitmaps. This can be used in lieu of specifying a single-colored bitmap as a background. In Java, this is equivalent to creating a colored rectangle and setting it as a background for a view. The <drawable> value tag in the values subdirectory supports this. These resource IDs are exposed in R.java as R.drawable.*. The XML node in the file is /resources/drawable. Android also supports rounded rectangles and gradient rectangles through XML files placed in /res/drawable with the root XML tag of <shape>. These resource IDs are also exposed in R.java as R.drawable.*. Each file name in this case translates to a unique drawable ID.
Arbitrary XML files	/res/xml/*.xml	Android allows arbitrary XML files as resources. These files are compiled by the AAPT compiler. These resource IDs are exposed in R.java as R.xml.*.
Arbitrary raw resources	/res/raw/*.*	Android allows arbitrary *noncompiled* binary or text files under this directory. Each file gets a unique resource ID. These resource IDs are exposed in R.java as R.raw.* .
Arbitrary raw assets	/assets/*.*/*.*	Android allows arbitrary files in arbitrary subdirectories starting at the /assets subdirectory. These are not really resources, just raw files. This directory, unlike the /res resources subdirectory, allows an arbitrary depth of subdirectories. These files do not generate any resource IDs. You have to use a relative pathname starting at and excluding /assets.

Each of the resources specified in this table is further elaborated in the following sections with XML and Java code snippets.

NOTE: Looking at the nature of ID generation, it appears—although we haven't seen it officially stated anywhere—that there are IDs generated based on file names if those XML files are anywhere but in the /res/values subdirectory. If they are in the values subdirectory, only the contents of the files are looked at to generate the IDs.

String Arrays

You can specify an array of strings as a resource in any file under the /res/values subdirectory. To do so, you use an XML node called string-array. This node is a child node of resources just like the string resource node. Listing 3–10 is an example of specifying an array in a resource file.

Listing 3–10. *Specifying String Arrays*

```
<resources ....>
......Other resources
<string-array name="test_array">
    <item>one</item>
    <item>two</item>
    <item>three</item>
</string-array>
......Other resources
</resources>
```

Once you have this string-array resource definition, you can retrieve this array in the Java code as shown in Listing 3–11.

Listing 3–11. *Specifying String Arrays*

```
//Get access to Resources object from an Activity
Resources res = your-activity.getResources();
String strings[] = res.getStringArray(R.array.test_array);

//Print strings
for (String s: strings)
{
    Log.d("example", s);
}
```

Plurals

The resource plurals is a set of strings. These strings are various ways of expressing a numerical quantity, such as how many eggs are in a nest. Consider an example:

```
There is 1 egg.
There are 2 eggs.
There are 0 eggs.
There are 100 eggs.
```

Notice how the sentences are identical for the numbers 2, 0, and 100. However, the sentence for 1 egg is different. Android allows you to represent this variation as a plurals resource. Listing 3–12 shows how you would represent these two variations based on quantity in a resource file.

Listing 3–12. *Specifying String Arrays*

```
<resources...>
<plurals name="eggs_in_a_nest_text">
    <item quantity="one">There is 1 egg</item>
    <item quantity="other">There are %d eggs</item>
</plurals>
</resources>
```

The two variations are represented as two different strings under one plural. Now you can use the Java code in Listing 3–13 to use this plural resource to print a string given a quantity. The first parameter to the `getQuantityString()` method is the `plurals` resource ID. The second parameter selects the string to be used. When the value of the quantity is 1, you use the string as is. When the value is not 1, you must supply a third parameter whose value is to be placed where `%d` is. You must always have at least three parameters if you use a formatting string in your `plurals` resource. The second parameter can be confusing; the only distinction in this parameter is whether its value is 1 or other than 1.

Listing 3–13. *Specifying String Arrays*

```
Resources res = your-activity.getResources();
String s1 = res.getQuantityString(R.plurals.eggs_in_a_nest_text, 0,0);
String s2 = res.getQuantityString(R.plurals.eggs_in_a_nest_text, 1,1);
String s3 = res.getQuantityString(R.plurals.eggs_in_a_nest_text, 2,2);
String s4 = res.getQuantityString(R.plurals.eggs_in_a_nest_text, 10,10);
```

Given this code, each quantity results in an appropriate string that is suitable for its plurality.

However, what other possibilities exist for the `quantity` attribute of the preceding `item` node? We strongly recommend that you read the source code of `Resources.java` and `PluralRules.java` in the Android source code distribution to truly understand this. Our research link in "Resources" at the end of this chapter has extracts from these source files.

The bottom line is that, for the `en` (English) locale, the only two possible values are `"one"` and `"other"`. This is true for all other languages as well, except for `cs` (Czech), in which case the values are `"one"` (for 1), `"few"` (for 2 to 4), and `"other"` for the rest.

More on String Resources

You covered string resources briefly in earlier sections. Let's revisit them to provide additional nuances, including HTML strings and how to substitute variables in string resources.

> **NOTE:** Most UI frameworks allow string resources. However, unlike other UI frameworks, Android offers the ability to quickly associate IDs with string resources through `R.java`, so using strings as resources is that much easier in Android.

You start by seeing how to define normal strings, quoted strings, HTML strings, and substitutable strings in an XML resource file (see Listing 3–14).

Listing 3–14. *XML Syntax for Defining String Resources*

```
<resources>
    <string name="simple_string">simple string</string>
    <string name="quoted_string">"quoted 'xyz' string"</string>
    <string name="double_quoted_string">\"double quotes\"</string>
    <string name="java_format_string">
           hello %2$s Java format string. %1$s again
     </string>
     <string name="tagged_string">
         Hello <b><i>Slanted Android</i></b>, You are bold.
     </string>
</resources>
```

This XML string resource file needs to be in the /res/values subdirectory. The name of the file is arbitrary.

Notice that quoted strings need to be either escaped or placed in alternate quotes. The string definitions also allow standard Java string-formatting sequences.

Android also allows child XML elements such as <b>, <i>, and other simple text-formatting HTML within the <string> node. You can use this compound HTML string to style the text before painting in a text view.

The Java examples in Listing 3–15 illustrate each usage.

Listing 3–15. *Using String Resources in Java Code*

```
//Read a simple string and set it in a text view
String simpleString = activity.getString(R.string.simple_string);
textView.setText(simpleString);

//Read a quoted string and set it in a text view
String quotedString = activity.getString(R.string.quoted_string);
textView.setText(quotedString);

//Read a double quoted string and set it in a text view
String doubleQuotedString = activity.getString(R.string.double_quoted_string);
textView.setText(doubleQuotedString);

//Read a Java format string
String javaFormatString = activity.getString(R.string.java_format_string);
//Convert the formatted string by passing in arguments
String substitutedString = String.format(javaFormatString, "Hello" , "Android");
//set the output in a text view
textView.setText(substitutedString);

//Read an html string from the resource and set it in a text view
String htmlTaggedString = activity.getString(R.string.tagged_string);
//Convert it to a text span so that it can be set in a text view
//android.text.Html class allows painting of "html" strings
//This is strictly an Android class and does not support all html tags
Spanned textSpan = android.text.Html.fromHtml(htmlTaggedString);
//Set it in a text view
textView.setText(textSpan);
```

Once you've defined the strings as resources, you can set them directly on a view such as a TextView in the XML layout definition for that TextView. Listing 3–16 shows an example where an HTML string is set as the text content of a TextView.

Listing 3–16. *Using String Resources in XML*

```
<TextView android:layout_width="fill_parent"
          android:layout_height="wrap_content"
          android:gravity="center_horizontal"
          android:text="@string/tagged_string"/>
```

The TextView automatically realizes that this string is an HTML string and honors its formatting accordingly, which is nice because you can quickly set attractive text in your views as part of the layout.

Color Resources

As you can with string resources, you can use reference identifiers to indirectly reference colors. Doing this enables Android to localize colors and apply themes. Once you've defined and identified colors in resource files, you can access them in Java code through their IDs. Whereas string-resource IDs are available under the *<your-package>*.R.string namespace, the color IDs are available under the *<your-package>*.R.color namespace.

Android also defines a base set of colors in its own resource files. These IDs, by extension, are accessible through the Android android.R.color namespace. Check out this URL to learn the color constants available in the android.R.color namespace:

http://code.google.com/android/reference/android/R.color.html

Listing 3–17 has some examples of specifying color in an XML resource file.

Listing 3–17. *XML Syntax for Defining Color Resources*

```
<resources>
    <color name="red">#f00</color>
    <color name="blue">#0000ff</color>
    <color name="green">#f0f0</color>
    <color name="main_back_ground_color">#ffffff00</color>
</resources>
```

The entries in Listing 3–17 need to be in a file residing in the /res/values subdirectory. The name of the file is arbitrary, meaning the file name can be anything you choose. Android reads all the files and then processes them and looks for individual nodes such as resources and color to figure out individual IDs.

Listing 3–18 shows an example of using a color resource in Java code.

Listing 3–18. *Color Resources in Java code*

```
int mainBackGroundColor
     = activity.getResources.getColor(R.color.main_back_ground_color);
```

Listing 3–19 shows how you can use a color resource in a view definition.

Listing 3–19. *Using Colors in View Definitions*

```
<TextView android:layout_width="fill_parent"
          android:layout_height="wrap_content"
          android:textColor="@color/red"
          android:text="Sample Text to Show Red Color"/>
```

Dimension Resources

Pixels, inches, and points are all examples of dimensions that can play a part in XML layouts or Java code. You can use these dimension resources to style and localize Android UIs without changing the source code.

Listing 3–20 shows how you can use dimension resources in XML.

Listing 3–20. *XML Syntax for Defining Dimension Resources*

```
<resources>
    <dimen name="mysize_in_pixels">1px</dimen>
    <dimen name="mysize_in_dp">5dp</dimen>
    <dimen name="medium_size">100sp</dimen>
</resources>
```

You can specify the dimensions in any of the following units:

- px: Pixels
- in: Inches
- mm: Millimeters
- pt: Points
- dp: Density-independent pixels based on a 160dpi (pixel density per inch) screen (dimensions adjust to screen density)
- sp: Scale-independent pixels (dimensions that allow for user sizing; helpful for use in fonts)

In Java, you need to access your Resources object instance to retrieve a dimension. You can do this by calling getResources on an activity object (see Listing 3–21). Once you have the Resources object, you can ask it to locate the dimension using the dimension ID (again, see Listing 3–21).

Listing 3–21. *Using Dimension Resources in Java Code*

```
float dimen = activity.getResources().getDimension(R.dimen.mysize_in_pixels);
```

> **NOTE:** The Java method call uses Dimension (full word) whereas the R.java namespace uses the shortened version dimen to represent *dimension*.

As in Java, the resource reference for a dimension in XML uses dimen as opposed to the full word *dimension* (see Listing 3–22).

Listing 3–22. *Using Dimension Resources in XML*

```
<TextView android:layout_width="fill_parent"
          android:layout_height="wrap_content"
          android:textSize="@dimen/medium_size"/>
```

Image Resources

Android generates resource IDs for image files placed in the /res/drawable subdirectory. The supported image types include .gif, .jpg, and .png. Each image file in this directory generates a unique ID from its base file name. If the image file name is sample_image.jpg, for example, then the resource ID generated is R.drawable.sample_image.

> **CAUTION:** You get an error if you have two file names with the same base file name. Also, subdirectories underneath /res/drawable are ignored. Any files placed under those subdirectories aren't read.

You can reference the images available in /res/drawable in other XML layout definitions, as shown in Listing 3–23.

Listing 3–23. *Using Image Resources in XML*

```
<Button
      android:id="@+id/button1"
      android:layout_width="fill_parent"
      android:layout_height="wrap_content"
      android:text="Dial"
      android:background="@drawable/sample_image"
/>
```

You can also retrieve an image programmatically using Java and set it yourself against a UI object like a button (see Listing 3–24).

Listing 3–24. *Using Image Resources in Java*

```
//Call getDrawable to get the image
BitmapDrawable d = activity.getResources().getDrawable(R.drawable.sample_image);

//You can use the drawable then to set the background
button.setBackgroundDrawable(d);

//or you can set the background directly from the Resource Id
button.setBackgroundResource(R.drawable.sample_image);
```

> **NOTE:** These background methods go all the way back to the View class. As a result, most of the UI controls have this background support.

Android also supports a special type of image called a *stretchable* image. This is a kind of .png where parts of the image can be specified as static and stretchable. Android provides a tool called the Draw 9-patch tool to specify these regions (you can read more about it at http://developer.android.com/guide/developing/tools/draw9patch.html).

Once the .png image is made available, you can use it like any other image. It comes in handy when used as a background for a button where the button has to stretch itself to accommodate the text.

Color-Drawable Resources

In Android, an image is one type of a drawable resource. Android supports another drawable resource called a *color-drawable resource*; it's essentially a colored rectangle.

> **CAUTION:** The Android documentation seems to suggest that rounded corners are possible, but we have not been successful in creating those. We have presented an alternate approach instead. The documentation also suggests that the instantiated Java class is PaintDrawable, but the code returns a ColorDrawable.

To define one of these color rectangles, you define an XML element by the node name of drawable in any XML file in the /res/values subdirectory. Listing 3–25 shows a couple of color-drawable resource examples.

Listing 3–25. *XML Syntax for Defining Color-Drawable Resources*

```
<resources>
    <drawable name="red_rectangle">#f00</drawable>
    <drawable name="blue_rectangle">#0000ff</drawable>
    <drawable name="green_rectangle">#f0f0</drawable>
</resources>
```

Listings 3–26 and 3–27 show how you can use a color-drawable resource in Java and XML, respectively.

Listing 3–26. *Using Color-Drawable Resources in Java Code*

```
// Get a drawable
ColorDrawable redDrawable = (ColorDrawable)
     activity.getResources().getDrawable(R.drawable.red_rectangle);

//Set it as a background to a text view
textView.setBackgroundDrawable(redDrawable);
```

Listing 3–27. *Using Color-Drawable Resources in XML Code*

```
<TextView android:layout_width="fill_parent"
          android:layout_height="wrap_content"
          android:textAlign="center"
          android:background="@drawable/red_rectangle"/>
```

To achieve the rounded corners in your Drawable, you can use the currently undocumented <shape> tag. However, this tag needs to reside in a file by itself in the /res/drawable directory. Listing 3–28 shows how you can use the <shape> tag to define a rounded rectangle in a file called /res/drawable/my_rounded_rectangle.xml.

Listing 3–28. *Defining a Rounded Rectangle*

```
<shape xmlns:android="http://schemas.android.com/apk/res/android">
    <solid android:color="#f0600000"/>
    <stroke android:width="3dp" color="#ffff8080"/>
    <corners android:radius="13dp" />
    <padding android:left="10dp" android:top="10dp"
        android:right="10dp" android:bottom="10dp" />
</shape>
```

You can then use this drawable resource as a background of the previous text view example, as shown in Listing 3–29.

Listing 3–29. *Using a Drawable from Java Code*

```
// Get a drawable
GradientDrawable roundedRectangle =
(GradientDrawable)
activity.getResources().getDrawable(R.drawable.my_rounded_rectangle);

//Set it as a background to a text view
textView.setBackgroundDrawable(roundedRectangle);
```

NOTE: It is not necessary to cast the returned base `Drawable` to a `GradientDrawable`, but it was done to show you that this `<shape>` tag becomes a `GradientDrawable`. This information is important because you can look up the Java API documentation for this class to know the XML tags it defines.

In the end, a bitmap image in the `drawable` subdirectory resolves to a `BitmapDrawable` class. A drawable resource value, such as one of the rectangles in Listing 3–29, resolves to a `ColorDrawable`. An XML file with a `<shape>` tag in it resolves to a `GradientDrawable`.

Working with Arbitrary XML Resource Files

In addition to the structured resources described so far, Android allows arbitrary XML files as resources. This approach extends the advantages of using resources to arbitrary XML files. This approach provides a quick way to reference these files based on their generated resource IDs. Second, the approach allows you to localize these resource XML files. Third, you can compile and store these XML files on the device efficiently.

XML files that need to be read in this fashion are stored under the `/res/xml` subdirectory. Listing 3–30 is an example XML file called `/res/xml/test.xml`.

Listing 3–30. *Example XML File*

```
<rootelem1>
   <subelem1>
      Hello World from an xml sub element
   </subelem1>
</rootelem1>
```

As it does with other Android XML resource files, the AAPT compiles this XML file before placing it in the application package. You need to use an instance of XmlPullParser if you want to parse these files. You can get an instance of the XmlPullParser implementation using the code in Listing 3–31 from any context (including activity).

Listing 3–31. *Reading an XML File*

```
Resources res = activity.getResources();
XmlResourceParser xpp = res.getXml(R.xml.test);
```

The returned XmlResourceParser is an instance of XmlPullParser, and it also implements java.util.AttributeSet. Listing 3–32 shows a more complete code snippet that reads the test.xml file.

Listing 3–32. *Using XmlPullParser*

```
private String getEventsFromAnXMLFile(Activity activity)
throws XmlPullParserException, IOException
{
   StringBuffer sb = new StringBuffer();
   Resources res = activity.getResources();
   XmlResourceParser xpp = res.getXml(R.xml.test);

   xpp.next();
   int eventType = xpp.getEventType();
    while (eventType != XmlPullParser.END_DOCUMENT)
    {
        if(eventType == XmlPullParser.START_DOCUMENT)
        {
           sb.append("******Start document");
        }
        else if(eventType == XmlPullParser.START_TAG)
        {
           sb.append("\nStart tag "+xpp.getName());
        }
        else if(eventType == XmlPullParser.END_TAG)
        {
           sb.append("\nEnd tag "+xpp.getName());
        }
        else if(eventType == XmlPullParser.TEXT)
        {
           sb.append("\nText "+xpp.getText());
        }
        eventType = xpp.next();
    }//eof-while
    sb.append("\n******End document");
    return sb.toString();
}//eof-function
```

In Listing 3–32, you can see how to get XmlPullParser, how to use XmlPullParser to navigate the XML elements in the XML document, and how to use additional methods of XmlPullParser to access the details of the XML elements. If you want to run this code, you must create an XML file as shown earlier and call the getEventsFromAnXMLFile function from any menu item or button click. It returns a string, which you can print out to the log stream using the Log.d debug method.

Working with Raw Resources

Android also allows raw files in addition to arbitrary XML files. These resources, placed in /res/raw, are raw files such as audio, video, or text files that require localization or references through resource IDs. Unlike the XML files placed in /res/xml, these files are not compiled but are moved to the application package as they are. However, each file has an identifier generated in R.java. If you were to place a text file at /res/raw/test.txt, you would be able to read that file using the code in Listing 3–33.

Listing 3–33. *Reading a Raw Resource*

```
String getStringFromRawFile(Activity activity)
   throws IOException
   {
      Resources r = activity.getResources();
      InputStream is = r.openRawResource(R.raw.test);
      String myText = convertStreamToString(is);
      is.close();
      return myText;
   }

   String convertStreamToString(InputStream is)
   throws IOException
   {
      ByteArrayOutputStream baos = new ByteArrayOutputStream();
      int i = is.read();
      while (i != -1)
      {
         baos.write(i);
         i = is.read();
      }
      return baos.toString();
   }
```

CAUTION: File names with duplicate base names generate a build error in the Eclipse ADT plug-in. This is the case for all resource IDs generated for resources that are based on files.

Working with Assets

Android offers one more directory where you can keep files to be included in the package: /assets. It's at the same level as /res, meaning it's not part of the /res subdirectories. The files in /assets do not generate IDs in R.java; you must specify the file path to read them. The file path is a relative path starting at /assets. You use the AssetManager class to access these files, as shown in Listing 3–34.

Listing 3–34. *Reading an Asset*

```
//Note: Exceptions are not shown in the code
String getStringFromAssetFile(Activity activity)
{
```

```
    AssetManager am = activity.getAssets();
    InputStream is = am.open("test.txt");
    String s = convertStreamToString(is);
    is.close();
    return s;
}
```

Reviewing the Resources Directory Structure

In summary, Listing 3–35 offers a quick look at the overall resources directory structure.

Listing 3–35. *Resource Directories*

```
/res/values/strings.xml
           /colors.xml
           /dimens.xml
           /attrs.xml
           /styles.xml
    /drawable/*.png
             /*.jpg
             /*.gif
             /*.9.png
    /anim/*.xml
    /layout/*.xml
    /raw/*.*
    /xml/*.xml
/assets/*.*/*.*
```

> **NOTE:** Because it's not under the `/res` directory, only the `/assets` directory can contain an arbitrary list of subdirectories. Every other directory can only have files at the level of that directory and no deeper. This is how `R.java` generates identifiers for those files.

Resources and Configuration Changes

Resources help with localization. For example, you can have a string value that changes based on the language locale of the user. Android resources generalize this idea to any configuration of the device, of which language is just one configuration choice. Another example of a configuration change is when a device is turned from a vertical position to a horizontal position. The vertical mode is called the *portrait mode* and the horizontal mode the *landscape mode*.

Android allows you to pick different sets of layouts based on this layout mode for the same resource ID. Android does this by using different directories for each configuration. An example is shown in Listing 3–36.

Listing 3–36. *Alternate Resource Directories*

```
\res\layout\main_layout.xml
\res\layout-port\main_layout.xml
\res\layout-land\main_layout.xml
```

Even though there are three separate layout files here, they all generate only one layout ID in R.java. This ID looks as follows:

```
R.layout.main_layout
```

However, when you retrieve the layout corresponding to this layout ID, you get the appropriate layout suitable for that device layout.

In this example the directory extensions -port and -land are called *configuration qualifiers*. These qualifiers are *case insensitive* and separated from the resource directory name with a hyphen (-). Resources that you specify in these configuration qualifier directories are called *alternate resources*. The resources in resource directories without the configuration qualifiers are called *default resources*.

Most of the available configuration qualifiers are listed next. Please note that new qualifiers may be added with newer APIs. Refer to the URL in the references section to see the latest set of resource qualifiers:

- mccAAA: AAA is the mobile country code.
- mncAAA: AAA is the carrier/network code.
- en-rUS: Language and region.
- sw<N>dp, w<N>dp, h<N>dp: Smallest width, available width, available height (since API 13).
- small, normal, large, xlarge: Screen size.
- long, notlong: Screen type.
- port, land: Portrait or landscape.
- car, desk: Type of docking.
- night, notnight: Night or day.
- ldpi, mdpi, hdpi, xhdpi, nodpi, tvdpi: Screen density.
- notouch, stylus, finger: Kind of screen.
- keysexposed, keyssoft, keyshidden: Kind of keyboard.
- nokeys, qwerty, 12key: Number of keys.
- navexposed, navhidden: Navigation keys hidden or exposed.
- nonav, dpad, trackball, wheel: Type of navigation device.
- v3, v4, v7: API level.

With these qualifiers, you can have resource directories such as those shown in Listing 3–37.

Listing 3–37. ***Additional Alternate Resource Directories***

```
\res\layout-mcc312-mnc222-en-rUS
\res\layout-ldpi
\res\layout-hdpi
```

```
\res\layout-car
```

You can discover your current locale by navigating to the Custom Locale application available on the device. The navigation path for this application is Home ➤ List of Applications ➤ Custom Locale.

Given a resource ID, Android uses an algorithm to pick up the right resource. You can refer to the URLs included in the "Reference URLs" section to understand more about these rules, but let's look at a few rules of thumb.

The primary rule is that the qualifiers listed earlier are in the order of precedence. Consider the directories in Listing 3–38.

Listing 3–38. *Layout File Variations*

```
\res\layout\main_layout.xml
\res\layout-port\main_layout.xml
\res\layout-en\main_layout.xml
```

In Listing 3–38, the layout file `main_layout.xml` is available in two additional variations. There is one variation for the language and one variation for the layout mode. Now, let's examine what layout file is picked up if you are viewing the device in portrait mode. Even though you are in portrait mode, Android picks the layout from the `layout-en` directory, because the language variation comes before the orientation variation in the list of configuration qualifiers. The SDK links mentioned in the "Reference URLs" section of this chapter list all the configuration qualifiers and their precedence order.

Let's look at the precedence rules further by experimenting with a few string resources. Please note that string resources are based on individual IDs, whereas layout resources are file-based. To test the configuration qualifier precedence with string resources, let's come up with five resource IDs that can participate in the following variations: `default`, `en`, `en_us`, `port`, and `en_port`. The five resource IDs follow:

- `teststring_all`: This ID is in all variations of the `values` directory including the default.
- `testport_port`: This ID is in the default and in only the `-port` variation.
- `t1_enport`: This ID is in the default and in the `-en` and `-port` variations.
- `t1_1_en_port`: This is in the default and only in the `-en-port` variation.
- `t2`: This is only in the default.

Listing 3–39 shows all the variations of the `values` directory.

Listing 3–39. *String Variations Based on Configuration*

```
// values/strings.xml
<resources xmlns="http://schemas.android.com/apk/res/android">
  <string name="teststring_all">teststring in root</string>
  <string name="testport_port">testport-port</string>
  <string name="t1_enport">t1 in root</string>
  <string name="t1_1_en_port">t1_1 in root</string>
  <string name="t2">t2 in root</string>
</resources>
```

```
// values-en/strings_en.xml
<resources xmlns="http://schemas.android.com/apk/res/android">
  <string name="teststring_all">teststring-en</string>
  <string name="t1_enport">t1_en</string>
  <string name="t1_1_en_port">t1_1_en</string>
</resources>

// values-en-rUS/strings_en_us.xml
<resources xmlns="http://schemas.android.com/apk/res/android">
  <string name="teststring_all">test-en-us</string>
</resources>

// values-port/strings_port.xml
<resources xmlns="http://schemas.android.com/apk/res/android">
  <string name="teststring_all">test-en-us-port</string>
  <string name="testport_port">testport-port</string>
  <string name="t1_enport">t1_port</string>
  <string name="t1_1_en_port">t1_1_port</string>
</resources>

// values-en-port/strings_en_port.xml
<resources xmlns="http://schemas.android.com/apk/res/android">
  <string name="teststring_all">test-en-port</string>
  <string name="t1_1_en_port">t1_1_en_port</string>
</resources>
```

Listing 3–40 shows the `R.java` file for these.

Listing 3–40. *R.java to Support String Variations*

```
public static final class string {
    public static final int teststring_all=0x7f050000;
    public static final int testport_port=0x7f050004;
    public static final int t1_enport=0x7f050001;
    public static final int t1_1_en_port=0x7f050002;
    public static final int t2=0x7f050003;
}
```

Right off the bat, you can see that even though you have a ton of strings defined, only five string resource IDs are generated. If you retrieve these string values, the behavior of each string retrieval is documented next (the configuration we tested with is `en_US` and portrait mode):

- `teststring_all`: This ID is in all five variations of the `values` directory. Because it is there in all variations, the variation from the `values-en-rUS` directory is picked up. Based on precedence rules, the specific language trumps the default, `en`, `port`, and `en-port` variations.
- `testport_port`: This ID is in the default and in only the `-port` variation. Because it is not in any value directory starting with `-en`, the `-port` takes precedence over the default, and the value from the `-port` variation is picked up. If this had been in one of the `-en` variations, the value would have been picked up from there.

- t1_enport: This ID is in three variations: default, -en, and -port. Because this is in -en and -port at the same time, the value from -en is picked up.
- t1_1_en_port: This is in four variations: default, -port, -en, and -en-port Because this is available in -en-port, it is picked up from -en-port, ignoring default, -en, and -port.
- t2: This is only in the default, so the value is picked up from default.

Android SDK has a more detailed algorithm that you can read up on. However, the example in this section gives you the essence of it. The key is to realize the precedence of one variation over the other. The following reference section provides a URL for this SDK link. It also includes a URL to download an importable project for this chapter; you can use this project to experiment with these configuration variations.

Reference URLs

As you learn about Android resources, you may want to keep the following reference URLs handy. In addition to the URLs, we list what you will gain from each of them:

- http://developer.android.com/guide/topics/resources/index.html: A roadmap to the documentation on resources.
- http://developer.android.com/guide/topics/resources/available-resources.html: Android documentation of various types of resources.
- http://developer.android.com/guide/topics/resources/providing-resources.html#AlternativeResources: A list of various configuration qualifiers provided by the latest Android SDK.
- http://developer.android.com/guide/practices/screens_support.html: Guidelines on how to design Android applications for multiple screen sizes.
- http://developer.android.com/reference/android/content/res/Resources.html: Various methods available to read resources.
- http://developer.android.com/reference/android/R.html: Resources as defined to the core Android platform.
- www.androidbook.com/item/3542: Our research on plurals, string arrays, and alternate resources, as well as links to other references.
- androidbook.com/proandroid4/projects: URL for a downloadable Eclipse project that demonstrates many concepts in this chapter. The name of the file is ProAndroid4_Ch03_TestResources.zip.

Summary

Let's conclude this chapter by quickly enumerating what you have learned about resources so far:

- You know the types of resources supported in Android.
- You know how to create resources in XML files.
- You know how resource IDs are generated and how to use them in Java code.
- You learned that resource ID generation is a convenient scheme that simplifies resource usage in Android.
- You learned how to work with XML resources, raw resources, and assets.
- You briefly touched on alternate resources.
- You saw how to define and use plurals and string arrays.
- You learned about resource-reference syntax.

Interview Questions

You can use the following questions as a guide to consolidate your understanding of this chapter:

1. How many resource types can you name?
2. What is `R.java`?
3. Why is `R.java` such a convenience for working with resources?
4. What is the connection between resource-reference syntax and allocating IDs for UI controls?
5. Is a file extension used in generating a resource ID?
6. What happens if two file-based resources differ only in their extension?
7. What are raw and XML resources, and how do they differ from assets?
8. Can you localize XML resources?
9. Can you localize assets?
10. Can you write down and explain resource-reference syntax?
11. Can you predeclare IDs for controls? If so, why?
12. What XML node is used to create an ID?

13. If you place files in XML and raw directories, does Android generate IDs for those files through `R.java`?
14. Does Android generate IDs for files in the asset directory?
15. What is the meaning of `one` and `other` in the `plurals` resource?
16. Can you use HTML strings in a string resource?
17. How can you display an HTML string in a text view?
18. How can you define a rectangle as a drawable?
19. How do you use a shape drawable?
20. What class do you use to read XML files from the `/res/xml` directory?
21. What is the primary class for dealing with XML files in Android?
22. What is the `AssetManager` class, and how do you get access to it?
23. What is the `Resources` class, and how do you get an instance of it?
24. Can you have arbitrary subdirectories under the `assets` folder?
25. Can you have subdirectories under `/res/xml` resource folder?
26. What are resource configuration qualifiers?

With that, let's turn our attention to content providers in the next chapter.

Chapter 4

Understanding Content Providers

Android uses a concept called *content providers* for abstracting data into services. This idea of content providers makes data sources look like REST-enabled data providers, such as web sites. In that sense, a content provider is a wrapper around data. A SQLite database on an Android device is an example of a data source that you can encapsulate into a content provider.

> **NOTE:** REST stands for REpresentational State Transfer. When you type a URL in a web browser and the web server responds with HTML, you have essentially performed a REST-based "query" operation on the web server. REST is also usually contrasted with (SOAP—Simple Object Access Protocol) web services. You can read more about REST at the following Wikipedia entry: `http://en.wikipedia.org/wiki/Representational_State_Transfer`.

To retrieve data from a content provider or save data into a content provider, you will need to use a set of REST-like URIs. For example, if you were to retrieve a set of books from a content provider that is an encapsulation of a book database, you would need to use a URI like this:

```
content://com.android.book.BookProvider/books
```

To retrieve a specific book from the book database (book 23), you would need to use a URI like this:

```
content://com.android.book.BookProvider/books/23
```

You will see in this chapter how these URIs translate to underlying database-access mechanisms. Any application on the device can make use of these URIs to access and manipulate data. As a consequence, content providers play a significant role in sharing data between applications.

Strictly speaking, a content provider's responsibilities consist more of an encapsulation mechanism than a data-access mechanism. You'll need an actual data-access mechanism such as SQLite or network access to get to the underlying data sources. So, content-provider abstraction is required only if you want to share data externally or between applications. For internal data access, an application can use any data storage/access mechanism that it deems suitable, such as the following:

- *Preferences*: A set of key/value pairs that you can persist to store application preferences
- *Files*: Files internal to applications, which you can store on a removable storage medium
- *SQLite*: SQLite databases, each of which is private to the package that creates that database
- *Network*: A mechanism that lets you retrieve or store data externally through the Internet via HTTP services

> **NOTE:** Despite the number of data-access mechanisms allowed in Android, this chapter focuses on SQLite and the content-provider abstraction because content providers form the basis of data sharing, which is much more common in the Android framework compared to other UI frameworks. We'll cover the network approach in Chapter 11 and the preferences mechanism in Chapter 9.

Exploring Android's Built-in Providers

Android comes with a number of built-in content providers, which are documented in the SDK's `android.provider` Java package. You can view the list of these providers here:

`http://developer.android.com/reference/android/provider/package-summary.html`

The providers include, for example, Contacts and Media Store. These SQLite databases typically have an extension of `.db` and are accessible only from the implementation package. Any access outside that package must go through the content-provider interface.

Exploring Databases on the Emulator and Available Devices

Because many content providers in Android use SQLite databases (`www.sqlite.org/`), you can use tools provided both by Android and by SQLite to examine the databases. Many of these tools reside in the `\android-sdk-install-directory\tools` subdirectory; others are in `\android-sdk-install-directory\platform-tools`.

NOTE: Refer to Chapter 2 for information on locating the `tools` directories and invoking a command window for different operating systems. This chapter, like most of the remaining chapters, gives examples primarily on Windows platforms. As you go through this section, in which we use a number of command-line tools, you can focus on the name of the executable or the batch file and not pay as much attention to the directory the tool is in. We covered how to set the path for the tools directories on various platforms in Chapter 2.

Android uses a command-line tool called Android Debug Bridge (adb), which is found here:

```
platform-tools\adb.exe
```

adb is a special tool in the Android toolkit that most other tools go through to get to the device. However, you must have an emulator running or an Android device connected for adb to work. You can find out whether you have running devices or emulators by typing this at the command line:

```
adb devices
```

If the emulator is not running, you can start it by typing this at the command line:

```
emulator.exe @avdname
```

The argument `@avdname` is the name of an Android Virtual Device (AVD). (Chapter 2 covered the need for android virtual devices and how to create them in Chapter 2.) To find out what virtual devices you already have, you can run the following command:

```
android    list    avd
```

This command will list the available AVDs. If you have developed and run any Android applications through Eclipse Android Development Tool (ADT), then you will have configured at least one virtual device. The preceding command will list at least that one.

Here is some example output of that `list` command. (Depending on where your `tools` directory is and also depending on the Android release, the following printout may vary as to the path or release numbers, such as `i:\android`.)

```
I:\android\tools>android list avd
Available Android Virtual Devices:
    Name: avd
    Path: I:\android\tools\..\avds\avd3
  Target: Google APIs (Google Inc.)
          Based on Android 1.5 (API level 3)
    Skin: HVGA
  Sdcard: 32M
---------
    Name: titanium
    Path: C:\Documents and Settings\Satya\.android\avd\titanium.avd
  Target: Android 1.5 (API level 3)
    Skin: HVGA
```

As indicated, AVDs are covered in detail in Chapter 2.

You can also start the emulator through the Eclipse ADT plug-in. This automatically happens when you choose a program to run or debug in the emulator. Once the emulator is up and running, you can test again for a list of running devices by typing this:

```
adb devices
```

Now you should see a printout that looks like this:

```
List of devices attached
emulator-5554 device
```

You can see the many options and commands that you can run with adb by typing this at the command line:

```
adb help
```

You can also visit the following URL for many of the runtime options for adb: `http://developer.android.com/guide/developing/tools/adb.html`.

You can use adb to open a shell on the connected device by typing this:

```
adb shell
```

> **NOTE:** This shell is a Unix `ash`, albeit with a limited command set. You can do `ls`, for example, but `find`, `grep`, and `awk` are not available in the shell.

You can see the available command set in the shell by typing this at the shell prompt:

```
#ls    /system/bin
```

The # sign is the prompt for the shell. For brevity, we will omit this prompt in the following examples. To see a list of root-level directories and files, you can type the following in the shell:

```
ls    -l
```

You'll need to access this directory to see the list of databases:

```
ls     /data/data
```

This directory contains the list of installed packages on the device. Let's look at an example by exploring the `com.android.providers.contacts` package:

```
ls   /data/data/com.android.providers.contacts/databases
```

This will list a database file called `contacts.db`, which is a SQLite database. (This file and path are still device and release dependent.)

> **NOTE:** In Android, databases may be created when they are accessed the first time. This means you may not see this file if you have never accessed the Contacts application.

If there were a `find` command in the included `ash`, you could look at all the `*.db` files. But there is no good way to do this with `ls` alone. The nearest thing you can do is this:

```
ls -R /data/data/*/databases
```

With this command, you will notice that the Android distribution has the following databases (again, a bit of caution; depending on your release, this list may vary):

```
alarms.db
contacts.db
downloads.db
internal.db
settings.db
mmssms.db
telephony.db
```

You can invoke `sqlite3` on one of these databases inside the adb shell by typing this:

```
sqlite3  /data/data/com.android.providers.contacts/databases/contacts.db
```

You can exit `sqlite3` by typing this:

```
sqlite>.exit
```

Notice that the prompt for adb is `#` and the prompt for `sqlite3` is `sqlite>`. You can read about the various `sqlite3` commands by visiting `www.sqlite.org/sqlite.html`. However, we will list a few important commands here so you don't have to make a trip to the Web. You can see a list of tables by typing

```
sqlite> .tables
```

This command is a shortcut for

```
SELECT name FROM sqlite_master
WHERE type IN ('table','view') AND name NOT LIKE 'sqlite_%'
UNION ALL
SELECT name FROM sqlite_temp_master
WHERE type IN ('table','view')
ORDER BY 1
```

As you probably guessed, the table `sqlite_master` is a master table that keeps track of tables and views in the database. The following command line prints out a `create` statement for a table called `people` in `contacts.db`:

```
.schema people
```

This is one way to get at the column names of a table in SQLite. This will also print out the column data types. While working with content providers, you should note these column types because access methods depend on them.

However, it is pretty tedious to manually parse through this long `create` statement just to learn the column names and their types. There is a workaround: you can pull `contacts.db` down to your local box and then examine the database using any number of GUI tools for SQLite version 3. You can issue the following command from your OS command prompt to pull down the contacts.db file:

```
adb pull  /data/data/com.android.providers.contacts/databases/contacts.db ➥
c:/somelocaldir/contacts.db
```

We used a free download of Sqliteman (http://sqliteman.com/), a GUI tool for SQLite databases, which seemed to work fine. We experienced a few crashes but otherwise found the tool completely usable for exploring Android SQLite databases.

Quick SQLite Primer

The following sample SQL statements could help you navigate through an SQLite database quickly:

```
//Set the column headers to show in the tool
sqlite>.headers on

//select all rows from a table
select * from table1;

//count the number of rows in a table
select count(*) from table1;

//select a specific set of columns
select col1, col2 from table1;

//Select distinct values in a column
select distinct col1 from table1;

//counting the distinct values
select count(col1) from (select distinct col1  from table1);

//group by
select count(*), col1 from table1 group by col1;

//regular inner join
select * from table1 t1, table2 t2
where t1.col1 = t2.col1;

//left outer join
//Give me everything in t1 even though there are no rows in t2
select * from table t1 left outer join table2 t2
on t1.col1 = t2.col1
where ....
```

Architecture of Content Providers

You now know how to explore existing content providers through Android and SQLite tools. Next, we'll examine some of the architectural elements of content providers and how these content providers relate to other data-access abstractions in the industry.

Overall, the content-provider approach has parallels to the following industry abstractions:

- Web sites
- REST
- Web services
- Stored procedures

Each content provider on a device registers itself like a web site with a string (akin to a domain name, but called an *authority*). This uniquely identifiable string forms the basis of a set of URIs that this content provider can offer. This is not unlike how a web site with a domain offers a number of URLs to expose its documents or content in general.

This authority registration occurs in the `AndroidManifest.xml` file. Here are two examples of how you can register providers in `AndroidManifest.xml`:

```
<provider android:name="SomeProvider"
        android:authorities="com.your-company.SomeProvider" />

<provider android:name="NotePadProvider"
   android:authorities="com.google.provider.NotePad"
/>
```

An authority is like a domain name for that content provider. Given the preceding authority registration, these providers will honor URLs starting with that authority prefix:

```
content://com.your-company.SomeProvider/
content://com.google.provider.NotePad/
```

You see that a content provider, like a web site, has a base domain name that acts as a starting URL.

> **NOTE:** The providers offered by Android may not carry a fully qualified authority name. It is recommended at the moment only for third-party content providers. This is why you sometimes see that content providers are referenced with a simple word such as `contacts` as opposed to `com.google.android.contacts` (in the case of a third-party provider).

Content providers also provide REST-like URLs to retrieve or manipulate data. For the preceding registration, the URI to identify a directory or a collection of notes in the `NotePadProvider` database is

```
content://com.google.provider.NotePad/Notes
```

The URI to identify a specific note is

```
content://com.google.provider.NotePad/Notes/#
```

where # is the `id` of a particular note. Here are some additional examples of URIs that some data providers accept:

```
content://media/internal/images
content://media/external/images
content://contacts/people/
content://contacts/people/23
```

Notice how these providers' media (`content://media`) and contacts (`content://contacts`) don't have a fully qualified structure. This is because these are not third-party providers and are controlled by Android.

Content providers exhibit characteristics of web services as well. A content provider, through its URIs, exposes internal data as a service. However, the output from the URL of a content provider is not typed data, as is the case for a SOAP-based web-service call. This output is more like a result set coming from a JDBC statement. Even there, the similarities to JDBC are conceptual. We don't want to give the impression that this is the same as a JDBC ResultSet.

The caller is expected to know the structure of the rows and columns that are returned. Also, as you will see in this chapter's "Structure of Android MIME Types" section, a content provider has a built-in mechanism that allows you to determine the Multipurpose Internet Mail Extensions (MIME) type of the data represented by this URI.

In addition to resembling web sites, REST, and web services, a content provider's URIs also resemble the names of stored procedures in a database. Stored procedures present service-based access to the underlying relational data. URIs are similar to stored procedures, because URI calls against a content provider return a cursor. However, content providers differ from stored procedures in that the input to a service call in a content provider is typically embedded in the URI itself.

We've provided these comparisons to give you an idea of the broader scope of content providers.

Structure of Android Content URIs

We compared a content provider to a web site because it responds to incoming URIs. So, to retrieve data from a content provider, all you have to do is invoke a URI. The retrieved data in the case of a content provider, however, is in the form of a set of rows and columns represented by an Android `cursor` object. In this context, we'll examine the structure of the URIs that you could use to retrieve data.

Content URIs in Android look similar to HTTP URIs, except that they start with `content` and have the general form

```
content://*/*/*
```

or

```
content://authority-name/path-segment1/path-segment2/etc…
```

Here's an example URI that identifies a note numbered 23 in a database of notes:

```
content://com.google.provider.NotePad/notes/23
```

After content:, the URI contains a unique identifier for the authority, which is used to locate the provider in the provider registry. In the preceding example, com.google.provider.NotePad is the authority portion of the URI.

/notes/23 is the path section of the URI that is specific to each provider. The notes and 23 portions of the path section are called *path segments*. It is the responsibility of the provider to document and interpret the path section and path segments of the URIs.

The developer of the content provider usually does this by declaring constants in a Java class or a Java interface in that provider's implementation Java package. Furthermore, the first portion of the path might point to a collection of objects. For example, /notes indicates a collection or a directory of notes, whereas /23 points to a specific note item.

Given this URI, a provider is expected to retrieve rows that the URI identifies. The provider is also expected to alter content at this URI using any of the state-change methods: insert, update, or delete.

Structure of Android MIME Types

Just as a web site returns a MIME type for a given URL (this allows browsers to invoke the right program to view the content), a content provider has an added responsibility to return the MIME type for a given URI. This allows flexibility of viewing data. Knowing what kind of data it is, you may have more than one program that knows how to handle that data. For example, if you have a text file on your hard drive, there are many editors that can display that text file. Depending on the OS, it may even give you an option of which editor to pick.

MIME types work in Android similarly to how they work in HTTP. You ask a provider for the MIME type of a given URI that it supports, and the provider returns a two-part string identifying its MIME type according to the standard web MIME conventions. You can find the MIME-type standard here:

```
http://tools.ietf.org/html/rfc2046
```

According to the MIME-type specification, a MIME type has two parts: a type and a subtype. Here are some examples of well-known MIME-type pairs:

```
text/html
text/css
text/xml
text/vnd.curl
application/pdf
application/rtf
application/vnd.ms-excel
```

You can see a complete list of registered types and subtypes at the Internet Assigned Numbers Authority (IANA) web site:

```
http://www.iana.org/assignments/media-types/
```

The primary registered content types are

```
application
audio
example
image
message
model
multipart
text
video
```

Each of these primary types has subtypes. But if a vendor has proprietary data formats, the subtype name begins with vnd. For example, Microsoft Excel spreadsheets are identified by the subtype `vnd.ms-excel`, whereas `pdf` is considered a nonvendor standard and is represented as such without any vendor-specific prefix.

Some subtypes start with `x-`; these are nonstandard subtypes that don't have to be registered. They're considered private values that are bilaterally defined between two collaborating agents. Here are a few examples:

```
application/x-tar
audio/x-aiff
video/x-msvideo
```

Android follows a similar convention to define MIME types. The vnd in Android MIME types indicates that these types and subtypes are nonstandard, vendor-specific forms. To provide uniqueness, Android further demarcates the types and subtypes with multiple parts similar to a domain specification. Furthermore, the Android MIME type for each content type has two forms: one for a specific record and one for multiple records.

For a single record, the MIME type looks like this:

```
vnd.android.cursor.item/vnd.yourcompanyname.contenttype
```

For a collection of records or rows, the MIME type looks like this:

```
vnd.android.cursor.dir/vnd.yourcompanyname.contenttype
```

Here are a couple of examples:

```
//One single note
vnd.android.cursor.item/vnd.google.note

//A collection or a directory of notes
vnd.android.cursor.dir/vnd.google.note
```

> **NOTE:** The implication here is that Android natively recognizes a directory of items and a single item. As a programmer, your flexibility is limited to the subtype. For example, things like list controls rely on what is returned from a cursor as one of these MIME main types.

MIME types are extensively used in Android, especially in intents, where the system figures out what activity to invoke based on the MIME type of data. MIME types are invariably derived from their URIs through content providers. You need to keep three things in mind when you work with MIME types:

- The type and subtype need to be unique for what they represent. The type is pretty much decided for you, as pointed out. It is primarily a directory of items or a single item. In the context of Android, these may not be as open as you might think.
- The type and subtype need to be preceded with vnd if they are not standard (which is usually the case when you talk about specific records).
- They are typically name-spaced for your specific need.

To reiterate this point, the primary MIME type for a collection of items returned through an Android `cursor` should always be `vnd.android.cursor.dir`, and the primary MIME type of a single item retrieved through an Android `cursor` should be `vnd.android.cursor.item`. You have more wiggle room when it comes to the subtype, as in `vnd.google.note`; after the `vnd.` part, you are free to subtype it with anything you'd like.

Reading Data Using URIs

Now you know that to retrieve data from a content provider, you need to use URIs supplied by that content provider. Because the URIs defined by a content provider are unique to that provider, it is important that these URIs are documented and available to programmers to see and then call. The providers that come with Android do this by defining constants representing these URI strings.

Consider these three URIs defined by helper classes in the Android SDK:

```
MediaStore.Images.Media.INTERNAL_CONTENT_URI
MediaStore.Images.Media.EXTERNAL_CONTENT_URI
ContactsContract.Contacts.CONTENT_URI
```

The equivalent textual URI strings would be as follows:

```
content://media/internal/images
content://media/external/images
content://com.android.contacts/contacts/
```

Given these URIs, the code to retrieve a single row of people from the Contacts provider looks like this:

```
Uri peopleBaseUri = ContactsContract.Contacts.CONTENT_URI;
Uri myPersonUri = Uri.withAppendedPath(peopleBaseUri, "23");

//Query for this record.
//managedQuery is a method on Activity class
Cursor cur = managedQuery(myPersonUri, null, null, null);
```

Notice how the `ContactsContract.Contacts.CONTENT_URI` is predefined as a constant in the `Contacts` class. We have named the variable `peopleBaseUri` to indicate that if your intention is to discover people, you go after the Contacts content URI. Of course, you can call this variable `contactsBaseUri` if you conceptually think of people as contacts.

> **NOTE:** Refer to Chapter 30 for a much more detailed exploration of the Contacts content provider. Also note that the Contacts API and its associated constants could change with each API release. This chapter has been tested using Android 2.2 (API 8) and above.

In this example, the code takes the root URI, adds a specific person ID to it, and makes a call to the `managedQuery` method.

As part of the query against this URI, it is possible to specify a sort order, the columns to select, and a `where` clause. These additional parameters are set to `null` in this example.

> **NOTE:** A content provider should list which columns it supports by implementing a set of interfaces or by listing the column names as constants. However, the class or interface that defines constants for columns should also make the column types clear through a column-naming convention, or comments or documentation, because there is no formal way to indicate the type of a column through constants.

Listing 4–1 shows how to retrieve a cursor with a specific list of columns from the Contacts content provider, based on the previous example.

Listing 4–1. *Retrieving a Cursor from a Content Provider*

```
//Use this interface to see the constants
import ContactsContract.Contacts;
...
// An array specifying which columns to return.
string[] projection = new string[] {
    Contacts._ID,
    Contacts.DISPLAY_NAME_PRIMARY
};

Uri mContactsUri = ContactsContract.Contacts.CONTENT_URI;

// Best way to retrieve a query; returns a managed query.
Cursor managedCursor = managedQuery( mContactsUri,
                        projection, //Which columns to return.
                        null,       // WHERE clause
                        Contacts.DISPLAY_NAME_PRIMARY + " ASC"); // Order-by clause.
```

Notice how a `projection` is merely an array of strings representing column names. So unless you know what these columns are, you'll find it difficult to create a `projection`. You should look for these column names in the same class that provides the URI, in this case the `Contacts` class. You can discover more about each of these columns by looking at the SDK documentation for the `android.provider.ContactsContract.Contacts` class, available at this URL:

`http://developer.android.com/reference/android/provider/ContactsContract.Contacts.html`

Let's revisit the cursor that is returned: it contains zero or more records. Column names, order, and type are provider specific. However, every row returned has a default column called _id representing a unique ID for that row.

Using the Android Cursor

Here are a few facts about an Android cursor:

- A cursor is a collection of rows.
- You need to use moveToFirst() before reading any data because the cursor starts off positioned before the first row.
- You need to know the column names.
- You need to know the column types.
- All field-access methods are based on column number, so you must convert the column name to a column number first.
- The cursor is random (you can move forward and backward, and you can jump).
- Because the cursor is random, you can ask it for a row count.

An Android cursor has a number of methods that allow you to navigate through it. Listing 4–2 shows how to check if a cursor is empty and how to walk through the cursor row by row when it is not empty.

Listing 4–2. *Navigating Through a Cursor Using a while Loop*

```
if (cur.moveToFirst() == false)
{
   //no rows empty cursor
   return;
}

//The cursor is already pointing to the first row
//let's access a few columns
int nameColumnIndex = cur.getColumnIndex(Contacts.DISPLAY_NAME_PRIMARY);
String name = cur.getString(nameColumnIndex);

//let's now see how we can loop through a cursor

while(cur.moveToNext())
{
   //cursor moved successfully
   //access fields
}
```

The assumption at the beginning of Listing 4–2 is that the cursor has been positioned before the first row. To position the cursor on the first row, we use the moveToFirst() method on the cursor object. This method returns false if the cursor is empty. We then use the moveToNext() method repetitively to walk through the cursor.

To help you learn where the cursor is, Android provides the following methods:

```
isBeforeFirst()
isAfterLast()
isClosed()
```

Using these methods, you can also use a for loop as in Listing 4–3 to navigate through the cursor instead of the while loop used in Listing 4–2.

Listing 4–3. *Navigating Through a Cursor Using a for Loop*

```
//Get your indexes first outside the for loop
int nameColumn = cur.getColumnIndex(Contacts.DISPLAY_NAME_PRIMARY);

//Walk the cursor now based on column indexes
for(cur.moveToFirst();!cur.isAfterLast();cur.moveToNext())
{
   String name = cur.getString(nameColumn);
}
```

The index order of columns seems to be a bit arbitrary. As a result, we advise you to explicitly get the indexes first from the cursor to avoid surprises. To find the number of rows in a cursor, Android provides a method on the cursor object called getCount().

Working with the where Clause

Content providers offer two ways of passing a where clause:

- Through the URI
- Through the combination of a string clause and a set of replaceable string-array arguments

We will cover both of these approaches through some sample code.

Passing a where Clause Through a URI

Imagine you want to retrieve a note whose ID is 23 from the Google notes database. You'd use the code in Listing 4–4 to retrieve a cursor containing one row corresponding to row 23 in the notes table.

Listing 4–4. *Passing SQL where Clauses Through the URI*

```
Activity someActivity;
//..initialize someActivity
String noteUri = "content://com.google.provider.NotePad/notes/23";
Cursor managedCursor = someActivity.managedQuery( noteUri,
                        projection, //Which columns to return.
                        null,       // WHERE clause
                        null); // Order-by clause.
```

We left the where clause argument of the managedQuery method null because, in this case, we assumed that the note provider is smart enough to figure out the id of the book we wanted. This id is embedded in the URI itself. We used the URI as a vehicle to pass the where clause. This becomes apparent when you notice how the notes provider implements the corresponding query method. Here is a code snippet from that query method:

```
//Retrieve a note id from the incoming uri that looks like
//content://.../notes/23
int noteId = uri.getPathSegments().get(1);

//ask a query builder to build a query
//specify a table name
queryBuilder.setTables(NOTES_TABLE_NAME);

//use the noteid to put a where clause
queryBuilder.appendWhere(Notes._ID + "=" + noteId);
```

Notice how the ID of a note is extracted from the URI. The `Uri` class representing the incoming argument `uri` has a method to extract the portions of a URI after the root `content://com.google.provider.NotePad`. These portions are called *path segments*; they're strings between / separators such as `/seg1/seg3/seg4/`, and they're indexed by their positions. For the URI here, the first path segment would be 23. We then used this ID of 23 to append to the `where` clause specified to the `QueryBuilder` class. In the end, the equivalent select statement would be

```
select * from notes where _id = 23
```

NOTE: The classes `Uri` and `UriMatcher` are used to identify URIs and extract parameters from them. (We'll cover `UriMatcher` further in the section "Using UriMatcher to Figure Out the URIs.") `SQLiteQueryBuilder` is a helper class in `android.database.sqlite` that allows you to construct SQL queries to be executed by `SQLiteDatabase` on a SQLite database instance.

Using Explicit where Clauses

Now that you have seen how to use a URI to send in a `where` clause, consider the other method by which Android lets us send a list of explicit columns and their corresponding values as a `where` clause. To explore this, let's take another look at the `managedQuery` method of the `Activity` class that we used in Listing 4–4. Here's its signature:

```
public final Cursor managedQuery(Uri uri,
   String[] projection,
   String selection,
   String[] selectionArgs,
   String sortOrder)
```

Notice the argument named `selection`, which is of type `String`. This selection string represents a filter (a `where` clause, essentially) declaring which rows to return, formatted as a SQL `where` clause (excluding the `WHERE` itself). Passing `null` will return all rows for the given URI. In the selection string you can include ?s, which will be replaced by the values from `selectionArgs` in the order that they appear in the selection. The values will be bound as `String`s.

Because you have two ways of specifying a where clause, you might find it difficult to determine how a provider has used these where clauses and which where clause takes precedence if both where clauses are utilized.

For example, you can query for a note whose ID is 23 using either of these two methods:

```
//URI method
managedQuery("content://com.google.provider.NotePad/notes/23"
,null
,null
,null
,null);
```

or

```
//explicit where clause
managedQuery("content://com.google.provider.NotePad/notes"
,null
,"_id=?"
,new String[] {23}
,null);
```

The convention is to use where clauses through URIs where applicable and use the explicit option as a special case.

Inserting Records

So far, we have talked about how to retrieve data from content providers using URIs. Now, let's turn our attention to inserts, updates, and deletes.

> **NOTE:** In explaining content providers so far, we have used examples from the Notepad application that Google's tutorials provide as the prototypical app. However, it is not necessary to be completely familiar with that application. Even if you haven't seen the application you should be able to follow along with the examples. We will, however, give you complete code for a sample provider later in this chapter.

Android uses a class called android.content.ContentValues to hold the values for a single record that is to be inserted. ContentValues is a dictionary of key/value pairs, much like column names and their values. You insert records by first populating a record into ContentValues and then asking android.content.ContentResolver to insert that record using a URI.

> **NOTE:** You need to locate ContentResolver, because at this level of abstraction, you are not asking a database to insert a record; instead, you are asking to insert a record into a provider identified by a URI. ContentResolver is responsible for resolving the URI reference to the right provider and then passing on the ContentValues object to that specific provider.

Here is an example of populating a single row of notes in ContentValues in preparation for an insert:

```
ContentValues values = new ContentValues();
values.put("title", "New note");
values.put("note","This is a new note");

//values object is now ready to be inserted
```

You can get a reference to ContentResolver by asking the Activity class:

```
ContentResolver contentResolver = activity.getContentResolver();
```

Now, all you need is a URI to tell ContentResolver to insert the row. These URIs are defined in a class corresponding to the Notes table. In the Notepad example, this URI is

```
Notepad.Notes.CONTENT_URI
```

We can take this URI and the ContentValues we have and make a call to insert the row:

```
Uri uri = contentResolver.insert(Notepad.Notes.CONTENT_URI, values);
```

This call returns a URI pointing to the newly inserted record. This returned URI would match the following structure:

```
Notepad.Notes.CONTENT_URI/new_id
```

Adding a File to a Content Provider

Occasionally, you might need to store a file in a database. The usual approach is to save the file to disk and then update the record in the database that points to the corresponding file name.

Android takes this protocol and automates it by defining a specific procedure for saving and retrieving these files. Android uses a convention where a reference to the file name is saved in a record with a reserved column name of _data.

When a record is inserted into that table, Android returns the URI to the caller. Once you save the record using this mechanism, you also need to follow it up by saving the file in that location. To do this, Android allows ContentResolver to take the URI of the database record and return a writable output stream. Behind the scenes, Android allocates an internal file and stores the reference to that file name in the _data field.

If you were to extend the Notepad example to store an image for a given note, you could create an additional column called _data and run an insert first to get a URI back. The following code demonstrates this part of the protocol:

```
ContentValues values = new ContentValues();
values.put("title", "New note");
values.put("note","This is a new note");

//Use a content resolver to insert the record
ContentResolver contentResolver = activity.getContentResolver();
Uri newUri = contentResolver.insert(Notepad.Notes.CONTENT_URI, values);
```

Once you have the URI of the record, the following code asks the ContentResolver to get a reference to the file output stream:

```
....
//Use the content resolver to get an output stream directly
//ContentResolver hides the access to the _data field where
//it stores the real file reference.
OutputStream outStream = activity.getContentResolver().openOutputStream(newUri);
someSourceBitmap.compress(Bitmap.CompressFormat.JPEG, 50, outStream);
outStream.close();
```

The code then uses that output stream to write to.

Updates and Deletes

So far, we have talked about queries and inserts; updates and deletes are fairly straightforward. Performing an update is similar to performing an insert, in which changed column values are passed through a ContentValues object. Here is the signature of an update method on the ContentResolver object:

```
int numberOfRowsUpdated =
activity.getContentResolver().update(
    Uri uri,
    ContentValues values,
    String whereClause,
    String[] selectionArgs )
```

The whereClause argument constrains the update to the pertinent rows. Similarly, the signature for the delete method is

```
int numberOfRowsDeleted =
activity.getContentResolver().delete(
    Uri uri,
    String whereClause,
    String[] selectionArgs )
```

Clearly, a delete method will not require the ContentValues argument because you will not need to specify the columns you want when you are deleting a record.

Almost all the calls from managedQuery and ContentResolver are directed eventually to the provider class. Knowing how a provider implements each of these methods gives us enough clues as to how those methods are used by a client. In the next section, we'll cover from scratch the implementation of an example content provider: called BookProvider.

Implementing Content Providers

We've discussed how to interact with a content provider for data needs but haven't yet discussed how to write a content provider. To write a content provider, you have to extend android.content.ContentProvider and implement the following key methods:

```
query
insert
update
delete
getType
```

You'll also need to set up a number of things before implementing them. We will illustrate all the details of a content-provider implementation by describing the steps you'll need to take:

1. Plan your database, URIs, column names, and so on, and create a metadata class that defines constants for all of these metadata elements.
2. Extend the abstract class `ContentProvider`.
3. Implement these methods: `query`, `insert`, `update`, `delete`, and `getType`.
4. Register the provider in the manifest file.

Planning a Database

To explore this topic, we'll create a database that contains a collection of books. The book database contains only one table called `books`, and its columns are `name`, `isbn`, and `author`. These column names fall under metadata. You'll define this sort of relevant metadata in a Java class. This metadata-bearing Java class `BookProviderMetaData` is shown in Listing 4–5. Some key elements of this metadata class are highlighted.

Listing 4–5. *Defining Metadata for Your Database: The BookProviderMetaData Class*

```
public class BookProviderMetaData
{
    public static final String AUTHORITY = "com.androidbook.provider.BookProvider";

    public static final String DATABASE_NAME = "book.db";
    public static final int DATABASE_VERSION = 1;
    public static final String BOOKS_TABLE_NAME = "books";

    private BookProviderMetaData() {}

    //inner class describing BookTable
    public static final class BookTableMetaData implements BaseColumns
    {
        private BookTableMetaData() {}
        public static final String TABLE_NAME = "books";

        //uri and MIME type definitions
        public static final Uri CONTENT_URI =
                          Uri.parse("content://" + AUTHORITY + "/books");

        public static final String CONTENT_TYPE =
                         "vnd.android.cursor.dir/vnd.androidbook.book";

        public static final String CONTENT_ITEM_TYPE =
                        "vnd.android.cursor.item/vnd.androidbook.book";
```

```
        public static final String DEFAULT_SORT_ORDER = "modified DESC";

        //Additional Columns start here.
        //string type
        public static final String BOOK_NAME = "name";

        //string type
        public static final String BOOK_ISBN = "isbn";

        //string type
        public static final String BOOK_AUTHOR = "author";

        //Integer from System.currentTimeMillis()
        public static final String CREATED_DATE = "created";

        //Integer from System.currentTimeMillis()
        public static final String MODIFIED_DATE = "modified";
    }
}
```

This BookProviderMetaData class starts by defining its authority to be com.androidbook.provider.BookProvider. We are going to use this string to register the provider in the Android manifest file. This string forms the front part of the URIs intended for this provider.

This class then proceeds to define its one table (books) as an inner BookTableMetaData class. The BookTableMetaData class then defines a URI for identifying a collection of books. Given the authority in the previous paragraph, the URI for a collection of books will look like this:

```
content://com.androidbook.provider.BookProvider/books
```

This URI is indicated by the constant

```
BookProviderMetaData.BookTableMetaData.CONTENT_URI
```

The BookTableMetaData class then proceeds to define the MIME types for a collection of books and a single book. The provider implementation will use these constants to return the MIME types for the incoming URIs.

BookTableMetaData then defines the set of columns: name, isbn, author, created (creation date), and modified (last-updated date).

> **NOTE:** You should point out your columns' data types through comments in the code.

The metadata class BookTableMetaData also inherits from the BaseColumns class that provides the standard _id field, which represents the row ID. With these metadata definitions in hand, we're ready to tackle the provider implementation.

Extending ContentProvider

Implementing our BookProvider sample content provider involves extending the ContentProvider class and overriding onCreate() to create the database and then implement the query, insert, update, delete, and getType methods. This section covers the setup and creation of the database, while the following sections deal with each of the individual methods: query, insert, update, delete, and getType. Listing 4–6 provides the complete source code for this class. Important subsections of this class are highlighted.

A query method requires the set of columns it needs to return. This is similar to a select clause that requires column names along with their as counterparts (sometimes called *synonyms*). Android uses a map object that it calls a projection map to represent these column names and their synonyms. We will need to set up this map so we can use it later in the query-method implementation. In the code for the provider implementation (see Listing 4–6), you will see this done up front as part of Project map setup.

Most of the methods we'll be implementing take a URI as an input. Although all the URIs that this content provider is able to respond to start with the same pattern, the tail ends of the URIs will be different—just like a web site. Each URI, although it starts the same, must be different to identify different data or documents. Here's an example:

```
Uri1: content://com.androidbook.provider.BookProvider/books
Uri2: content://com.androidbook.provider.BookProvider/books/12
```

The book provider needs to distinguish each of these URIs. This is a simple case. If our book provider had been housing more objects rather than just books, then there would be more URIs to identify those objects.

The provider implementation needs a mechanism to distinguish one URI from the other; Android uses a class called UriMatcher for this purpose. So we need to set up this object with all our URI variations. You will see this code in Listing 4–6 after the segment that creates a projection map. We'll further explain the UriMatcher class in the section "Using UriMatcher to Figure Out the URIs."

The code in Listing 4–6 then overrides the onCreate() method to facilitate the database creation. The source code then implements the insert(), query(), update(), getType(), and delete() methods. The code for all of this is presented in one listing, but we will explain each aspect in a separate subsection.

Listing 4–6. *Implementing the BookProvider Content Provider*

```
public class BookProvider extends ContentProvider
{
    //Logging helper tag. No significance to providers.
    private static final String TAG = "BookProvider";

    //Setup projection Map
    //Projection maps are similar to "as" (column alias) construct
    //in an sql statement where by you can rename the
    //columns.
    private static HashMap<String, String> sBooksProjectionMap;
```

```
static
{
    sBooksProjectionMap = new HashMap<String, String>();
    sBooksProjectionMap.put(BookTableMetaData._ID,
                            BookTableMetaData._ID);

    //name, isbn, author
    sBooksProjectionMap.put(BookTableMetaData.BOOK_NAME,
                            BookTableMetaData.BOOK_NAME);
    sBooksProjectionMap.put(BookTableMetaData.BOOK_ISBN,
                            BookTableMetaData.BOOK_ISBN);
    sBooksProjectionMap.put(BookTableMetaData.BOOK_AUTHOR,
                            BookTableMetaData.BOOK_AUTHOR);

    //created date, modified date
    sBooksProjectionMap.put(BookTableMetaData.CREATED_DATE,
                            BookTableMetaData.CREATED_DATE);
    sBooksProjectionMap.put(BookTableMetaData.MODIFIED_DATE,
                            BookTableMetaData.MODIFIED_DATE);
}

//Setup URIs
//Provide a mechanism to identify
//all the incoming uri patterns.
private static final UriMatcher sUriMatcher;
private static final int INCOMING_BOOK_COLLECTION_URI_INDICATOR = 1;
private static final int INCOMING_SINGLE_BOOK_URI_INDICATOR = 2;
static {
    sUriMatcher = new UriMatcher(UriMatcher.NO_MATCH);
    sUriMatcher.addURI(BookProviderMetaData.AUTHORITY, "books",
                    INCOMING_BOOK_COLLECTION_URI_INDICATOR);
    sUriMatcher.addURI(BookProviderMetaData.AUTHORITY, "books/#",
                    INCOMING_SINGLE_BOOK_URI_INDICATOR);

}

/**
 * Setup/Create Database
 * This class helps open, create, and upgrade the database file.
 */
private static class DatabaseHelper extends SQLiteOpenHelper {

    DatabaseHelper(Context context) {
        super(context,
            BookProviderMetaData.DATABASE_NAME,
            null,
            BookProviderMetaData.DATABASE_VERSION);
    }

    @Override
    public void onCreate(SQLiteDatabase db)
    {
        Log.d(TAG,"inner oncreate called");
        db.execSQL("CREATE TABLE " + BookTableMetaData.TABLE_NAME + " ("
                + BookTableMetaData._ID + " INTEGER PRIMARY KEY,"
                + BookTableMetaData.BOOK_NAME + " TEXT,"
                + BookTableMetaData.BOOK_ISBN + " TEXT,"
```

```
                + BookTableMetaData.BOOK_AUTHOR + " TEXT,"
                + BookTableMetaData.CREATED_DATE + " INTEGER,"
                + BookTableMetaData.MODIFIED_DATE + " INTEGER"
                + ");");
    }

    @Override
    public void onUpgrade(SQLiteDatabase db, int oldVersion, int newVersion)
    {
        Log.d(TAG,"inner onupgrade called");
        Log.w(TAG, "Upgrading database from version "
                + oldVersion + " to "
                + newVersion + ", which will destroy all old data");
        db.execSQL("DROP TABLE IF EXISTS " +
                 BookTableMetaData.TABLE_NAME);
        onCreate(db);
    }
}

private DatabaseHelper mOpenHelper;

//Component creation callback
@Override
public boolean onCreate()
{
    Log.d(TAG,"main onCreate called");
    mOpenHelper = new DatabaseHelper(getContext());
    return true;
}

@Override
public Cursor query(Uri uri, String[] projection, String selection,
        String[] selectionArgs,  String sortOrder)
{
    SQLiteQueryBuilder qb = new SQLiteQueryBuilder();

    switch (sUriMatcher.match(uri)) {
    case INCOMING_BOOK_COLLECTION_URI_INDICATOR:
        qb.setTables(BookTableMetaData.TABLE_NAME);
        qb.setProjectionMap(sBooksProjectionMap);
        break;

    case INCOMING_SINGLE_BOOK_URI_INDICATOR:
        qb.setTables(BookTableMetaData.TABLE_NAME);
        qb.setProjectionMap(sBooksProjectionMap);
        qb.appendWhere(BookTableMetaData._ID + "="
                    + uri.getPathSegments().get(1));
        break;

    default:
        throw new IllegalArgumentException("Unknown URI " + uri);
    }

    // If no sort order is specified use the default
    String orderBy;
    if (TextUtils.isEmpty(sortOrder)) {
        orderBy = BookTableMetaData.DEFAULT_SORT_ORDER;
```

```
        } else {
            orderBy = sortOrder;
        }

        // Get the database and run the query
        SQLiteDatabase db = mOpenHelper.getReadableDatabase();
        Cursor c = qb.query(db, projection, selection,
                    selectionArgs, null, null, orderBy);

        //example of getting a count
        int i = c.getCount();

        // Tell the cursor what uri to watch,
        // so it knows when its source data changes
        c.setNotificationUri(getContext().getContentResolver(), uri);
        return c;
    }

    @Override
    public String getType(Uri uri)
    {
        switch (sUriMatcher.match(uri)) {
        case INCOMING_BOOK_COLLECTION_URI_INDICATOR:
            return BookTableMetaData.CONTENT_TYPE;

        case INCOMING_SINGLE_BOOK_URI_INDICATOR:
            return BookTableMetaData.CONTENT_ITEM_TYPE;

        default:
            throw new IllegalArgumentException("Unknown URI " + uri);
        }
    }

    @Override
    public Uri insert(Uri uri, ContentValues initialValues)
    {
        // Validate the requested uri
        if (sUriMatcher.match(uri)
                != INCOMING_BOOK_COLLECTION_URI_INDICATOR)
        {
            throw new IllegalArgumentException("Unknown URI " + uri);
        }

        ContentValues values;
        if (initialValues != null) {
            values = new ContentValues(initialValues);
        } else {
            values = new ContentValues();
        }

        Long now = Long.valueOf(System.currentTimeMillis());

        // Make sure that the fields are all set
        if (values.containsKey(BookTableMetaData.CREATED_DATE) == false)
        {
            values.put(BookTableMetaData.CREATED_DATE, now);
        }
```

```
        if (values.containsKey(BookTableMetaData.MODIFIED_DATE) == false)
        {
            values.put(BookTableMetaData.MODIFIED_DATE, now);
        }

        if (values.containsKey(BookTableMetaData.BOOK_NAME) == false)
        {
            throw new SQLException(
               "Failed to insert row because Book Name is needed " + uri);
        }

        if (values.containsKey(BookTableMetaData.BOOK_ISBN) == false) {
            values.put(BookTableMetaData.BOOK_ISBN, "Unknown ISBN");
        }
        if (values.containsKey(BookTableMetaData.BOOK_AUTHOR) == false) {
            values.put(BookTableMetaData.BOOK_ISBN, "Unknown Author");
        }

        SQLiteDatabase db = mOpenHelper.getWritableDatabase();
        long rowId = db.insert(BookTableMetaData.TABLE_NAME,
                BookTableMetaData.BOOK_NAME, values);
        if (rowId > 0) {
            Uri insertedBookUri =
                ContentUris.withAppendedId(
                        BookTableMetaData.CONTENT_URI, rowId);
            getContext()
               .getContentResolver()
                    .notifyChange(insertedBookUri, null);

            return insertedBookUri;
        }

        throw new SQLException("Failed to insert row into " + uri);
    }

    @Override
    public int delete(Uri uri, String where, String[] whereArgs)
    {
        SQLiteDatabase db = mOpenHelper.getWritableDatabase();
        int count;
        switch (sUriMatcher.match(uri)) {
        case INCOMING_BOOK_COLLECTION_URI_INDICATOR:
            count = db.delete(BookTableMetaData.TABLE_NAME,
                    where, whereArgs);
            break;

        case INCOMING_SINGLE_BOOK_URI_INDICATOR:
            String rowId = uri.getPathSegments().get(1);
            count = db.delete(BookTableMetaData.TABLE_NAME,
                    BookTableMetaData._ID + "=" + rowId
                    + (!TextUtils.isEmpty(where) ? " AND (" + where + ')' : ""),
                    whereArgs);
            break;

        default:
            throw new IllegalArgumentException("Unknown URI " + uri);
```

```
        }

        getContext().getContentResolver().notifyChange(uri, null);
        return count;
    }

    @Override
    public int update(Uri uri, ContentValues values,
            String where, String[] whereArgs)
    {
        SQLiteDatabase db = mOpenHelper.getWritableDatabase();
        int count;
        switch (sUriMatcher.match(uri)) {
        case INCOMING_BOOK_COLLECTION_URI_INDICATOR:
            count = db.update(BookTableMetaData.TABLE_NAME,
                    values, where, whereArgs);
            break;

        case INCOMING_SINGLE_BOOK_URI_INDICATOR:
            String rowId = uri.getPathSegments().get(1);
            count = db.update(BookTableMetaData.TABLE_NAME,
                    values, BookTableMetaData._ID + "=" + rowId
                    + (!TextUtils.isEmpty(where) ? " AND (" + where + ')' : ""),
                    whereArgs);
            break;

        default:
            throw new IllegalArgumentException("Unknown URI " + uri);
        }

        getContext().getContentResolver().notifyChange(uri, null);
        return count;
    }
}
```

Fulfilling MIME-Type Contracts

The BookProvider content provider must also implement the getType() method to return a MIME type for a given URI. This method, like many other methods of a content provider, is overloaded with respect to the incoming URI. As a result, the first responsibility of the getType() method is to distinguish the type of the URI. Is it a collection of books or a single book?

As we pointed out in the previous section, we will use the UriMatcher to decipher this URI type. Depending on this URI, the BookTableMetaData class has defined the MIME-type constants to return for each URI. You can see the implementation for this method in Listing 4–6.

Implementing the Query Method

The query method in a content provider is responsible for returning a collection of rows depending on an incoming URI and a where clause.

Like the other methods, the query method uses UriMatcher to identify the URI type. If the URI type is a single-item type, the method retrieves the book ID from the incoming URI like this:

1. It extracts the path segments using getPathSegments().

2. It indexes into the URI to get the first path segment, which happens to be the book ID.

The query method then uses the projections that we created up front in Listing 4–6 to identify the return columns. In the end, query returns the cursor to the caller. Throughout this process, the query method uses the SQLiteQueryBuilder object to formulate and execute the query (see Listing 4–6).

Implementing an Insert Method

The insert method in a content provider is responsible for inserting a record into the underlying database and then returning a URI that points to the newly created record.

Like the other methods, insert uses UriMatcher to identify the URI type. The code first checks whether the URI indicates the proper collection-type URI. If not, the code throws an exception (see Listing 4–6).

The code then validates the optional and mandatory column parameters. The code can substitute default values for some columns if they are missing.

Next, the code uses a SQLiteDatabase object to insert the new record and returns the newly inserted ID. In the end, the code constructs the new URI using the returned ID from the database.

Implementing an Update Method

The update method in a content provider is responsible for updating a record (or records) based on the column values passed in, as well as the where clause that is passed in. The update method then returns the number of rows updated in the process.

Like the other methods, update uses UriMatcher to identify the URI type. If the URI type is a collection, the where clause is passed through so it can affect as many records as possible. If the URI type is a single-record type, then the book ID is extracted from the URI and specified as an additional where clause. In the end, the code returns the number of records updated (see Listing 4–6). Also notice how this notifyChange method enables you to announce to the world that the data at that URI has changed. Potentially, you can do the same in the insert method by saying that "…/books" has changed when a record is inserted.

Implementing a Delete Method

The delete method in a content provider is responsible for deleting a record (or records) based on the where clause that is passed in. The delete method then returns the number of rows deleted in the process.

Like the other methods, delete uses UriMatcher to identify the URI type. If the URI type is a collection type, the where clause is passed through so you can delete as many records as possible. If the where clause is null, all records will be deleted. If the URI type is a single-record type, the book ID is extracted from the URI and specified as an additional where clause. In the end, the code returns the number of records deleted (see Listing 4–6).

Using UriMatcher to Figure Out the URIs

We've mentioned the UriMatcher class several times now; let's look into it. Almost all methods in a content provider are overloaded with respect to the URI. For example, the same query() method is called whether you want to retrieve a single book or a list of multiple books. It is up to the method to know which type of URI is being requested. Android's UriMatcher utility class helps you identify the URI types.

Here's how it works. You tell an instance of UriMatcher what kind of URI patterns to expect. You will also associate a unique number with each pattern. Once these patterns are registered, you can then ask UriMatcher if the incoming URI matches a certain pattern.

As we've mentioned, our BookProvider content provider has two URI patterns: one for a collection of books and one for a single book. The code in Listing 4–7 registers both these patterns using UriMatcher. It allocates 1 for a collection of books and a 2 for a single book (the URI patterns themselves are defined in the metadata for the books table).

Listing 4–7. *Registering URI Patterns with* `UriMatcher`

```
private static final UriMatcher sUriMatcher;
//define ids for each uri type
private static final int INCOMING_BOOK_COLLECTION_URI_INDICATOR = 1;
private static final int INCOMING_SINGLE_BOOK_URI_INDICATOR = 2;

static {
    sUriMatcher = new UriMatcher(UriMatcher.NO_MATCH);
    //Register pattern for the books
    sUriMatcher.addURI(BookProviderMetaData.AUTHORITY
                        , "books"
                        , INCOMING_BOOK_COLLECTION_URI_INDICATOR);
    //Register pattern for a single book
    sUriMatcher.addURI(BookProviderMetaData.AUTHORITY
                       , "books/#",
                       INCOMING_SINGLE_BOOK_URI_INDICATOR);

}
```

Now that this registration is in place, you can see how UriMatcher plays a part in the query-method implementation:

```
switch (sUriMatcher.match(uri)) {
   case INCOMING_BOOK_COLLECTION_URI_INDICATOR:
   ......
   case INCOMING_SINGLE_BOOK_URI_INDICATOR:
   ......
   default:
      throw new IllegalArgumentException("Unknown URI " + uri);
}
```

Notice how the match method returns the same number that was registered earlier. The constructor of UriMatcher takes an integer to use for the root URI. UriMatcher returns this number if there are neither path segments nor authorities on the URL. UriMatcher also returns NO_MATCH when the patterns don't match. You can construct a UriMatcher with no root-matching code; in that case, Android initializes UriMatcher to NO_MATCH internally. So you could have written the code in Listing 4–7 as follows instead:

```
 static {
        sUriMatcher = new UriMatcher();
        sUriMatcher.addURI(BookProviderMetaData.AUTHORITY
                                  , "books"
                                  , INCOMING_BOOK_COLLECTION_URI_INDICATOR);

        sUriMatcher.addURI(BookProviderMetaData.AUTHORITY
                                 , "books/#",
                                 INCOMING_SINGLE_BOOK_URI_INDICATOR);
}
```

Using Projection Maps

A content provider acts like an intermediary between an abstract set of columns and a real set of columns in a database, yet these column sets might differ. While constructing queries, you must map between the where clause columns that a client specifies and the real database columns. You set up this *projection map* with the help of the SQLiteQueryBuilder class.

Here is what the Android SDK documentation says about the mapping method public void setProjectionMap(Map columnMap) available on the QueryBuilder class:

> *Sets the projection map for the query. The projection map maps from column names that the caller passes into query to database column names. This is useful for renaming columns as well as disambiguating column names when doing joins. For example you could map "name" to "people.name". If a projection map is set it must contain all column names the user may request, even if the key and value are the same.*

Here is how our BookProvider content provider sets up the projection map:

```
sBooksProjectionMap = new HashMap<String, String>();
sBooksProjectionMap.put(BookTableMetaData._ID, BookTableMetaData._ID);

//name, isbn, author
sBooksProjectionMap.put(BookTableMetaData.BOOK_NAME
                                            , BookTableMetaData.BOOK_NAME);
sBooksProjectionMap.put(BookTableMetaData.BOOK_ISBN
                                            , BookTableMetaData.BOOK_ISBN);
sBooksProjectionMap.put(BookTableMetaData.BOOK_AUTHOR
                                            , BookTableMetaData.BOOK_AUTHOR);

//created date, modified date
sBooksProjectionMap.put(BookTableMetaData.CREATED_DATE
                                            , BookTableMetaData.CREATED_DATE);
sBooksProjectionMap.put(BookTableMetaData.MODIFIED_DATE
                                            , BookTableMetaData.MODIFIED_DATE);
```

And then the query builder uses the variable sBooksProjectionMap like this:

```
queryBuilder.setTables(BookTableMetaData.TABLE_NAME);
queryBuilder.setProjectionMap(sBooksProjectionMap);
```

Registering the Provider

Finally, you must register the content provider in the Android.Manifest.xml file using the tag structure in Listing 4–8.

Listing 4–8. *Registering a Provider*

```
<provider android:name=".BookProvider"
   android:authorities="com.androidbook.provider.BookProvider"/>
```

Exercising the Book Provider

Now that we have a book provider, we are going to show you sample code to exercise that provider. The sample code includes adding a book, removing a book, getting a count of the books, and finally displaying all the books.

Keep in mind that these are code extracts from the sample project and will not compile, because they require additional dependency files. However, we feel this sample code is valuable in demonstrating the concepts we have explored.

At the end of this chapter, we have included a link to the downloadable sample project, which you can use in your Eclipse environment to compile and test.

Adding a Book

The code in Listing 4–9 inserts a new book into the book database.

Listing 4–9. *Exercising a Provider Insert*

```
public void addBook(Context context)
{
    String tag = "Exercise BookProvider";
    Log.d(tag,"Adding a book");
    ContentValues cv = new ContentValues();
    cv.put(BookProviderMetaData.BookTableMetaData.BOOK_NAME, "book1");
    cv.put(BookProviderMetaData.BookTableMetaData.BOOK_ISBN, "isbn-1");
    cv.put(BookProviderMetaData.BookTableMetaData.BOOK_AUTHOR, "author-1");

    ContentResolver cr = context.getContentResolver();
    Uri uri = BookProviderMetaData.BookTableMetaData.CONTENT_URI;
    Log.d(tag,"book insert uri:" + uri);
    Uri insertedUri = cr.insert(uri, cv);
    Log.d(tag,"inserted uri:" + insertedUri);
}
```

Removing a Book

The code in Listing 4–10 deletes the last record from the book database. See Listing 4–11 for an example of how the getCount() method in Listing 4–10 works.

Listing 4–10. *Exercising a Provider delete*

```
public void removeBook(Context context)
{
    String tag = "Exercise BookProvider";
    int i = getCount(context); //See the getCount function in Listing 4-11
    ContentResolver cr = context.getContentResolver();
    Uri uri = BookProviderMetaData.BookTableMetaData.CONTENT_URI;
    Uri delUri = Uri.withAppendedPath(uri, Integer.toString(i));
    Log.d(tag, "Del Uri:" + delUri);
    cr.delete(delUri, null, null);
    Log.d(tag, "New count:" + getCount(context));
}
```

Please note that this is a quick example to show how delete works with a URI. The algorithm to get the last URI may not be valid in all cases. However, it should work if you were to add five records and proceed to delete them one by one from the end. In a real case, you would want to display the records in a list and ask the user to pick one to delete, in which case you will know the exact URI of the record.

Getting a Count of the Books

The code in Listing 4–11 gets the database cursor and counts the number of records in the cursor.

Listing 4–11. *Counting the Records in a Table*

```
private int getCount(Context context)
{
    Uri uri = BookProviderMetaData.BookTableMetaData.CONTENT_URI;
    Activity a = (Activity)context;
    Cursor c = a.managedQuery(uri,
            null, //projection
            null, //selection string
            null, //selection args array of strings
            null); //sort order
    int numberOfRecords = c.getCount();
    c.close();
    return numberOfRecords;
}
```

Displaying the List of Books

The code in Listing 4–12 retrieves all the records in the book database.

Listing 4–12. *Displaying a List of Books*

```
public void showBooks(Context context)
{
    String tag = "Exercise BookProvider";
    Uri uri = BookProviderMetaData.BookTableMetaData.CONTENT_URI;
    Activity a = (Activity)context;
    Cursor c = a.managedQuery(uri,
            null, //projection
            null, //selection string
            null, //selection args array of strings
            null); //sort order

    int iname = c.getColumnIndex(
        BookProviderMetaData.BookTableMetaData.BOOK_NAME);

    int iisbn = c.getColumnIndex(
        BookProviderMetaData.BookTableMetaData.BOOK_ISBN);
    int iauthor = c.getColumnIndex(
         BookProviderMetaData.BookTableMetaData.BOOK_AUTHOR);

    //Report your indexes
    Log.d(tag,"name,isbn,author:" + iname + iisbn + iauthor);

    //walk through the rows based on indexes
    for(c.moveToFirst();!c.isAfterLast();c.moveToNext())
    {
        //Gather values
        String id = c.getString(1);
        String name = c.getString(iname);
        String isbn = c.getString(iisbn);
```

```
        String author = c.getString(iauthor);

        //Report or log the row
        StringBuffer cbuf = new StringBuffer(id);
        cbuf.append(",").append(name);
        cbuf.append(",").append(isbn);
        cbuf.append(",").append(author);
        Log.d(tag, cbuf.toString());
    }

    //Report how many rows have been read
    int numberOfRecords = c.getCount();
    Log.d(tag,"Num of Records:" + numberOfRecords);

    //Close the cursor
    //ideally this should be done in
    //a finally block.
    c.close();
}
```

Resources

Here are some additional Android resources that can help you with the topics covered in this chapter:

- http://developer.android.com/guide/topics/providers/content-providers.html: Android documentation on content providers.
- http://developer.android.com/reference/android/content/ContentProvider.html: API description for a ContentProvider, where you can learn about ContentProvider contracts.
- http://developer.android.com/reference/android/content/UriMatcher.html: Information that is useful for understanding UriMatcher.
- http://developer.android.com/reference/android/database/Cursor.html: Information that helps you read data from a content provider or a database directly.
- www.sqlite.org/sqlite.html: Home page of SQLite, where you can learn more about SQLite and download tools that you can use to work with SQLite databases.
- androidbook.com/proandroid4/projects: Downloadable test project for this chapter is accessible from this URL. The name of the zip file is ProAndroid4_Ch04_TestProvider.zip.

Summary

In this chapter, you have learned the following:

- What content providers are

- How to discover existing content-provider databases
- The nature of content URIs, MIME types, and content providers
- How to use SQLite to construct providers that respond to URIs
- How to expose data across process boundaries to various applications
- How to write a new content provider
- How to access an existing content provider
- How to use `URIMatcher` to aid the implementation of content providers

Interview Questions

The following questions should solidify your understanding of content providers as covered in this chapter:

1. How are content providers similar to web sites?
2. Can you name some built-in content providers?
3. What can you do with an adb tool?
4. What is an AVD?
5. How do you list available AVDs?
6. What are the names of some useful command-line tools in Android?
7. Where are your databases kept for content providers?
8. What is a good way to explore a content-provider database?
9. What is the `authority` property of a content provider?
10. Can the authority of a content provider be shortened?
11. What are MIME types, and how are they connected to content providers?
12. How does a programmer discover the URIs to access a content provider?
13. How do you use a content provider URI to access data?
14. How do you supply a `where` clause to a content provider query?
15. How do you walk through a cursor?
16. What is the role of the `ContentValues` class?
17. What is the role of the `ContentResolver` class?
18. What is the protocol to save a file in a content provider?
19. How does `URIMatcher` work, and what is it used for?

Chapter 5

Understanding Intents

Android introduced a concept called *intents* to invoke components. The list of components in Android includes activities (UI components), services (background code), broadcast receivers (code that responds to broadcast messages), and content providers (code that abstracts data).

Basics of Android Intents

Although an intent is easily understood as a mechanism to invoke components, Android folds multiple ideas into the concept of an *intent.* You can use intents to invoke external applications from your application. You can use intents to invoke internal or external components from your application. You can use intents to raise events so that others can respond in a manner similar to a publish-and-subscribe model. You can use intents to raise alarms.

> **NOTE:** What is an intent? The short answer may be that an intent is an action with its associated data payload.

At the simplest level, an intent is an action that you can tell Android to perform (or *invoke*). The action Android invokes depends on what is registered for that action. Imagine you've written the following activity:

```
public class BasicViewActivity extends Activity
{
    @Override
    public void onCreate(Bundle savedInstanceState)
    {
        super.onCreate(savedInstanceState);
        setContentView(R.layout.some_view);
    }
}//eof-class
```

The layout some_view needs to point to a valid layout file in the /res/layout directory. Android then allows you to register this activity in the manifest file of that application, making it available for other applications to invoke. The registration looks like this:

```
<activity android:name=".BasicViewActivity"
          android:label="Basic View Tests">
 <intent-filter>
   <action android:name="com.androidbook.intent.action.ShowBasicView"/>
   <category android:name="android.intent.category.DEFAULT" />
 </intent-filter>
</activity>
```

The registration here involves not only an activity but also an action that you can use to invoke that activity. The activity designer usually chooses a name for the action and specifies that action as part of an intent filter for this activity. As we go through the rest of the chapter, you will have a chance to learn more about these intent filters.

Now that you have specified the activity and its registration against an action, you can use an intent to invoke this BasicViewActivity:

```
public static void invokeMyApplication(Activity parentActivity)
{
   String actionName= "com.androidbook.intent.action.ShowBasicView";
   Intent intent = new Intent(actionName);
   parentActivity.startActivity(intent);
}
```

NOTE: The general convention for an action name is <*your-package-name*>.intent.action.*YOUR_ACTION_NAME*.

Once the BasicViewActivity is invoked, it has the ability to discover the intent that invoked it. Here is the BasicViewActivity code rewritten to retrieve the intent that invoked it:

```
public class BasicViewActivity extends Activity
{
    @Override
    public void onCreate(Bundle savedInstanceState)
    {
        super.onCreate(savedInstanceState);
        setContentView(R.layout.some_view);
        Intent intent = this.getIntent();
        if (intent == null)
        {
            Log.d("test tag", "This activity is invoked without an intent");
        }
    }
}//eof-class
```

Available Intents in Android

You can give intents a test run by invoking some of the applications that come with Android. The page at `http://developer.android.com/guide/appendix/g-app-intents.html` documents some of the available Google applications and the intents that invoke them.

> **NOTE:** This list may change depending on the Android release.

The set of available applications could include the following:

- A browser application to open a browser window
- An application to call a telephone number
- An application to present a phone dialer so the user can enter the numbers and make a call through the UI
- A mapping application to show the map of the world at a given latitude and longitude coordinate
- A detailed mapping application that can show Google street views

Listing 5–1 has the code to invoke these applications through their published intents.

Listing 5–1. *Exercising Android's Prefabricated Applications*

```
public class IntentsUtils
{
    public static void invokeWebBrowser(Activity activity)
    {
       Intent intent = new Intent(Intent.ACTION_VIEW);
       intent.setData(Uri.parse("http://www.google.com"));
       activity.startActivity(intent);
    }
    public static void invokeWebSearch(Activity activity)
    {
       Intent intent = new Intent(Intent.ACTION_WEB_SEARCH);
       intent.setData(Uri.parse("http://www.google.com"));
       activity.startActivity(intent);
    }
    public static void dial(Activity activity)
    {
       Intent intent = new Intent(Intent.ACTION_DIAL);
       activity.startActivity(intent);
    }

    public static void call(Activity activity)
    {
       Intent intent = new Intent(Intent.ACTION_CALL);
       intent.setData(Uri.parse("tel:555-555-5555"));
       activity.startActivity(intent);
    }
```

```
    public static void showMapAtLatLong(Activity activity)
    {
        Intent intent = new Intent(Intent.ACTION_VIEW);
        //geo:lat,long?z=zoomlevel&q=question-string
        intent.setData(Uri.parse("geo:0,0?z=4&q=business+near+city"));
        activity.startActivity(intent);
    }

    public static void tryOneOfThese(Activity activity)
    {
        IntentsUtils.invokeWebBrowser(activity);
    }
}
```

You will be able to exercise this code as long you have a simple activity with a menu item to invoke `tryOneOfThese(activity)`. Creating a simple menu is easy (see Listing 5–2).

Listing 5–2. *A Test Harness to Create a Simple Menu*

```
public class MainActivity extends Activity
{
    public void onCreate(Bundle savedInstanceState)    {
        super.onCreate(savedInstanceState);

        TextView tv = new TextView(this);
        tv.setText("Hello, Android. Say hello");
        setContentView(tv);
    }
    @Override
    public boolean onCreateOptionsMenu(Menu menu)    {
        super.onCreateOptionsMenu(menu);
        int base=Menu.FIRST; // value is 1
        MenuItem item1 = menu.add(base,base,base,"Test");
        return true;
    }

    @Override
    public boolean onOptionsItemSelected(MenuItem item)    {
        if (item.getItemId() == 1)        {
            IntentUtils.tryOneOfThese(this);
        }
        else {
          return super.onOptionsItemSelected(item);
        }
        return true;
    }
}
```

NOTE: See Chapter 2 for instructions on how to make an Android project out of these files, as well as how to compile and run it. You can also read the early parts of Chapter 7 to see more sample code relating to menus. Or you can download the sample Eclipse project dedicated for this chapter using the URL supplied at the end of this chapter. However, when you download the sample code, this basic activity may be slightly different, but the concept remains the same. In the download sample, we also load the menus from an XML file.

Exploring Intent Composition

Another sure way to further understand an intent is to see what an intent object contains. An intent has an action, data (represented by a data URI), a key/value map of extra data elements, and an explicit class name (called a *component name*). Almost all of these are optional as long as the intent carries at least one of these. We will explore each of these parts in turn.

> **NOTE:** When an intent carries a component name with it, it is called an *explicit* intent. When an intent doesn't carry a component name but relies on other parts such as action and data, it is called an *implicit* intent. As we go through the rest of the chapter, you will see that there are subtle differences between these two types of intents.

Intents and Data URIs

So far, we've covered the simplest of the intents, where all we need is the name of an action. The ACTION_DIAL activity in Listing 5–1 is one of these; to invoke the dialer, all we needed in that listing is the dialer's action and nothing else:

```
public static void dial(Activity activity)
{
   Intent intent = new Intent(Intent.ACTION_DIAL);
   activity.startActivity(intent);
}
```

Unlike ACTION_DIAL, the intent ACTION_CALL (again referring to Listing 5–1) that is used to make a call to a given phone number takes an additional parameter called Data. This parameter points to a URI, which, in turn, points to the phone number:

```
public static void call(Activity activity)
{
   Intent intent = new Intent(Intent.ACTION_CALL);
   intent.setData(Uri.parse("tel:555-555-5555"));
   activity.startActivity(intent);
}
```

The action portion of an intent is a string or a string constant, usually prefixed by the Java package name.

The data portion of an intent is not really data but a pointer to the data. This data portion is a string representing a URI. An intent's URI can contain arguments that can be inferred as data, just like a web site's URL.

The format of this URI could be specific to each activity that is invoked by that action. In this case, the CALL action decides what kind of data URI it would expect. From the URI, it extracts the telephone number.

NOTE: The invoked activity can also use the URI as a pointer to a data source, extract the data from the data source, and use that data instead. This would be the case for media such as audio, video, and images.

Generic Actions

The actions Intent.ACTION_CALL and Intent.ACTION_DIAL could easily lead us to the wrong assumption that there is a one-to-one relationship between an action and what it invokes. To disprove this, let's consider a counterexample from the IntentUtils code in Listing 5–1:

```
public static void invokeWebBrowser(Activity activity)
{
   Intent intent = new Intent(Intent.ACTION_VIEW);
   intent.setData(Uri.parse("http://www.google.com"));
   activity.startActivity(intent);
}
```

Note that the action is simply stated as ACTION_VIEW. How does Android know which activity to invoke in response to such a generic action name? In these cases, Android relies not only on the generic action name but also on the nature of the URI. Android looks at the scheme of the URI, which happens to be http, and questions all the registered activities to see which ones understand this scheme. Out of these, it inquires which ones can handle the VIEW and then invokes that activity. For this to work, the browser activity should have registered a VIEW intent against the data scheme of http. That intent declaration might look like this in the manifest file:

```
<activity......>
<intent-filter>
         <action android:name="android.intent.action.VIEW" />
        <data android:scheme="http"/>
        <data android:scheme="https"/>
</intent-filter>
</activity>
```

You can learn about more allowed attributes of the data node by looking at the XML definition for the data element of the intent filter at http://developer.android.com/guide/topics/manifest/data-element.html. The child elements or attributes of the data XML subnode of the intent filter node include these:

```
host
mimeType
path
pathPattern
pathPrefix
port
scheme
```

mimeType is one attribute you'll see used often. For example, the following intent filter for the activity that displays a list of notes indicates the MIME type as a directory of notes:

```
<intent-filter>
   <action android:name="android.intent.action.VIEW" />
   <data android:mimeType="vnd.android.cursor.dir/vnd.google.note" />
</intent-filter>
```

This intent filter declaration can be read as "Invoke this activity to view a collection of notes."

The screen that displays a single note, on the other hand, declares its intent filter using a MIME type indicating a single note item:

```
<intent-filter>
   <action android:name="android.intent.action.VIEW" />
   <data android:mimeType="vnd.android.cursor.item/vnd.google.note" />
</intent-filter>
```

This intent filter declaration can be read as "Invoke this activity to view a single of note."

Using Extra Information

In addition to its primary attributes of action and data, an intent can include an additional attribute called *extras*. An extra can provide more information to the component that receives the intent. The extra data is in the form of key/value pairs: the key name typically starts with the package name, and the value can be any fundamental data type or arbitrary object as long as it implements the `android.os.Parcelable` interface. This extra information is represented by an Android class called `android.os.Bundle`.

The following two methods on an `Intent` class provide access to the extra `Bundle`:

```
//Get the Bundle from an Intent
Bundle extraBundle = intent.getExtras();

// Place a bundle in an intent
Bundle anotherBundle = new Bundle();

//populate the bundle with key/value pairs
...
//and then set the bundle on the Intent
intent.putExtras(anotherBundle);
```

`getExtras` is straightforward: it returns the `Bundle` that the intent has. `putExtras` checks whether the intent currently has a bundle. If the intent already has a bundle, `putExtras` transfers the additional keys and values from the new bundle to the existing bundle. If the bundle doesn't exist, `putExtras` will create one and copy the key/value pairs from the new bundle to the created bundle.

> **NOTE:** `putExtras` replicates the incoming bundle rather than referencing it. So if you were to later change the incoming bundle, you wouldn't be changing the bundle inside the intent.

You can use a number of methods to add fundamental types to the bundle. Here are some of the methods that add simple data types to the extra data:

```
putExtra(String name, boolean value);
putExtra(String name, int value);
putExtra(String name, double value);
putExtra(String name, String value);
```

And here are some not-so-simple extras:

```
//simple array support
putExtra(String name, int[] values);
putExtra(String name, float[] values);

//Serializable objects
putExtra(String name, Serializable value);

//Parcelable support
putExtra(String name, Parcelable value);

//Add another bundle at a given key
//Bundles in bundles
putExtra(String name, Bundle value);

//Add bundles from another intent
//copy of bundles
putExtra(String name, Intent anotherIntent);

//Explicit Array List support
putIntegerArrayListExtra(String name, ArrayList arrayList);
putParcelableArrayListExtra(String name, ArrayList arrayList);
putStringArrayListExtra(String name, ArrayList arrayList);
```

On the receiving side, equivalent methods starting with `get` retrieve information from the extra bundle based on key names.

The `Intent` class defines extra key strings that go with certain actions. You can discover a number of these extra-information key constants at `http://developer.android.com/reference/android/content/Intent.html#EXTRA_ALARM_COUNT`.

Let's consider a couple of example extras listed at this URL that involve sending e-mails:

- `EXTRA_EMAIL`: You will use this string key to hold a set of e-mail addresses. The value of the key is `android.intent.extra.EMAIL`. It should point to a string array of textual e-mail addresses.
- `EXTRA_SUBJECT`: You will use this key to hold the subject of an e-mail message. The value of the key is `android.intent.extra.SUBJECT`. The key should point to a string that provides the subject.

Using Components to Directly Invoke an Activity

You've seen a couple of ways to start an activity using intents. You saw an explicit action start an activity, and you saw a generic action start an activity with the help of a data URI. Android also provides a more direct way to start an activity: you can specify the activity's ComponentName, which is an abstraction around an object's package name and class name. There are a number of methods available on the Intent class to specify a component:

```
setComponent(ComponentName name);
setClassName(String packageName, String classNameInThatPackage);
setClassName(Context context, String classNameInThatContext);
setClass(Context context, Class classObjectInThatContext);
```

Ultimately, they are all shortcuts for calling one method:

```
setComponent(ComponentName name);
```

ComponentName wraps a package name and a class name together. For example, the following code invokes the contacts activity that ships with the emulator:

```
Intent intent = new Intent();
intent.setComponent(new ComponentName(
     "com.android.contacts"
    ,"com.android.contacts.DialContactsEntryActivity");
startActivity(intent);
```

Notice that the package name and the class name are fully qualified and are used in turn to construct the ComponentName before passing it to the Intent class.

You can also use the class name directly without constructing a ComponentName. Consider the BasicViewActivity code snippet again:

```
public class BasicViewActivity extends Activity
{
    @Override
    public void onCreate(Bundle savedInstanceState)
    {
        super.onCreate(savedInstanceState);
        setContentView(R.layout.some_view);
    }
}//eof-class
```

Given this, you can use the following code to start this activity:

```
Intent directIntent = new Intent(activity, BasicViewActivity.class);
activity.start(directIntent);
```

If you want any type of intent to start an activity, however, you should register the activity in the AndroidManifest.xml file like this:

```
<activity android:name=".BasicViewActivity"
          android:label="Test Activity">
```

NOTE: No intent filters are necessary for invoking an activity directly through its class name or name. As explained earlier, this type of intent is called an *explicit intent*. Because an explicit intent specifies a fully qualified Android component to invoke, the additional parts of that intent are ignored while invoking that component.

Understanding Intent Categories

You can classify activities into categories so you can search for them based on a category name. For example, during startup, Android looks for activities whose category is marked as CATEGORY_LAUNCHER. It then picks up these activity names and icons and places them on the home screen to launch.

Here's another example: Android looks for an activity tagged as CATEGORY_HOME to show the home screen during startup. Similarly, CATEGORY_GADGET marks an activity as suitable for embedding or reuse inside another activity.

The format of the string for a category like CATEGORY_LAUNCHER follows the category definition convention:

```
android.intent.category.LAUNCHER
```

You will need to know these text strings for category definitions because activities register their categories in the AndroidManifest.xml file as part of their activity filter definitions. Here is an example:

```
<activity android:name=".HelloWorldActivity"
          android:label="@string/app_name">
    <intent-filter>
        <action android:name="android.intent.action.MAIN" />
        <category android:name="android.intent.category.LAUNCHER" />
    </intent-filter>
</activity>
```

NOTE: Activities might have certain capabilities that restrict them or enable them, such as whether you can embed them in a parent activity. These types of activity characteristics are declared through categories.

Let's take a quick look at some predefined Android categories and how to use them (see Table 5–1).

Table 5–1. *Activity Categories and Their Descriptions*

Category Name	Description
CATEGORY_DEFAULT	An activity can declare itself as a DEFAULT activity if it wants to be invoked by implicit intents. If you don't define this category for your activity, that activity will need to be invoked explicitly every time through its class name. This is why you see activities that get invoked through generic actions or other action names that use default category specification.
CATEGORY_BROWSABLE	An activity can declare itself as BROWSABLE by promising the browser that it will not violate browser security considerations when started.
CATEGORY_TAB	An activity of this type is embeddable in a tabbed parent activity.
CATEGORY_ALTERNATIVE	An activity can declare itself as an ALTERNATIVE activity for a certain type of data that you are viewing. These items normally show up as part of the options menu when you are looking at that document. For example, print view is considered an alternative to regular view.
CATEGORY_SELECTED_ALTERNATIVE	An activity can declare itself as an ALTERNATIVE activity for a certain type of data. This is similar to listing a series of possible editors for a text document or an HTML document.
CATEGORY_LAUNCHER	Assigning this category to an activity will allow it to be listed on the launcher screen.
CATEGORY_HOME	An activity of this type will be the home screen. Typically, there should be only one activity of this type. If there are more, the system will provide a prompt to pick one.
CATEGORY_PREFERENCE	This activity identifies an activity as a preference activity, so it will be shown as part of the preferences screen.
CATEGORY_GADGET	An activity of this type is embeddable in a parent activity.
CATEGORY_TEST	This is a test activity.
CATEGORY_EMBED	This category has been superseded by the GADGET category, but it's been kept for backward compatibility.

You can read the details of these activity categories at the following Android SDK URL for the Intent class:
http://developer.android.com/android/reference/android/content/Intent.html#CATEGORY_ALTERNATIVE.

When you use an intent to start an activity, you can specify the kind of activity to choose by specifying a category. Or you can search for activities that match a certain category. Here is an example to retrieve a set of main activities that match the category of CATEGORY_LAUNCHER:

```
Intent mainIntent = new Intent(Intent.ACTION_MAIN, null);
mainIntent.addCategory(Intent.CATEGORY_LAUNCHER);
```

```
PackageManager pm = getPackageManager();
List<ResolveInfo> list = pm.queryIntentActivities(mainIntent, 0);
```

PackageManager is a key class that allows you to discover activities that match certain intents without invoking them. You can cycle through the received activities and invoke them as you see fit, based on the ResolveInfo API. Here is an extension to the preceding code that walks through the list of activities and invokes one of the activities if it matches a name. In the code, we have a used an arbitrary name to test it:

```
for(ResolveInfo ri: list)
{
    //ri.activityInfo.
    Log.d("test",ri.toString());
    String packagename = ri.activityInfo.packageName;
    String classname = ri.activityInfo.name;
    Log.d("test", packagename + ":" + classname);
    if (classname.equals("com.ai.androidbook.resources.TestActivity"))
    {
        Intent ni = new Intent();
        ni.setClassName(packagename,classname);
        activity.startActivity(ni);
    }
}
```

You can also start an activity based purely on an intent category such as CATEGORY_LAUNCHER:

```
public static void invokeAMainApp(Activity activity)
{
   Intent mainIntent = new Intent(Intent.ACTION_MAIN, null);
   mainIntent.addCategory(Intent.CATEGORY_LAUNCHER);
   activity.startActivity(mainIntent);
}
```

More than one activity will match the intent, so which activity will Android pick? To resolve this, Android presents a Complete Action Using dialog that lists all the possible activities so you can choose one to run.

Here is another example of using an intent to go to a home page:

```
//Go to home screen
Intent mainIntent = new Intent(Intent.ACTION_MAIN, null);
mainIntent.addCategory(Intent.CATEGORY_HOME);
startActivity(mainIntent);
```

If you don't want to use Android's default home page, you can write your own and declare that activity to be of category HOME. In that case, the preceding code will give you an option to open your home activity because more than one home activity is registered now:

```
//Replace the home screen with yours
<intent-filter>
   <action android:value="android.intent.action.MAIN" />
   <category android:value="android.intent.category.HOME"/>
   <category android:value="android.intent.category.DEFAULT" />
</intent-filter>
```

Rules for Resolving Intents to Their Components

So far, we have discussed a number of aspects about intents. To recap, we talked about actions, data URIs, extra data, and, finally, categories. Given these aspects, Android uses multiple strategies to match intents to their target activities based on intent filters.

At the top of the hierarchy is the component name attached to an intent. If this is set, the intent is known as an *explicit intent*. For an explicit intent, only the component name matters; every other aspect or attribute of the intent is ignored. When a component name is not present on an intent, the intent is said to be an *implicit intent*. The rules for resolving targets for implicit intents are numerous.

The basic rule is that an incoming intent's action, category, and data characteristics *must match* (or present) those specified in the intent filter. An intent filter, unlike an intent, can specify multiple actions, categories, and data attributes. This means the same intent filter can satisfy multiple intents, which is to say that an activity can respond to many intents. However, the meaning of "match" differs among actions, data attributes, and categories. Let's look the matching criteria for each of the parts of an implicit intent.

Action

If an intent has an action on it, the intent filter must have that action as part of its action list or not have any actions at all. So if an intent filter *doesn't define an action*, that intent filter *is a match* for any incoming intent action.

If one or more actions are specified in the intent filter, at least one of the actions must match the incoming intent's action.

Data

If no data characteristics are specified in an intent filter, it does not match an incoming intent that carries any data or data attribute. This means it will only look for intents that have no data specified at all.

Lack of data and lack of action (in the filter) work the opposite. If there is no action in the filter, every thing is a match. If there is no data in the filter, every bit of data in the intent is a mismatch.

Data Type

For a data type to match, the incoming intent's data type must be one of the data types that is specified in the intent filter. The data type in the intent must be present in the intent filter.

The incoming intent's data type is determined in one of two ways. First, if the data URI is a content or file URI, the content provider or Android will figure out the type. The second way is to look at the explicit data type of the intent. For this to work, the incoming intent should not have a data URI set, because this is automatically taken care of when `setType` is called on the intent.

Android also allows its MIME type specification to have an asterisk (*) as its subtype to cover all possible subtypes.

Also, the data type is case sensitive.

Data Scheme

For a data scheme to match, the incoming intent data scheme must be one of those specified in the intent filter. In other words, the incoming data scheme must be present in the intent filter.

The incoming intent's scheme is the first part of the data URI. On an intent, there is no method to set the scheme. It is purely derived from the intent data URI that looks like `http://www.somesite.com/somepath`.

If the data scheme of the incoming intent URI is `content:` or `file:`, it is considered a match regardless of the intent filter scheme, domain, and path. According to the SDK, this is so because every component is expected to know how to read data from content or file URLs, which are essentially local. In other words, all components are expected to support these two types of URLs.

The scheme is also case sensitive.

Data Authority

If there are no authorities in the filter, you have a match for any incoming data URI authority (or domain name). If an authority is specified in the filter—for example, `www.somesite.com`—then one scheme and one authority should match the incoming intent's data URI.

For example, if we specify `www.somesite.com` as the authority in the intent filter and the scheme as `https`, the intent will fail to match `http://www.somesite.com/somepath` because `http` is not indicated as the supporting scheme.

The authority is case sensitive as well.

Data Path

No data paths in the intent filter means a match for any incoming data URI's path. If a path is specified in the filter—for example, `somepath`—one scheme, one authority, and one data path should match the incoming intent's data URI.

In other words scheme, authority, and path work together to validate an incoming intent URI such as `http://www.somesite.com/somepath`. So `path`, `authority`, and `scheme` work not in isolation but together.

The path, too, is case sensitive.

Intent Categories

Every category in the incoming intent must be present in the filter category list. Having more categories in the filter is OK. If a filter *doesn't have any categories*, it will match *only with an intent that doesn't have any* categories mentioned.

However, there is a caveat. Android treats all *implicit* intents passed to `startActivity()` as if they contained at least one category: `android.intent.category.DEFAULT`. The code in `startActivity()` will search only for those activities that have the `DEFAULT` category defined if the incoming intent is an implicit intent. So every activity that wants to be invoked through an implicit intent must include the default category in its filters.

Even if an activity doesn't have the default category in its intent filter, if you know its explicit component names, you will be able to start it like the launcher does. If you explicitly search for matching intents yourself without having a default category as a search criterion, you will be able to start those activities that way.

In that sense, this `DEFAULT` category is an artifact of the `startActivity()` implementation and not an inherent behavior of filters.

There is an additional wrinkle because Android states that the `DEFAULT` category is unnecessary if the activity is intended to be invoked only from launcher screens. So these activities tend to have only `MAIN` and `LAUNCHER` categories as part of their filters. However, the `DEFAULT` category can be optionally specified for these activities as well.

Exercising the ACTION_PICK

So far, we have exercised intents or actions that mainly invoke another activity without expecting a result back. Let's look at an action that is a bit more involved and returns a value after being invoked. `ACTION_PICK` is one such generic action.

The idea of `ACTION_PICK` is to start an activity that displays a list of items. The activity then should allow a user to pick one item from that list. Once the user picks the item, the activity should return the URI of the picked item to the caller. This allows reuse of the UI's functionality to select items of a certain type.

You should indicate the collection of items to choose from using a MIME type that points to an Android content cursor. The MIME type of this URI should look similar to the following:

```
vnd.android.cursor.dir/vnd.google.note
```

It is the responsibility of the activity to retrieve the data from the content provider based on the URI. This is also the reason that data should be encapsulated into content providers where possible.

For actions that return data like this, we cannot use `startActivity()`, because `startActivity()` does not return a result. `startActivity()` cannot return a result, because it opens the new activity as a modal dialog in a separate thread and leaves the main thread for attending events. In other words, `startActivity()` is an asynchronous call with no callbacks to indicate what happened in the invoked activity. If you want to return data, you can use a variation of `startActivity()` called `startActivityForResult()`, which comes with a callback.

Let's look at the signature of the `startActivityForResult()` method from the `Activity` class:

```
public void startActivityForResult(Intent intent, int requestCode)
```

This method launches an activity from which you would like a result. When this activity exits, the source activity's `onActivityResult()` method will be called with the given `requestCode`. The signature of this callback method is

```
protected void onActivityResult(int requestCode, int resultCode, Intent data)
```

`requestCode` is what you passed in to the `startActivityForResult()` method. The `resultCode` can be `RESULT_OK`, `RESULT_CANCELED`, or a custom code. The custom codes should start at `RESULT_FIRST_USER`. The `Intent` parameter contains any additional data that the invoked activity wants to return. In the case of `ACTION_PICK`, the returned data in the intent points to the data URI of a single item.

Listing 5–3 demonstrates invoking an activity that sends a result back.

> **NOTE:** The code in Listing 5–3 assumes that you have installed the NotePad sample project from the Android SDK distribution. We have included a link at the end of this chapter that gives you directions on how to download the NotePad sample if you don't have it in the SDK already.

Listing 5–3. *Returning Data After Invoking an Action*

```
public class SomeActivity extends Activity
{
.....
.....
public static void invokePick(Activity activity)
{
  Intent pickIntent = new Intent(Intent.ACTION_PICK);
  int requestCode = 1;
  pickIntent.setData(Uri.parse(
     "content://com.google.provider.NotePad/notes"));
  activity.startActivityForResult(pickIntent, requestCode);
}

protected void onActivityResult(int requestCode
    ,int resultCode
    ,Intent outputIntent)
{
  //This is to inform the parent class (Activity)
  //that the called activity has finished and the baseclass
  //can do the necessary clean up
  super.onActivityResult(requestCode, resultCode, outputIntent);
  parseResult(this, requestCode, resultCode, outputIntent);
}
public static void parseResult(Activity activity
    , int requestCode
```

```
    , int resultCode
    , Intent outputIntent)
{
    if (requestCode != 1)
    {
     Log.d("Test", "Some one else called this. not us");
            return;
    }
    if (resultCode != Activity.RESULT_OK)
    {
      Log.d(Test, "Result code is not ok:" + resultCode);
             return;
    }
    Log.d("Test", "Result code is ok:" + resultCode);
    Uri selectedUri = outputIntent.getData();
    Log.d("Test", "The output uri:" + selectedUri.toString());

    //Proceed to display the note
    outputIntent.setAction(Intent.ACTION_VIEW);
    startActivity(outputIntent);
}
```

The constants RESULT_OK, RESULT_CANCELED, and RESULT_FIRST_USER are all defined in the Activity class. The numerical values of these constants are

```
RESULT_OK = -1;
RESULT_CANCELED = 0;
RESULT_FIRST_USER = 1;
```

To make the PICK functionality work, the implementer that is responding should have code that explicitly addresses the needs of a PICK. Let's look at how this is done in the Google sample NotePad application. When the item is selected in the list of items, the intent that invoked the activity is checked to see whether it's a PICK intent. If it is, the data URI is set in a new intent and returned through setResult():

```
@Override
protected void onListItemClick(ListView l, View v, int position, long id) {
    Uri uri = ContentUris.withAppendedId(getIntent().getData(), id);

    String action = getIntent().getAction();
    if (Intent.ACTION_PICK.equals(action) ||
              Intent.ACTION_GET_CONTENT.equals(action))
   {
        // The caller is waiting for us to return a note selected by
        // the user.  They have clicked on one, so return it now.
        setResult(RESULT_OK, new Intent().setData(uri));
    } else {
        // Launch activity to view/edit the currently selected item
        startActivity(new Intent(Intent.ACTION_EDIT, uri));
    }
}
```

Exercising the GET_CONTENT Action

ACTION_GET_CONTENT is similar to ACTION_PICK. In the case of ACTION_PICK, you are specifying a URI that points to a collection of items, such as a collection of notes. You will expect the action to pick one of the notes and return it to the caller. In the case of ACTION_GET_CONTENT, you indicate to Android that you need an item of a particular MIME type. Android searches for either activities that can create one of those items or activities that can choose from an existing set of items that satisfy that MIME type.

Using ACTION_GET_CONTENT, you can pick a note from a collection of notes supported by the NotePad application using the following code:

```
public static void invokeGetContent(Activity activity)
{
       Intent pickIntent = new Intent(Intent.ACTION_GET_CONTENT);
       int requestCode = 2;
       pickIntent.setType("vnd.android.cursor.item/vnd.google.note");
       activity.startActivityForResult(pickIntent, requestCode);
}
```

Notice how the intent type is set to the MIME type of a single note. Contrast this with the ACTION_PICK code in the following snippet, where the input is a data URI:

```
public static void invokePick(Activity activity)
{
  Intent pickIntent = new Intent(Intent.ACTION_PICK);
  int requestCode = 1;
  pickIntent.setData(Uri.parse(
     "content://com.google.provider.NotePad/notes"));
  activity.startActivityForResult(pickIntent, requestCode);
}
```

For an activity to respond to ACTION_GET_CONTENT, the activity has to register an intent filter indicating that the activity can provide an item of that MIME type. Here is how the SDK's NotePad application accomplishes this:

```
<activity android:name="NotesList" android:label="@string/title_notes_list">
......
<intent-filter>
    <action android:name="android.intent.action.GET_CONTENT" />
    <category android:name="android.intent.category.DEFAULT" />
    <data android:mimeType="vnd.android.cursor.item/vnd.google.note" />
      </intent-filter>
......
</activity>
```

The rest of the code for responding to onActivityResult() is identical to the previous ACTION_PICK example. If there are multiple activities that can return the same MIME type, Android will show you the chooser dialog to let you pick an activity.

Introducing Pending Intents

Android has a variation on an intent called a *pending intent*. In this variation, Android allows a component to store an intent for future use in a location from which it can be invoked again. For example, in an alarm manager, you want to start a service when the alarm goes off. Android does this by creating a wrapper pending intent around a normal corresponding intent and storing it away so that even if the calling process dies off, the intent can be dispatched to its target. At the time of the pending intent creation, Android stores enough information about the originating process that security credentials can be checked at the time of dispatch or invocation.

Let's see how we can go about creating a pending intent:

```
Intent regularIntent;
PendingIntent pi = PendingIntent.getActivity(context, 0, regularIntent,...);
```

> **NOTE:** The second argument to the `PendingIntent.getActivity()` method is called `requestCode`, and in this example we are setting it to zero. This argument is used to distinguish two pending intents when their underlying intents are the same. This aspect is discussed in much more detail in Chapter 20, where we talk about pending intents in the context of alarm managers.

There are a couple of odd things here when it comes to the naming of the method `PendingActivity.getActivity()`. What is the role of an activity here? And why don't we call `create` for creating a pending intent but instead use `get`?

To understand the first point, we have to dig a bit into the usage of a regular intent. A regular intent can be used to start an activity or a service or invoke a broadcast receiver. (You will learn about services and broadcast receivers later in this book.) The nature of using an intent to call these different sorts of components is different. To accommodate this, an Android context (a superclass of `Activity`) provides three distinct methods:

```
startActivty(intent)
startService(intent)
sendBroadcast(intent)
```

Given these variations, if we were to store an intent to be reused later, how would Android know whether to start an activity, start a service, or start a broadcast receiver due to a broadcast? This is why we have to explicitly specify the purpose for which we are creating the pending intent when it's created, and it explains the following three separate methods:

```
PendingIntent.getActivity(context, 0, intent, ...)
PendingIntent.getService(context, 0, intent, ...)
PendingIntent.getBroadcast(context, 0, intent, ...)
```

Now to explain the "get" part. Android stores away intents and reuses them. If you ask for a pending intent using the same intent object twice, you get the same pending intent.

This becomes a bit clearer if you see the full signature of the `PendingIntent.getActivity()` method:

```
PendingIntent.getActivity(Context context, //originating context
     int requestCode, //1,2, 3, etc
     Intent intent, //original intent
     int flags ) //flags
```

If your goal is to get a different copy of the pending intent, you have to supply a different `requestCode`. This need is explained in much greater detail when we cover alarm managers in Chapter 20. Two intents are considered identical if their internal parts match except for the extra bundle. The extra bundle is allowed to differ and will not affect the uniqueness of intents. If you want to force uniqueness among two otherwise identical underlying intents, you can vary the request code argument value. This will make the pending intents unique even though the underlying intents are not.

The flags indicate what to do if there is an existing pending intent—whether to return a null, overwrite extras, and so on. See the following URL to see more detail about the possible flags:

`http://developer.android.com/reference/android/app/PendingIntent.html`

Usually, you can pass a zero for `requestCode` and flags to get the default behavior.

Resources

Here are some useful links to further strengthen your understanding of this chapter:

- `http://developer.android.com/reference/android/content/Intent.html`: Overview of intents, including well-known actions, extras, and so on.
- `http://developer.android.com/guide/appendix/g-app-intents.html`: Lists the intents for a set of Google applications. Here, you will see here how to invoke Browser, Map, Dialer, and Google Street View.
- `http://developer.android.com/reference/android/content/IntentFilter.html`: Talks about intent filters and is useful when you are registering intent filters.
- `http://developer.android.com/guide/topics/intents/intents-filters.html`: Goes into the resolution rules of intent filters.
- `http://developer.android.com/resources/samples/get.html`: URL where you can download the sample code for NotePad application. You need this sample project loaded to test some of the intents.
- `http://developer.android.com/resources/samples/NotePad/index.html`: Online source code for the NotePad application.
- `www.openintents.org/`: A web effort to collect open intents from various vendors.
- `www.androidbook.com/proandroid4/projects`: Downloadable test project for this chapter is available at this URL. The name of the ZIP file is `ProAndroid4_ch05_TestIntents.zip`.

Summary

This chapter has covered the following about intents:

- An implicit intent is a collection of actions, data URIs, and explicit data passed in as extras.
- An explicit intent is an intent that is directly tied to a class name with no regard to its implicit parts mentioned in previous parts.
- You use an intent to invoke activities or other components in Android.
- Components such as activities declare which intents they would like to respond to through intent filters.
- Resolution rules between intents and intent filters.
- How to start activities using intents.
- How to start activities that return results.
- The role of intent categories.
- The nuances of the default category.
- What pending intents are and how they are used.
- The uniqueness of pending intents.
- How to use `PICK` and `GET_CONTENT` actions.

Interview Questions

1. How can you use an intent to invoke an activity?
2. What are explicit intents and implicit intents?
3. What are constituent parts of an intent?
4. How do you send data through an intent to a receiving component?
5. Can you name the main components in an Android application?
6. Does the data portion of an intent contain data directly?
7. Should the action part of an intent directly refer to an activity or a component?
8. What additional portions of an intent are considered when the class name is specified explicitly in the intent?
9. What is the meaning of `action.MAIN`?
10. If you don't specify an action in an intent filter, does it mean your activity can respond to all actions?

11. If you don't specify data in your intent filter, what type of intents do you match?
12. Why is it necessary to have a default category for your activity in your intent filter?
13. Does your launcher activity need a default category?
14. How can you call an activity that can return a result to the caller?
15. What is the quickest way to invoke an activity?
16. What is the difference between `action_pick` and `action_get_content`?

Chapter 6

Building User Interfaces and Using Controls

Thus far, we have covered the fundamentals of Android but have not touched the user interface (UI). In this chapter, we are going to discuss user interfaces and controls. We will begin by discussing the general philosophy of UI development in Android, and then we'll describe many of the UI controls that ship with the Android SDK. These are the building blocks of the interfaces you'll create. We will also discuss view adapters and layout managers. View adapters are used to provide data to the controls that show sets of data, whether those sets come from arrays, databases, or other data sources. As the name suggests, a layout manager governs where controls appear on the screen. Along the way, we'll also cover styles and themes, which help to encapsulate control-appearance attributes for easier setup and maintenance.

By the end of this chapter, you'll have a solid understanding of how to lay out UI controls into screens and populate them with data.

UI Development in Android

UI development in Android is fun. It's fun because it's relatively easy. With Android, we have a simple-to-understand framework with a limited set of out-of-the-box controls. The available screen area is generally limited. Android also takes care of a lot of the heavy lifting normally associated to designing and building quality UIs. This, combined with the fact that the user usually wants to do one specific action, allows us to easily build a good UI to deliver a good user experience.

The Android SDK ships with a host of controls that you can use to build UIs for your application. Similar to other SDKs, the Android SDK provides text fields, buttons, lists, grids, and so on. In addition, Android provides a collection of controls that are appropriate for mobile devices.

At the heart of the common controls are two classes: `android.view.View` and `android.view.ViewGroup`. As the name of the first class suggests, the `View` class represents a general-purpose `View` object. The common controls in Android ultimately extend the `View` class. `ViewGroup` is also a view, but it contains other views too. `ViewGroup` is the base class for a list of layout classes. Android, like Swing, uses the concept of *layouts* to manage how controls are laid out within a container view. Using layouts, as we'll see, makes it easy for us to control the position and orientation of the controls in our UIs.

You can choose from several approaches to build UIs in Android. You can construct UIs entirely in code. You can also define UIs in XML. You can even combine the two—define the UI in XML and then refer to it, and modify it, in code. To demonstrate this, in this chapter we are going to build a simple UI using each of these three approaches.

Before we get started, let's define some nomenclature. In this book and other Android literature, you will find the terms *view*, *control*, *widget*, *container*, and *layout* in discussions regarding UI development. If you are new to Android programming or UI development in general, you might not be familiar with these terms. We'll briefly describe them before we get started (see Table 6–1).

Table 6–1. *UI Nomenclature*

Term	Description
View, widget, control	Each of these represents a UI element. Examples include a button, a grid, a list, a window, a dialog box, and so on. The terms *view*, *widget*, and *control* are used interchangeably in this chapter.
Container	This is a view used to contain other views. For example, a grid can be considered a container because it contains cells, each of which is a view.
Layout	This is a visual arrangement of containers and views and can include other layouts.

Figure 6–1 shows a screenshot of the application that we are going to build. Next to the screenshot is the layout hierarchy of the controls and containers in the application.

Figure 6–1. *The UI and layout of an activity*

We will refer to this layout hierarchy as we discuss the sample programs. For now, know that the application has one activity. The UI for the activity is composed of three containers: a container that contains a person's name, a container that contains the address, and an outer parent container for the child containers.

Building a UI Completely in Code

The first example, Listing 6–1, demonstrates how to build the UI entirely in code. To try this, create a new Android project with an activity named `MainActivity` and then copy the code from Listing 6–1 into your `MainActivity` class.

> **NOTE:** We will give you a URL at the end of the chapter that you can use to download projects from this chapter. This will allow you to import these projects into Eclipse directly instead of copying and pasting code.

Listing 6–1. *Creating a Simple User Interface Entirely in Code*

```
package com.androidbook.controls;
import android.app.Activity;
import android.os.Bundle;
import android.view.ViewGroup.LayoutParams;
import android.widget.LinearLayout;
import android.widget.TextView;
public class MainActivity extends Activity
{
    private LinearLayout nameContainer;

    private LinearLayout addressContainer;

    private LinearLayout parentContainer;

    /** Called when the activity is first created. */
    @Override
    public void onCreate(Bundle savedInstanceState)
    {
        super.onCreate(savedInstanceState);

        createNameContainer();

        createAddressContainer();

        createParentContainer();

        setContentView(parentContainer);
    }

    private void createNameContainer()
    {
        nameContainer = new LinearLayout(this);

        nameContainer.setLayoutParams(new LayoutParams(LayoutParams.FILL_PARENT,
                LayoutParams.WRAP_CONTENT));
        nameContainer.setOrientation(LinearLayout.HORIZONTAL);

        TextView nameLbl = new TextView(this);
        nameLbl.setText("Name: ");

        TextView nameValue = new TextView(this);
```

```
        nameValue.setText("John Doe");

        nameContainer.addView(nameLbl);
        nameContainer.addView(nameValue);
    }

    private void createAddressContainer()
    {
        addressContainer = new LinearLayout(this);

        addressContainer.setLayoutParams(new LayoutParams(LayoutParams.FILL_PARENT,
                LayoutParams.WRAP_CONTENT));
        addressContainer.setOrientation(LinearLayout.VERTICAL);

        TextView addrLbl = new TextView(this);
        addrLbl.setText("Address:");

        TextView addrValue = new TextView(this);
        addrValue.setText("911 Hollywood Blvd");

        addressContainer.addView(addrLbl);
        addressContainer.addView(addrValue);
    }

    private void createParentContainer()
    {
        parentContainer = new LinearLayout(this);

        parentContainer.setLayoutParams(new LayoutParams(LayoutParams.FILL_PARENT,
                LayoutParams.FILL_PARENT));
        parentContainer.setOrientation(LinearLayout.VERTICAL);

        parentContainer.addView(nameContainer);
        parentContainer.addView(addressContainer);
    }
}
```

As shown in Listing 6–1, the activity contains three `LinearLayout` objects. As we mentioned earlier, layout objects contain logic to position objects within a portion of the screen. A `LinearLayout`, for example, knows how to lay out controls either vertically or horizontally. Layout objects can contain any type of view—even other layouts.

The `nameContainer` object contains two `TextView` controls: one for the label `Name:` and the other to hold the actual name (such as John Doe). The `addressContainer` also contains two `TextView` controls. The difference between the two containers is that the `nameContainer` is laid out horizontally and the `addressContainer` is laid out vertically. Both of these containers live within the `parentContainer`, which is the root view of the activity. After the containers have been built, the activity sets the content of the view to the root view by calling `setContentView(parentContainer)`. When it comes time to render the UI of the activity, the root view is called to render itself. The root view then calls its children to render themselves, and the child controls call their children, and so on, until the entire UI is rendered.

As shown in Listing 6–1, we have several `LinearLayout` controls. Two of them are laid out vertically, and one is laid out horizontally. The `nameContainer` is laid out horizontally.

This means the two TextView controls appear side by side horizontally. The addressContainer is laid out vertically, which means the two TextView controls are stacked one on top of the other. The parentContainer is also laid out vertically, which is why the nameContainer appears above the addressContainer. Note a subtle difference between the two vertically laid-out containers, addressContainer and parentContainer. parentContainer is set to take up the entire width and height of the screen:

```
parentContainer.setLayoutParams(new LayoutParams(LayoutParams.FILL_PARENT,
        LayoutParams.FILL_PARENT));
```

And addressContainer wraps its content vertically:

```
addressContainer.setLayoutParams(new LayoutParams(LayoutParams.FILL_PARENT,
        LayoutParams.WRAP_CONTENT));
```

Said another way, WRAP_CONTENT means the view should take just the space it needs in that dimension and no more, up to what the containing view will allow. For the addressContainer, this means the container will take two lines vertically, because that's all it needs.

Building a UI Completely in XML

Now let's build the same UI in XML (see Listing 6–2). Recall from Chapter 3 that XML layout files are stored under the resources (/res/) directory in a folder called layout. To try this example, create a new Android project in Eclipse. By default, you will get an XML layout file named main.xml, located under the res/layout folder. Double-click main.xml to see the contents. Eclipse will display a visual editor for your layout file. You probably have a string at the top of the view that says "Hello World, MainActivity!" or something like that. Click the main.xml tab at the bottom of the view to see the XML of the main.xml file. This reveals a LinearLayout and a TextView control. Using either the Layout or main.xml tab, or both, re-create Listing 6–2 in the main.xml file. Save it.

Listing 6–2. *Creating a User Interface Entirely in XML*

```
<?xml version="1.0" encoding="utf-8"?>
<LinearLayout xmlns:android="http://schemas.android.com/apk/res/android"
    android:orientation="vertical" android:layout_width="fill_parent"
    android:layout_height="fill_parent">
    <!-- NAME CONTAINER -->
    <LinearLayout xmlns:android="http://schemas.android.com/apk/res/android"
        android:orientation="horizontal" android:layout_width="fill_parent"
        android:layout_height="wrap_content">

            <TextView  android:layout_width="wrap_content"
        android:layout_height="wrap_content" android:text="Name:" />

            <TextView android:layout_width="wrap_content"
        android:layout_height="wrap_content" android:text="John Doe" />

    </LinearLayout>

    <!-- ADDRESS CONTAINER -->
    <LinearLayout xmlns:android="http://schemas.android.com/apk/res/android"
        android:orientation="vertical" android:layout_width="fill_parent"
```

```
        android:layout_height="wrap_content">

            <TextView android:layout_width="fill_parent"
        android:layout_height="wrap_content" android:text="Address:" />

            <TextView android:layout_width="fill_parent"
        android:layout_height="wrap_content" android:text="911 Hollywood Blvd." />
    </LinearLayout>

</LinearLayout>
```

Under your new project's `src` directory, there is a default `.java` file containing an `Activity` class definition. Double-click that file to see its contents. Notice the statement `setContentView(R.layout.main)`. The XML snippet shown in Listing 6–2, combined with a call to `setContentView(R.layout.main)`, will render the same UI as before when we generated it completely in code. The XML file is self-explanatory, but note that we have three container views defined. The first `LinearLayout` is the equivalent of our parent container. This container sets its orientation to vertical by setting the corresponding property like this: `android:orientation="vertical"`. The parent container contains two `LinearLayout` containers, which represent `nameContainer` and `addressContainer`.

Running this application will produce the same UI as our previous example application. The labels and values will be displayed as shown in Figure 6–1.

Building a UI in XML with Code

Listing 6–2 is a contrived example. It doesn't make any sense to hard-code the values of the `TextView` controls in the XML layout. Ideally, we should design our UIs in XML and then reference the controls from code. This approach enables us to bind dynamic data to the controls defined at design time. In fact, this is the recommended approach. It is fairly easy to build layouts in XML and then use code to populate the dynamic data.

Listing 6–3 shows the same UI with slightly different XML. This XML assigns IDs to the `TextView` controls so that we can refer to them in code.

Listing 6–3. ***Creating a User Interface in XML with IDs***

```
<?xml version="1.0" encoding="utf-8"?>
<LinearLayout xmlns:android="http://schemas.android.com/apk/res/android"
    android:orientation="vertical" android:layout_width="fill_parent"
    android:layout_height="fill_parent">
    <!-- NAME CONTAINER -->
    <LinearLayout xmlns:android="http://schemas.android.com/apk/res/android"
        android:orientation="horizontal" android:layout_width="fill_parent"
        android:layout_height="wrap_content">

            <TextView android:layout_width="wrap_content"
        android:layout_height="wrap_content" android:text="@string/name_text" />

            <TextView android:id="@+id/nameValue"
        android:layout_width="wrap_content" android:layout_height="wrap_content" />

    </LinearLayout>
```

```
    <!-- ADDRESS CONTAINER -->
    <LinearLayout xmlns:android="http://schemas.android.com/apk/res/android"
        android:orientation="vertical" android:layout_width="fill_parent"
        android:layout_height="wrap_content">

            <TextView android:layout_width="fill_parent"
        android:layout_height="wrap_content" android:text="@string/addr_text" />

            <TextView android:id="@+id/addrValue"
        android:layout_width="fill_parent" android:layout_height="wrap_content" />
    </LinearLayout>

</LinearLayout>
```

In addition to adding the IDs to the TextView controls that we want to populate from code, we also have label TextView controls that we're populating with text from our strings resource file. These are the TextViews without IDs that have an android:text attribute. As you may recall from Chapter 3, the actual strings for these TextViews will come from our strings.xml file in the /res/values folder. Listing 6–4 shows what our strings.xml file might look like.

Listing 6–4. *strings.xml File for Listing 6–3*

```
<?xml version="1.0" encoding="utf-8"?>
<resources>
    <string name="app_name">Common Controls</string>
    <string name="name_text">Name:</string>
    <string name="addr_text">Address:</string>
</resources>;
```

The code in Listing 6–5 demonstrates how you can obtain references to the controls defined in the XML to set their properties. You might put this into your onCreate() method for your activity.

Listing 6–5. *Referring to Controls in Resources at Runtime*

```
setContentView(R.layout.main);

TextView nameValue = (TextView)findViewById(R.id.nameValue);
nameValue.setText("John Doe");
TextView addrValue = (TextView)findViewById(R.id.addrValue);
addrValue.setText("911 Hollywood Blvd.");
```

The code in Listing 6–5 is straightforward, but note that we load the resource by calling setContentView(R.layout.main) before calling findViewById()—we cannot get references to views if they have not been loaded yet.

The developers of Android have done a nice job of making just about every aspect of a control settable via XML or code. It's usually a good idea to set the control's attributes in the XML layout file rather than using code. However, there will be lots of times when you need to use code, such as setting a value to be displayed to the user.

FILL_PARENT vs. MATCH_PARENT

The constant FILL_PARENT was deprecated in Android 2.2 and replaced with MATCH_PARENT. This was strictly a name change, though. The value of this constant is still -1. Similarly, for XML layouts, fill_parent was replaced with match_parent. So what value do you use? Instead of FILL_PARENT or MATCH_PARENT, you could simply use the value -1, and you'd be fine. However, this isn't very easy to read, and you don't have an equivalent unnamed value to use with your XML layouts. There's a better way.

Depending on which Android APIs you need to use in your application, you can either build your application against a version of Android before 2.2 and rely on forward compatibility or build your application against version 2.2 or later of Android and set minSdkVersion to the lowest version of Android your application will run on. For example, if you only need APIs that existed in Android 1.6, build against Android 1.6, and use FILL_PARENT and fill_parent. Your application should run with no problems in all later versions of Android including 2.2 and beyond. If you need APIs from Android 2.2 or later, go ahead and build against that version of Android, use MATCH_PARENT and match_parent, and set minSdkVersion to something older: for example, 4 (for Android 1.6). You can still deploy an Android application built in Android 2.2 to an older version of Android, but you'll have to be careful about the classes and/or methods that aren't in the earlier releases of the Android SDK. There are ways around this, such as using reflection or creating wrapper classes to handle differences in Android versions. We get into those advanced topics in Chapter 12.

Understanding Android's Common Controls

We will now start our discussion of the common controls in the Android SDK. We'll start with text controls and then cover buttons, check boxes, radio buttons, lists, grids, date and time controls, and a map-view control. We will also talk about layout controls.

Text Controls

Text controls are likely to be the first type of control that you'll work with in Android. Android has a complete but not overwhelming set of text controls. In this section, we are going to discuss the TextView, EditText, AutoCompleteTextView, and MultiCompleteTextView controls. Figure 6–2 shows the controls in action.

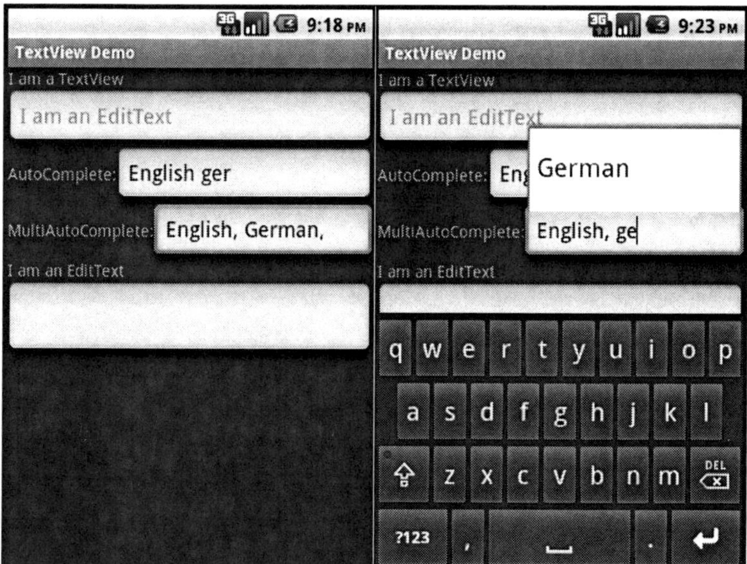

Figure 6–2. *Text controls in Android*

TextView

You've already seen a simple XML specification for a `TextView` control, in Listing 6–3, and how to handle `TextViews` in code in Listing 6–4. Notice how we specified the ID, width, height, and value of the text in XML and how we set the value using the `setText()` method. The `TextView` control knows how to display text but does not allow editing. This might lead you to conclude that the control is essentially a dummy label. Not true. The `TextView` control has a few interesting properties that make it very handy. If you know that the content of the `TextView` is going to contain a web URL or an e-mail address, for example, you can set the `autoLink` property to `email|web`, and the control will find and highlight any e-mail addresses and URLs. Moreover, when the user clicks on one of these highlighted items, the system will take care of launching the e-mail application with the e-mail address, or a browser with the URL. In XML, this attribute would be inside the `TextView` tag and would look something like this:

```
<TextView   ...     android:autoLink="email|web"    ...    />
```

You specify a pipe-delimited set of values including `web`, `email`, `phone`, or `map`, or use `none` (the default) or `all`. If you want to set autoLink behavior in code instead of using XML, the corresponding method call is `setAutoLinkMask()`. You would pass it an `int` representing the combination of values sort of like before, such as `Linkify.EMAIL_ADDRESSES|Linkify.WEB_ADDRESSES`. To achieve this functionality, `TextView` is utilizing the `android.text.util.Linkify` class. Listing 6–6 shows an example of auto-linking with code.

Listing 6–6. *Using Linkify on Text in a TextView*

```
TextView tv =(TextView)this.findViewById(R.id.tv);
tv.setAutoLinkMask(Linkify.ALL);
tv.setText("Please visit my website, http://www.androidbook.com
or email me at davemac327@gmail.com.");
```

Notice that we set the auto-link options on our TextView before we set the text. This is important because setting the auto-link options after setting the text won't affect the existing text. Because we're using code to add hyperlinks to our text, our XML for the TextView in Listing 6–6 does not require any special attributes and can look as simple as this:

```
<TextView android:id="@+id/tv" android:layout_width="wrap_content"
  android:layout_height="wrap_content"/>
```

If you want to, you can invoke the static addLinks() method of the Linkify class to find and add links to the content of any TextView or any Spannable on demand. Instead of using setAutoLinkMask(), we could have done the following *after* setting the text:

```
Linkify.addLinks(tv, Linkify.ALL);
```

Clicking a link will cause the default intent to be called for that action. For example, clicking a web URL will launch the browser with the URL. Clicking a phone number will launch the phone dialer, and so on. The Linkify class can perform this work right out of the box.

Linkify can also detect custom patterns you want to look for, decide whether they are a match for something you decide needs to be clickable, and set up how to fire an intent to make a click turn into some sort of action. We won't go into those details here, but know that these things can be done.

There are many more features of TextView to explore, from font attributes to minLines and maxLines and many more. These are fairly self-explanatory, and you are encouraged to experiment to see how you might be able to use them. Although you should keep in mind that some functionality in the TextView class is not applicable to a read-only field, the functionality is there for the subclasses of TextView, one of which we will cover next.

EditText

The EditText control is a subclass of TextView. As suggested by the name, the EditText control allows for text editing. EditText is not as powerful as the text-editing controls that you find on the Internet, but users of Android-based devices probably won't type documents—they'll type a couple paragraphs at most. Therefore, the class has limited but appropriate functionality and may even surprise you. For example, one of the most significant properties of an EditText is the inputType. You can set the inputType property to textAutoCorrect have the control correct common misspellings. You can set it to textCapWords to have the control capitalize words. Other options expect only phone numbers or passwords.

There are older, now deprecated, ways of specifying capitalization, multiline text, and other features. If these are specified without an inputType property, they can be read; but if inputType is specified, these older properties are ignored.

The old default behavior of the EditText control is to display text on one line and expand as needed. In other words, if the user types past the first line, another line will appear, and so on. You could, however, force the user to a single line by setting the singleLine property to true. In this case, the user will have to continue typing on the same line. With inputType, if you don't specify textMultiLine, the EditText will default to single-line only. So if you want the old default behavior of multiline typing, you need to specify inputType with textMultiLine.

One of the nice features of EditText is that you can specify hint text. This text will be displayed slightly faded and disappears as soon as the user starts to type text. The purpose of the hint is to let the user know what is expected in this field, without the user having to select and erase default text. In XML, this attribute is android:hint="your hint text here" or android:hint="@string/your_hint_name", where your_hint_name is a resource name of a string to be found in /res/values/strings.xml. In code, you would call the setHint() method with either a CharSequence or a resource ID.

AutoCompleteTextView

The AutoCompleteTextView control is a TextView with auto-complete functionality. In other words, as the user types in the TextView, the control can display suggestions for selection. Listing 6–7 demonstrates the AutoCompleteTextView control with XML and with the corresponding code.

Listing 6–7. *Using an AutoCompleteTextView Control*

```
<AutoCompleteTextView android:id="@+id/actv"
    android:layout_width="fill_parent"  android:layout_height="wrap_content" />


AutoCompleteTextView actv = (AutoCompleteTextView) this.findViewById(R.id.actv);

ArrayAdapter<String> aa = new ArrayAdapter<String>(this,
                android.R.layout.simple_dropdown_item_1line,
                new String[] {"English", "Hebrew", "Hindi", "Spanish",
                "German", "Greek" });

actv.setAdapter(aa);
```

The AutoCompleteTextView control shown in Listing 6–7 suggests a language to the user. For example, if the user types **en**, the control suggests English. If the user types **gr**, the control recommends Greek, and so on.

If you have used a suggestion control or a similar auto-complete control, you know that controls like this have two parts: a text-view control and a control that displays the suggestion(s). That's the general concept. To use a control like this, you have to create the control, create the list of suggestions, tell the control the list of suggestions, and possibly tell the control how to display the suggestions. Alternatively, you could create a second control for the suggestions and then associate the two controls.

Android has made this simple, as is evident from Listing 6–7. To use an `AutoCompleteTextView`, you can define the control in your layout file and reference it in your activity. You then create an adapter class that holds the suggestions and define the ID of the control that will show the suggestion (in this case, a simple list item). In Listing 6–7, the second parameter to the `ArrayAdapter` tells the adapter to use a simple list item to show the suggestion. The final step is to associate the adapter with the `AutoCompleteTextView`, which you do using the `setAdapter()` method. Don't worry about the adapter for the moment; we'll cover those later in this chapter.

MultiAutoCompleteTextView

If you have played with the `AutoCompleteTextView` control, you know that the control offers suggestions only for the *entire* text in the text view. In other words, if you type a sentence, you don't get suggestions for each word. That's where `MultiAutoCompleteTextView` comes in. You can use the `MultiAutoCompleteTextView` to provide suggestions as the user types. For example, Figure 6–2 shows that the user typed the word **English** followed by a comma, and then **Ge**, at which point the control suggested **German**. If the user were to continue, the control would offer additional suggestions.

Using the `MultiAutoCompleteTextView` is like using the `AutoCompleteTextView`. The difference is that you have to tell the control where to start suggesting again. For example, in Figure 6–2, you can see that the control can offer suggestions at the beginning of the sentence and after it sees a comma. The `MultiAutoCompleteTextView` control requires that you give it a tokenizer that can parse the sentence and tell it whether to start suggesting again. Listing 6–8 demonstrates using the `MultiAutoCompleteTextView` control with the XML and then the Java code.

Listing 6–8. *Using the MultiAutoCompleteTextView Control*

```
<MultiAutoCompleteTextView android:id="@+id/mactv"
    android:layout_width="fill_parent"  android:layout_height="wrap_content" />


MultiAutoCompleteTextView mactv = (MultiAutoCompleteTextView) this
                .findViewById(R.id.mactv);
ArrayAdapter<String> aa2 = new ArrayAdapter<String>(this,
                android.R.layout.simple_dropdown_item_1line,
new String[] {"English", "Hebrew", "Hindi", "Spanish", "German", "Greek" });

mactv.setAdapter(aa2);

mactv.setTokenizer(new MultiAutoCompleteTextView.CommaTokenizer());
```

The only significant differences between Listings 6–7 and 6–8 are the use of `MultiAutoCompleteTextView` and the call to the `setTokenizer()` method. Because of the `CommaTokenizer` in this case, after a comma is typed into the `EditText` field, the field will again make suggestions using the array of strings. Any other characters typed in will not trigger the field to make suggestions. So even if you were to type **French Spani**, the partial word *Spani* would not trigger the suggestion because it did not follow a comma. Android provides another tokenizer for e-mail addresses called Rfc822Tokenizer. You can always create your own tokenizer if you want to.

Button Controls

Buttons are common in any widget toolkit, and Android is no exception. Android offers the typical set of buttons as well as a few extras. In this section, we will discuss three types of button controls: the basic button, the image button, and the toggle button. Figure 6–3 shows a UI with these controls. The button at the top is the basic button, the middle button is an image button, and the last one is a toggle button.

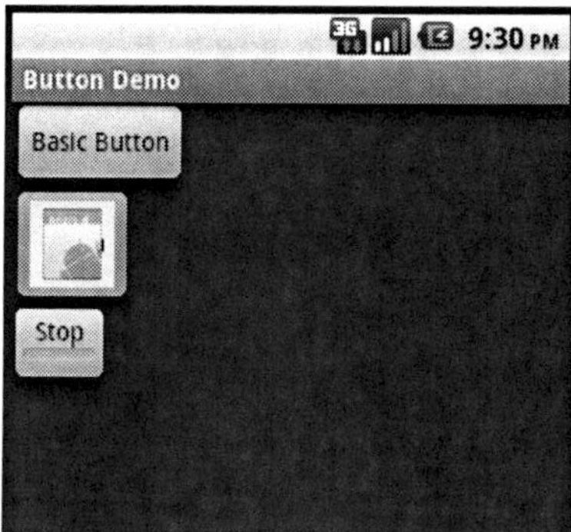

Figure 6–3. *Android button controls*

Let's get started with the basic button.

The Button Control

The basic button class in Android is android.widget.Button. There's not much to this type of button, beyond how you use it to handle click events. Listing 6–9 shows a fragment of an XML layout for the Button control, plus some Java that we might set up in the onCreate() method of our activity. Our basic button would look like the top button in Figure 6–3.

Listing 6–9. *Handling Click Events on a Button*

```
<Button android:id="@+id/button1"
    android:text="@string/basicBtnLabel"
    android:layout_width="fill_parent"
    android:layout_height="wrap_content" />

Button button1 = (Button)this.findViewById(R.id.button1);
button1.setOnClickListener(new OnClickListener()
{
     public void onClick(View v)
     {
        Intent intent = new Intent(Intent.ACTION_VIEW,
```

```
                                Uri.parse("http://www.androidbook.com"));
        startActivity(intent);
    }
});
```

Listing 6–9 shows how to register for a button-click event. You register for the on-click event by calling the `setOnClickListener()` method with an `OnClickListener`. In Listing 6–9, an anonymous listener is created on the fly to handle click events for `button1`. When the button is clicked, the `onClick()` method of the listener is called and, in this case, launches the browser to our web site.

Since Android SDK 1.6, there is an easier way to set up a click handler for your button or buttons. Listing 6–10 shows the XML for a `Button` where you specify an attribute for the handler, plus the Java code that is the click handler.

Listing 6–10. *Setting Up a Click Handler for a Button*

```
<Button   ...   android:onClick="myClickHandler"   ...  />

    public void myClickHandler(View target) {
        switch(target.getId()) {
        case R.id.button1:
        ...
```

The handler method will be called with target set to the `View` object representing the button that was clicked. Notice how the `switch` statement in the click handler method uses the resource IDs of the buttons to select the logic to run. Using this method means you won't have to explicitly create each `Button` object in your code, and you can reuse the same method across multiple buttons. This makes things easier to understand and maintain. This works with the other button types as well.

The ImageButton Control

Android provides an image button via `android.widget.ImageButton`. Using an image button is similar to using the basic button (see Listing 6–11). Our image button would look like the middle button in Figure 6–3.

Listing 6–11. *Using an `ImageButton`*

```
<ImageButton android:id="@+id/imageButton2"
    android:layout_width="wrap_content" android:layout_height="wrap_content"
    android:onClick="myClickHandler"
    android:src="@drawable/icon"  />

ImageButton imageButton2 = (ImageButton)this.findViewById(R.id.imageButton2);
imageButton2.setImageResource(R.drawable.icon);
```

Here we've created the image button in XML and set the button's image from a drawable resource. The image file for the button must exist under `/res/drawable`. In our case, we're simply reusing the Android icon for the button. We also show in Listing 6–11 how you can set the button's image dynamically by calling `setImageResource()` method on the button and passing it a resource ID. Note that you only need to do one or the other. You don't need to specify the button image in both the XML file and in code.

One of the nice features of an image button is that you can specify a transparent background for the button. The result will be a clickable image that acts like a button but can look like whatever you want it to look like. Just set `android:background="@null"` for the image button.

Because your image may be something very different than a standard button, you can customize how the button looks in the two other states it can be in when used in your UI. Besides appearing as normal, buttons can have focus, and they can be pressed. Having *focus* simply means the button is currently where events will go. You can direct focus to a button using the arrow keys on the keypad or D-pad, for example. *Pressed* means that the button's appearance changes when it has been pressed but before the user has let go. To tell Android what the three images are for our button, and which one is which, we set up a selector. This is a simple XML file that resides in the `/res/drawable` folder of our project. This is somewhat counterintuitive, because this is an XML file and not an image file, yet that is where the selector file must go. The content of a selector file will look like Listing 6–12.

Listing 6–12. *Using a Selector with an ImageButton*

```
<?xml version="1.0" encoding="utf-8"?>
    <selector xmlns:android="http://schemas.android.com/apk/res/android">
     <item android:state_pressed="true"
           android:drawable="@drawable/button_pressed" /> <!-- pressed -->
     <item android:state_focused="true"
           android:drawable="@drawable/button_focused" /> <!-- focused -->
     <item android:drawable="@drawable/icon" /> <!-- default -->
    </selector>
```

There are several things to note about the selector file. First, you do not specify a `<resources>` tag as in values XML files. Second, the order of the button images is important. Android will test each item in the selector, in order, to see if it matches. Therefore, you want the normal image to be last so it is used only if the button is not pressed and if the button does not have focus. If the normal image was listed first, it would always match and be selected even if the button is pressed or has focus. Of course, the drawables you refer to must exist in the `/res/drawables` folder. In the definition of your button in the layout XML file, you want to set the `android:src` property to the selector XML file as if it were a regular drawable, like so:

```
<Button   ...    android:src="@drawable/imagebuttonselector"    ...   />
```

The ToggleButton Control

The `ToggleButton` control, like a check box or a radio button, is a two-state button. This button can be in either the On or Off state. As shown in Figure 6–3, the `ToggleButton`'s default behavior is to show a green bar when in the On state and a grayed-out bar when in the Off state. Moreover, the default behavior also sets the button's text to On when it's in the On state and Off when it's in the Off state. You can modify the text for the `ToggleButton` if On/Off is not appropriate for your application. For example, if you have a background process that you want to start and stop via a `ToggleButton`, you could set the button's text to Stop and Run by using android:textOn and android:textOff properties.

Listing 6–13 shows an example. Our toggle button is the bottom button in Figure 6–3, and it is in the On position, so the label on the button says Stop.

Listing 6–13. *The Android ToggleButton*

```
<ToggleButton android:id="@+id/cctglBtn"
        android:layout_width="wrap_content"
        android:layout_height="wrap_content"
        android:text="Toggle Button"
        android:textOn="Stop"
        android:textOff="Run"/>
```

Because ToggleButtons have on and off text as separate attributes, the android:text attribute of a ToggleButton is not really used. It's available because it has been inherited (from TextView), but in this case, you don't need to use it.

The CheckBox Control

The CheckBox control is another two-state button that allows the user to toggle its state. The difference is that, for many situations, the users don't view it as a button that invokes immediate action. From Android's point of view, however, it is a button, and you can do anything with a check box that you can do with a button.

In Android, you can create a check box by creating an instance of android.widget.CheckBox. See Listing 6–14 and Figure 6–4.

Listing 6–14. *Creating Check Boxes*

```
<LinearLayout xmlns:android="http://schemas.android.com/apk/res/android"
        android:orientation="vertical" android:layout_width="fill_parent"
        android:layout_height="fill_parent">

<CheckBox android:id="@+id/chickenCB"  android:text="Chicken" android:checked="true"
    android:layout_width="wrap_content" android:layout_height="wrap_content" />

<CheckBox android:id="@+id/fishCB"  android:text="Fish"
    android:layout_width="wrap_content" android:layout_height="wrap_content" />

<CheckBox android:id="@+id/steakCB"  android:text="Steak" android:checked="true"
    android:layout_width="wrap_content" android:layout_height="wrap_content" />

</LinearLayout>
```

Figure 6–4. *Using the* `CheckBox` *control*

You manage the state of a check box by calling `setChecked()` or `toggle()`. You can obtain the state by calling `isChecked()`.

If you need to implement specific logic when a check box is checked or unchecked, you can register for the `on-checked` event by calling `setOnCheckedChangeListener()` with an implementation of the `OnCheckedChangeListener` interface. You'll then have to implement the `onCheckedChanged()` method, which will be called when the check box is checked or unchecked. Listing 6–15 show some code that deals with a `CheckBox`.

Listing 6–15. *Using Check Boxes in Code*

```
public class CheckBoxActivity extends Activity {
        /** Called when the activity is first created. */
        @Override
        public void onCreate(Bundle savedInstanceState) {
            super.onCreate(savedInstanceState);
            setContentView(R.layout.checkbox);

            CheckBox fishCB = (CheckBox)findViewById(R.id.fishCB);

            if(fishCB.isChecked())
                fishCB.toggle();     // flips the checkbox to unchecked if it was
checked

            fishCB.setOnCheckedChangeListener(
                        new CompoundButton.OnCheckedChangeListener() {

                @Override
                public void onCheckedChanged(CompoundButton arg0, boolean isChecked) {
                    Log.v("CheckBoxActivity", "The fish checkbox is now "
                            + (isChecked?"checked":"not checked"));
                }});
        }
}
```

The nice part of setting up the `OnCheckedChangeListener` is that you are passed the new state of the `CheckBox` button. You could instead use the `OnClickListener` technique as we used with basic buttons. When the `onClick()` method is called, you would need to determine the new state of the button by casting it appropriately and then calling `isChecked()` on it. Listing 6–16 shows what this code might look like if we added `android:onClick="myClickHandler"` to the XML definition of our `CheckBox` buttons.

Listing 6–16. *Using Check Boxes in Code with* `android:onClick`

```
public void myClickHandler(View view) {
    switch(view.getId()) {
    case R.id.steakCB:
        Log.v("CheckBoxActivity", "The steak checkbox is now " +
                (((CheckBox)view).isChecked()?"checked":"not checked"));
    }
}
```

The RadioButton Control

`RadioButton` controls are an integral part of any UI toolkit. A radio button gives the users several choices and forces them to select a single item. To enforce this single-selection model, radio buttons generally belong to a group, and each group is forced to have only one item selected at a time.

To create a group of radio buttons in Android, first create a `RadioGroup`, and then populate the group with radio buttons. Listing 6–17 and Figure 6–5 show an example.

Listing 6–17. *Using Android* `RadioButton` *Widgets*

```
<LinearLayout xmlns:android="http://schemas.android.com/apk/res/android"
        android:orientation="vertical" android:layout_width="fill_parent"
        android:layout_height="fill_parent">

<RadioGroup     android:id="@+id/rBtnGrp" android:layout_width="wrap_content"
            android:layout_height="wrap_content"  android:orientation="vertical" >

    <RadioButton     android:id="@+id/chRBtn" android:text="Chicken"
            android:layout_width="wrap_content"  android:layout_height="wrap_content"/>

    <RadioButton  android:id="@+id/fishRBtn" android:text="Fish" android:checked="true"
            android:layout_width="wrap_content"  android:layout_height="wrap_content"/>

    <RadioButton android:id="@+id/stkRBtn" android:text="Steak"
            android:layout_width="wrap_content"  android:layout_height="wrap_content"/>

</RadioGroup>

</LinearLayout>
```

In Android, you implement a radio group using `android.widget.RadioGroup` and a radio button using `android.widget.RadioButton`.

Figure 6–5. *Using radio buttons*

Note that the radio buttons within the radio group are, by default, unchecked to begin with, although you can set one to checked in the XML definition, as we did with Fish in Listing 6–17. To set one of the radio buttons to the checked state programmatically, you can obtain a reference to the radio button and call `setChecked()`:

```
RadioButton steakBtn = (RadioButton)this.findViewById(R.id.stkRBtn);
steakBtn.setChecked(true);
```

You can also use the `toggle()` method to toggle the state of the radio button. As with the `CheckBox` control, you will be notified of `on-checked` or `on-unchecked` events if you call the `setOnCheckedChangeListener()` with an implementation of the `OnCheckedChangeListener` interface. There is a slight difference here, though. This is a different class than before. This time, it's technically the `RadioGroup.OnCheckedChangeListener` class, whereas before it was the `CompoundButton.OnCheckedChangeListener` class.

The `RadioGroup` can also contain views other than the radio button. For example, Listing 6–18 adds a `TextView` after the last radio button. Also note that the first radio button (`anotherRadBtn`) lies outside the radio group.

Listing 6–18. *A `RadioGroup` with More Than Just `RadioButtons`*

```
<LinearLayout xmlns:android="http://schemas.android.com/apk/res/android"
        android:orientation="vertical"  android:layout_width="fill_parent"
        android:layout_height="fill_parent">

<RadioButton android:id="@+id/anotherRadBtn"  android:text="Outside"
          android:layout_width="wrap_content"  android:layout_height="wrap_content"/>

<RadioGroup android:id="@+id/radGrp"
          android:layout_width="wrap_content"  android:layout_height="wrap_content">

      <RadioButton android:id="@+id/chRBtn"  android:text="Chicken"
          android:layout_width="wrap_content"  android:layout_height="wrap_content"/>

      <RadioButton android:id="@+id/fishRBtn"  android:text="Fish"
          android:layout_width="wrap_content"  android:layout_height="wrap_content"/>

      <RadioButton android:id="@+id/stkRBtn"  android:text="Steak"
          android:layout_width="wrap_content"  android:layout_height="wrap_content"/>

      <TextView android:text="My Favorite"
          android:layout_width="wrap_content"  android:layout_height="wrap_content"/>

</RadioGroup>
</LinearLayout>
```

Listing 6–18 shows that you can have non-`RadioButton` controls inside a radio group. You should also know that the radio group can only enforce single-selection on the radio buttons in its own container. That is, the radio button with ID `anotherRadBtn` will not be affected by the radio group shown in Listing 6–18 because it is not one of the group's children.

You can manipulate the RadioGroup programmatically. For example, you can obtain a reference to a radio group and add a radio button (or other type of control). Listing 6–19 demonstrates this concept.

Listing 6–19. *Adding a RadioButton to a RadioGroup in Code*

```
RadioGroup radGrp = (RadioGroup)findViewById(R.id.radGrp);
RadioButton newRadioBtn = new RadioButton(this);
newRadioBtn.setText("Pork");
radGrp.addView(newRadioBtn);
```

Once a user has checked a radio button within a radio group, the user cannot uncheck it by clicking it again. The only way to clear all radio buttons in a radio group is to call the clearCheck() method on the RadioGroup programmatically.

Of course, you want to do something interesting with the RadioGroup. You probably don't want to poll each RadioButton to determine whether it's checked. Fortunately, the RadioGroup has several methods to help you out. We demonstrate those with Listing 6–20. The XML for this code is in Listing 6–18.

Listing 6–20. *Using a RadioGroup Programmatically*

```
public class RadioGroupActivity extends Activity {
        protected static final String TAG = "RadioGroupActivity";

        /** Called when the activity is first created. */
        @Override
        public void onCreate(Bundle savedInstanceState) {
            super.onCreate(savedInstanceState);
            setContentView(R.layout.radiogroup);

            RadioGroup radGrp = (RadioGroup)findViewById(R.id.radGrp);

            int checkedRadioButtonId = radGrp.getCheckedRadioButtonId();

            radGrp.setOnCheckedChangeListener(new RadioGroup.OnCheckedChangeListener() {
                @Override
                public void onCheckedChanged(RadioGroup arg0, int id) {
                    switch(id) {
                    case -1:
                        Log.v(TAG, "Choices cleared!");
                        break;
                    case R.id.chRBtn:
                        Log.v(TAG, "Chose Chicken");
                        break;
                    case R.id.fishRBtn:
                        Log.v(TAG, "Chose Fish");
                        break;
                    case R.id.stkRBtn:
                        Log.v(TAG, "Chose Steak");
                        break;
                    default:
                        Log.v(TAG, "Huh?");
                        break;
                    }
                }});
        }
}
```

We can always get the currently checked RadioButton using getCheckedRadioButtonId(), which returns the resource ID of the checked item or –1 if nothing is checked (possible if there's no default and the user hasn't chosen an option yet). We showed this in our onCreate() method previously, but in reality, you'd want to use it at the appropriate time to read the user's current choice. We can also set up a listener to be notified immediately when the user chooses one of the RadioButtons. Notice that the onCheckedChanged() method takes a RadioGroup parameter, allowing you to use the same OnCheckedChangeListener for multiple RadioGroups. You may have noticed the switch option of –1. This can also occur if the RadioGroup is cleared through code using clearCheck().

The ImageView Control

One of the basic controls we haven't covered yet is the ImageView control. This is used to display an image, where the image can come from a file, a content provider, or a resource such as a drawable. You can even specify just a color, and the ImageView will display that color. Listing 6–21 shows some XML examples of ImageViews, followed by some code that shows how to create an ImageView.

Listing 6–21. *ImageViews in XML and in Code*

```
<ImageView android:id="@+id/image1"
  android:layout_width="wrap_content"  android:layout_height="wrap_content"
  android:src="@drawable/icon" />

<ImageView android:id="@+id/image2"
  android:layout_width="125dip"  android:layout_height="25dip"
  android:src="#555555" />

<ImageView android:id="@+id/image3"
  android:layout_width="wrap_content"  android:layout_height="wrap_content" />

<ImageView android:id="@+id/image4"
  android:layout_width="wrap_content"  android:layout_height="wrap_content"
  android:src="@drawable/manatee02"
  android:scaleType="centerInside"
  android:maxWidth="35dip"  android:maxHeight="50dip"
  />


  ImageView imgView = (ImageView)findViewById(R.id.image3);

  imgView.setImageResource( R.drawable.icon );

  imgView.setImageBitmap(BitmapFactory.decodeResource(
              this.getResources(), R.drawable.manatee14) );

  imgView.setImageDrawable(
              Drawable.createFromPath("/mnt/sdcard/dave2.jpg") );

  imgView.setImageURI(Uri.parse("file://mnt/sdcard/dave2.jpg"));
```

In this example, we have four images defined in XML. The first is simply the icon for our application. The second is a gray bar that is wider than it is tall. The third definition does not specify an image source in the XML, but we associate an ID with this one (image3) that we can use from our code to set the image. The fourth image is another of our drawable image files where we not only specify the source of the image file but also set the maximum dimensions of the image on the screen and define what to do if the image is larger than our maximum size. In this case, we tell the ImageView to center and scale the image so it fits inside the size we specified.

In the Java code of Listing 6–21 we show several ways to set the image of image3. We first of course must get a reference to the ImageView by finding it using its resource ID. The first setter method, setImageResource(), simply uses the image's resource ID to locate the image file to supply the image for our ImageView. The second setter uses the BitmapFactory to read in an image resource into a Bitmap object and then sets the ImageView to that Bitmap. Note that we could have done some modifications to the Bitmap before applying it to our ImageView, but in our case, we used it as is. In addition, the BitmapFactory has several methods of creating a Bitmap, including from a byte array and an InputStream. You could use the InputStream method to read an image from a web server, create the Bitmap image, and then set the ImageView from there.

The third setting uses a Drawable for our image source. In this case, we're showing the source of the image coming from the SD card. You'll need to put some sort of image file out on the SD card with the proper name for this to work for you. Similar to BitmapFactory, the Drawable class has a few different ways to construct Drawables, including from an XML stream.

The final setter method takes the URI of an image file and uses that as the image source. For this last call, don't think that you can use any image URI as the source. This method is really only intended to be used for local images on the device, not for images that you might find through HTTP. To use Internet-based images as the source for your ImageView, you'd most likely use BitmapFactory and an InputStream.

Date and Time Controls

Date and time controls are common in many widget toolkits. Android offers several date- and time-based controls, some of which we'll discuss in this section. Specifically, we are going to introduce the DatePicker, TimePicker, DigitalClock, and AnalogClock controls.

The DatePicker and TimePicker Controls

As the names suggest, you use the DatePicker control to select a date and the TimePicker control to pick a time. Listing 6–22 and Figure 6–6 show examples of these controls.

Listing 6–22. *The `DatePicker` and `TimePicker` Controls in XML*

```
<LinearLayout xmlns:android="http://schemas.android.com/apk/res/android"
        android:orientation="vertical"
        android:layout_width="fill_parent"
        android:layout_height="fill_parent">

  <TextView android:id="@+id/dateDefault"
    android:layout_width="fill_parent" android:layout_height="wrap_content" />

  <DatePicker android:id="@+id/datePicker"
    android:layout_width="wrap_content" android:layout_height="wrap_content" />

  <TextView android:id="@+id/timeDefault"
    android:layout_width="fill_parent" android:layout_height="wrap_content" />

  <TimePicker android:id="@+id/timePicker"
    android:layout_width="wrap_content" android:layout_height="wrap_content" />

</LinearLayout>
```

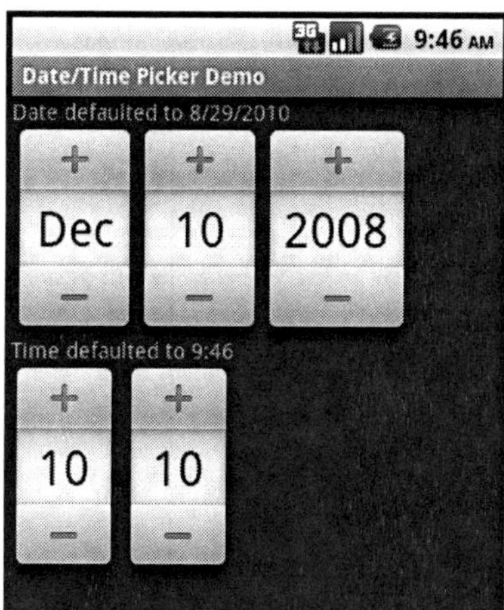

Figure 6–6. *The `DatePicker` and `TimePicker` UIs*

If you look at the XML layout, you can see that defining these controls is easy. As with any other control in the Android toolkit, you can access the controls programmatically to initialize them or to retrieve data from them. For example, you can initialize these controls as shown in Listing 6–23.

Listing 6–23. *Initializing the `DatePicker` and `TimePicker` with Date and Time, Respectively*

```
    public void onCreate(Bundle savedInstanceState) {
        super.onCreate(savedInstanceState);
        setContentView(R.layout.datetimepicker);

        TextView dateDefault = (TextView)findViewById(R.id.dateDefault);
        TextView timeDefault = (TextView)findViewById(R.id.timeDefault);

        DatePicker dp = (DatePicker)this.findViewById(R.id.datePicker);
        // The month, and just the month, is zero-based. Add 1 for display.
```

```
        dateDefault.setText("Date defaulted to " + (dp.getMonth() + 1) + "/" +
                dp.getDayOfMonth() + "/" + dp.getYear());
        // And here, subtract 1 from December (12) to set it to December
        dp.init(2008, 11, 10, null);

        TimePicker tp = (TimePicker)this.findViewById(R.id.timePicker);

        java.util.Formatter timeF = new java.util.Formatter();
        timeF.format("Time defaulted to %d:%02d", tp.getCurrentHour(),
                        tp.getCurrentMinute());
        timeDefault.setText(timeF.toString());

        tp.setIs24HourView(true);
        tp.setCurrentHour(new Integer(10));
        tp.setCurrentMinute(new Integer(10));
    }
}
```

Listing 6–23 sets the date on the `DatePicker` to December 10, 2008. Note that for the month, the internal value is zero-based, which means that January is 0 and December is 11. For the `TimePicker`, the number of hours and minutes is set to 10. Note also that this control supports 24–hour view. If you do not set values for these controls, the default values will be the current date and time as known to the device.

Finally, note that Android offers versions of these controls as modal windows, such as `DatePickerDialog` and `TimePickerDialog`. These controls are useful if you want to display the control to the user and force the user to make a selection. We'll cover dialogs in more detail in Chapter 8.

The DigitalClock and AnalogClock Controls

Android also offers `DigitalClock` and `AnalogClock` controls (see Figure 6–7).

Figure 6–7. *Using the `AnalogClock` and `DigitalClock`*

As shown, the digital clock supports seconds in addition to hours and minutes. The analog clock in Android is a two-handed clock, with one hand for the hour indicator and the other hand for the minute indicator. To add these to your layout, use the XML as shown in Listing 6–24.

Listing 6–24. Adding a *DigitalClock* or an *AnalogClock* in XML

```
<DigitalClock
  android:layout_width="wrap_content" android:layout_height="wrap_content" />

<AnalogClock
  android:layout_width="wrap_content" android:layout_height="wrap_content" />
```

These two controls are really just for displaying the current time, as they don't let you modify the date or time. In other words, they are controls whose only capability is to display the current time. Thus, if you want to change the date or time, you'll need to stick to the DatePicker/TimePicker or DatePickerDialog/TimePickerDialog. The nice part about these two clocks, though, is that they will update themselves without you having to do anything. That is, the seconds tick away in the DigitalClock, and the hands move on the AnalogClock without anything extra from us.

The MapView Control

The com.google.android.maps.MapView control can display a map. You can instantiate this control either via XML layout or code, but the activity that uses it must extend MapActivity. MapActivity takes care of multithreading requests to load a map, perform caching, and so on.

Listing 6–25 shows an example instantiation of a MapView.

Listing 6–25. *Creating a MapView Control via XML Layout*

```
<LinearLayout xmlns:android="http://schemas.android.com/apk/res/android"
        android:orientation="vertical" android:layout_width="fill_parent"
        android:layout_height="fill_parent">

    <com.google.android.maps.MapView
        android:layout_width="fill_parent"
        android:layout_height="fill_parent"
        android:enabled="true"
        android:clickable="true"
        android:apiKey="myAPIKey"
        />

</LinearLayout>
```

We'll discuss the MapView control in detail in Chapter 17, when we discuss location-based services. This is also where you'll learn how to obtain your own mapping API key.

Understanding Adapters

Before we get into the details of list controls of Android, we need to talk about adapters. List controls are used to display collections of data. But instead of using a single type of control to manage both the display and the data, Android separates these two responsibilities into list controls and adapters. List controls are classes that extend android.widget.AdapterView and include ListView, GridView, Spinner, and Gallery (see Figure 6–8).

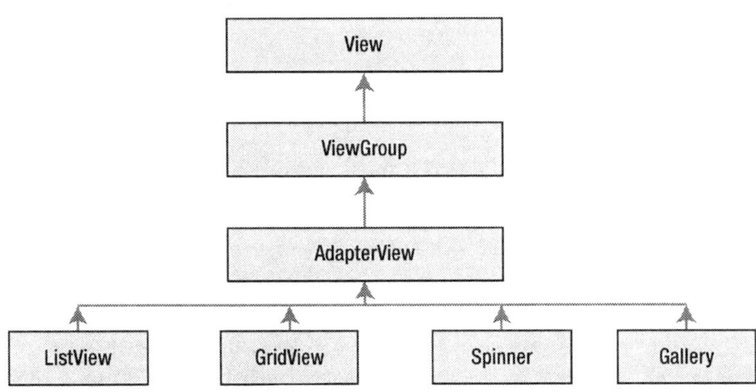

Figure 6–8. *`AdapterView` class hierarchy*

`AdapterView` itself extends `android.widget.ViewGroup`, which means that `ListView`, `GridView`, and so on are container controls. In other words, list controls contain collections of child views. The purpose of an adapter is to manage the data for an `AdapterView` and to provide the child views for it. Let's see how this works by examining the `SimpleCursorAdapter`.

Getting to Know SimpleCursorAdapter

The `SimpleCursorAdapter` is depicted in Figure 6–9.

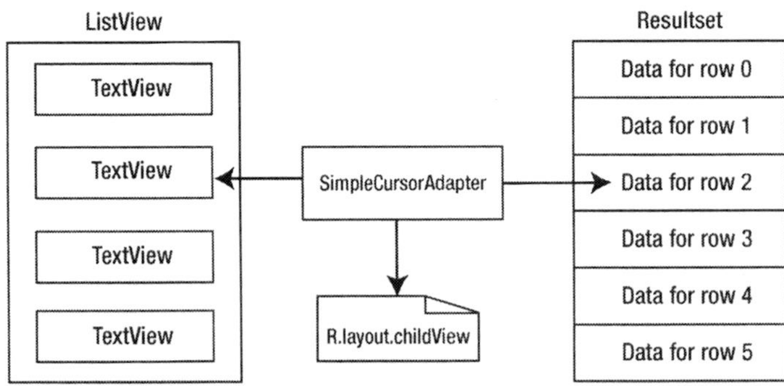

Figure 6–9. *The `SimpleCursorAdapter`*

This is a very important picture to understand. On the left side is the `AdapterView`; in this example, it is a `ListView` made up of `TextView` children. On the right side is the data; in this example, it's represented as a result set of data rows that came from a query against a content provider.

To map the data rows to the `ListView`, the `SimpleCursorAdapter` needs to have a child layout resource ID. The child layout must describe the layout for each of the data elements from the right side that should be displayed on the left side. A layout in this case is just like the layouts we've been working with for our activities, but it only needs to specify the layout of a single row of our `ListView`. For example, if you have a result set of information from the Contacts content provider, and you only want to display each contact name in your ListView, you would need to provide a layout to describe what the name field should look like. If you wanted to display the name and an image from the result set in each row of the ListView, your layout must say how to display the name and the image.

This does not mean you must provide a layout specification for every field in your result set, nor does it mean you must have a piece of data in your result set for everything you want to include in each row of the ListView. For example, we'll show you in a bit how you can have check boxes in your ListView for selecting rows, and those check boxes don't need to be set from data in a result set. We'll also show you how to get to data in the result set that is not part of the ListView. And although we've just talked about ListViews, TextViews, cursors, and result sets, please keep in mind that the adapter concept is more general than this. The left side can be a gallery, and the right side can be a simple array of images. But let's keep things fairly simple for now and look at SimpleCursorAdapter in more detail.

The constructor of SimpleCursorAdapter looks like this:

```
SimpleCursorAdapter(Context context, int childLayout, Cursor c, String[] from, int[] to)
```

This adapter converts a row from the cursor to a child view for the container control. The definition of the child view is defined in an XML resource (childLayout parameter). Note that because a row in the cursor might have many columns, you tell the SimpleCursorAdapter which columns you want to select from the row by specifying an array of column names (using the from parameter).

Similarly, because each column you select must be mapped to a View in the layout, you must specify the IDs in the to parameter. There's a one-to-one mapping between the column you select and a View that displays the data in the column, so the from and to parameter arrays must have the same number of elements. As we mentioned before, the child view could contain other types of views; they don't have to be TextViews. You could use an ImageView, for example.

There is a careful collaboration going on between the ListView and our adapter. When the ListView wants to display a row of data, it calls the getView() method of the adapter, passing in the position to specify the row of data to be displayed. The adapter responds by building the appropriate child view using the layout that was set in the adapter's constructor and by pulling the data from the appropriate record in the result set. The ListView, therefore, doesn't have to deal with how the data exists on the adapter side; it only needs to call for child views as needed. This is a critical point, because it means our ListView doesn't necessarily need to create every child view for every data row. It really only needs to have as many child views as are necessary for what's visible in the display window. If only ten rows are being displayed, technically the ListView only needs to have ten child layouts instantiated, even if there are hundreds of records in our result set. In reality, more than ten child layouts get instantiated, because Android usually keeps extras on hand to make it faster to bring a new row to visibility. The conclusion you should reach is that the child views managed by the ListView can be recycled. We'll talk more about that a little later.

Figure 6–9 reveals some flexibility in using adapters. Because the list control uses an adapter, you can substitute various types of adapters based on your data and child view. For example, if you are not going to populate an AdapterView from a content provider or database, you don't have to use the SimpleCursorAdapter. You can opt for an even "simpler" adapter—the ArrayAdapter.

Getting to Know ArrayAdapter

The ArrayAdapter is the simplest of the adapters in Android. It specifically targets list controls and assumes that TextView controls represent the list items (the child views). Creating a new ArrayAdapter can look as simple as this:

```
ArrayAdapter<String> adapter = new ArrayAdapter<String>(this,
                android.R.layout.simple_list_item_1,
                new string[]{"Dave","Satya","Dylan"});
```

We still pass the context (this) and a childLayout resource ID. But instead of passing a from array of data field specifications, we pass in an array of strings as the actual data. We don't pass a cursor or a to array of View resource IDs. The assumption here is that our child layout consists of a single TextView, and that's what the ArrayAdapter will use as the destination for the strings that are in our data array.

Now we're going to introduce a nice shortcut for the childLayout resource ID. Instead of creating our own layout file for the list items, we can take advantage of predefined layouts in Android. Notice that the prefix on the resource for the child layout resource ID is android.. Instead of looking in our local /res directory, Android looks in its own. You can browse to this folder by navigating to the Android SDK folder and looking under platforms/<android-version>/data/res/layout. There you'll find simple_list_item_1.xml and can see inside that it defines a simple TextView. That TextView is what our ArrayAdapter will use to create a view (in its getView() method) to give to the ListView. Feel free to browse through these folders to find predefined layouts for all sorts of uses. We'll be using more of these later.

ArrayAdapter has other constructors. If the childLayout is not a simple TextView, you can pass in the row layout resource ID plus the resource ID of the TextView to receive the data. When you don't have a ready-made array of strings to pass in, you can use the createFromResource() method. Listing 6–26 shows an example in which we create an ArrayAdapter for a spinner.

Listing 6–26. *Creating an ArrayAdapter from a String-Resource File*

```
<Spinner android:id="@+id/spinner"
    android:layout_width="wrap_content"  android:layout_height="wrap_content" />


Spinner spinner = (Spinner) findViewById(R.id.spinner);

ArrayAdapter<CharSequence> adapter = ArrayAdapter.createFromResource(this,
        R.array.planets, android.R.layout.simple_spinner_item);

adapter.setDropDownViewResource(android.R.layout.simple_spinner_dropdown_item);
```

```
spinner.setAdapter(adapter);


<?xml version="1.0" encoding="utf-8"?>
<!-- This file is /res/values/planets.xml -->
<resources>
  <string-array name="planets">
    <item>Mercury</item>
    <item>Venus</item>
    <item>Earth</item>
    <item>Mars</item>
    <item>Jupiter</item>
    <item>Saturn</item>
    <item>Uranus</item>
    <item>Neptune</item>
  </string-array>
</resources>
```

Listing 6–26 has three parts. The first part is the XML layout for a spinner. The second Java part shows how you can create an `ArrayAdapter` whose data source is defined in a string resource file. Using this method allows you to not only externalize the contents of the list to an XML file but also use localized versions. We'll talk about spinners a little later, but for now, know that a spinner has a view to show the currently selected value, plus a list view to show the values that can be selected from. It's basically a drop-down menu. The third part of Listing 6–26 is the XML resource file called `/res/values/planets.xml`, which is read in to initialize the `ArrayAdapter`.

Worth mentioning is that the `ArrayAdapter` allows for dynamic modifications to the underlying data. For example, the `add()` method will append a new value on the end of the array. The `insert()` method will add a new value at a specified position within the array. And `remove()` takes an object out of the array. You can also call `sort()` to reorder the array. Of course, once you've done this, the data array is out of sync with the `ListView`, so that's when you call the `notifyDataSetChanged()` method of the adapter. This method will resync the `ListView` with the adapter.

The following list summarizes the adapters that Android provides:

- `ArrayAdapter<T>`: This is an adapter on top of a generic array of arbitrary objects. It's meant to be used with a `ListView`.
- `CursorAdapter`: This adapter, also meant to be used in a `ListView`, provides data to the list via a cursor.
- `SimpleAdapter`: As the name suggests, this adapter is a simple adapter. It is generally used to populate a list with static data (possibly from resources).
- `ResourceCursorAdapter`: This adapter extends `CursorAdapter` and knows how to create views from resources.
- `SimpleCursorAdapter`: This adapter extends `ResourceCursorAdapter` and creates `TextView`/`ImageView` views from the columns in the cursor. The views are defined in resources.

We've covered enough of adapters to start showing you some real examples of working with adapters and list controls (also known as AdapterViews). Let's get to it.

Using Adapters with AdapterViews

Now that you've been introduced to adapters, it is time to put them to work for us, providing data for list controls. In this section, we're going to first cover the basic list control, the ListView. Then, we'll describe how to create your own custom adapter, and finally, we'll describe the other types of list controls: GridViews, spinners, and the gallery.

The Basic List Control: ListView

The ListView control displays a list of items vertically. That is, if we've got a list of items to view and the number of items extends beyond what we can currently see in the display, we can scroll to see the rest of the items. You generally use a ListView by writing a new activity that extends android.app.ListActivity. ListActivity contains a ListView, and you set the data for the ListView by calling the setListAdapter() method.

As we described previously, adapters link list controls to the data and help prepare the child views for the list control. Items in a ListView can be clicked to take immediate action or selected to act on the set of selected items later. We're going to start really simple and then add functionality as we go.

Displaying Values in a ListView

Figure 6–10 shows a ListView control in its simplest form.

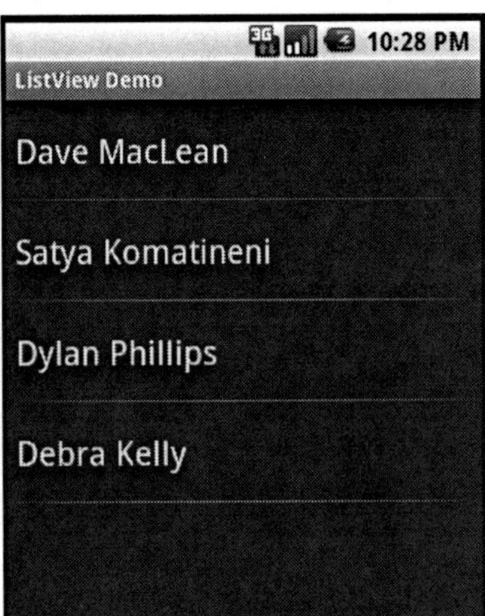

Figure 6–10. *Using the ListView control*

For this exercise, we will fill the entire screen with the ListView, so we don't even need to specify a ListView in our main layout XML file. Listing 6–27 shows the Java code for our ListActivity.

Listing 6–27. *Adding Items to a ListView*

```
public class ListViewActivity extends ListActivity
{
    @Override
    protected void onCreate(Bundle savedInstanceState)
    {
        super.onCreate(savedInstanceState);

        Cursor c = managedQuery(Contacts.CONTENT_URI,
                        null, null, null, Contacts.DISPLAY_NAME + " ASC");

        String[] cols = new String[] {Contacts.DISPLAY_NAME};
        int[]    views = new int[]   {android.R.id.text1};

        SimpleCursorAdapter adapter = new SimpleCursorAdapter(this,
                        android.R.layout.simple_list_item_1,
                        c, cols, views);
        this.setListAdapter(adapter);
    }
}
```

Listing 6–27 creates a ListView control populated with the list of contacts on the device. In our example, we query the device for the list of contacts. For demonstration purposes, we're selecting all fields from Contacts (using the first null parameter in the managedQuery() method), and we're sorting on the Contacts.DISPLAY_NAME field (using the final parameter in the managedQuery() method). We then create a projection (cols) to select only the names of the contacts for our ListView—a projection defines the columns that we are interested in. Next, we provide the corresponding resource ID array (views) to map the name column (Contacts.DISPLAY_NAME) to a TextView control (android.R.id.text1). After that, we create a cursor adapter and set the list's adapter. The adapter class has the smarts to take the rows in the data source and pull out the name of each contact to populate the UI.

There's one more thing we need to do to make this work. Because this demonstration is accessing the phone's Contacts database, we need to ask permission to do so. This security topic will be covered in more detail in Chapter 14, so for now, we'll just walk you through getting our ListView to show up. Double-click the AndroidManifest.xml file for this project, and click the Permissions tab. Click the Add button, choose Uses Permission, and click OK. Scroll down the Name list until you get to android.permission.READ_CONTACTS. Your Eclipse window should look like the one shown in Figure 6–11. Then, save the AndroidManifest.xml file. Now you can run this application in the emulator. You might need to add some contacts using the Contacts application before any names will show up in this example application.

You'll notice that the onCreate() method does not set the content view of the activity. Instead, because the base class ListActivity contains a ListView already, it just needs to provide the data for the ListView. We've used a couple of shortcuts in this example, the first being that we've taken advantage of our ListActivity supplying the main layout. We're also using an Android-provided layout for our child view (resource ID `android.R.layout.simple_list_item_1`), which contains an Android-provided `TextView` (resource ID `android.R.id.text1`). All in all, pretty simple to set up.

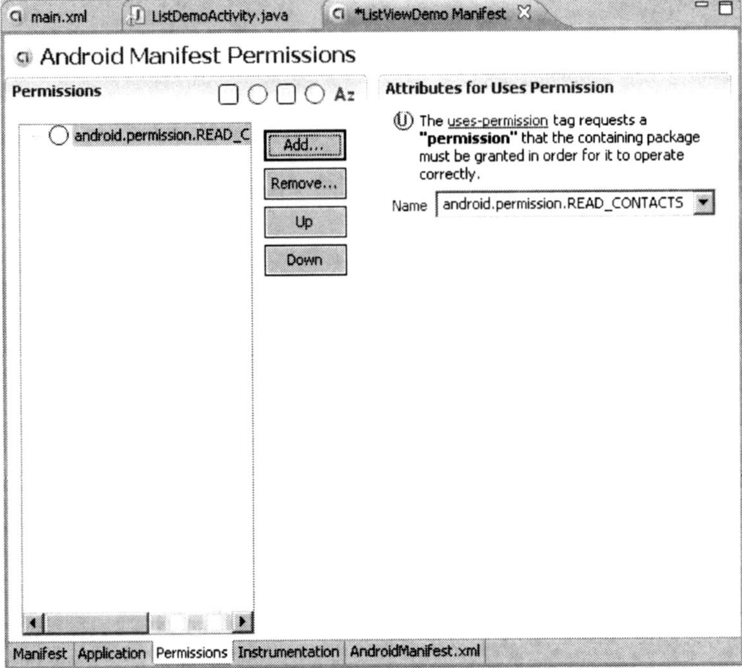

Figure 6–11. *Modifying `AndroidManifest.xml` so our application will run*

Clickable Items in a ListView

Of course, when you run this example, you'll see that you're able to scroll up and down the list to see all your contact names, but that's about it. What if we want to do something a little more interesting with this example, like launch the Contact application when a user clicks one of the items in our `ListView`? Listing 6–28 shows a modification to our example to accept user input.

Listing 6–28. *Accepting User Input on a `ListView`*

```
public class ListViewActivity2 extends ListActivity implements OnItemClickListener
{
    @Override
    protected void onCreate(Bundle savedInstanceState)
    {
        super.onCreate(savedInstanceState);

        ListView lv = getListView();

        Cursor c = managedQuery(Contacts.CONTENT_URI,
                null, null, null, Contacts.DISPLAY_NAME + " ASC");

        String[] cols = new String[] {Contacts.DISPLAY_NAME};
```

```
        int[]   views = new int[]   {android.R.id.text1};

        SimpleCursorAdapter adapter = new SimpleCursorAdapter(this,
                android.R.layout.simple_list_item_1,
                c, cols, views);
        this.setListAdapter(adapter);
        lv.setOnItemClickListener(this);
    }

    @Override
    public void onItemClick(AdapterView<?> adView, View target, int position, long id) {
        Log.v("ListViewActivity", "in onItemClick with " + ((TextView) target).getText()
+
                ". Position = " + position + ". Id = " + id);
        Uri selectedPerson = ContentUris.withAppendedId(
                Contacts.CONTENT_URI, id);
        Intent intent = new Intent(Intent.ACTION_VIEW, selectedPerson);
        startActivity(intent);
    }
}
```

Our activity is now implementing the OnItemClickListener interface, which means we'll receive a callback when the user clicks something in our ListView. As you can see by our onItemClick() method, we get a lot of information about what was clicked, including the view receiving the click, the position of the clicked item in the ListView, and the ID of the item according to our adapter. Because we know that our ListView is made up of TextViews, we assume that we received a TextView and cast accordingly before calling the getText() method to retrieve the contact's name. The position value represents where this item is in relation to the overall list of items in the ListView, and it's zero-based. Therefore, the first item in the list is at position 0.

The ID value depends entirely on the adapter and the source of the data. In our example, we happen to be querying the Contacts content provider, so the ID according to this adapter is the _ID of the record from the content provider. But your data source in other situations may not be from a content provider, so you should not think that you can always create a URI as we've done in this example. If we were using an ArrayAdapter that had read its values from a resource XML file, the ID given to us is very likely the position of the value in the data array and could, in fact, be exactly the same as the position value.

When we discussed ArrayAdapters before, we mentioned the notifyDataSetChanged() method to have the adapter update the ListView if the data has changed. Try this little experiment with our current example. Click one of your contacts, which should launch the Contacts application. Now, edit the contact by changing the name of the contact; click Done, and click the Back button so you're back to our example application. You should see that the name of that contact in your ListView has automatically been updated. How cool is that? Through the SimpleCursorAdapter and the Contacts content provider, our ListView has been updated for us. With ArrayAdapters, however, you will need to invoke the notifyDataSetChanged() method yourself.

That was pretty easy to do. We generated our own ListView of contact names, and by clicking a name, we launched the Contacts application for the selected person. But what if we want to select a bunch of names first and then do something with the subset of people? For the next example application, we're going to modify the layout of a list item to include a check box, and we're going to add a button to the UI to then act on the subset of selected items.

Adding Other Controls with a ListView

If you want additional controls in your main layout, you can provide your own layout XML file, put in a `ListView`, and add other desired controls. For example, you could add a button below the `ListView` in the UI to submit an action on the selected items, as shown in Figure 6–12.

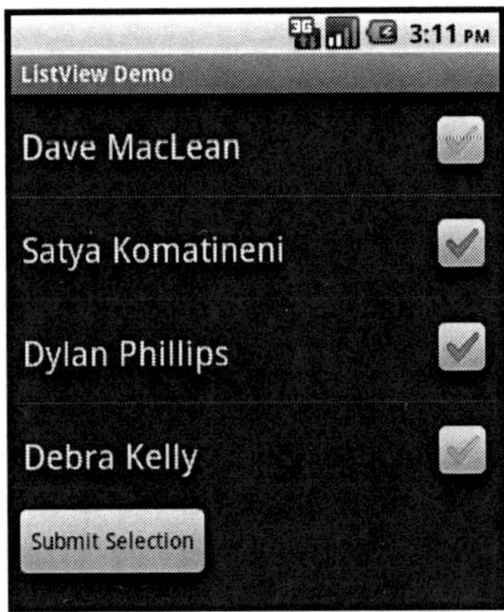

Figure 6–12. *An additional button that lets the user submit the selected item(s)*

The main layout for this example is in Listing 6–29, and it contains the UI definition of the activity—the `ListView` and the `Button`.

Listing 6–29. *Overriding the `ListView` Referenced by `ListActivity`*

```
<?xml version="1.0" encoding="utf-8"?>
<!-- This file is at /res/layout/list.xml -->
<LinearLayout xmlns:android="http://schemas.android.com/apk/res/android"
    android:orientation="vertical"
    android:layout_width="fill_parent"  android:layout_height="fill_parent">

    <ListView android:id="@android:id/list"
            android:layout_width="fill_parent"  android:layout_height="0dip"
            android:layout_weight="1" />

    <Button android:id="@+id/btn" android:onClick="doClick"
           android:layout_width="wrap_content"  android:layout_height="wrap_content"
           android:text="Submit Selection" />

</LinearLayout>
```

Notice the specification of the ID for the `ListView`. We've had to use `@android:id/list` because the `ListActivity` expects to find a `ListView` in our layout with this name. If we had relied on the default `ListView` that `ListActivity` would have created for us, it would have this ID.

The other thing to note is the way we have to specify the height of the `ListView` in `LinearLayout`. We want our button to appear on the screen at all times no matter how many items are in our `ListView`, and we don't want to be scrolling all the way to the bottom of the page just to find the button. To accomplish this, we set the `layout_height` to 0 and then use `layout_weight` to say that this control should take up all available room from the parent container. This trick allows room for the button and retains our ability to scroll the `ListView`. We'll talk more about layouts and weights later in this chapter.

The activity implementation would then look like Listing 6–30.

Listing 6–30. *Reading User Input from the `ListActivity`*

```
public class ListViewActivity3 extends ListActivity
{
    private static final String TAG = "ListViewActivity3";
    private ListView lv = null;
    private Cursor cursor = null;
    private int idCol = -1;
    private int nameCol = -1;
    private int timesContactedCol = -1;

    @Override
    protected void onCreate(Bundle savedInstanceState)
    {
        super.onCreate(savedInstanceState);
        setContentView(R.layout.list);

        lv = getListView();

        cursor = managedQuery(Contacts.CONTENT_URI,
                        null, null, null, Contacts.DISPLAY_NAME + " ASC");

        String[] cols = new String[]{Contacts.DISPLAY_NAME};
        idCol = cursor.getColumnIndex(Contacts._ID);
        nameCol = cursor.getColumnIndex(Contacts.DISPLAY_NAME);
        timesContactedCol = cursor.getColumnIndex(Contacts.TIMES_CONTACTED);

        int[] views = new int[]{android.R.id.text1};

        SimpleCursorAdapter adapter = new SimpleCursorAdapter(this,
                android.R.layout.simple_list_item_multiple_choice,
                cursor, cols, views);

        this.setListAdapter(adapter);

        lv.setChoiceMode(ListView.CHOICE_MODE_MULTIPLE);
    }

    public void doClick(View view) {
        int count=lv.getCount();
        SparseBooleanArray viewItems = lv.getCheckedItemPositions();
        for(int i=0; i<count; i++) {
```

```
            if(viewItems.get(i)) {
                // CursorWrapper cw = (CursorWrapper) lv.getItemAtPosition(i);
                cursor.moveToPosition(i);
                long id = cursor.getLong(idCol);
                String name = cursor.getString(nameCol);
                int timesContacted = cursor.getInt(timesContactedCol);
                Log.v(TAG, name + " is checked. Times contacted = " + timesContacted +
                        ". Position = " + i + ". Id = " + id);
            }
        }
    }
}
```

Now, we're back to calling setContentView() to set the UI for the activity. And within the setup of the adapter, we're passing another of the Android-provided views for a ListView line item (android.R.layout.simple_list_item_multiple_choice), which results in each row having a TextView and a CheckBox. If you look inside this layout file, you will see another subclass of TextView, this one called CheckedTextView. This special type of TextView is intended for use with ListViews. See, we told you there were some interesting things in that Android layout folder! You will see that the ID of the CheckedTextView is text1, which is what we needed to pass in our views array to the constructor of the SimpleCursorAdapter.

Because we want the user to be able to select our rows, we set the choice mode to CHOICE_MODE_MULTIPLE. By default, the choice mode is CHOICE_MODE_NONE. The other possible value is CHOICE_MODE_SINGLE. If you want to use that choice mode for this example, you would want to use a different layout, most likely android.R.layout.simple_list_item_single_choice.

In this example, we've implemented a basic button that calls the doClick() method of our activity. To keep things simple, we just want to write out to LogCat the names of the items that were checked by the user. The good news is that the solution is pretty easy; the bad news is that Android has evolved so the best solution depends on which version of Android you're targeting. The ListView solution we've shown here has worked since Android 1 (although we took the Android 1.6 shortcut on the button callback). That is, the getCheckedItemPositions() method is old, but it still works. The return value is an array that can tell you whether an item has been checked. So, we iterate through the array. viewItems.get(i) will return true if the corresponding row in our ListView has been checked. Our data is accessible through the cursor. So instead of looking up data in the ListView, we look up data in the cursor. The ListView will tell us is where in the adapter to look.

When we get a position number from the ListView that has been checked, we can use the cursor's moveToPosition() method to prepare to read the data. There's another method that does nearly the same thing, the getItemAtPosition() method of the ListView. In our case, the object returned from getItemAtPosition() would turn out to be a CursorWrapper object. As we said before, in other situations, we might get some other type of object. It's only because we're working with a content provider that we would get a CursorWrapper here. You have to understand your data source and your adapter to know what to expect.

We can then use our Cursor (or CursorWrapper if we went with that) to retrieve the data that is connected to our ListView row. Notice how, in our example, we can retrieve not only the name of the contact but notes as well, even though we never mapped notes to the ListView. When we set up the cursor for our adapter, we selected all available fields. In practice, you won't need all fields, so you should restrict your query to just the fields you're going to use. But this is a case where we query for more fields than we need for display in the ListView, so we can get easy access to the other fields in our button callback.

Another Way to Read Selections from a ListView

Android 1.6 introduced another method for retrieving a list of the checked rows from a ListView: getCheckItemIds(). Then, in Android 2.2, this method was deprecated and replaced with getCheckedItemIds(). It was a subtle name change, but the way you use the method is basically the same. Also, the way you deal with contacts changed in Android 2.2. For our next example, we'll use Android 2.2 features to show how our example might look. Listing 6–31 shows the Java code. For the XML layout of list.xml, we can continue to use the file in Listing 6–29.

Listing 6–31. *Another Way of Reading User Input From the ListActivity*

```
public class ListViewActivity4 extends ListActivity
{
    private static final String TAG = "ListViewActivity4";
    private SimpleCursorAdapter adapter = null;
    private ListView lv = null;

    @Override
    protected void onCreate(Bundle savedInstanceState)
    {
        super.onCreate(savedInstanceState);
        setContentView(R.layout.list);

        lv = getListView();

        String[] projection = new String[] { Contacts._ID,
                Contacts.DISPLAY_NAME};
        Cursor c = managedQuery(Contacts.CONTENT_URI,
                        projection, null, null, Contacts.DISPLAY_NAME);

        String[] cols = new String[] { Contacts.DISPLAY_NAME};
        int[]    views = new int[]      {android.R.id.text1};

        adapter = new SimpleCursorAdapter(this,
                android.R.layout.simple_list_item_multiple_choice,
                c, cols, views);

        this.setListAdapter(adapter);

        lv.setChoiceMode(ListView.CHOICE_MODE_MULTIPLE);
    }

    public void doClick(View view) {
```

```
        if(!adapter.hasStableIds()) {
            Log.v(TAG, "Data is not stable");
            return;
        }
        long[] viewItems = lv.getCheckedItemIds();
        for(int i=0; i<viewItems.length; i++) {
            Uri selectedPerson = ContentUris.withAppendedId(
                    Contacts.CONTENT_URI, viewItems[i]);

            Log.v(TAG, selectedPerson.toString() + " is checked.");
        }
    }
}
```

In this example application, when we click the button, our callback calls the method `getCheckedItemIds()`. Whereas in our last example, we got an array of positions of the checked items in the `ListView`, this time we get an array of IDs of the records from the adapter that have been checked in the `ListView`. We can bypass the `ListView` and the cursor now, because the IDs can be used with the content provider to take whatever action we desire. In our example, we simply construct a URI that represents the specific record from the Contacts content provider, and we write that URI to LogCat. We could have operated on the data using the content provider directly. This technique works equally well using the older Contacts content provider and the Android 1.6 `getCheckItemIds()` method.

Something else we've done differently in this example is to only select a couple of columns when we created our cursor. This is the normal practice because you do not want to read more data than is necessary. The last thing to point out from this example is that the method `getCheckedItemIds()` requires that the underlying data in the adapter is stable. Therefore, it is highly recommended that you call `hasStableIds()` on the adapter before calling `getCheckedItemIds()` on the `ListView`. In our example, we took a shortcut and simply logged the fact and returned. In reality, you'd want to do something more intelligent, like maybe initiate a background thread to do retries and display a dialog indicating that you're doing processing.

We've shown you how to work with `ListViews` from a variety of scenarios. We've shown that adapters do a lot of the work to support a `ListView`. Next, we'll cover the other types of list controls, starting with the `GridView`.

The GridView Control

Most widget toolkits offer one or more grid-based controls. Android has a `GridView` control that can display data in the form of a grid. Note that although we use the term *data* here, the contents of the grid can be text, images, and so on.

The `GridView` control displays information in a grid. The usage pattern for the `GridView` is to define the grid in the XML layout (see Listing 6–32) and then bind the data to the grid using an `android.widget.ListAdapter`. Don't forget to add the `uses-permission` tag to the `AndroidManifest.xml` file to make this example work.

Listing 6–32. *Definition of a* `GridView` *in an XML Layout and Associated Java Code*

```
<?xml version="1.0" encoding="utf-8"?>
<!-- This file is at /res/layout/gridview.xml -->
<GridView xmlns:android="http://schemas.android.com/apk/res/android"
    android:id="@+id/gridview"
    android:layout_width="fill_parent"
    android:layout_height="fill_parent"
    android:padding="10px"
    android:verticalSpacing="10px"
    android:horizontalSpacing="10px"
    android:numColumns="auto_fit"
    android:columnWidth="100px"
    android:stretchMode="columnWidth"
    android:gravity="center"
    />


public class GridViewActivity extends Activity
{
    @Override
    protected void onCreate(Bundle savedInstanceState)
    {
        super.onCreate(savedInstanceState);
        setContentView(R.layout.gridview);

        GridView gv = (GridView)findViewById(R.id.gridview);

        Cursor c = managedQuery(Contacts.CONTENT_URI,
                        null, null, null, Contacts.DISPLAY_NAME);

        String[] cols = new String[] {Contacts.DISPLAY_NAME};
        int[]    views = new int[]     {android.R.id.text1};

        SimpleCursorAdapter adapter = new SimpleCursorAdapter(this,
                android.R.layout.simple_list_item_1,
                c, cols, views);

        gv.setAdapter(adapter);
    }
}
```

Listing 6–32 defines a simple `GridView` in an XML layout. The grid is then loaded into the activity's content view. The generated UI is shown in Figure 6–13.

Figure 6–13. *A `GridView` populated with contact information*

The grid shown in Figure 6–13 displays the names of the contacts on the device. We have decided to show a `TextView` with the contact names, but you could easily generate a grid filled with images or other controls. We've again taken advantage of predefined layouts in Android. In fact, this example looks very much like Listing 6–27 except for a few important differences. First, our `GridViewActivity` extends `Activity`, not `ListActivity`. Second, we must call `setContentView()` to set the layout for our `GridView`; there are no default views to fall back on. And finally, to set the adapter, we call `setAdapter()` on the `GridView` object instead of calling `setListAdapter()` on `Activity`.

You've no doubt noticed that the adapter used by the grid is a `ListAdapter`. Lists are generally one-dimensional, whereas grids are two-dimensional. We can conclude, then, that the grid actually displays list-oriented data. And it turns out that the list is displayed by rows. That is, the list goes across the first row, then across the second row, and so on.

As before, we have a list control that works with an adapter to handle the data management, and the generation of the child views. The same techniques we used before should work just fine with `GridView`s. One exception relates to making selections: there is no way to specify multiple choices in a `GridView`, as we did in Listing 6–30.

The Spinner Control

The `Spinner` control is like a drop-down menu. It is typically used to select from a relatively short list of choices. If the choice list is too long for the display, a scrollbar is automatically added for you. You can instantiate a `Spinner` via XML layout as simply as this:

```
<Spinner
    android:id="@+id/spinner"  android:prompt="@string/spinnerprompt"
    android:layout_width="wrap_content"  android:layout_height="wrap_content" />
```

Although a spinner is technically a list control, it will appear to you more like a simple TextView control. In other words, only one value will be displayed when the spinner is at rest. The purpose of the spinner is to allow the user to choose from a set of predetermined values: when the user clicks the small arrow, a list is displayed, and the user is expected to pick a new value. Populating this list is done in the same way as the other list controls: with an adapter.

Because a spinner is often used like a drop-down menu, it is common to see the adapter get the list choices from a resource file. An example that sets up a spinner using a resource file is shown in Listing 6–33. Notice the new attribute called android:prompt for setting a prompt at the top of the list to choose from. The actual text for our spinner prompt is in our /res/values/strings.xml file. As you should expect, the Spinner class has a method for setting the prompt in code as well.

Listing 6–33. *Code to Create a Spinner from a Resource File*

```
public class SpinnerActivity extends Activity {
    /** Called when the activity is first created. */
    @Override
    public void onCreate(Bundle savedInstanceState) {
        super.onCreate(savedInstanceState);
        setContentView(R.layout.spinner);

        Spinner spinner = (Spinner)findViewById(R.id.spinner);

        ArrayAdapter<CharSequence> adapter = ArrayAdapter.createFromResource(this,
                R.array.planets, android.R.layout.simple_spinner_item);

        adapter.setDropDownViewResource(android.R.layout.simple_spinner_dropdown_item);

        spinner.setAdapter(adapter);
    }
}
```

You may recall seeing the planets.xml file in Listing 6–26. We show in this example how a Spinner control is created; the adapter is set up and then associated to the spinner. See Figure 6–14 for what this looks like in action.

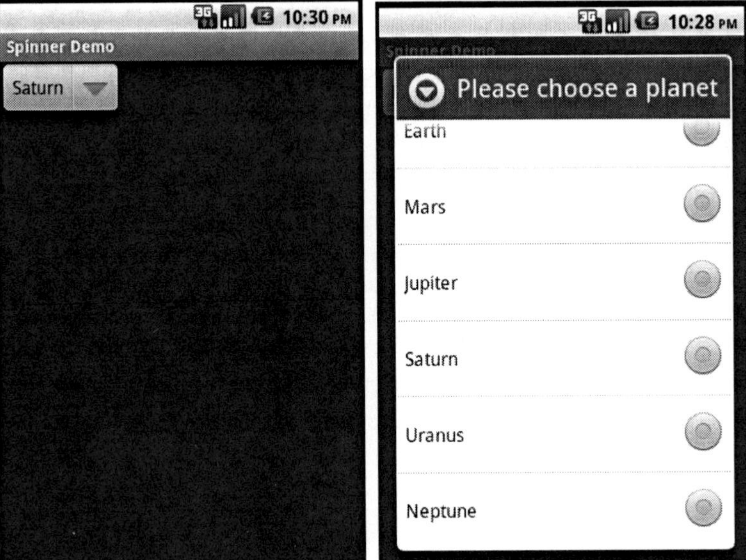

Figure 6–14. *A spinner for choosing a planet*

One of the differences from our earlier list controls is that we've got an extra layout to contend with when working with a spinner. The left side of Figure 6–14 shows the normal mode of a spinner, where the current selection is shown. In this case, the current selection is Saturn. Next to the word is a downward-pointing arrow indicating that this control is a spinner and can be used to pop up a list to select a different value. The first layout, supplied as a parameter to the `ArrayAdapter.createFromResource()` method, defines how the spinner looks in normal mode. On the right side of Figure 6–14, we show the spinner in the pop-up list mode, waiting for the user to choose a new value. The layout for this list is set using the `setDropDownViewResource()` method. Again in this example, we're using Android-provided layouts for these two needs, so if you want to inspect the definition of either of these layouts, you can visit the Android `res/layout` folder. And of course, you can specify your own layout definition for either of these to get the effect you want.

The Gallery Control

The `Gallery` control is a horizontally scrollable list control that always focuses at the center of the list. This control generally functions as a photo gallery in touch mode. You can instantiate a `Gallery` via either XML layout or code:

```
<Gallery
    android:id="@+id/gallery"
    android:layout_width="fill_parent"
    android:layout_height="wrap_content"
/>
```

The Gallery control is typically used to display images, so your adapter is likely going to be specialized for images. We'll show you a custom image adapter in next section on custom adapters. Visually, a Gallery looks like Figure 6–15.

Figure 6–15. *A gallery with images of manatees*

Creating Custom Adapters

Standard adapters in Android are easy to use, but they have some limitations. To address this, Android provides an abstract class called BaseAdapter that you can extend if you need a custom adapter. You would use a custom adapter if you had special data-management needs or if you wanted more control over how to display child views. You might also use a custom adapter to improve performance by using caching techniques. We're going to show you how to build a custom adapter next.

Listing 6–34 shows what the XML layout and the Java code could look like for a custom Adapter. For this next example, our adapter is going to deal with images of manatees, so we'll call it ManateeAdapter. We're going to create it inside of an activity as well.

Listing 6–34. *Our Custom Adapter: ManateeAdapter*

```
<?xml version="1.0" encoding="utf-8"?>
<!-- This file is at /res/layout/gridviewcustom.xml -->
<GridView xmlns:android="http://schemas.android.com/apk/res/android"
    android:id="@+id/gridview"
    android:layout_width="fill_parent"
    android:layout_height="fill_parent"
    android:padding="10dip"
    android:verticalSpacing="10dip"
    android:horizontalSpacing="10dip"
    android:numColumns="auto_fit"
    android:gravity="center"
```

```
    />

public class GridViewCustomAdapter extends Activity
{
    @Override
    protected void onCreate(Bundle savedInstanceState)
    {
        super.onCreate(savedInstanceState);
        setContentView(R.layout.gridviewcustom);

        GridView gv = (GridView)findViewById(R.id.gridview);

        ManateeAdapter adapter = new ManateeAdapter(this);

        gv.setAdapter(adapter);
    }

    public static class ManateeAdapter extends BaseAdapter {
        private static final String TAG = "ManateeAdapter";
        private static int convertViewCounter = 0;
        private Context mContext;
        private LayoutInflater mInflater;

        static class ViewHolder {
            ImageView image;
        }

        private int[] manatees = {
                R.drawable.manatee00, R.drawable.manatee01, R.drawable.manatee02,
                R.drawable.manatee03, R.drawable.manatee04, R.drawable.manatee05,
                R.drawable.manatee06, R.drawable.manatee07, R.drawable.manatee08,
                R.drawable.manatee09, R.drawable.manatee10, R.drawable.manatee11,
                R.drawable.manatee12, R.drawable.manatee13, R.drawable.manatee14,
                R.drawable.manatee15, R.drawable.manatee16, R.drawable.manatee17,
                R.drawable.manatee18, R.drawable.manatee19, R.drawable.manatee20,
                R.drawable.manatee21, R.drawable.manatee22, R.drawable.manatee23,
                R.drawable.manatee24, R.drawable.manatee25, R.drawable.manatee26,
                R.drawable.manatee27, R.drawable.manatee28, R.drawable.manatee29,
                R.drawable.manatee30, R.drawable.manatee31, R.drawable.manatee32,
                R.drawable.manatee33 };

        private Bitmap[] manateeImages = new Bitmap[manatees.length];
        private Bitmap[] manateeThumbs = new Bitmap[manatees.length];

        public ManateeAdapter(Context context) {
            Log.v(TAG, "Constructing ManateeAdapter");
            this.mContext = context;
            mInflater = LayoutInflater.from(context);

            for(int i=0; i<manatees.length; i++) {
                manateeImages[i] = BitmapFactory.decodeResource(
                        context.getResources(), manatees[i]);
                manateeThumbs[i] = Bitmap.createScaledBitmap(manateeImages[i],
                        100, 100, false);
            }
        }
```

```
        @Override
        public int getCount() {
            Log.v(TAG, "in getCount()");
            return manatees.length;
        }

        public int getViewTypeCount() {
            Log.v(TAG, "in getViewTypeCount()");
            return 1;
        }

        public int getItemViewType(int position) {
            Log.v(TAG, "in getItemViewType() for position " + position);
            return 0;
        }

        @Override
        public View getView(int position, View convertView, ViewGroup parent) {
            ViewHolder holder;

            Log.v(TAG, "in getView for position " + position +
                    ", convertView is " +
                    ((convertView == null)?"null":"being recycled"));

            if (convertView == null) {
                convertView = mInflater.inflate(R.layout.gridimage, null);
                convertViewCounter++;
                Log.v(TAG, convertViewCounter + " convertViews have been created");

                holder = new ViewHolder();
                holder.image = (ImageView) convertView.findViewById(R.id.gridImageView);

                convertView.setTag(holder);
            } else {
                holder = (ViewHolder) convertView.getTag();
            }

            holder.image.setImageBitmap( manateeThumbs[position] );

            return convertView;
        }

        @Override
        public Object getItem(int position) {
            Log.v(TAG, "in getItem() for position " + position);
            return manateeImages[position];
        }

        @Override
        public long getItemId(int position) {
            Log.v(TAG, "in getItemId() for position " + position);
            return position;
        }
    }
}
```

When you run this application, you should see a display that looks like Figure 6–16.

Figure 6–16. *A `GridView` with images of manatees*

There is a lot to explain in this example, even though it looks relatively simple. We'll start with our `Activity` class, which looks a lot like the ones we've been working with throughout this section of the chapter. There's a main layout from `gridviewcustom.xml`, which contains just a `GridView` definition. We need to get a reference to the `GridView` from inside the layout, so we define and set `gv`. We instantiate our `ManateeAdapter`, passing it our context, and we set the adapter on our `GridView`. This is pretty standard stuff so far, although you've no doubt noticed that our custom adapter doesn't use nearly as many parameters as pre-defined adapters when being created. This is mainly because we're in complete control over this particular adapter, and we're using it with only this application. If we were making this adapter more general, we would most likely be setting more parameters. But let's keep going.

Our job inside an adapter is to manage the passing of data into Android `View` objects. The `View` objects will be used by the list control (a `GridView` in this case). The data comes from some data source. In the earlier examples, the data came via a cursor object that was passed into the adapter. In our custom case here, our adapter knows all about the data and where it comes from. The list control will ask for things so it knows how to build the UI. It is also kind enough to pass in views for recycling when it has a view it no longer needs. It may seem a bit strange to think that our adapter must know how to construct views, but in the end, it all makes sense.

When we instantiate our custom adapter `ManateeAdapter`, it is customary to pass in the context and for the adapter to hold onto it. It is often very useful to have it available when needed. The second thing we want to do in our adapter is to hang onto the inflater. This will help performance when we need to create a new view to return to the list control. The third thing that is typical in an adapter is to create a `ViewHolder` object, to contain the `View` objects for the data we are managing. For this example, we are simply storing an ImageView, but if we had additional fields to deal with, we would add them into the definition of ViewHolder. For example, if we had a ListView where each row contained an ImageView and two TextViews, our ViewHolder would have an ImageView and two TextViews.

Because we're dealing with images of manatees in this adapter, we set up an array of their resource IDs to be used during construction to create bitmaps. We also define an array of bitmaps to use as our data list.

As you can see from our ManateeAdapter constructor, we save the context, create and hang onto an inflater, and then, we iterate through the image resource IDs and build an array of bitmaps. This bitmap array will be our data.

As you learned previously, setting the adapter will cause our GridView to call methods on the adapter to set itself up with data to display. For example, gv will call the adapter's getCount() method to determine how many objects there are for displaying. It will also call the getViewTypeCount() method to determine how many different types of views could be displayed within the GridView. For our purposes in this example, we set this to 1. However, if we had a ListView and wanted to put separators in between regular rows of data, we would have two types of views and would need to return 2 from getViewTypeCount(). You could have as many different view types as you like, as long as you appropriately return the correct count from this method. Related to this method is getItemViewType(). We just said that we could have more than one type of view to return from the adapter, but to keep things simpler, getItemViewType() needs to return only an integer value to indicate which of our view types is at a particular position in the data. Therefore, if we had two types of views to return, getItemViewType() would need to return either 0 or 1 to indicate which type. If we have three types of views, this method needs to return 0, 1, or 2.

If our adapter is dealing with separators in a ListView, it must treat the separators as data. That means there is a position in the data that is taken up by a separator. When getView() is called by a list control to retrieve the appropriate view for that position, getView() will need to return a separator as a view instead of regular data as a view. And when asked in getItemViewType() for the view type for that position, we need to return the appropriate integer value that we've decided matches that view type. The other thing you should do if using separators is to implement the isEnabled() method. This should return true for list items and false for separators because separators should not be selectable or clickable.

The most interesting method in ManateeAdapter is the getView() method call. Once gv has determined how many items are available, it starts to ask for the data. Now, we can talk about recycling views. A list control can only show as many child views on the display as will fit. That means there's no point in calling getView() for every piece of data in the adapter; it only makes sense to call getView() for as many items as can be displayed. As gv gets child views back from the adapter, it is determining how many will fit on the display. When the display is full of child views, gv can stop calling getView().

If you look at LogCat after starting this example application, you will see the various calls, but you will also see that getView() stops being called before all images have been requested. If you start scrolling up and down the GridView, you will see more calls to getView() in LogCat, and you will notice that, once we've created a certain number of child views, getView() is being called with convertView set to something, not null. This means we're now recycling child views—and that's very good for performance.

If we get a nonnull convertView value from gv in getView(), it means gv is recycling that view. By reusing the view passed in, we avoid having to inflate an XML layout, and we avoid having to find the ImageView. By linking a ViewHolder object to the View that we return, we can be much faster at recycling the view the next time it comes back to us. All we have to do in getView() is reacquire the ViewHolder and assign the right data into the view.

For this example, we wanted to show that the data placed into the view is not necessarily exactly what exists in the data. The createScaledBitmap() method is creating a smaller version of the data for display purposes. The point is that our list control does not call the getItem() method. This method would be called by our other code that wants to do something with the data if the user acts on the list control. Once again, for any adapter, it is very important that you understand what it is doing. You don't necessarily want to rely on data in the view from the list control, as created by getView() in the adapter. Sometimes, you will need to call the adapter's getItem() method to get the actual data to be operated on. And sometimes, as we did in the earlier ListView examples, you'll want to go to a cursor for the data. It all depends on the adapter and where the data is ultimately coming from. Although we used the createScaledBitmap() method in our example, Android 2.2 introduced another class that might have been helpful here: ThumbnailUtils. This class has some static methods for generating thumbnail images from bitmaps and videos.

The last thing to point out from this example is the getItemId() method call. In our earlier examples with ListViews and contacts, the item ID was the _ID value from the content provider. For this example, we don't really need to use anything other than position for the item ID. The point of item IDs is to provide a mechanism to refer to the data separately from its position. This is especially true when the data has a life away from this adapter, as is the case with our contacts. When we have this kind of direct control over the data, as we do with our images of manatees, and we understand how to get to the actual data in our application, it is a common shortcut to simply use position as the item ID. This is particularly true in our case, because we don't even allow adding or removal of data.

Other Controls in Android

There are many, many controls in Android that you can use. We've covered quite a few so far, and more will be covered in later chapters (such as MapView in Chapter 22 and VideoView and MediaController in Chapter 24). You will find that the other controls, because they're all descended from View, have a lot in common what the ones we've covered here. For now, we'll just mention a few of the controls you might want to explore further on your own.

ScrollView is a control for setting up a View container with a vertical scrollbar. This is useful when you have too much to fit onto a single screen. See this chapter's "References" section for a link to a blog post from Romain Guy on how to use this.

The ProgressBar and RatingBar controls are like sliders. The first shows the progress of some operation visually (perhaps a file download or music playing), and the second shows a rating scale of stars.

The Chronometer control is a timer that counts up. There's a CountDownTimer class if you want something to help you display a countdown timer, but it's not a View class.

New in Android 4.0 is the Switch control, which functions like a ToggleButton but visually has a side-to-side presentation. Also new in 4.0 is the Space view, a lightweight view that can be used in layouts to more easily create spaces between other views.

WebView is a very special view for displaying HTML. It can do a lot more than that, including handling cookies and JavaScript and linking to Java code in your application. But before you go implementing a web browser inside your application, you should carefully consider invoking the on-device web browser to let it do all that heavy lifting.

That completes our introduction of controls in this chapter. We'll now move on to styles and themes for modifying the look and feel of our controls and then to layouts for arranging our controls on screens.

Styles and Themes

Android provides several ways to alter the style of views in your application. We'll first cover using markup tags in strings and then how to use spannables to change specific visual attributes of text. But what if you want to control how things look using a common specification for several views or across an entire activity or application? We'll discuss Android styles and themes to show you how.

Using Styles

Sometimes, you want to highlight or style a portion of the View's content. You can do this statically or dynamically. Statically, you can apply markup directly to the strings in your string resources, as shown here:

```
<string name="styledText"><i>Static</i> style in a <b>TextView</b>.</string>
```

You can then reference it in your XML or from code. Note that you can use the following HTML tags with string resources: <i>, <b>, and <u> for italics, bold, and underlined, respectively, as well as <sup> (superscript), <sub> (subscript), <strike> (strikethrough), <big>, <small>, and <monospace>. You can even nest these to get, for example, small superscripts. This works not just in TextViews but also in other views, like buttons. Figure 6–17 shows what styled and themed text looks like, using many of the examples in this section.

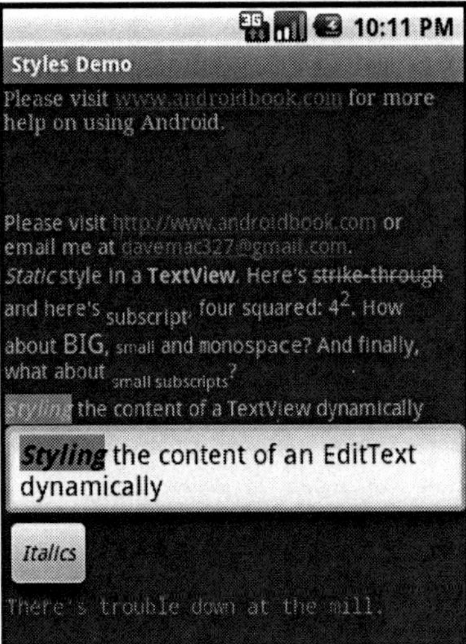

Figure 6–17. *Examples of styles and themes*

Styling a `TextView` control's content programmatically requires a little additional work but allows for much more flexibility (see Listing 6–35), because you can style it at runtime. This flexibility can only be applied to a spannable, though, which is how `EditText` normally manages the internal text, whereas `TextView` does not normally use `Spannable`. `Spannable` is basically a `String` that you can apply styles to. To get a `TextView` to store text as a spannable, you can call `setText()` this way:

```
tv.setText("This text is stored in a Spannable", TextView.BufferType.SPANNABLE);
```

Then, when you call `tv.getText()`, you'll get a spannable.

As shown in Listing 6–35, you can get the content of the `EditText` (as a `Spannable` object) and then set styles for portions of the text. The code in the listing sets the text styling to bold and italics and sets the background to red. You can use all the styling options as we have with the HTML tags as described previously, and then some.

Listing 6–35. *Applying Styles Dynamically to the Content of an* `EditText`

```
EditText et =(EditText)this.findViewById(R.id.et);
et.setText("Styling the content of an EditText dynamically");
Spannable spn = (Spannable) et.getText();
spn.setSpan(new BackgroundColorSpan(Color.RED), 0, 7,
            Spannable.SPAN_EXCLUSIVE_EXCLUSIVE);
spn.setSpan(new StyleSpan(android.graphics.Typeface.BOLD_ITALIC),
            0, 7, Spannable.SPAN_EXCLUSIVE_EXCLUSIVE);
```

These two techniques for styling only work on the one view they're applied to. Android provides a style mechanism to define a common style to be reused across views, as well as a theme mechanism, which basically applies a style to an entire activity or the entire application. To begin with, we need to talk about styles.

A *style* is a collection of `View` attributes that is given a name so you can refer to that collection by its name and assign that style by name to views. For example, Listing 6–36 shows a resource XML file, saved in /res/values, that we could use for all error messages.

Listing 6–36. *Defining a Style to Be Used Across Many Views*

```
<?xml version="1.0" encoding="utf-8"?>
<resources>
    <style name="ErrorText">
        <item name="android:layout_width">fill_parent</item>
        <item name="android:layout_height">wrap_content</item>
        <item name="android:textColor">#FF0000</item>
        <item name="android:typeface">monospace</item>
    </style>
</resources>
```

The size of the view is defined as well as the font color (red) and typeface. Notice how the name attribute of the item tag is the XML attribute name we used in our layout XML files, and the value of the item tag no longer requires double quotes. We can now use this style for an error TextView, as shown in Listing 6–37.

Listing 6–37. *Using a Style in a View*

```
<TextView  android:id="@+id/errorText"
    style="@style/ErrorText"
    android:text="No errors at this time"
    />
```

It is important to note that the attribute name for a style in this View definition does not start with android:. Watch out for this, because everything seems to use android: except the style. When you've got many views in your application that share a style, changing that style in one place is much simpler; you only need to modify the style's attributes in the one resource file. You can, of course, create many different styles for various controls. Buttons could share a common style, for example, that's different from the common style for text in menus.

One really nice aspect of styles is that you can set up a hierarchy of them. We could define a new style for really bad error messages and base it on the style of ErrorText. Listing 6–38 shows how this might look.

Listing 6–38. *Defining a Style from a Parent Style*

```
<?xml version="1.0" encoding="utf-8"?>
<resources>
    <style name="ErrorText.Danger" >
        <item name="android:textStyle">bold</item>
    </style>
</resources>
```

This example shows that we can simply name our child style using the parent style as a prefix to the new style name. Therefore, ErrorText.Danger is a child of ErrorText and inherits the style attributes of the parent. It then adds a new attribute for textStyle. This can be repeated again and again to create a whole tree of styles.

As was the case for adapter layouts, Android provides a large set of styles that we can use. To specify an Android-provided style, use syntax like this:

```
style="@android:style/TextAppearance"
```

This style sets the default style for text in Android. To locate the master Android styles.xml file, visit the Android SDK/platforms/<android-version>/data/res/values/ folder. Inside this file, you will find quite a few styles that are ready-made for you to use or extend. Here's a word of caution about extending the Android-provided styles: the previous method of using a prefix won't work with Android-provided styles. Instead, you must use the parent attribute of the style tag, like this:

```
<style name="CustomTextAppearance" parent="@android:style/TextAppearance">
    <item  ... your extensions go here ...    />
</style>
```

You don't always have to pull in an entire style on your view. You could choose to borrow just a part of the style instead. For example, if you want to set the color of the text in your TextView to a system style color, you could do the following:

```
<EditText id="@+id/et2"
    android:layout_width="fill_parent"  android:layout_height="wrap_content"
    android:textColor="?android:textColorSecondary"
    android:text="@string/hello_world" />
```

Notice that in this example, the name of the textColor attribute value starts with the ? character instead of the @ character. The ? character is used so Android knows to look for a style value in the current theme. Because we see ?android, we look in the Android system theme for this style value.

Using Themes

One problem with styles is that you need to add an attribute specification of style="@style/..." to every view definition that you want it to apply to. If you have some style elements you want applied across an entire activity, or across the whole application, you should use a theme instead. A *theme* is really just a style applied broadly; but in terms of defining a theme, it's exactly like a style. In fact, themes and styles are fairly interchangeable: you can extend a theme into a style or refer to a style as a theme. Typically, only the names give a hint as to whether a style is intended to be used as a style or a theme.

To specify a theme for an activity or an application, you would add an attribute to the <activity> or <application> tag in the AndroidManifest.xml file for your project. The code might look like one of these:

```
<activity android:theme="@style/MyActivityTheme">
<application android:theme="@style/MyApplicationTheme">
<application android:theme="@android:style/Theme.NoTitleBar">
```

You can find the Android-provided themes in the same folder as the Android-provided styles, with the themes in a file called themes.xml. When you look inside the themes file, you will see a large set of styles defined, with names that start with Theme. You will also notice that within the Android-provided themes and styles, there is a lot of extending going on, which is why you end up with styles called Theme.Dialog.AppError", for example.

This concludes our discussion of the Android control set. As we mentioned in the beginning of the chapter, building UIs in Android requires you to master two things: the control set and the layout managers. In the next section, we are going to discuss the Android layout managers.

Understanding Layout Managers

Android offers a collection of view classes that act as containers for views. These container classes are called *layouts* (or *layout managers*), and each implements a specific strategy to manage the size and position of its children. For example, the `LinearLayout` class lays out its children either horizontally or vertically, one after the other. All layout managers derive from the `View` class, therefore you can nest layout managers inside of one another.

The layout managers that ship with the Android SDK are defined in Table 6–2.

Table 6–2. *Android Layout Managers*

Layout Manager	Description
`LinearLayout`	Organizes its children either horizontally or vertically
`TableLayout`	Organizes its children in tabular form
`RelativeLayout`	Organizes its children relative to one another or to the parent
`FrameLayout`	Allows you to dynamically change the control(s) in the layout
`GridLayout`	Organizes its children in a grid arrangement

We will discuss these layout managers in the sections that follow. The layout manager called `AbsoluteLayout` has been deprecated and will not be covered in this book.

The LinearLayout Layout Manager

The `LinearLayout` layout manager is the most basic. This layout manager organizes its children either horizontally or vertically based on the value of the `orientation` property. We've used `LinearLayout` in several of our examples so far. Listing 6–39 shows `LinearLayout` with a horizontal configuration.

Listing 6–39. *LinearLayout with a Horizontal Configuration*

```
<LinearLayout xmlns:android="http://schemas.android.com/apk/res/android"
    android:orientation="horizontal"
    android:layout_width="fill_parent"  android:layout_height="wrap_content">

    <!-- add children here-->

</LinearLayout>
```

You can create a vertically oriented `LinearLayout` by setting the value of `orientation` to `vertical`. Because layout managers can be nested, you could, for example, construct a vertical layout manager that contained horizontal layout managers to create a fill-in form, where each row had a label next to an `EditText` control. Each row would be its own horizontal layout, but the rows as a collection would be organized vertically.

Understanding Weight and Gravity

The `orientation` attribute is the first important attribute recognized by the `LinearLayout` layout manager. Other important properties that can affect size and position of child controls are weight and gravity.

You use *weight* to assign size importance to a control relative to the other controls in the container. Suppose a container has three controls: one has a weight of 1, whereas the others have a weight of 0. In this case, the control whose weight equals 1 will consume the empty space in the container. *Gravity* is essentially alignment. For example, if you want to align a label's text to the right, you would set its gravity to `right`. There are quite a few possible values for `gravity`, including `left`, `center`, `right`, `top`, `bottom`, `center_vertical`, `clip_horizontal`, and others. See the web pages in the "References" section for details on these and the other values of gravity.

> **NOTE:** Layout managers extend `android.widget.ViewGroup`, as do many control-based container classes such as `ListView`. Although the layout managers and control-based containers extend the same class, the layout manager classes strictly deal with the sizing and position of controls and not user interaction with child controls. For example, compare `LinearLayout` to the `ListView` control. On the screen, they look similar in that both can organize children vertically. But the `ListView` control provides APIs for the user to make selections, whereas `LinearLayout` does not. In other words, the control-based container (`ListView`) supports user interaction with the items in the container, whereas the layout manager (`LinearLayout`) addresses sizing and positioning only.

Now let's look at an example involving the weight and gravity properties (see Figure 6–18).

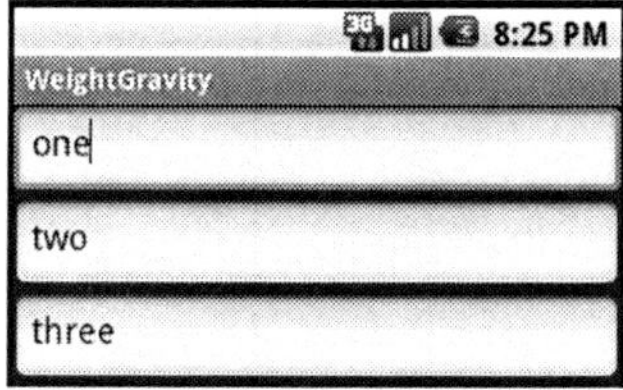

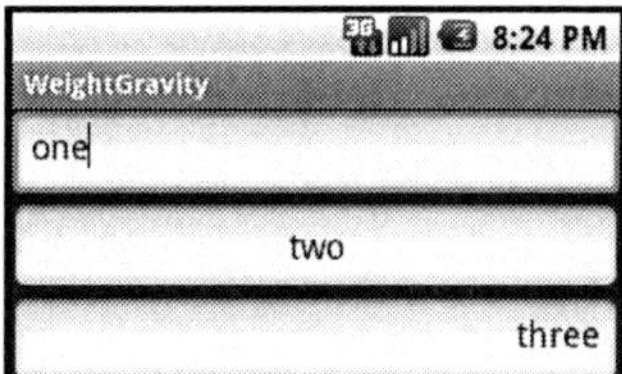

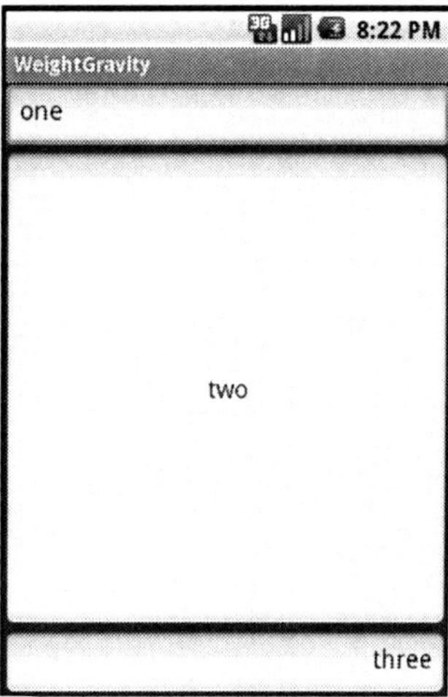

Figure 6–18. *Using the `LinearLayout` layout manager*

Figure 6–18 shows three UIs that utilize `LinearLayout`, with different weight and gravity settings. The UI on the left uses the default settings for weight and gravity. The XML layout for this first UI is shown in Listing 6–40.

Listing 6–40. *Three Text Fields Arranged Vertically in a LinearLayout, Using Default Values for Weight and Gravity*

```
<LinearLayout xmlns:android="http://schemas.android.com/apk/res/android"
    android:orientation="vertical" android:layout_width="fill_parent"
    android:layout_height="fill_parent">

    <EditText android:layout_width="fill_parent"
        android:layout_height="wrap_content"
        android:text="one"/>
    <EditText android:layout_width="fill_parent"
        android:layout_height="wrap_content"
        android:text="two"/>
    <EditText android:layout_width="fill_parent"
        android:layout_height="wrap_content"
        android:text="three"/>
</LinearLayout>
```

The UI in the center of Figure 6–18 uses the default value for weight but sets `android:gravity` for the controls in the container to `left`, `center`, and `right`, respectively. The last example sets the `android:layout_weight` attribute of the center component to 1.0 and leaves the others to the default value of 0.0 (see Listing 6–41). By setting the weight attribute to 1.0 for the middle component and leaving the weight attributes for the other two components at 0.0, we are specifying that the center component should take up all the remaining whitespace in the container and that the other two components should remain at their ideal size.

Similarly, if you want two of the three controls in the container to share the remaining whitespace among them, you would set the weight to 1.0 for those two and leave the third one at 0.0. Finally, if you want the three components to share the space equally, you'd set all of their weight values to 1.0. Doing this would expand each text field equally.

Listing 6–41. *`LinearLayout` with Weight Configurations*

```
<LinearLayout xmlns:android="http://schemas.android.com/apk/res/android"
    android:orientation="vertical" android:layout_width="fill_parent"
    android:layout_height="fill_parent">

    <EditText android:layout_width="fill_parent" android:layout_weight="0.0"
    android:layout_height="wrap_content" android:text="one"
    android:gravity="left"/>

    <EditText android:layout_width="fill_parent" android:layout_weight="1.0"
    android:layout_height="wrap_content" android:text="two"
    android:gravity="center"/>

    <EditText android:layout_width="fill_parent" android:layout_weight="0.0"
    android:layout_height="wrap_content" android:text="three"
    android:gravity="right"
    />
</LinearLayout>
```

android:gravity vs. android:layout_gravity

Note that Android defines two similar gravity attributes: `android:gravity` and `android:layout_gravity`. Here's the difference: `android:gravity` is a setting used by the view, whereas `android:layout_gravity` is used by the container (`android.view.ViewGroup`). For example, you can set `android:gravity` to `center` to have the text in the `EditText` centered within the control. Similarly, you can align an `EditText` to the far right of a `LinearLayout` (the container) by setting `android:layout_gravity="right"`. See Figure 6–19 and Listing 6–42.

Figure 6–19. *Applying gravity settings*

Listing 6–42. *Understanding the Difference Between `android:gravity` and `android:layout_gravity`*

```
<LinearLayout xmlns:android="http://schemas.android.com/apk/res/android"
    android:orientation="vertical" android:layout_width="fill_parent"
    android:layout_height="fill_parent">

    <EditText android:layout_width="wrap_content" android:gravity="center"
    android:layout_height="wrap_content" android:text="one"
 android:layout_gravity="right"/>
</LinearLayout>
```

As shown in Figure 6–19, the text is centered in the `EditText`, which is aligned to the right of the `LinearLayout`.

The TableLayout Layout Manager

The `TableLayout` layout manager is an extension of `LinearLayout`. This layout manager structures its child controls into rows and columns. Listing 6–43 shows an example.

Listing 6–43. *A Simple* `TableLayout`

```
<?xml version="1.0" encoding="utf-8"?>
<TableLayout xmlns:android="http://schemas.android.com/apk/res/android"
    android:layout_width="fill_parent"  android:layout_height="fill_parent">

  <TableRow>
    <TextView android:text="First Name:"
        android:layout_width="wrap_content"  android:layout_height="wrap_content" />

    <EditText android:text="Edgar"
        android:layout_width="wrap_content"  android:layout_height="wrap_content" />
  </TableRow>

  <TableRow>
    <TextView android:text="Last Name:"
        android:layout_width="wrap_content"  android:layout_height="wrap_content" />

    <EditText android:text="Poe"
        android:layout_width="wrap_content"  android:layout_height="wrap_content" />
  </TableRow>

</TableLayout>
```

To use this layout manager, you create an instance of `TableLayout` and place `TableRow` elements within it. These `TableRow` elements contain the controls of the table. The UI for Listing 6–43 is shown in Figure 6–20.

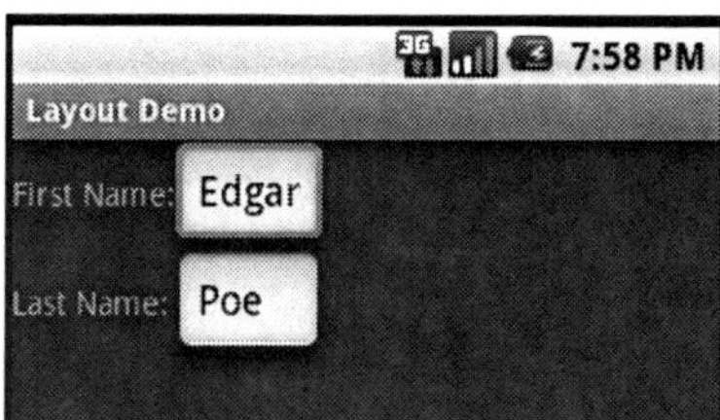

Figure 6–20. *The* `TableLayout` *layout manager*

Because the contents of a `TableLayout` are defined by rows as opposed to columns, Android determines the number of columns in the table by finding the row with the most cells. For example, Listing 6–44 creates a table with two rows where one row has two cells and the other has three cells (see Figure 6–21). In this case, Android creates a table with two rows and three columns. The last column of the first row is an empty cell.

Listing 6–44. *An Irregular Table Definition*

```
<TableLayout xmlns:android="http://schemas.android.com/apk/res/android"
    android:layout_width="fill_parent"  android:layout_height="fill_parent">

  <TableRow>
    <TextView android:text="First Name:"
        android:layout_width="wrap_content"  android:layout_height="wrap_content" />

    <EditText android:text="Edgar"
        android:layout_width="wrap_content"  android:layout_height="wrap_content" />
  </TableRow>

  <TableRow>
    <TextView android:text="Last Name:"
        android:layout_width="wrap_content"  android:layout_height="wrap_content" />

    <EditText android:text="Allen"
        android:layout_width="wrap_content"  android:layout_height="wrap_content" />

    <EditText android:text="Poe"
        android:layout_width="wrap_content"  android:layout_height="wrap_content" />
  </TableRow>

</TableLayout>
```

Figure 6–21. *An irregular* `TableLayout`

In Listings 6–43 and 6–44, we populated the `TableLayout` with `TableRow` elements. Although this is the usual pattern, you can place any `android.widget.View` as a child of the table. For example, Listing 6–45 creates a table where the first row is an `EditText` (see Figure 6–22).

Listing 6–45. *Using an* `EditText` *Instead of a* `TableRow`

```
<?xml version="1.0" encoding="utf-8"?>
<TableLayout xmlns:android="http://schemas.android.com/apk/res/android"
    android:layout_width="fill_parent"  android:layout_height="fill_parent"
    android:stretchColumns="0,1,2" >

  <EditText android:text="Fullname:"
        android:layout_width="wrap_content"  android:layout_height="wrap_content" />

  <TableRow>
    <TextView android:text="Edgar"
        android:layout_width="wrap_content"  android:layout_height="wrap_content" />

    <TextView android:text="Allen"
```

```
        android:layout_width="wrap_content"  android:layout_height="wrap_content" />

    <TextView android:text="Poe"
        android:layout_width="wrap_content"  android:layout_height="wrap_content" />
  </TableRow>

</TableLayout>
```

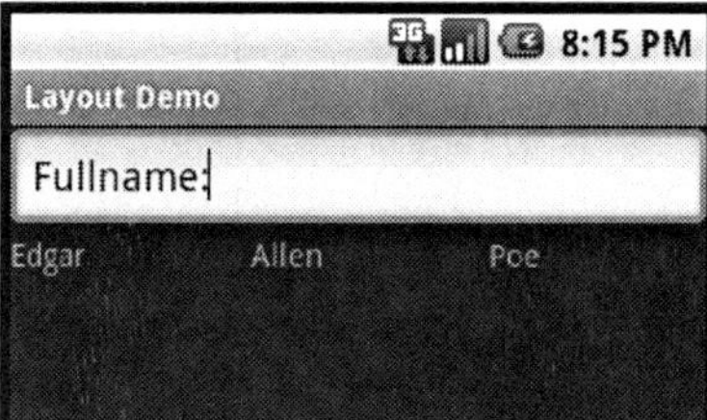

Figure 6–22. *`EditText` as a child of a `TableLayout`*

The UI for Listing 6–45 is shown in Figure 6–22. Notice that the `EditText` takes up the entire width of the screen, even though we have not specified this in the XML layout. That's because children of `TableLayout` always span the entire row. In other words, children of `TableLayout` can specify `android:layout_width="wrap_content"` (as we did with `EditText`), but it won't affect actual layout—they are forced to accept `fill_parent`. They can, however, set `android:layout_height`.

Because the content of a table is not always known at design time, `TableLayout` offers several attributes that can help you control the layout of a table. For example, Listing 6–45 sets the `android:stretchColumns` property on the `TableLayout` to `"0,1,2"`. This gives a hint to the `TableLayout` that columns 0, 1, and 2 can be stretched if required, based on the contents of the table. If we had not used `stretchColumns` in Listing 6–45, we would have seen `"EdgarAllenPoe"` all squished together. Technically, the second row takes up the entire width, but without `stretchColumns`, the three `TextViews` would not spread across.

Similarly, you can set `android:shrinkColumns` to wrap the content of a column or columns if other columns require more space. You can also set `android:collapseColumns` to make columns invisible. Note that columns are identified with a zero-based indexing scheme.

`TableLayout` also offers `android:layout_span`. You can use this property to have a cell span multiple columns. This field is similar to the HTML `colspan` property.

At times, you might also need to provide spacing within the contents of a cell or a control. The Android SDK supports this via `android:padding` and its siblings. `android:padding` lets you control the space between a view's outer boundary and its content (see Listing 6–46).

Listing 6–46. *Using `android:padding`*

```
<LinearLayout xmlns:android="http://schemas.android.com/apk/res/android"
    android:orientation="vertical" android:layout_width="fill_parent"
    android:layout_height="fill_parent">
```

```
    <EditText android:text="one"
    android:layout_width="wrap_content"  android:layout_height="wrap_content"
    android:padding="40px" />
</LinearLayout>
```

Listing 6–46 sets the padding to 40px. This creates 40 pixels of whitespace between the EditText control's outer boundary and the text displayed within it. Figure 6–23 shows the same EditText with two different padding values. The UI on the left does not set any padding, and the one on the right sets android:padding="40px".

Figure 6–23. *Utilizing padding*

android:padding sets the padding for all sides: left, right, top, and bottom. You can control the padding for each side by using android:leftPadding, android:rightPadding, android:topPadding, and android:bottomPadding.

Android also defines android:layout_margin, which is similar to android:padding. In fact, android:padding/android:layout_margin is analogous to android:gravity/android:layout_gravity, but one is for a view, and the other is for a container.

Finally, the padding value is always set as a dimension type, usually dp, px, or sp. We covered the dimension types in Chapter 3.

The RelativeLayout Layout Manager

Another interesting layout manager is RelativeLayout. As the name suggests, this layout manager implements a policy where the controls in the container are laid out relative to either the container or another control in the container. Listing 6–47 and Figure 6–24 show an example.

Listing 6–47. *Using a RelativeLayout Layout Manager*

```
<RelativeLayout xmlns:android="http://schemas.android.com/apk/res/android"
        android:layout_width="fill_parent"
        android:layout_height="wrap_content">

<TextView android:id="@+id/userNameLbl"
        android:layout_width="fill_parent"  android:layout_height="wrap_content"
        android:text="Username: "
        android:layout_alignParentTop="true" />

<EditText android:id="@+id/userNameText"
        android:layout_width="fill_parent"  android:layout_height="wrap_content"
        android:layout_toRightOf="@id/userNameLbl" />
```

```
<TextView android:id="@+id/pwdLbl"
        android:layout_width="wrap_content"  android:layout_height="wrap_content"
        android:layout_below="@id/userNameText"
        android:text="Password: " />

<EditText android:id="@+id/pwdText"
        android:layout_width="fill_parent"  android:layout_height="wrap_content"
        android:layout_toRightOf="@id/pwdLbl"
        android:layout_below="@id/userNameText" />

<TextView android:id="@+id/pwdCriteria"
        android:layout_width="fill_parent"  android:layout_height="wrap_content"
        android:layout_below="@id/pwdText"
        android:text="Password Criteria... " />

<TextView android:id="@+id/disclaimerLbl"
        android:layout_width="fill_parent"  android:layout_height="wrap_content"
        android:layout_alignParentBottom="true"
        android:text="Use at your own risk... " />

</RelativeLayout>
```

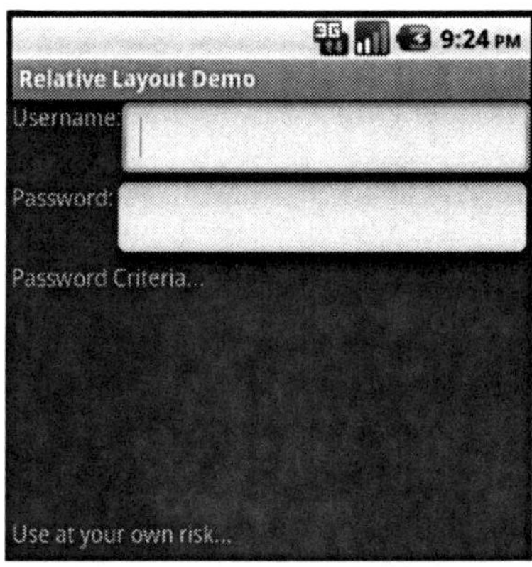

Figure 6–24. *A UI laid out using the `RelativeLayout` layout manager*

As shown, the UI looks like a simple login form. The username label is pinned to the top of the container, because we set `android:layout_alignParentTop` to `true`. Similarly, the Username input field is positioned below the Username label because we set `android:layout_below`. The Password label appears below the Username label, and the Password input field appears below the Password label. The disclaimer label is pinned to the bottom of the container because we set `android:layout_alignParentBottom` to `true`.

Besides these three layout attributes, you can also specify `layout_above`, `layout_toRightOf`, `layout_toLeftOf`, `layout_centerInParent`, and several more. Working with `RelativeLayout` is fun due to its simplicity. In fact, once you start using it, it'll become your favorite layout manager—you'll find yourself going back to it over and over again.

Support in ADT for Designing RelativeLayouts

As the Android Development Tools (ADT) evolve for Eclipse, more and more features become available. In the early days, the visual editor for layouts was not very good. But lately it's much better. In particular, the tool's ability to visually lay out UIs using `RelativeLayout` has greatly improved. You should definitely check out this tool in Eclipse if you use `RelativeLayouts`.

The FrameLayout Layout Manager

The layout managers that we've discussed so far implement various layout strategies. In other words, each one has a specific way that it positions and orients its children on the screen. With these layout managers, you can have many controls on the screen at one time, each taking up a portion of the screen. Android also offers a layout manager that is mainly used to display a single item: `FrameLayout`. You mainly use this utility layout class to dynamically display a single view, but you can populate it with many items, setting one to visible while the others are invisible. Listing 6–48 demonstrates using `FrameLayout`.

Listing 6–48. *Populating* `FrameLayout`

```
<?xml version="1.0" encoding="utf-8"?>
<FrameLayout xmlns:android="http://schemas.android.com/apk/res/android"
    android:id="@+id/frmLayout"
    android:layout_width="fill_parent"  android:layout_height="fill_parent">

     <ImageView
        android:id="@+id/oneImgView" android:src="@drawable/one"
        android:scaleType="fitCenter"
        android:layout_width="fill_parent"  android:layout_height="fill_parent"/>
    <ImageView
        android:id="@+id/twoImgView" android:src="@drawable/two"
        android:scaleType="fitCenter"
        android:layout_width="fill_parent"  android:layout_height="fill_parent"
        android:visibility="gone" />

</FrameLayout>


public class FrameLayoutActivity extends Activity{
    private ImageView one = null;
    private ImageView two = null;
    @Override
    protected void onCreate(Bundle savedInstanceState) {
        super.onCreate(savedInstanceState);
        setContentView(R.layout.listing6_48);

        one = (ImageView)findViewById(R.id.oneImgView);
        two = (ImageView)findViewById(R.id.twoImgView);

        one.setOnClickListener(new OnClickListener(){
```

```
            public void onClick(View view) {
                two.setVisibility(View.VISIBLE);

                view.setVisibility(View.GONE);
            }});

        two.setOnClickListener(new OnClickListener(){

            public void onClick(View view) {
                one.setVisibility(View.VISIBLE);

                view.setVisibility(View.GONE);
            }});
    }
}
```

Listing 6–48 shows the layout file as well as the `onCreate()` method of the activity. The idea of the demonstration is to load two `ImageView` objects in the `FrameLayout`, with only one of the `ImageView` objects visible at a time. In the UI, when the user clicks the visible image, we hide one image and show the other one.

Look at Listing 6–48 more closely now, starting with the layout. You can see that we define a `FrameLayout` with two `ImageView` objects (an `ImageView` is a control that knows how to display images). Notice that the second `ImageView`'s visibility is set to `gone`, making the control invisible. Now, look at the `onCreate()` method. In the `onCreate()` method, we register listeners to click events on the `ImageView` objects. In the click handler, we hide one `ImageView` and show the other.

As we said earlier, you generally use `FrameLayout` when you need to dynamically set the content of a view to a single control. Although this is the general practice, the control will accept many children, as we demonstrated. Listing 6–48 adds two controls to the layout but has one of the controls visible at a time. `FrameLayout`, however, does not force you to have only one control visible at a time. If you add many controls to the layout, `FrameLayout` will simply stack the controls, one on top of the other, with the last one on top. This can create an interesting UI. For example, Figure 6–25 shows a `FrameLayout` control with two `ImageView` objects that are visible. You can see that the controls are stacked, and that the top one is partially covering the image behind it.

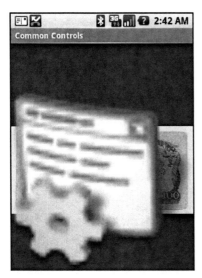

Figure 6–25. *FrameLayout with two ImageView objects*

Another interesting aspect of the FrameLayout is that if you add more than one control to the layout, the size of the layout is computed as the size of the largest item in the container. In Figure 6–25, the top image is actually much smaller than the image behind it, but because the size of the layout is computed based on the largest control, the image on top is stretched.

Also note that if you put many controls inside a FrameLayout with one or more of them invisible to start, you might want to consider using setMeasureAllChildren(true) on your FrameLayout. Because the largest child dictates the layout size, you'll have a problem if the largest child is invisible to begin with: when it becomes visible, it is only partially visible. To ensure that all items are rendered properly, call setMeasureAllChildren() and pass it a value of true. The equivalent XML attribute for FrameLayout is android:measureAllChildren="true".

The GridLayout Layout Manager

Android 4.0 brought with it a new layout manager called GridLayout. As you might expect, it lays out views in a grid pattern of rows and columns, somewhat like TableLayout. However, it's easier to use than TableLayout. With a GridLayout, you can specify a row and column value for a view, and that's where it goes in the grid. This means you don't need to specify a view for every cell, just those that you want to hold a view. Views can span multiple grid cells. You can even put more than one view into the same grid cell.

When laying out views, you must not use the weight attribute, because it does not work in child views of a GridLayout. You can use the layout_gravity attribute instead. Other interesting attributes you can use with GridLayout child views include layout_column and layout_columnSpan to specify the left-most column and the number of columns the view takes up, respectively. Similarly, there are layout_row and layout_rowSpan attributes. Interestingly, you do not need to specify layout_height and layout_width for GridLayout child views; they default to WRAP_CONTENT.

Customizing the Layout for Various Device Configurations

By now, you know very well that Android offers a host of layout managers that help you build UIs. If you've played around with the layout managers we've discussed, you know that you can combine the layout managers in various ways to obtain the look and feel you want. But even with all the layout managers, building UIs—and getting them right—can be a challenge. This is especially true for mobile devices. Users and manufacturers of mobile devices are getting more and more sophisticated, and that makes the developer's job even more challenging.

One of the challenges is building a UI for an application that displays in various screen configurations. For example, what would your UI look like if your application were displayed in portrait versus landscape mode? If you haven't run into this yet, your mind is probably racing right now, wondering how to deal with this common scenario. Interestingly, and fortunately, Android provides some support for this use case.

Here's how it works: when building a layout, Android will find and load layouts from specific folders based on the configuration of the device. A device can be in one of three configurations: portrait, landscape, or square (square is rare). To provide different layouts for the various configurations, you have to create specific folders for each configuration from which Android will load the appropriate layout. As you know, the default layout folder is located at res/layout. To support portrait display, create a folder called res/layout-port. For landscape, create a folder called res/layout-land. And for a square, create one called res/layout-square.

A good question at this point is, "With these three folders, do I need the default layout folder (res/layout)?" Generally, yes. Android's resource-resolution logic looks in the configuration-specific directory first. If Android doesn't find a resource there, it goes to the default layout directory. Therefore, you should place default layout definitions in res/layout and the customized versions in the configuration-specific folders.

Another trick is to use the <include /> tag in a layout file. This allows you to create common chunks of layout code (for example, in the default layout directory) and include them in layouts defined in layout-port and layout-land. An include tag might look like this:

```
<include layout="@layout/common_chunk1" />
```

If the concept of include interests you, you should also check out the <merge /> tag and the ViewStub class in the Android API. These give you even more flexibility when organizing layouts, without duplicating views.

Note that the Android SDK does not offer any APIs for you to programmatically specify which configuration to load—the system simply selects the folder based on the configuration of the device. You can, however, set the orientation of the device in code, for example, using the following:

```
import android.content.pm.ActivityInfo;
...
setRequestedOrientation(ActivityInfo.SCREEN_ORIENTATION_LANDSCAPE);
```

This forces your application to appear on the device in landscape mode. Go ahead and try it in one of your earlier projects. Add the code to your `onCreate()` method of an activity, run it in the emulator, and see your application sideways.

The layout is not the only resource that is configuration driven, and other qualifiers of the device configuration are taken into account when finding the resource to use. The entire contents of the `res` folder can have variations for each configuration. For example, to have different drawables loaded for each configuration, create folders for `drawable-port`, `drawable-land`, and `drawable-square`, in addition to the default `drawable`. Remember that in your code, you still only refer to the resource as `R.resource_type.name` without any qualifiers. For example, if you have lots of different variations of your layout file `main.xml` in several different qualified resource directories, your code will still refer to `R.layout.main`. Android takes care of finding the appropriate `main.xml` for you.

You can refer back to Chapter 3 for more on the basics of resources. We'll be covering this concept of configuration changes in more detail in Chapter 12. This concludes our discussion of building UIs.

References

Here are some helpful references to topics you may wish to explore further:

- `www.androidbook.com/proandroid4/projects`: A list of downloadable projects related to this book. For this chapter, look for a ZIP file called `ProAndroid4_Ch06_Controls.zip`. This ZIP file contains all projects from this chapter, listed in separate root directories. There is also a `README.TXT` file that describes exactly how to import projects into Eclipse from one of these ZIP files.
- `http://developer.android.com/reference/android/widget/LinearLayout.html#attr_android:gravity`: Reference page describing different values for gravity when used with a `LinearLayout`.
- `www.curious-creature.org/2010/08/15/scrollviews-handy-trick`: A blog post by Romain Guy (of the Android team) that explains how to use the `ScrollView` properly.
- `http://developer.android.com/resources/articles/index.html`: Several "Layout Tricks" technical articles that are well worth reading. They get into performance aspects of designing and building UIs in Android. Look for other articles in this list related to building UIs.

Summary

Let's conclude this chapter by quickly enumerating what you have learned about building user interfaces:

- How XML resources define UI appearances, and how code fills in the data
- The three main types of layouts and when to use each
- Views supported in Android and how to define them both in XML and via code
- The main list controls and how to use adapters to populate the data
- Styles and themes you can use to manage the look and feel of your application from a common set of resources

Interview Questions

You can use the following questions as a guide to test your understanding of this chapter.

1. Can a `TextView` receive typed input from a user?
2. When would you use a `Spannable` with an `EditText` field?
3. How can you customize the way an `ImageButton` looks during the press and release actions?
4. Does an adapter always need to get data from a content provider?
5. Can a `ListView` contain more than one view?
6. Must a `ListView` always contain at least one `TextView`?
7. What defines the layout for a `ListView` item, and where do you find one of these?
8. How is a `Spinner` different from the other list controls?
9. To display a list of names from a database table in a row-and-column format, is it better to use a `TableLayout` or a `GridView`?
10. When do you use an `AbsoluteLayout` in your user interface?
11. What does `android:layout_weight` do?
12. What's the difference between `android:gravity` and `android:layout_gravity`?

13. Is it possible to force an activity to be displayed in a specific orientation?
14. With which layout manager do you use `android:layout_toRightOf`?
15. Under what circumstances is the ID from a list adapter the same value as the position?

In the next chapter, we'll take user interface development further—we are going to discuss menus.

Chapter 7

Working with Menus

Android SDK has extensive support for menus. It supports regular menus, submenus, context menus, icon menus, secondary menus, and alternative menus. Android 3.0, in addition, introduced the action bar, which integrates well with menu items. The action bar and menu interaction are covered in Chapter 10. Now Android 4.0 has introduced pop-up menus: menus that can be invoked at any time based on a button click or any other UI event.

In Android, menus, much like other resources, can be represented as both Java objects and entries in XML files. Android generates resource IDs for each of the loaded menu items. This chapter covers the XML menu resources in detail as well.

Being resources, all menu items take advantage of auto-generated resource IDs.

Understanding Android Menus

The key class in Android menu support is `android.view.Menu`. Every activity in Android is associated with one menu object of this type. The menu object then contains a number of menu items and submenus.

Menu items are represented by `android.view.MenuItem`. Submenus are represented by `android.view.SubMenu`. These relationships are graphically represented in Figure 7–1. Strictly speaking, this is not a class diagram but a structural diagram designed to help you visualize the relationships between various menu-related classes and functions.

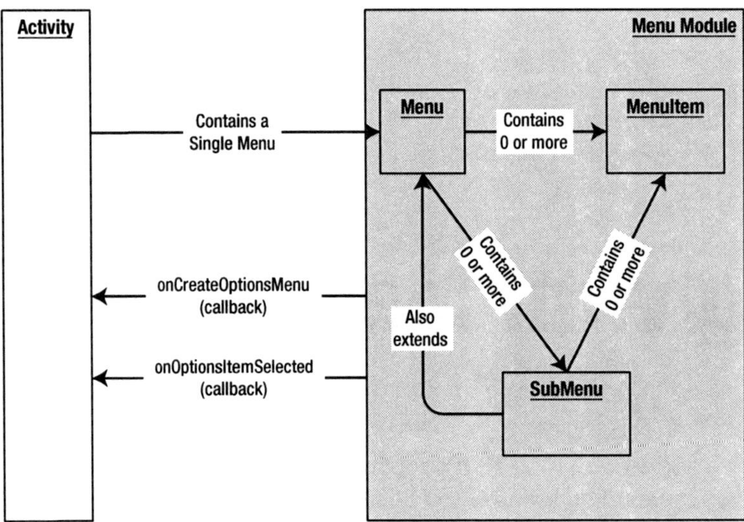

Figure 7–1. *Structure of Android menu-related classes*

Figure 7–1 illustrates that a `Menu` object contains a set of menu items. A menu item carries the following attributes:

- *Name:* A string title
- *Menu item ID:* An integer
- *Group ID:* An integer representing which group this item should be part of
- *Sort order:* An integer identifying the order of this menu item when it is displayed in the menu

The name and menu item ID attributes are self explanatory.

You can group menu items together by assigning each one a group ID. Multiple menu items that carry the same group ID are considered part of the same group.

The sort-order attribute demands a bit of coverage. If one menu item carries an order number of 4 and another menu item carries a order number of 6, the first menu item will appear above the second menu item in the menu. Some of these menu item sort-order-number ranges are reserved for certain kinds of menus. These are called menu *categories*. The available menu categories are as follows:

- *Secondary:* Secondary menu items, which are considered less important (and are less frequently used) than others, start at 0x30000 and are defined by the constant `Menu.CATEGORY_SECONDARY`. Other types of menu categories—such as system menus, alternative menus, and container menus—have different order-number ranges.

- *System:* The sort-order numbers for system menu items start at 0x20000 and are defined by the constant `Menu.CATEGORY_SYSTEM`. This sort-order range is reserved for menu items added by the Android system. As of 4.0 we haven't seen any system menu items added to applications. For example, on a Microsoft Windows platform, Close, Refresh, and so on could be considered system menu items that are applicable to all activities. We haven't seen them yet on the Android platform, but we expect they may be considered in the future as the platform expands beyond phones.
- *Alternative:* The alternative menu item sort-order range starts at 0x40000. These menu items are defined by the constant `Menu.CATEGORY_ALTERNATIVE`. They're usually contributed by external applications that provide alternative ways to deal with the data that is under consideration.
- *Container:* Container menu items have a sort-order range starting at 0x10000 and are defined by the constant `Menu.CATEGORY_CONTAINER`. In Android, the parents of views, such as layouts, are considered containers. The documentation is not clear about whether this category pertains to layouts, but it is as good a guess as any. Most likely, container-related menu items can be placed in this range. Reordering a set of views in a grid could be considered a container menu item.

These menu categories are primarily designed to make one set of menu items more important than another. However, there is no stipulation not to use these category integers as starting numbers for menu item IDs to keep them unique. So you can use these category integers as starting numbers for a certain class of menu item IDs. It is also possible to pass this menu category starting numbers as the menu item group ID.

Usually, menu item ID uniqueness is not a problem, because system menus are not added at the moment into your application menus as of Android 4.0. The mechanism to invoke alternative menus is different (alternative menus are covered later in the chapter), where their IDs are not passed to your menu handler. Most of the time, the menus are declared in XML files where Android generates the unique IDs for you. So these category ranges come into play only when you're deciding on the menu display order or, sometimes, when you're indicating that an entire group of menu items belongs to a particular category (in which case the group ID matches the category ID).

Figure 7–1 also shows two callback methods that you can use to create and respond to menu items: `onCreateOptionsMenu()` and `onOptionsItemSelected()`. You will cover these next.

Creating a Menu

In the Android SDK, you don't need to create a menu object from scratch. Because an activity is associated with a single menu, Android creates this single menu for that activity and passes it to the `onCreateOptionsMenu()` callback method of the activity class. (As the name of the method indicates, menus in Android are also known as *options menus*.)

Prior to SDK 3.0, `onCreateOptionsMenu()` is called the first time an activity's options menu is accessed. Starting with 3.0, this method is called as part of activity creation. This change is due to the fact that the action bar is always present in an activity. A menu item that you create in this method for the options menu may sit in an action bar. Because an action bar is always visible (unlike the options menu), the action bar must know its menu items from the beginning. So Android cannot wait until the user opens an options menu to call the `onCreateOptionsMenu()` method.

This callback menu setup method allows you to populate the single passed-in menu with a set of menu items (see Listing 7–1).

Listing 7–1. *Signature for the* `onCreateOptionsMenu` *Method*

```
@Override
public boolean onCreateOptionsMenu(Menu menu)
{
    // populate menu items
    .....
   ...return true;
}
```

Once the menu items are populated, the code should return `true` to make the menu visible. If this method returns `false`, the menu is invisible. The code in Listing 7–2 shows how to add three menu items using a single group ID along with incremental menu item IDs and order IDs.

Listing 7–2. *Adding Menu Items*

```
@Override
public boolean onCreateOptionsMenu(Menu menu)
{
   //call the base class to include system menus
   super.onCreateOptionsMenu(menu);

   menu.add(0            // Group
         ,1              // item id
         ,0              //order
         ,"append");     // title

   menu.add(0,2,1,"item2");
   menu.add(0,3,2,"clear");

   //It is important to return true to see the menu
   return true;
}
```

You should also call the base-class implementation of this method to give the system an opportunity to populate the menu with system menu items (no system menu items are defined so far). To keep these system menu items separate from other kinds of menu items, Android adds the system menu items with a sort order starting at 0x20000. (As we mentioned before, the constant `Menu.CATEGORY_SYSTEM` defines the starting sort order ID for these system menu items. In all releases so far, Android has not added any system menus.)

The arguments to create the menu item are explained in Listing 7–2. The last argument is the name or title of the menu item. Instead of free text, you can use a string resource through the `R.java` constants file. The group, menu item, and order IDs are all optional; you can use `Menu.NONE` if you don't want to specify any of them.

Working with Menu Groups

Now, let's look at how to work with menu groups. Listing 7–3 shows how you add two groups of menus: Group 1 and Group 2.

Listing 7–3. *Using Group IDs to Create Menu Groups*

```
@Override
public boolean onCreateOptionsMenu(Menu menu)
{
   //Group 1
   int group1 = 1;
   menu.add(group1,1,1,"g1.item1");
   menu.add(group1,2,2,"g1.item2");

   //Group 2
   int group2 = 2;
   menu.add(group2,3,3,"g2.item1");
   menu.add(group2,4,4,"g2.item2");

   return true; // it is important to return true
}
```

Notice how the menu item IDs and the order IDs are independent of the groups. What good is a group, then? Well, Android provides a set of methods on the `android.view.Menu` class that are based on group IDs. You can manipulate a group's menu items using these methods:

```
removeGroup(id)
setGroupCheckable(id, checkable, exclusive)
setGroupEnabled(id,boolean enabled)
setGroupVisible(id,visible)
```

`removeGroup()` removes all menu items from that group, given the group ID. You can enable or disable menu items in a given group using the `setGroupEnabled` method(). Similarly, you can control the visibility of a group of menu items using `setGroupVisible()`.

`setGroupCheckable()` is interesting. You can use this method to show a check mark on a menu item when that menu item is selected. When applied to a group, it enables this functionality for all menu items within that group. If this method's exclusive flag is set, only one menu item within that group is allowed to go into a checked state. The other menu items remain unchecked.

You now know how to populate an activity's main menu with a set of menu items and group them according to their nature. Next, you see how to respond to these menu items.

Responding to Menu Items

There are multiple ways of responding to menu item clicks in Android. You can use the onOptionsItemSelected() method of the activity class; you can use stand-alone listeners, or you can use intents. You will cover each of these techniques in this section.

Responding to Menu Items through onOptionsItemSelected

When a menu item is clicked, Android calls the onOptionsItemSelected() callback method on the Activity class (see Listing 7–4).

Listing 7–4. *Signature and Body of the onOptionsItemSelected Method*

```
@Override
public boolean onOptionsItemSelected(MenuItem item)
{
   switch(item.getItemId()) {
      .....
   //for items handled
   return true;

   //for the rest
   ...return super.onOptionsItemSelected(item);
   }
}
```

The key pattern here is to examine the menu item ID through the getItemId() method of the MenuItem class and do what's necessary. If onOptionsItemSelected() handles a menu item, it returns true. The menu event will not be further propagated. For the menu item callbacks that onOptionsItemSelected() doesn't deal with, onOptionsItemSelected() should call the parent method through super.onOptionsItemSelected(). The default implementation of the onOptionsItemSelected() method returns false so that the normal processing can take place. Normal processing includes alternative means of invoking responses for a menu click, such as invoking a listener directly that can be directly tied to the menu item.

Responding to Menu Items Through Listeners

You usually respond to menus by overriding onOptionsItemSelected(); this is the recommended technique for better performance. However, a menu item allows you to register a listener that could be used as a callback. A listener implies object creation and a registry of the listener. So this is the overhead that the *performance* refers to in the first sentence of this paragraph. However, you may choose to give more importance to reuse and clarity, in which case listeners provide flexibility.

This approach is a two-step process. In the first step, you implement the `OnMenuClickListener` interface. Then, you take an instance of this implementation and pass it to the menu item. When the menu item is clicked, the menu item calls the `onMenuItemClick()` method of the `OnMenuClickListener` interface (see Listing 7–5).

Listing 7–5. *Using a Listener as a Callback for a Menu Item Click*

```
//Step 1
public class MyResponse implements OnMenuClickListener
{
   //some local variable to work on
   //...
   //Some constructors
   @override
   boolean onMenuItemClick(MenuItem item)
   {
      //do your thing
      return true;
   }
}

//Step 2
MyResponse myResponse = new MyResponse(...);
menuItem.setOnMenuItemClickListener(myResponse);
...
```

The `onMenuItemClick()` method is called when the menu item has been invoked. This code executes as soon as the menu item is clicked, even before the `onOptionsItemSelected()` method is called. If `onMenuItemClick()` returns `true`, no other callbacks are executed—including the `onOptionsItemSelected()` callback method. This means that the listener code takes precedence over the `onOptionsItemSelected()` method.

Using an Intent to Respond to Menu Items

You can also associate a menu item with an intent by using the `MenuItem`'s method `setIntent(intent)`. By default, a menu item has no intent associated with it. But when an intent *is* associated with a menu item, and nothing else handles the menu item, then the default behavior is to invoke the intent using `startActivity(intent)`. For this to work, all the handlers—especially the `onOptionsItemSelected()` method—should call the parent class's `onOptionsItemSelected()` method for those items that are not handled. Or you could look at it this way: the system gives `onOptionsItemSelected()` an opportunity to handle menu items first (followed by the listener, of course). This is assuming there is no listener directly associated with that menu item; if there is, then the listener overrides the rest.

If you don't override the `onOptionsItemSelected()` method, the base class in the Android framework does what's necessary to invoke the intent on the menu item. But if you do override this method and you're not interested in this menu item, you must call the parent method, which, in turn, facilitates the intent invocation. So here's the bottom line: either don't override the onOptionsItemSelected() method, or override it and invoke the parent for the menu items that you are not handling.

> **NOTE:** The Menu class defines a few convenience constants, one of which is Menu.FIRST. You can use this as a baseline number for menu IDs and other menu-related sequential numbers.

Secondary menu items, as mentioned earlier, start at 0x30000 and are defined by the constant Menu.CATEGORY_SECONDARY. Their sort-order IDs are higher than regular menu items, so they appear after the regular menu items in a menu. Note that the sort order is the only thing that distinguishes a secondary menu item from a regular menu item. In all other aspects, a secondary menu item works and behaves like any other menu item.

Now that you have covered the basics of menu support in Android, Figure 7–2 shows a screen shot of an activity that displays a menu. This section refers to this picture as it explains the concepts.

Figure 7–2. *Sample Menus application*

Working with Other Menu Types

So far you've covered some of the simpler, although quite functional, menu types. As you walk through the SDK, you see that Android also supports icon menus, submenus, context menus, and alternative menus. Out of these, alternative menus are unique to Android. This section covers all of these menu types.

Expanded Menus

If an application has more menu items than it can display on the main screen, Android shows the More menu item to allow the user to see the rest. This menu, called an *expanded menu*, appears automatically when there are too many menu items to display in the limited amount of space. You can see this More menu item in Figure 7–2.

Working with Icon Menus

Android supports not only text but also images or icons as part of its menu repertoire. You can also see this in Figure 7–2. You can use icons to represent menu items instead of and in addition to text.

Note a few limitations when it comes to using icon menus. First, you can't use icon menus for expanded menus. This restriction may be lifted in the future, depending on device size and SDK support. Larger devices may allow this functionality, whereas smaller devices may keep the restriction.

Second, icon menu items do not support menu item check marks.

Third, if the text in an icon menu item is too long, it's truncated after a certain number of characters, depending on the size of the display. (This last limitation applies to text-based menu items also.)

Creating an icon menu item is straightforward. You create a regular text-based menu item as before, and then you use the `setIcon()` method on the `MenuItem` class to set the image. You need to use the image's resource ID, so you must generate it first by placing the image or icon in the `/res/drawable` directory. For example, if the icon's file name is `balloons`, then the resource ID is `R.drawable.balloons`.

Listing 7–6 demonstrates how to add an icon to a menu item

Listing 7–6. ***Attaching an Icon to a Menu Item***

```
//add a menu item and remember it so that you can use it
//subsequently to set the icon on it.
MenuItem item = menu.add(...);
item.setIcon(R.drawable.balloons);
```

As you add menu items to the menu, you rarely need to keep a local variable returned by the `menu.add` method. But in this case, you need to remember the returned object so you can add the icon to the menu item. The code in this example also demonstrates that the type returned by the `menu.add()` method is `MenuItem`.

The icon shows as long as the menu item is displayed on the main application screen. If it's displayed as part of the expanded menu, the icon doesn't show, just the text. The menu item displaying an image of balloons in Figure 7–2 is an example of an icon menu item.

Working with Submenus

Let's take a look at Android's submenus now. Figure 7–1 points out the structural relationship of a SubMenu to a Menu and a MenuItem. A Menu object can have multiple SubMenu objects. Each SubMenu object is added to the Menu object through a call to the Menu.addSubMenu() method (see Listing 7–7). You add menu items to a submenu the same way that you add menu items to a menu. This is because SubMenu is also derived from a Menu object. However, you cannot add additional submenus to a submenu.

Listing 7–7. *Adding Submenus*

```
private void addSubMenu(Menu menu)
{
   //Secondary items are shown just like everything else
   int base=Menu.FIRST + 100;
   SubMenu sm = menu.addSubMenu(base,base+1,Menu.NONE,"submenu");
   sm.add(base,base+2,base+2,"sub item1");
   sm.add(base,base+3,base+3,"sub item2");
   sm.add(base,base+4,base+4,"sub item3");

    //submenu item icons are not supported
   item1.setIcon(R.drawable.icon48x48_2);

   //the following is ok however
   sm.setIcon(R.drawable.icon48x48_1);

   //This will result in runtime exception
    //sm.addSubMenu("try this");
}
```

NOTE: SubMenu, as a subclass of the Menu object, continues to carry the addSubMenu() method. The compiler won't complain if you add a submenu to another submenu, but you'll get a runtime exception if you try to do it.

The Android SDK documentation also suggests that submenus do not support icon menu items. When you add an icon to a menu item and then add that menu item to a submenu, the menu item ignores that icon, even if you don't see a compile-time or runtime error. However, the submenu itself can have an icon.

Working with Context Menus

Users of desktop programs are familiar with context menus. In Windows applications, for example, you can access a context menu by right-clicking a UI element. Android supports the same idea of context menus through an action called a *long click*. A long click is a mouse click held down slightly longer than usual on any Android view.

On handheld devices such as cell phones, mouse clicks are implemented in a number of ways, depending on the navigation mechanism. If your phone has a wheel to move the cursor, a press of the wheel serves as the mouse click. Or if the device has a touch pad, a tap or a press is equivalent to a mouse click. Or you might have a set of arrow buttons for movement and a selection button in the middle; clicking that button is equivalent to clicking the mouse. Regardless of how a mouse click is implemented on your device, if you hold the mouse click a bit longer, you realize the long click.

A context menu differs structurally from the standard options menu that you've been examining (see Figure 7–3). Context menus have nuances that options menus don't have.

Figure 7–3 shows that a context menu is represented as a `ContextMenu` class in the Android menu architecture. Just like a `Menu`, a `ContextMenu` can contain a number of menu items. You use the same set of `Menu` methods to add menu items to the context menu.

The difference between a `Menu` and a `ContextMenu` boils down to the ownership of the menu in question. An activity owns a regular options menu, whereas a view owns a context menu. This is to be expected, because the long clicks that activate context menus apply to the *view* being clicked. So an activity can have only one options menu but many context menus. Because an activity can contain multiple views, and each view can have its own context menu, an activity can have as many context menus as there are views.

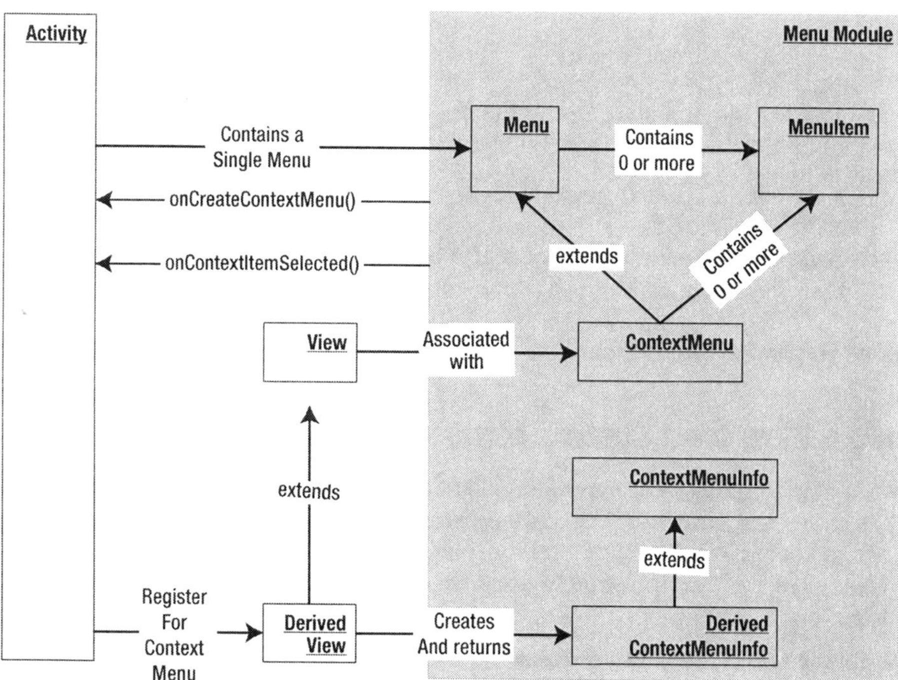

Figure 7–3. *Activities, views, and context menus*

Although a context menu is owned by a view, the method to populate context menus resides in the `Activity` class. This method is called `activity.onCreateContextMenu()`, and its role resembles that of the `activity.onCreateOptionsMenu()` method. This callback method also carries with it (as an argument to the method) the view for which the context menu items are to be populated.

There is another notable wrinkle to the context menu. Whereas the `onCreateOptionsMenu()` method is automatically called for every activity, this is not the case with `onCreateContextMenu()`. A view in an activity does not *have* to own a context menu. You can have three views in your activity, for example, but perhaps you want to enable context menus for only one view and not the others. If you want a particular view to own a context menu, you must register that view with its activity specifically for the purpose of owning a context menu. You do this through the `activity.registerForContextMenu(view)` method, which is discussed in the section "Registering a View for a Context Menu."

Now note the `ContextMenuInfo` class shown in Figure 7–3. An object of this type is passed to the `onCreateContextMenu()` method. This is one way for the view to pass additional information to this method. For a view to do this, it needs to override the `getContextViewInfo()` method and return a derived class of `ContextMenuInfo` with additional methods to represent the additional information. You might want to look at the source code for `android.view.View` to fully understand this interaction.

> **NOTE:** Per the Android SDK documentation, context menus do not support shortcuts, icons, or submenus.

Now that you know the general structure of the context menus, let's look at some sample code that demonstrates each of the steps to implement a context menu:

1. Register a view for a context menu in an activity's `onCreate()` method.
2. Populate the context menu using `onCreateContextMenu()`. You must complete step 1 before this callback method is invoked by Android.
3. Respond to context menu clicks.

Registering a View for a Context Menu

The first step in implementing a context menu is to register a view for the context menu in an activity's `onCreate()` method. You can register a `TextView` for a context menu by using the code in Listing 7–8. You first find the `TextView` and then call `registerForContextMenu()` on the activity using the `TextView` as an argument. This sets up the `TextView` for context menus.

Listing 7–8. ***Registering a `TextView` for a Context Menu***

```
@Override
public void onCreate(Bundle savedInstanceState) {
    super.onCreate(savedInstanceState);
    setContentView(R.layout.main);

    TextView tv = (TextView)this.findViewById(R.id.textViewId);
    registerForContextMenu(tv);
}
```

Populating a Context Menu

Once a view like the TextView in this example is registered for context menus, Android calls the onCreateContextMenu() method with this view as the argument. This is where you can populate the context menu items for that context menu. The onCreateContextMenu() callback method provides three arguments to work with.

The first argument is a preconstructed ContextMenu object, the second is the view (such as the TextView) that generated the callback, and the third is the ContextMenuInfo class that you saw briefly in the discussion of Figure 7–3. For a lot of simple cases, you can just ignore the ContextMenuInfo object. However, some views may pass extra information through this object. In those cases, you need to cast the ContextMenuInfo class to a subclass and then use the additional methods to retrieve the additional information.

Some examples of classes derived from ContextMenuInfo include AdapterContextMenuInfo and ExpandableContextMenuInfo. Views that are tied to database cursors in Android use the AdapterContextMenuInfo class to pass the row ID within that view for which the context menu is being displayed. In a sense, you can use this class to further clarify the object underneath the mouse click, even within a given view.

Listing 7–9 demonstrates the onCreateContextMenu() method.

Listing 7–9. *The onCreateContextMenu() Method*

```
@Override
public void onCreateContextMenu(ContextMenu menu, View v, ContextMenuInfo menuInfo)
{
        menu.setHeaderTitle("Sample Context Menu");
        menu.add(200, 200, 200, "item1");
}
```

Responding to Context Menu Items

The third step in the implementation of a context menu is responding to context menu clicks. The mechanism of responding to context menus is similar to the mechanism of responding to options menus. Android provides a callback method similar to onOptionsItemSelected() called onContextItemSelected(). This method, like its counterpart, is also available on the Activity class. Listing 7–10 demonstrates onContextItemSelected().

Listing 7–10. *Responding to Context Menus*

```
@Override
 public boolean onContextItemSelected(MenuItem item)
{
     if (item.getitemId() = some-menu-item-id)
     {
        //handle this menu item
        return true;
     }
... other exception processing
}
```

Working with Alternative Menus

So far, you have learned to create and work with menus, submenus, and context menus. Android introduces a new concept called *alternative menus*, which allow alternative menu items to be part of menus, submenus, and context menus. Alternative menus allow multiple applications on Android to assist one another.

Specifically, alternative menus allow one application to include menus from another application. When the alternative menus are chosen, the target application or activity is launched with a URL to the data needed by that activity. The invoked activity then uses the data URL from the intent that is passed. To understand alternative menus well, you must first understand content providers, content URIs, content MIME types, and intents (see Chapters 4 and 5).

The general idea here is this: imagine you are writing a screen to display some data. Most likely, this screen is an activity. On this activity, you have an options menu that allows you to manipulate or work with the data in a number of ways. Also assume for a moment that you are working with a document or a note that is identified by a URI and a corresponding MIME type. What you want to do as a programmer is anticipate that the device will eventually contain more programs that will know how to work with this data or display this data. You want to give this new set of programs an opportunity to display their menu items as part of the menu that you are constructing for this activity.

To attach alternative menu items to a menu, follow these steps while setting up the menu in the `onCreateOptionsMenu()` method:

1. Create an intent whose data URI is set to the data URI that you are showing at the moment.
2. Set the category of the intent as `CATEGORY_ALTERNATIVE`.
3. Search for activities that allow operations on data supported by this type of URI.
4. Add intents that can invoke those activities as menu items to the menu.

These steps tell you a lot about the nature of Android applications, so let's examine each one. As you know now, attaching the alternative menu items to the menu happens in the `onCreateOptionsMenu()` method:

```
@Override public boolean onCreateOptionsMenu(Menu menu)
{
}
```

Let's now figure out what code makes up this function. You first need to know the URI for the data you may be working on in this activity. You can get the URI like this:

```
this.getIntent().getData()
```

This works because the `Activity` class has a method called `getIntent()` that returns the data URI for which this activity is invoked. This invoked activity might be the main activity invoked by the main menu; in that case, it might not have an intent, and the `getIntent()`method returns `null`. In your code, you have to guard against this situation.

If you happen to know the URI of the data regardless of how the activity is invoked, you can use that URI directly as well.

Your goal now is to find out the other programs that know how to work with this kind of data. You do this search using an intent as an argument. Here's the code to construct that intent:

```
Intent criteriaIntent = new Intent(null, getIntent().getData());
intent.addCategory(Intent.CATEGORY_ALTERNATIVE);
```

Once you construct the intent, you also add a category of actions that you are interested in. Specifically, you are interested only in activities that can be invoked as part of an alternative menu. You are ready now to tell the Menu object to search for matching activities and add them as menu options (see Listing 7–11).

Listing 7–11. *Populating a Menu with Alternative Menu Items*

```
// Search for, and populate the menu with matching Activities.

//You can use the following defined constant
//to serve as a starting point for a number
//of unique ids.
int menuItemGroupId = Menu.CATEGORY_ALTERNATIVE;
int startingMenuItemId = Menu.CATEGORY_ALTERNATIVE;
int startingMenuItemOrderId = Menu.CATEGORY_ALTERNATIVE;

menu.addIntentOptions(
     menuItemGroupId,             // Group
     startingMenuItemId,          // Starting menu item ID for all items.
     startingMenuItemOrderId,     // Starting order id for each of the menus
     this.getComponentName(),     // Name of the activity class displaying
                                  // the menu--here, it's this class.
                                    //variable "this" points to activity
     null,                          // No specifics.
     criteriaIntent,                // Previously created intent that
                                    // describes our requirements.
     0,                             // No flags.
     null);                         // returned menu items
```

The method addIntentOptions() on the Menu class is responsible for looking up the activities that match an intent's URI and category attributes. Then, the method adds these activities to the menu under the right group with the appropriate menu item and sort order IDs. The first three arguments deal with this aspect of the method's responsibility. In Listing 7–11, you start with the Menu.CATEGORY_ALTERNATIVE as the group under which the new menu items will be added. You also use this same constant as the starting point for the menu item and order IDs.

The next argument points to the fully qualified component name of the activity that this menu is part of. The code uses a method from the Activity class called getComponentName(). A *component name* is simply the name of the package and the name of the class, and this component name is needed because when a new menu item is added, that menu item will need to invoke the target activity. To do that, the system needs the source activity that started the target activity. The next argument is an array of intents that you should use as a filter on the returned intents. You use null in the example.

The next argument points to criteriaIntent, which you just constructed. This is the search criteria you want to use. The argument after that is a flag such as Menu.FLAG_APPEND_TO_GROUP to indicate whether to append to the set of existing menu items in this group or replace them. The default value is 0, which indicates that the menu items in the menu group should be replaced.

The last argument in Listing 7–11 is an array of menu items that are added. You could use these added menu item references if you want to manipulate them in some manner after adding them.

All of this is well and good. But a few questions remain unanswered. For example, what will be the names of the added menu items? The Android documentation is silent about this, so we snooped around the source code to see what this function is actually doing behind the scenes (refer to Chapter 1 to see how to get to Android's source code).

As it turns out, the Menu class is only an interface, so you can't see any implementation source code for it. The class that implements the Menu interface is called MenuBuilder. Listing 7–12 shows the source code of a relevant method, addIntentOptions(), from the MenuBuilder class (we're providing the code for your reference; we won't explain it line by line).

Listing 7–12. *MenuBuilder.addIntentOptions() Method*

```
public int addIntentOptions(int group, int id, int categoryOrder,
                            ComponentName caller,
                            Intent[] specifics,
                            Intent intent, int flags,
                            MenuItem[] outSpecificItems)
{
    PackageManager pm = mContext.getPackageManager();
    final List<ResolveInfo> lri =
            pm.queryIntentActivityOptions(caller, specifics, intent, 0);
    final int N = lri != null ? lri.size() : 0;

    if ((flags & FLAG_APPEND_TO_GROUP) == 0) {
        removeGroup(group);
    }

    for (int i=0; i<N; i++) {
        final ResolveInfo ri = lri.get(i);
        Intent rintent = new Intent(
            ri.specificIndex < 0 ? intent : specifics[ri.specificIndex]);
        rintent.setComponent(new ComponentName(
                ri.activityInfo.applicationInfo.packageName,
                ri.activityInfo.name));
 final MenuItem item = add(group, id, categoryOrder,
                        ri.loadLabel(pm));
        item.setIntent(rintent);
        if (outSpecificItems != null && ri.specificIndex >= 0) {
            outSpecificItems[ri.specificIndex] = item;
        }
    }
```

```
    return N;
}
```

Note the line in Listing 7–12 highlighted in bold; this portion of the code constructs a menu item. The code delegates the work of figuring out a menu title to the `ResolveInfo` class. The source code of the `ResolveInfo` class shows that the intent filter that declared this intent should have a title associated with it. Listing 7–13 shows an example of an intent filter definition.

Listing 7–13. *An Intent Filter's Label*

```
<intent-filter android:label="Menu Title ">
     .......
      <category android:name="android.intent.category.ALTERNATE" />
     <data android:mimeType="some type data" />
</intent-filter>
```

The `label` value of the intent filter ends up serving as the menu name. You can look at the Android NotePad example to see this behavior.

Dynamic Menus

So far, we've talked about static menus—you set them up once, and they don't change dynamically according to what's onscreen. If you want to create dynamic menus, use the `onPrepareOptionsMenu()` method that Android provides on an activity class. This method resembles `onCreateOptionsMenu()` except that it is called every time a menu is invoked. You should use `onPrepareOptionsMenu()` if you want to disable some menu items or menu groups based on what you are displaying. For 3.0 and above, you have to explicitly call a new provisioned method called `invalidateOptionsMenu()`, which in turn invokes the `onPrepareOptionsMenu()`. You can call this method any time something changes in your application state that would require a change to the menu.

Loading Menus Through XML Files

Up until this point, we've created all our menus programmatically. This is not the most convenient way to create menus, because for every menu, you have to provide several IDs and define constants for each of those IDs. No doubt this is tedious.

Instead, you can define menus through XML files, which is possible in Android because menus are also resources. The XML approach to menu creation offers several advantages, such as the ability to name menus, order them automatically, and give them IDs. You can also get localization support for the menu text.

Follow these steps to work with XML-based menus:

1. Define an XML file with menu tags.

2. Place the file in the `/res/menu` subdirectory. The name of the file is arbitrary, and you can have as many files as you want. Android automatically generates a resource ID for this menu file.

3. Use the resource ID for the menu file to load the XML file into the menu.

4. Respond to the menu items using the resource IDs generated for each menu item.

The following sections talk about each of these steps and provide corresponding code snippets.

Structure of an XML Menu Resource File

First, let's look at an XML file with menu definitions (see Listing 7–14). All menu files start with the same high-level menu tag followed by a series of group tags. Each of these group tags corresponds to the menu item group you talked about at the beginning of the chapter. You can specify an ID for the group using the @+id approach. Each menu group has a series of menu items with their menu item IDs tied to symbolic names. You can refer to the Android SDK documentation for all the possible arguments for these XML tags.

Listing 7–14. *An XML File with Menu Definitions*

```
<menu xmlns:android="http://schemas.android.com/apk/res/android">
    <!-- This group uses the default category. -->
    <group android:id="@+id/menuGroup_Main">

        <item android:id="@+id/menu_testPick"
            android:orderInCategory="5"
            android:title="Test Pick" />
        <item android:id="@+id/menu_testGetContent"
            android:orderInCategory="5"
            android:title="Test Get Content" />
        <item android:id="@+id/menu_clear"
            android:orderInCategory="10"
            android:title="clear" />
        <item android:id="@+id/menu_dial"
            android:orderInCategory="7"
            android:title="dial" />
        <item android:id="@+id/menu_test"
            android:orderInCategory="4"
            android:title="@+string/test" />
        <item android:id="@+id/menu_show_browser"
            android:orderInCategory="5"
            android:title="show browser" />
    </group>
</menu>
```

The menu XML file in Listing 7–14 has one group. Based on the resource ID definition @+id/menuGroup_main, this group is automatically assigned a resource ID called menuGroup_main in the R.java resource ID file. Similarly, all the child menu items are allocated menu item IDs based on their symbolic resource ID definitions in this XML file.

Inflating XML Menu Resource Files

Let's assume that the name of this XML file is my_menu.xml. You need to place this file in the /res/menu subdirectory. Placing the file in /res/menu automatically generates a resource ID called R.menu.my_menu.

Now, let's look at how you can use this menu resource ID to populate the options menu. Android provides a class called android.view.MenuInflater to populate Menu objects from XML files. You use an instance of this MenuInflater to make use of the R.menu.my_menu resource ID to populate a menu object. This is shown in Listing 7–15.

Listing 7–15. *Using the Menu Inflater*

```
@Override
public boolean onCreateOptionsMenu(Menu menu)
{
   MenuInflater inflater = getMenuInflater(); //from activity
   inflater.inflate(R.menu.my_menu, menu);

   //It is important to return true to see the menu
   return true;

}
```

In this code, you first get the MenuInflater from the Activity class and then tell it to inflate the menu XML file into the menu directly.

Responding to XML-Based Menu Items

You respond to XML menu items the way you respond to menus created programmatically, but with a small difference. As before, you handle the menu items in the onOptionsItemSelected() callback method. This time, you have some help from Android's resources (see Chapter 3 for details on resources). As mentioned in the section "Structure of an XML Menu Resource File," Android not only generates a resource ID for the XML file but also generates the necessary menu item IDs to help you distinguish between the menu items. This is an advantage in terms of responding to the menu items because you don't have to explicitly create and manage their menu item IDs.

> **NOTE:** The resource type to identify the ID of a menu item (R.id.some_menu_item_id) is different from the resource type to identify the menu itself (R.menu.some_menu_file_id).

To further elaborate on this, in the case of XML menus, you don't have to define constants for these IDs and you don't have to worry about their uniqueness because resource ID generation takes care of that. The code in Listing 7–16 illustrates.

Listing 7–16. *Responding to Menu Items from an XML Menu Resource File*

```
private void onOptionsItemSelected (MenuItem item)
{
   if (item.getItemId() == R.id.menu_clear)
   {
        //do something
   }
   else if (item.getItemId() == R.id.menu_dial)
   {
        //do something
   }
    ......etc
 }
```

Notice how the menu item names from the XML menu resource file have automatically generated menu item IDs in the `R.id` space.

Starting in SDK 3.0, you can use the `android:onClick` attribute of a menu item to directly indicate the name of a method in an activity that is attached to this menu. This activity method is then called with the menu item object as the sole input. This feature is only available in 3.0 and above. Listing 7–17 shows an example.

Listing 7–17. *Specifying a Menu Callback Method in an XML Menu Resource File*

```
<item android:id="... "
        android:onClick="a-method-name-in-your-activity"
       ...
</item>
```

Pop-up Menus in 4.0

Android 3.0 introduced another type of menu called a pop-up menu. SDK 4.0 enhanced this slightly by adding a couple of utility methods (for example, `PopupMenu.inflate`) to the `PopupMenu` class. (See the `PopupMenu` API documentation to learn about these methods. Listing 7–19 also draws attention to this difference.)

A pop-up menu can be invoked against any view in response to a UI event. An example of a UI event is a button click or a click on an image view. Figure 7–4 shows a pop-up menu invoked against a view.

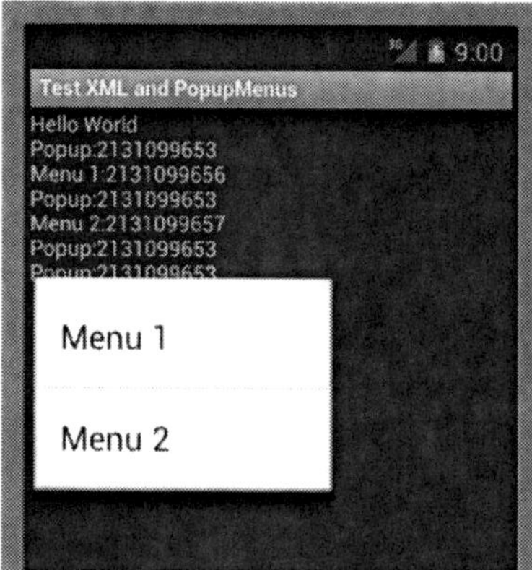

Figure 7–4. *Pop-up menu attached to a text view*

To create a pop-up menu like the one in Figure 7–4, start with a regular XML menu file as shown in Listing 7–18.

Listing 7–18. *A Sample XML File for a Pop-up Menu*

```
<menu xmlns:android="http://schemas.android.com/apk/res/android">
    <!-- This group uses the default category. -->
    <group android:id="@+id/menuGroup_Popup">
        <item android:id="@+id/popup_menu_1"
            android:title="Menu 1" />

        <item android:id="@+id/popup_menu_2"
            android:title="Menu 2" />
    </group>
</menu>
```

Assuming the code in Listing 7–18 is in a file called popup_menu.xml, you can then use the Java code in Listing 7–19 to load this menu XML as a pop-up menu.

Listing 7–19. *Working with a Pop-up Menu*

```
//Other activity code goes here...
//Invoke the following method to show a popup menu
private void showPopupMenu()
{
    //Get hold of a view to anchor the popup
    //getTextView() can be any method that returns a view
    TextView tv = getTextView();

    //instantiate a popup menu
    //the var "this" stands for activity
    PopupMenu popup = new PopupMenu(this, tv);

    //the following code for 3.0 sdk
    //popup.getMenuInflater().inflate(
    //        R.menu.popup_menu, popup.getMenu());

    //Or in sdk 4.0
    popup.inflate(R.menu.popup_menu);
    popup.setOnMenuItemClickListener(new PopupMenu.OnMenuItemClickListener()
```

```
            {
                public boolean onMenuItemClick(MenuItem item)
                {
                    //some local method to log that item
                    //See the sample project to see how this method works
                    appendMenuItemText(item);
                    return true;
                }
            }
    );
    popup.show();
}
```

As you can see, a pop-up menu behaves much like an options menu. The key differences are as follows:

- A pop-up menu is used on demand, whereas an options menu is always available.
- A pop-up menu is anchored to a view, whereas an options menu belong to the entire activity.
- A pop-up menu uses its own menu item callback, whereas the options menu uses the `onOptionsItemSelected()` callback on the activity.

A Brief Introduction to Additional XML Menu Tags

As you construct your menu XML files, you need to know the various XML menu tags that are possible. You can look these up online or quickly get this information by examining the API demonstrations that come with the Android SDK. These Android API demonstrations include a series of menus that help you explore all aspects of Android programming. If you look at the `/res/menu` subdirectory of the API demos project, you find a number of XML menu samples. The following sections briefly cover some key tags borrowed from there.

Group Category Tag

In an XML file, you can specify the category of a group by using the `menuCategory` tag:

```
<group android:id="@+id/some_group_id "
        android:menuCategory="secondary">
```

Checkable Behavior Tags

You can use the `checkableBehavior` tag to control checkable behavior at a group level:

```
<group android:id="@+id/noncheckable_group"
       android:checkableBehavior="none">
```

You can use the `checked` tag to control checkable behavior at an item level:

```
<item android:id=".."
      android:title="..."
      android:checked="true" />
```

Tags to Simulate a Submenu

A submenu is represented as a menu element under a menu item:

```
<item android:title="All without group">
       <menu>
                  <item...>
       </menu>
</item>
```

Menu Icon Tag

You can use the icon tag to associate an image with a menu item:

```
<item android:id=".. "
       android:icon="@drawable/some-file" />
```

Menu Enabling/Disabling Tag

You can enable and disable a menu item using the enabled tag:

```
<item android:id=".. "
        android:enabled="true"
        android:icon="@drawable/some-file" />
```

Menu Item Shortcuts

You can set a shortcut for a menu item using the alphabeticShortcut tag:

```
<item android:id="... "
       android:alphabeticShortcut="a"
       ...
  </item>
```

Menu Visibility

You can control a menu item's visibility using the visible flag:

```
<item android:id="... "
        android:visible="true"
       ...
</item>
```

Resources

As you learn about and work with Android menus, you may want to keep the following URLs handy:

- `http://developer.android.com/guide/topics/ui/menus.html`: The primary document from Google describing how to work with menus. As SDKs advance, this is a good place to check if things have changed since this chapter was written.
- `http://developer.android.com/guide/topics/resources/menu-resource.html`: Information about various XML tags you can use in a menu resource.
- `http://androidbook.com/proandroid4/projects`: Project download URL for this book. The downloadable project ZIP file for this chapter is `ProAndroid4_ch07_TestMenus.zip`. This project uses all the menus covered in this chapter. The screen shot in Figure 7–2 is taken from a running example of this project.

Summary

This chapter has covered the following:

- How to create and respond to regular menus
- How to create and respond to context menus
- How to change an existing menu to respond to changing context in your application
- How to work with alternative menus
- How to create, load, and respond to XML-based menu resources
- How to work with pop-up menus

Interview Questions

The following questions should consolidate what you have learned in this chapter:

1. What types of menus are available in the Android SDK?
2. What are some of the significant attributes of a menu item?
3. What callback method is called to create menu items?
4. What callback method is called to respond to menu items?
5. Is the order of a menu item impacted by its group ID?

6. How do you attach a listener to a menu item click?
7. How do you use an intent to respond to a menu item?
8. Can you start a service via an intent from a menu item?
9. What are expanded menus?
10. How do you add an image to a menu item?
11. Do images appear when a menu is expanded?
12. What is a long click in Android?
13. From an ownership perspective, what is the difference between a menu and a context menu?
14. What callback method is used to create a context menu?
15. What callback method is used to respond to a context menu?
16. Is `OnCreateContextMenu` called for every view?
17. How do you trigger the creation of a context menu?
18. What are the three arguments that are passed to the `onCreateContextMenu()` callback method?
19. How do you add alternative menus to your application?
20. How can you dynamically change a menu every time it is invoked or when certain data changes in your application?
21. Where do menu XML files go?
22. Do you need a separate menu XML file for each menu?
23. How do you identify a menu item ID using `R.java`?
24. How do you attach a menu from an XML file to an activity?
25. Do menu IDs and menu item IDs share the same resource type?
26. What is the role of a menu inflater class?
27. How do you get an instance of a menu inflater class?

Chapter 8

Fragments for Tablets and More

So far, we've explored several bits and pieces of an Android application, and you've run some simple applications tailored to a smartphone-sized screen. All you had to think about was how to lay out the UI controls on the screen for an activity, and how one activity flowed to the next, and so on. For the first two major releases of Android, small screens were it. Then came the Android tablets: devices with screen sizes of 10". And that complicated things. Why? Because now there was so much screen real estate that a simple activity had a hard time filling a screen while at the same time keeping to a single function. It no longer made sense to have an e-mail application that showed only headers in one activity (filling the screen), and a separate activity to show an individual e-mail (also filling the screen). With that much room to work with, an application could show a list of e-mail headers down the left side of the screen and the selected e-mail contents on the right side of the screen. Could it be done in a single activity with a single layout? Well, yes, but you couldn't reuse that activity or layout for any of the smaller-screen devices.

One of the core classes introduced in Android 3.0 was the `Fragment` class, especially designed to help developers manage application functionality so it would provide great usability as well as lots of reuse. This chapter will introduce you to the fragment, what it is, how it fits into an application's architecture, and how to use it. Fragments make a lot of interesting things possible that were difficult before. At about the same time, Google released a fragment SDK that works on old Androids. So even if you weren't interested in writing applications for tablets, you may have found that fragments made your life easier on non-tablet devices. Now, with Android 4.0, it's easier than ever to write great applications for smartphones and tablets and even TVs and other devices.

Let's get started with Android fragments.

What Is a Fragment?

This first section will explain what a fragment is and what it does. But first, let's set the stage to see why we need fragments. As you learned earlier, an Android application on small-screen devices uses activities to show data and functionality to a user, and each activity has a fairly simple, well-defined purpose. For example, an activity might show the user a list of contacts from their address book. Another activity might allow the user to type an e-mail. The Android application is the series of these activities grouped together to achieve a larger purpose, such as managing an e-mail account via the reading and sending of messages. This is fine for a small-screen device, but when the user's screen is very large (10 inches or larger), there's room on the screen to do more than just one simple thing. An application might want to let the user view the list of e-mails in their inbox and at the same time show the currently selected e-mail text in another window. Or an application might want to show a list of contacts and at the same time show the currently selected contact in a detail view.

As an Android developer, you know that this functionality could be accomplished by defining yet another layout for the xlarge screen with `ListViews` and layouts and all sorts of other views. And by "yet another layout" we mean layouts in addition to those you've probably already defined for the smaller screens. Of course, you'll want to have separate layouts for the portrait case as well as the landscape case. And with the size of an xlarge screen, this could mean quite a few views for all the labels and fields and images and so on that you'll need to lay out and then provide code for. If only there were a way to group these view objects together and consolidate the logic for them, so that chunks of an application could be reused across screen sizes and devices, minimizing how much work a developer has to do to maintain their application. And that is why we have fragments.

One way to think of a fragment is as a sub-activity. And in fact, the semantics of a fragment are a lot like an activity. A fragment can have a view hierarchy associated with it, and it has a lifecycle much like an activity's lifecycle. Fragments can even respond to the Back button like activities do. If you were thinking, "If only I could put multiple activities together on a tablet's screen at the same time," then you're on the right track. But because it would be too messy to have more than one activity of an application active at the same time on a tablet screen, fragments were created to implement basically that thought. This means fragments are contained within an activity. Fragments can only exist within the context of an activity; you can't use a fragment without an activity. Fragments can coexist with other elements of an activity, which means you do *not* need to convert the entire user interface of your activity to use fragments. You can create an activity's layout as before and only use a fragment for one piece of the user interface.

Fragments are not like activities, however, when it comes to saving state and restoring it later. The fragments framework provides several features to make saving and restoring fragments much simpler than the work you need to do on activities.

How you decide when to use a fragment depends on a few considerations, which are discussed next.

When to Use Fragments

One of the primary reasons to use a fragment is so you can reuse a chunk of user interface and functionality across devices and screen sizes. This is especially true with tablets. Think of how much can happen when the screen is as large as a tablet's. It's more like a desktop than a phone, and many of your desktop applications have a multipane user interface. As described earlier, you can have a list and a detail view of the selected item on screen at the same time. This is easy to picture in a landscape orientation with the list on the left and the details on the right. But what if the user rotates the device to portrait mode so that now the screen is taller than it is wide? Perhaps you now want the list to be in the top portion of the screen and the details in the bottom portion. But what if this application is running on a small screen and there's just no room for the two portions to be on the screen at the same time? Wouldn't you want the separate activities for the list and for the details to be able to share the logic you've built into these portions for a large screen? We hope you answered yes. Fragments can help with that. Figure 8–1 makes this a little clearer.

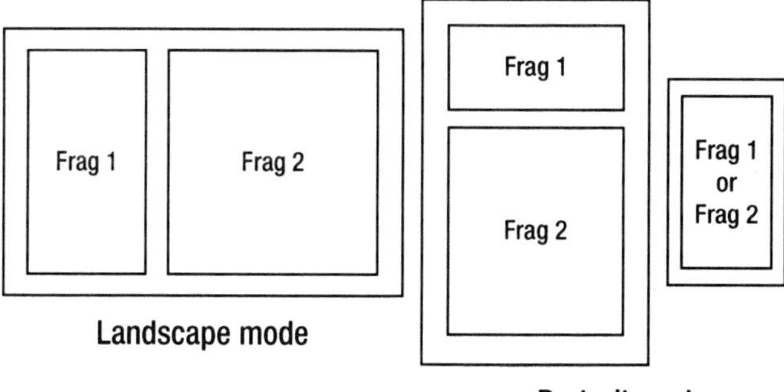

Figure 8–1. *Fragments used for a tablet UI and for a smartphone UI*

In landscape mode, two fragments may sit nicely side by side. In portrait mode, we might be able to put one fragment above the other. But if we're trying to run the same application on a device with a smaller screen, we might need to show either fragment 1 or fragment 2 but not both at the same time. If we tried to manage all these scenarios with layouts, we'd be creating quite a few, which means difficulty trying to keep everything correct across many separate layouts. When using fragments, our layouts become simple; they're dealing with the fragments themselves, not the internal structure of each fragment. Each fragment will have its own layout that can be reused across many configurations.

Let's go back to the rotating orientation example. If you've had to code for orientation changes of an activity, you know that it can be a real pain to save the current state of the activity and to restore the state once the activity has been re-created. Wouldn't it be nice if your activity had chunks that could be easily retained across orientation changes, so you could avoid all the tearing down and re-creating every time the orientation changed? Of course it would. Fragments can help with that.

Now imagine that a user is in your activity, and they've been doing some work. And imagine that the user interface has changed within the same activity, and the user wants to go back a step, or two, or three. In an old-style activity, pressing the Back button will take the user out of the activity entirely. With fragments, the Back button can step backward through a stack of fragments while staying inside the current activity.

Next, think about an activity's user interface when a big chunk of content changes; you'd like to make the transition look smooth, like a polished application. Fragments can do that, too.

Now that you have some idea of what a fragment is and why you'd want to use one, let's dig a little deeper into the structure of a fragment.

The Structure of a Fragment

As mentioned, a fragment is like a sub-activity: it has a fairly specific purpose and almost always displays a user interface. But where an activity is subclassed below `Context`, a fragment is extended from `Object` in package `android.app`. A fragment is *not* an extension of `Activity`. Like activities, however, you will always extend `Fragment` (or one of its subclasses) so you can override its behavior.

A fragment can have a view hierarchy to engage with a user. This view hierarchy is like any other view hierarchy in that it can be created (inflated) from an XML layout specification or created in code. The view hierarchy needs to be attached to the view hierarchy of the surrounding activity if it is to be seen by the user, which you'll get to shortly. The view objects that make up a fragment's view hierarchy are the same sorts of views that are used elsewhere in Android. So everything you know about views applies to fragments as well.

Besides the view hierarchy, a fragment has a bundle that serves as its initialization arguments. Similar to an activity, a fragment can be saved and later restored automatically by the system. When the system restores a fragment, it calls the default constructor (with no arguments) and then restores this bundle of arguments to the newly created fragment. Subsequent callbacks on the fragment have access to these arguments and can use them to get the fragment back to its previous state. For this reason, it is imperative that you

- Ensure that there's a default constructor for your fragment class.
- Add a bundle of arguments as soon as you create a new fragment so these subsequent methods can properly set up your fragment, and so the system can restore your fragment properly when necessary.

An activity can have multiple fragments in play at one time; and if a fragment has been switched out with another fragment, the fragment-switching transaction can be saved on a back stack. The back stack is managed by the fragment manager tied to the activity. The back stack is how the Back button behavior is managed. The fragment manager is discussed later in this chapter. What you need to know here is that a fragment knows which activity it is tied to, and from there it can get to its fragment manager. A fragment can also get to the activity's resources through its activity.

Because a fragment can be managed, it has some identifying information about itself, including a tag and an ID. These identifiers can be used to find this fragment later, which helps with reuse.

Also similar to an activity, a fragment can save state into a bundle object when the fragment is being re-created, and this bundle object gets given back to the fragment's `onCreate()` callback. This saved bundle is also passed to `onInflate()`, `onCreateView()`, and `onActivityCreated()`. Note that this is not the same bundle as the one attached as initialization arguments. This bundle is one in which you are likely to store the current state of the fragment, not the values that should be used to initialize it.

A Fragment's Lifecycle

Before you start using fragments in sample applications, you need understand the lifecycle of a fragment. Why? A fragment's lifecycle is more complicated than an activity's lifecycle, and it's very important to understand *when* you can do things with fragments. Figure 8–2 shows the lifecycle of a fragment.

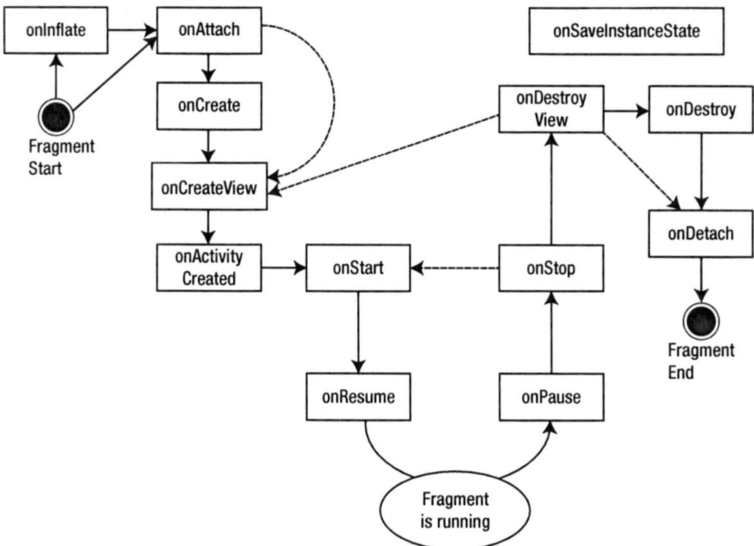

Figure 8–2. *Lifecycle of a fragment*

If you compare this to Figure 2-15 (the lifecycle for an activity), you'll notice several differences, due mostly to the interaction required between an activity and a fragment. A fragment is very dependent on the activity in which it lives and can go through multiple steps while its activity goes through one.

At the very beginning, a fragment is instantiated. It now exists as an object in memory. The first thing that is likely to happen is that initialization arguments will be added to your fragment object. This is definitely true in the situation where the system is re-creating your fragment from a saved state. When the system is restoring a fragment from a saved state, the default constructor is invoked, followed by the attachment of the initialization arguments bundle. If you are doing the creation of the fragment in code, a nice pattern to use is that in Listing 8–1, which shows a factory type of instantiator within the `MyFragment` class definition.

Listing 8–1. *Instantiating a Fragment Using a Static Factory Method*

```
public static MyFragment newInstance(int index) {
    MyFragment f = new MyFragment();
    Bundle args = new Bundle();
    args.putInt("index", index);
    f.setArguments(args);
    return f;
}
```

From the client's point of view, they get a new instance by calling the static `newInstance()` method with a single argument. They get the instantiated object back, and the initialization argument has been set on this fragment in the arguments bundle. If this fragment is saved and reconstructed later, the system will go through a very similar process of calling the default constructor and then reattaching the initialization arguments. For your particular case, you would define the signature of your `newInstance()` method (or methods) to take the appropriate number and type of arguments, and then build the arguments bundle appropriately. This is all you want your `newInstance()` method to do. The callbacks that follow will take care of the rest of the setup of your fragment.

The onInflate() Callback

The next thing that *could* happen is layout view inflation. If your fragment is defined by a `<fragment>` tag in a layout that is being inflated (typically when an activity has called `setContentView()` for its main layout), your fragment's `onInflate()` callback is called. This passes in the activity just mentioned, an `AttributeSet` with the attributes from the `<fragment>` tag, and a saved bundle. The saved bundle is the one with the saved state values in it, put there by `onSaveInstanceState()`. The expectation of `onInflate()` is that you'll read attribute values and save them for later use. At this stage in the fragment's life, it's too early to actually do anything with the user interface. The fragment is not even associated to its activity yet. But that's the next event to occur to your fragment.

The onAttach() Callback

The `onAttach()` callback is invoked after your fragment is associated with its activity. The activity reference is passed to you if you want to use it. You can at least use the activity to determine information about your enclosing activity. You can also use the activity as a context to do other operations. One thing to note is that the `Fragment` class has a `getActivity()` method that will always return the attached activity for your fragment should you need it. Keep in mind that all during this lifecycle, the initialization arguments bundle is available to you from the fragment's `getArguments()` method. However, once the fragment is attached to its activity, you can't call `setArguments()` again. So you can't add to the initialization arguments except in the very beginning.

The onCreate() Callback

Next up is the `onCreate()` callback. Although this is similar to the activity's `onCreate()`, the difference is that you should not put code in here that relies on the existence of the activity's view hierarchy. Your fragment may be associated to its activity by now, but you haven't yet been notified that the activity's `onCreate()` has finished. That's coming up. This callback gets the saved state bundle passed in, if there is one. This callback is about as early as possible to create a background thread to get data that this fragment will need. Your fragment code is running on the UI thread, and you don't want to do disk I/O or network accesses on the UI thread. In fact, it makes a lot of sense to fire off a background thread to get things ready. Your background thread is where blocking calls should be. You'll need to hook up with the data later, perhaps using a handler or some other technique.

> **NOTE:** One of the ways to load data in a background thread is to use the `Loader` class. We didn't have room in the book to cover this.

The onCreateView() Callback

The next callback is `onCreateView()`. The expectation here is that you will return a view hierarchy for this fragment. The arguments passed in to this callback include a `LayoutInflater` (which you can use to inflate a layout for this fragment), a `ViewGroup` parent (called *container* in Listing 8–2), and the saved bundle if one exists. It is very important to note that you should not attach the view hierarchy to the `ViewGroup` parent passed in. That association will happen automatically later. You will very likely get exceptions if you attach the fragment's view hierarchy to the parent in this callback—or at least odd and unexpected application behavior.

The parent is provided so you can use it with the `inflate()` method of the `LayoutInflater`. If the parent container value is null, that means this particular fragment won't be viewed because there's no view hierarchy for it to attach to. In this case, you can simply return null from here. Remember that there may be fragments floating around in your application that aren't being displayed. Listing 8–2 shows a sample of what you might want to do in this method.

Listing 8–2. *Creating a Fragment View Hierarchy in onCreateView()*

```
@Override
public View onCreateView(LayoutInflater inflater,
                ViewGroup container, Bundle savedInstanceState) {
        If(container == null)
            return null;

        View v = inflater.inflate(R.layout.details, container, false);
        TextView text1 = (TextView) v.findViewById(R.id.text1);
        text1.setText(myDataSet[ getPosition() ] );
        return v;
}
```

Here you see how you can access a layout XML file that is just for this fragment and inflate it to a view that you return to the caller. There are several advantages to this approach. You could always construct the view hierarchy in code, but by inflating a layout XML file, you're taking advantage of the system's resource-finding logic. Depending on which configuration the device is in, or for that matter which device you're on, the appropriate layout XML file will be chosen. You can then access a particular view within the layout—in this case, the `text1` `TextView` field—to do what you want with. To repeat a very important point: do not attach the fragment's view to the container parent in this callback. You can see in Listing 8–2 that you use a container in the call to `inflate()`, but you also pass `false` for the `attachToRoot` parameter.

The onActivityCreated() Callback

You're now getting close to the point where the user can interact with your fragment. The next callback is `onActivityCreated()`. This is called after the activity has completed its `onCreate()` callback. You can now trust that the activity's view hierarchy, including your own view hierarchy if you returned one earlier, is ready and available. This is where you can do final tweaks to the user interface before the user sees it. This could be especially important if this activity and its fragments are being re-created from a saved state. It's also where you can be sure that any other fragment for this activity has been attached to your activity.

The onStart() Callback

The next callback in your fragment lifecycle is `onStart()`. Now your fragment is visible to the user. But you haven't started interacting with the user just yet. This callback is tied to the activity's `onStart()`. As such, whereas previously you may have put your logic into the activity's `onStart()`, now you're more likely to put your logic into the fragment's `onStart()`, because that is also where the user interface components are.

The onResume() Callback

The last callback before the user can interact with your fragment is `onResume()`. This callback is tied to the activity's `onResume()`. When this callback returns, the user is free to interact with this fragment. For example, if you have a camera preview in your fragment, you would probably enable it in the fragment's `onResume()`.

So now you've reached the point where the app is busily making the user happy. And then the user decides to get out of your app, either by Back'ing out, or by pressing the Home button, or by launching some other application. The next sequence, similar to what happens with an activity, goes in the opposite direction of setting up the fragment for interaction.

The onPause() Callback

The first undo callback on a fragment is `onPause()`. This callback is tied to the activity's `onPause()`; just as with an activity, if you have a media player in your fragment or some other shared object, you could pause it, stop it, or give it back via your `onPause()` method. The same good-citizen rules apply here: you don't want to be playing audio if the user is taking a phone call.

The onSaveInstanceState() Callback

Similar to activities, fragments have an opportunity to save state for later reconstruction. This callback passes in a `Bundle` object to be used as the container for whatever state information you want to hang onto. This is the saved-state bundle passed to the callbacks covered earlier. To prevent memory problems, be careful about what you save into this bundle. Only save what you need. If you need to keep a reference to another fragment, save its tag instead of trying to save the other fragment.

Although you may see this method usually called right after `onPause()`, the activity to which this fragment belongs calls it when it feels that the fragment's state should be saved. This can occur any time before `onDestroy()`.

The onStop() Callback

The next undo callback is `onStop()`. This one is tied to the activity's `onStop()` and serves a purpose similar to an activity's `onStop()`. A fragment that has been stopped could go straight back to the `onStart()` callback, which then leads to `onResume()`.

The onDestroyView() Callback

If your fragment is on its way to being killed off or saved, the next callback in the undo direction is `onDestroyView()`. This will be called after the view hierarchy you created on your `onCreateView()` callback earlier has been detached from your fragment.

The onDestroy() Callback

Next up is `onDestroy()`. This is called when the fragment is no longer in use. Note that it is still attached to the activity and is still findable, but it can't do much.

The onDetach() Callback

The final callback in a fragment's lifecycle is `onDetach()`. Once this is invoked, the fragment is not tied to its activity, it does not have a view hierarchy anymore, and all its resources should have been released.

Using setRetainInstance()

You may have noticed the dotted lines in the diagram in Figure 8–2. One of the cool features of a fragment is that you can specify that you don't want the fragment completely destroyed if the activity is being re-created and therefore your fragments will be coming back also. Therefore, fragment comes with a method called `setRetainInstance()`, which takes a boolean parameter to tell it "Yes; I want you to hang around when my activity restarts" or "No; go away, and I'll create a new fragment from scratch." The best place to call `setRetainInstance()` is in the `onCreate()` callback of a fragment.

If the parameter is `true`, that means you want to keep your fragment object in memory and not start over from scratch. However, if your activity is going away and being re-created, you'll have to detach your fragment from this activity and attach it to the new one. The bottom line is that if the retain instance value is `true`, you won't actually destroy your fragment instance, and therefore you won't need to create a new one on the other side. All other callbacks will be invoked, however. The dotted lines on the diagram mean you would skip the `onDestroy()` callback on the way out, and you'd skip the `onCreate()` callback when your fragment is being re-attached to your new activity. Because an activity is re-created most likely for configuration changes, your fragment callbacks should probably assume that the configuration has changed, and therefore should take appropriate action. This would include inflating the layout to create a new view hierarchy in `onCreateView()`, for example. The code provided in Listing 8–2 would take care of that as it is written. If you choose to use the retain-instance feature, you may decide not to put some of your initialization logic in `onCreate()` because it won't always get called the way the other callbacks will.

Sample Fragment App Showing the Lifecycle

There's nothing like seeing a real example to get an appreciation for a concept. You'll create a sample application that has been instrumented so you can see all these callbacks in action. You're going to work with a sample application that uses a list of Shakespearean titles in one fragment; when the user clicks one of the titles, some text from that play will appear in a separate fragment. This sample application will work in both landscape and portrait modes on a tablet. Then you'll configure it to run as if on a smaller screen so you can see how to separate the text fragment into an activity. You'll start with the XML layout of your activity in landscape mode in Listing 8–3, which will look like Figure 8–3 when it runs.

> **NOTE:** At the end of the chapter is the URL you can use to download the projects in this chapter. This will allow you to import these projects into Eclipse directly.

Listing 8–3. *Your Activity's Layout XML for Landscape Mode*

```
<?xml version="1.0" encoding="utf-8"?>
<!-- This file is res/layout-land/main.xml -->
<LinearLayout xmlns:android="http://schemas.android.com/apk/res/android"
        android:orientation="horizontal"
        android:layout_width="match_parent"
        android:layout_height="match_parent">

    <fragment class="com.androidbook.fragments.bard.TitlesFragment"
            android:id="@+id/titles" android:layout_weight="1"
            android:layout_width="0px"
            android:layout_height="match_parent" />

    <FrameLayout
            android:id="@+id/details" android:layout_weight="2"
            android:layout_width="0px"
            android:layout_height="match_parent" />

</LinearLayout>
```

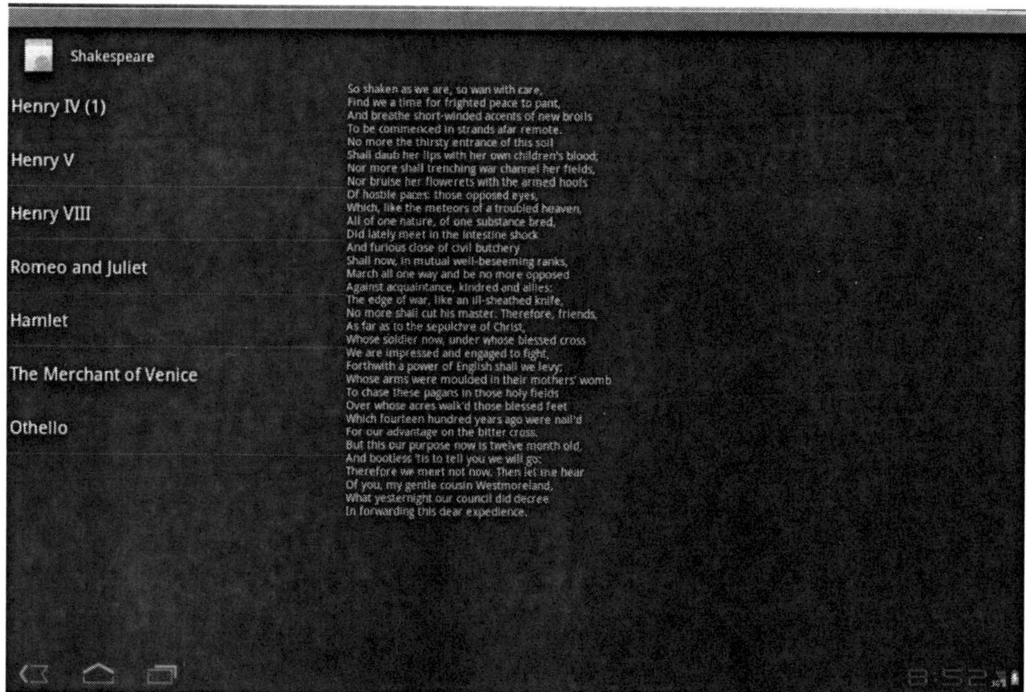

Figure 8–3. *The user interface of your sample fragment application*

This layout looks like a lot of other layouts you've seen throughout the book, horizontally left to right with two main objects. There's a special new tag, though, called `<fragment>`, and this tag has a new attribute called `class`. Keep in mind that a fragment is not a view, so the layout XML is a little different for a fragment than it is for everything else. The other thing to keep in mind is that the `<fragment>` tag is just a placeholder in this layout. You should not put child tags under `<fragment>` in a layout XML file.

The other attributes for a fragment look familiar and serve a purpose similar to that for a view. The fragment tag's `class` attribute specifies your extended class for the titles of your application. That is, you must extend one of the Android `Fragment` classes to implement your logic, and the `<fragment>` tag must know the name of your extended class. A fragment has its own view hierarchy that will be created later by the fragment itself. The next tag is a `FrameLayout`—not another `<fragment>` tag. Why is that? We'll explain in more detail later, but for now, you should be aware that you're going to be doing some transitions on the text, swapping out one fragment with another. You use the `FrameLayout` as the view container to hold the current text fragment. With your titles fragment, you have one—and only one—fragment to worry about: no swapping and no transitions. For the area that displays the Shakespearean text, you'll have several fragments.

The `MainActivity` Java code is in Listing 8–4. Actually, the listing only shows the interesting code. The code is instrumented with logging messages so you can see what's going on through LogCat. Please review the source code files from the web site to see all of it.

Listing 8–4. *Interesting Source Code from* `MainActivity`

```
public boolean isMultiPane() {
    return getResources().getConfiguration().orientation
            == Configuration.ORIENTATION_LANDSCAPE;
}

/**
 * Helper function to show the details of a selected item, either by
 * displaying a fragment in-place in the current UI, or starting a
 * whole new activity in which it is displayed.
 */
public void showDetails(int index) {
    Log.v(TAG, "in MainActivity showDetails(" + index + ")");

    if (isMultiPane()) {
        // Check what fragment is shown, replace if needed.
        DetailsFragment details = (DetailsFragment)
                getFragmentManager().findFragmentById(R.id.details);
        if ( (details == null) ||
             (details.getShownIndex() != index) ) {
            // Make new fragment to show this selection.
            details = DetailsFragment.newInstance(index);

            // Execute a transaction, replacing any existing
            // fragment with this one inside the frame.
            Log.v(TAG, "about to run FragmentTransaction...");
            FragmentTransaction ft
                    = getFragmentManager().beginTransaction();
```

```
            ft.setTransition(
                    FragmentTransaction.TRANSIT_FRAGMENT_FADE);
            //ft.addToBackStack("details");
            ft.replace(R.id.details, details);
            ft.commit();
        }

    } else {
        // Otherwise you need to launch a new activity to display
        // the dialog fragment with selected text.
        Intent intent = new Intent();
        intent.setClass(this, DetailsActivity.class);
        intent.putExtra("index", index);
        startActivity(intent);
    }
}
```

This is a very simple activity to write. To determine multipane mode (that is, whether you need to use fragments side by side), you just use the orientation of the device. If you're in landscape mode, you're multipane; if you're in portrait mode, you're not. The helper method `showDetails()` is there to figure out how to show the text when a title is selected. The index is the position of the title in the title list. If you're in multipane mode, you're going to use a fragment to show the text. You're calling this fragment a `DetailsFragment`, and you use a factory-type method to create one with the index. The interesting code for the `DetailsFragment` class is shown in Listing 8–5. As we did before in `TitlesFragment`, the various callbacks of `DetailsFragment` have logging added so we can watch what happens via LogCat. You'll come back to your `showDetails()` method later.

Listing 8–5. *Source Code for* `DetailsFragment`

```
public class DetailsFragment extends Fragment {

    private int mIndex = 0;

    public static DetailsFragment newInstance(int index) {
        Log.v(MainActivity.TAG, "in DetailsFragment newInstance(" +
                                index + ")");

        DetailsFragment df = new DetailsFragment();

        // Supply index input as an argument.
        Bundle args = new Bundle();
        args.putInt("index", index);
        df.setArguments(args);
        return df;
    }

    public static DetailsFragment newInstance(Bundle bundle) {
        int index = bundle.getInt("index", 0);
        return newInstance(index);
    }

    @Override
    public void onCreate(Bundle myBundle) {
        Log.v(MainActivity.TAG,
```

```
                "in DetailsFragment onCreate. Bundle contains:");
        if(myBundle != null) {
            for(String key : myBundle.keySet()) {
                Log.v(MainActivity.TAG, "    " + key);
            }
        }
        else {
            Log.v(MainActivity.TAG, "    myBundle is null");
        }
        super.onCreate(myBundle);

        mIndex = getArguments().getInt("index", 0);
    }

    public int getShownIndex() {
        return mIndex;
    }

    @Override
    public View onCreateView(LayoutInflater inflater,
            ViewGroup container, Bundle savedInstanceState) {
        Log.v(MainActivity.TAG,
                "in DetailsFragment onCreateView. container = " +
                container);

        // Don't tie this fragment to anything through the inflater.
        // Android takes care of attaching fragments for us. The
        // container is only passed in so you can know about the
        // container where this View hierarchy is going to go.
        View v = inflater.inflate(R.layout.details, container, false);
        TextView text1 = (TextView) v.findViewById(R.id.text1);
        text1.setText(Shakespeare.DIALOGUE[ mIndex ] );
        return v;
    }
}
```

The DetailsFragment class is actually fairly simple as well. Now you can see how to instantiate this fragment. It's important to point out that you're instantiating this fragment in code because your layout defines the ViewGroup container (a FrameLayout) that your details fragment is going to go into. Because the fragment is not itself defined in the layout XML for the activity, as your titles fragment was, you need to instantiate your details fragments in code.

To create a new details fragment , you use your newInstance() method. As discussed earlier, this factory method invokes the default constructor and then sets the arguments bundle with the value of index. Once newInstance() has run, your details fragment can retrieve the value of index in any of its callbacks by referring to the arguments bundle via getArguments(). For your convenience, in onCreate() you can save the index value from the arguments bundle to a member field in your DetailsFragment class.

You might wonder why you didn't simply set the mIndex value in newInstance(). The reason is that Android will, behind the scenes, re-create your fragment using the default constructor. Then it sets the arguments bundle to what it was before. Android won't use your newInstance() method, so the only reliable way to ensure that mIndex is set is to read the value from the arguments bundle and set it in onCreate(). The convenience method getShownIndex() retrieves the value of that index. Now the only method left to describe in the details fragment is onCreateView(). And this is very simple, too.

The purpose of onCreateView() is to return the view hierarchy for your fragment. Remember that based on your configuration, you could want all kinds of different layouts for this fragment. Therefore, the most common thing to do is utilize a layout XML file for your fragment. In your sample application, you specify the layout for the fragment to be details.xml using the resource R.layout.details. The XML for details.xml is in Listing 8–6.

Listing 8–6. *The details.xml Layout File for the Details Fragment*

```
<?xml version="1.0" encoding="utf-8"?>
<!-- This file is res/layout/details.xml -->
<LinearLayout
  xmlns:android="http://schemas.android.com/apk/res/android"
  android:layout_width="match_parent"
  android:layout_height="match_parent">
  <ScrollView android:id="@+id/scroller"
      android:layout_width="match_parent"
      android:layout_height="match_parent">
    <TextView android:id="@+id/text1"
        android:layout_width="match_parent"
        android:layout_height="match_parent" />
  </ScrollView>
</LinearLayout>
```

For your sample application, you can use the exact same layout file for details whether you're in landscape mode or in portrait mode. This layout is not for the activity, it's just for your fragment to display the text. Because it could be considered the default layout, you can store it in the /res/layout directory and it will be found and used even if you're in landscape mode. When Android goes looking for the details XML file, it tries the specific directories that closely match the device's configuration, but it will end up in the /res/layout directory if it can't find the details.xml file in any of the other places. Of course, if you want to have a different layout for your fragment in landscape mode, you could define a separate details.xml layout file and store it under /res/layout-land. Feel free to experiment with different details.xml files.

When your details fragment's onCreateView() is called, you will simply grab the appropriate details.xml layout file, inflate it, and set the text to the text from the Shakespeare class. We won't include the entire Java code for Shakespeare here, but a portion is in Listing 8–7 so you understand how it was done. For the complete source, access the project download files, as described in the "References" section at the end of this chapter.

Listing 8–7. *Source Code for Shakespeare*

```
public class Shakespeare {
    public static String TITLES[] = {
            "Henry IV (1)",
            "Henry V",
            "Henry VIII",
            "Romeo and Juliet",
```

```
            "Hamlet",
            "The Merchant of Venice",
            "Othello"
    };
    public static String DIALOGUE[] = {
        "So shaken as we are, so wan with care,\n...
... and so on ...
```

So now your details fragment view hierarchy contains the text from the selected title. Your details fragment is ready to go. And you can return to the showDetails() method to talk about FragmentTransactions.

FragmentTransactions and the Fragment Back Stack

The code in showDetails() that pulls in your new details fragment (shown again in Listing 8–8) looks rather simple, but there's a lot going on here. It's worth spending some time to explain what is happening and why. If your activity is in multipane mode, you want to show the details in a fragment next to the title list. You may already be showing details, which means you may have a details fragment visible to the user. Either way, the resource ID R.id.details is for the FrameLayout for your activity, as shown in Listing 8–3. If you have a details fragment sitting in the layout because you didn't assign any other ID to it, it will have this ID. Therefore, to find out if there's a details fragment in the layout, you can ask the fragment manager using findFragmentById(). This will return null if the frame layout is empty or will give you the current details fragment. You can then decide if you need to place a new details fragment in the layout, either because the layout is empty or because there's a details fragment for some other title. Once you make the determination to create and use a new details fragment, you invoke the factory method to create a new instance of a details fragment. Now you can put this new fragment into place for the user to see.

Listing 8–8. *Fragment Transaction Example*

```
public void showDetails(int index) {
    Log.v(TAG, "in MainActivity showDetails(" + index + ")");

    if (isMultiPane()) {
        // Check what fragment is shown, replace if needed.
        DetailsFragment details = (DetailsFragment)
                getFragmentManager().findFragmentById(R.id.details);
        if (details == null || details.getShownIndex() != index) {
            // Make new fragment to show this selection.
            details = DetailsFragment.newInstance(index);

            // Execute a transaction, replacing any existing
            // fragment with this one inside the frame.
            Log.v(TAG, "about to run FragmentTransaction...");
            FragmentTransaction ft
                    = getFragmentManager().beginTransaction();
            ft.setTransition(
                    FragmentTransaction.TRANSIT_FRAGMENT_FADE);
            //ft.addToBackStack("details");
            ft.replace(R.id.details, details);
            ft.commit();
```

```
        }
            // The rest was left out to save space.
}
```

A key concept to understand is that a fragment must live inside a view container, also known as a *view group*. The ViewGroup class includes such things as layouts and their derived classes. FrameLayout is a good choice as the container for the details fragment in the main.xml layout file of your activity. A FrameLayout is simple, and all you need is a simple container for your fragment, without the extra baggage that comes with other types of layouts. The FrameLayout is where your details fragment is going to go. If you had instead specified another <fragment> tag in the activity's layout file instead of a FrameLayout, you would not be able to do the swapping that you want to do.

The FragmentTransaction is what you use to do your swapping. You tell the fragment transaction that you want to replace whatever is in your frame layout with your new details fragment. You could have avoided all this by locating the resource ID of the details TextView and just setting the text of it to the new text for the new Shakespeare title. But there's another side to fragments that explains why you use FragmentTransactions.

As you know, activities are arranged in a stack, and as you get deeper and deeper into an application, it's not uncommon to have a stack of several activities going at once. When you press the Back button, the topmost activity goes away, and you are returned to the activity below, which resumes for you. This can continue all the way down until you're at the home screen again.

This was fine when an activity was just single-purpose, but now that an activity can have several fragments going at once, and because you can go deeper into your application without leaving the topmost activity, Android really needed to extend the Back button stack concept to include fragments as well. In fact, fragments demand this even more. When there are several fragments interacting with each other at the same time in an activity, and there's a transition to new content across several fragments at once, pressing the Back button should cause each of the fragments to roll back one step *together*. To ensure that each fragment properly participates in the rollback, a FragmentTransaction is created and managed to perform that coordination.

Be aware that a back stack for fragments is not required within an activity. You can code your application to let the Back button work at the activity level and not at the fragment level at all. If there's no back stack for your fragments, pressing the Back button will pop the current activity off the stack and return the user to whatever was underneath. If you choose to take advantage of the back stack for fragments, you will want to uncomment in Listing 8–8 the line that says ft.addToBackStack("details"). For this particular case, you've hardcoded the tag parameter to be the string "details". This tag should be an appropriate string name that represents the state of the fragments at the time of the transaction. You will be able to interrogate the back stack in code using the tag value to delete entries, as well as pop entries off. You will want meaningful tags on these transactions to be able to find the appropriate ones later.

Fragment Transaction Transitions and Animations

One of the very nice things about fragment transactions is that you can perform transitions from an old fragment to a new fragment using transitions and animations. These are not like the animations coming later, in Chapter 21. These are much simpler and do not require in-depth graphics knowledge. Let's use a fragment transaction transition to add special effects when you swap out the old details fragment with a new details fragment. This can add polish to your application, making the switch from the old to the new fragment look smooth.

One method to accomplish this is `setTransition()`, as shown in Listing 8–8. However, there are a few different transitions available. You used a fade in your example, but you can also use the `setCustomAnimations()` method to describe other special effects, such as sliding one fragment out to the right as another slides in from the left. The custom animations use the new object animation definitions, not the old ones. The old anim XML files use tags such as `<translate>`, whereas the new XML files use `<objectAnimator>`. The old standard XML files are located in the `/data/res/anim` directory under the appropriate Android SDK `platforms` directory (such as `platforms/android-11` for Honeycomb). There are some new XML files located in the `/data/res/animator` directory here, too. Your code could be something like

```
ft.setCustomAnimations(android.R.animator.fade_in, android.R.animator.fade_out);
```

which will cause the new fragment to fade in as the old fragment fades out. The first parameter applies to the fragment entering, and the second parameter applies to the fragment exiting. Feel free to explore the Android animator directory for more stock animations. If you'd like to create your own, there's section on the object animator later in this chapter to help you. The other very important bit of knowledge you need is that the transition calls need to come before the `replace()` call; otherwise, they will have no effect.

Using the object animator for special effects on fragments can be a fun way to do transitions. There are two other methods on `FragmentTransaction` you should know about: `hide()` and `show()`. Both of these methods take a fragment as a parameter, and they do exactly what you'd expect. For a fragment in the fragment manager associated to a view container, the methods simply hide or show the fragment in the user interface. The fragment does not get removed from the fragment manager in the process, but it certainly must be tied into a view container in order to affect its visibility. If a fragment does not have a view hierarchy, or if its view hierarchy is not tied into the displayed view hierarchy, then these methods won't do anything.

Once you've specified the special effects for your fragment transaction, you have to tell it the main work that you want done. In your case, you're replacing whatever is in the frame layout with your new details fragment. That's where the `replace()` method comes in. This is equivalent to calling `remove()` for any fragments that are already in the frame layout and then `add()` for your new details fragment, which means you could just call `remove()` or `add()` as needed instead.

The final action you must take when working with a fragment transaction is to commit it. The commit() method does not cause things to happen immediately but rather schedules the work for when the UI thread is ready to do it.

Now you should understand why you need to go to so much trouble to change the content in a simple fragment. It's not just that you want to change the text; you might want a special graphics effect during the transition. You may also want to save the transition details in a fragment transaction that you can reverse later. That last point may be confusing, so we'll clarify.

This is not a transaction in the truest sense of the word. When you pop fragment transactions off the back stack, you are not undoing all the data changes that may have taken place. If data changed within your activity, for example, as you created fragment transactions on the back stack, pressing the Back button does not cause the activity data changes to revert back to their previous values. You are merely stepping back through the user interface views the way you came in, just as you do with activities, but in this case it's for fragments. Because of the way fragments are saved and restored, the inner state of a fragment that has been restored from a saved state will depend on what values you saved with the fragment and how you manage to restore them. So your fragments may look the same as they did previously but your activity will not, unless you take steps to restore activity state when you restore fragments.

In your example, you're only working with one view container and bringing in one details fragment. If your user interface were more complicated, you could manipulate other fragments within the fragment transaction. What you are actually doing is beginning the transaction, replacing any existing fragment in your details frame layout with your new details fragment, specifying a fade-in animation, and committing the transaction. You commented out the part where this transaction is added to the back stack, but you could certainly uncomment it to take part in the back stack.

The FragmentManager

The FragmentManager is a component that takes care of the fragments belonging to an activity. This includes fragments on the back stack and fragments that may just be hanging around. We'll explain.

Fragments should only be created within the context of an activity. This occurs either through the inflation of an activity's layout XML or through direct instantiation using code like that in Listing 8–1. When instantiated through code, a fragment usually gets attached to the activity using a fragment transaction. In either case, the FragmentManager class is used to access and manage these fragments for an activity.

You use the getFragmentManager() method on either an activity or an attached fragment to retrieve a fragment manager. You saw in Listing 8–8 that a fragment manager is where you get a fragment transaction. Besides getting a fragment transaction, you can also get a fragment using the fragment's ID, its tag, or a combination of bundle and key.

For this, the getter methods include `findFragmentById()`, `findFragmentByTag()`, and `getFragment()`. The `getFragment()` method would be used in conjunction with `putFragment()`, which also takes a bundle, a key, and the fragment to be put. The bundle is most likely going to be the `savedState` bundle, and `putFragment()` will be used in the `onSaveInstanceState()` callback to save the state of the current activity (or another fragment). The `getFragment()` method would probably be called in `onCreate()` to correspond to `putFragment()`, although for a fragment, the bundle is available to the other callback methods, as described earlier.

Obviously, you can't use the `getFragmentManager()` method on a fragment that has not been attached to an activity yet. But it's also true that you can attach a fragment to an activity without making it visible to the user yet. If you do this, you should associate a `String` tag to the fragment so you can get to it in the future. You'd most likely use this method of `FragmentTransaction` to do this:

```
public FragmentTransaction add (Fragment fragment, String tag)
```

In fact, you can have a fragment that does not exhibit a view hierarchy. This might be done to encapsulate certain logic together such that it could be attached to an activity, yet still retain some autonomy from the activity's lifecycle and from other fragments. When an activity goes through a re-create cycle due to a device-configuration change, this non-UI fragment could remain largely intact while the activity goes away and comes back again. This would be a good candidate for the `setRetainInstance()` option.

The fragment back stack is also the domain of the fragment manager. Whereas a fragment transaction is used to put fragments onto the back stack, the fragment manager can take fragments off the back stack. This is usually done using the fragment's ID or tag, but it can be done based on position in the back stack or just to pop the top-most fragment.

Finally, the fragment manager has methods for some debugging features, such as turning on debugging messages to LogCat using `enableDebugLogging()` or dumping the current state of the fragment manager to a stream using `dump()`. Note that you turned on fragment manager debugging in the `onCreate()` method of your activity in Listing 8–4.

Caution When Referencing Fragments

It's time to revisit the earlier discussion of the fragment's lifecycle and the arguments and saved-state bundles. Android could save one of your fragments at many different times. This means that at the moment your application wants to retrieve that fragment, it's possible that it is not in memory. For this reason, we caution you *not* to think that a variable reference to a fragment is going to remain valid for a long time. If fragments are being replaced in a container view using fragment transactions, any reference to the old fragment is now pointing to a fragment that is possibly on the back stack. Or a fragment may get detached from the activity's view hierarchy during an application configuration change such as a screen rotation. Be careful.

If you're going to hold onto a reference to a fragment, be aware of when it could get saved away; when you need to find it again, use one of the getter methods of the fragment manager. If you want to hang onto a fragment reference, such as when an activity is going through a configuration change, you can use the putFragment() method with the appropriate bundle. In the case of both activities and fragments, the appropriate bundle is the savedState bundle that is used in onSaveInstanceState() and that reappears in onCreate() (or, in the case of fragments, the other early callbacks of the fragment's lifecycle). You will probably never store a direct fragment reference into the arguments bundle of a fragment; if you're tempted to do so, please think very carefully about it first.

The other way you can get to a specific fragment is by querying for it using a known tag or known ID. The getter methods described previously will allow retrieval of fragments from the fragment manager this way, which means you have the option of just remembering the tag or ID of a fragment so that you can retrieve it from the fragment manager using one of those values, as opposed to using putFragment() and getFragment().

Saving Fragment State

Another interesting class was introduced in Android 3.2: Fragment.SavedState. Using the saveFragmentInstanceState() method of FragmentManager, you can pass this method a fragment, and it returns an object representing the state of that fragment. You can then use that object when initializing a fragment, using Fragment's setInitialSavedState() method. Chapter 12 discusses this in more detail.

ListFragments and <fragment>

There are still a few more things to cover to make your sample application complete. The first is the TitlesFragment class. This is the one that is created via the layout.xml file of your main activity. The <fragment> tag serves as your placeholder for where this fragment will go and does not define what the view hierarchy will look like for this fragment. The code for your TitlesFragment is in Listing 8–9. TitlesFragment displays the list of titles for your application.

Listing 8–9. *TitlesFragment Java Code*

```
public class TitlesFragment extends ListFragment {
    private MainActivity myActivity = null;
    int mCurCheckPosition = 0;

    @Override
    public void onAttach(Activity myActivity) {
        Log.v(MainActivity.TAG,
            "in TitlesFragment onAttach; activity is: " + myActivity);
        super.onAttach(myActivity);
        this.myActivity = (MainActivity)myActivity;
    }


    @Override
    public void onActivityCreated(Bundle savedState) {
```

```
        Log.v(MainActivity.TAG,
            "in TitlesFragment onActivityCreated. savedState contains:");
        if(savedState != null) {
            for(String key : savedState.keySet()) {
                Log.v(MainActivity.TAG, "    " + key);
            }
        }
        else {
            Log.v(MainActivity.TAG, "    savedState is null");
        }
        super.onActivityCreated(savedState);

        // Populate list with your static array of titles.
        setListAdapter(new ArrayAdapter<String>(getActivity(),
                android.R.layout.simple_list_item_1,
                Shakespeare.TITLES));

        if (savedState != null) {
            // Restore last state for checked position.
            mCurCheckPosition = savedState.getInt("curChoice", 0);
        }

        // Get your ListFragment's ListView and update it
        ListView lv = getListView();
        lv.setChoiceMode(ListView.CHOICE_MODE_SINGLE);
        lv.setSelection(mCurCheckPosition);

        // Activity is created, fragments are available
        // Go ahead and populate the details fragment
        myActivity.showDetails(mCurCheckPosition);
    }
    @Override
    public void onSaveInstanceState(Bundle outState) {
        Log.v(MainActivity.TAG, "in TitlesFragment onSaveInstanceState");
        super.onSaveInstanceState(outState);
        outState.putInt("curChoice", mCurCheckPosition);
    }

    @Override
    public void onListItemClick(ListView l, View v, int pos, long id) {
        Log.v(MainActivity.TAG,
            "in TitlesFragment onListItemClick. pos = "
            + pos);
        myActivity.showDetails(pos);
        mCurCheckPosition = pos;
    }

    @Override
    public void onDetach() {
        Log.v(MainActivity.TAG, "in TitlesFragment onDetach");
        super.onDetach();
        myActivity = null;
    }
}
```

Unlike DetailsFragment, for this fragment you don't do anything in the onCreateView() callback. This is because you're extending the ListFragment class, which contains a

ListView already. The default onCreateView() for a ListFragment creates this ListView for you and returns it. It's not until onActivityCreated() that you do any real application logic. By this time in your application, you can be sure that the activity's view hierarchy, plus this fragment's, has been created. The resource ID for that ListView is android.R.id.list1, but you can always call getListView() if you need to get a reference to it, which you do in onActivityCreated(). However, because a ListFragment is not the same as a ListView, do not attach the adapter to the ListView directly. You must use the ListFragment's setListAdapter() method instead. Because the activity's view hierarchy is set up, you're safe going back into the activity to do the showDetails() call.

At this point in your sample activity's life, you've added a list adapter to your list view, you've restored the current position (if you came back from a restore, due perhaps to a configuration change), and you've asked the activity (in showDetails()) to set the text to correspond to the selected Shakespearean title.

Your TitlesFragment class also has a listener on the list so when the user clicks another title, the onListItemClick() callback is called, and you switch the text to correspond to that title, again using the showDetails() method.

Another difference between this fragment and the earlier details fragment is that when this fragment is being destroyed and re-created, you save state in a bundle (the value of the current position in the list), and you read it back in onCreate(). Unlike the details fragments that get swapped in and out of the FrameLayout on your activity's layout, there is just one titles fragment to think about. So when there is a configuration change and your titles fragment is going through a save-and-restore operation, you want to remember where you were. With the details fragments, you can re-create them without having to remember the previous state.

Invoking a Separate Activity When Needed

There's a piece of code we haven't talked about yet, and that is in showDetails() when you're in portrait mode and the details fragment won't fit properly on the same page as the titles fragment. You're going to pretend that's the case even though it really isn't on a tablet screen. Since fragments got made available for the older Android releases via the compatibility library, you are able to use fragments on phones as well as tablets, which means the scenario described here could be common. If the screen real estate won't permit feasible viewing of a fragment that would otherwise be shown alongside the other fragments, you will need to launch a separate activity to show the user interface of that fragment. For your sample application, you implement a details activity; the code is in Listing 8–10.

Listing 8–10. *Showing a New Activity When a Fragment Doesn't Fit*

```
public class DetailsActivity extends Activity {

    @Override
    public void onCreate(Bundle savedInstanceState) {
        Log.v(MainActivity.TAG, "in DetailsActivity onCreate");
        super.onCreate(savedInstanceState);
```

```
        if (getResources().getConfiguration().orientation
                == Configuration.ORIENTATION_LANDSCAPE) {
            // If the screen is now in landscape mode, it means
            // that your MainActivity is being shown with both
            // the titles and the text, so this activity is
            // no longer needed. Bail out and let the MainActivity
            // do all the work.
            finish();
            return;
        }

        if(getIntent() != null) {
            // This is another way to instantiate a details
            // fragment.
            DetailsFragment details =
                DetailsFragment.newInstance(getIntent().getExtras());

            getFragmentManager().beginTransaction()
                .add(android.R.id.content, details)
                .commit();
        }
    }
}
```

There are several interesting aspects to this code. For one thing, it is really easy to implement. You make a simple determination of the device's orientation, and as long as you're in portrait mode, you set up a new details fragment within this details activity. If you're in landscape mode, your `MainActivity` is able to display both the titles fragment and the details fragment, so there is no reason to be displaying this activity at all. You may wonder why you would ever launch this activity if you're in landscape mode, and the answer is, you wouldn't. However, once this activity has been started in portrait mode, if the user rotates the device to landscape mode, this details activity will get restarted due to the configuration change. So now the activity is starting up, and it's in landscape mode. At that moment, it makes sense to finish this activity and let the `MainActivity` take over and do all the work.

Another interesting aspect about this details activity is that you never set the root content view using `setContentView()`. So how does the user interface get created? If you look carefully at the `add()` method call on the fragment transaction, you will see that the view container to which you add the fragment is specified as the resource `android.R.id.content`. This is the top-level view container for an activity, and therefore when you attach your fragment view hierarchy to this container, your fragment view hierarchy becomes the only view hierarchy for the activity. You used the very same `DetailsFragment` class as before with the other `newInstance()` method to create the fragment (the one that takes a bundle as a parameter), then you simply attached it to the top of the activity's view hierarchy. This causes the fragment to be displayed within this new activity.

From the user's point of view, they are now looking at just the details fragment view, which is the text from the Shakespearean play. If the user wants to select a different title, they press the Back button, which pops this activity to reveal your main activity (with the titles fragment only). The other choice for the user is to rotate the device to get back to landscape mode. Then your details activity will call finish() and go away, revealing the also-rotated main activity underneath.

When the device is in portrait mode, if you're not showing the details fragment in your main activity, you should have a separate main.xml layout file for portrait mode like the one in Listing 8–11.

Listing 8–11. *The Layout for a Portrait Main Activity*

```
<?xml version="1.0" encoding="utf-8"?>
<!-- This file is res/layout/main.xml -->
<LinearLayout xmlns:android="http://schemas.android.com/apk/res/android"
        android:orientation="vertical"
        android:layout_width="match_parent"
        android:layout_height="match_parent">

    <fragment class="com.androidbook.fragments.bard.TitlesFragment"
            android:id="@+id/titles"
            android:layout_width="match_parent"
            android:layout_height="match_parent" />

</LinearLayout>
```

Of course, you could make this layout whatever you want it to be. For your purposes here, you simply make it show the titles fragment by itself. It's very nice that your titles fragment class doesn't need to include much code to deal with the device reconfiguration.

Take a moment to view this application's manifest file. In it you find the main activity with a category of LAUNCHER so that it will appear in the device's list of apps. Then you have the separate DetailsActivity with a category of DEFAULT. This allows you to start the details activity from code but will not show the details activity as an app in the App list.

Persistence of Fragments

When you play with this sample application, make sure you rotate the device (pressing Ctrl+F11 rotates the device in the emulator). You will see that the device rotates, and the fragments rotate right along with it. If you watch the LogCat messages, you will see a lot of them for this application. In particular, during a device rotation, pay careful attention to the messages about fragments; not only does the activity get destroyed and re-created, but the fragments do also.

So far, you only wrote a tiny bit of code on the titles fragment to remember the current position in the titles list across restarts. You didn't do anything in the details fragment code to handle reconfigurations, and that's because you didn't need to. Android will take care of hanging onto the fragments that are in the fragment manager, saving them away, then restoring them when the activity is being re-created. You should realize that the fragments you get back after the reconfiguration is complete are very likely not the same fragments in memory that you had before. These fragments have been reconstructed for you. Android saved the arguments bundle and the knowledge of which type of fragment it was, and it stored the saved-state bundles for each fragment that contain saved-state information about the fragment to use to restore it on the other side.

The LogCat messages show you the fragments going through their lifecycles in sync with the activity. You will see that your details fragment gets re-created, but your `newInstance()` method does not get called again. Instead, Android uses the default constructor, attaches the arguments bundle to it, and then starts calling the callbacks on the fragment. This is why it is so important not to do anything fancy in the `newInstance()` method: when the fragment gets re-created, it won't do it through `newInstance()`.

You should also appreciate by now that you've been able to reuse your fragments in a few different places. The titles fragment was used in two different layouts, but if you look at the titles fragment code, it doesn't worry about the attributes of each layout. You could make the layouts rather different from each other, and the titles fragment code would look the same. The same can be said of the details fragment. It was used in your main landscape layout and within the details activity all by itself. Again, the layout for the details fragment could have been very different between the two, and the code of the details fragment would be the same. The code of the details activity was very simple, also.

So far, you've explored two of the fragment types: the base `Fragment` class and the `ListFragment` subclass. `Fragment` has other subclasses: the `DialogFragment`, `PreferenceFragment`, and `WebViewFragment`. We'll cover `DialogFragment` and `PreferenceFragment` in Chapters 9 and 13, respectively.

Communications with Fragments

Because the fragment manager knows about all fragments attached to the current activity, the activity or any fragment in that activity can ask for any other fragment using the getter methods described earlier. Once the fragment reference has been obtained, the activity or fragment could cast the reference appropriately and then call methods directly on that activity or fragment. This would cause your fragments to have more knowledge about the other fragments than might normally be desired, but don't forget that you're running this application on a mobile device, so cutting corners can sometimes be justified. A code snippet is provided in Listing 8–12 to show how one fragment might communicate directly with another fragment.

Listing 8–12. *Direct Fragment-to-Fragment Communication*

```
FragmentOther fragOther =
        (FragmentOther)getFragmentManager().findFragmentByTag("other");
fragOther.callCustomMethod( arg1, arg2 );
```

In Listing 8–12, the current fragment has direct knowledge of the class of the other fragment and also which methods exist on that class. This may be okay because these fragments are part of one application, and it can be easier to simply accept the fact that some fragments will know about other fragments. We'll show you a cleaner way to communicate between fragments in the `DialogFragment` sample application in Chapter 9.

Using startActivity() and setTargetFragment()

A feature of fragments that is very much like activities is the ability of a fragment to start an activity. Fragment has a startActivity() method and startActivityForResult() method. These work just like the ones for activities; when a result is passed back, it will cause the onActivityResult() callback to fire on the fragment that started the activity.

There's another communication mechanism you should know about. When one fragment wants to start another fragment, there is a feature that lets the calling fragment set its identity with the called fragment. Listing 8–13 shows an example of what it might look like.

Listing 8–13. *Fragment-to-Target-Fragment Setup*

```
mCalledFragment = new CalledFragment();
mCalledFragment.setTargetFragment(this, 0);
fm.beginTransaction().add(mCalledFragment, "work").commit();
```

With these few lines, you've created a new CalledFragment object, set the target fragment on the called fragment to the current fragment, and added the called fragment to the fragment manager and activity using a fragment transaction. When the called fragment starts to run, it will be able to call getTargetFragment(), which will return a reference to the calling fragment. With this reference, the called fragment could invoke methods on the calling fragment or even access view components directly. For example, in Listing 8–14, the called fragment could set text in the UI of the calling fragment directly.

Listing 8–14. *Target Fragment-to-Fragment Communication*

```
TextView tv = (TextView)
    getTargetFragment().getView().findViewById(R.id.text1);
tv.setText("Set from the called fragment");
```

Custom Animations with ObjectAnimator

Earlier, we exposed you to a little custom animations on fragments. You used a custom animation to fade out the current details fragment while you faded in the new details fragment. We also told you that the stock animations under the Android SDK were few, and some don't even work. This section will help you understand how to create your own custom animations so you can do interesting transitions between old fragments and new fragments.

The mechanism for implementing custom animations on fragments is the ObjectAnimator class. This is actually a generic feature in Android that can be applied to View objects and not just fragments. You're only going to worry about fragments in this section, but the principles here can apply to other objects as well. An *object animator* is a device that takes an object and animates it from a "from" state to a "to" state over a period of time. The period of time is defined in the animator in milliseconds. There is a routine that defines how the animation behaves over that period of time; these routines are called *interpolators*.

If you imagine the transition from the "from" state to the "to" state as a straight line, the interpolator defines where along that straight line the transition will be at any point during the time period. One of the simplest interpolators is the linear interpolator; it divides the straight line into equal chunks and steps evenly through those chunks for the duration of the time period. The effect is that the object moves at a constant speed from the "from" to the "to" with no acceleration at the beginning and no deceleration at the end.

The default interpolator is `accelerate_decelerate`, which adds a smooth accelerated beginning and a smooth decelerated end. What's really interesting is that the interpolator could go past the "to" point on that line and then come back. This is what the overshoot interpolator does. There's another interpolator called bounce that goes from "from" to "to," but when it first gets to the "to" point, it bounces back towards "from" a few times before finally settling to rest on the "to" point.

An interpolator acts on a dimension of the object. For the `fade_in` and `fade_out` animators you used earlier, the dimension was the fragment's alpha (that is, the amount of transparency of the object). The `fade_in` animator took the alpha dimension from zero (0) to one (1). The `fade_out` animator took the alpha dimension of the other fragment from one (1) to zero (0). One fragment went from invisible to completely visible, while the other went from completely visible to invisible.

Behind the scenes, the object animator is finding the root view of the fragment and applying repeated calls to the `setAlpha()` method, changing the parameter value over the time period a little in each call. The frequency of the repeated calls depends on the interpolator. The linear interpolator makes regular calls at regular intervals in time. The `accelerate_decelerate` interpolator starts out setting the parameter values smaller at first per unit of time and then makes the parameter values larger, creating the effect of an acceleration. It then does the opposite at the other end, making the object appear to decelerate on its dimension.

Dimensions can be many of the values that are settable and gettable on a `View`. In fact, reflection is used by the object animator to work on the view being manipulated. If you specify that you want to animate rotation, the object animator will call the `setRotation()` method on the object (or object's view). The animator takes a "from" and a "to" value and uses them to animate the object from "from" to "to." If the "from" value is not specified, a getter method will be determined and used to get the current value from the object. Let's see how this applies to your fragments.

The only method in the `FragmentTransaction` class that specifies a custom animation is the `setCustomAnimations()` method, which takes two resource ID parameters:

- The first parameter specifies an animator resource for the fragment entering the view container.
- The second specifies an animator resource for the fragment exiting the view container.

These two animators do not need to be related, but it's probably best visually to pair them. In other words, if you're fading one fragment out, fade the other fragment in. Or if you're sliding one fragment out to the right, slide the other fragment in from the left.

Animator resources can be found in the Android SDK folder under the appropriate platform, and then under /data/res/animator. This is where you will find the fade_in.xml and fade_out.xml that you used earlier. Or you could create your own. If you decide to create your own, it would be best to use your project's /res/animator directory, creating it manually if you need to. For an example of a simple local animator XML file (slide_in_left.xml), refer to Listing 8–15.

Listing 8–15. *A Custom Animator to Slide In from the Left*

```
<?xml version="1.0" encoding="utf-8" ?>
<objectAnimator xmlns:android="http://schemas.android.com/apk/res/android"
    android:interpolator="@android:interpolator/accelerate_decelerate"
    android:valueFrom="-1280"
    android:valueTo="0"
    android:valueType="floatType"
    android:propertyName="x"
    android:duration="2000" />
```

This resource file uses the new (in Android 3.0) objectAnimator tag. The basic structure of this file should look familiar to you. It is a bunch of android: attributes to indicate what you want to do. For object animator, there are several things that you need to specify. The first one is the interpolator. The types that are available to you are listed in android.R.interpolator. Using your knowledge of resource names, the interpolator attribute resolves to a file in the Android SDK under the appropriate platform, in /data/res/interpolator, with a filename of accelerate_decelerate.xml.

The android:propertyName attribute specifies the dimension on which you want to animate. In this case, you want to animate on the x dimension. If you investigate the setX() method on a View, you will find that it takes a float value as a parameter, and that is why the android:valueType attribute is set to floatType. The android:duration value is set to 2000, which means 2 seconds. This is probably too slow for a real production app, but it helps you to see what's happening as it happens. Finally, the android:valueFrom and android:valueTo attributes have values of -1280 and 0, respectively. These are chosen because you want the fragment to be at 0 when the animation is done. That is, you want the fragment to be visible to the user with its left edge on the left edge of the view container when the animation stops. Because you want to have the effect of the fragment sliding in from the left, you want it to start from off to the left, and -1280 seems like a big enough number to make that happen. As you might expect, an animator resource file that slides out to the right would look very similar to the one in Listing 8–15, except that valueFrom would be 0 and valueTo would be some large positive number, such as 1280.

Most of the time, you will find that the dimension you're interested in animating is a floatType, although there may be times when you pick an intType. Just look at the type of the parameter that the setter requires. This is where the object animator gets really powerful. In fact, it does not care where the setter method came from. That means you could add your own dimension to an object, and the object animator can animate it for you. All you need to do is supply the setter method and then set the attributes in a resource file; the object animator will do the rest. One caveat here is that if you do not specify a valueFrom attribute in your XML, the object animator will use a getter method to determine the starting value for the object. The getter method must return the appropriate type for the dimension in question.

You might also be interested in animating more than one dimension at a time. For this, you can use the <set> tag to enclose more than one <objectAnimator> tag. Listing 8–16 shows an animator resource file (slide_out_down.xml) that animates along y at the same time it animates on alpha.

Listing 8–16. *A Custom Animator that Animates on Y and Alpha*

```
<?xml version="1.0" encoding="utf-8" ?>
<set xmlns:android="http://schemas.android.com/apk/res/android">
<objectAnimator
    android:interpolator="@android:interpolator/accelerate_cubic"
    android:valueFrom="0"
    android:valueTo="1280"
    android:valueType="floatType"
    android:propertyName="y"
    android:duration="2000" />
<objectAnimator
    android:interpolator="@android:interpolator/accelerate_cubic"
    android:valueFrom="1"
    android:valueTo="0"
    android:valueType="floatType"
    android:propertyName="alpha"
    android:duration="2000" />
</set>
```

The <set> tag corresponds to the AnimatorSet class in Android; however, in XML, <set> has only one attribute, and that is android:ordering. The allowed attribute values are together, the default, which causes the enclosed object animators to run in parallel; and sequential, which causes the object animators to run one after the other in the order in which they are declared in the XML file.

References

Here are some helpful references to topics you may wish to explore further:

- www.androidbook.com/proandroid4/projects: A list of downloadable projects related to this book. The file called ProAndroid4_Ch08_Fragments.zip contains all projects from this chapter, listed in separate root directories. There is also a README.TXT file that describes exactly how to import projects into Eclipse from one of these zip files. It includes some projects that use the Fragment Compatibility SDK for older Androids as well.
- ApiDemos: A project within the Android SDK samples. This project includes several example applications that use fragments, and it should help you to understand how to use them.
- http://developer.android.com/guide/topics/fundamentals/fragments.html: The Android Developer's Guide page to fragments.

- http://android-developers.blogspot.com/2011/02/android-30-fragments-api.html: The Android blog post that introduced fragments.
- http://android-developers.blogspot.com/2011/02/animation-in-honeycomb.html: The Android blog post that introduced the new animations framework and object animators.

Summary

This chapter introduced the Fragment class and its related classes for the manager, transactions, and subclasses. This is a summary of what's been covered in this chapter:

- The Fragment class, what it does, and how to use it.
- Why fragments cannot be used without being attached to one and only one activity.
- That although fragments can be instantiated with a static factory method such as newInstance(), you must always have a default constructor and a way to save initialization values into an initialization arguments bundle.
- The lifecycle of a fragment and how it is intertwined with the lifecycle of the activity that owns the fragment.
- FragmentManager and its features.
- Managing device configurations using fragments.
- Combining fragments into a single activity, or splitting them between multiple activities.
- Using fragment transactions to change what's displayed to a user, and animating those transitions using cool effects.
- New behaviors that are possible with the Back button when using fragments.
- Using the <fragment> tag in a layout.
- Using a FrameLayout as a placeholder for a fragment when you want to use transitions.
- ListFragment and how to use an adapter to populate the data (very much like a ListView).
- Launching a new activity when a fragment can't fit onto the current screen, and how to adjust when a configuration change makes it possible to see multiple fragments again.
- Communicating between fragments, and between a fragment and its activity.

Interview Questions

Here are some questions to ask yourself to ensure that you have a good understanding of fragments in Android:

1. What is the parent class of `Fragment`?
2. How is a fragment like an activity?
3. How is a fragment different from an activity?
4. What is a back stack used for?
5. Can you have an application that does not use a back stack?
6. True or false: rolling back a `FragmentTransaction` restores an application to its prior state.
7. How does a `<fragment>` tag specify which fragment to use?
8. Do you attach the data adapter to the `ListFragment`, or to the `ListView` within the `ListFragment`?
9. What are the different ways to find an existing fragment running in your application?
10. What is an interpolator, and what is it used for?
11. List some of the view dimensions on which you can use an animator.
12. Can you use an animator on a custom dimension of a class, and if so, what does the animator require for this to work?

Chapter 9

Working with Dialogs

The Android SDK offers extensive support for dialogs. A dialog is a smaller window that pops up in front of the current window to show an urgent message, to prompt the user for a piece of input, or to show some sort of status like the progress of a download. The user is generally expected to interact with the dialog and then return to the window underneath to continue with the application. Technically, Android allows a dialog fragment to also be embedded within an activity's layout, and we'll cover that as well.

Dialogs that are explicitly supported in Android include the alert, prompt, pick-list, single-choice, multiple-choice, progress, time-picker, and date-picker dialogs. (This list could vary depending on the Android release.) Android also supports custom dialogs for other needs. The primary purpose of this chapter is not to cover every single one of these dialogs but to cover the underlying architecture of Android dialogs with a sample application. From there you should be able to use any of the Android dialogs.

It's important to note that Android 3.0 added dialogs based on fragments. The expectation from Google is that developers will only use fragment dialogs, even in the versions of Android before 3.0. This can be done with the fragment-compatibility library. For this reason, this chapter focuses on `DialogFragment`.

Using Dialogs in Android

Dialogs in Android are asynchronous, which provides flexibility. However, if you are accustomed to a programming framework where dialogs are primarily synchronous (such as Microsoft Windows, or JavaScript dialogs in web pages), you might find asynchronous dialogs a bit unintuitive. With a synchronous dialog, the line of code after the dialog is shown does not run until the dialog has been dismissed. This means the next line of code could interrogate which button was pressed, or what text was typed into the dialog. In Android however, dialogs are asynchronous. As soon as the dialog has been shown, the next line of code runs, even though the user hasn't touched the dialog yet. Your application has deal with this fact by implementing callbacks from the dialog, to allow the application to be notified of user interaction with the dialog.

This also means your application has the ability to dismiss the dialog from code, which is powerful. If the dialog is displaying a busy message because your application is doing something, as soon as your application has completed that task, it can dismiss the dialog from code.

Understanding Dialog Fragments

In this section, you learn how to use dialog fragments to present a simple alert dialog and a custom dialog that is used to collect prompt text.

DialogFragment Basics

Before we show you working examples of a prompt dialog and an alert dialog, we would like to cover the high-level idea of dialog fragments. Dialog-related functionality uses a class called `DialogFragment`. A `DialogFragment` is derived from the class `Fragment` and behaves much like a fragment. You will then use the `DialogFragment` as the base class for your dialogs. Once you have a derived dialog from this class such as

```
public class MyDialogFragment extends DialogFragment { ... }
```

you can then show this dialog fragment `MyDialogFragment` as a dialog using a fragment transaction. Listing 9–1 shows a code snippet to do this.

> **NOTE:** We provide a link to a downloadable project at the end of this chapter in the "References" section. You can use this download to experiment with the code and the concepts presented in this chapter.

Listing 9–1. *Showing a Dialog Fragment*

```
SomeActivity
{
    //....other activity functions
    public void showDialog()
    {
        //construct MyDialogFragment
        MyDialogFragment mdf = MyDialogFragment.newInstance(arg1,arg2);
        FragmentManager fm = getFragmentManager();
        FragmentTransaction ft = fm.beginTransaction();
        mdf.show(ft,"my-dialog-tag");
    }
    //....other activity functions
}
```

From Listing 9–1, the steps to show a dialog fragment are as follows:

1. Create a dialog fragment.
2. Get a fragment transaction.
3. Show the dialog using the fragment transaction from step 2.

Let's talk about each of these steps.

Constructing a Dialog Fragment

A dialog fragment being a fragment, the same rules and regulations apply when constructing a dialog fragment. The recommended pattern is to use a factory method such as newInstance() as you did before. Inside that newInstance() method, you use the default constructor for your dialog fragment, and then you add an arguments bundle that contains your passed-in parameters. You don't want to do other work inside this method because you must make sure that what you do here is the same as what Android does when it restores your dialog fragment from a saved state. And all that Android does is call the default constructor and re-create the arguments bundle on it.

Overriding onCreateView

When you inherit from a dialog fragment, you need to override one of two methods to provide the view hierarchy for your dialog. The first option is to override onCreateView() and return a view. The second option is to override onCreateDialog() and return a dialog (like the one constructed by an AlertDialog.Builder, which we'll get to shortly).

Listing 9–2 shows an example of overriding the onCreateView().

Listing 9–2. *Overriding onCreateView() of a DialogFragment*

```
MyDialogFragment
{
    .....other functions
    public View onCreateView(LayoutInflater inflater,
            ViewGroup container, Bundle savedInstanceState)
    {
        //Create a view by inflating desired layout
        View v =
            inflater.inflate(R.layout.prompt_dialog, container, false);

        //you can locate a view and set values
        TextView tv = (TextView)v.findViewById(R.id.promptmessage);
        tv.setText(this.getPrompt());

        //You can set callbacks on buttons
        Button dismissBtn = (Button)v.findViewById(R.id.btn_dismiss);
        dismissBtn.setOnClickListener(this);

        Button saveBtn = (Button)v.findViewById(R.id.btn_save);
        saveBtn.setOnClickListener(this);
        return v;
    }
    .....other functions
}
```

In Listing 9–2, you are loading a view identified by a layout. Then you look for two buttons and set up callbacks on them. This is very similar to how you created the details fragment in the previous chapter. However, unlike the earlier fragments, a dialog fragment has another way to create the view hierarchy.

Overriding onCreateDialog

As an alternate to supplying a view in onCreateView(), you can override onCreateDialog() and supply a dialog instance. Listing 9–3 supplies sample code for this approach.

Listing 9–3. *Overriding onCreateDialog() of a DialogFragment*

```
MyDialogFragment
{
    .....other functions
    @Override
    public Dialog onCreateDialog(Bundle icicle)
    {
        AlertDialog.Builder b = new AlertDialog.Builder(getActivity())
          .setTitle("My Dialog Title")
          .setPositiveButton("Ok", this)
          .setNegativeButton("Cancel", this)
          .setMessage(this.getMessage());
        return b.create();
    }
    .....other functions
}
```

In this example, you use the alert dialog builder to create a dialog object to return. This works well for simple dialogs. The first option of overriding onCreateView() is equally easy and provides much more flexibility.

AlertDialog.Builder is actually a carryover from pre-3.0 Android. This is one of the old ways to create a dialog, and it's still available to you to create dialogs within DialogFragments. As you can see, it's fairly easy to build a dialog by calling the various methods available, as we've done here.

Displaying a Dialog Fragment

Once you have a dialog fragment constructed, you need a fragment transaction to show it. Like all other fragments, operations on dialog fragments are conducted through fragment transactions.

The show() method on a dialog fragment takes a fragment transaction as an input. You can see this in Listing 9–1. The show() method uses the fragment transaction to add this dialog to the activity and then commits the fragment transaction. However, the show() method does not add the transaction to the back stack. If you want to do this, you need to add this transaction to the back stack first and then pass it to the show() method. The show() method of a dialog fragment has the following signatures:

```
public int show(FragmentTransaction transaction, String tag)
public int show(FragmentManager manager, String tag)
```

The first show() method displays the dialog by adding this fragment to the passed-in transaction with the specified tag. This method then returns the identifier of the committed transaction.

The second show() method automates getting a transaction from the transaction manager. This is a shortcut method. However, when you use this second method, you don't have an option to add the transaction to the back stack. If you want that control, you need to use the first method. The second method could be used if you wanted to simply display the dialog, and you had no other reason to work with a fragment transaction at that time.

A nice thing about a dialog being a fragment is that the underlying fragment manager does the basic state management. For example, even if the device rotates when a dialog is being displayed, the dialog is reproduced without you performing any state management.

The dialog fragment also offers methods to control the frame in which the dialog's view is displayed, such as the title and the appearance of the frame. Refer to the DialogFragment class documentation to see more of these options; this URL is provided at the end of this chapter.

Dismissing a Dialog Fragment

There are two ways you can dismiss a dialog fragment. The first is to explicitly call the dismiss() method on the dialog fragment in response to a button or some action on the dialog view, as shown in Listing 9–4.

Listing 9–4. *Calling dismiss()*

```
if (someview.getId() == R.id.btn_dismiss)
{
    //use some callbacks to advise clients
    //of this dialog that it is being dismissed
    //and call dismiss
    dismiss();
    return;
}
```

The dialog fragment's dismiss() method removes the fragment from the fragment manager and then commits that transaction. If there is a back stack for this dialog fragment, then the dismiss() pops the current dialog out of the transaction stack and presents the previous fragment transaction state. Whether there is a back stack or not, calling dismiss() results in calling the standard dialog fragment destroy callbacks, including onDismiss().

One thing to note is that you can't rely on onDismiss() to conclude that a dismiss() has been called by your code. This is because onDismiss() is also called when a device configuration changes and hence is not a good indicator of what the user did to the dialog itself. If the dialog is being displayed when the user rotates the device, the dialog fragment sees onDismiss() called even though the user did not press a button in the dialog. Instead, you should always rely on explicit button clicks on the dialog view.

If the user presses the Back button while the dialog fragment is displayed, this causes the onCancel() callback to fire on the dialog fragment. By default, Android makes the dialog fragment go away, so you don't need to call dismiss() on the fragment yourself. But if you want the calling activity to be notified that the dialog has been cancelled, you need to invoke logic from within onCancel() to make that happen. This is a difference between onCancel() and onDismiss() with dialog fragments. With onDismiss(), you can't be sure exactly what happened that caused the onDismiss() callback to fire. You might also have noticed that a dialog fragment does not have a cancel() method, just dismiss(); but as we said, when a dialog fragment is being cancelled by pressing the Back button, Android takes care of cancelling/dismissing it for you.

The other way to dismiss a dialog fragment is to present another dialog fragment. The way you dismiss the current dialog and present the new one is slightly different than just dismissing the current dialog. Listing 9–5 shows an example.

Listing 9–5. *Setting Up a Dialog for a Back Stack*

```
if (someview.getId() == R.id.btn_invoke_another_dialog)
{
    Activity act = getActivity();
    FragmentManager fm = act.getFragmentManager();
    FragmentTransaction ft = fm.beginTransaction();
    ft.remove(this);

    ft.addToBackStack(null);
    //null represents no name for the back stack transaction

    HelpDialogFragment hdf =
        HelpDialogFragment.newInstance(R.string.helptext);
    hdf.show(ft, "HELP");
    return;
}
```

Within a single transaction, you're removing the current dialog fragment and adding the new dialog fragment. This has the effect of making the current dialog disappear visually and making the new dialog appear. If the user presses the Back button, because you've saved this transaction on the back stack, the new dialog is dismissed and the previous dialog is displayed. This is a handy way of displaying a help dialog, for example.

Implications of a Dialog Dismiss

When you add any fragment to a fragment manager, the fragment manager does the state management for that fragment. This means when a device configuration changes (for example, the device rotates), the activity is restarted and the fragments are also restarted. You saw this earlier when you rotated the device while running the Shakespeare sample application.

A device-configuration change doesn't affect dialogs because they are also managed by the fragment manager. But the implicit behavior of show() and dismiss() means you can easily lose track of a dialog fragment if you're not careful. The show() method automatically adds the fragment to the fragment manager; the dismiss() method automatically removes the fragment from the fragment manager. You may have a direct pointer to a dialog fragment before you start showing the fragment. But you can't add this fragment to the fragment manager and later call show(), because a fragment can only be added once to the fragment manager. You may plan to retrieve this pointer through restore of the activity. However, if you show and dismiss this dialog, this fragment is implicitly removed from the fragment manager, thereby denying that fragment's ability to be restored and repointed (because the fragment manager doesn't know this fragment exists after it is removed).

If you want to keep the state of a dialog after it is dismissed, you need to maintain the state outside of the dialog either in the parent activity or in a non-dialog fragment that hangs around for a longer time.

DialogFragment Sample Application

In this section, you create a sample application that demonstrates these concepts of a dialog fragment. You also examine communication between a fragment and the activity that contains it. To make it all happen, you need five Java files:

- MainActivity.java: The main activity of your application. It displays a simple view with help text in it and a menu from which dialogs can be started.
- PromptDialogFragment.java: An example of a dialog fragment that defines its own layout in XML and allows input from the user. It has three buttons: Save, Dismiss (cancel), and Help.
- AlertDialogFragment.java: An example of a dialog fragment that uses the AlertBuilder class to create a dialog within this fragment. This is the old-school way of creating a dialog.
- HelpDialogFragment.java: A very simple fragment that displays a help message from the application's resources. The specific help message is identified when a help dialog object is created. This help fragment can be shown from both the main activity and the prompt dialog fragment.
- OnDialogDoneListener.java: An interface that you require your activity to implement in order to get messages back from the fragments. Using an interface means your fragments don't need to know much about the calling activity, except that it must have implemented this interface. This helps encapsulate functionality where it belongs. From the activity's point of view, it has a common way to receive information back from fragments without needing to know too much about them.

There are three layouts for this application: for the main activity, for the prompt dialog fragment, and for the help dialog fragment. Note that you don't need a layout for the alert dialog fragment because the AlertBuilder takes care of that layout for you internally. When you're done, the application looks like Figure 9–1.

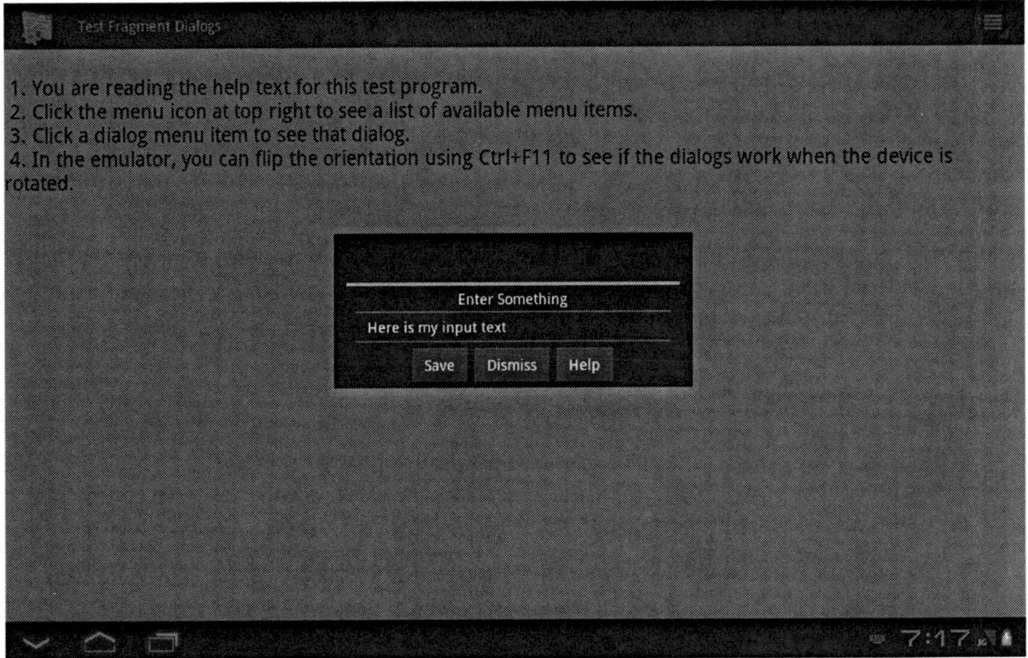

Figure 9–1. *The user interface for the dialog fragment sample application*

Dialog Sample: MainActivity

Let's get to the source code. Listing 9–6 shows pieces of the main activity that has to do with dialogs. To see the entire source code, you can download it from the book's web site (see the "References" section).

Listing 9–6. *The Main Activity for Dialog Fragments*

```
public class MainActivity extends Activity
implements OnDialogDoneListener
{
    public static final String LOGTAG = "DialogFragmentDemo";
    public static final String ALERT_DIALOG_TAG = "ALERT_DIALOG_TAG";
    public static final String HELP_DIALOG_TAG = "HELP_DIALOG_TAG";
    public static final String PROMPT_DIALOG_TAG = "PROMPT_DIALOG_TAG";
    public static final String EMBED_DIALOG_TAG = "EMBED_DIALOG_TAG";

    private void testPromptDialog()
    {
        FragmentTransaction ft = getFragmentManager().beginTransaction();

        PromptDialogFragment pdf =
            PromptDialogFragment.newInstance("Enter Something");

        pdf.show(ft, PROMPT_DIALOG_TAG);
    }

    private void testAlertDialog()
```

```
    {
        FragmentTransaction ft = getFragmentManager().beginTransaction();

        AlertDialogFragment adf =
            AlertDialogFragment.newInstance("Alert Message");

        adf.show(ft, ALERT_DIALOG_TAG);
    }

    private void testHelpDialog()
    {
        FragmentTransaction ft = getFragmentManager().beginTransaction();

        HelpDialogFragment hdf =
            HelpDialogFragment.newInstance(R.string.help_text);

        hdf.show(ft, HELP_DIALOG_TAG);
    }

    private void testEmbedDialog()
    {
        FragmentTransaction ft = getFragmentManager().beginTransaction();

        PromptDialogFragment pdf =
            PromptDialogFragment.newInstance(
                 "Enter Something (Embedded)");

        ft.add(R.id.embeddedDialog, pdf, EMBED_DIALOG_TAG);
        ft.commit();
    }

    public void onDialogDone(String tag, boolean cancelled,
                             CharSequence message) {
        String s = tag + " responds with: " + message;
        if(cancelled)
            s = tag + " was cancelled by the user";
        Toast.makeText(this, s, Toast.LENGTH_LONG).show();
        Log.v(LOGTAG, s);
    }
}
```

The code for the main activity is very straightforward. You display a simple page of text and set up a menu. Each menu item invokes an activity method, and each method does basically the same thing: gets a fragment transaction, creates a new fragment, and shows the fragment. Note that each fragment has a unique tag that's used with the fragment transaction. This tag becomes associated with the fragment in the fragment manager, so you can locate these fragments later by tag name. The fragment can also determine its own tag value with the `getTag()` method on `Fragment`.

The last method definition in the main activity is `onDialogDone()`, which is a callback that is part of the `OnDialogDoneListener` interface that your activity is implementing. As you can see, the callback supplies a tag of the fragment that is calling you, a boolean value indicating whether the dialog fragment was cancelled, and a message. For your purposes, you merely want to log the information to LogCat; you also show it to the user using `Toast`. Toast will be covered later in this chapter.

Dialog Sample: OnDialogDoneListener

So that you can know when a dialog has gone away, create a listener interface that your dialog callers implement. The code of the interface is in Listing 9–7.

Listing 9–7. *The Listener Interface*

```
// This file is OnDialogDoneListener.java
/*
 * An interface implemented typically by an activity
 * so that a dialog can report back
 * on what happened.
 */
public interface OnDialogDoneListener {
  public void onDialogDone(String tag, boolean cancelled,
                           CharSequence message);
}
```

This is a very simple interface, as you can see. You choose only one callback for this interface, which the activity must implement. Your fragments don't need to know the specifics of the calling activity, only that the calling activity must implement the `OnDialogDoneListener` interface; therefore the fragments can call this callback to communicate with the calling activity. Depending on what the fragment is doing, there could be multiple callbacks in the interface. For this sample application, you're showing the interface separately from the fragment class definitions. For easier management of code, you could embed the fragment listener interface inside of the fragment class definition itself, thus making it easier to keep the listener and the fragment in sync with each other.

Dialog Sample: PromptDialogFragment

Now let's look at your first fragment, `PromptDialogFragment`, whose layout and Java code are shown in Listing 9–8.

Listing 9–8. *The `PromptDialogFragment` Layout and Java Code*

```
<?xml version="1.0" encoding="utf-8"?>
<!-- This file is /res/layout/prompt_dialog.xml -->
<LinearLayout xmlns:android="http://schemas.android.com/apk/res/android"
    android:orientation="vertical" android:padding="4dip"
    android:gravity="center_horizontal"
    android:layout_width="match_parent"
    android:layout_height="match_parent">

    <TextView
        android:id="@+id/promptmessage"
        android:layout_height="wrap_content"
        android:layout_width="wrap_content"
        android:layout_marginLeft="20dip"
        android:layout_marginRight="20dip"
        android:text="Enter Text"
        android:layout_weight="1"
        android:layout_gravity="center_vertical|center_horizontal"
```

```
        android:textAppearance="?android:attr/textAppearanceMedium"
        android:gravity="top|center_horizontal" />

    <EditText
        android:id="@+id/inputtext"
        android:layout_height="wrap_content"
        android:layout_width="400dip"
        android:layout_marginLeft="20dip"
        android:layout_marginRight="20dip"
        android:scrollHorizontally="true"
        android:autoText="false"
        android:capitalize="none"
        android:gravity="fill_horizontal"
        android:textAppearance="?android:attr/textAppearanceMedium" />

    <LinearLayout
        android:orientation="horizontal"
        android:layout_width="wrap_content"
        android:layout_height="wrap_content">

      <Button android:id="@+id/btn_save"
        android:layout_width="wrap_content"
        android:layout_height="wrap_content"
        android:layout_weight="0"
        android:text="Save">
      </Button>

      <Button android:id="@+id/btn_dismiss"
        android:layout_width="wrap_content"
        android:layout_height="wrap_content"
        android:layout_weight="0"
        android:text="Dismiss">
      </Button>

      <Button android:id="@+id/btn_help"
        android:layout_width="wrap_content"
        android:layout_height="wrap_content"
        android:layout_weight="0"
        android:text="Help">
      </Button>

    </LinearLayout>
</LinearLayout>


// This file is PromptDialogFragment.java
public class PromptDialogFragment
extends DialogFragment
implements View.OnClickListener
{
    private EditText et;

    public static PromptDialogFragment
    newInstance(String prompt)
    {
        PromptDialogFragment pdf = new PromptDialogFragment();
        Bundle bundle = new Bundle();
```

```
        bundle.putString("prompt",prompt);
        pdf.setArguments(bundle);

        return pdf;
    }

    @Override
    public void onAttach(Activity act) {
        // If the activity you're being attached to has
        // not implemented the OnDialogDoneListener
        // interface, the following line will throw a
        // ClassCastException. This is the earliest you
        // can test if you have a well-behaved activity.
        try {
            OnDialogDoneListener test = (OnDialogDoneListener)act;
        }
        catch(ClassCastException cce) {
            // Here is where we fail gracefully.
            Log.e(MainActivity.LOGTAG, "Activity is not listening");
        }
        super.onAttach(act);
    }

    @Override
    public void onCreate(Bundle icicle)
    {
        super.onCreate(icicle);
        this.setCancelable(true);
        int style = DialogFragment.STYLE_NORMAL, theme = 0;
        setStyle(style,theme);
    }

    public View onCreateView(LayoutInflater inflater,
            ViewGroup container, Bundle icicle)
    {
        View v = inflater.inflate(R.layout.prompt_dialog, container,
                                  false);

        TextView tv = (TextView)v.findViewById(R.id.promptmessage);
        tv.setText(getArguments().getString("prompt"));

        Button dismissBtn = (Button)v.findViewById(R.id.btn_dismiss);
        dismissBtn.setOnClickListener(this);

        Button saveBtn = (Button)v.findViewById(R.id.btn_save);
        saveBtn.setOnClickListener(this);

        Button helpBtn = (Button)v.findViewById(R.id.btn_help);
        helpBtn.setOnClickListener(this);

        et = (EditText)v.findViewById(R.id.inputtext);
        if(icicle != null)
            et.setText(icicle.getCharSequence("input"));
        return v;
    }

    @Override
```

```
    public void onSaveInstanceState(Bundle icicle) {
        super.onSaveInstanceState(icicle);
        icicle.putCharSequence("input", et.getText());
    }

    public void onClick(View v)
    {
        OnDialogDoneListener act = (OnDialogDoneListener)getActivity();
        if (v.getId() == R.id.btn_save)
        {
            TextView tv =
                (TextView)getView().findViewById(R.id.inputtext);
            act.onDialogDone(this.getTag(), false, tv.getText());
            dismiss();
            return;
        }
        if (v.getId() == R.id.btn_dismiss)
        {
            act.onDialogDone(this.getTag(), true, null);
            dismiss();
            return;
        }
        if (v.getId() == R.id.btn_help)
        {
            FragmentTransaction ft =
                getFragmentManager().beginTransaction();
            ft.remove(this);

            // in this case, you want to show the help text, but
            // come back to the previous dialog when you're done
            ft.addToBackStack(null);
            //null represents no name for the back stack transaction

            HelpDialogFragment hdf =
                    HelpDialogFragment.newInstance(R.string.help1);
            hdf.show(ft, MainActivity.HELP_DIALOG_TAG);
            return;
        }
    }
}
```

This prompt dialog layout looks like many you've seen previously. There is a `TextView` to serve as the prompt; an `EditText` to take the user's input; and three buttons for saving the input, dismissing (cancelling) the dialog fragment, and popping a help dialog.

The `PromptDialogFragment` Java code starts out looking just like your earlier fragments. You have a `newInstance()` static method to create new objects, and within this method you call the default constructor, build an arguments bundle, and attach it to your new object. Next, you have something new in the `onAttach()` callback. You want to make sure the activity you just got attached to has implemented the `OnDialogDoneListener` interface. In order to test that, you cast the activity passed in to the `OnDialogDoneListener` interface. Here's that code again:

```
try {
    OnDialogDoneListener test = (OnDialogDoneListener)act;
}
```

```
catch(ClassCastException cce) {
    // Here is where we fail gracefully.
    Log.e(MainActivity.LOGTAG, "Activity is not listening");
}
```

If the activity does not implement this interface, a ClassCastException is thrown. You could handle this exception and deal with it more gracefully, but this example keeps the code as simple as possible.

Next up is the onCreate() callback. As is common with fragments, you don't build your user interface here, but you can set the dialog style. This is unique to dialog fragments. You can set both the style and the theme yourself, or you can set just style and use a theme value of zero (0) to let the system choose an appropriate theme for you. Here's that code again:

```
int style = DialogFragment.STYLE_NORMAL, theme = 0;
setStyle(style,theme);
```

In onCreateView() you create the view hierarchy for your dialog fragment. Just like other fragments, you do not attach your view hierarchy to the view container passed in (that is, by setting the attachToRoot parameter to false). You then proceed to set up the button callbacks, and you set the dialog prompt text to the prompt that was passed originally to newInstance(). Finally, you check to see whether any values are being passed in through the saved state bundle (icicle). This would indicate that your fragment is being re-created, most likely due to a configuration change, and it's possible that the user has already typed some text. If so, you need to populate the EditText with what the user has done so far. Remember that because your configuration has changed, the actual view object in memory is not the same as before, so you must locate it and set the text accordingly. The very next callback is onSaveInstanceState(); it's where you save any current text typed by the user into the saved state bundle.

The onCancel() and onDismiss() callbacks are not shown because all they do is logging; you be able to see when these callbacks fire during the fragment's lifecycle.

The final callback in the prompt dialog fragment is for the buttons. Once again, you grab a reference to your enclosing activity and cast it to the interface you expect the activity to have implemented. If the user pressed the Save button, you grab the text as entered and call the interface's callback onDialogDone(). This callback takes the tag name of this fragment, a boolean indicating whether this dialog fragment was cancelled, and a message, which in this case is the text typed by the user. Here it is again from the MainActivity at the end of Listing 9–6:

```
public void onDialogDone(String tag, boolean cancelled,
                         CharSequence message) {
    String s = tag + " responds with: " + message;
    if(cancelled)
        s = tag + " was cancelled by the user";
    Toast.makeText(this, s, Toast.LENGTH_LONG).show();
    Log.v(LOGTAG, s);
}
```

You then call dismiss() to get rid of the dialog fragment. Remember that dismiss() not only makes the fragment go away visually, but also pops the fragment out of the fragment manager so it is no longer available to you. If the button pressed is Dismiss, you again call the interface callback, this time with no message, and then you call `dismiss()`. And finally, if the user pressed the Help button, you don't want to lose the prompt dialog fragment, so you do something a little different. We described this earlier. In order to remember the prompt dialog fragment so you can come back to it later, you need to create a fragment transaction to remove the prompt dialog fragment and add the help dialog fragment with the `show()` method; this needs to go onto the back stack. Notice, too, how the help dialog fragment is created with a reference to a resource ID. This means your help dialog fragment can be used with any help text available to your application.

Dialog Sample: HelpDialogFragment

We'll show the code for the help dialog fragment shortly, but we'll describe the operation now. You created a fragment transaction to go from the prompt dialog fragment to the help dialog fragment, and you placed that fragment transaction on the back stack. This has the effect of making the prompt dialog fragment disappear from view, but it's still accessible through the fragment manager and the back stack. The new help dialog fragment appears in its place and allows the user to read the help text. When the user dismisses the help dialog fragment, the fragment back stack entry is popped, with the effect of the help dialog fragment being dismissed (both visually and from the fragment manager) and the prompt dialog fragment restored to view. This is a pretty easy way to make all this happen. It is very simple yet very powerful; it even works if the user rotates the device while these dialogs are being displayed.

Download and look at the source code of the `HelpDialogFragment.java` file and its layout (`help_dialog.xml`). The point of this dialog fragment is to display help text. The layout is a `TextView` and a Close button. The Java code should be starting to look familiar to you. There's a `newInstance()` method to create a new help dialog fragment, an `onCreate()` method to set the style and theme, and an `onCreateView()` method to build the view hierarchy. In this particular case, you want to locate a string resource to populate the `TextView`, so you access the resources through the activity and choose the resource ID that was passed in to `newInstance()`. Finally, `onCreateView()` sets up a button-click handler to capture the clicks of the Close button. In this case, you don't need to do anything interesting at the time of dismissal.

This fragment is called two ways: from the activity and from the prompt dialog fragment. When this help dialog fragment is shown from the main activity, dismissing it simply pops the fragment off the top and reveals the main activity underneath. When this help dialog fragment is shown from the prompt dialog fragment, because the help dialog fragment was part of a fragment transaction on the back stack, dismissing it causes the fragment transaction to be rolled back, which pops the help dialog fragment but restores the prompt dialog fragment. The user sees the prompt dialog fragment reappear.

Dialog Sample: AlertDialogFragment

We have one last dialog fragment to show you in this sample application: the alert dialog fragment. Although you could create an alert dialog fragment in a way similar to the help dialog fragment, you can also create a dialog fragment using the old `AlertBuilder` framework that has worked for many releases of Android. Listing 9–9 shows the source code of the alert dialog fragment.

Listing 9–9. *The `AlertDialogFragment` Java Code*

```
public class AlertDialogFragment
extends DialogFragment
implements DialogInterface.OnClickListener
{
    public static AlertDialogFragment
    newInstance(String message)
    {
        AlertDialogFragment adf = new AlertDialogFragment();
        Bundle bundle = new Bundle();
        bundle.putString("alert-message",message);
        adf.setArguments(bundle);

        return adf;
    }

    @Override
    public void onCreate(Bundle savedInstanceState)
    {
        super.onCreate(savedInstanceState);
        this.setCancelable(true);
        int style = DialogFragment.STYLE_NORMAL, theme = 0;
        setStyle(style,theme);
    }

    @Override
    public Dialog onCreateDialog(Bundle savedInstanceState)
    {
        AlertDialog.Builder b =
            new AlertDialog.Builder(getActivity())
           .setTitle("Alert!!")
           .setPositiveButton("Ok", this)
           .setNegativeButton("Cancel", this)
           .setMessage(this.getArguments().getString("alert-message"));
        return b.create();
    }

    public void onClick(DialogInterface dialog, int which)
    {
        OnDialogDoneListener act = (OnDialogDoneListener) getActivity();
        boolean cancelled = false;
        if (which == AlertDialog.BUTTON_NEGATIVE)
        {
            cancelled = true;
        }
        act.onDialogDone(getTag(), cancelled, "Alert dismissed");
```

```
    }
}
```

You don't need a layout for this one because the AlertBuilder takes care of that for you. Note that this dialog fragment starts out like any other, but instead of an onCreateView() callback, you have a onCreateDialog() callback. You either implement onCreateView() or onCreateDialog() but not both. The return from onCreateDialog() is not a view; it's a dialog. Of interest here is that to get parameters for the dialog, you should be accessing your arguments bundle. In this example application, you only do this for the alert message, but you could access other parameters through the arguments bundle as well.

Notice also that with this type of dialog fragment, you need your fragment class to implement the DialogInterface.OnClickListener, which means your dialog fragment must implement the onClick() callback. This callback is fired when the user acts on the embedded dialog. Once again, you get a reference to the dialog that fired and an indication of which button was pressed. As before, you should be careful not to depend on an onDismiss() because this could fire when there is a device configuration change.

Dialog Sample: Embedded Dialogs

There's one more feature of a DialogFragment that you may have noticed. In the main layout for the application, under the text, is a FrameLayout that can be used to hold a dialog. In the application's menu, the last item causes a fragment transaction to add a new instance of a PromptDialogFragment to the main screen. Without any modifications, the dialog fragment can be displayed embedded in the main layout, and it functions as you would expect.

One thing that is different about this technique is that the code to show the embedded dialog is not the same as the code to do a pop-up dialog. The embedded dialog code looks like this:

```
ft.add(R.id.embeddedDialog, pdf, EMBED_DIALOG_TAG);
ft.commit();
```

This looks just the same as in Chapter 8, when we displayed a fragment in a FrameLayout. This time, however, you make sure to pass in a tag name, which is used when the dialog fragment notifies your activity of the user's input.

Dialog Sample: Observations

When you run this sample application, make sure you try all the menu options in different orientations of the device. Rotate the device while the dialog fragments are displayed. You should be pleased to see that the dialogs go with the rotations; you do not need to worry about a lot of code to manage the saving and restoring of fragments due to configuration changes.

The other thing we hope you appreciate is the ease with which you can communicate between the fragments and the activity. Of course, the activity has references, or can get references, to all the available fragments, so it can access methods exposed by the fragments themselves. This isn't the only way to communicate between fragments and the activity. You can always use the getter methods on the fragment manager to retrieve an instance of a managed fragment, and then cast that reference appropriately and call a method on that fragment directly. You can even do this from within another fragment. The degree to which you isolate your fragments from each other with interfaces and through activities, or build in dependencies with fragment-to-fragment communication, is based on how complex your application is and how much reuse you want to achieve.

Working with Toast

The chapter began by indicating how alert messages are commonly used for debugging JavaScript on error pages. If you are pressed to use a similar approach for infrequent debug messages, you can use the `Toast` object in Android.

A `Toast` is like an alert dialog that has a message and displays for a certain amount of time and then goes away. It does not have any buttons. So it can be said that it is a transient alert message. It's called `Toast` because it pops up like toast out of a toaster.

Listing 9–10 shows an example of how you can show a message using `Toast`.

Listing 9–10. ***Using `Toast` for Debugging***

```
//Create a function to wrap a message as a toast
//show the toast
public void reportToast(String message)
{
    String s = MainActivity.LOGTAG + ":" + message;
    Toast.makeText(activity, s, Toast.LENGTH_SHORT).show();
}
```

The `makeText()` method in Listing 9–10 can take not only an activity but any context object, such as the one passed to a broadcast receiver or a service, for example. This extends the use of `Toast` outside of activities.

Dialog Fragments for Older Android

Although the hope is that all older phones will upgrade to Ice Cream Sandwich and therefore be able to take advantage of all the wonderfulness of dialog fragments, the reality is that some older models will stay on Android 1.6, 2.1, and 2.2. To provide dialog fragment support for these phones, Google provides the Fragment Compatibility library. For dialog fragments, you use the same `DialogFragment` class discussed earlier; just make sure you include the compatibility jar file in your application.

References

- www.androidbook.com/proandroid4/projects: This chapter's test project. The name of the ZIP file is ProAndroid4_ch09_Dialogs.zip. The download includes examples of the date- and time-picker dialog.
- http://developer.android.com/guide/topics/ui/dialogs.html: Android SDK document that provides an excellent introduction to working with Android dialogs. You will find here an explanation of how to use managed dialogs and various examples of available dialogs.
- http://developer.android.com/reference/android/content/DialogInterface.html: The many constants defined for dialogs.
- http://developer.android.com/reference/android/app/Dialog.html: A number of methods available on a Dialog object.
- http://developer.android.com/reference/android/app/AlertDialog.Builder.html: API documentation for the AlertDialog builder class.
- http://developer.android.com/reference/android/app/ProgressDialog.html: API documentation for ProgressDialog.
- http://developer.android.com/reference/android/app/DatePickerDialog.html: API documentation for DatePickerDialog.
- http://developer.android.com/reference/android/app/TimePickerDialog.html: API documentation for TimePickerDialog.
- http://developer.android.com/resources/tutorials/views/hello-datepicker.html: An Android tutorial for using the date-picker dialog.
- http://developer.android.com/resources/tutorials/views/hello-timepicker.html: An Android tutorial for using the time-picker dialog.

Summary

This chapter discussed asynchronous dialogs and how to use dialog fragments, including the following topics:

- What a dialog is and why you use one
- The asynchronous nature of a dialog in Android
- The three steps of getting a dialog to display on the screen
- Creating a fragment
- Two methods for how a dialog fragment can create a view hierarchy
- How a fragment transaction is involved in displaying a dialog fragment, and how to get one

- What happens when the user presses the Back button while viewing a dialog fragment
- The back stack, and managing dialog fragments
- What happens when a button on a dialog fragment is clicked, and how you deal with it
- A clean way to communicate back to the calling activity from a dialog fragment
- How one dialog fragment can call another dialog fragment and still get back to the previous dialog fragment
- The `Toast` class and how it can be used as a simple alert pop-up

Interview Questions

The following questions should help you solidify your understanding of this chapter's discussion of dialogs:

1. Can your code read the user's input from a prompt dialog fragment right after the dialog fragment has been shown? Why or why not?
2. Which method of `DialogFragment` do you use to build a view hierarchy?
3. Which method do you use to return a ready-to-go dialog?
4. What method of a dialog fragment do you use to display the dialog?
5. What are the two ways this method can be called? Which one is the simplest to call?
6. Can you use `onDismiss()` to tell you when the user has clicked a button in a dialog fragment? Why or why not?
7. Can a dialog fragment show another dialog fragment? If so, could the new dialog fragment display yet another dialog fragment with no issues?
8. Where do you set the style and theme of a dialog?
9. What are the time-duration choices for a `Toast` message?
10. How do you get a dialog fragment to remember what the user typed in during a configuration change such as a device rotation?

Chapter 10

Exploring ActionBar

ActionBar was introduced in the Android 3.0 SDK for tablets and is now available for phones as well in 4.0. It allows you to customize the title bar of an activity. Prior to the 3.0 SDK release, the title bar of an activity merely contained the title of an activity. Android ActionBar is modeled similar to the menu/title bar of a web browser.

> **NOTE:** This chapter refers to both ActionBar and action bar. ActionBar refers to the actual class, and *action bar* means the concept.

The 3.0 SDK is optimized and available only for tablets. This means the action bar API is not available for phones that run Android versions prior to 4.0. With the 4.0 SDK, the phone and tablet aspects of the SDK are merged to provide a uniform API.

A key goal of the action bar design is to make the frequently used actions easily available to the user without searching through option menus or context menus.

> **NOTE:** In the current computer technology literature, the convenient access to actions is fashionably called *affordance*, which refers to the ability to conveniently discover/invoke actions. We include a few reference URLs on affordance at the end of the chapter.

As you go through this chapter, you learn the following about an action bar:

- An action bar is owned by an activity and follows its lifecycle.
- An action bar can take one of three forms: tabbed action bar, list action bar, or standard action bar. You see how these various action bars look and behave in each of the modes.
- You learn how tabbed listeners allow you to interact with a tabbed action bar.

- You see how spinner adapters and list listeners are used to interact with the list action bar.
- You learn how the Home icon of an action bar interacts with the menu infrastructure.
- You see how icon menu items can be shown and reacted to on the action bar real estate.
- You see how to place a custom search widget in the action bar

You explore these concepts by planning three different activities. Each activity sports an action bar in a different mode. This gives you an opportunity to examine the behavior of the action bar in each mode. But first, let's take a quick look at visual aspects of an action bar.

Anatomy of an ActionBar

Figure 10–1 shows a typical action bar in tabbed navigation mode.

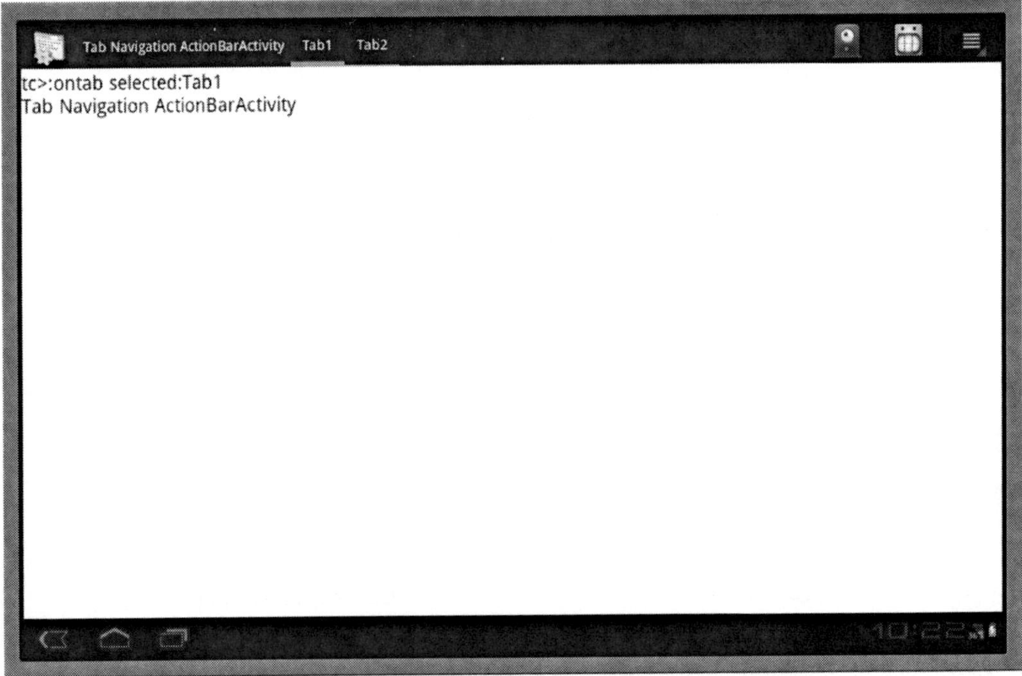

Figure 10–1. *An activity with a tabbed action bar*

This screenshot is taken from the actual working example that is presented later in the chapter. The action bar in Figure 10–1 has five parts. These parts are as follows (from left to right):

- *Home Icon area:* The icon at upper left on the action bar is sometimes called the Home icon. This is similar to a web site navigation context, where clicking the Home icon takes you to a starting point. When you transfer the user to the home activity, don't start a new home activity; instead, transfer to it by using an intent flag that clears the stack of all activities on top of the home activity. You see later that clicking this Home icon sends a callback to the option menu with menu ID `android.R.id.home`.
- *Title area:* The Title area displays the title for the action bar.
- *Tabs area:* The Tabs area is where the action bar paints the list of tabs specified. The content of this area is variable. If the action bar navigation mode is tabs, then tabs are shown here. If the mode is list-navigation mode, then a navigable list of drop-down items is shown. In standard mode, this area is ignored and left empty.
- *Action Icon area:* Following the Tabs area, the Action Icon area shows some of the option menu items as icons. You see how to choose which option menus are displayed as action icons in the example later.
- *Menu Icon area:* Last is the Menu Icon area. It is a single standard menu icon. When you click this menu icon, you see the expanded menu. This expanded menu looks different or shows up in a different location depending on the size of the Android device. You can also attach a search view as if it is an action icon of the menu. We will cover this later in the chapter.

In addition to the action bar, the activity in Figure 10–1 is showing a debug text view to which a number of actions are logged. These actions may be a result of clicking the tabs or the Home icon or the action menus or the actual option menus.

Let's look at how to implement the three types of action bar activities mentioned earlier: the tabbed action bar, the list action bar, and the standard action bar. You've seen the tabbed action bar as the visual example of an action bar, so you start with the implementation of a tabbed action bar.

Tabbed Navigation Action Bar Activity

Although you are planning three different activities, each with its own type of action bar, these example activities have a lot of common functionality:

- All of these activities have the same debug text view so you can monitor the actions as they are invoked.
- All of these activities have the same Home icon.

- All of these activities have a title.
- All of these activities have the same action icons.
- All of these activities have the same Options menu.

The primary difference between these activities is that each configures the action bar differently. In the example, you encapsulate the common behavior in a base class and allow each of the derived activities, including this tabbed action bar activity, to configure the action bar.

It is difficult to explain these common files without the context of at least one action bar activity. So you look first at these common files and how the tabbed action bar activity uses them. Then the other two status bar activities can be added to this project with fewer files.

Following is a list of files that are needed for this tabbed action bar exercise. These files include both the common files and the files specific to the tabbed action bar. The list seems numerous because you are encapsulating the common behavior into base classes. This reduces the number of files for later examples. Listing numbers are indicated for each of the files:

- `DebugActivity.java`: Base class activity that allows for a debug text view as shown in Figure 10–1 (Listing 10–2).
- `BaseActionBarActivity.java`: Derived from `DebugActivity` and allows for common navigation (such as responding to common actions including switching between the three activities) (Listing 10–3).
- `IReportBack.java`: An interface that works as a communication vehicle between the debug activity and the various listeners of the action bar (Listing 10–1).
- `BaseListener.java`: Base listener class that works with the `DebugActivity` and the various actions that gets invoked from the action bar. Acts as a base class for both tab listeners and list navigation listeners (Listing 10–4).
- `TabNavigationActionBarActivity.java`: inherits from `BaseActionBarActivity.java` and configures the action bar as a tabbed action bar. Most of the code pertaining to the tabbed action bar is in this class (Listing 10–6).
- `TabListener.java`: Required to add a tab to the tabbed action bar. This where you respond to tab clicks. In your case this simply logs a message to the debug view through the `BaseListener` (Listing 10–5).

- AndroidManifest.xml: Where activities are defined to be invoked (Listing 10–13).
- Layout/main.xml: Layout file for the DebugActivity. Because all the three status bar activities inherit this base DebugActivity, they all share this layout file (Listing 10–7).
- menu/menu.xml: A set of menu items to test the menu interaction with the action bar. The menu file is also shared across all the derived status bar activities (Listing 10–9).

Implementing Base Activity Classes

A number of the base classes use the IReportBack interface. It is introduced in Listing 10–1.

Listing 10–1. *IReportBack.java*

```
//IReportBack.java
package com.androidbook.actionbar;

public interface IReportBack
{
   public void reportBack(String tag, String message);
   public void reportTransient(String tag, String message);
}
```

A class that implements this interface takes a message and reports it on a screen, like a debug message. This is done through the reportBack() method. The method reportTransient() does the same thing except it uses a Toast to report that message to the user.

In this example, the class that implements IReportBack is DebugActivity. This allows DebugActivity to pass itself around without exposing all of its internals. The source code for DebugActivity is presented in Listing 10–2.

Listing 10–2. *DebugActivity with a Debug Text View*

```
//DebugActivity.java
package com.androidbook.actionbar;
//
//Use CTRL-SHIFT-O to import dependencies
//
public abstract class DebugActivity
extends Activity
implements IReportBack
{
   //Derived classes needs first
   protected abstract boolean
   onMenuItemSelected(MenuItem item);

   //private variables set by constructor
   private static String tag=null;
```

```
private int menuId = 0;
private int layoutid = 0;
private int debugTextViewId = 0;

public DebugActivity(int inMenuId,
      int inLayoutId,
      int inDebugTextViewId,
      String inTag)
{
   tag = inTag;
   menuId = inMenuId;
   layoutid = inLayoutId;
   debugTextViewId = inDebugTextViewId;

}
 @Override
 protected void onCreate(Bundle savedInstanceState) {
     super.onCreate(savedInstanceState);
     setContentView(this.layoutid);

     //You need the following to be able to scroll
     //the text view.
     TextView tv = this.getTextView();
     tv.setMovementMethod(
       ScrollingMovementMethod.getInstance());
 }
 @Override
 public boolean onCreateOptionsMenu(Menu menu){
    super.onCreateOptionsMenu(menu);
    MenuInflater inflater = getMenuInflater();
    inflater.inflate(menuId, menu);
    return true;
 }
 @Override
 public boolean onOptionsItemSelected(MenuItem item){
    appendMenuItemText(item);
    if (item.getItemId() == R.id.menu_da_clear){
       this.emptyText();
       return true;
    }
    boolean b = onMenuItemSelected(item);
    if (b == true)
    {
       return true;
    }
    return super.onOptionsItemSelected(item);
 }
 protected TextView getTextView(){
     return
     (TextView)this.findViewById(this.debugTextViewId);
 }
 protected void appendMenuItemText(MenuItem menuItem){
    String title = menuItem.getTitle().toString();
    appendText("MenuItem:" + title);
 }
 protected void emptyText(){
    TextView tv = getTextView();
```

```
        tv.setText("");
    }
    protected void appendText(String s){
        TextView tv = getTextView();
        tv.setText(s + "\n" + tv.getText());
        Log.d(tag,s);
    }
   public void reportBack(String tag, String message)
   {
      this.appendText(tag + ":" + message);
      Log.d(tag,message);
   }
   public void reportTransient(String tag, String message)
   {
      String s = tag + ":" + message;
      Toast mToast =
        Toast.makeText(this, s, Toast.LENGTH_SHORT);
      mToast.show();
      reportBack(tag,message);
      Log.d(tag,message);
   }
}//eof-class
```

The primary goal of this base activity class is to present an activity with a debug text view in it. This text view is used to log messages coming from the `reportBack()` method. You use this activity as the base activity for all of the action bar activities.

Assigning Uniform Behavior for the Action Bar

You have more opportunities to refactor the code from the derived activities into another level of a base class called `BaseActionBarActivity`.

The primary goal of this refactoring class is to provide a common behavior in response to the menu items. These menu items are there to switch between the three activities that represent three different action bar modes. Once switched, you can test that particular action bar activity.

This activity is presented in Listing 10–3.

Listing 10–3. *A Common Base Class for Action Bar Enabled Activities*

```
// BaseActionBarActivity.java
package com.androidbook.actionbar;
//
//Use CTRL-SHIFT-O to import dependencies
//
public abstract class BaseActionBarActivity
extends DebugActivity
{
    private String tag=null;
    public BaseActionBarActivity(String inTag)
    {
        super(R.menu.menu,      //Provides a common menu
            R.layout.main,      //Provides a common layout
            R.id.textViewId,    //Text view for the base debug activity
            inTag);             //Debug tag for logging
```

```
        tag = inTag;
    }
    @Override
    public void onCreate(Bundle savedInstanceState)
    {
        super.onCreate(savedInstanceState);
        TextView tv = this.getTextView();
        tv.setText(tag);
    }
    protected boolean onMenuItemSelected(MenuItem item)
    {
        //Responding to Home Icon
        if (item.getItemId() == android.R.id.home) {
            this.reportBack(tag,"Home Pressed");
            return true;
        }

        //Common behavior to invoke sibling activities
        if (item.getItemId() == R.id.menu_invoke_tabnav){
            if (getNavMode() ==
              ActionBar.NAVIGATION_MODE_TABS)
            {
                this.reportBack(tag,
                  "You are already in tab nav");
            }
            else {
                this.invokeTabNav();
            }
            return true;
        }
        if (item.getItemId() == R.id.menu_invoke_listnav){
            if (getNavMode() ==
            ActionBar.NAVIGATION_MODE_LIST)
            {
                this.reportBack(tag,
                "You are already in list nav");
            }
            else{
                this.invokeListNav();
            }
            return true;
        }
        if (item.getItemId() == R.id.menu_invoke_standardnav){
            if (getNavMode() ==
            ActionBar.NAVIGATION_MODE_STANDARD)
            {
                this.reportBack(tag,
                "You are already in standard nav");
            }
            else{
                this.invokeStandardNav();
            }
            return true;
        }
        return false;
    }
    private int getNavMode(){
```

```
        ActionBar bar = this.getActionBar();
        return bar.getNavigationMode();
    }
    private void invokeTabNav(){
        Intent i = new Intent(this,
          TabNavigationActionBarActivity.class);
        startActivity(i);
    }

    //Uncomment the following method bodies
    //as you implement these additional activities

    private void invokeListNav(){
        //Intent i = new Intent(this,
        //  ListNavigationActionBarActivity.class);
        //startActivity(i);
    }
    private void invokeStandardNav(){
        //Intent i = new Intent(this,
        //  StandardNavigationActionBarActivity.class);
        //startActivity(i);
    }
}//eof-class
```

If you notice the code responding to menu items in Listing 10–3, you see that you are checking if the current activity is also the one that is being switched to. If it is, you log a message and don't switch the current activity.

This base action bar activity also simplifies the derived action bar navigation activities including the tabbed navigation action bar activity.

Implementing the Tabbed Listener

Before you are able to work with a tabbed action bar, you need a tabbed listener. A tabbed listener allows you to respond to the click events on the tabs. You derive your tabbed listener from a base listener that allows you to log tab actions. Listing 10–4 shows the base listener that uses the IReportBack for logging.

Listing 10–4. *A Common Listener for Action Bar Enabled Activities*

```
//BaseListener.java
package com.androidbook.actionbar;
//
//Use CTRL-SHIFT-O to import dependencies
//
public class BaseListener
{
   protected IReportBack mReportTo;
   protected Context mContext;
   public BaseListener(Context ctx, IReportBack target)
   {
      mReportTo = target;
      mContext = ctx;
   }
}
```

This base class holds a reference to an implementation of IReportBack and also the activity that can be used as a context. In this case, the DebugActivity from Listing 10–2 is the implementer of IReportBack and also plays the role of the context.

Now that you have a base listener, Listing 10–5 shows the tabbed listener.

Listing 10–5. *Tab Listener to Respond to Tab Actions*

```
// TabListener.java
package com.androidbook.actionbar;
//
//Use CTRL-SHIFT-O to import dependencies
//
public class TabListener extends BaseListener
implements ActionBar.TabListener
{
    private static String tag = "tc>";
    public TabListener(Context ctx,
                IReportBack target)
    {
        super(ctx, target);
    }
    public void onTabReselected(Tab tab,
                  FragmentTransaction ft)
    {
        this.mReportTo.reportBack(tag,
          "ontab re selected:" + tab.getText());
    }
    public void onTabSelected(Tab tab,
               FragmentTransaction ft)
    {
        this.mReportTo.reportBack(tag,
          "ontab selected:" + tab.getText());
    }
    public void onTabUnselected(Tab tab,
                 FragmentTransaction ft)
    {
        this.mReportTo.reportBack(tag,
          "ontab un selected:" + tab.getText());
    }
}
```

This tabbed listener documents the callbacks from the action bar tabs to the debug text view of Figure 10–1.

Implementing the Tabbed Action Bar Activity

With the tabbed listener in place, you can finally construct the tabbed navigation activity. This is presented in Listing 10–6.

Listing 10–6. *Tab-Navigation Enabled Action Bar Activity*

```
// TabNavigationActionBarActivity.java
package com.androidbook.actionbar;
//
//Use CTRL-SHIFT-O to import dependencies
```

```
//
public class TabNavigationActionBarActivity
extends BaseActionBarActivity
{
    private static String tag =
      "Tab Navigation ActionBarActivity";
    public TabNavigationActionBarActivity()
    {
        super(tag);
    }
    @Override
    public void onCreate(Bundle savedInstanceState)
    {
        super.onCreate(savedInstanceState);
        workwithTabbedActionBar();
    }

    public void workwithTabbedActionBar()
    {
        ActionBar bar = this.getActionBar();
        bar.setTitle(tag);
        bar.setNavigationMode(
          ActionBar.NAVIGATION_MODE_TABS);

        TabListener tl = new TabListener(this,this);

        Tab tab1 = bar.newTab();
        tab1.setText("Tab1");
        tab1.setTabListener(tl);
        bar.addTab(tab1);

        Tab tab2 = bar.newTab();
        tab2.setText("Tab2");
        tab2.setTabListener(tl);
        bar.addTab(tab2);
    }
}//eof-class
```

You now look at the code for this activity (Listing 10–6) in the following subsections, which draw attention to each aspect of working with a tabbed action bar. You start with getting access to the action bar belonging to an activity.

Obtaining an Action Bar Instance

In Listing 10–6, notice that the code that controls the action bar is pretty simple. You get access to the action bar of an activity by calling `getActionbar()` on the activity. Here is that line of code again:

```
ActionBar bar = this.getActionBar();
```

As this snippet of code shows, action bar is a property of the activity, and does not cross activity boundaries. In other words, one cannot use an action bar to control or influence multiple activities.

Action Bar Navigation Modes

In Listing 10–6, once you obtain the action bar for an activity, you set its navigation mode to ActionBar.NAVIGATION_MODE_TABS:

```
bar.setNavigationMode(
  ActionBar.NAVIGATION_MODE_TABS);
```

The other two possible action bar navigation modes are

- ActionBar.NAVIGATION_MODE_LIST
- ActionBar.NAVIGATION_MODE_STANDARD

Once you set the tabbed navigation mode, you have a number of tab-related methods to work with in the API of the ActionBar class. Listing 10–6 use these tab-related APIs to add two tabs to the action bar. You also use the tabbed listener from Listing 10–5 to initialize the tabs.

Here is a quick code snippet from Listing 10–6 that shows how a tab is added to the action bar:

```
Tab tab1 = bar.newTab();
tab1.setText("Tab1");
tab1.setTabListener(tl);
bar.addTab(tab1);
```

If you forget to call the setTabListener() method on a tab that is added to the action bar, you get a runtime error indicating that a listener is needed.

Scrollable Debug Text View Layout

As the tabs of the action bar are clicked, the tab listeners are set up in such a way that debug messages are sent to the debug text view. Listing 10–7 shows the layout file for the DebugActivity, which in turn contains the debug text view.

Listing 10–7. *Debug Activity Text View Layout File*

```
<?xml version="1.0" encoding="utf-8"?>
<!-- /res/layout/main.xml -->
<LinearLayout
xmlns:android="http://schemas.android.com/apk/res/android"
    android:orientation="vertical"
    android:layout_width="fill_parent"
    android:layout_height="fill_parent"
    android:gravity="fill"
    >
<TextView android:id="@+id/textViewId"
    android:layout_width="fill_parent"
    android:layout_height="fill_parent"
    android:background="@android:color/white"
    android:text="Initial Text Message"
    android:textColor="@android:color/black"
    android:textSize="25sp"
```

```
    android:scrollbars="vertical"
    android:scrollbarStyle="insideOverlay"
    android:scrollbarSize="25dip"
    android:scrollbarFadeDuration="0"
    />
</LinearLayout>
```

There are a few things worth noting about this layout. You set the background color of the text view to white. This lets you capture screens in brighter light. The text size is also set to a large font to aid screen capture.

You also set up the text view so that it is enabled for scrolling. Although typically layouts use `ScrollView`, a text view is already enabled for scrolling by itself. In addition to enabling the scrolling properties in the XML file for the text view, you need to call the `setMovementMethod()` method on the text view as shown in Listing 10–8.

Listing 10–8. *Enabling Text View for Scrolling*

```
TextView tv = this.getTextView();
tv.setMovementMethod(
     ScrollingMovementMethod.getInstance());
```

This code is extracted from the `DebugActivity` (Listing 10–2).

As the text view is scrolled, notice that the scrollbar appears and then fades away. This is not a good indicator if there is text beyond visible range. You can tell the scrollbar to stay by setting the fade duration to 0. See Listing 10–7 for how to set this parameter.

Action Bar and Menu Interaction

This example also demonstrates how menus interact with the action bar. So, you need to set up a menu file. This file is presented in Listing 10–9.

Listing 10–9. *Menu XML File for This Project*

```
<!-- /res/menu/menu.xml -->
<menu
xmlns:android="http://schemas.android.com/apk/res/android">
    <!-- This group uses the default category. -->
    <group android:id="@+id/menuGroup_Main">

        <item android:id="@+id/menu_action_icon1"
            android:title="Action Icon1"
            android:icon="@drawable/creep001"
            android:showAsAction="ifRoom"/>

        <item android:id="@+id/menu_action_icon2"
            android:title="Action Icon2"
            android:icon="@drawable/creep002"
            android:showAsAction="ifRoom"/>

        <item android:id="@+id/menu_icon_test"
            android:title="Icon Test"
            android:icon="@drawable/creep003"/>

        <item android:id="@+id/menu_invoke_listnav"
```

```
            android:title="Invoke List Nav"
            />
        <item android:id="@+id/menu_invoke_standardnav"
            android:title="Invoke Standard Nav"
            />
        <item android:id="@+id/menu_invoke_tabnav"
            android:title="Invoke Tab Nav"
            />
        <item android:id="@+id/menu_da_clear"
            android:title="clear" />
    </group>
</menu>
```

> **NOTE:** The menu XML file in Listing 10–9 uses 3 icons (creep001, 002, and 003) from `www.androidicons.com`. As per the web site, these icons are under Creative Commons License 3.0.

The following section talks about this menu in a bit more detail.

Displaying the Menu

In releases 2.3 and earlier, devices often had an explicit menu button. In 3.0, the emulator doesn't show physical Home, Back, and Menu buttons. These may still be available on some devices.

As shown in Figure 10–2, the Back and Home buttons are now soft buttons available at the bottom of the screen. However, the Menu button is shown in the context of an application, specifically as part of the action bar in the upper-right corner.

Figure 10–2 shows what the menu looks like when it is expanded.

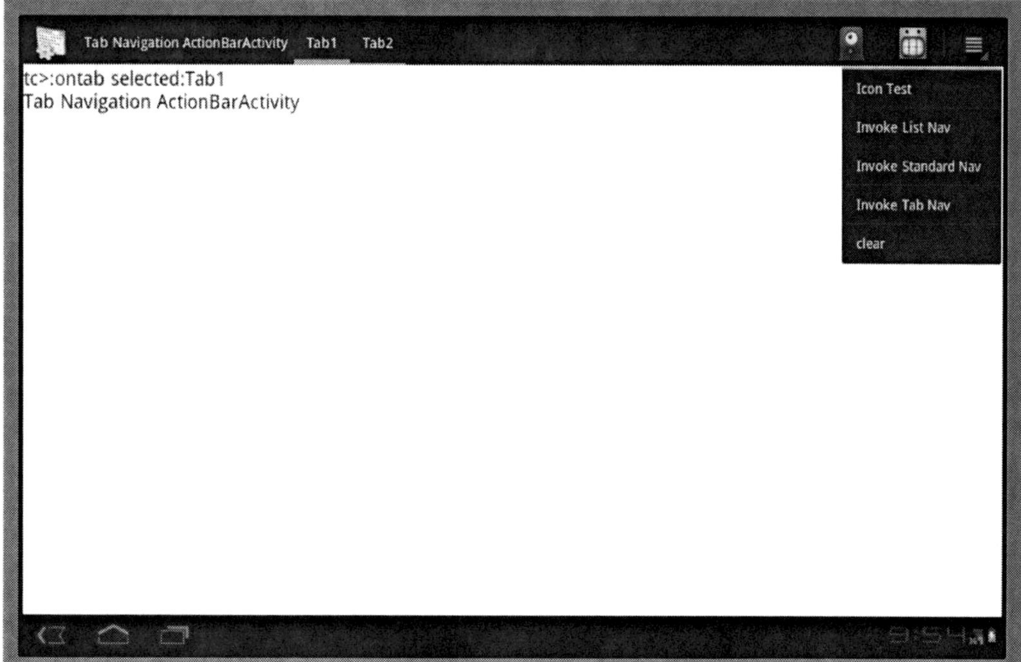

Figure 10–2. *An activity with a tabbed action bar and expanded menu*

One thing of note is that a menu bar may not show the icons for menu items. You should not rely on icons for menu items being shown in all cases.

Menu Items as Actions

As indicated at the beginning of the chapter, you can assign some of the menu items to appear directly on the action bar. These menu items are indicated with the tag `showAsAction`. You can see this tag in Listing 10–9 in the menu XML file. This tag line is extracted and shown again in Listing 10–10.

Listing 10–10. *Menu Item Attribute for* `showAsAction`

```
android:showAsAction="ifRoom"
```

The other possible values for this XML tag are as follows:

- `always`
- `never`
- `withText`

You can also accomplish the same affect with a Java API available on the `MenuItem` class:

```
menuItem.setShowAsAction(int actionEnum)
```

The values for the actionEnum are as follows:

- SHOW_AS_ACTION_ALWAYS
- SHOW_AS_ACTION_IF_ROOM
- SHOW_AS_ACTION_NEVER
- SHOW_AS_ACTION_WITH_TEXT

Because these actions are merely menu items, they behave as such and call the onOptionsItemSelected() callback method of the activity class.

Finally, the example uses a number of icons. You can replace these icons with some of your own, or you can download the project for this chapter using the URL at the end of this chapter.

Android Manifest File

Listing 10–11 shows the manifest file for this project so far.

Listing 10–11. *AndroidManifest.xml*

```
<?xml version="1.0" encoding="utf-8"?>
<manifest xmlns:android="http://schemas.android.com/apk/res/android"
      package="com.androidbook.actionbar"
      android:versionCode="1"
      android:versionName="1.0.0">
    <application android:icon="@drawable/icon"
        android:label="ActionBars Demo App">
        <activity android:name=".TabNavigationActionBarActivity"
                  android:label="Action Bar Demonstration: TabNav">
            <intent-filter>
                <action android:name="android.intent.action.MAIN" />
                <category android:name="android.intent.category.LAUNCHER" />
            </intent-filter>
        </activity>
    </application>
    <uses-sdk android:minSdkVersion="11" />
</manifest>
```

The minSDKVersion need to point to 11, the API number for release 3.0.

Examining the Tabbed Action Bar Activity

When you compile these files and run the application, you see the tabbed action bar, as shown in Figure 10–1. Then if you click the menu icon to the right, you see the menu of the application, as expanded in Figure 10–2.

The application is designed in such a way that any action on the action bar is logged to the debug text view. While you are running this application, you can test the following:

- If you click the Home icon, you see a message logged to the debug screen indicating that the Home button is clicked.

- If you click tab1, you see a message that tab1 is reselected.
- If you click tab2, you see two messages: one indicates that tab1 is losing focus and the other that tab2 is clicked. These messages are provisioned through the tab listener in Listing 10–5.
- If you click the action buttons on the right side, you see that their corresponding menu items are invoked and debug messages logged to the debug view.
- If you expand the menu, you see that there are menu items to invoke other activities, which demonstrate the rest of the action bar modes. However, you need to wait until the other activities are developed later in the chapter. Until then, just notice that those items are invoked and debug messages logged.

This concludes your implementation of the tabbed action bar activity as well as setting up the base framework so that coding the other two activities is much simpler. Let's move onto the list navigation mode action bar.

List Navigation Action Bar Activity

Because your base classes are carrying the most of the work, it is fairly easy to implement and test the list action bar navigation activity. You need the following additional files to implement this activity:

- `SimpleSpinnerArrayAdapter.java`: Needed to set up the list navigation bar along with the listener. This class provides the rows required by a drop-down navigation list (Listing 10–12).
- `ListListener.java`: Acts as a listener to the list navigation activity. This class needs to be passed to the action bar when setting it up as a list action bar (Listing 10–13).
- `ListNavigationActionBarActivity.java`: Implements the list navigation action bar activity (Listing 10–14).

Once you have these three new files, you need to update the following two files:

- `BaseActionBarActivity.java`: Uncomment the invocation of the list action bar activity (Listing 10–3).
- `AndroidManifest.xml`: Define the new list navigation action bar activity in the manifest file (Listing 10–11).

Creating a Spinner Adapter

To be able to initialize the action bar with list navigation mode, you need the following two things:

- A spinner adapter that can tell the list navigation what the list of navigation text is
- A list navigation listener so that when one of the list items is picked you can get a call back

Listing 10–12 presents the `SimpleSpinnerArrayAdapter` that implements the `SpinnerAdapter` interface. As stated earlier, the goal of this class is to give a list of items to show.

Listing 10–12. *Creating a Spinner Adapter for List Navigation*

```
//SimpleSpinnerArrayAdapter.java
package com.androidbook.actionbar;
//
//Use CTRL-SHIFT-O to import dependencies
//
public class SimpleSpinnerArrayAdapter
extends ArrayAdapter<String>
implements SpinnerAdapter
{
    public SimpleSpinnerArrayAdapter(Context ctx)
    {
        super(ctx,
          android.R.layout.simple_spinner_item,
          new String[]{"one","two"});

        this.setDropDownViewResource(
          android.R.layout.simple_spinner_dropdown_item);
    }
    public View getDropDownView(
      int position, View convertView, ViewGroup parent)
    {
        return super.getDropDownView(
          position, convertView, parent);
    }
}
```

There is no SDK class that directly implements the `SpinnerAdapter` interface required by list navigation. So, you derive this class from an `ArrayAdapter` and provide a simple implementation for the `SpinnerAdapter`. At the end of the chapter is a reference URL on spinner adapters for further reading. Let's move on now to the list navigation listener.

Creating a List Listener

This is a simple class implementing the `ActionBar.OnNavigationListener`. Listing 10–13 shows the code for this class.

Listing 10–13. *Creating a List Listener for List Navigation*

```
//ListListener.java
package com.androidbook.actionbar;
//
//Use CTRL-SHIFT-O to import dependencies
//
public class ListListener
extends BaseListener
implements ActionBar.OnNavigationListener
{
    public ListListener(
    Context ctx, IReportBack target)
    {
        super(ctx, target);
    }
    public boolean onNavigationItemSelected(
    int itemPosition, long itemId)
    {
        this.mReportTo.reportBack(
          "list listener","ItemPostion:" + itemPosition);
        return true;
    }
}
```

Like the tabbed listener in Listing 10–5, you inherit from your `BaseListener` so that you can log events to the debug text view through the `IReportBack` interface.

Setting Up a List Action Bar

You now have what you require to set up a list navigation action bar. The source code for the list navigation action bar activity is shown in Listing 10–14. This class is very similar to the tabbed activity you coded earlier.

Listing 10–14. *List Navigation Action Bar Activity*

```
// ListNavigationActionBarActivity.java
package com.androidbook.actionbar;
//
//Use CTRL-SHIFT-O to import dependencies
//
public class ListNavigationActionBarActivity
extends BaseActionBarActivity
{
    private static String tag=
      "List Navigation ActionBarActivity";

    public ListNavigationActionBarActivity()
    {
        super(tag);
    }
    @Override
    public void onCreate(Bundle savedInstanceState)
    {
        super.onCreate(savedInstanceState);
        workwithListActionBar();
```

```
    }
    public void workwithListActionBar()
    {
        ActionBar bar = this.getActionBar();
        bar.setTitle(tag);
        bar.setNavigationMode(ActionBar.NAVIGATION_MODE_LIST);
        bar.setListNavigationCallbacks(
           new SimpleSpinnerArrayAdapter(this),
           new ListListener(this,this));
    }
}//eof-class
```

The important code is highlighted in Listing 10–14. The code is quite simple: you take a spinner adapter and a list listener and set them as list navigation callbacks on the action bar.

Making Changes to BaseActionBarActivity

With this list navigation action bar activity (Listing 10–14) available, you can go back and change the BaseActionBarActivity so that the menu item intended for ListNavigationActionBarActivity invokes this activity. When uncommented, the corresponding function in Listing 10–3 looks like the extracted and uncommented code in Listing 10–15.

Listing 10–15. *Code to Uncomment for Invoking the List Navigation Action Bar Activity*

```
private void invokeListNav(){
     Intent i = new Intent(this,
        ListNavigationActionBarActivity.class);
     startActivity(i);
}
```

After you uncomment this, the menu item and the code are wired to invoke this list navigation action bar activity.

Making Changes to AndroidManifest.xml

Before you can invoke the activity, you need to register it in the Android manifest file. Add the code in Listing 10–16 to the Android manifest file in Listing 10–11 to complete the activity registration.

Listing 10–16. *Registering the List Navigation Action Bar Activity*

```
<activity android:name=".ListNavigationActionBarActivity"
        android:label="Action Bar Demonstration: ListNav">
</activity>
```

Examining the List Action Bar Activity

When you compile the files covered so far (the new and changed files mentioned at the beginning of this section about the list navigation action bar) and run the application, you see the list action bar shown in Figure 10–3.

Figure 10–3. *An activity with a list navigation action bar*

In Figure 10–3, you can see the unexpanded list next to the title of the activity. This is the same place the SDK puts the tabs when the action bar mode is tab navigation. If you click the One item, the list expands and allows you to choose an option, as shown in Figure 10–4.

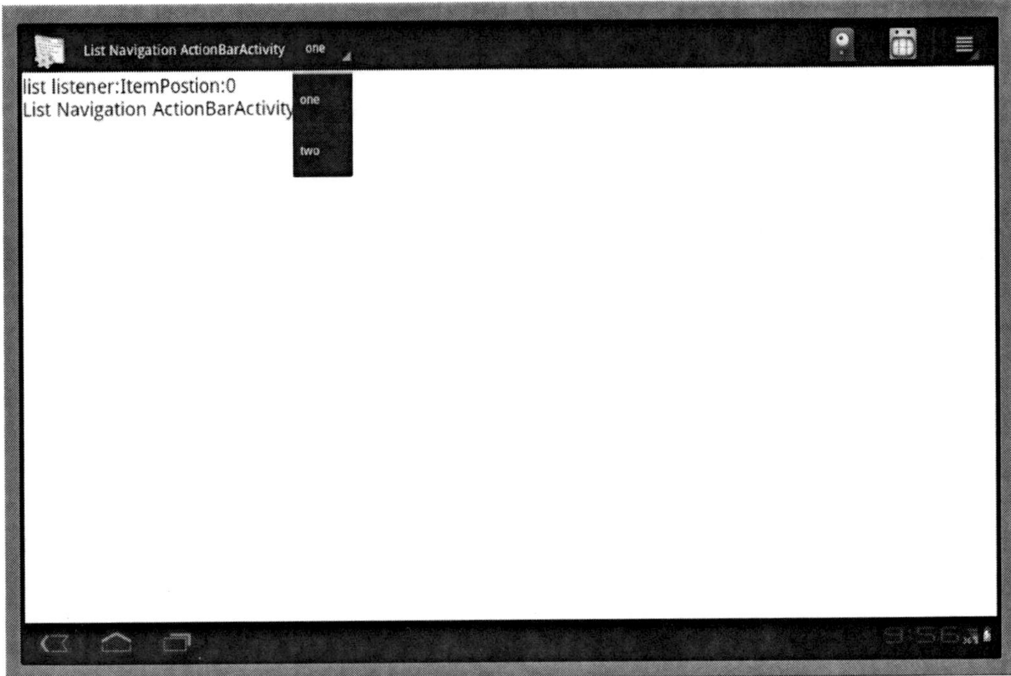

Figure 10–4. *An activity with an opened navigation list*

When you compare this activity to the activity in Figures 10–1 and 10–2, they look very similar, except that in one case you have tabs and in the other case you have a list to navigate. The motif of these two activities illustrates an important parallel to the way web sites are designed.

> **NOTE:** In a web site, there may be a number of web pages, but each page displays a uniform look and feel through master pages. In this simpler case, you use the base class to accomplish this effect.

Although you have used multiple activities to showcase action bars, the action bars seem to be more applicable to orchestrate fragments on a single activity. However, should you need to work with multiple activities, you can use this base-class pattern to provide that master page design pattern.

The behavior of this list navigation activity is very much like the one for the tabbed activity in the previous section. The difference is what happens when you click the list items. Each time you choose a list item, you see a call back to the list listener, and the list listener sends a message to the debug text view.

Now that you have two activities available, the menu items allow you to switch between the tabbed activity and the list activity.

Let's move on to the simpler standard action bar activity.

Standard Navigation Action Bar Activity

In this section, you examine the nature of a standard navigation action bar. You set up an activity and set its action bar navigation mode as standard. You then see what the standard navigation looks like and examine its behavior.

As in the case of `ListNavigationActionBarActivity`, because your base classes are carrying most of the work, it is easy to implement and test the standard action bar navigation activity. You need the following additional file:

- `StandardNavigationActionBarActivity.java`: This is the implementation file for configuring the action bar as a standard navigation mode action bar (Listing 10–17).

You also need to update the following two files:

- `BaseActionBarActivity.java`: Uncomment the invocation of the standard action bar activity in response to a menu item (see Listing 10–18 for changes and Listing 10–3 for the original file).

- `AndroidManifest.xml`: Define this new activity in the manifest file (see Listing 10–19 for this activity's definition so you can add this to the main Android manifest file in Listing 10–11).

Setting up the Standard Navigation Action Bar Activity

You used tabbed listeners while setting up the tabbed action bar and list listeners for setting up the list navigation action bar. For a standard action bar, there are no listeners other than the menu callbacks. The menu callbacks don't need to be specially set up because they are hooked up automatically by the SDK. As a result, it is quite easy to set up the action bar in the standard navigation mode.

Listing 10–17 presents the source code for the standard navigation action bar activity.

Listing 10–17. *Standard Navigation Action Bar Activity*

```
//StandardNavigationActionBarActivity.java
package com.androidbook.actionbar;
//
//Use CTRL-SHIFT-O to import dependencies
//
public class StandardNavigationActionBarActivity
extends BaseActionBarActivity
{
    private static String tag=
      "Standard Navigation ActionBarActivity";
    public StandardNavigationActionBarActivity()
    {
```

```
        super(tag);
    }
    @Override
    public void onCreate(Bundle savedInstanceState)
    {
        super.onCreate(savedInstanceState);
        workwithStandardActionBar();
    }

    public void workwithStandardActionBar()
    {
        ActionBar bar = this.getActionBar();
        bar.setTitle(tag);
        bar.setNavigationMode(ActionBar.NAVIGATION_MODE_STANDARD);
        //test to see what happens if you were to attach tabs
        attachTabs(bar);
    }
    public void attachTabs(ActionBar bar)
    {
        TabListener tl = new TabListener(this,this);

        Tab tab1 = bar.newTab();
        tab1.setText("Tab1");
        tab1.setTabListener(tl);
        bar.addTab(tab1);

        Tab tab2 = bar.newTab();
        tab2.setText("Tab2");
        tab2.setTabListener(tl);
        bar.addTab(tab2);
    }
}//eof-class
```

The only thing necessary to set up an action bar as a standard navigation action bar is to set its navigation mode as such. That portion of the code is highlighted in Listing 10–17.

> **NOTE:** Listing 10–17 also includes code to see what happens if you add tabs while the mode is standard navigation. Testing shows that these tabs do not cause any runtime errors but are ignored by the framework.

Before you see what the standard action bar looks like, you need to make a couple of changes to existing files.

Making Changes to BaseActionBarActivity

Once the standard navigation action bar activity (Listing 10–17) is available, you can go back and change the `BaseActionBarActivity` (Listing 10–3) so that the menu item intended for `StandardNavigationActionBarActivity` invokes this activity. When uncommented, the corresponding function in Listing 10–3 looks like the code in Listing 10–18.

Listing 10–18. *Section to Uncomment to Invoke the Standard Navigation Action Bar Activity*

```
private void invokeStandardNav(){
    Intent i = new Intent(this,
      StandardNavigationActionBarActivity.class);
    startActivity(i);
}
```

After you uncomment this, the menu item and the code are already wired to invoke the `StandardNavigationActionBarActivity`.

Making Changes to AndroidManifest.xml

Before you can invoke this activity, you need to register this activity in the Android manifest file. Add the lines from Listing 10–19 to the Android manifest file in Listing 10–11 to complete the activity registration.

Listing 10–19. *Registering the Standard Navigation Action Bar Activity*

```
<activity android:name=".StandardNavigationActionBarActivity "
        android:label="Action Bar Demonstration: Standard Nav">
</activity>
```

Examining the Standard Action Bar activity

When you compile the files covered so far (and listed in the section “Standard Navigation Action Bar Activity”) and run the application, the application opens with the tabbed activity as the first activity (Figure 10–1). If you click the menu item, you see the activity from Figure 10–2. And if you choose the menu item Invoke Standard Nav, you now see the standard navigation action bar activity shown in Figure 10–5.

Figure 10–5. *An activity with a standard navigation action bar*

The first thing you notice in Figure 10–5 is that this action bar is missing the area that was previously dedicated to either a tab or a list navigation. If you click the action buttons on the right, they write their invocation to the debug text view. Click the Home button: this also writes its invocation signature to the debug text view. At the end of these three clicks, the debug text view looks like Figure 10–6.

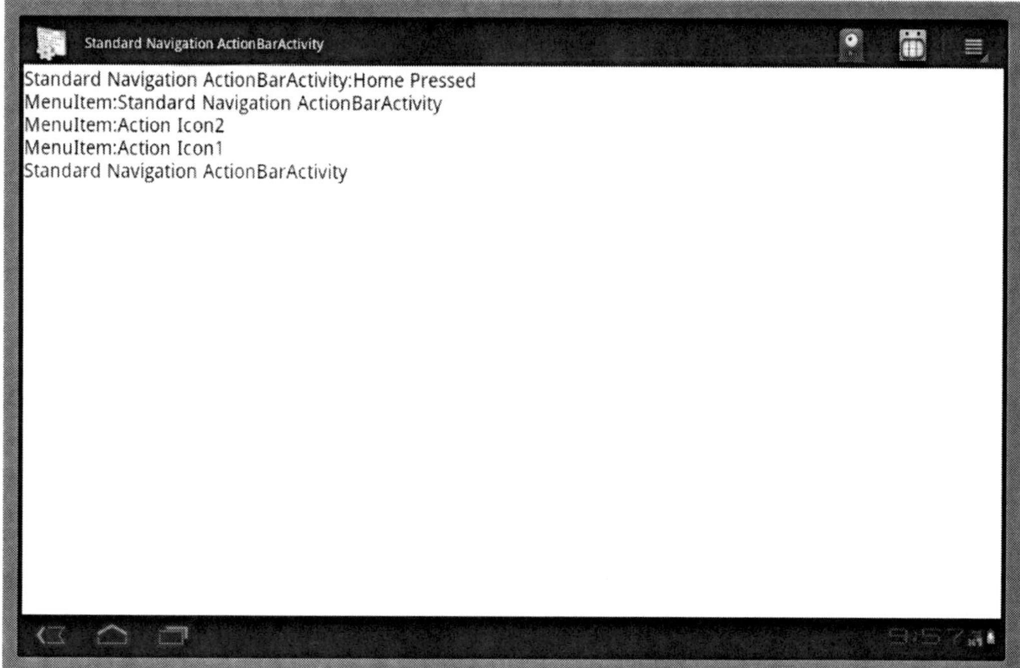

Figure 10–6. *Responding to events from an action bar*

Action Bar and Search View

In Android 4.0, because the action bar is available on phones, there is an increasing interest in using it as a search facility. This section shows how to use a search widget in the action bar.

We will provide code snippets that you can use to modify the project you have seen so far to include a search widget. Although we present only snippets, you can see the full code in the downloadable project for this chapter.

A search view widget is a search box that fits between your tabs and the menu icons in the action bar, as shown in Figure 10–7.

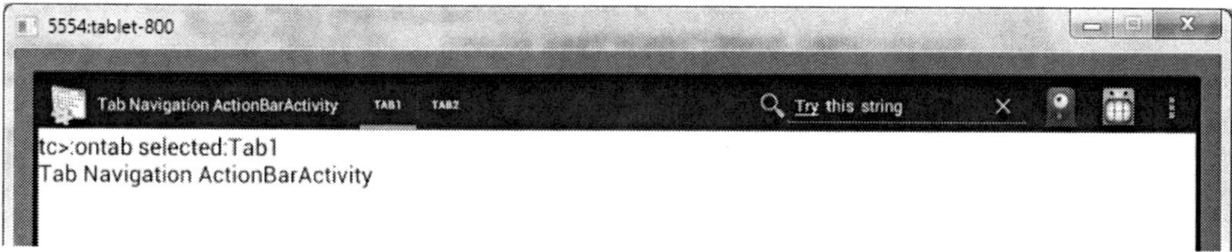

Figure 10–7. *An action bar search view*

You need to do the following to use search in your action bar:

1. Define a menu item pointing to a search view provided by the SDK. You also need an activity into which you can load this menu. This is often called the *search invoker activity*.
2. Create another activity that can take the query from the search view in step 1 and provide results. This is often called the *search results activity*.
3. Create an XML file that allows you to customize the search view widget. This file is often called `searchable.xml` and resides in the `res/xml` subdirectory.
4. Declare the search results activity in the manifest file. This definition needs to point to the XML file defined in step 3.
5. In your menu setup for the search invoker activity, indicate that the search view needs to target the search results activity from step 2.

We will provide code snippets for each of these steps. As mentioned earlier, the complete code is available in the downloadable project. In fact, when you run the project for this chapter, the search view is visible on all the action bars presented in the previous sections of this chapter: tab, list, and standard.

Defining a Search View Widget as a Menu Item

To define a search view to appear in the action bar of your activity, you need to define a menu item in one of your menu XML files, as shown in Listing 10–20.

Listing 10–20. *Search View Menu Item Definition*

```
<item android:id="@+id/menu_search"
    android:title="Search"
    android:showAsAction="ifRoom"
    android:actionViewClass="android.widget.SearchView"
    />
```

The key element in Listing 10–20 is the `actionViewClass` attribute pointing to `android.widget.SearchView`. You saw the other attributes earlier in the chapter when you declared your normal menu items to appear as action icons in the action bar.

Casting a Search Results Activity

To enable search in your application, you need an activity that can respond to a search query. This can be like any other activity. An example is shown in Listing 10–21.

Listing 10–21. *Search Results Activity*

```
public class SearchResultsActivity
{
    private static String tag="Search Results Activity";
```

```
    @Override
    public void onCreate(Bundle savedInstanceState)
    {
        super.onCreate(savedInstanceState);
        final Intent queryIntent = getIntent();
        doSearchQuery(queryIntent);
    }
    @Override
    public void onNewIntent(final Intent newIntent)
    {
        super.onNewIntent(newIntent);
        final Intent queryIntent = getIntent();
        doSearchQuery(queryIntent);
    }
    private void doSearchQuery(final Intent queryIntent)
    {
        final String queryAction = queryIntent.getAction();
        if (!(Intent.ACTION_SEARCH.equals(queryAction)))
        {
            Log.d(tag,"intent NOT for search");
            return;
        }
        final String queryString =
            queryIntent.getStringExtra(SearchManager.QUERY);
        Log.d(tag, queryString);
    }
}//eof-class
```

A few notes about the search results activity in Listing 10–21:

- The activity checks to see whether the action that invoked it is initiated by search.
- This activity could have been newly created or just brought to the top. In the latter case, this activity needs to do something identical to the `oncreate()` method in its `onNewIntent()` method as well
- If this activity is invoked by search, it retrieves the query string using an extra parameter called `SearchManager.QUERY`. Then the activity logs what that string is. In a real scenario, you would use that string to return matching results.

Customizing Search Through a Searchable XML File

As indicated in the earlier steps, let's look at the XML file that customizes the search experience; see Listing 10–22.

Listing 10–22. *Searchable XML File*

```
<!-- /res/xml/searchable.xml -->
<searchable xmlns:android="http://schemas.android.com/apk/res/android"
    android:label="@string/search_label"
    android:hint="@string/search_hint"
/>
```

The hint attribute will appear on the search view widget as a hint that disappears when you start typing. The label doesn't play a significant role in the action bar. However, when you use the same search results activity in a search dialog, the dialog has the label defined here.

You can learn more about searchable XML attributes at the following URL:

`http://developer.android.com/guide/topics/search/searchable-config.html`

Defining the Search Results Activity in the Manifest File

Now let's see how to tie this XML file to the search results activity. This is done in the manifest file as part of defining the search results activity: see Listing 10–23. Notice the metadata definition pointing to the searchable XML file resource.

Listing 10–23. *Tying an Activity to Its `Searchable.xml`*

```
<activity android:name=".SearchResultsActivity"
   android:label="Search Results">
     <intent-filter>
        <action android:name="android.intent.action.SEARCH"/>
      </intent-filter>
     <meta-data android:name="android.app.searchable"
                      android:resource="@xml/searchable"/>
</activity>
```

Identifying the Search Target for the Search View Widget

So far, you have the search view in your action bar, and you have the activity that can respond to search. You need to tie together these two pieces using Java code. You do this in the `onCreateOptions()` callback of the search-invoking activity as part of setting up your menu. The function in Listing 10–24 can be called from `onCreateOptions()` to link the search view widget and the search results activity.

Listing 10–24. *Tying the Search View Widget to the Search Results Activity*

```
private void setupSearchView(Menu menu)
{
    //Locate the search view widget
    SearchView searchView =
        (SearchView) menu.findItem(R.id.menu_search).getActionView();
    if (searchView == null)
    {
        this.reportBack(tag, "Failed to get search view");
        return;
    }

    //setup searchview
    SearchManager searchManager =
        (SearchManager) getSystemService(Context.SEARCH_SERVICE);
    ComponentName cn =
        new ComponentName(this,SearchResultsActivity.class);
    SearchableInfo info =
        searchManager.getSearchableInfo(cn);
```

```
    if (info == null)
    {
        this.reportBack(tag, "Failed to get search info");
        return;
    }

    searchView.setSearchableInfo(info);

    // Do not iconify the widget; expand it by default
    searchView.setIconifiedByDefault(false);
}
```

You can test the code in this section by modifying your project, or you can download the project for this chapter. When you do, you will see the search view as part of the action bar, as shown in Figure 10–7.

The Action Bar and Fragments

This chapter has shown how to use the action bar in association with activities. The action bar is generally recommended for use with fragments when you're dealing with tablets. Fragments are covered in Chapter 8. The same principles laid out here apply to fragments as well.

Because fragments are inside an activity, and an activity owns the action bar, you don't need the abstraction of a base class to ensure the same action bar for each activity. All fragments share the same activity, so they also share the same action bar. The solution is simpler.

References

The following URLs were helpful as we researched material for this chapter. The list also includes further reading material. In addition, the final URL allows you to download a zip file of the project of this chapter:

- *The Design of Everyday Things* by Donald A Norman. This book appropriated a previous idea from visual perception called *affordance* for human-computer interaction (HCI). This term is being increasingly used in Android UI literature. The action bar discussed in this chapter is touted as one of the key UI affordances.
- `http://en.wikipedia.org/wiki/Affordance`: Wikipedia reference that discusses UI affordances.
- `www.androidbook.com/item/3624`: Our research about the Android action bar, including a list of further references, sample code, links to examples, and UI figures representing various action bar modes.

- http://developer.android.com/reference/android/app/ActionBar.html: API URL for the ActionBar class.
- www.androidbook.com/item/3627: "Using Spinner Adapter." To set up list navigation mode, you need to understand how drop-down lists and spinners work. This brief article shows a few samples and reference links on how to use spinners in Android.
- www.androidbook.com/item/3885: Explains how search works, to help you utilize the action bar to its full extent.
- www.androidicons.com: Web site from which a couple of the icons used in this chapter are borrowed. These icons are under Creative Commons License 3.0.
- www.androidbook.com/item/3302: "Pleasing Android Layouts." A few notes and sample code for simple layouts.
- http://developer.android.com/reference/android/view/MenuItem.html: API for the MenuItem class. You find here documentation for attaching menu items as action icons on the action bar.
- http://developer.android.com/guide/topics/resources/menu-resource.html: XML elements available for defining menu items as action bar icons.
- www.androidbook.com/proandroid4/projects: Download site for this chapter's test project. The name of the zip file is ProAndroid4_ch10_TestActionbar.zip.

Summary

In summary, we have covered the following in this chapter:

- The look and feel of action bar
- The three types of action bars
- A common framework for an action bar to coordinate between multiple activities
- Accommodating a search view in an action bar

Interview Questions

Here is a list of questions to review what you have learned in this chapter:

1. What is an action bar, and what are its key visual components?
2. What are three types of action bars?

3. How do you show a menu item as an icon on the action bar?
4. What is `R.id.home`?
5. How do you design a pattern so that all your activities can share a common action/menu bar?
6. How do you get an instance of an action bar?
7. How do you add tabs to an action bar?
8. How do you respond to tabs being clicked?
9. Can a single action bar control multiple activities?
10. What are the possible values when a menu item is shown in the action bar?
11. Why is `SpinnerAdapter` important to a list navigation list?
12. How do you pass a list of items that are clickable to a list navigation tab?
13. How do fragments simplify an action bar pattern?
14. What artifacts do you need to provision a search widget in the action bar?

Chapter 11

Advanced Debugging and Analysis

At this point in your Android learning, you've got a few applications under your belt, and you may have encountered some unexpected behavior from them. We'd like to take some time in this chapter to explore the various ways you can debug an application, so you can look inside your application and find out what's going on. Although there are ways to do this without Eclipse and the Android plug-ins, that's what we're going to use here.

The Eclipse Debug perspective is the standard one that comes with Eclipse, and is not specific to Android programming. However, we want to make you aware of what can be done with it. The DDMS perspective has quite a few very useful features to help you debug your applications. These include Devices view (for seeing what you're connected to), Emulator Control (for sending in telephone calls, SMS messages, and GPS coordinates), File Explorer (for viewing/transferring files on the device), Threads, Heap, and Allocation Tracker (for seeing inside your application). We'll also dive into the Hierarchy View perspective so you can traverse the actual view structure of a running application. Then we'll touch on Traceview which makes it much easier to analyze a dump file from an application. Finally, we'll cover the `StrictMode` class, which can be used to trap policy violations to catch design errors that can cause poor user experiences.

Enabling Advanced Debugging

When you are testing in an emulator, the Eclipse Android Development Tools (ADT) plug-ins take care of setting everything up for you so you are able to use all the tools we're about to describe.

There are two things you need to know about debugging applications on a real device. The first is that the application must be set to be debuggable. This involves adding `android:debuggable="true"` to the `<application>` tag in the `AndroidManifest.xml` file.

Fortunately, ADT sets this for properly for you. When you're creating debug builds for the emulator or deploying directly from Eclipse to a device, this attribute is set to `true` by ADT. When you export your application to create a production version of it, ADT knows not to set `debuggable` to `true`. Note that if you set it yourself in `AndroidManifest.xml`, it stays set no matter what. The second thing to know is that the device must be put into USB Debug mode. To find this setting, go to the device's Settings screen, choose Application, and then choose Development. Make sure Enable USB Debugging is selected.

The Debug Perspective

Although LogCat is very useful for watching log messages, you'll definitely want to have more control and more information about your application as it runs. Debugging in Eclipse is fairly easy and also described in detail in many places on the Internet. As such, we're not going to go into great detail here but will list some of the useful features you get:

- Setting breakpoints in your code so execution stops there when the application runs
- Inspecting variables
- Stepping over and into lines of code
- Attaching the debugger to an application that's already running
- Disconnecting from an application you were attached to
- Viewing stack traces
- Viewing a list of threads
- Viewing LogCat

Figure 11–1 shows a sample screen layout of what you can do with the Debug perspective.

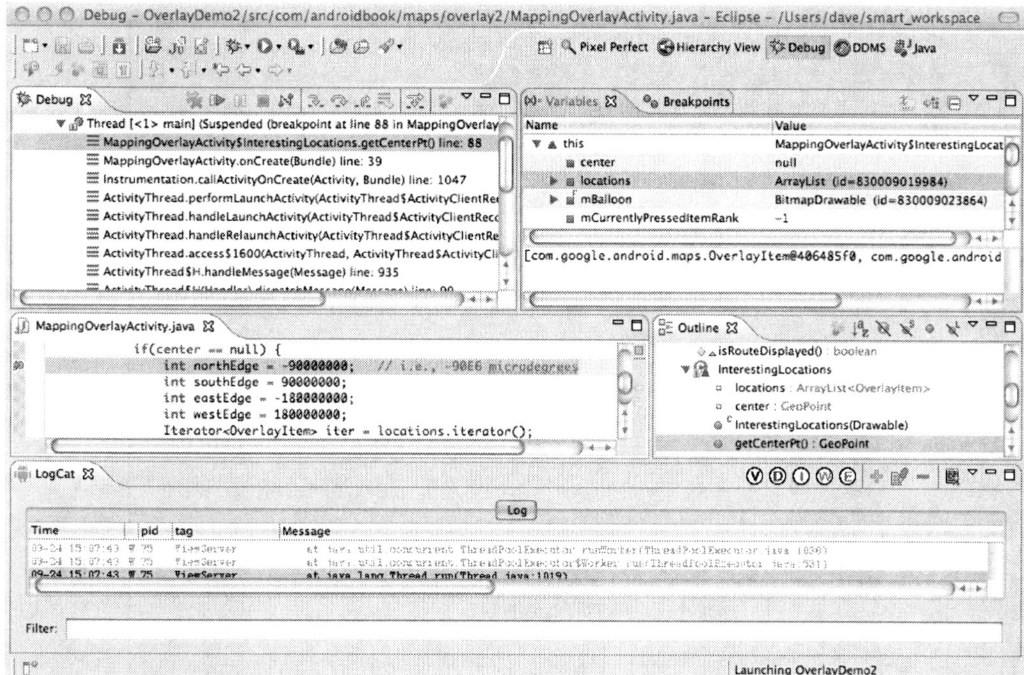

Figure 11–1. *The Debug perspective*

You can start debugging an application from the Java perspective (where you write the code) by right-clicking it and selecting Debug As ➤ Android Application; this will launch the application. You may need to switch to the Debug perspective to do the debugging.

The DDMS Perspective

DDMS stands for Dalvik Debug Monitor Server. This perspective gives you insight into the applications that are running on the emulator or device, allowing you to watch threads and memory, as well as gather statistics as the applications run. Figure 11–2 shows how it might look on your workstation. Although we'll use the term *device* for the remainder of this section, we mean device or emulator.

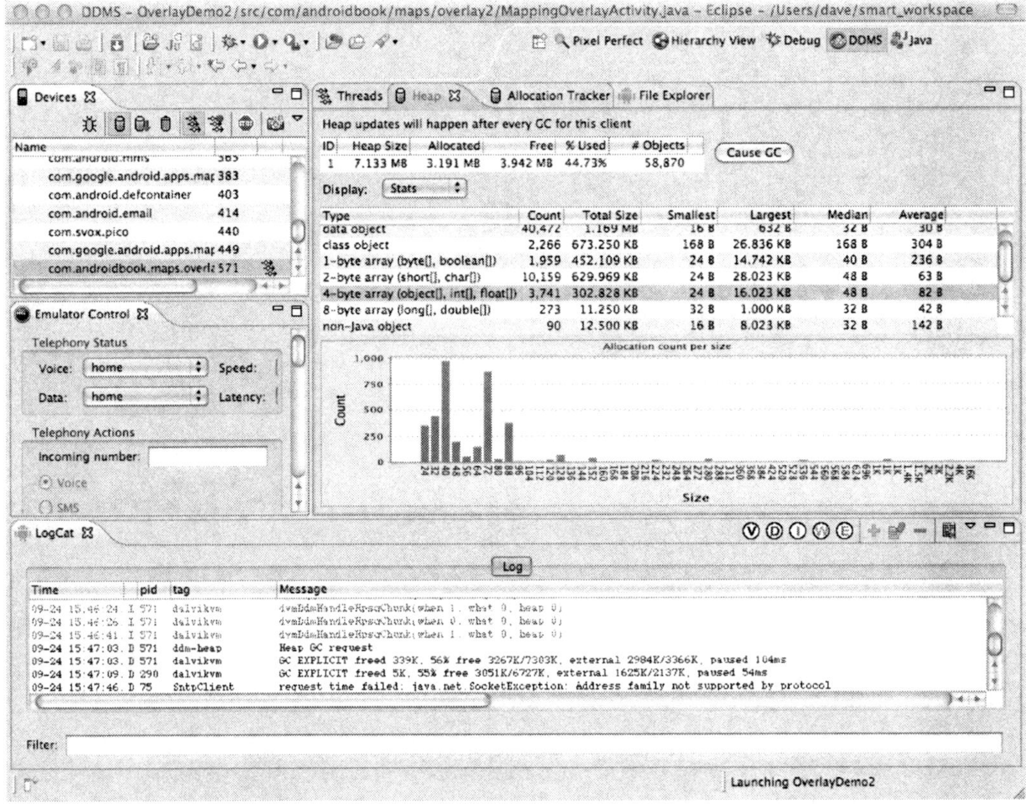

Figure 11–2. *The DDMS perspective*

In the upper-left corner of Figure 11–2, notice the Devices view. This will show you all devices connected to your workstation (you can have more than one device or emulator connected simultaneously) and, if you expand the device, all applications that are available for debugging. In our particular case, we're viewing an emulator, so the stock applications appear available for debugging (although we don't have source code). On a real device, you may see only a few applications, if any. Don't forget that if you're debugging a production application on a real device, you may need to tweak the `AndroidManifest.xml` file to set `android:debuggable` to `true`.

Within the Devices view, there are buttons to start debugging an application, as well as buttons to update the heap, dump a Heap and CPU Profiling Agent (HPROF) file, cause a garbage collection (GC), update the threads list, start method profiling, stop the process, or take a picture of the device's screen. We'll go into more details on each of these.

The little green bug button starts debugging on the selected application. Clicking it will take you to the Debug perspective we just described. What's nice about this option is that you can attach the debugger to a running application. You can get an application to a state where you want debugging to start, then select it and click this button. Now, as you continue to exercise the application, the breakpoints will cause execution to stop, and you can inspect variables and step through the code.

The next button is used to initiate inspection of the memory heap of a running process. You want your apps to use as little memory as feasible and not allocate memory too frequently. Similar to the debug button, you select the application you want to inspect and then click the Update Heap button. You'll want to pick only applications that you're actively debugging. In the Heap tab to the right (as shown in Figure 11–2), you can now click the Cause GC button to collect the information about memory in the heap. The summary results are displayed, with detailed results below. Then, for each type and size of allocated memory, you can see additional details about how memory is being used.

The Dump HPROF File button will do just that: give you an HPROF file. If you have installed the Eclipse Memory Analyzer (MAT) plug-in, this file will get processed and results will be displayed. This can be a powerful way to look for memory leaks. By default, an HPROF file will get opened in Eclipse. There is a Preferences setting for this under Android ➤ DDMS, where you can opt to save to a file instead.

The Update Threads button will populate the Threads tab to the right with the current set of threads from the selected application. This is a great way to watch as threads get created and destroyed, and to get some idea of what's happening at the thread level within your application. Below the list of threads, you can see where a thread is by following what looks like a stack trace (objects, source-code file reference, and line number).

The next button, Start Method Profiling, allows you to collect information about the methods within your application, including number of calls and timing information. You click this button, interact with your application, and then click the button again (it toggles between Start and Stop Method Profiling). When you click Stop Method Profiling, Eclipse will switch to the Traceview view, which we'll cover in the next section of this chapter.

The Stop button (which looks like a stop sign) allows you to stop the selected process. This is a hard application stop, not like pressing the Back button, which would only affect an activity. In this case, the application goes away.

The button that looks like a camera will capture the current state of the device screen, regardless of which application is selected in the Device view. You can then refresh the image, rotate it, save it, or copy it.

Finally, there is a menu next to the camera button that has all of the button features, and in addition a menu item called Reset adb. This one will restart the adb server that is talking to devices, in case things get out of sync and you can no longer see a device. This should (in effect) refresh the list of devices in this view. The other way you can reset the adb server is to use the following pair of commands in a tools window:

```
adb kill-server
adb start-server
```

At right in Figure 11–2 , there is a tab called Allocation Tracker. This allows you to start tracking individual memory allocations. After you click Start Tracking, you would exercise your application and then click Get Allocations. The list of memory allocations in that time period will be displayed, and you can click a specific allocation to see where it came from (class, method, source-code file reference, and line number). The Stop Tracking button is there so you can reset and start over.

DDMS also has a File Explorer and an Emulator Control so you can simulate incoming phone calls, SMS messages or GPS coordinates. The File Explorer allows you to browse through the file system on the device and even push or pull files between the device and your workstation. We'll talk more about the File Explorer in Chapter 24 in the section on SD Cards. The Emulator Control will be used in Chapters 22 and 23.

The Hierarchy View Perspective

In this perspective, you connect to a running instance of your application in an emulator (not on a real device), and you can then explore the views of the application, their structure, and their properties. You start by selecting which application you want. When it's selected and read, the view hierarchy will be displayed in various ways, as shown in Figure 11–3.

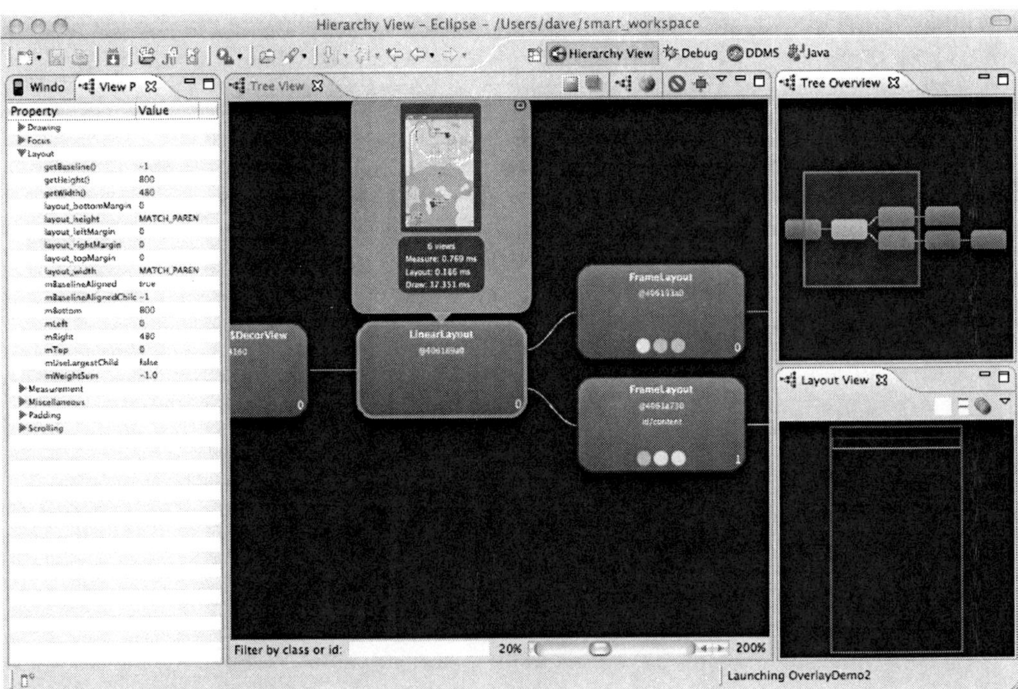

Figure 11–3. *The Hierarchy View perspective*

You can navigate around the structure, checking properties and making sure you don't have more views than you need. For example, if you have many nested layouts, you could probably replace these with a single `RelativeLayout`.

You probably noticed the three colored balls in the views in the center window. These correspond to a rating of that view's performance in terms of measuring, laying out, and drawing the view (which would include enclosed views). The colors are relative, so a red ball doesn't necessarily mean something is wrong, but it certainly means you should investigate.

Also notice the selected view and the information above it. Not only does it include a capture of the image of that view, but it also shows the absolute times for measuring, laying out, and drawing that view. These are valuable numbers for determining if you truly need to dig into that view and make improvements. Besides collapsing layouts as mentioned previously, you could alter how views are initialized and how much work is done for drawing. If your code is creating lots of objects, you might be able to reuse objects instead, to avoid that overhead. Use background threads or `AsyncTasks` or other techniques to do work that can take a long time.

Pixel Perfect View

Similar to the Hierarchy View, you can take the current screen image and bring it up in the Pixel Perfect View. This Eclipse plug-in gives you a zoomed-in image viewer, which allows you to see down to individual pixels and their associated color. What's interesting about this feature is that you can overlay another graphic (like a screen mockup) and compare it to the current screen. If you need to reproduce a specific look, this is a great way to see how well you're doing.

Traceview

We showed you earlier how you can collect statistics about the execution of methods in your application. Using DDMS, you can perform method profiling, after which the Traceview window will be shown with the results. Figure 11–4 shows what this looks like.

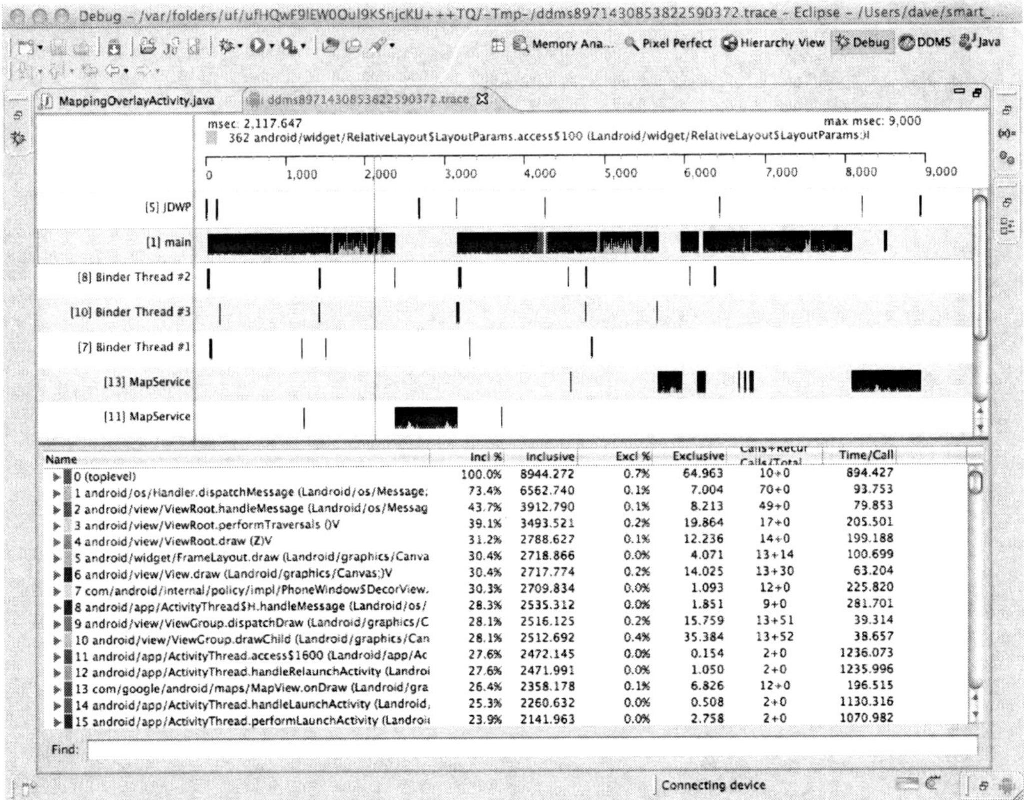

Figure 11–4. *Traceview*

Using the technique shown earlier, you get the results for all methods in the application. You can also get more specific tracing information for your Android application by using the `android.os.Debug` class, which provides a start-tracing method (`Debug.startMethodTracing("basename")`) and a stop-tracing method (`Debug.stopMethodTracing()`). Android will create a trace file on the device's SD card with the file name `basename.trace`. You put the start and stop code around what you want to trace, thus limiting how much data is collected into the trace file. You can then copy the trace file to your workstation and view the tracer output using the `traceview` tool included in the Android SDK tools directory, with the trace filename as the only argument to `traceview`. Chapter 24 has extensive coverage of the SD card and how to pull files from it.

You'll notice that the results of the analysis indicate what's being called, how often, and how much time is spent in each method. The breakdown is by thread, with color coding. Use this screen to look for methods that are taking up too much time or are being called too many times.

The adb Command

There are several other debugging tools you can use from a command line (or tools window). The Android Debug Bridge (adb) command allows you to install, update, and remove applications. You can start a shell on the emulator or device, and from there you can run the subset of Linux commands that Android provides. For example, you can browse the file system, list processes, read the log, and even connect to SQLite databases and execute SQL commands. We covered the SQLite commands in Chapter 4. As an example, the following command (in a tools window) will create a shell on an emulator:

```
adb -e shell
```

Notice the `-e` to specify an emulator. You would use `-d` if you were connecting to a device. Within an emulator shell, you will have elevated Linux privileges, whereas on a real device, you won't. This means you can poke around in SQLite databases within the emulator, but you won't be able to do so on a real device, even if it's your application!

Typing `adb` with no arguments will show you all of the available capabilities of the `adb` command.

The Emulator Console

Another powerful debugging technique is to run the Emulator Console, which obviously only works with the emulator. To get started once the emulator is up and running, you'd type the following in a tools window

```
telnet localhost port#
```

where `port#` is where the emulator is listening. The port# is typically displayed in the emulator window title and is often a value such as 5554. Once the emulator console has launched, you type commands to simulate GPS events, SMS messages, and even battery and network status changes. See the "References" section at the end of this chapter for a link to the Emulator Console commands and their usage.

StrictMode

Android 2.3 introduced a new debugging feature called `StrictMode`, and according to Google, this feature was used to make hundreds of improvements to the Google applications available for Android. So what does it do? It will report violations of policies related to threads and related to the virtual machine. If a policy violation is detected, you will get an alert, and that alert will include a stack trace to show you where your application was when the violation occurred. You can force a crash with the alert, or you can just log the alert and let your application carry on.

StrictMode Policies

There are two types of policies currently available with StrictMode. The first policy relates to threads and is intended mostly to run against the main thread (also known as the UI thread). It is not good practice to do disk reads and writes from the main thread, nor is it good practice to perform network accesses from the main thread. Google has added StrictMode hooks into the disk and network code; if you enable StrictMode for one of your threads, and that thread performs disk or network access, you can be alerted. You get to choose which aspects of the ThreadPolicy you want to alert on, and you get to choose the alert method. Some of the violations you can look for include custom slow calls, disk reads, disk writes, and network accesses. For alerts, you can choose to write to LogCat, display a dialog, flash the screen, write to the DropBox log file, or crash the application. The most common choices are to write to LogCat or to crash the application. Listing 11–1 shows a sample of what it takes to set up StrictMode for thread policies.

Listing 11–1. *Setting StrictMode's ThreadPolicy*

```
StrictMode.setThreadPolicy(new StrictMode.ThreadPolicy.Builder()
    .detectDiskReads()
    .detectDiskWrites()
    .detectNetwork()
    .penaltyLog()
    .build());
```

Note that the Builder class makes it really easy to set up StrictMode. The Builder methods that define the policy all return a reference to the Builder object, so these methods can be chained together as shown in Listing 11–1. The last method call, build(), returns a ThreadPolicy object that is the argument expected by the setThreadPolicy() method of StrictMode. Note that setThreadPolicy() is a static method, so you don't need to actually instantiate a StrictMode object. Internally, setThreadPolicy() uses the current thread for the policy, so subsequent thread actions will be evaluated against the ThreadPolicy and alerted as necessary. In this sample code, the policy is defined to alert on disk reads, disk writes, and network accesses with messages to LogCat. Instead of the specific detect methods, you could use the detectAll() method instead. You can also use different or additional penalty methods. For instance, you could use penaltyDeath() to cause the application to crash once it has written StrictMode alert messages to LogCat (as a result of the penaltyLog() method call).

Because you enable StrictMode on a thread, once you've enabled it, you don't need to keep enabling it. Therefore, you could enable StrictMode at the beginning of your main activity's onCreate() method, which runs on the main thread, and it would then be enabled for everything that happens on that main thread. Depending on what sorts of violations you want to look for, the first activity may be soon enough to enable StrictMode. You could also enable it in your application by extending the Application class and adding StrictMode setup to the application's onCreate() method. Anything that runs on a thread could conceivably set up StrictMode, but you certainly don't need to call the setup code from everywhere; once is enough.

Similar to ThreadPolicy, StrictMode has a VmPolicy. VmPolicy can check for memory leaks if a SQLite object is finalized before it has been closed, or if any Closeable object is finalized before it has been closed. A VmPolicy is created via a similar Builder class, as shown in Listing 11–2. One difference between a VmPolicy and a ThreadPolicy is that a VmPolicy can't alert via a dialog.

Listing 11–2. *Setting StrictMode's VmPolicy*

```
StrictMode.setVmPolicy(new StrictMode.VmPolicy.Builder()
    .detectLeakedSqlLiteObjects()
    .penaltyLog()
    .penaltyDeath()
    .build());
```

Turning Off StrictMode

Because the setup happens on a thread, StrictMode will find violations even as control flows from object to object to object. When a violation occurs, you may be surprised to realize that the code is running on the main thread, but the stack trace is there to help you follow along to uncover how it happened. You can then take steps to resolve the issue by moving that code to its own background thread. Or you could decide that it's okay to leave things the way they are. It's up to you. Of course, you will probably want to turn off StrictMode when your application goes to production; you don't want it crashing on your users because of an alert.

There are a couple of ways to go about turning off StrictMode for a production application. The most straightforward way is to remove the calls, but that makes it more difficult to continue to do development on the application. You could always define an application-level boolean and test it before calling the StrictMode code. Setting the value of the boolean to false just before you release your application to the world would effectively disable StrictMode. A more elegant method is to take advantage of the application's debug mode, as defined in AndroidManifest.xml. One of the attributes for the <application> tag in this file is android:debuggable. This value can be set to true when you want to debug an application, and it results in the ApplicationInfo object getting a flag set, which you can then read in code. Listing 11–3 shows how you might use this to your advantage, so that when the application is in debug mode, StrictMode is active (and when the application is not in debug mode, StrictMode is not active).

Listing 11–3. *Setting StrictMode Only for Debugging*

```
// Return if this application is not in debug mode
ApplicationInfo appInfo = context.getApplicationInfo();
int appFlags = appInfo.flags;
if ((appFlags & ApplicationInfo.FLAG_DEBUGGABLE) != 0) {
    // Do StrictMode setup here
}
```

Remember that ADT will set this attribute to true when launching a development version of an application in the emulator or on a device, which would therefore enable StrictMode in the previous code. When you're exporting your application to create a production version, ADT will set it to false.

All of this is well and good, but it doesn't work on Android prior to version 2.3. To use StrictMode explicitly, you have to deploy to an environment running Android 2.3 or later. If you deploy to anything older than 2.3, you're going to get verify errors because this class just doesn't exist prior to Android 2.3.

StrictMode with Old Android Versions

To use StrictMode with older versions of Android (prior to 2.3), you can utilize reflection techniques so you can indirectly invoke StrictMode methods if they are available and fail gracefully if they are not. The simplest thing you can do is shown in Listing 11–4; you invoke a special method created just for dealing with older versions of Android.

Listing 11–4. *Using StrictMode with Reflection*

```
try {
   Class sMode = Class.forName("android.os.StrictMode");
   Method enableDefaults = sMode.getMethod("enableDefaults");
   enableDefaults.invoke(null);
}
catch(Exception e) {
    // StrictMode not supported on this device, punt
    Log.v("StrictMode", "... not supported. Skipping...");
}
```

This determines whether the StrictMode class exists and, if it does, invokes the enableDefaults() method on it. If StrictMode is not found, your catch block is invoked with a ClassNotFoundException. You shouldn't get any exceptions if StrictMode does exist, because enableDefaults() is one of its methods. The enableDefaults() method sets up StrictMode to detect everything and to write any violations to LogCat. Because this StrictMode method you're calling is a static method, you specify null as the first argument when you invoke it.

There may be times when you don't want all violations to be reported. It's perfectly fine to set up StrictMode on threads other than the main thread, and that's when you might choose to alert on less than everything. For example, you may be fine with doing disk reads on the thread you're monitoring. If this is the case, you can either not call detectDiskReads() on the Builder, or you could call detectAll() and then permitDiskReads() on the Builder. There are similar permit methods for the other policy options.

If you want to use StrictMode but your application runs on Android versions prior to 2.3, is there a way? Of course there is! If StrictMode is not available for your application, a VerifyError will be thrown if you try to access it. If you wrap StrictMode in a class and then catch the error, you can ignore when StrictMode is not available and get it when it is. Listing 11–5 shows a sample StrictModeWrapper class that you can add to your application, and Listing 11–6 shows what the code inside your application would look like to set up StrictMode.

Listing 11–5. *Using `StrictMode` on Pre-2.3 Android*

```
public class StrictModeWrapper {
    public static void init(Context context) {
        // check if android:debuggable is set to true
        int appFlags = context.getApplicationInfo().flags;
        if ((appFlags & ApplicationInfo.FLAG_DEBUGGABLE) != 0) {
            StrictMode.setThreadPolicy(
              new StrictMode.ThreadPolicy.Builder()
                .detectDiskReads()
                .detectDiskWrites()
                .detectNetwork()
                .penaltyLog()
                .build());
            StrictMode.setVmPolicy(
              new StrictMode.VmPolicy.Builder()
                .detectLeakedSqlLiteObjects()
                .penaltyLog()
                .penaltyDeath()
                .build());
        }
    }
}
```

You can see how this is just like your code from before, except that you're combining everything you've learned so far. And finally, in order to set up `StrictMode` from your application, you only need to add the code shown in Listing 11–6.

Listing 11–6. *Invoking `StrictMode` with Pre-2.3 Android*

```
try {
    StrictModeWrapper.init(this);
}
catch(Throwable throwable) {
    Log.v("StrictMode", "... is not available. Punting...");
}
```

Note that `this` is the local context of whatever object you're in, such as from within the `onCreate()` method of your main activity. The code in Listing 11–6 will work on any release of Android. This second method of conditionally invoking `StrictMode` gives you more control, because it's easier to call all the methods you want and skip features you don't want. The earlier technique only used `enableDefaults()`, so reflection on one method is not too difficult.

StrictMode Exercise

As a reader exercise, go into Eclipse and make a copy of one of the applications you've developed so far. You'll have to pick a build target of 2.3 or later so it will find the `StrictMode` class. But set `minSDKVersion` to something below 2.3. Then add a new class under the `src` folder using the code in Listing 11–5. Within the `onCreate()` method of an activity that launches first, add code like that in Listing 11–6; run the program on a pre-2.3 version of Android in the emulator, and then on Android 2.3 or later in the emulator. When `StrictMode` is not available, you should see LogCat messages that indicate `StrictMode` is not present, but the application should continue to run well. When

`StrictMode` is available, you might see the occasional violation messages in LogCat as you use your application. If you try this on a pre-2.3 sample application such as NotePad, you will very likely see policy violations.

References

Here are some helpful references to topics you may wish to explore further:

- `http://developer.android.com/guide/developing/tools/index.html`: Developer documentation for the Android debugging tools described previously.
- `http://developer.android.com/guide/developing/devices/emulator.html#console`: Syntax and usage for the Emulator Console commands. This allows you to use a command-line interface to simulate events for an application running in the emulator.
- `www.eclipse.org/mat/`: The Eclipse project called Memory Analyzer (MAT). You can use this plug-in to read HPROF files as collected by that DDMS feature.

Summary

Here is what has been covered in this chapter:

- How to set up Eclipse and your device for debugging.
- The Debug perspective, which lets you stop an application to inspect variable values and also step through code line by line.
- The DDMS perspective, which has quite a few tools for investigating threads, memory, and method calls, as well as taking snapshots of the screen and generating events to send to the emulator.
- Resetting the adb server from DDMS and from the command line.
- The Hierarchy view, which exposes the view structure of a running application and includes metrics to help you tune and troubleshoot the application.
- Traceview, which shows the methods that are called while an application is running, as well as statistics to help you identify problem methods that need attention for a better user experience.
- The `adb` command, which you can use to log in to a device and look around.
- The Emulator Console, which is a great way to talk to an emulator from a command line. Imagine the scripting possibilities.

- `StrictMode`, a special class for verifying that your application is not doing non-recommended things like disk or network I/O from the main thread.

Interview Questions

Here are some questions you can ask yourself to solidify your understanding of this topic:

1. True or false: if you want to debug an application, you must explicitly set the `android:debuggable` attribute to `true` in the `<application>` tag of the `AndroidManifest.xml` file.
2. Name four things you can do with your application while using the Eclipse Debug perspective.
3. Is it possible to connect more than one device and/or emulator to Eclipse at the same time? If so, where do you select which application you want to work with?
4. Which DDMS feature do you use to get statistics for an application's current memory allocations?
5. How do you determine how many threads are running in your application?
6. How do you find out the number of times a particular method is called in your application, and what the time of execution is within that method?
7. Where do you go to capture a picture of a device's screen?
8. What Eclipse perspective is used to analyze the structure of an application's views?
9. What do the three colored balls mean in this perspective? Does yellow mean you have a big problem? Does red?
10. If you see a yellow or red ball and want to know how bad the situation is, what should you do to see the actual numeric metric values?
11. If you want to look at method profiles, but you don't want to see all the methods for the entire application, what do you do?
12. How do you create a Linux shell inside a running emulator?
13. Can you also do this on a real device? If so, are there any limitations to what you can do on a real device?
14. How do you figure out the port number of an emulator so you can connect to it using the Emulator Console?
15. What two main things does `StrictMode` check for?

Chapter 12

Responding to Configuration Changes

We've covered a fair bit of material so far, and now seems like a good time to cover configuration changes. When an application is running on a device, and the device's configuration changes (for example, is rotated 90 degrees), your application needs to respond accordingly. The new configuration will most likely look different from the previous configuration. For example, switching from portrait to landscape mode means the screen went from being tall and narrow to being short and wide. The UI elements (buttons, text, lists, and so on) will need to be rearranged, resized, or even removed to accommodate the new configuration.

In Android, a configuration change causes the current activity to go away and be re-created. The application itself keeps on running, but it has the opportunity to change how the activity is displayed in response to the configuration change.

Be aware that configuration changes can take on many forms, not just device rotation. If a device gets connected to a dock, that's also a configuration change. So is changing the language of the device. Whatever the new configuration is, as long as you've designed your activity for that configuration, Android takes care of most everything to transition to it, giving the user a seamless experience.

This chapter will take you through the process of a configuration change, from the perspectives of both activities and fragments. We'll show you how to design your application for those transitions and how to avoid traps that could cause your application to crash or misbehave.

The Configuration Change Process

The Android operating system keeps track of the current configuration of the device it's running on. Configuration includes lots of factors, and new ones get added all the time. For example, if a device is plugged into a docking station, that represents a change in the device configuration. When a configuration change is detected by Android, callbacks are invoked in running applications to tell them a change is occurring, so an application can properly respond to the change. We'll discuss those callbacks a little later, but for now let's refresh your memory with regard to resources.

One of the great features of Android is that resources get selected for your activity based on the current configuration of the device. You don't need to write code to figure out which configuration is active; you just access resources by name, and Android gets the appropriate resources for you. If the device is in portrait mode and your application requests a layout, you get the portrait layout. If the device is in landscape mode, you get the landscape layout. The code just requests a layout without specifying which one it should get. This is powerful because as new configuration factors get introduced, or new values for configuration factors, the code stays the same. All a developer needs to do is decide if new resources need to be created, and create them as necessary for the new configuration. Then, when the application goes through a configuration change, Android provides the new resources to the application, and everything continues to function as desired.

Because of a great desire to keep things simple, Android destroys the current activity when the configuration changes, and creates a new one in its place. This might seem rather harsh, but it's not. It would be a bigger challenge to take a running activity and figure out which parts would stay the same and which would not, and then only work with the pieces that need to change.

An activity that's about to be destroyed is properly notified first, giving you a chance to save anything that needs to be saved. When the new activity gets created, it has the opportunity to restore state using data from the previous activity. For a good user experience, obviously you do not want this save and restore to take very long.

It's fairly easy to save any data that you need saved and then let Android throw away the rest and start over, as long as the design of the application and its activities is such that activities don't contain a lot of non-UI stuff that would take a long time to re-create. Therein lies the secret to successful configuration change design: do not put "stuff" inside an activity that cannot be easily re-created during a configuration change.

Keep in mind that our application is not being destroyed, so anything that is in the application context, and not a part of our current activity, will still be there for the new activity. Singletons will still be available, as well as any background threads we might have spun off to do work for our application. Any databases or content providers that we were working with will also still be around. Taking advantage of these makes configuration changes quick and painless. Keep data and business logic outside of activities if you can.

The configuration change process is somewhat similar between activities and fragments. When an activity is being destroyed and re-created, the fragments within that activity get destroyed and re-created. What we need to worry about then is state information about our fragments and activity, such as data currently being displayed to the user, or internal values that we want to preserve. We will save what we want to keep, and pick it up again on the other side when the fragments and activities are being re-created. The larger chunks of data are better off in objects separate from our fragments and activities, so they'll naturally hang around during configuration changes.

The Destroy/Create Cycle of Activities

There are three callbacks to be aware of when dealing with configuration changes in activities:

- onSaveInstanceState()
- onCreate()
- onRestoreInstanceState()

The first is the callback that Android will invoke when it detects that a configuration change is happening. The activity has a chance to save state that it wants to restore when the new activity gets created at the end of the configuration change. The onSaveInstanceState() callback will be called prior to the call to onStop(). Whatever state exists can be accessed and saved into a Bundle object. This object will get passed in to both of the other callbacks (onCreate() and onRestoreInstanceState()) when the activity is re-created. You only need to put logic in one or the other to restore your activity's state.

The default onSaveInstanceState() callback does some nice things for you. For example, it goes through the currently active view hierarchy and saves the values for each view that has an android:id. This means if you have an EditText view that has received some user input, that input will be available on the other side of the activity destroy/create cycle to populate the EditText before the user gets control back. You do not need to go through and save this state yourself. If you do override onSaveInstanceState(), be sure to call super.onSaveInstanceState() with the bundle object so it can take care of this for you. It's not the views that are saved, only the attributes of their state that should persist across the destroy/create boundary.

To save data in the bundle object, use methods such as putInt() for integers and putString() for strings. There are quite a few methods in the android.os.Bundle class; you are not limited to integers and strings. For example, putParcelable() can be used to save complex objects. Each put is used with a string key, and you will retrieve the value later using the same key used to put the value in. A sample onSaveInstanceState() might look like Listing 12–1.

Listing 12–1. *Sample onSaveInstanceState()*

```
@Override
public void onSaveInstanceState(Bundle icicle) {
    super.onSaveInstanceState(icicle);
    icicle.putInt("counter", 1);
}
```

Sometimes the bundle is called icicle because it represents a small frozen piece of an activity. In this sample, you only save one value, and it has a key of counter. You could save more values by simply adding more put statements to this callback. The counter value in this example is somewhat temporary because if the application is completely destroyed, the current value will be lost. This could happen if the user turned off their device, for example. In Chapter 13, you'll learn about ways to save values more permanently. This instance state is only meant to hang onto values while the application is running this time. Do not use this mechanism for state that is important to keep for a longer term.

To restore activity state, you access the bundle object to retrieve values that you believe are there. Again, you use methods of the Bundle class such as getInt() and getString() with the appropriate key passed to tell which value you want back. If the key does not exist in the Bundle, a value of 0 or null is passed back (depending on the type of the object being requested). Or you can provide a default value in the appropriate getter method. Listing 12–2 shows a sample onRestoreInstanceState() callback.

Listing 12–2. *Sample onRestoreInstanceState()*

```
@Override
public void onRestoreInstanceState(Bundle icicle) {
    super.onRestoreInstanceState(icicle);
    int someInt = icicle.getInt("counter", -1);
    // Now go do something with someInt to restore the
    // state of the activity. -1 is the default if no
    // value was found.
}
```

It's up to you whether you restore state in onCreate() or in onRestoreInstanceState(). Many applications will restore state in onCreate() because that is where a lot of initialization is done. One reason to separate the two would be if you're creating an activity class that could be extended. The developers doing the extending might find it easier to just override onRestoreInstanceState() with the code to restore state, as compared to having to override all of onCreate().

What's very important to note here is that you need to be very concerned with references to activities and views and other objects that need to be garbage-collected when the current activity is fully destroyed. If you put something into the saved bundle that refers back to the activity being destroyed, that activity can't be garbage collected. This is very likely a memory leak that could grow and grow until your application crashes. Objects to avoid in bundles include Drawables, Adapters, Views, and anything else that is tied to the activity context. Instead of putting a Drawable into the bundle, serialize the bitmap. Or better yet, manage the bitmaps outside of the activity and fragment instead of inside. Add some sort of reference to the bitmap to the bundle. When it comes time to re-create any Drawables for the new fragment, use the reference to access the outside bitmaps to regenerate your Drawables.

The Destroy/Create Cycle of Fragments

The destroy/create cycle for fragments is very similar to that of activities. A fragment in the process of being destroyed and re-created will have its onSaveInstanceState() callback called, allowing the fragment to save values in a Bundle object for later. One difference is that four fragment callbacks receive this Bundle object when a fragment is being re-created: `onInflate()`, `onCreate()`, `onCreateView()`, and `onActivityCreated()`. This gives us lots of opportunities to rebuild the internal state of our reconstructed fragment from its previous state.

Android only guarantees that `onSaveInstanceState()` will be called for a fragment sometime before `onDestroy()`. That means the view hierarchy may or may not be attached when `onSaveInstanceState()` is called. Therefore, don't count on traversing the view hierarchy inside of `onSaveInstanceState()`. For example, if the fragment is on the fragment backstack, no UI will be showing, so no view hierarchy will exist. This is OK of course because if no UI is showing, there is no need to attempt to capture the current values of views to save them. You need to check if a view exists before trying to save its current value, and not consider it an error if the view does not exist.

Just like with activities, be careful not to include items in the bundle object that refer to an activity or to a fragment that might not exist later when this fragment is being re-created. Keep the size of the bundle as small as possible, and as much as possible store long-lasting data outside of activities and fragments and simply refer to it from your activities and fragments. Then your destroy/create cycles will go that much faster, you'll be much less likely to create a memory leak, and your activity and fragment code should be easier to maintain.

Using FragmentManager to Save Fragment State

Fragments have another way to save state, in addition to, or instead of, Android notifying the fragments that their state should be saved. The `FragmentManager` class has a `saveFragmentInstanceState()` method that can be called to generate an object of the class `Fragment.SavedState`. The methods mentioned in the previous sections for saving state do so within the internals of Android. While we know that the state is being saved, we do not have any direct access to it. This method of saving state gives you an object that represents the saved state of a fragment and allows you to control if and when a fragment is created from that state.

The way to use a `Fragment.SavedState` object to restore a fragment is through the `setInitialSavedState()` method of the `Fragment` class. In Chapter 8, you learned that it is best to create new fragments using a static factory method (for example, `newInstance()`). Within this method, you saw how a default constructor is called and then an arguments bundle is attached. You could instead call the `setInitialSavedState()` method to set it up for restoration to a previous state.

There are a few caveats you should know about this method of saving fragment state:

- The fragment to be saved must currently be attached to the fragment manager.
- A new fragment created using this saved state must be the same class type as the fragment it was created from.
- The saved state cannot contain dependencies on other fragments. Other fragments may not exist when the saved fragment is re-created.

Using setRetainInstance on a Fragment

A fragment can avoid being destroyed and re-created on a configuration change. If the `setRetainInstance()` method is called with an argument of `true`, the fragment will be retained in the application when its activity is being destroyed and re-created. The fragment's `onDestroy()` callback will not be called, nor will `onCreate()`. The `onDetach()` callback will be called because the fragment must be detached from the activity that's going away, and `onAttach()` and `onActivityCreated()` will also be called for the same reason because the fragment is attached to a new activity. This only works for fragments that are not on the backstack. It is especially useful for fragments that do not have a UI.

Deprecated Configuration Change Methods

A couple of methods on `Activity` have been deprecated, so you should no longer use them:

- `getLastNonConfigurationInstance()`
- `onRetainNonConfigurationInstance()`

These methods previously allowed you to save an arbitrary object from an activity that was being destroyed, to be passed to the next instance of the activity that was being created. Although they were useful, you should now use the methods described earlier instead to manage data between instances of activities in the destroy/create cycle.

References

Here are some helpful references to topics you may wish to explore further:

- `www.androidbook.com/proandroid4/projects`: A list of downloadable projects related to this book. For this chapter, look for a ZIP file called `ProAndroid4_Ch12_ConfigChanges.zip`. This ZIP file contains all the projects from this chapter, listed in separate root directories. There is also a `README.TXT` file that describes exactly how to import projects into Eclipse from one of these ZIP files.
- `http://developer.android.com/guide/topics/fundamentals/activities.html#SavingActivityState`: The Android Developer's Guide, which discusses saving and restoring state.

Summary

Let's conclude this chapter by quickly enumerating what you have learned about handling configuration changes:

- Activities get destroyed and re-created during configuration changes.

- Avoid putting lots of data and logic into activities so configuration changes occur quickly.
- Let Android provide the appropriate resources.
- Use singletons to hold data outside of activities to make it easier to destroy and re-create activities during configuration changes.
- Take advantage of the default `onSaveInstanceState()` callback to save UI state on views with `android:ids`.
- If a fragment can survive with no issues across an activity destroy and create cycle, use `setRetainInstance()` to tell Android it doesn't need to destroy and create the fragment.

Interview Questions

You can use the following questions as a guide to consolidate your understanding of this chapter:

1. True or false. All configuration changes are due to device rotation.
2. Configuration changes are made much easier due to what basic Android feature?
3. Activities are notified about a configuration change through what callback method?
4. What does the default configuration change callback method do?
5. What classes of objects should not be saved in a bundle object when saving state?
6. What sort of state information should not be saved in a bundle object for configuration changes?
7. What type of fragment would be most appropriate for the `setRetainInstance(true)` method call?

Chapter 13

Working with Preferences and Saving State

Android offers a robust and flexible framework for dealing with preferences. *And by Preferences*, we mean those feature choices that a user makes and saves to customize an application to their liking. For example, if the user wants a notification via a ringtone or vibration or not at all, that is a preference the user saves; the application remembers the choice until the user changes it. Android provides simple APIs that hide the reading and persisting of preferences. It also provides prebuilt user interfaces that you can use to let the user make preference selections. Because of the power built in to the Android preferences framework, we can also use preferences for more general-purpose storing of application state, to allow our application to pick up where it left off for example, should our application go away and come back later. As another example, a game's high scores could be stored as preferences, although you'll want to use your own UI to display them.

Before Android 3.0, preferences were managed a certain way, but then things changed. With the extra real estate of a tablet screen, preferences can be visually arranged much more nicely than on a phone's screen. Although the foundational bits of preferences (the different types of preferences) stayed the same, the way they are displayed has changed quite a bit. This chapter covers the foundational aspects of preferences and shows how pre-3.0 preferences are displayed. The chapter ends with coverage of `PreferenceFragment` and the new capabilities of `PreferenceActivity`.

Exploring the Preferences Framework

Before we dig into Android's preferences framework, let's establish a scenario that would require the use of preferences and then explore how we would go about addressing it. Suppose you are writing an application that provides a facility to search for airline flights. Moreover, suppose that the application's default setting is to display flights based on the lowest cost, but the user can set a preference to always sort flights by the least number of stops or by a specific airline. How would you go about doing that?

Understanding ListPreference

Obviously, you would have to provide a UI for the user to view the list of sort options. The list would contain radio buttons for each option, and the default (or current) selection would be preselected. To solve this problem with the Android preferences framework requires very little work. First, you would create a preferences XML file to describe the preference, and then you use a prebuilt activity class that knows how to show and persist preferences. Listing 13–1 shows the details.

> **NOTE:** We will give you a URL at the end of the chapter that you can use to download projects from this chapter. This will allow you to import these projects into Eclipse directly.

Listing 13–1. *The Flight-Options Preferences XML File and Associated Activity Class*

```
<?xml version="1.0" encoding="utf-8"?>
<!-- This file is /res/xml/flightoptions.xml -->
<PreferenceScreen
        xmlns:android="http://schemas.android.com/apk/res/android"
    android:key="flight_option_preference"
    android:title="@string/prefTitle"
    android:summary="@string/prefSummary">

  <ListPreference
    android:key="@string/selected_flight_sort_option"
    android:title="@string/listTitle"
    android:summary="@string/listSummary"
    android:entries="@array/flight_sort_options"
    android:entryValues="@array/flight_sort_options_values"
    android:dialogTitle="@string/dialogTitle"
    android:defaultValue="@string/flight_sort_option_default_value" />

</PreferenceScreen>


public class FlightPreferenceActivity extends PreferenceActivity
{
    @Override
    protected void onCreate(Bundle savedInstanceState) {
        super.onCreate(savedInstanceState);
        addPreferencesFromResource(R.xml.flightoptions);
    }
}
```

Listing 13–1 contains an XML fragment that represents the flight-option preference setting. The listing also contains an activity class that loads the preferences XML file. Let's start with the XML. Android provides an end-to-end preferences framework. This means the framework lets you define your preferences, display the setting(s) to the user, and persist the user's selection to the data store. You define your preferences in XML under `/res/xml/`. To show preferences to the user, you write an activity class that extends a predefined Android class called `android.preference.PreferenceActivity` and use the addPreferencesFromResource() method to add the resource to the activity's resource collection. The framework takes care of the rest (displaying and persisting).

In this flight scenario, you create a file called flightoptions.xml at /res/xml/flightoptions.xml. You then create an activity class called FlightPreferenceActivity that extends the android.preference.PreferenceActivity class. Next, you call addPreferencesFromResource(), passing in R.xml.flightoptions. Note that the preference resource XML points to several string resources. To ensure compilation, you need to add several string resources to your project. We will show you how to do that shortly. For now, have a look at the UI generated by Listing 13–1 (see Figure 13–1).

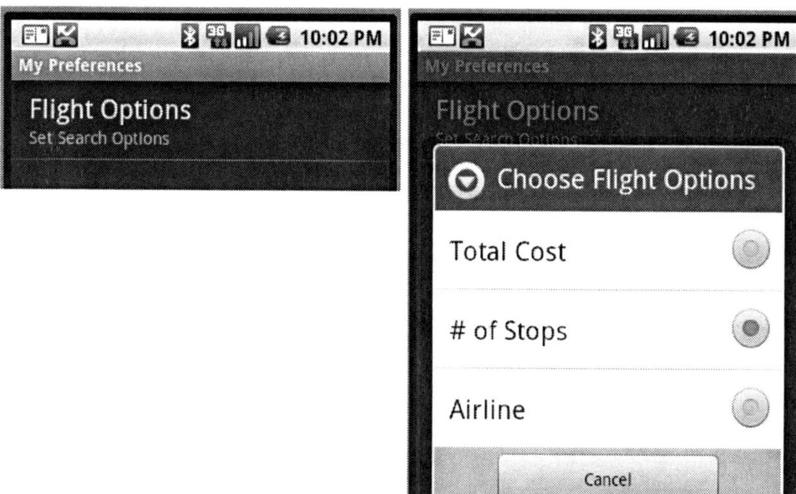

Figure 13–1. *The flight-options preference UI*

Figure 13–1 contains two views. The view on the left is called a *preference screen*, and the UI on the right is a *list preference*. When the user selects Flight Options, the Choose Flight Options view appears as a modal dialog with radio buttons for each option. The user selects an option, which immediately saves that option and closes the view. When the user returns to the options screen, the view reflects the saved selection from before.

The XML code in Listing 13–1 defines PreferenceScreen and then creates ListPreference as a child. For PreferenceScreen, you set three properties: key, title, and summary. key is a string you can use to refer to the item programmatically (similar to how you use android:id); title is the screen's title (Flight Options); and summary is a description of the screen's purpose, shown below the title in a smaller font (Set Search Options, in this case). For the list preference, you set key, title, and summary, as well as attributes for entries, entryValues, dialogTitle, and defaultValue. Table 13–1 summarizes these attributes.

Table 13–1. *A Few Attributes of* `android.preference.ListPreference`

Attribute	Description
android:key	A name or key for the option (such as selected_flight_sort_option).
android:title	The title of the option.
android:summary	A short summary of the option.
android:entries	The array of text items that the option can be set to.
android:entryValues	The key, or value, for each item. Note that each item has some text and a value. The text is defined by entries, and the values are defined by entryValues.
android:dialogTitle	The title of the dialog—used if the view is shown as a modal dialog.
android:defaultValue	The default value of the option from the list of items.

Listing 13–2 contains the source of several other files for the example, which we'll be talking about soon.

Listing 13–2. *Other Files from Our Example*

```
<?xml version="1.0" encoding="utf-8"?>
<!-- This file is /res/values/arrays.xml -->
<resources>
<string-array name="flight_sort_options">
    <item>Total Cost</item>
    <item># of Stops</item>
    <item>Airline</item>
</string-array>
<string-array name="flight_sort_options_values">
    <item>0</item>
    <item>1</item>
    <item>2</item>
</string-array>
</resources>


<?xml version="1.0" encoding="utf-8"?>
<!-- This file is /res/values/strings.xml -->
<resources>
    <string name="app_name">Preferences Demo</string>
    <string name="prefTitle">My Preferences</string>
    <string name="prefSummary">Set Flight Option Preferences</string>
    <string name="flight_sort_option_default_value">1</string>
    <string name="dialogTitle">Choose Flight Options</string>
    <string name="listSummary">Set Search Options</string>
    <string name="listTitle">Flight Options</string>
    <string name="selected_flight_sort_option">
        selected_flight_sort_option</string>
```

```
    <string name="menu_prefs_title">Settings</string>
</resources>


// This file is MainActivity.java
public class MainActivity extends Activity {
    private TextView tv = null;
    private Resources resources;

    /** Called when the activity is first created. */
    @Override
    public void onCreate(Bundle savedInstanceState) {
        super.onCreate(savedInstanceState);
        setContentView(R.layout.main);

        resources = this.getResources();

        tv = (TextView)findViewById(R.id.text1);

        setOptionText();
    }

    @Override
    public boolean onCreateOptionsMenu(Menu menu)
    {
       MenuInflater inflater = getMenuInflater();
       inflater.inflate(R.menu.mainmenu, menu);
       return true;
    }

    @Override
    public boolean onOptionsItemSelected (MenuItem item)
    {
       if (item.getItemId() == R.id.menu_prefs)
       {
           // Launch to our preferences screen.
           Intent intent = new Intent()
                   .setClass(this,
              com.androidbook.preferences.sample.FlightPreferenceActivity.class);
           this.startActivityForResult(intent, 0);
       }
       return true;
    }

    @Override
    public void onActivityResult(int reqCode, int resCode, Intent data)
    {
        super.onActivityResult(reqCode, resCode, data);
        setOptionText();
    }

    private void setOptionText()
    {
        SharedPreferences prefs =
                PreferenceManager.getDefaultSharedPreferences(this);
//      This is the other way to get to the shared preferences:
//      SharedPreferences prefs = getSharedPreferences(
```

```
//              "com.androidbook.preferences.sample_preferences", 0);
        String option = prefs.getString(
                resources.getString(R.string.selected_flight_sort_option),
                resources.getString(R.string.flight_sort_option_default_value));
        String[] optionText = resources.getStringArray(R.array.flight_sort_options);

        tv.setText("option value is " + option + " (" +
                optionText[Integer.parseInt(option)] + ")");
    }
}
```

When you run this application, you will first see a simple text message that says "option value is 1 (# of Stops)." Click the Menu button and then Settings to get to the `PreferenceActivity`. Click the back arrow when you're finished, and you will see any changes to the option text immediately.

The first file to talk about is `/res/values/arrays.xml`. This file contains the two string arrays that we need to implement the option choices. The first array holds the text to be displayed, and the second holds the values that we'll get back in our method calls plus the value that gets stored in the preferences XML file. For our purposes, we chose to use array index values 0, 1, and 2 for `flight_sort_options_values`. We could use any value that helps us run the application. If our option was numeric in nature (for example, a countdown timer starting value), then we could have used values such as 60, 120, 300, and so on. The values don't need to be numeric at all as long as they make sense to the developer; the user doesn't see these values unless you choose to expose them. The user only sees the text from the first string array `flight_sort_options`.

As we said earlier, the Android framework also takes care of persisting preferences. For example, when the user selects a sort option, Android stores the selection in an XML file within the application's `/data` directory on the device (see Figure 13–2).

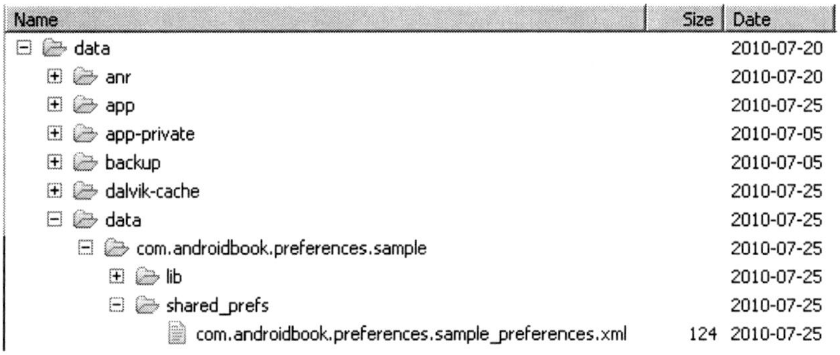

Name	Size	Date
data		2010-07-20
anr		2010-07-20
app		2010-07-25
app-private		2010-07-05
backup		2010-07-05
dalvik-cache		2010-07-25
data		2010-07-25
com.androidbook.preferences.sample		2010-07-25
lib		2010-07-25
shared_prefs		2010-07-25
com.androidbook.preferences.sample_preferences.xml	124	2010-07-25

Figure 13–2. *Path to an application's saved preferences*

> **NOTE:** You will only be able to inspect shared preferences files in the emulator. On a real device, the shared preferences files are not readable due to Android security.

The actual file path is /data/data/*[PACKAGE_NAME]*/shared_prefs/*[PACKAGE_NAME]*_preferences.xml. Listing 13–3 shows the com.androidbook.preferences.sample_preferences.xml file for our example.

Listing 13–3. *Saved Preferences for Our Example*

```
<?xml version='1.0' encoding='utf-8' standalone='yes' ?>
<map>
    <string name="selected_flight_sort_option">1</string>
</map>
```

You can see that for a list preference, the preferences framework persists the selected item's value using the list's key attribute. Note also that the selected item's *value* is stored—not the text. A word of caution here: because the preferences XML file is storing only the value and not the text, should you ever upgrade your application and change the text of the options or add items to the string arrays, any value stored in the preferences XML file should still line up with the appropriate text after the upgrade. The preferences XML file is kept during the application upgrade. If the preferences XML file had a "1" in it, and that meant "# of Stops" before the upgrade, it should still mean "# of Stops" after the upgrade.

The next file we are interested in is /res/values/strings.xml. We added several strings for our titles, summaries, and menu items. There are two strings to pay particular attention to. The first is flight_sort_option_default_value. We set the default value to 1 to represent "# of Stops" in our example. It is usually a good idea to choose a default value for each option. If you don't choose a default value and no value has yet been chosen, the methods that return the value of the option will return null. Your code would have to deal with null values in this case. The other interesting string is selected_flight_sort_option. Strictly speaking, the user is not going to see this string, so we don't need to put it inside strings.xml to provide alternate text for other languages. However, because this string value is a key used in the method call to retrieve the value, by creating an ID out of it, we can ensure at compile time that we didn't make a typographical error on the key's name.

Next up is the source code for our MainActivity. This is a basic activity that gets a reference to the preferences and a handle to a TextView and then calls a method to read the current value of our option to set it into the TextView. The layout for our application is not shown here but is simply a TextView to display a message about the current preference setting. Next, We set up our menu and the menu callback. Within the menu callback, we launch an Intent for the FlightPreferenceActivity. Launching an intent for our preferences is the best way to get to the preferences screen. You could use a menu or use a button to fire the intent. We'll not repeat this code for later examples, but you would do the same thing with them, except that you use the appropriate activity class name. When the preferences Intent returns to us, we call the setOptionText() method to update our TextView.

There are two ways to get a handle to the preferences:

- The easiest is what we show in the example: that is, to call `PreferenceManager.getDefaultSharedPreferences(this)`. The `this` argument is the context for finding the default shared preferences, and the method will use the package name of `this` to determine the file name and location of the preferences file, which happens to be the one created by our `PreferenceActivity`, because they share the same package name.

- The other way to get a handle to a preferences file is to use the `getSharedPreferences()` method call, passing in a file name argument as well as a mode argument. In Listing 13–2, we show this way, but it's been commented out. Notice that you only specify the base part of the file name, not the path and not the file name extension. The mode argument controls permissions to our XML preferences file. In our preceding example, the mode argument wouldn't affect anything because the file is only created within the `PreferenceActivity`, which sets the default permissions of `MODE_PRIVATE` (zero). We'll discuss the mode argument later, in the sections on saving state.

In most cases, you'll use the first method of locating the preferences. However, if you had multiple users for your application on a device, and each user managed their own preferences, you'd need to use the second option to keep the users' preferences separate from each other.

Inside of `setOptionText()`, with a reference to the preferences, you call the appropriate methods to retrieve the preference values. In our example, we call `getString()`, because we know we're retrieving a string value from the preferences. The first argument is the string value of the option key. We noted before that using an ID ensures that we haven't made any typographical errors while building our application. We could also have simply used the string `"selected_flight_sort_option"` for the first argument, but then it's up to you to ensure that this string is exactly the same as other parts of your code where the key value is used. For the second argument, you specify a default value in case the value can't be found in the preferences XML file. When your application runs for the very first time, you don't have a preferences XML file; so without specifying a value for the second argument, you'll always get `null` the first time. This is true even though you've specified a default value for the option in the `ListPreference` specification in `flightoptions.xml`. In our example, we set a default value in XML, and we used a resource ID to do it, so the code in `setOptionText()` can be used to read the value of the resource ID for the default value. Note that if we had not used an ID for the default value, it would be a lot tougher to read it directly from the `ListPreference`. By sharing a resource ID between the XML and our code, we have only one place in which to change the default value (that is, in `strings.xml`).

In addition to displaying the value of the preference, we also display the text of the preference. We're taking a shortcut in our example, because we used array indices for the values in `flight_sort_options_values`. By simply converting the value to an `int`, we know which string to read from `flight_sort_options`. Had we used some other set of values for flight_sort_options_values, we would need to determine the index of the element that is our preference and then turn around and use that index to grab the text of our preference from flight_sort_options.

Because we now have two activities in our application, we need two activity tags in AndroidManifest.xml. The first one is a standard activity of category LAUNCHER. The second one is for a PreferenceActivity, so we set the action name according to convention for intents, and we set the category to PREFERENCE as shown in Listing 13–4. We probably don't want the PreferenceActivity showing up on the Android page with all our other applications, which is why we chose not to use LAUNCHER for it. You would need to make similar changes to AndroidManifest.xml if you were to add other preferences screens.

Listing 13–4. *PreferenceActivity entry in AndroidManifest.xml*

```
        <activity android:name=".FlightPreferenceActivity"
                  android:label="@string/prefTitle">
            <intent-filter>
                <action android:name=
 "com.androidbook.preferences.sample.intent.action.FlightPreferences" />
                <category
                    android:name="android.intent.category.PREFERENCE" />
            </intent-filter>
        </activity>
```

We showed one way to read a default value for a preference in code. Android provides another way that is a bit more elegant. In onCreate(), we could have done the following instead:

```
PreferenceManager.setDefaultValues(this, R.xml.flightoptions, false);
```

Then, in setOptionText(), we could have done this to read the option value:

```
String option = prefs.getString(
    resources.getString(R.string.selected_flight_sort_option), null);
```

The first call will use flightoptions.xml to find the default values and generate the preferences XML file for us using the default values. If we already have an instance of the SharedPreferences object in memory, it will update that too. The second call will then find a value for selected_flight_sort_option, because we took care of loading defaults first.

After running this code the first time, if you look in the shared_prefs folder, you will see the preferences XML file even if the preferences screen has not yet been invoked. You will also see another file called _has_set_default_values.xml. This tells your application that the preferences XML file has already been created with the default values. The third argument to setDefaultValues()—that is, false—indicates that you only want the defaults set in the preferences XML file if it hasn't been done before. If you choose true instead, you'll always reset the preferences XML file with default values. Android remembers this information through the existence of this new XML file. If the user has selected new preference values, and you choose false for the third argument, the user preferences won't be overwritten the next time this code runs. Notice that now we don't need to provide a default value in the getString() method call, because we should always get a value from the preferences XML file.

If you need a reference to the preferences from inside of an activity that extends PreferenceActivity, you could do it this way:

```
SharedPreferences prefs = getPreferenceManager().getDefaultSharedPreferences(this);
```

We showed you how to use the ListPreference view; now, let's examine some other UI elements within the Android preferences framework. Namely, let's talk about the CheckBoxPreference view and the EditTextPreference view.

Understanding CheckBoxPreference

You saw that the ListPreference preference displays a list as its UI element. Similarly, the CheckBoxPreference preference displays a check-box widget as its UI element.

To extend the flight-search example application, suppose you want to let the user set the list of columns to see in the result set. This preference displays the available columns and allows the user to choose the desired columns by marking the corresponding check boxes. The user interface for this example is shown in Figure 13–3, and the preferences XML file is shown in Listing 13–5.

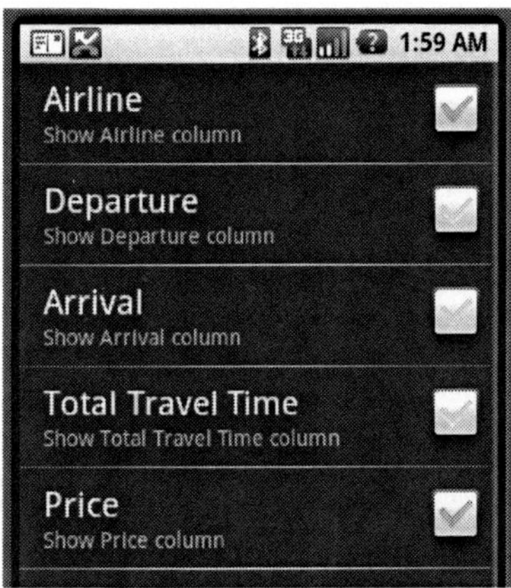

Figure 13–3. *The user interface for the check-box preference*

Listing 13–5. *Using CheckBoxPreference*

```
<?xml version="1.0" encoding="utf-8"?>
<!-- This file is /res/xml/chkbox.xml -->
    <PreferenceScreen
        xmlns:android="http://schemas.android.com/apk/res/android"
                android:key="flight_columns_pref"
                android:title="Flight Search Preferences"
                android:summary="Set Columns for Search Results">
        <CheckBoxPreference
                android:key="show_airline_column_pref"
                android:title="Airline"
                android:summary="Show Airline column" />
```

```
        <CheckBoxPreference
                android:key="show_departure_column_pref"
                android:title="Departure"
                android:summary="Show Departure column" />
        <CheckBoxPreference
                android:key="show_arrival_column_pref"
                android:title="Arrival"
                android:summary="Show Arrival column" />
         <CheckBoxPreference
                android:key="show_total_travel_time_column_pref"
                android:title="Total Travel Time"
                android:summary="Show Total Travel Time column" />
        <CheckBoxPreference
                android:key="show_price_column_pref"
                android:title="Price"
                android:summary="Show Price column" />

</PreferenceScreen>


public class CheckBoxPreferenceActivity extends PreferenceActivity
{
    @Override
    protected void onCreate(Bundle savedInstanceState) {
        super.onCreate(savedInstanceState);
        addPreferencesFromResource(R.xml.chkbox);
    }
}
```

Listing 13–5 shows the preferences XML file, chkbox.xml, and a simple activity class that loads it using addPreferencesFromResource(). As you can see, the UI has five check boxes, each of which is represented by a CheckBoxPreference node in the preferences XML file. Each of the check boxes also has a key, which—as you would expect—is ultimately used to persist the state of the UI element when it comes time to save the selected preference. With CheckBoxPreference, the state of the preference is saved when the user sets the state. In other words, when the user checks or unchecks the preference control, its state is saved. Listing 13–6 shows the preference data store for this example.

Listing 13–6. *The Preferences Data Store for the Check Box Preference*

```
<?xml version='1.0' encoding='utf-8' standalone='yes' ?>
<map>
    <boolean name="show_total_travel_time_column_pref" value="false" />
    <boolean name="show_price_column_pref" value="true" />
    <boolean name="show_arrival_column_pref" value="false" />
    <boolean name="show_airline_column_pref" value="true" />
    <boolean name="show_departure_column_pref" value="false" />
</map>
```

Again, you can see that each preference is saved through its key attribute. The data type of the CheckBoxPreference is a boolean, which contains a value of either true or false: true to indicate the preference is selected, and false to indicate otherwise. To read the value of one of the check-box preferences, you would get access to the shared preferences and call the getBoolean() method, passing the key of the preference:

```
boolean option = prefs.getBoolean("show_price_column_pref", false);
```

One other useful feature of CheckBoxPreference is that you can set different summary text depending on whether it's checked. The XML attributes are summaryOn and summaryOff. Now, let's have a look at the EditTextPreference.

Understanding EditTextPreference

The preferences framework also provides a free-form text preference called EditTextPreference. This preference allows you to capture raw text rather than ask the user to make a selection. To demonstrate this, let's assume you have an application that generates Java code for the user. One of the preference settings of this application might be the default package name to use for the generated classes. Here, you want to display a text field to the user for setting the package name for the generated classes. Figure 13–4 shows the UI, and Listing 13–7 shows the XML.

Figure 13–4. *Using the EditTextPreference*

Listing 13–7. *An Example of an EditTextPreference*

```
<?xml version="1.0" encoding="utf-8"?>
<!-- This file is /res/xml/packagepref.xml -->
<PreferenceScreen
        xmlns:android="http://schemas.android.com/apk/res/android"
                android:key="package_name_screen"
                android:title="Package Name"
                android:summary="Set package name">

        <EditTextPreference
                android:key="package_name_preference"
                android:title="Set Package Name"
                android:summary="Set the package name for generated code"
                android:dialogTitle="Package Name" />

</PreferenceScreen>

public class EditTextPreferenceActivity extends PreferenceActivity{
```

```
    @Override
    protected void onCreate(Bundle savedInstanceState) {
        super.onCreate(savedInstanceState);

        addPreferencesFromResource(R.xml.packagepref);
    }
}
```

You can see that Listing 13–6 defines PreferenceScreen with a single EditTextPreference instance as a child. The generated UI for the listing features the PreferenceScreen on the left and the EditTextPreference on the right (see Figure 13–4). When Set Package Name is selected, the user is presented with a dialog to input the package name. When the OK button is clicked, the preference is saved to the preference store.

As with the other preferences, you can obtain the EditTextPreference from your activity class by using the preference's key. Once you have the EditTextPreference, you can manipulate the actual EditText by calling getEditText()—if, for example, you want to apply validation, preprocessing, or post-processing on the value the user types in the text field. To get the text of the EditTextPreference, just use the getText() method.

Understanding RingtonePreference and MultiSelectListPreference

There is another preference called RingtonePreference, but we won't cover that here. It follows the same rules as the others and isn't used much. And finally, a preference called MultiSelectListPreference was introduced in Android 3.0. The concept is somewhat similar to a ListPreference, but instead of only being able to select one item in the list, the user can select several or none. Unfortunately, the implementation is buggy as of this writing. For example, the values array does not seem to be used—only the entries array. This means the XML preferences file contains the entry strings and not the corresponding values as the ListPreference does. It is also a mystery how to set the default values. For more information, see http://code.google.com/p/android/issues/detail?id=15966.

Until this type of preference is fixed, you're better off creating a set of CheckBoxPreferences. Figure 13–3 shows what a MultiSelectListPreference looks like (more or less).

Organizing Preferences

The preferences framework provides some support for you to organize your preferences into categories. If you have a lot of preferences, for example, you can build a view that shows high-level categories of preferences. Users could then drill down into each category to view and manage preferences specific to that group.

Using PreferenceCategory

You can implement something like this in one of two ways. You can introduce nested `PreferenceScreen` elements within the root `PreferenceScreen`, or you can use `PreferenceCategory` elements to get a similar result. Figure 13–5 and Listing 13–8 show how to implement the first technique, grouping preferences by using nested `PreferenceScreen` elements.

The view on the left in Figure 13–5 displays options for two preference screens, one with the title Meats and the other with the title Vegetables. Clicking a group takes you to the preferences within that group. Listing 13–8 shows how to create nested screens.

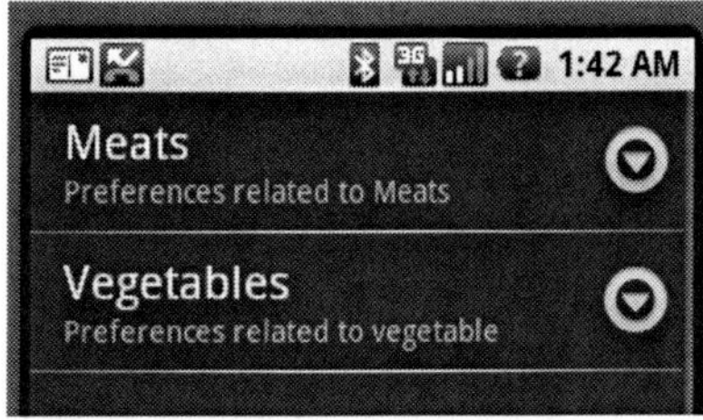

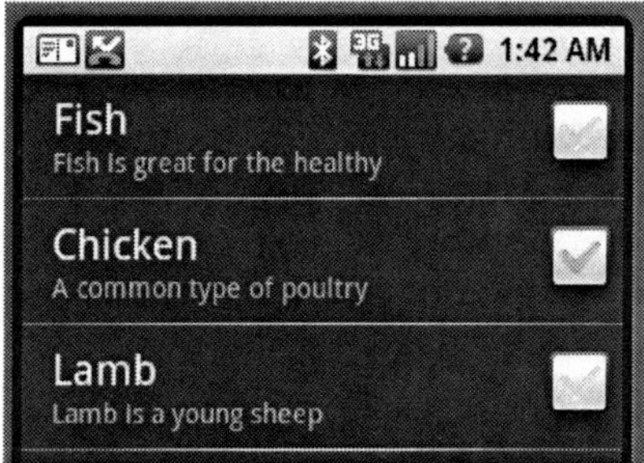

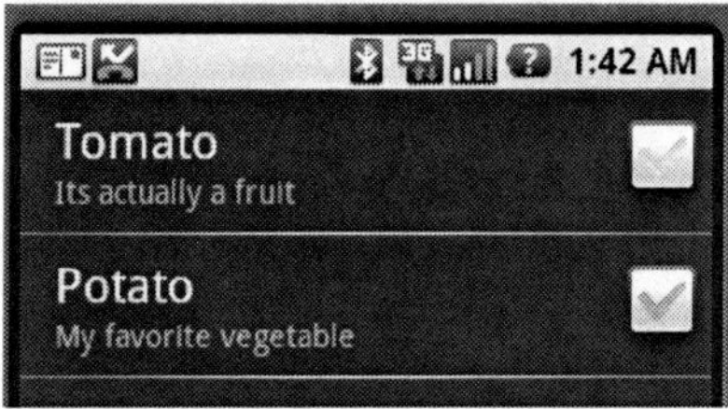

Figure 13–5. *Creating groups of preferences by nesting `PreferenceScreen` elements*

Listing 13–8. *Nesting `PreferenceScreen` Elements to Organize Preferences*

```
<?xml version="1.0" encoding="utf-8"?>
<PreferenceScreen
        xmlns:android="http://schemas.android.com/apk/res/android"
                android:key="using_categories_in_root_screen"
                android:title="Categories"
                android:summary="Using Preference Categories">

    <PreferenceScreen
        xmlns:android="http://schemas.android.com/apk/res/android"
                android:key="meats_screen"
                android:title="Meats"
                android:summary="Preferences related to meats">

        <CheckBoxPreference
                android:key="fish_selection_pref"
                android:title="Fish"
                android:summary="Fish is healthy" />
```

```
        <CheckBoxPreference
                android:key="chicken_selection_pref"
                android:title="Chicken"
                android:summary="A common type of poultry" />
        <CheckBoxPreference
                android:key="lamb_selection_pref"
                android:title="Lamb"
                android:summary="A young sheep" />

    </PreferenceScreen>
    <PreferenceScreen
        xmlns:android="http://schemas.android.com/apk/res/android"
                android:key="vegi_screen"
                android:title="Vegetables"
                android:summary="Preferences related to vegetables">
        <CheckBoxPreference
                android:key="tomato_selection_pref"
                android:title="Tomato "
                android:summary="It's actually a fruit" />
        <CheckBoxPreference
                android:key="potato_selection_pref"
                android:title="Potato"
                android:summary="My favorite vegetable" />

    </PreferenceScreen>

</PreferenceScreen>
```

You create the groups in Figure 13–5 by nesting `PreferenceScreen` elements within the root `PreferenceScreen`. Organizing preferences this way is useful if you have a lot of preferences and you're concerned about having the users scroll to find the preference they are looking for. If you don't have a lot of preferences but still want to provide high-level categories for your preferences, you can use `PreferenceCategory`, which is the second technique we mentioned. Figure 13–6 and Listing 13–9 show the details.

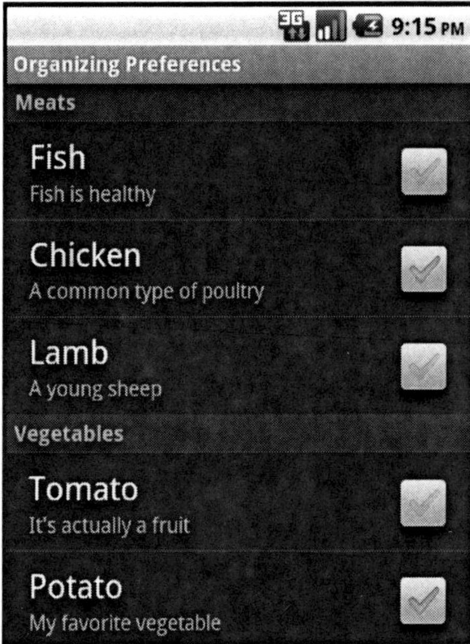

Figure 13–6. *Using* `PreferenceCategory` *to organize preferences*

Figure 13–6 shows the same groups we used in our previous example, but now organized with preference categories. The only difference between the XML in Listing 13–9 and the XML in Listing 13–8 is that you create a `PreferenceCategory` for the nested screens rather than nest `PreferenceScreen` elements.

Listing 13–9. *Creating Categories of Preferences*

```
<?xml version="1.0" encoding="utf-8"?>
<PreferenceScreen
        xmlns:android="http://schemas.android.com/apk/res/android"
                android:key="using_categories_in_root_screen"
                android:title="Categories"
                android:summary="Using Preference Categories">

    <PreferenceCategory
        xmlns:android="http://schemas.android.com/apk/res/android"
                android:key="meats_category"
                android:title="Meats"
                android:summary="Preferences related to meats">

        <CheckBoxPreference
                android:key="fish_selection_pref"
                android:title="Fish"
                android:summary="Fish is healthy" />
        <CheckBoxPreference
                android:key="chicken_selection_pref"
                android:title="Chicken"
                android:summary="A common type of poultry" />
        <CheckBoxPreference
                android:key="lamb_selection_pref"
```

```
            android:title="Lamb"
            android:summary="A young sheep" />

    </PreferenceCategory>
    <PreferenceCategory
        xmlns:android="http://schemas.android.com/apk/res/android"
            android:key="vegi_category"
            android:title="Vegetables"
            android:summary="Preferences related to vegetables">
        <CheckBoxPreference
            android:key="tomato_selection_pref"
            android:title="Tomato "
            android:summary="It's actually a fruit" />
        <CheckBoxPreference
            android:key="potato_selection_pref"
            android:title="Potato"
            android:summary="My favorite vegetable" />

    </PreferenceCategory>

</PreferenceScreen>
```

Creating Child Preferences with Dependency

Another way to organize preferences is to use a preference dependency. This creates a parent-child relationship between preferences. For example, we might have a preference that turns on alerts; and if alerts are on, there might be several other alert-related preferences to choose from. If the main alerts preference is off, the other preferences are not relevant and should be disabled. Listing 13–10 shows the XML, and Figure 13–7 shows what it looks like.

Listing 13–10. *Preference Dependency in XML*

```
<PreferenceScreen>

    <PreferenceCategory
          android:title="Alerts">

        <CheckBoxPreference
            android:key="alert_email"
            android:title="Send email?" />

        <EditTextPreference
            android:key="alert_email_address"
            android:layout="?android:attr/preferenceLayoutChild"
            android:title="Email Address"
            android:dependency="alert_email" />

    </PreferenceCategory>

</PreferenceScreen>
```

Figure 13–7. *Preference dependency*

Preferences with Headers

With the introduction of Android 3.0, we got another way to organize preferences. You see this on tablets under the main Settings app. Because tablet screen real estate offers much more room than a smartphone does, it makes sense to display more preference information at the same time. To accomplish this, we use preference headers. Take a look at Figure 13–8 to see what we mean.

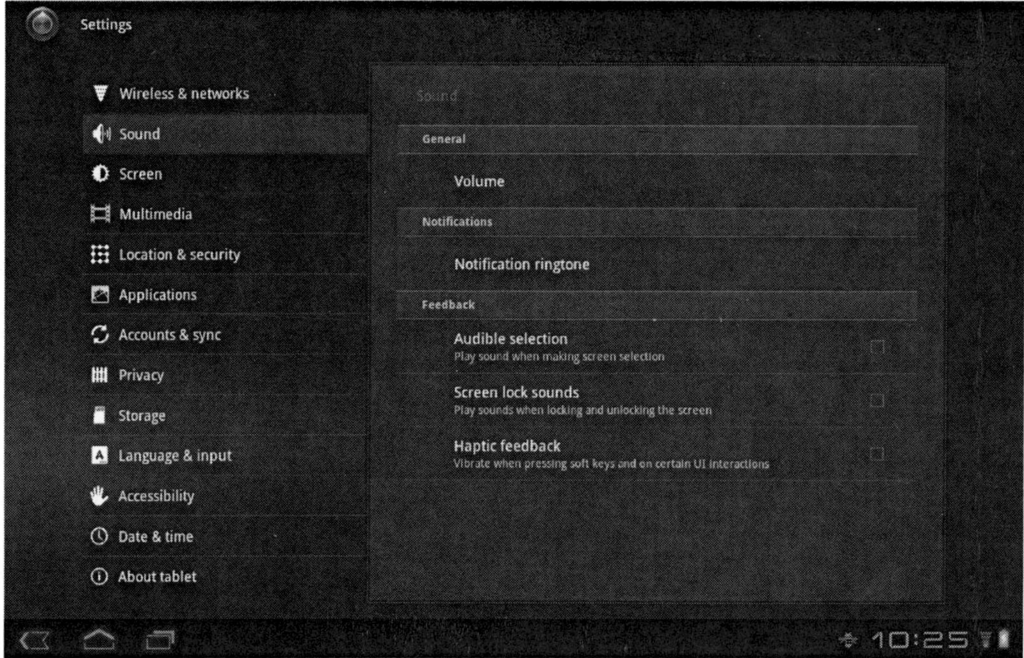

Figure 13–8. *Main Settings page with preference headers*

Notice that headers appear down the left side, like a vertical tab bar. As you click each item on the left, the screen to the right displays the preferences for that item. In Figure 13–8, Sound is chosen, and the sound preferences are displayed at right. The right side is a `PreferenceScreen` object, and this setup uses fragments. Obviously, we need to do something different than what has been discussed so far in this chapter.

The big change from Android 3.0 was the addition of headers to `PreferenceActivity`. This also means using a new callback within `PreferenceActivity` to do the headers setup. Now, when you extend PreferenceActivity, you'll want to implement this method:

```
public void onBuildHeaders(List<Header> target) {
    loadHeadersFromResource(R.xml.preferences, target);
}
```

The preferences.xml file contains some new tags that look like this:

```
<preference-headers
        xmlns:android="http://schemas.android.com/apk/res/android">
    <header android:fragment="com.example.PrefActivity$Prefs1Fragment"
            android:icon="@drawable/ic_settings_sound"
            android:title="Sound"
            android:summary="Your sound preferences" />
    ...
```

Each header tag points to a class that extends PreferenceFragment. In the example just given, the XML specifies an icon, the title, and summary text (which acts like a subtitle). Prefs1Fragment is an inner class of PreferenceActivity that could look something like this:

```
public static class Prefs1Fragment extends PreferenceFragment {
    @Override
    public void onCreate(Bundle savedInstanceState) {
        super.onCreate(savedInstanceState);
        addPreferencesFromResource(R.xml.sound_preferences);
    }
}
```

All this inner class needs to do is pull in the appropriate preferences XML file, as shown. That preferences XML file contains the types of preference specifications we covered earlier, such as ListPreference, CheckBoxPreference, PreferenceCategory, and so on. What's very nice is that Android takes care of doing the right thing when the screen configuration changes and when the preferences are displayed on a small screen. Headers behave like old preferences when the screen is too small to display both headers and the preference screen to the right. That is, you only see the headers; and when you click a header, you then see only the appropriate preference screen.

Manipulating Preferences Programmatically

It goes without saying that you might need to access the actual preference controls programmatically. For example, what if you need to provide the entries and entryValues for the ListPreference at runtime? You can define and access preference controls similarly to the way you define and access controls in layout files and activities. For example, to access the list preference defined in Listing 13–1, you would call the findPreference() method of PreferenceActivity, passing the preference's key (note the similarity to findViewById()). You would next cast the control to ListPreference and then go about manipulating the control. For example, if you want to set the entries of the ListPreference view, call the setEntries() method, and so on. Listing 13–11 shows what this might look like with a simple example of using code to set up the preference.

Of course, you could also create the entire PreferenceScreen starting with PreferenceManager.createPreferenceScreen().

Listing 13–11. ***Setting ListPreference Values Programmatically***

```
public class FlightPreferenceActivity extends PreferenceActivity
{
    @Override
    protected void onCreate(Bundle savedInstanceState) {
        super.onCreate(savedInstanceState);
        addPreferencesFromResource(R.xml.flightoptions);

        ListPreference listpref = (ListPreference) findPreference(
                                   "selected_flight_sort_option");

        listpref.setEntryValues(new String[] {"0","1","2"});
        listpref.setEntries(new String[] {"Food", "Lounge", "Frequent Flier Program"});
    }
}
```

Saving State with Preferences

Preferences are great for allowing users to customize applications to their liking, but we can use the Android preference framework for more than that. When your application needs to keep track of some data between invocations of the application, preferences are one way to accomplish the task. We've already talked about content providers for maintaining data. We could use custom files on the SD card. We can also use shared preference files and code.

The Activity class has a getPreferences(int mode) method. This, in reality, simply calls getSharedPreferences() with the class name of the activity as the tag plus the mode as passed in. The result is an activity-specific shared preferences file that you can use to store data about this activity across invocations. A simple example of how you could use this is shown in Listing 13–12.

Listing 13–12. ***Using Preferences to Save State for an Activity***

```
final String INITIALIZED = "initialized";
SharedPreferences myPrefs = getPreferences(MODE_PRIVATE);

boolean hasPreferences = myPrefs.getBoolean(INITIALIZED, false);
if(hasPreferences) {
        Log.v("Preferences", "We've been called before");
        // Read other values as desired from preferences file...
        someString = myPrefs.getString("someString", "");
}
else {
        Log.v("Preferences", "First time ever being called");
        // Set up initial values for what will end up
        // in the preferences file
        someString = "some default value";
}

// Later when ready to write out values
Editor editor = myPrefs.edit();
```

```
editor.putBoolean(INITIALIZED, true);
editor.putString("someString", someString);
// Write other values as desired
editor.commit();
```

What this code does is acquire a reference to preferences for our activity class and check for the existence of a boolean "preference" called initialized. We write "preference" in double quotation marks because this value is not something the user is going to see or set; it's merely a value that we want to store in a shared preferences file for use next time. If we get a value, the shared preferences file exists, so our application must have been called before. We could then read other values out of the shared preferences file. For example, someString could be an activity variable that should be set from the last time this activity ran or set to the default value if this is the first time.

To write values to the shared preferences file, we must first get a preferences Editor. We can then put values into preferences and commit those changes when we're finished. Note that, behind the scenes, Android is managing a SharedPreferences object that is truly shared. Ideally, there is never more than one Editor active at a time. But it is very important to call the commit() method so that the SharedPreferences object and the shared preferences XML file get updated. In the example, we write out the value of someString to be used the next time this activity runs.

You can access, write, and commit values anytime to your preferences file. Possible uses for this include writing out high scores for a game or recording when the application was last run. You can also use the getSharedPreferences() call with different names to manage separate sets of preferences, all within the same application or even the same activity.

We've used MODE_PRIVATE for mode in our examples thus far. The other possible values of mode are MODE_WORLD_READABLE and MODE_WORLD_WRITEABLE. These modes are used when creating the shared preferences XML file to set the file permissions accordingly. Because the shared preferences files are stored within your application's data directory and therefore are not accessible to other applications, you only need to use MODE_PRIVATE.

Using DialogPreference

So far, you've seen how to use the out-of-the-box capabilities of the preferences framework, but what if you want to create a custom preference? What if you want something like the slider of the Brightness preference under Screen Settings? This is where DialogPreference comes in. DialogPreference is the parent class of EditTextPreference and ListPreference. The behavior is a dialog that pops up, displays choices to the user, and is closed with a button or via the Back button. But you can extend DialogPreference to set up your own custom preference. Within your extended class, you provide your own layout, your own click handlers, and custom code in onDialogClosed() to write the data for your preference to the shared preferences file.

Reference

Here is a helpful reference to a topic you may wish to explore further:

- `www.androidbook.com/proandroid4/projects`: A list of downloadable projects related to this book. For this chapter, look for the file `ProAndroid4_Ch13_Preferences.zip`. This ZIP file contains all the projects from this chapter, listed in separate root directories. There is also a `README.TXT` file that describes how to import projects into Eclipse from one of these ZIP files.

Summary

This chapter talked about managing preferences in Android:

- Types of preferences available
- Reading the current values of preferences into your application
- Setting default values from embedded code and by writing the default values from the XML file to the saved preferences file
- Organizing preferences into groups, and defining dependencies between preferences
- Programmatically manipulating preferences
- Using the preferences framework to save and restore information from an activity across invocations
- Creating a custom preference

Interview Questions

Here are some questions you can ask yourself to solidify your understanding of this material:

1. Name five different types of preferences.
2. What attribute of a preference is used to store the selected value?
3. How many shared preference files can one application have?
4. Which source directory is the typical place to put preference definition files?
5. Where can you find a shared preferences file for an application with the package name `com.androidbook.myapp`?

6. When is a shared preferences file first created: when the application is installed, or some time later? If later, when?
7. What are some considerations if you are upgrading your application and there are new preferences?
8. If you want to retrieve the current value of an `int` preference, what class is used, and which method?
9. What is the category name that should be used in `AndroidManifest.xml` for a `PreferenceActivity`?
10. What two main tags are used in a preference XML file to set up headers?
11. Can you save a value in a shared preferences file without ever showing it to a user in a preferences screen?
12. What is so important about the `commit()` method of the `Editor` class?

Chapter 14

Exploring Security and Permissions

In this chapter, we are going to talk about Android's application-security model, which is a fundamental part of the Android platform. In Android, security spans all phases of the application life cycle—from design-time policy considerations to runtime boundary checks. You'll learn Android's security architecture and understand how to design secure applications.

Let's get started with the Android security model.

Understanding the Android Security Model

In this first section, we're going to cover security during the deployment and execution of the application. With respect to deployment, Android applications have to be signed with a digital certificate in order for you to install them onto a device. With respect to execution, Android runs each application within a separate process, each of which has a unique and permanent user ID (assigned at install time). This places a boundary around the process and prevents one application from having direct access to another's data. Moreover, Android defines a declarative permission model that protects sensitive features (such as the contact list).

In the next several sections, we are going to discuss these topics. But before we get started, let's provide an overview of some of the security concepts that we'll refer to later.

Overview of Security Concepts

Android requires that applications be signed with a digital certificate. One of the benefits of this requirement is that an application cannot be updated with a version that was not published by the original author. If we publish an application, for example, then you cannot update our application with your version (unless, of course, you somehow obtain our certificate). That said, what does it mean for an application to be signed? And what is the process of signing an application?

You sign an application with a digital certificate. A *digital certificate* is an artifact that contains information about you, such as your company name, address, and so on. A few important attributes of a digital certificate include its signature and public/private key. A public/private key is also called a *key pair*. Note that although you use digital certificates here to sign `.apk` files, you can also use them for other purposes (such as encrypted communication). You can obtain a digital certificate from a trusted certificate authority (CA) and you can also generate one yourself using tools such as the `keytool`, which we'll discuss shortly. Digital certificates are stored in keystores. A *keystore* contains a list of digital certificates, each of which has an alias that you can use to refer to it in the keystore.

Signing an Android application requires three things: a digital certificate, an `.apk` file, and a utility that knows how to apply a digital signature to the `.apk` file. As you'll see, we use a free utility that is part of the Java Development Kit (JDK) distribution called the `jarsigner`. This utility is a command-line tool that knows how to sign a `.jar` file using a digital certificate.

Now, let's move on and talk about how you can sign an `.apk` file with a digital certificate.

Signing Applications for Deployment

To install an Android application onto a device, you first need to sign the Android package (`.apk` file) using a digital certificate. The certificate, however, can be self-signed—you do not need to purchase a certificate from a certificate authority such as VeriSign.

Signing your application for deployment involves three steps. The first step is to generate a certificate using `keytool` (or a similar tool). The second step involves using the `jarsigner` tool to sign the `.apk` file with the generated certificate. The third step aligns portions of your application on memory boundaries for more efficient memory usage when running on a device. Note that during development, the ADT plug-in for Eclipse takes care of everything for you: signing your `.apk` file and doing the memory alignment, before deploying onto the emulator or a device.

Generating a Self-Signed Certificate Using the Keytool

The `keytool` utility manages a database of private keys and their corresponding X.509 certificates (a standard for digital certificates). This utility ships with the JDK and resides under the JDK `bin` directory. If you followed the instructions in Chapter 2 regarding changing your `PATH`, the JDK `bin` directory should already be in your `PATH`.

In this section, we'll show you how to generate a keystore with a single entry, which you'll later use to sign an Android .apk file. To generate a keystore entry, do the following:

1. Create a folder to hold the keystore, such as c:\android\release\.
2. Open a tools window, and execute the keytool utility with the parameters shown in Listing 14–1 (see Chapter 2 for details of what we mean by a *tools window*).

Listing 14–1. *Generating a Keystore Entry Using the keytool Utility*

```
keytool -genkey -v -keystore "c:\android\release\release.keystore"
-alias androidbook -storepass paxxword -keypass paxxword -keyalg RSA
-validity 14000
```

All of the arguments passed to the keytool are summarized in Table 14–1.

Table 14–1. *Arguments Passed to the keytool Utility*

Argument	Description
genkey	Tells keytool to generate a public/private key pair.
v	Tells keytool to emit verbose output during key generation.
keystore	Path to the keystore database (in this case, a file). The file will be created if necessary.
alias	Unique name for the keystore entry. This alias is used later to refer to the keystore entry.
storepass	Password for the keystore.
keypass	Password used to access the private key.
keyalg	Algorithm.
validity	Validity period.

keytool will prompt you for the passwords listed in Table 14–1 if you do not provide them on the command line. If you are not the sole user of your computer, it would be safer to not specify -storepass and -keypass on the command line, but rather to type them in when prompted by keytool.

The command in Listing 14–1 will generate a keystore database file in your keystore folder. The database will be a file named release.keystore. The validity of the entry will be 14,000 days (or approximately 38 years)—which is a long time from now. You should understand the reason for this. The Android documentation recommends that you specify a validity period long enough to surpass the entire lifespan of the application, which will include many updates to the application. It recommends that the validity be at least 25 years. If you plan to publish the application on Android Market, your certificate will need to be valid through at least October 22, 2033. Android Market checks each application when uploaded to make sure it will be valid at least until then.

Because your certificate in any application update must match the certificate you used the first time, make sure you safeguard your keystore file! If you lose it and you can't re-create it, you won't be able to update your application, and you'll have to issue a whole new application instead.

Going back to the `keytool`, the argument `alias` is a unique name given to the entry in the keystore database; you will use this name later to refer to the entry. When you run the `keytool` command in Listing 14–1, `keytool` will ask you a few questions (see Figure 14–1) and then generates the keystore database and entry. Note that in Figure 14–1, because the command was executed from within the `c:\android\release` directory, it was not necessary to specify the full pathname to the `release.keystore` file.

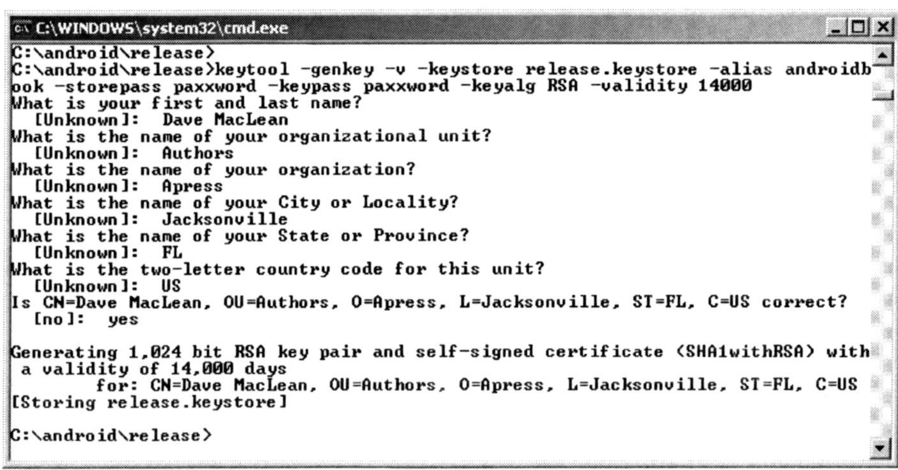

C:\WINDOWS\system32\cmd.exe
C:\android\release>
C:\android\release>keytool -genkey -v -keystore release.keystore -alias androidbook -storepass paxxword -keypass paxxword -keyalg RSA -validity 14000
What is your first and last name?
[Unknown]: Dave MacLean
What is the name of your organizational unit?
[Unknown]: Authors
What is the name of your organization?
[Unknown]: Apress
What is the name of your City or Locality?
[Unknown]: Jacksonville
What is the name of your State or Province?
[Unknown]: FL
What is the two-letter country code for this unit?
[Unknown]: US
Is CN=Dave MacLean, OU=Authors, O=Apress, L=Jacksonville, ST=FL, C=US correct?
[no]: yes
Generating 1,024 bit RSA key pair and self-signed certificate (SHA1withRSA) with a validity of 14,000 days
for: CN=Dave MacLean, OU=Authors, O=Apress, L=Jacksonville, ST=FL, C=US
[Storing release.keystore]
C:\android\release>

Figure 14–1. *Additional questions asked by* `keytool`

Once you have a keystore file for your production certificates, you can reuse this file to add more certificates. Just use `keytool` again, and specify your existing keystore file.

The Debug Keystore and the Development Certificate

We mentioned that the ADT plug-in for Eclipse takes care of setting up a development keystore for you. However, the default certificate used for signing during development cannot be used for production deployment onto a real device. This is partly because the ADT-generated development certificate is only valid for 365 days, which clearly does not get you past October 22, 2033. So what happens on the three hundred sixty-sixth day of development? You'll get a build error. Your existing applications should still run, but to build a new version of an application, you need to generate a new certificate. The easiest way to do this is to delete the existing debug.keystore file, and as soon as it is needed again, the ADT will generate a new file and certificate valid for another 365 days.

To find your debug.keystore file, open the Preferences screen of Eclipse and go to Android ➤ Build. The debug certificate's location will be displayed in the Default Debug Keystore field, as shown in Figure 14–2 (see Chapter 2 if you have trouble finding the Preferences menu).

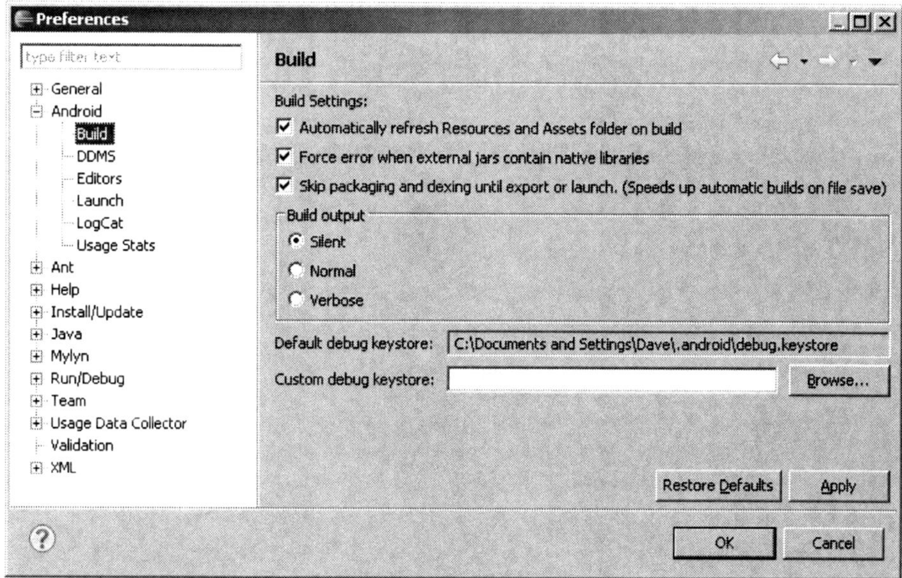

Figure 14–2. *The debug certificate's location*

Of course, now that you've got a new development certificate, you cannot update your existing applications in Android Virtual Devices (AVDs) or on devices using a new development certificate. Eclipse will provide messages in the Console telling you to uninstall the existing application first using adb, which you can certainly do. If you have a lot of your applications installed onto an AVD, you may feel it is easier to simply re-create the AVD, so it does not contain any of your applications and you can start fresh. To avoid this problem a year from now, you could generate your own debug.keystore file with whatever validity period you desire. Obviously, it needs to have the same file name and be in the same directory as the file that ADT would create. The certificate alias is androiddebugkey, and the storepass and keypass are both "android". ADT sets the first and last name on the certificate as "Android Debug", the organizational unit as "Android", and the two-letter country code as "US". You can leave the organization, city, and state values as "Unknown".

If you acquired a map-api key from Google using the old debug certificate, you will need to get a new map-api key to match the new debug certificate. We'll cover map-api keys in Chapter 22.

Now that you have a digital certificate that you can use to sign your production .apk file, you need to use the jarsigner tool to do the signing. Here's how to do that.

Using the Jarsigner Tool to Sign the .apk File

The keytool utility described in the previous section created a digital certificate, which is one of the parameters for the jarsigner tool. The other parameter for jarsigner is the actual Android package to be signed. To generate an Android package, you need to use the Export Unsigned Application Package utility in the ADT plug-in for Eclipse. You access the utility by right-clicking an Android project in Eclipse, selecting Android Tools, and selecting Export Unsigned Application Package. Running the Export Unsigned Application Package utility will generate an .apk file that will not be signed with the debug certificate.

To see how this works, run the Export Unsigned Application Package utility on one of your Android projects, and store the generated .apk file somewhere. For this example, we'll use the keystore folder we created earlier and generate an .apk file called c:\android\release\myappraw.apk.

With the .apk file and the keystore entry, run the jarsigner tool to sign the .apk file (see Listing 14–2). Use the full path names to your keystore file and .apk file as appropriate when you run this.

Listing 14–2. *Using jarsigner to Sign the .apk File*

```
jarsigner -keystore "PATH TO YOUR release.keystore FILE" -storepass paxxword
-keypass paxxword "PATH TO YOUR RAW APK FILE" androidbook
```

To sign the .apk file, you pass the location of the keystore, the keystore password, the private-key password, the path to the .apk file, and the alias for the keystore entry. The jarsigner will then sign the .apk file with the digital certificate from the keystore entry. To run the jarsigner tool, you will need to either open a tools window (as explained in Chapter 2) or open a command or Terminal window and either navigate to the JDK bin directory or ensure that your JDK bin directory is on the system path. For security reasons, it is safer to leave off the password arguments to the command and simply let jarsigner prompt you as necessary for passwords. Figure 14–3 shows what the jarsigner tool invocation looks like. You may have noticed that jarsigner prompted for only one password in Figure 14–3. Jarsigner figures out not to ask for the keypass password when the storepass and keypass are the same. Strictly speaking, the jarsigner command in Listing 14–2 only needs -keypass if it has a different password than -storepass.

As we pointed out earlier, Android requires that an application be signed with a digital signature to prevent a malicious programmer from updating your application with their version. For this to work, Android requires that updates to an application be signed with the same signature as the original. If you sign the application with a different signature, Android treats them as two different applications. So we remind you again, be careful with your keystore file so it's available to you later when you need to provide an update to your application.

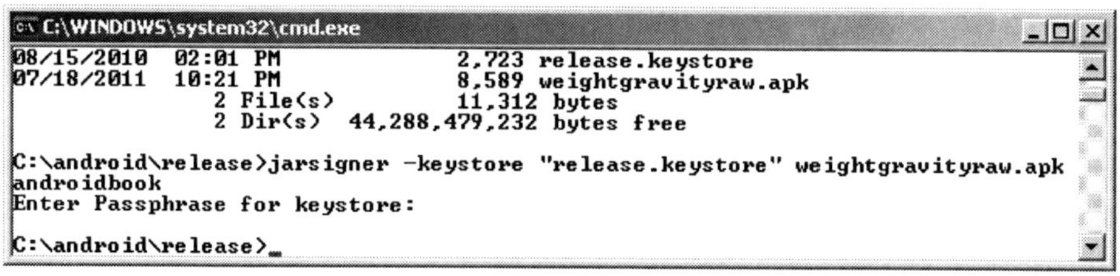

```
C:\WINDOWS\system32\cmd.exe
08/15/2010  02:01 PM             2,723 release.keystore
07/18/2011  10:21 PM             8,589 weightgravityraw.apk
               2 File(s)         11,312 bytes
               2 Dir(s)  44,288,479,232 bytes free

C:\android\release>jarsigner -keystore "release.keystore" weightgravityraw.apk
androidbook
Enter Passphrase for keystore:

C:\android\release>_
```

Figure 14–3. *Using* `jarsigner`

Aligning Your Application with zipalign

You want your application to be as memory efficient as possible when running on a device. If your application contains uncompressed data (perhaps certain image types or data files) at runtime, Android can map this data straight into memory using the `mmap()` call. For this to work, though, the data must be aligned on a 4-byte memory boundary. The CPUs in Android devices are 32-bit processors, and 32 bits equals 4 bytes. The `mmap()` call makes the data in your `.apk` file look like memory, but if the data is not aligned on a 4-byte boundary, it can't do that and extra copying of data must occur at runtime. The `zipalign` tool, found in the Android SDK tools directory, looks through your application and moves slightly any uncompressed data not already on a 4-byte memory boundary to a 4-byte memory boundary. You may see the file size of your application increase slightly but not significantly. To perform an alignment on your `.apk` file, use this command in a tools window (see also Figure 14–4):

```
zipalign -v 4 infile.apk outfile.apk
```

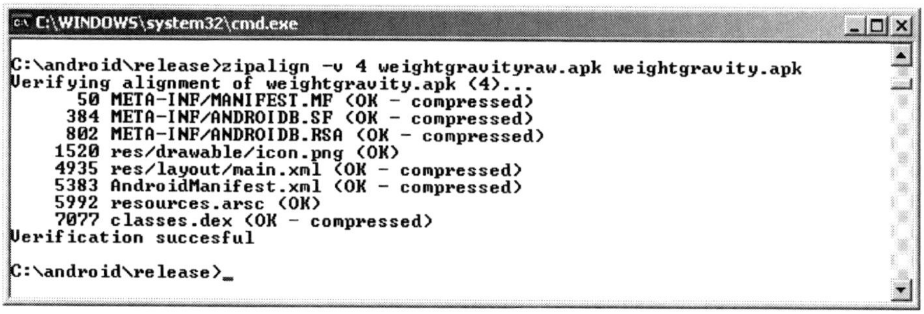

```
C:\WINDOWS\system32\cmd.exe
C:\android\release>zipalign -v 4 weightgravityraw.apk weightgravity.apk
Verifying alignment of weightgravity.apk (4)...
      50 META-INF/MANIFEST.MF (OK - compressed)
     384 META-INF/ANDROIDB.SF (OK - compressed)
     802 META-INF/ANDROIDB.RSA (OK - compressed)
    1520 res/drawable/icon.png (OK)
    4935 res/layout/main.xml (OK - compressed)
    5383 AndroidManifest.xml (OK - compressed)
    5992 resources.arsc (OK)
    7077 classes.dex (OK - compressed)
Verification succesful

C:\android\release>_
```

Figure 14–4. *Using* `zipalign`

Note that `zipalign` does not modify the input file, so this is why we chose to use "`raw`" as part of our file name when exporting from Eclipse. Now, our output file has an appropriate name for deployment. If you need to overwrite an existing `outfile.apk` file, you can use the `-f` option. Also note that `zipalign` performs a verification of the alignment when you create your aligned file. To verify that an existing file is properly aligned, use `zipalign` in the following way:

```
zipalign -c -v 4 filename.apk
```

It is very important that you align *after* signing; otherwise, signing could cause things to go back out of alignment. This does not mean your application would crash, but it could use more memory than it needs to.

Using the Export Wizard

In Eclipse, you may have noticed a menu choice under Android Tools called Export Signed Application Package. This launches what is called the *export wizard*, and it does all of the previous steps for you, prompting only for the path to your keystore file, key alias, the passwords and the name of your output `.apk` file. It will even create a new keystore or new key if you need one. You may find it easier to use the wizard, or you may prefer to script the steps yourself to operate on an exported unsigned application package. Now that you know how each works, you can decide which is better for you.

Manually Installing Apps

Once you have signed and aligned an `.apk` file, you can install it onto the emulator manually using the `adb` tool. As an exercise, start the emulator. One way to do this, which we haven't discussed yet, is to go to the Window menu of Eclipse and select Android SDK and AVD Manager. A window will be displayed, showing your available AVDs. Select the one you want to use for your emulator, and click the Start button. The emulator will start without copying over any of your development projects from Eclipse. Now, open a tools window, and run the `adb` tool with the `install` command:

```
adb install "PATH TO APK FILE GOES HERE"
```

This may fail for a couple of reasons, but the most likely are that the debug version of your application was already installed on the emulator, giving you a certificate error, or the release version of your application was already installed on the emulator, giving you an "already exists" error. In the first case, you can uninstall the debug application with this command:

```
adb uninstall packagename
```

Note that the argument to uninstall is the application's package name and not the `.apk` file name. The package name is defined in the `AndroidManifest.xml` file of the installed application.

For the second case, you can use this command, where `-r` says to reinstall the application while keeping its data on the device (or emulator):

```
adb install -r "PATH TO APK FILE GOES HERE"
```

Now, let's see how signing affects the process of updating an application.

Installing Updates to an Application and Signing

Earlier, we mentioned that a certificate has an expiration date and that Google recommends you set expiration dates far into the future, to account for a lot of application updates. That said, what happens if the certificate does expire? Would Android still run the application? Fortunately, yes—Android tests the certificate's expiration only at install time. Once your application is installed, it will continue to run even if the certificate expires.

But what about updates? Unfortunately, you will not be able to update the application once the certificate expires. In other words, as Google suggests, you need to make sure the life of the certificate is long enough to support the entire life of the application. If a certificate does expire, Android will not install an update to the application. The only choice left will be for you to create another application—an application with a different package name—and sign it with a new certificate. So as you can see, it is critical for you to consider the expiration date of the certificate when you generate it.

Now that you understand security with respect to deployment and installation, let's move on to runtime security in Android.

Performing Runtime Security Checks

Runtime security in Android happens at the process and operation levels. At the process level, Android prevents one application from directly accessing another application's data. It does this by running each application within a different process and under a unique and permanent user ID. At the operational level, Android defines a list of protected features and resources. For your application to access this information, you have to add one or more permission requests to your `AndroidManifest.xml` file. You can also define custom permissions with your application.

In the sections that follow, we will talk about process-boundary security and how to declare and use predefined permissions. We will also discuss creating custom permissions and enforcing them within your application. Let's start by dissecting Android security at the process boundary.

Understanding Security at the Process Boundary

Unlike your desktop environment, where most of the applications run under the same user ID, each Android application generally runs under its own unique ID. By running each application under a different ID, Android creates an isolation boundary around each process. This prevents one application from directly accessing another application's data.

Although each process has a boundary around it, data sharing between applications is obviously possible but has to be explicit. In other words, to get data from another application, you have to go through the components of that application. For example, you can query a content provider of another application, you can invoke an activity in another application, or—as you'll see in Chapter 15—you can communicate with a service of another application. All of these facilities provide methods for you to share information between applications, but they do so in an explicit manner because you don't directly access the underlying database, files, and so on.

Android's security at the process boundary is clear and simple. Things get interesting when we start talking about protecting resources (such as contact data), features (such as the device's camera), and our own components. To provide this protection, Android defines a permission scheme. Let's dissect that now.

Declaring and Using Permissions

Android defines a permission scheme meant to protect resources and features on the device. For example, applications, by default, cannot access the contacts list, make phone calls, and so on. To protect the user from malicious applications, Android requires applications to request permissions if they need to use a protected feature or resource. As you'll see shortly, permission requests go in the manifest file. At install time, the APK installer either grants or denies the requested permissions based on the signature of the `.apk` file and/or feedback from the user. If permission is not granted, any attempt to execute or access the associated feature will result in a permission failure.

Table 14–2 shows some commonly used features and the permissions they require. Although you are not yet familiar with all the features listed, you will learn about them later (either in this chapter or in subsequent chapters).

Table 14–2. *Features and Resources and the Permissions They Require*

Feature/Resource	Required Permission	Description
Camera	`android.permission.CAMERA`	Enables you to access the device's camera.
Internet	`android.permission.INTERNET`	Enables you to make a network connection.
User's contact data	`android.permission.READ_CONTACTS` `android.permission.WRITE_CONTACTS`	Enables you to read from or write to the user's contact data.
User's calendar data	`android.permission.READ_CALENDAR` `android.permission.WRITE_CALENDAR`	Enables you to read from or write to the user's calendar data.
Recording audio	`android.permission.RECORD_AUDIO`	Enables you to record audio.
Wi-Fi location information	`android.permission.ACCESS_COARSE_LOCATION`	Enables you to access coarse-grained location information from Wi-Fi and cell towers.
GPS location information	`android.permission.ACCESS_FINE_LOCATION`	Enables you to access fine-grained location information. This includes GPS location information. It is also sufficient for Wi-Fi and cell towers.

Feature/Resource	Required Permission	Description
Battery information	android.permission.BATTERY_STATS	Enables you to obtain battery-state information.
Bluetooth	android.permission.BLUETOOTH	Enables you to connect to paired Bluetooth devices.

For a complete list of permissions, see the following URL:

http://developer.android.com/reference/android/Manifest.permission.html

Application developers can request permissions by adding entries to the AndroidManifest.xml file. For example, Listing 14–3 asks to access the camera on the device, to read the list of contacts, and to read the calendar.

Listing 14–3. *Permissions in AndroidManifest.xml*

```
<manifest … >
    <application>
        …
    </application>
    <uses-permission android:name="android.permission.CAMERA" />
    <uses-permission android:name="android.permission.READ_CONTACTS"/>
    <uses-permission android:name="android.permission.READ_CALENDAR" />
</manifest>
```

Note that you can either hard-code permissions in the AndroidManifest.xml file or use the manifest editor. The manifest editor is wired up to launch when you open (double-click) the manifest file. The manifest editor contains a drop-down list that has all of the permissions preloaded to prevent you from making a mistake. As shown in Figure 14–5, you can access the permissions list by selecting the Permissions tab in the manifest editor.

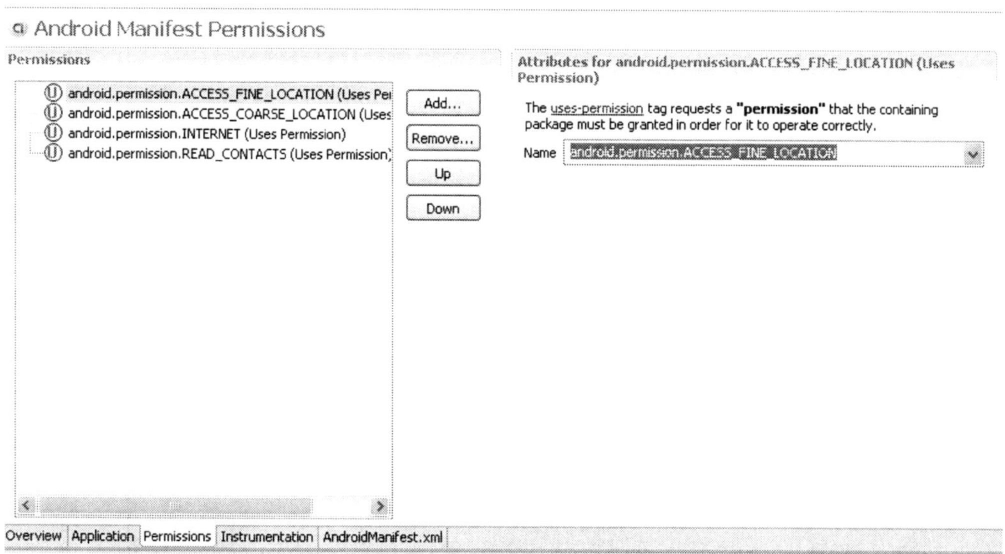

Figure 14–5. *The Android manifest editor tool in Eclipse*

You now know that Android defines a set of permissions that protects a set of features and resources. Similarly, you can define and enforce custom permissions with your application. Let's see how that works.

Understanding and Using Custom Permissions

Android allows you to define custom permissions with your application. For example, if you wanted to prevent certain users from starting one of the activities in your application, you could do that by defining a custom permission. To use custom permissions, you first declare them in your `AndroidManifest.xml` file. Once you've defined a permission, you can then refer to it as part of your component definition. We'll show you how this works.

Let's create an application containing an activity that not everyone is allowed to start. Instead, to start the activity, a user must have a specific permission. Once you have the application with a privileged activity, you can write a client that knows how to call the activity.

> **NOTE:** We will give you a URL at the end of the chapter that you can use to download projects from this chapter. This will allow you to import these projects into Eclipse directly.

First, create the project with the custom permission and activity. Open the Eclipse IDE, and select New ➤ New Project ➤ Android Project. This will open the New Android Project dialog box. Enter **CustomPermission** as the project name, select the Create New Project in Workspace radio button, and select the Use Default Location check box. Enter **Custom Permission** as the application name, **com.cust.perm** as the package name, and **CustPermMainActivity** as the activity name, and select a Build Target. Click the Finish button to create the project. The generated project will have the activity you just created, which will serve as the default (main) activity. Let's also create a *privileged activity*—an activity that requires a special permission. In the Eclipse IDE, go to the `com.cust.perm` package, create a class named `PrivActivity` whose superclass is `android.app.Activity`, and copy the code shown in Listing 14–4.

Listing 14–4. *The `PrivActivity` Class*

```
package com.cust.perm;

import android.app.Activity;
import android.os.Bundle;
import android.view.ViewGroup.LayoutParams;
import android.widget.LinearLayout;
import android.widget.TextView;

public class PrivActivity extends Activity
{
    @Override
    public void onCreate(Bundle savedInstanceState) {
        super.onCreate(savedInstanceState);
        LinearLayout view = new LinearLayout(this);
```

```
        view.setLayoutParams(new LayoutParams(
                LayoutParams.FILL_PARENT, LayoutParams.WRAP_CONTENT));
        view.setOrientation(LinearLayout.HORIZONTAL);

        TextView nameLbl = new TextView(this);

        nameLbl.setText("Hello from PrivActivity");
        view.addView(nameLbl);

        setContentView(view);
    }
}
```

As you can see, PrivActivity does not do anything miraculous. We just want to show you how to protect this activity with a permission and then call it from a client. If the client succeeds, you'll see the text "Hello from PrivActivity" on the screen. Now that you have an activity you want to protect, you can create the permission for it.

To create a custom permission, you have to define it in the AndroidManifest.xml file. The easiest way to do this is to use the manifest editor. Double-click the AndroidManifest.xml file, and select the Permissions tab. In the Permissions window, click the Add button, choose Permission, and click the OK button. The manifest editor will create an empty new permission for you. Populate the new permission by setting its attributes as shown in Figure 14–6. Fill in the fields on the right side, and if the label on the left still says just Permission, click it: it should update with the name from the right side.

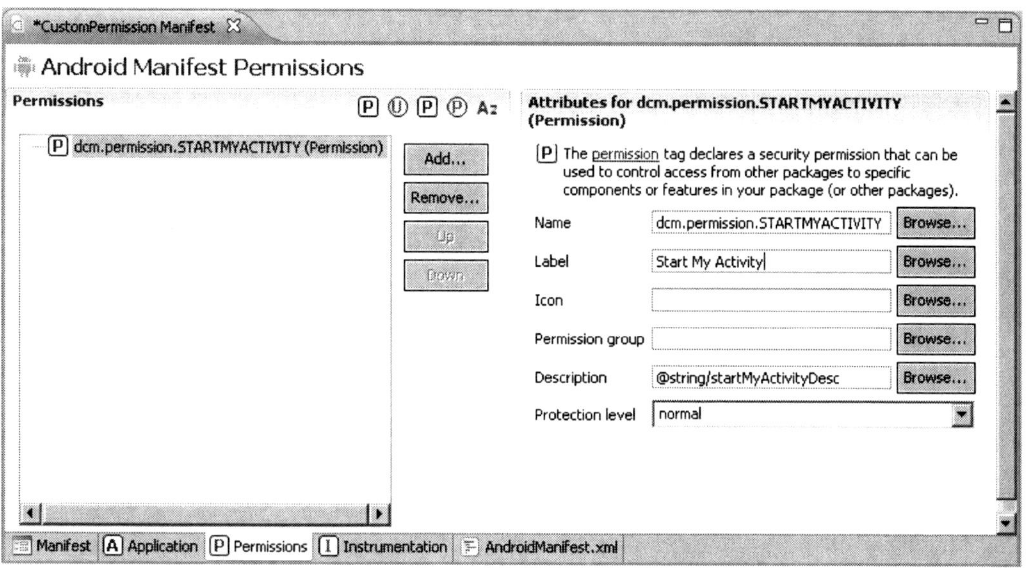

Figure 14–6. *Declaring a custom permission using the manifest editor*

As shown in Figure 14–6, each permission has a name, a label, an icon, a permission group, a description, and a protection level. Table 14–3 defines these properties.

Now you have a custom permission. Next, you want to tell the system that the `PrivActivity` activity should be launched only by applications that have the `dcm.permission.STARTMYACTIVITY` permission. You can set a required permission on an activity by adding the `android:permission` attribute to the activity definition in the `AndroidManifest.xml` file. For you to be able to launch the activity, you'll also need to add an intent-filter to the activity. Update your `AndroidManifest.xml` file with the content from Listing 14–5.

Table 14–3. *Attributes of a Permission*

Attribute	Required?	Description
`android:name`	Yes	Name of the permission. You should generally follow the Android naming scheme (`*.permission.*`).

`android:protectionLevel`	Yes	Defines the potential for risk associated with the permission. Must be one of the following values:
		`normal`
		`dangerous`
		`signature`
		`signatureOrSystem`
		Depending on the protection level, the system might take different action when determining whether to grant the permission. `normal` signals that the permission is low risk and will not harm the system, the user, or other applications. `dangerous` signals that the permission is high risk, and that the system will likely require input from the user before granting this permission. `signature` tells Android that the permission should be granted only to applications that have been signed with the same digital signature as the application that declared the permission. `signatureOrSystem` tells Android to grant the permission to applications with the same signature or to the Android package classes. This protection level is for very special cases involving multiple vendors needing to share features through the system image.

`android:permissionGroup`	No	You can place permissions into a group, but for custom permissions, you should avoid setting this property. If you really want to set this property, use this instead: `android.permission-group.SYSTEM_TOOLS`
`android:label`	No	Although it's not required, use this property to provide a short description of the permission.
`android:description`	No	Although it's not required, you should use this property to provide a more useful description of what the permission is for and what it protects.
`android:icon`	No	Permissions can be associated with an icon from your resources (such as `@drawable/myicon`).

Listing 14–5. *The* `AndroidManifest.xml` *File for the Custom-Permission Project*

```
<?xml version="1.0" encoding="utf-8"?>
<manifest xmlns:android="http://schemas.android.com/apk/res/android"
      package="com.cust.perm"
      android:versionCode="1"
      android:versionName="1.0.0">
    <application android:icon="@drawable/icon"
                 android:label="@string/app_name">
        <activity android:name=".CustPermMainActivity"
                  android:label="@string/app_name">
            <intent-filter>
              <action android:name="android.intent.action.MAIN" />
              <category android:name="android.intent.category.LAUNCHER"/>
            </intent-filter>
        </activity>
        <activity android:name="PrivActivity"
              android:permission="dcm.permission.STARTMYACTIVITY">
            <intent-filter>
                <action android:name="android.intent.action.MAIN" />
            </intent-filter>
        </activity>
    </application>

    <permission
        android:protectionLevel="normal"
        android:label="Start My Activity"
        android:description="@string/startMyActivityDesc"
        android:name="dcm.permission.STARTMYACTIVITY" />

    <uses-sdk android:minSdkVersion="4" />
</manifest>
```

Listing 14–5 requires that you add a string constant named `startMyActivityDesc` to your string resources. To ensure compilation of Listing 14–5, add the following string resource to the `res/values/strings.xml` file:

```
<string name="startMyActivityDesc">Allows starting my activity</string>
```

Now, run the project in the emulator. Although the main activity does not do anything, you want the application installed on the emulator before you write a client for the privileged activity.

Let's write a client for the privileged activity. In the Eclipse IDE, choose New ➤ Project ➤ Android Project. Enter **ClientOfCustomPermission** as the project name, select the Create New Project in Workspace radio button, and select the Use Default Location check box. Set the application name to **Client Of Custom Permission**, the package name to **com.client.cust.perm**, and the activity name to **ClientCustPermMainActivity**, and select a Build Target. Click the Finish button to create the project.

Next, you want to write an activity that displays a button you can click to call the privileged activity. Copy the layout shown in Listing 14–6 to the `main.xml` file in the project you just created.

Listing 14–6. *Main.xml File for the Client Project*

```
<?xml version="1.0" encoding="utf-8"?>
<LinearLayout xmlns:android="http://schemas.android.com/apk/res/android"
    android:orientation="vertical"
    android:layout_width="fill_parent"
    android:layout_height="fill_parent"  >

    <Button android:id="@+id/btn"    android:text="Launch PrivActivity"
    android:layout_width="wrap_content"
    android:layout_height="wrap_content"
    android:onClick="doClick"  />
</LinearLayout>
```

As you can see, the XML layout file defines a single button whose text reads Launch PrivActivity. Now, let's write an activity that will handle the button-click event and launch the privileged activity. Copy the code from Listing 14–7 to your `ClientCustPermMainActivity` class.

Listing 14–7. *The Modified `ClientCustPermMainActivity` Activity*

```
package com.client.cust.perm;
// This file is ClientCustPermMainActivity.java

import android.app.Activity;
import android.content.Intent;
import android.os.Bundle;
import android.view.View;

public class ClientCustPermMainActivity extends Activity {
    @Override
    public void onCreate(Bundle savedInstanceState) {
        super.onCreate(savedInstanceState);
        setContentView(R.layout.main);
    }

    public void doClick(View view) {
                Intent intent = new Intent();
                intent.setClassName("com.cust.perm","com.cust.perm.PrivActivity");
                startActivity(intent);
    }
}
```

As shown in Listing 14–7, when the button is invoked, you create a new intent and then set the class name of the activity you want to launch. In this case, you want to launch `com.cust.perm.PrivActivity` in the `com.cust.perm` package.

The only thing missing at this point is a `uses-permission` entry, which you add to the manifest file to tell the Android runtime that you need `dcm.permission.STARTMYACTIVITY` to run. Replace your client project's manifest file with that shown in Listing 14–8.

Listing 14–8. *The Client Manifest File*

```
<?xml version="1.0" encoding="utf-8"?>
<manifest xmlns:android="http://schemas.android.com/apk/res/android"
      package="com.client.cust.perm"
      android:versionCode="1"
      android:versionName="1.0.0">
```

```
    <application android:icon="@drawable/icon"
                 android:label="@string/app_name">
        <activity android:name=".ClientCustPermMainActivity"
                  android:label="@string/app_name">
            <intent-filter>
              <action android:name="android.intent.action.MAIN" />
              <category android:name="android.intent.category.LAUNCHER" />
            </intent-filter>
        </activity>
    </application>

    <uses-permission android:name="dcm.permission.STARTMYACTIVITY" />
    <uses-sdk android:minSdkVersion="4" />
</manifest>
```

As shown in Listing 14–8, we added a `uses-permission` entry to request the custom permission required to start the `PrivActivity` we implemented in the custom-permission project.

With that, you should be able to deploy the client project to the emulator and then click the Launch PrivActivity button. When the button is invoked, you should see the text "Hello from PrivActivity."

After you successfully call the privileged activity, remove the `uses-permission` entry from your client project's manifest file and redeploy the project to the emulator. Once it's deployed, confirm that you get an error when you invoke the button to launch the privileged activity. Note that LogCat will display a permission-denial exception.

Now you know how custom permissions work in Android. Obviously, custom permissions are not limited to activities. In fact, you can apply both predefined and custom permissions to Android's other types of components as well. We'll explore an important one next: URI permissions.

Understanding and Using URI Permissions

Content providers (discussed in Chapter 4) often need to control access at a finer level than all or nothing. Fortunately, Android provides a mechanism for this. Think about e-mail attachments. The attachment may need to be read by another activity to display it. But the other activity should not get access to all of the e-mail data and does not need access even to all attachments. This is where URI permissions come in.

Passing URI Permissions in Intents

When invoking another activity and passing a URI, your application can specify that it is granting permissions to the URI being passed. But before your application can do this, it needs permission itself to the URI, and the URI content provider must cooperate and allow the granting of permissions to another activity. The code to invoke an activity with granting of permissions looks like Listing 14–9, which is actually from the Android Email program, where it is launching an activity to view an e-mail attachment.

Listing 14–9. *Code to Launch an Activity with Granting of Permission*

```
try {
    Intent intent = new Intent(Intent.ACTION_VIEW);
    intent.setData(contentUri);
    intent.addFlags(Intent.FLAG_GRANT_READ_URI_PERMISSION);
    startActivity(intent);
} catch (ActivityNotFoundException e) {
    mHandler.attachmentViewError();
    // TODO: Add a proper warning message (and lots of upstream cleanup to prevent
    // it from happening) in the next release.
}
```

The attachment is specified by `contentUri`. Notice how the intent is created with the action `Intent.ACTION_VIEW`, and the data is set using `setData()`. The flag is set to grant read permission of the attachment to whatever activity will match on the intent. This is where the content provider comes into play. Just because an activity has read permission to content doesn't mean it can pass along that permission to some other activity that does not have the permission already. The content provider must allow it as well. As Android finds a matching intent filter on an activity, it consults with the content provider to make sure that permissions can be granted. In essence, the content provider is being asked to allow access to this new activity to the content specified by the URI. If the content provider refuses, then a `SecurityException` is thrown, and the operation fails. In Listing 14–9, this particular application is not checking for a `SecurityException`, because the developer is not expecting any refusals to grant permission. That's because the attachment content provider is part of the Email application! There is a possibility though that no activity can be found to handle the attachment, so that is the only exception being watched for.

In the case where the activity being called to process the URI already has permission to access that URI, the content provider does not get to deny access. That is, the calling activity can grant permission, and if the activity on the receiving end of the intent already has the necessary permissions for `contentURI`, the called activity will be allowed to proceed with no problems.

In addition to `Intent.FLAG_GRANT_READ_URI_PERMISSION`, there is a flag for write permissions: `Intent.FLAG_GRANT_WRITE_URI_PERMISSION`. It is possible to specify both in an `Intent`. Also, these flags can apply to services and `BroadcastReceivers` as well as activities because they can receive intents too.

Specifying URI Permissions in Content Providers

So how does a content provider specify URI permissions? It does so in the `AndroidManifest.xml` file in one of two ways:

- In the `<provider>` tag, the `android:grantUriPermissions` attribute can be set to either `true` or `false`. If `true`, any content from this content provider can be granted. If `false`, the second way of specifying URI permissions can happen, or the content provider can decide not to let anyone else grant permissions.

- Specify permissions with child tags of `<provider>`. The child tag is `<grant-uri-permission>`, and you can have more than one within `<provider>`. `<grant-uri-permission>` has three possible attributes:
  - Using the `android:path` attribute, you can specify a complete path which will then have permissions that are grantable.
  - Similarly, `android:pathPrefix` specifies the beginning of a URI path
  - `android:pathPattern` allows wildcards (the asterisk, *, character) to specify a path.

As we stated before, the granting entity must also have appropriate permissions to the content before being allowed to grant them to some other entity. Content providers have additional ways of controlling access to their content, through the `android:readPermission` attribute of the `<provider>` tag, the `android:writePermission` attribute, and the `android:permission` attribute (a convenient way to specify both read and write permissions with one permission `String` value). The value for any of these three attributes is a `String` that represents the permission a caller must have in order to read or write with this content provider. Before an activity could grant read permission to a content URI, that activity must have read permission first, as specified by either the `android:readPermission` attribute or the `android:permission` attribute. The entity wanting these permissions would declare them in their manifest file with the `<uses-permissions>` tag.

References

Here are some helpful references to topics you may wish to explore further:

- `www.androidbook.com/proandroid4/projects`: A list of downloadable projects related to this book. For this chapter, look for a zip file called `ProAndroid4_Ch14_Security.zip`. This zip file contains all projects from this chapter, listed in separate root directories. There is also a `README.TXT` file that describes exactly how to import projects into Eclipse from one of these zip files.
- `http://developer.android.com/guide/topics/security/security.html`: The *Android Developer's Guide* section "Security and Permissions." It provides an overview with links to lots of reference pages.
- `http://developer.android.com/guide/publishing/app-signing.html`: The *Android Developer's Guide* section "Signing Your Applications."
- `http://android.git.kernel.org/?p=platform/packages/apps/Email.git;a=blob_plain;f=src/com/android/email/activity/MessageView.java`: The source code from the stock Android Email application, where a `FLAG_GRANT_READ_URI_PERMISSION` is used. You can see how the Android team implements URI permissions by browsing in the source code for this application.

Summary

This security chapter covered the following topics:

- Unique application user IDs that help separate apps from each other to protect processing and data
- Digital certificates and their use in signing Android applications
- That an application can only be updated if the update is signed with the same digital certificate as the original
- Managing certificates in a keystore using `keytool`
- Running `jarsigner` to apply a certificate to an application `.apk` file
- `zipalign` and memory boundaries
- The Eclipse plugin wizard takes care of generating the apk, applying the certificate and `zipalign`-ing for you.
- Manually installing apps onto devices and emulators
- Permissions that applications can declare and use
- URI permissions and how content providers use them

Interview Questions

Ask yourself the following questions to solidify your understanding of this chapter:

1. For how long must a certificate be valid to be able to deploy an application to Android Market?
2. What other Android artifact is tied to an application's digital certificate?
3. Which tool is used to create or view a digital certificate?
4. Which tool is used to sign an application with a digital certificate?
5. What does `zipalign` do to an application?
6. Could `zipalign` increase the size of your application by a lot? Why or why not?
7. If a certificate expires for an application that has already been installed, does that application stop working on the device?
8. In which file are the permission declarations for an application stored?
9. What must happen first before an application can grant a URI permission to another activity?

Chapter 15

Building and Consuming Services

The Android Platform provides a complete software stack. This means you get an operating system and middleware, as well as working applications (such as a phone dialer). Alongside all of this, you have an SDK that you can use to write applications for the platform. Thus far, we've seen that we can build applications that directly interact with the user through a user interface. We have not, however, discussed background services or the possibilities of building components that run in the background.

In this chapter, we are going to focus on building and consuming services in Android. First we'll discuss consuming HTTP services, and then we'll cover a nice way to do simple background tasks, and finally we'll discuss interprocess communication—that is, communication between applications on the same device.

Consuming HTTP Services

Android applications and mobile applications in general are small apps with a lot of functionality. One of the ways that mobile apps deliver such rich functionality on such a small device is that they pull information from various sources. For example, most Android smartphones come with the Maps application, which provides sophisticated mapping functionality. We, however, know that the application is integrated with the Google Maps API and other services, which provide most of the sophistication.

That said, it is likely that the applications you write will also leverage information from other applications and APIs. A common integration strategy is to use HTTP. For example, you might have a Java servlet available on the Internet that provides services you want to leverage from one of your Android applications. How do you do that with Android? Interestingly, the Android SDK ships with a variation of Apache's `HttpClient` (`http://hc.apache.org/httpclient-3.x/`), which is universally used within the J2EE space. The Android version has been modified for Android, but the APIs are very similar to the APIs in the J2EE version.

The Apache HttpClient is a comprehensive HTTP client. Although it offers full support for the HTTP protocol, you will likely utilize only HTTP GET and POST. In this section, we will discuss using the HttpClient to make HTTP GET and HTTP POST calls.

Using the HttpClient for HTTP GET Requests

Here's one of the general patterns for using the HttpClient:

1. Create an HttpClient (or get an existing reference).
2. Instantiate a new HTTP method, such as PostMethod or GetMethod.
3. Set HTTP parameter names/values.
4. Execute the HTTP call using the HttpClient.
5. Process the HTTP response.

Listing 15–1 shows how to execute an HTTP GET using the HttpClient.

> **NOTE:** We give you a URL at the end of the chapter that you can use to download projects from this chapter. This will allow you to import these projects into your Eclipse directly. Also, because the code attempts to use the Internet, you will need to add android.permission.INTERNET to your manifest file when making HTTP calls using the HttpClient.

Listing 15–1. *Using HttpClient and HttpGet: HttpGetDemo.java*

```
import java.io.BufferedReader;
import java.io.IOException;
import java.io.InputStreamReader;
import org.apache.http.HttpResponse;
import org.apache.http.client.HttpClient;
import org.apache.http.client.methods.HttpGet;
import org.apache.http.impl.client.DefaultHttpClient;
import android.app.Activity;
import android.os.Bundle;

public class HttpGetDemo extends Activity {
    /** Called when the activity is first created. */
    @Override
    public void onCreate(Bundle savedInstanceState) {
        super.onCreate(savedInstanceState);
        setContentView(R.layout.main);

        BufferedReader in = null;
        try {

            HttpClient client = new DefaultHttpClient();
            HttpGet request = new HttpGet("http://code.google.com/android/");
            HttpResponse response = client.execute(request);

            in = new BufferedReader(
                    new InputStreamReader(
```

```
                    response.getEntity().getContent()));

            StringBuffer sb = new StringBuffer("");
            String line = "";
            String NL = System.getProperty("line.separator");
            while ((line = in.readLine()) != null) {
                sb.append(line + NL);
            }
            in.close();

            String page = sb.toString();
            System.out.println(page);
        } catch (Exception e) {
            e.printStackTrace();
        } finally {
            if (in != null) {
                try {
                    in.close();
                } catch (IOException e) {
                    e.printStackTrace();
                }
            }
        }
    }
}
```

The `HttpClient` provides abstractions for the various HTTP request types, such as `HttpGet`, `HttpPost`, and so on. Listing 15–1 uses the `HttpClient` to get the contents of the `http://code.google.com/android/` URL. The actual HTTP request is executed with the call to `client.execute()`. After executing the request, the code reads the entire response into a string object. Note that the `BufferedReader` is closed in the `finally` block, which also closes the underlying HTTP connection.

For our example we embedded the HTTP logic inside of an activity, but we don't need to be within the context of an activity to use `HttpClient`. You can use it from within the context of any Android component or use it as part of a stand-alone class. In fact, you really shouldn't use HttpClient directly within an activity, because a web call could take a while to complete and cause the activity to be force-closed. We'll cover that topic later in this chapter. For now we're going to cheat a little so we can focus on how to make `HttpClient` calls.

The code in Listing 15–1 executes an HTTP request without passing any HTTP parameters to the server. You can pass name/value parameters as part of the request by appending name/value pairs to the URL, as shown in Listing 15–2.

Listing 15–2. *Adding Parameters to an HTTP GET Request*

```
HttpGet request = new HttpGet("http://somehost/WS2/Upload.aspx?one=valueGoesHere");
client.execute(request);
```

When you execute an HTTP GET, the parameters (names and values) of the request are passed as part of the URL. Passing parameters this way has some limitations. Namely, the length of a URL should be kept below 2,048 characters. If you have more than this amount of data to submit, you should use HTTP POST instead. The POST method is more flexible and passes parameters as part of the request body.

Using the HttpClient for HTTP POST Requests (a Multipart Example)

Making an HTTP POST call is very similar to making an HTTP GET call (see Listing 15–3).

Listing 15–3. ***Making an HTTP POST Request with the*** `HttpClient`

```
HttpClient client = new DefaultHttpClient();
HttpPost request = new HttpPost(
        "http://192.165.13.37/services/doSomething.do");
List<NameValuePair> postParameters = new ArrayList<NameValuePair>();
postParameters.add(new BasicNameValuePair("first",
        "param value one"));
postParameters.add(new BasicNameValuePair("issuenum", "10317"));
postParameters.add(new BasicNameValuePair("username", "dave"));
UrlEncodedFormEntity formEntity = new UrlEncodedFormEntity(
        postParameters);
request.setEntity(formEntity);
HttpResponse response = client.execute(request);
```

The code in Listing 15–3 would replace the three lines in Listing 15–1 where the `HttpGet` is used. Everything else could stay the same. To make an HTTP POST call with the `HttpClient`, you have to call the `execute()` method of the `HttpClient` with an instance of `HttpPost`. When making HTTP POST calls, you generally pass URL-encoded name/value form parameters as part of the HTTP request. To do this with the `HttpClient`, you have to create a list that contains instances of `NameValuePair` objects and then wrap that list with a `UrlEncodedFormEntity` object. The `NameValuePair` wraps a name/value combination, and the `UrlEncodedFormEntity` class knows how to encode a list of `NameValuePair` objects suitable for HTTP calls (generally POST calls). After you create a `UrlEncodedFormEntity`, you can set the entity type of the `HttpPost` to the `UrlEncodedFormEntity` and then execute the request.

In Listing 15–3, we created an `HttpClient` and then instantiated the `HttpPost` with the URL of the HTTP endpoint. Next, we created a list of `NameValuePair` objects and populated it with several name/value parameters. We then created a `UrlEncodedFormEntity` instance, passing the list of `NameValuePair` objects to its constructor. Finally, we called the `setEntity()` method of the POST request and then executed the request using the `HttpClient` instance.

HTTP POST is actually much more powerful than this. With an HTTP POST, we can pass simple name/value parameters, as shown in Listing 15–3, as well as complex parameters such as files. HTTP POST supports another request-body format known as a *multipart POST*. With this type of POST, you can send name/value parameters as before, along with arbitrary files. Unfortunately, the version of `HttpClient` shipped with Android does not directly support multipart POST. To do multipart POST calls, you need to get three additional Apache open source projects: Apache Commons IO, Mime4j, and HttpMime. You can download these projects from the following web sites:

- *Commons IO*: `http://commons.apache.org/io/`
- *Mime4j*: `http://james.apache.org/mime4j/`
- *HttpMime*: `http://hc.apache.org/downloads.cgi` (inside of HttpClient)

Alternatively, you can visit this site to download all of the required `.jar` files to do multipart POST with Android:

`http://www.apress.com/book/view/1430226595`

Listing 15–4 demonstrates a multipart POST using Android.

Listing 15–4. ***Making a Multipart POST Call***

```
import java.io.ByteArrayInputStream;
import java.io.InputStream;
import org.apache.commons.io.IOUtils;
import org.apache.http.HttpResponse;
import org.apache.http.client.HttpClient;
import org.apache.http.client.methods.HttpPost;
import org.apache.http.entity.mime.MultipartEntity;
import org.apache.http.entity.mime.content.InputStreamBody;
import org.apache.http.entity.mime.content.StringBody;
import org.apache.http.impl.client.DefaultHttpClient;

import android.app.Activity;

public class TestMultipartPost extends Activity
{
    public void executeMultipartPost() throws Exception
    {
        try {
            InputStream is = this.getAssets().open("data.xml");
            HttpClient httpClient = new DefaultHttpClient();
            HttpPost postRequest =
              new HttpPost("http://mysomewebserver.com/services/doSomething.do");

            byte[] data = IOUtils.toByteArray(is);

            InputStreamBody isb = new InputStreamBody(new
                    ByteArrayInputStream(data), "uploadedFile");
            StringBody sb1 = new StringBody("some text goes here");
            StringBody sb2 = new StringBody("some text goes here too");

            MultipartEntity multipartContent = new MultipartEntity();
            multipartContent.addPart("uploadedFile", isb);
            multipartContent.addPart("one", sb1);
            multipartContent.addPart("two", sb2);

            postRequest.setEntity(multipartContent);
            HttpResponse response =httpClient.execute(postRequest);
            response.getEntity().getContent().close();
        } catch (Throwable e)
        {
            // handle exception here
        }
    }
}
```

> **NOTE:** The multipart example uses several `.jar` files that are not included as part of the Android runtime. To ensure that the `.jar` files will be packaged as part of your `.apk` file, you need to add them as external `.jar` files in Eclipse. To do this, right-click your project in Eclipse, select Properties, choose Java Build Path, select the Libraries tab, and then select Add External JARs.
>
> Following these steps will make the `.jar` files available during compile time as well as runtime.

To execute a multipart POST, you need to create an `HttpPost` and call its `setEntity()` method with a `MultipartEntity` instance (rather than the `UrlEncodedFormEntity` we created for the name/value parameter form post). `MultipartEntity` represents the body of a multipart POST request. As shown, you create an instance of a `MultipartEntity` and then call the `addPart()` method with each part. Listing 15–4 adds three parts to the request: two string parts and an XML file.

Finally, if you are building an application that requires you to pass a multipart POST to a web resource, you'll likely have to debug the solution using a dummy implementation of the service on your local workstation. When you're running applications on your local workstation, normally you can access the local machine by using `localhost` or IP address `127.0.0.1`. With Android applications, however, you will not be able to use `localhost` (or `127.0.0.1`) because the emulator will be its own `localhost`. You don't want to point this client to a service on the Android device, you want to point to your workstation. To refer to your development workstation from the application running in the emulator, you'll have to use your workstation's IP address. (Refer back to Chapter 2 if you need help figuring out what your workstation's IP address is.) You will need to modify Listing 15–4 by substituting the IP address with the IP address of your workstation.

SOAP, JSON, and XML Parsers

What about SOAP? There are lots of SOAP-based web services on the Internet, but to date, Google has not provided direct support in Android for calling SOAP web services. Google instead prefers REST-like web services, seemingly to reduce the amount of computing required on the client device. However, the tradeoff is that the developer must do more work to send data and to parse the returned data. Ideally, you will have some options for how you can interact with your web services. Some developers have used the kSOAP2 developer kit to build SOAP clients for Android. We won't be covering that approach, but it's out there if you're interested.

> **NOTE:** The original kSOAP2 source is located here: `http://ksoap2.sourceforge.net/`. The open source community has (thankfully!) contributed a version of kSOAP2 for Android, and you can find out more about it here: `http://code.google.com/p/ksoap2-android/`.

One approach that's been used successfully is to implement your own services on the Internet, which can talk SOAP (or whatever) to the destination service. Then your Android application only needs to talk to your services, and you now have complete control. If the destination services change, you might be able to handle that without having to update and release a new version of your application. You'd only have to update the services on your server. A side benefit of this approach is that you could more easily implement a paid subscription model for your application. If a user lets their subscription lapse, you can turn them off at your server.

Android *does* have support for JavaScript Object Notation (JSON). This is a fairly common method of packaging data between a web server and a client. The JSON parsing classes make it very easy to unpack data from a response so your application can act on it.

Android also has a couple of XML parsers that you can use to interpret the responses from the HTTP calls. The main one (`XMLPullParser`) was covered in Chapter 3.

Dealing with Exceptions

Dealing with exceptions is part of any program, but software that makes use of external services (such as HTTP services) must pay additional attention to exceptions because the potential for errors is magnified. There are several types of exceptions that you can expect while making use of HTTP services. These are transport exceptions, protocol exceptions, and timeouts. You should understand when these exceptions could occur.

Transport exceptions can occur due to a number of reasons, but the most likely scenario with a mobile device is poor network connectivity. Protocol exceptions are exceptions at the HTTP protocol layer. These include authentication errors, invalid cookies, and so on. You can expect to see protocol exceptions if, for example, you have to supply login credentials as part of your HTTP request but fail to do so. Timeouts, with respect to HTTP calls, come in two flavors: connection timeouts and socket timeouts. A connection timeout can occur if the `HttpClient` is not able to connect to the HTTP server—if, for example, the server is not available. A socket timeout can occur if the `HttpClient` fails to receive a response within a defined time period. In other words, the `HttpClient` was able to connect to the server, but the server failed to return a response within the allocated time limit.

Now that you understand the types of exceptions that might occur, how do you deal with them? Fortunately, the `HttpClient` is a robust framework that takes most of the burden off your shoulders. In fact, the only exception types that you'll have to worry about are the ones that you'll be able to manage easily. The `HttpClient` takes care of transport exceptions by detecting transport issues and retrying requests (which works very well with this type of exception). Protocol exceptions are exceptions that can generally be flushed out during development. Timeouts are the most likely exceptions that you'll have to deal with. A simple and effective approach to dealing with both types of timeouts—connection timeouts and socket timeouts—is to wrap the `execute()` method of your HTTP request with a `try/catch` and then retry if a failure occurs. This is demonstrated in Listing 15–5.

Listing 15–5. *Implementing a Simple Retry Technique to Deal with Timeouts*

```
import java.io.BufferedReader;
import java.io.IOException;
import java.io.InputStreamReader;
import java.net.URI;

import org.apache.http.HttpResponse;
import org.apache.http.client.HttpClient;
import org.apache.http.client.methods.HttpGet;
import org.apache.http.impl.client.DefaultHttpClient;

public class TestHttpGet {

    public String executeHttpGetWithRetry() throws Exception {
        int retry = 3;

        int count = 0;
        while (count < retry) {
            count += 1;
            try {
                String response = executeHttpGet();
                /**
                 * if we get here, that means we were successful and we
                 * can stop.
                 */
                return response;
            } catch (Exception e) {
                /**
                 * if we have exhausted our retry limit
                 */
                if (count < retry) {
                    /**
                     * we have retries remaining, so log the message
                     * and go again.
                     */
                    System.out.println(e.getMessage());
                } else {
                    System.out.println("all retries failed");
                    throw e;
                }
            }
        }
        return null;
    }

    public String executeHttpGet() throws Exception {
        BufferedReader in = null;
        try {
            HttpClient client = new DefaultHttpClient();
            HttpGet request = new
                    HttpGet("http://code.google.com/android/");
            HttpResponse response = client.execute(request);
            in = new BufferedReader(
                        new InputStreamReader(
                               response.getEntity().getContent()));

            StringBuffer sb = new StringBuffer("");
```

```
            String line = "";
            String NL = System.getProperty("line.separator");
            while ((line = in.readLine()) != null) {
                sb.append(line + NL);
            }
            in.close();

            String result = sb.toString();
            return result;
        } finally {
            if (in != null) {
                try {
                    in.close();
                } catch (IOException e) {
                    e.printStackTrace();
                }
            }
        }
    }
}
```

The code in Listing 15–5 shows how you can implement a simple retry technique to recover from timeouts when making HTTP calls. The listing shows two methods: one that executes an HTTP GET (`executeHttpGet()`), and another that wraps this method with the retry logic (`executeHttpGetWithRetry()`). The logic is very simple. We set the number of retries we want to attempt to 3, and then we enter a `while` loop. Within the loop, we execute the request. Note that the request is wrapped with a `try/catch` block, and in the `catch` block we check whether we have exhausted the number of retry attempts.

When using the `HttpClient` as part of a real-world application, you need to pay some attention to multithreading issues that might come up. Let's delve into these now.

Addressing Multithreading Issues

The examples we've shown so far created a new `HttpClient` for each request. In practice, however, you should probably create one `HttpClient` for the entire application and use that for all of your HTTP communication. With one `HttpClient` servicing all of your HTTP requests, you should also pay attention to multithreading issues that could surface if you make simultaneous requests through the same `HttpClient`. Fortunately, the `HttpClient` provides facilities that make this easy—all you have to do is create the `DefaultHttpClient` using a `ThreadSafeClientConnManager`, as shown in Listing 15–6.

Listing 15–6. *Creating an `HttpClient` for Multithreading: `CustomHttpClient.java`*

```
import org.apache.http.HttpVersion;
import org.apache.http.client.HttpClient;
import org.apache.http.conn.ClientConnectionManager;
import org.apache.http.conn.params.ConnManagerParams;
import org.apache.http.conn.scheme.PlainSocketFactory;
import org.apache.http.conn.scheme.Scheme;
import org.apache.http.conn.scheme.SchemeRegistry;
import org.apache.http.conn.ssl.SSLSocketFactory;
import org.apache.http.impl.client.DefaultHttpClient;
import org.apache.http.impl.conn.tsccm.ThreadSafeClientConnManager;
```

```
import org.apache.http.params.BasicHttpParams;
import org.apache.http.params.HttpConnectionParams;
import org.apache.http.params.HttpParams;
import org.apache.http.params.HttpProtocolParams;
import org.apache.http.protocol.HTTP;

public class CustomHttpClient {
    private static HttpClient customHttpClient;

    /** A private Constructor prevents instantiation */
    private CustomHttpClient() {
    }

    public static synchronized HttpClient getHttpClient() {
        if (customHttpClient == null) {
            HttpParams params = new BasicHttpParams();
            HttpProtocolParams.setVersion(params, HttpVersion.HTTP_1_1);
            HttpProtocolParams.setContentCharset(params,
                    HTTP.DEFAULT_CONTENT_CHARSET);
            HttpProtocolParams.setUseExpectContinue(params, true);
            HttpProtocolParams.setUserAgent(params,
"Mozilla/5.0 (Linux; U; Android 2.2.1; en-us; Nexus One Build/FRG83) AppleWebKit/533.1
(KHTML, like Gecko) Version/4.0 Mobile Safari/533.1"
            );

            ConnManagerParams.setTimeout(params, 1000);

            HttpConnectionParams.setConnectionTimeout(params, 5000);
            HttpConnectionParams.setSoTimeout(params, 10000);

            SchemeRegistry schReg = new SchemeRegistry();
            schReg.register(new Scheme("http",
                            PlainSocketFactory.getSocketFactory(), 80));
            schReg.register(new Scheme("https",
                            SSLSocketFactory.getSocketFactory(), 443));
            ClientConnectionManager conMgr = new
                            ThreadSafeClientConnManager(params,schReg);

            customHttpClient = new DefaultHttpClient(conMgr, params);
        }
        return customHttpClient;
    }

    public Object clone() throws CloneNotSupportedException {
        throw new CloneNotSupportedException();
    }
}
```

If your application needs to make more than a few HTTP calls, you should create an `HttpClient` that services all your HTTP requests. The simplest way to do this is to create a singleton class that can be accessed from anywhere in the application, as we've shown here. This is a fairly standard Java pattern in which we synchronize access to a getter method, and that getter method returns the one and only `HttpClient` object for the singleton, creating it the first time as necessary.

Now, take a look at the getHttpClient() method of CustomHttpClient. This method is responsible for creating our singleton HttpClient. We set some basic parameters, some timeout values, and the schemes that our HttpClient will support (that is, HTTP and HTTPS). Notice that when we instantiate the DefaultHttpClient(), we pass in a ClientConnectionManager. The ClientConnectionManager is responsible for managing HTTP connections for the HttpClient. Because we want to use a single HttpClient for all the HTTP requests (requests that could overlap if we're using threads), we create a ThreadSafeClientConnManager.

We also show you a simpler way of collecting the response from the HTTP request, using a BasicResponseHandler. The code for our activity that uses our CustomHttpClient is in Listing 15–7.

Listing 15–7. *Using Our CustomHttpClient: HttpActivity.java*

```
import java.io.IOException;
import org.apache.http.client.HttpClient;
import org.apache.http.client.methods.HttpGet;
import org.apache.http.impl.client.BasicResponseHandler;
import org.apache.http.params.HttpConnectionParams;
import org.apache.http.params.HttpParams;
import android.app.Activity;
import android.os.Bundle;
import android.util.Log;

public class HttpActivity extends Activity
{
    private HttpClient httpClient;
    @Override
    public void onCreate(Bundle savedInstanceState)
    {
        super.onCreate(savedInstanceState);
        setContentView(R.layout.main);

        httpClient = CustomHttpClient.getHttpClient();
        getHttpContent();
    }

    public void getHttpContent()
    {
        try {
            HttpGet request = new HttpGet("http://www.google.com/");
            String page = httpClient.execute(request,
                    new BasicResponseHandler());
            System.out.println(page);
        } catch (IOException e) {
            // covers:
            //      ClientProtocolException
            //      ConnectTimeoutException
            //      ConnectionPoolTimeoutException
            //      SocketTimeoutException
            e.printStackTrace();
        }
    }
}
```

For this sample application, we do a simple HTTP get of the Google home page. We also use a BasicResponseHandler object to take care of rendering the page as a big String, which we then write out to LogCat. As you can see, adding a BasicResponseHandler to the execute() method is very easy to do.

You may be tempted to take advantage of the fact that each Android application has an associated Application object. By default, if you don't define a custom application object, Android uses android.app.Application. Here's the interesting thing about the application object: there will always be exactly one application object for your application, and all of your components can access it (using the global context object). It is possible to extend the Application class and add functionality such as our CustomHttpClient. However, in our case there is really no reason to do this within the Application class itself, and you will be much better off not messing with the Application class when you can simply create a separate singleton class to handle this type of need.

Fun with Timeouts

There are other terrific advantages to setting up a single HttpClient for our application to use. We can modify the properties of it in one place, and everyone can take advantage of it. For example, if we want to set up common timeout values for our HTTP calls, we can do that when creating our HttpClient by calling the appropriate setter functions against our HttpParams object. Please refer to Listing 15–6 and the getHttpClient() method. Notice that there are three timeouts we can play with. The first is a timeout for the connection manager, and it defines how long we should wait to get a connection out of the connection pool managed by the connection manager. In our example, we set this to 1 second. About the only time we might ever have to wait is if all connections from the pool are in use. The second timeout value defines how long we should wait to make a connection over the network to the server on the other end. Here, we used a value of 2 seconds. And lastly, we set a socket timeout value to 4 seconds to define how long we should wait to get data back for our request.

Corresponding to the three timeouts described previously, we could get these three exceptions: ConnectionPoolTimeoutException, ConnectTimeoutException, or SocketTimeoutException. All three of these exceptions are subclasses of IOException, which we've used in our HttpActivity instead of catching each subclass exception separately.

If you investigate each of the parameter-setting classes that we used in getHttpClient(), you might discover even more parameters that you would find useful.

We've described for you how to set up a common pool of HTTP connections for use across your application. And the implication is that every time you need to use a connection, the various settings will apply to your particular needs. But what if you want different settings for a particular message? Thankfully, there's an easy way to do that as well. We showed you how to use an HttpGet or an HttpPost object to describe the request to be made across the network. In a similar way to how we set HttpParams on our HttpClient, you can set HttpParams on both HttpGet and HttpPost objects. The settings you apply at the message level will override the settings at the HttpClient level without changing the HttpClient settings. Listing 15–8 shows what this might look like if we wanted to have a socket timeout of 1 minute instead of 4 seconds for one particular request. You would use these lines in place of the lines in the try block of getHttpContent() in Listing 15–7.

Listing 15–8. *Overriding the Socket Timeout at the Request Level*

```
HttpGet request = new HttpGet("http://www.google.com/");
HttpParams params = request.getParams();
HttpConnectionParams.setSoTimeout(params, 60000);    // 1 minute
request.setParams(params);
String page = httpClient.execute(request,
                    new BasicResponseHandler());
System.out.println(page);
```

Using the HttpURLConnection

Android provides another way to deal with HTTP services, and that is using the java.net.HttpURLConnection class. This is not unlike the HttpClient classes we've just covered, but HttpURLConnection tends to require more statements to get things done. On the other hand, this class is much smaller and lightweight than HttpClient. Starting with the Gingerbread release, it is also fairly stable, so you should consider it for apps on more recent devices when you just need basic HTTP features and you want a compact application.

Using the AndroidHttpClient

Android 2.2 introduced a new subclass of HttpClient called AndroidHttpClient. The idea behind this class is to make things easier for the developer of Android apps by providing default values and logic appropriate for Android apps. For example, the timeout values for the connection and the socket (that is, operation) default to 20 seconds each. The connection manager defaults to the ThreadSafeClientConnManager. For the most part, it is interchangeable with the HttpClient we used in the previous examples. There are a few differences, though, that you should be aware of:

- To create an AndroidHttpClient, you invoke the static newInstance() method of the AndroidHttpClient class, like this:

```
AndroidHttpClient httpClient = AndroidHttpClient.newInstance("my-http-agent-string");
```

- Notice that the parameter to the newInstance() method is an HTTP agent string. In Android's default browser, you might see a string such as the following, but you can use whatever you want:

```
Mozilla/5.0 (Linux; U; Android 2.1; en-us; ADR6200 Build/ERD79) AppleWebKit/530.17
(KHTML, like Gecko) Version/ 4.0 Mobile Safari/530.17
```

- When execute() is called on this client, you must be in a thread separate from the main UI thread. This means that you'll get an exception if you simply attempt to replace our previous HttpClient with an AndroidHttpClient. It is bad practice to make HTTP calls from the main UI thread, so AndroidHttpClient won't let you. We'll be covering threading issues in the next section.
- You must call close() on the AndroidHttpClient instance when you are done with it. This is so memory can be freed up properly.
- There are some handy static methods for dealing with compressed responses from a server, including
  - modifyRequestToAcceptGzipResponse(HttpRequest request)
  - getCompressedEntity(byte[] data, ContentResolver resolver)
  - getUngzippedContent(HttpEntity entity)

Once you've acquired an instance of the AndroidHttpClient, you cannot modify any parameter settings in it, nor can you add any parameter settings to it (such as the HTTP protocol version, for example). Your options are to override settings within the HttpGet object as shown previously or to not use the AndroidHttpClient.

This concludes our discussion of using HTTP services with the HttpClient. In the sections that follow, we will turn our focus to another interesting part of the Android Platform: writing background/long-running services. Although not immediately obvious, the processes of making HTTP calls and writing Android services are linked in that you will do a lot of integration from within Android services. Take, for example, a simple mail-client application. On an Android device, this type of application will likely be composed of two pieces: one that will provide the UI to the user, and another to poll for mail messages. The polling will likely have to be done within a background service. The component that polls for new messages will be an Android service, which will in turn use the HttpClient to perform the work.

NOTE: For a great tutorial on using HttpClient and these other concepts, please check out the Apache site at http://hc.apache.org/httpcomponents-client-ga/tutorial/html/.

Using Background Threads (AsyncTask)

So far in our examples, we've been using the main thread of the activity to do our HTTP calls. While we may get lucky and get fast response times to every call, our network connection and the Internet are not always so speedy. Because the main thread of an activity is used mainly to process events from the user (button clicks and so on) and perform updates to the user interface, we should use a background thread to do work that could take a while. Android forces us into this position because if the main thread does not handle something within 5 seconds, an Application Not Responding (ANR) condition will be triggered, which ruins the user's experience by displaying a nasty dialog box asking the user to confirm that the current application should be terminated (also known as a *force close*). We'll get into the details of the main thread and the 5-second time limit in Chapter 17, but for now, just know that we can't tie up the main thread for long.

If all you want to do is some computing, with no updates required to the user interface, you could use a simple Thread object to offload some processing from the main thread. This technique won't work, however, if you need to do updates to the user interface. And that's because the Android user interface toolkit is not thread safe, so it should always be updated only from the main thread.

If you intend to update the user interface in any way as a result of your background thread, you should seriously consider using an AsyncTask. The AsyncTask provides a convenient way of backgrounding some processing that wishes to update the user interface. The AsyncTask takes care of creating a background thread for us where the work will get done, as well as providing callbacks that will run on the main thread to allow easy access to the user interface element (that is, views). The callbacks can fire before, during, and after our background thread has run.

For example, consider the problem of grabbing an image from a network server to display in our application. Perhaps the image needs to be created on the fly. We cannot guarantee how long it will take the image to be returned to us, so we really need to use a background thread for the job.

Listing 15–9 shows a simple implementation of an AsyncTask that will do the job for us. We'll talk about that and then show you the layout file and the Java code for an activity that can call this AsyncTask.

Listing 15–9. *AsyncTask for Downloading an Image: DownloadImageTask.java*

```
import java.io.IOException;
import org.apache.http.HttpResponse;
import org.apache.http.client.HttpClient;
import org.apache.http.client.methods.HttpGet;
import org.apache.http.params.BasicHttpParams;
import org.apache.http.params.HttpConnectionParams;
import org.apache.http.params.HttpParams;
import org.apache.http.util.EntityUtils;
import android.app.Activity;
import android.content.Context;
import android.graphics.Bitmap;
import android.graphics.BitmapFactory;
import android.os.AsyncTask;
import android.util.Log;
import android.widget.ImageView;
import android.widget.TextView;

public class DownloadImageTask extends AsyncTask<String, Integer, Bitmap> {
    private Context mContext;

    DownloadImageTask(Context context) {
        mContext = context;
    }
```

```
protected void onPreExecute() {
    // We could do some setup work here before doInBackground() runs
}

protected Bitmap doInBackground(String... urls) {
    Log.v("doInBackground", "doing download of image");
    return downloadImage(urls);
}

protected void onProgressUpdate(Integer... progress) {
    TextView mText = (TextView)
            ((Activity) mContext).findViewById(R.id.text);
    mText.setText("Progress so far: " + progress[0]);
}

protected void onPostExecute(Bitmap result) {
    if(result != null) {
        ImageView mImage = (ImageView)
            ((Activity) mContext).findViewById(R.id.image);
        mImage.setImageBitmap(result);
    }
    else {
        TextView errorMsg = (TextView)
            ((Activity) mContext).findViewById(R.id.errorMsg);
        errorMsg.setText(
            "Problem downloading image. Please try again later.");
    }
}

private Bitmap downloadImage(String... urls)
{
  HttpClient httpClient = CustomHttpClient.getHttpClient();
  try {
    HttpGet request = new HttpGet(urls[0]);
    HttpParams params = new BasicHttpParams();
    HttpConnectionParams.setSoTimeout(params, 60000);   // 1 minute
    request.setParams(params);

    publishProgress(25);

    HttpResponse response = httpClient.execute(request);

    publishProgress(50);

    byte[] image = EntityUtils.toByteArray(response.getEntity());

    publishProgress(75);

    Bitmap mBitmap = BitmapFactory.decodeByteArray(
                        image, 0, image.length);

    publishProgress(100);

    return mBitmap;
  } catch (IOException e) {
    // covers:
    //      ClientProtocolException
```

```
        //      ConnectTimeoutException
        //      ConnectionPoolTimeoutException
        //      SocketTimeoutException
        e.printStackTrace();
      }
      return null;
    }
}
```

Because AsyncTask is abstract, you need to customize it by extending it, which we do with the class DownloadImageTask. We're going to use a constructor that takes a reference to the calling context, which, in our case, will be the calling activity. We'll use that context to get to the activity's views. We'll also reuse the CustomHttpClient class from before.

There are four steps to an AsyncTask:

1. Do any setup work in the onPreExecute() method. This method executes on the main thread.
2. Run a background thread with doInBackground(). Thread creation is all handled for us behind the scenes. This code runs in a separate background thread.
3. Update progress using publishProgress() and onProgressUpdate(). publishProgress() gets called from within the code of doInBackground(), while onProgressUpdate() is executed in the main thread as a result of the call to publishProgress(). With these two methods, the backgrounded thread is able to communicate with the main thread while it is executing, so status updates can be made in the user interface before the backgrounded thread has completed its work.
4. Update the user interface in onPostExecute() with the results. This method executes in the main thread.

Steps 1 and 3 are optional. In our example, we chose not to do any initialization in onPreExecute(), but we did utilize the progress updating as in step 3. The main work of the background thread is done in the downloadImage() method called from doInBackground(). The downloadImage() method takes a URL and uses our HttpClient to execute an HttpGet request and response. Notice that we're now able to set a timeout of 60 seconds without worrying about getting any ANRs. You can see in the code where the progress is updated during the steps of setting up the HttpClient connection, executing the HTTP request, converting the image response to a byte array, and then building a Bitmap object from it. When downloadImage() returns back to doInBackground() and doInBackground() returns, Android takes care of taking our return value and passing it to onPostExecute(). Once the Bitmap has been passed to onPostExecute(), it is safe to update our ImageView with it, because onPostExecute() runs on the main thread of our activity. But what if we got some sort of exception while doing the download? If we do not get an image back from our HTTP call but get an exception instead, our Bitmap will be null. We can detect that fact in onPostExecute()

and display an error message instead of attempting to set the ImageView to a Bitmap. Of course, we could take other action if we know that our download failed.

Please keep in mind that the only code that does not run on the main thread is the code from doInBackground(). So be careful not to work with the UI within the doInBackground() method, because that is where you could get into trouble. Do not, for instance, call methods from doInBackground() that modify the UI elements. Only touch UI elements in onPreExecute(), onProgressUpdate(), and onPostExecute().

Let's fill out our latest example with the layout XML file and the Java code for our activity in Listings 15–10 and 15–11 respectively.

Listing 15–10. *Layout for Calling Our AsyncTask: /res/layout/main.xml*

```
<?xml version="1.0" encoding="utf-8"?>
<LinearLayout xmlns:android="http://schemas.android.com/apk/res/android"
    android:layout_width="fill_parent"
    android:layout_height="fill_parent"
    android:orientation="vertical"
    >
<LinearLayout
    android:layout_width="fill_parent"
    android:layout_height="wrap_content"
    android:orientation="horizontal"
    >
  <Button android:id="@+id/button"  android:text="Get Image"
    android:layout_width="wrap_content"
    android:layout_height="wrap_content"
    android:onClick="doClick"
    />
  <TextView android:id="@+id/text"
    android:layout_width="wrap_content"
    android:layout_height="wrap_content"
    />
</LinearLayout>
<TextView android:id="@+id/errorMsg"  android:textColor="#ff0000"
    android:layout_width="wrap_content"
    android:layout_height="wrap_content"
    />
<ImageView  android:id="@+id/image"
    android:layout_width="fill_parent"  android:layout_height="0dip"
    android:layout_weight="1" />
</LinearLayout>
```

Listing 15–11. *Activity for Calling Our AsyncTask: HttpActivity.java*

```
import android.app.Activity;
import android.os.AsyncTask;
import android.os.Bundle;
import android.util.Log;
import android.view.View;

public class HttpActivity extends Activity {
    private DownloadImageTask diTask;

    @Override
    public void onCreate(Bundle savedInstanceState)
```

```
    {
        super.onCreate(savedInstanceState);
        setContentView(R.layout.main);
    }

    public void doClick(View view) {
        if(diTask != null) {
            AsyncTask.Status diStatus = diTask.getStatus();
            Log.v("doClick", "diTask status is " + diStatus);
            if(diStatus != AsyncTask.Status.FINISHED) {
                Log.v("doClick", "... no need to start a new task");
                return;
            }
            // Since diStatus must be FINISHED, we can try again.
        }
        diTask = new DownloadImageTask(this);

diTask.execute("http://chart.apis.google.com/chart?&cht=p&chs=460x250&chd=t:15.3,20.3,0.
2,59.7,4.5&chl=Android%201.5%7CAndroid%201.6%7COther*%7CAndroid%202.1%7CAndroid%202.2&ch
co=c4df9b,6fad0c");
    }
}
```

When you run this sample and click the button, you should see a display like Figure 15–1.

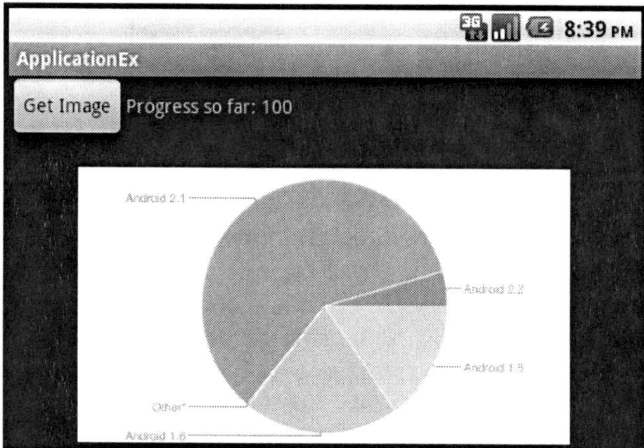

Figure 15–1. *Using* `AsyncTask` *to download an image (the Android device chart as of August 2, 2010)*

The layout is pretty straightforward. We have a button with a text message next to it. This text will be our progress message. Underneath that, we have room for an error message, whose text will be colored red. And finally, we have a place for our image.

Within our button callback method `doClick()`, we need to instantiate a new instance of our customized `AsyncTask` class and call the `execute()` method. This is the pattern you would use also. Instantiate an extension of `AsyncTask`, and call the `execute()` method. For our example, we're calling a Google charts service that takes data values and label names and creates a chart image for us, returning it as a PNG image. But before we launch our task, we really should check to see if a task is already running. If the user double-clicks the button, we could end up with two backgrounded tasks. Fortunately, the AsyncTask class allows us to check its status. If doTask is not null, there's a possibility that we have a running task. So we check the status of our AsyncTask. If it's anything but FINISHED, the task is either RUNNING or PENDING and about to run. Therefore, we only want to drop through and create a new AsyncTask if we have a task and it is already FINISHED. Of course, if the previous AsyncTask was able to successfully download the image, we might not want to download it again. But for our example, we'll go ahead and get it again.

While our sample application runs, you should notice the progress message updating after pressing the button, and then the image appears. The button goes from a pressed state back to the normal state *before* the progress message starts to change. This is an important observation, because it means our main thread has returned to managing the user interface while our download is underway.

Just for fun, go into the URL string for the Google charts call, and make a change that will cause an error situation. Now, run the application again. You should see a result similar to Figure 15–2.

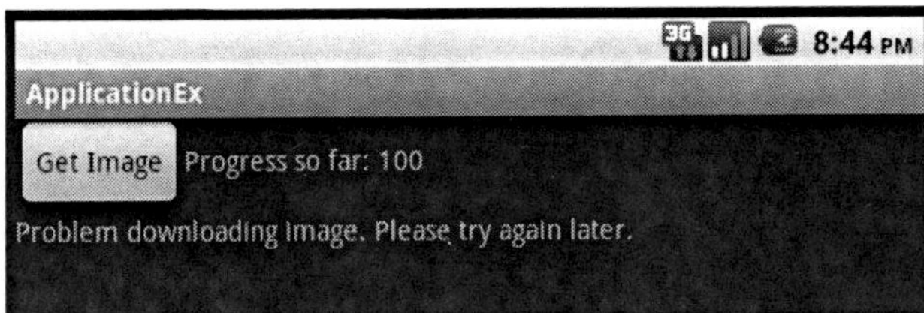

Figure 15–2. *Communicating exceptions back to the user interface with* AsyncTask

Here are a few more things to know about AsyncTasks. Once we've instantiated our extension of AsyncTask and launched the execute() method, our main thread goes back to executing. But we still have a reference to the task and can operate on it from the main thread. For example, we could call cancel() to kill it. We could call isCancelled() to see if it's been cancelled. We might want to modify our logic in onPostExecute() to deal with those cancellations. And AsyncTask has two forms of get() where we could get the result from doInBackground() instead of letting onPostExecute() do our work. One form of get() blocks, while the other uses a timeout value to prevent the calling thread from waiting too long.

An AsyncTask can only be run once. Therefore, if you do keep a reference to an AsyncTask, do not call execute() more than once on it. You will get an exception if you do. You are free to create new instances of your AsyncTask, but each of those instances can only be executed once. That's why we create a new DownloadImageTask every time we need one.We'll have more on AsyncTask in Chapter 18, where you'll see some advanced concepts. For now, we want to show you a special case where AsyncTask is embedded to help with downloading files in the background.

Getting Files Using DownloadManager

Under certain circumstances, your application may need to download a large file to the device. Because this can take a while, and because the procedure can be standardized, Android 2.3 introduced a special class just to manage this type of operation: DownloadManager. The purpose of the DownloadManager is to satisfy a DownloadManager request by using a background thread to download a large file to a local location on the device. It is possible to configure the DownloadManager to provide a notification of the download to the user.

In our next sample application, we use the DownloadManager to pull down one of the Android SDK ZIP files. This sample project will have the following files:

- res/layout/main.xml (Listing 15–12)
- MainActivity.java (Listing 15–13)
- AndroidManifest.xml (Listing 15–14)

Listing 15–12. *Using DownloadManager: /res/layout/main.xml*

```
<?xml version="1.0" encoding="utf-8"?>
<LinearLayout xmlns:android="http://schemas.android.com/apk/res/android"
    android:orientation="vertical"
    android:layout_width="fill_parent"
    android:layout_height="fill_parent" >
  <Button android:onClick="doClick" android:text="Start"
    android:layout_width="wrap_content"
    android:layout_height="wrap_content" />
  <TextView  android:id="@+id/tv"
    android:layout_width="fill_parent"
    android:layout_height="wrap_content" />
</LinearLayout>
```

Our layout is a simple one with a button and a text view. The button will cause the download to start, and we'll display some messages in the text view to indicate the beginning and end of the download. The user interface looks like Figure 15–3.

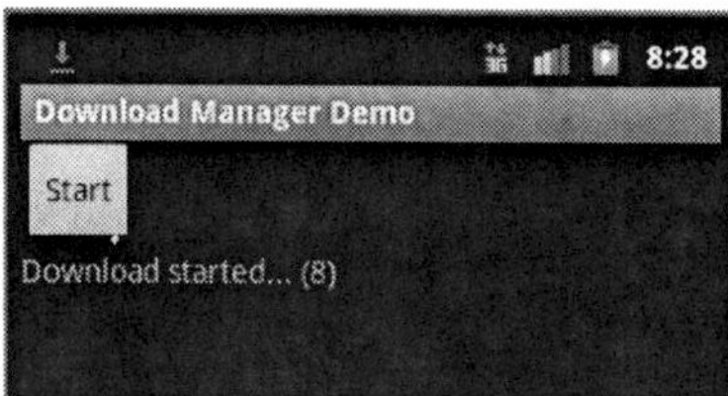

Figure 15–3. *User interface of our DownloadManagerDemo sample application*

Our next listing has the Java code for this application.

Listing 15–13. *Using DownloadManager: MainActivity.java*

```
import android.app.Activity;
import android.app.DownloadManager;
import android.content.BroadcastReceiver;
```

```
import android.content.Context;
import android.content.Intent;
import android.content.IntentFilter;
import android.net.Uri;
import android.os.Bundle;
import android.util.Log;
import android.view.View;
import android.widget.TextView;

public class MainActivity extends Activity {
    protected static final String TAG = "DownloadMgr";
    private DownloadManager dMgr;
    private TextView tv;
    private long downloadId;

    /** Called when the activity is first created. */
    @Override
    public void onCreate(Bundle savedInstanceState) {
        super.onCreate(savedInstanceState);
        setContentView(R.layout.main);

        tv = (TextView)findViewById(R.id.tv);
    }

    @Override
    protected void onResume() {
        super.onResume();
        dMgr = (DownloadManager) getSystemService(DOWNLOAD_SERVICE);
    }

    public void doClick(View view) {
        DownloadManager.Request dmReq = new DownloadManager.Request(
            Uri.parse(
                "http://dl-ssl.google.com/android/repository/" +
                "platform-tools_r01-linux.zip"));
        dmReq.setTitle("Platform Tools");
        dmReq.setDescription("Download for Linux");
        dmReq.setAllowedNetworkTypes(DownloadManager.Request.NETWORK_MOBILE);

        IntentFilter filter = new
IntentFilter(DownloadManager.ACTION_DOWNLOAD_COMPLETE);
        registerReceiver(mReceiver, filter);

        downloadId = dMgr.enqueue(dmReq);

        tv.setText("Download started... (" + downloadId + ")");
    }

    public BroadcastReceiver mReceiver = new BroadcastReceiver() {
        public void onReceive(Context context, Intent intent) {
            Bundle extras = intent.getExtras();
            long doneDownloadId =
                extras.getLong(DownloadManager.EXTRA_DOWNLOAD_ID);
            tv.setText(tv.getText() + "\nDownload finished (" +
                doneDownloadId + ")");
            if(downloadId == doneDownloadId)
                Log.v(TAG, "Our download has completed.");
```

```
        }
    };

    @Override
    protected void onPause() {
        super.onPause();
        unregisterReceiver(mReceiver);
        dMgr = null;
    }
}
```

The code for this application is very straightforward. First we initialize our main view, and then we get a reference to the text view. Within our `onResume()` method, we get a reference to the `DOWNLOAD_SERVICE` service. Note that we de-reference this service in `onPause()`. Our button click method `doClick()` creates a new `DownloadManager.Request` object using the path to the ZIP file we want to download. We also set the title, description, and allowed network type for the download. There are a few more options to choose from; see the online Android documentation for details.

Simply for demonstration purposes, we chose to use the mobile network for downloading, but you can also choose just WiFi (using `NETWORK_WIFI` instead of `NETWORK_MOBILE`) or you can OR the two values together to allow either. By default both networks are allowed for download, which means for our sample application we only want to use the mobile network for downloading, even if WiFi is available.

Once we've set up our request object, we create a filter for a broadcast receiver and we register it. We'll get to our broadcast receiver code shortly. By registering the receiver, we'll be notified when any download has completed. This means we need to keep track of the ID of our request, which is returned when we call `enqueue()` on our `DownloadManager`. Finally, we update the status message in our UI to indicate that a download has started.

For this application to work, we need to specify a couple of permissions, as shown in our `AndroidManifest.xml` file in Listing 15–14, to allow our application to access the Internet and to be able to write the file to the SD card. What's strange about Android 2.3 is that if you don't specify the permissions as indicated in Listing 15–14, you'll get an error message in LogCat that complains about not having the `ACCESS_ALL_DOWNLOADS` permission, which you don't even need for this example. So make sure you set the two permissions as indicated.

Listing 15–14. *Using `DownloadManager`: `AndroidManifest.xml`*

```
<?xml version="1.0" encoding="utf-8"?>
<manifest xmlns:android="http://schemas.android.com/apk/res/android"
      package="com.androidbook.services.download"
      android:versionCode="1"
      android:versionName="1.0">
    <application android:icon="@drawable/icon"
                 android:label="@string/app_name">
        <activity android:name=".MainActivity"
                  android:label="@string/app_name">
            <intent-filter>
              <action android:name="android.intent.action.MAIN" />
              <category android:name="android.intent.category.LAUNCHER" />
```

```
            </intent-filter>
        </activity>

    </application>
    <uses-sdk android:minSdkVersion="10" />

<uses-permission android:name="android.permission.INTERNET" />
<uses-permission android:name="android.permission.WRITE_EXTERNAL_STORAGE" />
</manifest>
```

When you run this application, it should show the button. Clicking the button will initiate the download operation and show the message as in Figure 15–3. Notice that there's a download icon in the notification bar in the upper-left corner of the screen. If you were to drag down on the download icon, you would see a notification window that looks like Figure 15–4.

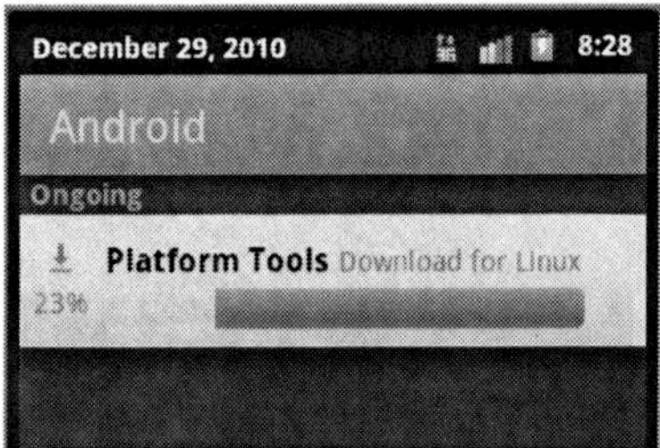

Figure 15–4. *Downloads in the Notification List*

The notification is the download occurring in the background. Once the download has completed, this notification item will be cleared, and we'll see an additional message in our application, as shown in Figure 15–5.

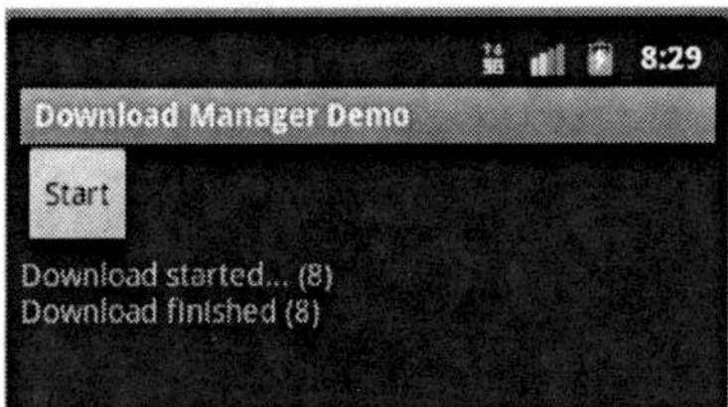

Figure 15–5. *Application shows download is complete*

In our broadcast receiver, we interrogate the intent to find out if the download that completed was ours or not. If it is, we update our status message in the UI, and that's all we're doing. Remember that we cannot do much processing in our broadcast receiver because we must return from `onReceive()` quickly. For example, we could instead invoke a service to process the file that was downloaded. Within that service, we could call something like Listing 15–15 to get to the file contents.

Listing 15–15. *Reading a Downloaded File*

```
try {
    ParcelFileDescriptor pfd =
            dMgr.openDownloadedFile(doneDownloadId);
    // Now we have a read-only handle to the downloaded file
    // Proceed to read the file...
} catch (FileNotFoundException e) {
    e.printStackTrace();
}
```

One way to locate the downloaded file is to use the DownloadManager service, and we need to specify the download ID to get the appropriate file. This was shown in Listing 15–15. The DownloadManager class takes care of resolving the download ID to the actual file. While our example downloads a file to the public area on the SD card, you can in fact download a file to the application's private data area on the SD card, using one of the setDestination*() methods of DownloadManager.Request.

DownloadManager has its own application that you can also access to see downloaded files. From the application menu on the Android device or emulator, look for the icon as shown in Figure 15–6.

Figure 15–6. *The Downloads application icon*

You can use the Downloads application to also get to downloaded files. Go ahead and try it now. When you launch the Downloads application, you'll see a screen that looks like Figure 15–7. Actually, it won't have the menu along the bottom until you click in a check box to select a specific download, as we did before taking the screen shot.

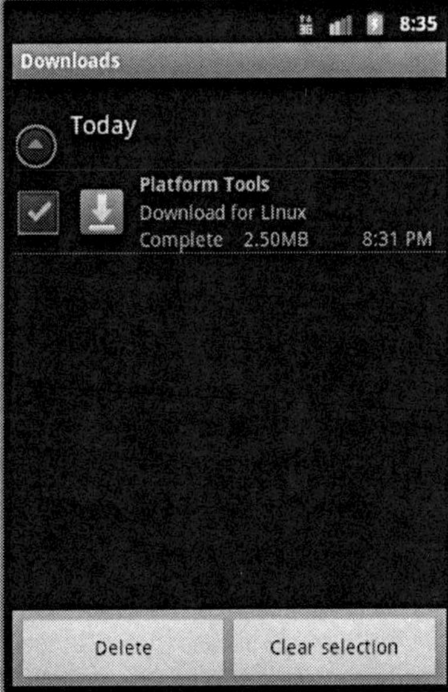

Figure 15–7. *The Downloads application*

The DownloadManager contains a content provider for the download file information. The Downloads application is simply accessing this content provider to show the list of available downloads to the user. This means you can also interrogate the content provider within your application to get information about downloads. To do that, you would use a DownloadManager.Query object and DownloadManager's query() method. There aren't very many options for searching, however. You can search by download ID (one or more), or you can search by download status. The result from the query() method is a Cursor object, which can be used to interrogate rows from the DownloadManager content provider. The columns available are listed in the documentation for DownloadManager and include things like the local Uri of the downloaded file, the number of bytes, the media type of the file, the download status, and several others. When you access the content provider in this way, you need to add the ACCESS_ALL_DOWNLOADS permission to your AndroidManifest.xml file.

Finally, you can use DownloadManager's remove() method to cancel a download, although this does not remove the file if it has been downloaded.

We've shown you how to operate with HTTP-based services, and we showed you how to manage the interface to those services using a special class called AsyncTask. The typical use case of an AsyncTask is some operation that will last for a time, but not for a long time, and where the conclusion of that operation should directly affect the user interface in some way. But what if we wanted to run some background processing that lasted longer than a short while, or what if we wanted to invoke some non-UI functionality that exists in another application? For these needs, Android provides services. We will discuss them next.

Using Android Services

Android supports the concept of services. *Services* are components that run in the background, without a user interface. You can think of these components as similar to Windows services or Unix daemons. Similar to these types of services, Android services can be always available but don't have to be actively doing something. More important, Android services can have life cycles separate from activities. When an activity pauses, stops, or gets destroyed, there may be some processing that you want to continue. Services are good for that too.

Android supports two types of services: local services and remote services. A *local service* is a service that is only accessible to the application that is hosting it, and it is not accessible from other applications running on the device. Generally, these types of services simply support the application that is hosting the service. A *remote service* is accessible from other applications on the device in addition to the application hosting the service. Remote services define themselves to clients using Android Interface Definition Language (AIDL). We're going to talk about both of these types of services, although in the next few chapters, we're going deep into local services. Therefore, we will introduce them here but not spend that much time on them. We'll cover remote services in more detail in this chapter.

Understanding Services in Android

The Android `Service` class is a wrapper of sorts for code that has service-like behavior. Unlike the `AsyncTask` we covered earlier, a `Service` object does not create its own threads automatically. For a `Service` object to use threads, the developer must make it happen. This means that without adding threading to a service, the code of the service will run on the main thread. If our service is performing operations that will complete quickly, this won't be a problem. If our service might run for a while, we definitely want to involve threading. Keep in mind there is nothing wrong with using `AsyncTasks` to do threading within services.

Android supports the concept of a service for two reasons:

- First, to allow you to implement background tasks easily
- Second, to allow you to do interprocess communication between applications running on the same device

These two reasons correspond to the two types of services that Android supports: local services and remote services. An example of the first case might be a local service implemented as part of an e-mail application. The service could handle the sending of a new e-mail to the e-mail server, complete with attachments and retries. As this could take a while to complete, a service is a nice way of wrapping up that functionality so the main thread can kick it off and get back to the user. Plus, if the e-mail activity goes away, you still want the sent e-mails to be delivered. An example of the second case, as we'll see later, is a language translation application. Suppose you have several applications running on a device, and you need a service to accept text that needs to be translated from one language to another. Rather than repeat the logic in every application, you could write a remote translation service and have the applications talk to the service.

There are some important differences between local services and remote services. Specifically, if a service is strictly used by the components in the same process, the clients must start the service by calling `Context.startService()`. This type of service is a local service, because its purpose is, generally, to run background tasks for the application that is hosting the service. If the service supports the `onBind()` method, it's a remote service that can be called via interprocess communication (`Context.bindService()`). We also call remote services *AIDL-supporting services* because clients communicate with the service using AIDL.

Although the interface of `android.app.Service` supports both local and remote services, it's not a good idea to provide one implementation of a service to support both types. The reason for this is that each type of service has a predefined life cycle; mixing the two, although allowed, can cause errors.

Now, we can begin a detailed examination of the two types of services. We will start by talking about local services and then discuss remote services (AIDL-supporting services). As mentioned before, local services are services that are called only by the application that hosts them. Remote services are services that support a remote procedure call (RPC) mechanism. These services allow external clients, on the same device, to connect to the service and use its facilities.

> **NOTE:** The second type of service in Android is known by several names: remote service, AIDL-supporting service, AIDL service, external service, and RPC service. These terms all refer to the same type of service—one that's meant to be accessed remotely by other applications running on the device.

Understanding Local Services

Local services are services that are started via `Context.startService()`. Once started, these types of services will continue to run until a client calls `Context.stopService()` on the service or the service itself calls `stopSelf()`. Note that when `Context.startService()` is called and the service has not already been created, the system will instantiate the service and call the service's `onStartCommand()` method. Keep in mind that calling `Context.startService()` after the service has been started (that is, while it exists) will not result in another instance of the service, but will reinvoke the running service's `onStartCommand()` method. Here are a couple of examples of local services:

- A service to monitor sensor data from the device and do analysis, issuing alerts if a certain condition is realized. This service might run constantly.

- A task-executor service that lets your application's activities submit jobs and queue them for processing. This service might only run for the duration of the operation to submit the job.

Listing 15–16 demonstrates a local service by implementing a service that executes background tasks. We'll end up with four artifacts required to create and consume the service: `BackgroundService.java` (the service itself), `main.xml` (a layout file for the activity), `MainActivity.java` (an activity class to call the service), and `AndroidManifest.xml`. Listing 15–16 only contains `BackgroundService.java`. We'll dissect this code first and then move on to the other three. This implementation requires Android 2.0 or later.

Listing 15–16. *Implementing a Local Service:* `BackgroundService.java`

```
import android.app.Notification;
import android.app.NotificationManager;
import android.app.PendingIntent;
import android.app.Service;
import android.content.Intent;
import android.os.IBinder;
import android.util.Log;

public class BackgroundService extends Service
{
    private static final String TAG = "BackgroundService";
    private NotificationManager notificationMgr;
    private ThreadGroup myThreads = new ThreadGroup("ServiceWorker");

    @Override
    public void onCreate() {
        super.onCreate();

        Log.v(TAG, "in onCreate()");
        notificationMgr =(NotificationManager)getSystemService(
               NOTIFICATION_SERVICE);
        displayNotificationMessage("Background Service is running");
    }

    @Override
    public int onStartCommand(Intent intent, int flags, int startId) {
        super.onStartCommand(intent, flags, startId);

        int counter = intent.getExtras().getInt("counter");
        Log.v(TAG, "in onStartCommand(), counter = " + counter +
                ", startId = " + startId);

        new Thread(myThreads, new ServiceWorker(counter),
            "BackgroundService")
                .start();

        return START_STICKY;
    }

    class ServiceWorker implements Runnable
    {
        private int counter = -1;
```

```
        public ServiceWorker(int counter) {
            this.counter = counter;
        }

        public void run() {
            final String TAG2 = "ServiceWorker:" +
                Thread.currentThread().getId();
            // do background processing here... we'll just sleep...
            try {
                Log.v(TAG2, "sleeping for 10 seconds. counter = " +
                    counter);
                Thread.sleep(10000);
                Log.v(TAG2, "... waking up");
            } catch (InterruptedException e) {
                Log.v(TAG2, "... sleep interrupted");
            }
        }
    }

    @Override
    public void onDestroy()
    {
        Log.v(TAG, "in onDestroy(). Interrupting threads and cancelling notifications");
        myThreads.interrupt();
        notificationMgr.cancelAll();
        super.onDestroy();
    }

    @Override
    public IBinder onBind(Intent intent) {
        Log.v(TAG, "in onBind()");
        return null;
    }

    private void displayNotificationMessage(String message)
    {
        Notification notification =
            new Notification(R.drawable.emo_im_winking,
                message, System.currentTimeMillis());

        notification.flags = Notification.FLAG_NO_CLEAR;

        PendingIntent contentIntent =
            PendingIntent.getActivity(this, 0,
                new Intent(this, MainActivity.class), 0);

        notification.setLatestEventInfo(this, TAG, message,
            contentIntent);

        notificationMgr.notify(0, notification);
    }
}
```

The structure of a `Service` object is somewhat similar to an activity. There is an `onCreate()` method where you can do initialization, and an `onDestroy()` where you do cleanup. Prior to Android 2.0, a service had an `onStart()` method, and since 2.0 it's called `onStartCommand()`. The difference between the two is the addition of a `flags` parameter, which is used to specify to the service that an intent is being redelivered or that the service should restart. We're using the onStartCommand() version for our example. Services don't pause or resume the way activities do so we don't use onPause() or onResume() methods. Because this is a local service, we won't be binding to it, but because Service requires an implementation of the onBind() method, we provide one that simply returns null.

Going back to our onCreate() method, we don't need to do much except to notify the user that this service has been created. We do this using the NotificationManager. You've probably noticed the notification bar at the top left of an Android screen. By pulling down on this, the user can view messages of importance, and by clicking notifications can act on the notifications, which usually means returning to some activity related to the notification. With services, because they can be running, or at least existing, in the background without a visible activity, there has to be some way for the user to get back in touch with the service, perhaps to turn it off. Therefore, we create a Notification object, populate it with a PendingIntent, which will get us back to our control activity, and post it. This all happens in the displayNotificationMessage() method. One more thing we really need to do is set a flag on our Notification object so the user can't clear it from the list. We really need that Notification to exist as long as our service exists, so we set Notification.FLAG_NO_CLEAR to keep it in the Notifications list until we clear it ourselves from our service's onDestroy() method. The method we used in onDestroy() to clear our notification is cancelAll() on the NotificationManager.

There's another thing you need to have for this example to work. You'll need to create a drawable named emo_im_winking and place it within your project's drawable folder. A good source of drawables for this demonstration purpose is to look under the Android platform folder at Android SDK/platforms/<version>/data/res/drawable, where <version> is the version you're interested in. Unfortunately, you can't refer to Android system drawables from your code the way you can with layouts, so you'll need to copy what you want over to your project's drawables folder. If you choose a different drawable file for your example, just go ahead and rename the resource ID in the constructor for the Notification.

When an intent is sent into our service using startService(), onCreate() is called if necessary, and our onStartCommand() method is called to receive the caller's intent. In our case, we're not going to do anything special with it, except to unpack the counter and use it to start a background thread. In a real-world service, we would expect any data to be passed to us via the intent, and this could include URIs for example. Notice the use of a ThreadGroup when creating the Thread. This will prove to be useful later when we want to get rid of our background threads. Also notice the startId parameter. This is set for us by Android and is a unique identifier of the service calls since this service was started.

Our ServiceWorker class is a typical runnable and is where the work happens for our service. In our particular case, we're simply logging some messages and sleeping. We're also catching any interruptions and logging them. One thing we're not doing is manipulating the user interface. We're not updating any views for example. Because we're not on the main thread anymore, we cannot touch the UI directly. There are ways for our ServiceWorker to effect changes in the user interface, and we'll get into those details in the next few chapters.

The last item to pay attention to in our BackgroundService is the onDestroy() method. This is where we perform the cleanup. For our example, we want to get rid of the threads we created earlier, if any are still around. If we don't do this, they could simply hang around and take up memory. Second, we want to get rid of our notification message. Because our service is going away, there's no longer any need for the user to get to the activity to get rid of it. In a real-world application, however, we might want to keep our workers working. If our service is sending e-mails, we certainly don't want to simply kill off the threads. Our example is overly simple, because we imply through the use of the interrupt() method that you can easily kill off background threads. In reality, however, the most you can do is interrupt. This won't necessarily kill off a thread, though. There are deprecated methods for killing threads, but you should not use these. They can cause memory and stability problems for you and your users. Interrupting works in our example, because we're doing sleeps, which can be interrupted.

It's worthwhile taking a look at the ThreadGroup class because it provides ways for you to get access to your threads. We created a single ThreadGroup object within our service and then used that when creating our individual threads. Within our onDestroy() method of the service, we simply interrupt() on the ThreadGroup, and it issues an interrupt to each thread in the ThreadGroup.

So there you have the makings of a simple local service. Before we show you the code for our activity, Listing 15–17 shows the XML layout file for our user interface.

Listing 15–17. *Implementing a Local Service:* *main.xml*

```
<?xml version="1.0" encoding="utf-8"?>
<!-- This file is /res/layout/main.xml -->
<LinearLayout xmlns:android="http://schemas.android.com/apk/res/android"
    android:orientation="vertical"
    android:layout_width="fill_parent"
    android:layout_height="fill_parent"
    >
<Button  android:id="@+id/startBtn"
    android:layout_width="wrap_content"
    android:layout_height="wrap_content"
    android:text="Start Service"  android:onClick="doClick" />
<Button  android:id="@+id/stopBtn"
    android:layout_width="wrap_content"
    android:layout_height="wrap_content"
    android:text="Stop Service"  android:onClick="doClick" />

</LinearLayout>
```

We're going to show two buttons on the user interface, one to do startService() and the other to do stopService(). We could have chosen to use a ToggleButton, but then you would not be able to call startService() multiple times in a row. This is an important point. There is not a one-to-one relationship between startService() and stopService(). When stopService() is called, the service object will be destroyed, and all threads created from all startService() calls should also go away. For our example, we require a minSdkVersion of 5 because we're using the newer onStartCommand()

instead of the older `onStart()`. Therefore, we can also take advantage of the `android:onClick` attribute of the `Button` tag in our layout XML file. Now, let's look at the code for our activity in Listing 15–18.

Listing 15–18. *Implementing a Local Service:* `MainActivity.java`

```
// MainActivity.java
import android.app.Activity;
import android.content.Intent;
import android.os.Bundle;
import android.util.Log;
import android.view.View;

public class MainActivity extends Activity
{
    private static final String TAG = "MainActivity";
    private int counter = 1;

    @Override
    public void onCreate(Bundle savedInstanceState)
    {
        super.onCreate(savedInstanceState);
        setContentView(R.layout.main);
    }

    public void doClick(View view) {
        switch(view.getId()) {
        case R.id.startBtn:
            Log.v(TAG, "Starting service... counter = " + counter);
            Intent intent = new Intent(MainActivity.this,
                    BackgroundService.class);
            intent.putExtra("counter", counter++);
            startService(intent);
            break;
        case R.id.stopBtn:
            stopService();
        }
    }

    private void stopService() {
        Log.v(TAG, "Stopping service...");
        if(stopService(new Intent(MainActivity.this,
                    BackgroundService.class)))
            Log.v(TAG, "stopService was successful");
        else
            Log.v(TAG, "stopService was unsuccessful");
    }

    @Override
    public void onDestroy()
    {
        stopService();
        super.onDestroy();
    }
}
```

Our MainActivity looks a lot like other activities you've seen. There's a simple onCreate() to set up our user interface from the main.xml layout file. There's a doClick() method to handle the button callbacks. In our example, we're calling startService() when the Start Service button is pressed, and we're calling stopService() when the Stop Service button is pressed. When we start the service, we want to pass in some data, which we do via the intent. We chose to pass the data in the Extras bundle, but we could have added it using setData() if we had a URI. When we stop the service, we check to see the return result. It should normally be true, but if the service was not running, we could get a return of false. Last, when our activity dies, we want to stop the service, so we also stop the service in our onDestroy() method. There's one more item to discuss, and that's the AndroidManifest.xml file, which we show in Listing 15–19.

Listing 15–19. *Implementing a Local Service: AndroidManifest.xml*

```
<?xml version="1.0" encoding="utf-8"?>
<manifest xmlns:android="http://schemas.android.com/apk/res/android"
      package="com.androidbook.services.simplelocal"
      android:versionCode="1"
      android:versionName="1.0">
    <application android:icon="@drawable/icon"
            android:label="@string/app_name">
        <activity android:name=".MainActivity"
                android:label="@string/app_name"
                android:launchMode="singleTop" >
            <intent-filter>
              <action android:name="android.intent.action.MAIN" />
              <category android:name="android.intent.category.LAUNCHER" />
            </intent-filter>
        </activity>
        <service android:name="BackgroundService"/>
    </application>
    <uses-sdk android:minSdkVersion="5" />

</manifest>
```

In addition to our regular <activity> tags in the manifest file, we now have a <service> tag. Because this is a local service that we're calling explicitly using the class name, we don't need to put much into the <service> tag. All that is required is the name of our service. But there is one other thing to point out about this manifest file. Our service creates a notification so that the user can get back to our MainActivity if, for example, the user pressed the Home key on MainActivity without stopping the service.

The MainActivity is still there; it's just not visible. One way to get back to the MainActivity is to click the notification that our service created. What we don't want to have happen is for a new MainActivity to be created in addition to our existing, invisible MainActivity. To prevent this from happening, we set an attribute in our manifest file for MainActivity called android:launchMode, and we set it to singleTop. This will help ensure that the existing invisible MainActivity will be brought forward and displayed, rather than creating another MainActivity.

When you run this application, you will see our two buttons. By clicking the Start Service button, you will be instantiating the service and calling onStartCommand(). Our code logs several messages to LogCat, so you can follow along. Go ahead and click Start Service several times in a row, even quickly. You will see threads created to handle each request. You'll also notice that the value of counter is passed along through to each `ServiceWorker` thread. When you press the Stop Service button, our service will go away, and you'll see the log messages from our `MainActivity`'s `stopService()` method, from our `BackgroundService`'s `onDestroy()` method, and possibly from `ServiceWorker` threads if they got interrupted.

You should also notice the notification message when the service has been started. With the service running, go ahead and press the Back button from our `MainActivity` and notice that the notification message disappears. This means our service has gone away also. To restart our `MainActivity`, click Start Service to get the service going again. Now, press the Home button. Our `MainActivity` disappears from view, but the notification remains, meaning our service is still in existence. Go ahead and click the notification, and you'll again see our `MainActivity`.

Note that our example uses an activity to interface with the service, but any component in your application can use the service. This includes other services, activities, generic classes, and so on. Also note that our service does not stop itself; it relies on the activity to do that for it. There are some methods available to a service to allow the service to stop itself, namely `stopSelf()` and `stopSelfResult()`.

Our `BackgroundService` is a typical example of a service that is used by the components of the application that is hosting the service. In other words, the application that is running the service is also the only consumer. Because the service does not support clients from outside its process, the service is a local service. And because it's a local service, as opposed to a remote service, it returns `null` in the `bind()` method. Therefore, the only way to bind to this service is to call `Context.startService()`. The critical methods of a local service are `onCreate()`, `onStartCommand()`, `stop*()`, and `onDestroy()`.

There's another option with a local service, and that is for the case where you'll only have one instance of the service with one background thread. In this case, in the `onCreate()` method of the `BackgroundService`, we could create a thread that does the service's heavy lifting. We could create and start the thread in `onCreate()` rather than `onStartCommand()`. We could do this because `onCreate()` is called only once, and we want the thread to be created only once during the life of the service. One thing we wouldn't have in `onCreate()`, though, is the content of the intent passed by `startService()`. If we need that, we might as well use the pattern as described previously, and we'd just know that `onStartCommand()` should only be called once.

This concludes our introduction to local services. Remember that we'll get into more details of local services in subsequent chapters. Let's move on to AIDL services—the more complicated type of service.

Understanding AIDL Services

In the previous section, we showed you how to write an Android service that is consumed by the application that hosts the service. Now, we are going to show you how to build a service that can be consumed by other processes via remote procedure call (RPC). As with many other RPC-based solutions, in Android you need an interface definition language (IDL) to define the interface that will be exposed to clients. In the Android world, this IDL is called Android Interface Definition Language (AIDL). To build a remote service, you do the following:

1. Write an AIDL file that defines your interface to clients. The AIDL file uses Java syntax and has an `.aidl` extension. Use the same package name inside your AIDL file as the package for your Android project.
2. Add the AIDL file to your Eclipse project under the `src` directory. The Android Eclipse plug-in will call the AIDL compiler to generate a Java interface from the AIDL file (the AIDL compiler is called as part of the build process).
3. Implement a service, and return the interface from the `onBind()` method.
4. Add the service configuration to your `AndroidManifest.xml` file. The sections that follow show you how to execute each step.

Defining a Service Interface in AIDL

To demonstrate an example of a remote service, we are going to write a stock-quoter service. This service will provide a method that takes a ticker symbol and returns the stock value. To write a remote service in Android, the first step is to define the service interface definition in an AIDL file. Listing 15–20 shows the AIDL definition of `IStockQuoteService`. This file goes into the same place as a regular Java file would for your `StockQuoteService` project.

Listing 15–20. *The AIDL Definition of the Stock-Quoter Service*

```
// This file is IStockQuoteService.aidl
package com.androidbook.services.stockquoteservice;
interface IStockQuoteService
{
        double getQuote(String ticker);
}
```

The `IStockQuoteService` accepts the stock-ticker symbol as a string and returns the current stock value as a double. When you create the AIDL file, the Android Eclipse plug-in runs the AIDL compiler to process your AIDL file (as part of the build process). If your AIDL file compiles successfully, the compiler generates a Java interface suitable for RPC communication. Note that the generated file will be in the package named in your AIDL file—`com.androidbook.services.stockquoteservice`, in this case.

Listing 15–21 shows the generated Java file for our `IStockQuoteService` interface. The generated file will be put into the gen folder of our Eclipse project.

Listing 15–21. *The Compiler-Generated Java File*

```
/*
 * This file is auto-generated.  DO NOT MODIFY.
 * Original file: C:\\android\\StockQuoteService\\src\\com\\androidbook\\➥
services\\stockquoteservice\\IStockQuoteService.aidl
 */
package com.androidbook.services.stockquoteservice;
import java.lang.String;
import android.os.RemoteException;
import android.os.IBinder;
import android.os.IInterface;
import android.os.Binder;
import android.os.Parcel;
public interface IStockQuoteService extends android.os.IInterface
{
/** Local-side IPC implementation stub class. */
public static abstract class Stub extends android.os.Binder implements ➥
com.androidbook.services.stockquoteservice.IStockQuoteService
{
private static final java.lang.String DESCRIPTOR = ➥
"com.androidbook.services.stockquoteservice.IStockQuoteService";
/** Construct the stub at attach it to the interface. */
public Stub()
{
this.attachInterface(this, DESCRIPTOR);
}
/**
 * Cast an IBinder object into an IStockQuoteService interface,
 * generating a proxy if needed.
 */
public static com.androidbook.services.stockquoteservice.IStockQuoteService ➥
asInterface(android.os.IBinder obj)
{
if ((obj==null)) {
return null;
}
android.os.IInterface iin = (android.os.IInterface)obj.queryLocalInterface(DESCRIPTOR);
if (((iin!=null)&&(iin instanceof
com.androidbook.services.stockquoteservice.IStockQuoteService))) {
return ((com.androidbook.services.stockquoteservice.IStockQuoteService)iin);
}
return ((com.androidbook.services.stockquoteservice.IStockQuoteService)iin);
}
return new
com.androidbook.services.stockquoteservice.IStockQuoteService.Stub.Proxy(obj);
}
public android.os.IBinder asBinder()
{
return this;
}
@Override public boolean onTransact(int code, android.os.Parcel data,➥
     android.os.Parcel reply, int flags) throws android.os.RemoteException
{
switch (code)
{
case INTERFACE_TRANSACTION:
```

```
{
reply.writeString(DESCRIPTOR);
return true;
}
case TRANSACTION_getQuote:
{
data.enforceInterface(DESCRIPTOR);
java.lang.String _arg0;
_arg0 = data.readString();
double _result = this.getQuote(_arg0);
reply.writeNoException();
reply.writeDouble(_result);
return true;
}
}
return super.onTransact(code, data, reply, flags);
}
private static class Proxy implements
        com.androidbook.services.stockquoteservice.IStockQuoteService
{
private android.os.IBinder mRemote;
Proxy(android.os.IBinder remote)
{
mRemote = remote;
}
public android.os.IBinder asBinder()
{
return mRemote;
}
public java.lang.String getInterfaceDescriptor()
{
return DESCRIPTOR;
}
public double getQuote(java.lang.String ticker) throws android.os.RemoteException
{
android.os.Parcel _data = android.os.Parcel.obtain();
android.os.Parcel _reply = android.os.Parcel.obtain();
double _result;
try {
_data.writeInterfaceToken(DESCRIPTOR);
_data.writeString(ticker);
mRemote.transact(Stub.TRANSACTION_getQuote, _data, _reply, 0);
_reply.readException();
_result = _reply.readDouble();
}
finally {
_reply.recycle();
_data.recycle();
}
return _result;
}
}
static final int TRANSACTION_getQuote = (IBinder.FIRST_CALL_TRANSACTION + 0);
}
public double getQuote(java.lang.String ticker) throws android.os.RemoteException;
}
```

Note the following important points regarding the generated classes:

- The interface we defined in the AIDL file is implemented as an interface in the generated code (that is, there is an interface named IStockQuoteService).
- A static final abstract class named Stub extends android.os.Binder and implements IStockQuoteService. Note that the class is an abstract class.
- An inner class named Proxy implements the IStockQuoteService that proxies the Stub class.
- The AIDL file must reside in the package where the generated files are supposed to be (as specified in the AIDL file's package declaration).

Now, let's move on and implement the AIDL interface in a service class.

Implementing an AIDL Interface

In the previous section, we defined an AIDL file for a stock-quoter service and generated the binding file. Now, we are going to provide an implementation of that service. To implement the service's interface, we need to write a class that extends android.app.Service and implements the IStockQuoteService interface. The class we are going to write we'll call StockQuoteService. To expose the service to clients, our StockQuoteService will need to provide an implementation of the onBind() method, and we'll need to add some configuration information to the AndroidManifest.xml file. Listing 15–22 shows an implementation of the IStockQuoteService interface. This file also goes into the src folder of the StockQuoteService project.

Listing 15–22. *The IStockQuoteService Service Implementation*

```
// StockQuoteService.java
import android.app.Service;
import android.content.Intent;
import android.os.IBinder;
import android.os.RemoteException;
import android.util.Log;

public class StockQuoteService extends Service
{
    private static final String TAG = "StockQuoteService";
    public class StockQuoteServiceImpl extends IStockQuoteService.Stub
    {
        @Override
        public double getQuote(String ticker) throws RemoteException
        {
            Log.v(TAG, "getQuote() called for " + ticker);
            return 20.0;
        }
    }
```

```
    @Override
    public void onCreate() {
        super.onCreate();
        Log.v(TAG, "onCreate() called");
    }

    @Override
    public void onDestroy()
    {
        super.onDestroy();
        Log.v(TAG, "onDestroy() called");
    }

    @Override
    public IBinder onBind(Intent intent)
    {
        Log.v(TAG, "onBind() called");
        return new StockQuoteServiceImpl();
    }
}
```

The `StockQuoteService.java` class in Listing 15–22 resembles the local `BackgroundService` we created earlier, but without the `NotificationManager`. The important difference is that we now implement the `onBind()` method. Recall that the `Stub` class generated from the AIDL file was an abstract class and that it implemented the `IStockQuoteService` interface. In our implementation of the service, we have an inner class that extends the `Stub` class called `StockQuoteServiceImpl`. This class serves as the remote-service implementation, and an instance of this class is returned from the `onBind()` method. With that, we have a functional AIDL service, although external clients cannot connect to it yet.

To expose the service to clients, we need to add a service declaration in the `AndroidManifest.xml` file, and this time, we need an intent filter to expose the service. Listing 15–23 shows the service declaration for the `StockQuoteService`. The `<service>` tag is a child of the `<application>` tag.

Listing 15–23. *Manifest Declaration for the `IStockQuoteService`*

```
<?xml version="1.0" encoding="utf-8"?>
<manifest xmlns:android="http://schemas.android.com/apk/res/android"
      package="com.androidbook.services.stockquoteservice"
      android:versionCode="1"
      android:versionName="1.0">
    <application android:icon="@drawable/icon"
        android:label="@string/app_name">
      <service android:name="StockQuoteService">
        <intent-filter>
                <action android:name=

"com.androidbook.services.stockquoteservice.IStockQuoteService" />
        </intent-filter>
      </service>
    </application>
    <uses-sdk android:minSdkVersion="4" />
</manifest>
```

As with all services, we define the service we want to expose with a `<service>` tag. For an AIDL service, we also need to add an `<intent-filter>` with an `<action>` entry for the service interface we want to expose.

With this in place, we have everything we need to deploy the service. When you are ready to deploy the service application from Eclipse, just go ahead and choose Run As the way you would for any other application. Eclipse will comment in the Console that this application has no Launcher, but it will deploy the app anyway, which is what we want. Let's now look at how we would call the service from another application (on the same device, of course).

Calling the Service from a Client Application

When a client talks to a service, there must be a protocol or contract between the two. With Android, the contract is in our AIDL file. So the first step in consuming a service is to take the service's AIDL file and copy it to your client project. When you copy the AIDL file to the client project, the AIDL compiler creates the same interface-definition file that was created when the service was implemented (in the service-implementation project). This exposes to the client all of the methods, parameters, and return types on the service. Let's create a new project and copy the AIDL file:

1. Create a new Android project named `StockQuoteClient`. Use a different package name, such as `com.androidbook.stockquoteclient`. Use `MainActivity` for the Create Activity field.
2. Create a new Java package in this project named `com.androidbook.services.stockquoteservice` in the `src` directory.
3. Copy the `IStockQuoteService.aidl` file from the `StockQuoteService` project to this new package. Note that after you copy the file to the project, the AIDL compiler will generate the associated Java file.

The service interface that you regenerate serves as the contract between the client and the service. The next step is to get a reference to the service so we can call the `getQuote()` method. With remote services, we have to call the `bindService()` method rather than the `startService()` method. Listing 15–24 shows an activity class that acts as a client of the `IStockQuoteService` service. Listing 15–25 contains the layout file for the activity.

Listing 15–24 shows our `MainActivity.java` file. Realize that the package name of the client activity is not that important—you can put the activity in any package you'd like. However, the AIDL artifacts that you create are package-sensitive because the AIDL compiler generates code from the contents of the AIDL file.

Listing 15–24. *A Client of the* `IStockQuoteService` *Service*

```
// This file is MainActivity.java
import com.androidbook.services.stockquoteservice.IStockQuoteService;
import android.app.Activity;
import android.content.ComponentName;
import android.content.Context;
```

```
import android.content.Intent;
import android.content.ServiceConnection;
import android.os.Bundle;
import android.os.IBinder;
import android.os.RemoteException;
import android.util.Log;
import android.view.View;
import android.widget.Button;
import android.widget.Toast;
import android.widget.ToggleButton;

public class MainActivity extends Activity {
    private static final String TAG = "StockQuoteClient";
    private IStockQuoteService stockService = null;
    private ToggleButton bindBtn;
    private Button callBtn;

    /** Called when the activity is first created. */
    @Override
    public void onCreate(Bundle savedInstanceState) {
        super.onCreate(savedInstanceState);
        setContentView(R.layout.main);

        bindBtn = (ToggleButton)findViewById(R.id.bindBtn);
        callBtn = (Button)findViewById(R.id.callBtn);
    }

    public void doClick(View view) {
        switch(view.getId()) {
        case R.id.bindBtn:
            if(((ToggleButton) view).isChecked()) {
                bindService(new Intent(
                    IStockQuoteService.class.getName()),
                    serConn, Context.BIND_AUTO_CREATE);
            }
            else {
                unbindService(serConn);
                callBtn.setEnabled(false);
            }
            break;
        case R.id.callBtn:
            callService();
            break;
        }
    }

    private void callService() {
        try {
            double val = stockService.getQuote("ANDROID");
            Toast.makeText(MainActivity.this,
                    "Value from service is " + val,
                    Toast.LENGTH_SHORT).show();
        } catch (RemoteException ee) {
            Log.e("MainActivity", ee.getMessage(), ee);
        }
    }
```

```
    private ServiceConnection serConn = new ServiceConnection() {

        @Override
        public void onServiceConnected(ComponentName name,
            IBinder service)
        {
            Log.v(TAG, "onServiceConnected() called");
            stockService = IStockQuoteService.Stub.asInterface(service);
            bindBtn.setChecked(true);
            callBtn.setEnabled(true);
        }

        @Override
        public void onServiceDisconnected(ComponentName name) {
            Log.v(TAG, "onServiceDisconnected() called");
            bindBtn.setChecked(false);
            callBtn.setEnabled(false);
            stockService = null;
        }
    };

    protected void onDestroy() {
        Log.v(TAG, "onDestroy() called");
        if(callBtn.isEnabled())
            unbindService(serConn);
        super.onDestroy();
    }
}
```

The activity displays our layout and grabs a reference to the Call Service button so we can properly enable it when the service is running and disable it when the service is stopped. When the user clicks the Bind button, the activity calls the `bindService()` method. Similarly, when the user clicks UnBind, the activity calls the `unbindService()` method. Notice that three parameters are passed to the `bindService()` method: the name of the AIDL service, a `ServiceConnection` instance, and a flag to autocreate the service.

Listing 15–25. ***The*** `IStockQuoteService` ***Service Client Layout***

```
<?xml version="1.0" encoding="utf-8"?>
<!-- This file is /res/layout/main.xml -->
<LinearLayout xmlns:android="http://schemas.android.com/apk/res/android"
    android:orientation="vertical"
    android:layout_width="fill_parent"
    android:layout_height="fill_parent" >

<ToggleButton android:id="@+id/bindBtn"
    android:layout_width="wrap_content"
    android:layout_height="wrap_content"
    android:textOff="Bind"  android:textOn="Unbind"
    android:onClick="doClick" />

<Button android:id="@+id/callBtn"
    android:layout_width="wrap_content"
    android:layout_height="wrap_content"
    android:text="Call Service"  android:enabled=”false”
    android:onClick="doClick" />
</LinearLayout>
```

With an AIDL service, you need to provide an implementation of the `ServiceConnection` interface. This interface defines two methods: one called by the system when a connection to the service has been established and one called when the connection to the service has been destroyed. In our activity implementation, we define a private anonymous member that implements the `ServiceConnection` for the `IStockQuoteService`. When we call the `bindService()` method, we pass in the reference to this member. When the connection to the service is established, the `onServiceConnected()` callback is invoked, and we then obtain a reference to the `IStockQuoteService` using the `Stub` and enable the Call Service button.

Note that the `bindService()` call is an asynchronous call. It is asynchronous because the process or service might not be running and thus might have to be created or started. And we cannot wait on the main thread for the service to start. Because `bindService()` is asynchronous, the platform provides the `ServiceConnection` callback, so we know when the service has been started and when the service is no longer available.

Please notice the `onServiceDisconnected()` callback. This does *not* get invoked when we unbind from the service. It is only invoked if the service crashes. If it does, we should not think that we're still connected, and we might need to reinvoke the `bindService()` call. That is why we change the status of our buttons in the UI when this callback is invoked. But notice we said "we might need to reinvoke the `bindService()` call." Android could restart our service for us and invoke our `onServiceConnected()` callback. You can try this yourself by running the client, binding to the service, and using DDMS to do a Stop on the Stock Quote Service application.

When you run this example, watch the log messages in LogCat to get a feel for what is going on behind the scenes.

Now you know how to create and consume an AIDL interface. Before we move on and complicate matters further, let's review what it takes to build a simple local service versus an AIDL service. A local service is a service that does not support `onBind()`—it returns `null` from `onBind()`. This type of service is accessible only to the components of the application that is hosting the service. You call local services by calling `startService()`.

On the other hand, an AIDL service is a service that can be consumed both by components within the same process and by those that exist in other applications. This type of service defines a contract between itself and its clients in an AIDL file. The service implements the AIDL contract, and clients bind to the AIDL definition. The service implements the contract by returning an implementation of the AIDL interface from the `onBind()` method. Clients bind to an AIDL service by calling `bindService()`, and they disconnect from the service by calling `unbindService()`.

In our service examples thus far, we have strictly dealt with passing simple Java primitive types. Android services actually support passing complex types, too. This is very useful, especially for AIDL services, because you might have an open-ended number of parameters that you want to pass to a service, and it's unreasonable to pass them all as simple primitives. It makes more sense to package them as complex types and then pass them to the service.

Let's see how we can pass complex types to services.

Passing Complex Types to Services

Passing complex types to and from services requires more work than passing Java primitive types. Before embarking on this work, you should get an idea of AIDL's support for nonprimitive types:

- AIDL supports `String` and `CharSequence`.
- AIDL allows you to pass other AIDL interfaces, but you need to have an `import` statement for each AIDL interface you reference (even if the referenced AIDL interface is in the same package).
- AIDL allows you to pass complex types that implement the `android.os.Parcelable` interface. You need to have an `import` statement in your AIDL file for these types.
- AIDL supports `java.util.List` and `java.util.Map`, with a few restrictions. The allowable data types for the items in the collection include Java primitive, `String`, `CharSequence`, and `android.os.Parcelable`. You do not need `import` statements for `List` or `Map`, but you do need them for the `Parcelables`.
- Nonprimitive types, other than `String`, require a directional indicator. Directional indicators include `in`, `out`, and `inout`. `in` means the value is set by the client; `out` means the value is set by the service; and `inout` means both the client and service set the value.

The `Parcelable` interface tells the Android runtime how to serialize and deserialize objects during the marshalling and unmarshalling process. Listing 15–26 shows a `Person` class that implements the `Parcelable` interface.

Listing 15–26. *Implementing the `Parcelable` Interface*

```
// This file is Person.java
package com.androidbook.services.stock2;
import android.os.Parcel;
import android.os.Parcelable;

public class Person implements Parcelable {
    private int age;
    private String name;
    public static final Parcelable.Creator<Person> CREATOR =
        new Parcelable.Creator<Person>()
    {
        public Person createFromParcel(Parcel in) {
            return new Person(in);
        }

        public Person[] newArray(int size) {
            return new Person[size];
        }
    };
```

```
    public Person() {
    }

    private Person(Parcel in) {
        readFromParcel(in);
    }

    @Override
    public int describeContents() {
        return 0;
    }

    @Override
    public void writeToParcel(Parcel out, int flags) {
        out.writeInt(age);
        out.writeString(name);
    }

    public void readFromParcel(Parcel in) {
        age = in.readInt();
        name = in.readString();
    }

    public int getAge() {
        return age;
    }

    public void setAge(int age) {
        this.age = age;
    }

    public String getName() {
        return name;
    }

    public void setName(String name) {
        this.name = name;
    }
}
```

To get started on implementing this, create a new Android Project in Eclipse called `StockQuoteService2`. For Create Activity, use a name of `MainActivity`, and use a package of `com.androidbook.services.stock2`. Then add the `Person.java` file from Listing 15–26 to the `com.androidbook.services.stock2` package of our new project.

The `Parcelable` interface defines the contract for hydration and dehydration of objects during the marshalling/unmarshalling process. Underlying the `Parcelable` interface is the `Parcel` container object. The `Parcel` class is a fast serialization/deserialization mechanism specially designed for interprocess communication within Android. The class provides methods that you use to flatten your members to the container and to expand the members back from the container. To properly implement an object for interprocess communication, we have to do the following:

1. Implement the Parcelable interface. This means that you implement writeToParcel() and readFromParcel(). The write method will write the object to the parcel, and the read method will read the object from the parcel. Note that the order in which you write properties must be the same as the order in which you read them.
2. Add a static final property to the class with the name CREATOR. The property needs to implement the android.os.Parcelable.Creator<T> interface.
3. Provide a constructor for the Parcelable that knows how to create the object from the Parcel.
4. Define a Parcelable class in an .aidl file that matches the .java file containing the complex type. The AIDL compiler will look for this file when compiling your AIDL files. An example of a Person.aidl file is shown in Listing 15–27. This file should be in the same place as Person.java.

NOTE: Seeing Parcelable might have triggered the question, why is Android not using the built-in Java serialization mechanism? It turns out that the Android team came to the conclusion that the serialization in Java is far too slow to satisfy Android's interprocess-communication requirements. So the team built the Parcelable solution. The Parcelable approach requires that you explicitly serialize the members of your class, but in the end, you get a much faster serialization of your objects.

Also realize that Android provides two mechanisms that allow you to pass data to another process. The first is to pass a bundle to an activity using an intent, and the second is to pass a Parcelable to a service. These two mechanisms are not interchangeable and should not be confused. That is, the Parcelable is not meant to be passed to an activity. If you want to start an activity and pass it some data, use a Bundle. Parcelable is meant to be used only as part of an AIDL definition.

Listing 15–27. *An Example of a Person.aidl File*

```
// This file is Person.aidl
package com.androidbook.services.stock2;
parcelable Person;
```

You will need an .aidl file for each Parcelable in your project. In this case, we have just one Parcelable, which is Person. You may notice that you don't get a Person.java file created in the gen folder. This is to be expected. We already have this file from when we created it previously.

Now, let's use the Person class in a remote service. To keep things simple, we will modify our IStockQuoteService to take an input parameter of type Person. The idea is that clients will pass a Person to the service to tell the service who is requesting the quote. The new IStockQuoteService.aidl looks like Listing 15–28.

Listing 15–28. *Passing Parcelables to Services*

```
// This file is IStockQuoteService.aidl
package com.androidbook.services.stock2;
import com.androidbook.services.stock2.Person;

interface IStockQuoteService
{
    String getQuote(in String ticker,in Person requester);
}
```

The getQuote() method now accepts two parameters: the stock's ticker symbol and a Person object to specify who is making the request. Note that we have directional indicators on the parameters because the parameters include nonprimitive types and that we have an import statement for the Person class. The Person class is also in the same package as the service definition (com.androidbook.services.stock2).

The service implementation now looks like Listing 15–29, with the layout in Listing 15–30.

Listing 15–29. *The StockQuoteService2 Implementation*

```
package com.androidbook.services.stock2;
// This file is StockQuoteService2.java

import android.app.Notification;
import android.app.NotificationManager;
import android.app.PendingIntent;
import android.app.Service;
import android.content.Intent;
import android.os.IBinder;
import android.os.RemoteException;

public class StockQuoteService2 extends Service
{
    private NotificationManager notificationMgr;

    public class StockQuoteServiceImpl extends IStockQuoteService.Stub
    {
        public String getQuote(String ticker, Person requester)
                throws RemoteException {
            return "Hello " + requester.getName() +
                "! Quote for " + ticker + " is 20.0";
        }
    }

    @Override
    public void onCreate() {
        super.onCreate();

        notificationMgr =
          (NotificationManager)getSystemService(NOTIFICATION_SERVICE);

        displayNotificationMessage(
```

```
                "onCreate() called in StockQuoteService2");
    }

    @Override
    public void onDestroy()
    {
        displayNotificationMessage(
              "onDestroy() called in StockQuoteService2");
        // Clear all notifications from this service
        notificationMgr.cancelAll();
        super.onDestroy();
    }

    @Override
    public IBinder onBind(Intent intent)
    {
        displayNotificationMessage(
              "onBind() called in StockQuoteService2");
        return new StockQuoteServiceImpl();
    }

    private void displayNotificationMessage(String message)
    {
        Notification notification =
            new Notification(R.drawable.emo_im_happy,
                message, System.currentTimeMillis());

        PendingIntent contentIntent =
            PendingIntent.getActivity(this, 0,
                new Intent(this, MainActivity.class), 0);

        notification.setLatestEventInfo(this,
            "StockQuoteService2", message,
            contentIntent);

        notification.flags = Notification.FLAG_NO_CLEAR;

        notificationMgr.notify(R.id.app_notification_id, notification);
    }
}
```

Listing 15–30. *The* `StockQuoteService2` *Layout*

```
<?xml version="1.0" encoding="utf-8"?>
<!-- This file is /res/layout/main.xml -->
<LinearLayout xmlns:android="http://schemas.android.com/apk/res/android"
    android:orientation="vertical"
    android:layout_width="fill_parent"
    android:layout_height="fill_parent" >
<TextView
    android:layout_width="fill_parent"
    android:layout_height="wrap_content"
    android:text="This is where the service could ask for help." />
</LinearLayout>
```

The differences between this implementation and the previous one are that we brought back the notifications, and we now return the stock value as a string and not a double. The string returned to the user contains the name of the requester from the `Person` object, which demonstrates that we read the value sent from the client and that the Person object was passed correctly to the service.

There are a few other things that need to be done to make this work:

1. Find the emo_im_happy.png image file from under Android SDK/platforms/android-2.1/data/res/drawable-mdpi, and copy it to the /res/drawable directory of our project. Or change the name of the resource in the code, and put whatever image you want in the drawables folder.

2. Add a new <item type="id" name="app_notification_id"/> tag to the /res/values/strings.xml file

3. We need to modify the application in the AndroidManifest.xml file as shown in Listing 15–31.

Listing 15–31. *Modified* <application> *in* AndroidManifest.xml *File for* StockQuoteService2

```
<?xml version="1.0" encoding="utf-8"?>
<manifest xmlns:android="http://schemas.android.com/apk/res/android"
      package="com.androidbook.services.stock2"
      android:versionCode="1"
      android:versionName="1.0">
    <application android:icon="@drawable/icon"
            android:label="@string/app_name">
        <activity android:name=".MainActivity"
                android:label="@string/app_name"
                android:launchMode="singleTop" >
            <intent-filter>
                <action android:name="android.intent.action.MAIN" />
            </intent-filter>
        </activity>
      <service android:name="StockQuoteService2">
        <intent-filter>
            <action android:name="com.androidbook.services.stock2.IStockQuoteService" />
        </intent-filter>
      </service>
    </application>
    <uses-sdk android:minSdkVersion="7" />
</manifest>
```

While it is OK to use the dot notation for our android:name=".MainActivity" attribute, it is not OK to use dot notation inside of our <action> tag inside the service's <intent-filter> tag. We need to spell it out; otherwise, our client will not find the service specification.

Last, we'll use the default MainActivity.java file that simply displays a basic layout with a simple message. We showed you earlier how to launch to the activity from a notification. This activity would serve that purpose also in real life, but for this example, we'll keep that part simple. Now that we have our service implementation, let's create a new Android project called StockQuoteClient2. Use com.dave for the package and MainActivity for the activity name. To implement a client that passes the Person object to the service, we need to copy everything that the client needs from the service project to the client project. In our previous example, all we needed was the IStockQuoteService.aidl file. We also need to copy the Person.java and Person.aidl files, because the Person object is now part of the interface. After you copy these three files to the client project, modify main.xml according to Listing 15–32, and modify MainActivity.java according to Listing 15–33. Or simply import this project from the source code on our web site.

Listing 15–32. *Updated* main.xml *for* StockQuoteClient2

```
<?xml version="1.0" encoding="utf-8"?>
<!-- This file is /res/layout/main.xml -->
<LinearLayout xmlns:android="http://schemas.android.com/apk/res/android"
    android:orientation="vertical"
    android:layout_width="fill_parent"
    android:layout_height="fill_parent" >

<ToggleButton android:id="@+id/bindBtn"
    android:layout_width="wrap_content"
    android:layout_height="wrap_content"
    android:textOff="Bind"  android:textOn="Unbind"
    android:onClick="doClick" />

<Button android:id="@+id/callBtn"
    android:layout_width="wrap_content"
    android:layout_height="wrap_content"
    android:text="Call Service" android:enabled="false"
    android:onClick="doClick" />
</LinearLayout>>
```

Listing 15–33. *Calling the Service with a* Parcelable

```
package com.dave;
// This file is MainActivity.java
import android.app.Activity;
import android.content.ComponentName;
import android.content.Context;
import android.content.Intent;
import android.content.ServiceConnection;
import android.os.Bundle;
import android.os.IBinder;
import android.os.RemoteException;
import android.util.Log;
import android.view.View;
import android.widget.Button;
import android.widget.Toast;
import android.widget.ToggleButton;

import com.androidbook.services.stock2.IStockQuoteService;
import com.androidbook.services.stock2.Person;

public class MainActivity extends Activity {

    protected static final String TAG = "StockQuoteClient2";
    private IStockQuoteService stockService = null;
    private ToggleButton bindBtn;
    private Button callBtn;
```

```
    /** Called when the activity is first created. */
    @Override
    public void onCreate(Bundle savedInstanceState) {
        super.onCreate(savedInstanceState);
        setContentView(R.layout.main);

        bindBtn = (ToggleButton)findViewById(R.id.bindBtn);
        callBtn = (Button)findViewById(R.id.callBtn);
    }

    public void doClick(View view) {
        switch(view.getId()) {
        case R.id.bindBtn:
            if(((ToggleButton) view).isChecked()) {
                bindService(new Intent(
                    IStockQuoteService.class.getName()),
                    serConn, Context.BIND_AUTO_CREATE);
            }
            else {
                unbindService(serConn);
                callBtn.setEnabled(false);
            }
            break;
        case R.id.callBtn:
            callService();
            break;
        }
    }

    private void callService() {
        try {
            Person person = new Person();
            person.setAge(47);
            person.setName("Dave");
            String response = stockService.getQuote("ANDROID", person);
            Toast.makeText(MainActivity.this,
                        "Value from service is "+response,
                        Toast.LENGTH_SHORT).show();
        } catch (RemoteException ee) {
            Log.e("MainActivity", ee.getMessage(), ee);
        }
    }

    private ServiceConnection serConn = new ServiceConnection() {

        @Override
        public void onServiceConnected(ComponentName name,
            IBinder service)
        {
            Log.v(TAG, "onServiceConnected() called");
            stockService = IStockQuoteService.Stub.asInterface(service);
            bindBtn.setChecked(true);
            callBtn.setEnabled(true);
        }

        @Override
        public void onServiceDisconnected(ComponentName name) {
```

```
            Log.v(TAG, "onServiceDisconnected() called");
            bindBtn.setChecked(false);
            callBtn.setEnabled(false);
            stockService = null;
        }
    };

    protected void onDestroy() {
        if(callBtn.isEnabled())
            unbindService(serConn);
        super.onDestroy();
    }
}
```

This is now ready to run. Remember to send over the service to the emulator before you send over the client to run. The user interface should look like Figure 15–8.

Figure 15–8. *User interface of* `StockQuoteClient2`

Let's take a look at what we've got. As before, we bind to our service, and then we can invoke a service method. The `onServiceConnected()` method is where we get told that our service is running, so we can then enable the Call Service button so the button can invoke the `callService()` method. As shown, we create a new `Person` object and set its `Age` and `Name` properties. We then execute the service and display the result from the service call. The result looks like Figure 15–9.

Figure 15–9. *Result from calling the service with a* `Parcelable`

Notice that when the service is called, you get a notification in the status bar. This is coming from the service itself. We briefly touched on Notifications earlier as a way for a service to communicate to the user. Normally, services are in the background and do not display any sort of UI. But what if a service needs to interact with the user? While it's tempting to think that a service can invoke an activity, a service should *never* invoke an activity directly. A service should instead create a notification, and the notification should be how the user gets to the desired activity. This was shown in our last exercise. We defined a simple layout and activity implementation for our service. When we created the notification within the service, we set the activity in the notification. The user can click the notification, and it will take the user to our activity that is part of this service. This will allow the user to interact with the service.

Notifications are saved so that you can get to them by pulling up the menu on the Android Home page and clicking Notifications. A user can also drag down from the notification icon in the status bar to see them. Note the use of the `setLatestEventInfo()` method call and the fact that we reuse the same ID for every message. This combination means that we are updating the one and only notification every time, rather than creating new notification entries. Therefore, if you go to the Notifications screen in Android after clicking Bind, Call Again, and Unbind a few times, you will only see one message in Notifications, and it will be the last one sent by the `BackgroundService`. If we used different IDs, we could have multiple notification messages, and we could update each one separately. Notifications can also be set with additional user "prompts" such as sound, lights, and/or vibration.

It is also useful to see the artifacts of the service project and the client that calls it (see Figure 15–10).

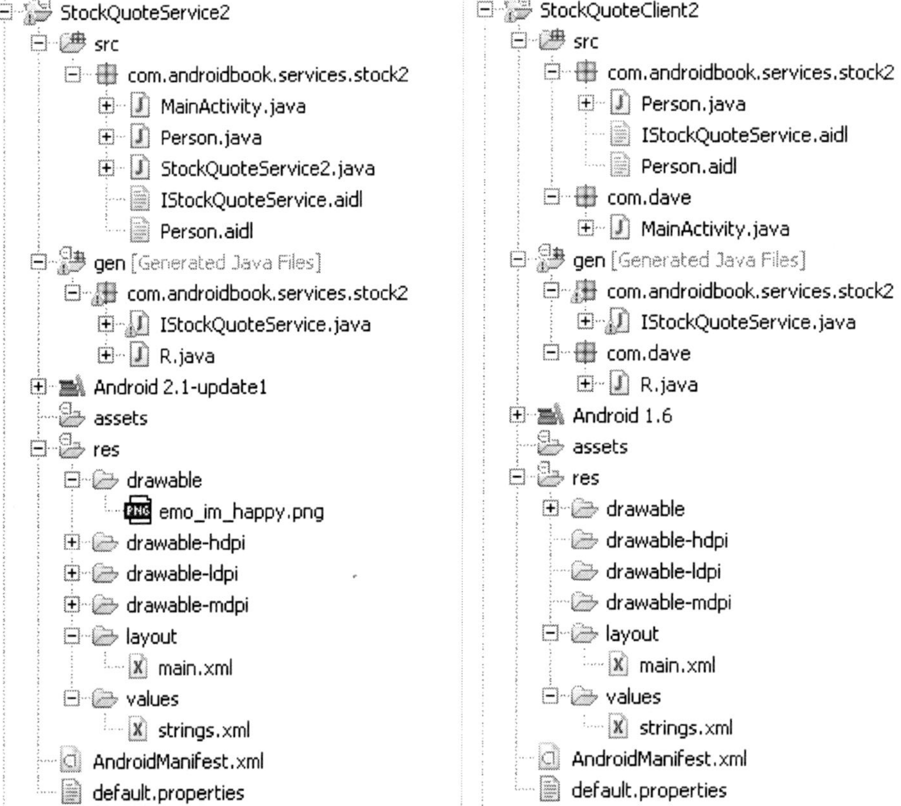

Figure 15–10. *The artifacts of the service and the client*

Figure 15–10 shows the Eclipse project artifacts for the service (left) and the client (right). Note that the contract between the client and the service consists of the AIDL artifacts and the `Parcelable` objects exchanged between the two parties. This is the reason that we see `Person.java`, `IStockQuoteService.aidl`, and `Person.aidl` on both sides. Because the AIDL compiler generates the Java interface, stub, proxy, and so on from the AIDL artifacts, the build process creates the `IStockQuoteService.java` file on the client side when we copy the contract artifacts to the client project.

Now you know how to exchange complex types between services and clients. Let's briefly touch on another important aspect of calling services: synchronous versus asynchronous service invocation.

All of the calls that you make on services are synchronous. This brings up the obvious question, do you need to implement all of your service calls in a worker thread? Not necessarily. On most other platforms, it's common for a client to use a service that is a complete black box, so the client would have to take appropriate precautions when making service calls. With Android, you will likely know what is in the service (generally because you wrote the service yourself), so you can make an informed decision. If you know that the method you are calling is doing a lot of heavy lifting, you should consider using a secondary thread to make the call. If you are sure that the method does not have any bottlenecks, you can safely make the call on the UI thread. If you conclude that it's best to make the service call within a worker thread, you can create the thread and then call the service. You can then communicate the result to the UI thread.

References

Here are some helpful references to topics you may wish to explore further:

- `www.androidbook.com/proandroid4/projects`: A list of downloadable projects related to this book. For this chapter, look for a ZIP file called `ProAndroid4_Ch15_Services.zip`. This ZIP file contains all projects from this chapter, listed in separate root directories. There is also a `README.TXT` file that describes exactly how to import projects into Eclipse from one of these ZIP files.
- `http://hc.apache.org/httpcomponents-client-ga/tutorial/html/`: Great tutorials on using the `HttpClient` classes, including authentication and the use of cookies.

Summary

This chapter was all about services, specifically:

- We talked about consuming external HTTP services using the Apache `HttpClient`.
- With regard to using the `HttpClient`, we showed you how to do HTTP GET calls and HTTP POST calls.
- We also showed you how to do multipart POSTs.
- You learned that SOAP can be done from Android, but it's not the preferred way to call web services.

- We talked about how you could setup an Internet proxy to manage a SOAP service on your application's behalf from a server somewhere, so your application can use RESTful services to your proxy and keep the application simpler.
- We then covered exception handling and the likely types of exceptions that your application is likely to experience (timeouts mostly).
- You saw how to use the `ThreadSafeClientConnManager` to share a common `HttpClient` inside your application.
- You learned how to check and set timeout values for connections to the network.
- We covered a couple of options for making connections to web services, including `HttpURLConnection` and `AndroidHttpClient`.
- Since network connections should never be made from the main UI thread, we showed you how to use the `AsyncTask` to perform operations in the background.
- We covered another special class for background tasks: the `DownloadManager`. This handles downloading of files on background threads.
- We then moved on to the formal Android Services topics.
- We explained the difference between local services and remote services. Local services are services that are consumed by the components (such as activities) in the same process as the service. Remote services are services whose clients are outside the process hosting the services.
- You learned that even though a service is meant to be on a separate thread, it is still up to the developer to create and manage the background threads associated with services.
- You discovered how to start and stop local services, and how to create and bind to a remote service.
- You saw how the `NotificationManager` is used to track running services.
- We covered how to pass data to a service, using Parcelables for the complex types.

Interview Questions

Here are some questions you can ask yourself to solidify your understanding of this topic:

1. Why is it bad practice to call a web service from the main UI thread?

2. Name the timeouts that you can query and set for a web connection?
3. What methods are available in Android for parsing XML?
4. What sorts of exceptions might you get while calling a web service? What are some of the ways that you could deal with each?
5. What is the lightweight class for dealing with HTTP connections?
6. What are the four callbacks for an `AsyncTask`? Which ones run on the main UI thread? Which ones are optional?
7. How many times can an instance of an `AsyncTask` be run?
8. How do you force `DownloadManager` to only download via a WiFi network?
9. Does an Android Service provide its own background thread?
10. What are reasons for using services in an Android application?
11. Can you bind to a local service?
12. What is one reason for giving a notification to the `NotificationManager` from a service?
13. What are `ThreadGroups` good for?
14. Can a Parcelable be used to send data to an activity? Why or why not?
15. Why is it not a good idea to launch an activity from a service?

Chapter 16

Exploring Packages

In the book thus far, we have covered the basics of the Android platform. These chapters detailed the happy path through Android. In the next few chapters, starting with this one (Chapters 16, 17, 18, and 19), we will cover the next grain of detail around the core of Android.

We will start that exploration by looking under the hood of Android packages, the Android package signing process, sharing data between packages, and Android library projects. You will understand the Linux process context in which an .apk file runs. You will see how multiple .apk files can share data and resources given that context.

Although you were introduced to signing Android package files in Chapter 14, in this chapter, you will learn the meaning, implication, and use of signed JAR files. In the context of data sharing, we will also look at Android library projects to see how they work and if they could be used for resource and code sharing.

Let's start this discussion by going back to the basics of an .apk file, as it forms the basis for an Android process.

Packages and Processes

As you have witnessed in previous chapters, when you develop an application in Android, you end up with an .apk file. You then sign this .apk file and deploy it to the device. Let's learn a little bit more about Android packages.

Details of a Package Specification

Each .apk file is uniquely identified by its root package name, which is specified in its manifest file. Here is an example of a package definition that we will be using for this chapter (the package name is highlighted):

```
<manifest xmlns:android="http://schemas.android.com/apk/res/android"
      package="com.androidbook.library.testlibraryapp"
      ...>
      ...rest of the xml nodes
</manifest>
```

If you were the developer of this package and signed it and installed on the device, no one else other than you can update this package. **The package name is tied to the signature with which it is signed**. Subsequently, a developer with a different signature cannot sign and install a package with the same fully qualified Java package name.

Translating the Package Name to a Process Name

Android uses the package name as the name of the process under which to run the components of this package. Android also allocates a unique user ID for this package process to run under. This allocated user ID is essentially an ID for the underlying Linux OS. You can discover this information by looking at the details of the installed package.

Listing Installed Packages

On the emulator, you can see a list of installed applications by navigating to the package browser using the path Home ➤ Applications ➤ Dev Tools ➤ Package Browser. (Note that you may or may not find a similar package browser on a real device. This could also change based on the Android release.)

For instance, on a 2.3 device (we tested this on an LG Revolution), you can see the list of installed applications by going to Settings ➤ Applications ➤ Manage Applications. This path leads to an icon that allows you to uninstall the application and hence its package.

Once you see the list of packages, you can highlight a package for a particular application such as, say, a browser, and click/touch it. This will bring up a package detail screen that looks like Figure 16–1.

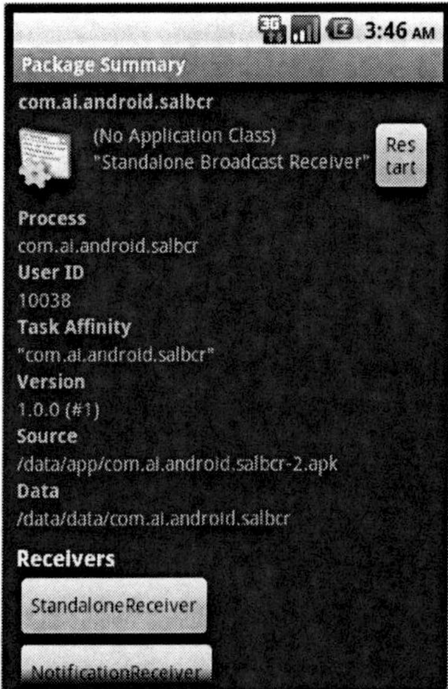

Figure 16–1. *Android package details*

Figure 16–1 shows the name of the process as indicated by the Java package name in the manifest file and the unique user ID allocated to this package. In the case of the browser, the manifest file would have indicated its package name as `com.android.browser` (reflected by the attribute process in Figure 16–1).

Any resources created by this process or package will be secured under that Linux user ID. This screen also lists the components inside this package. Examples of components are activities, services, and broadcast receivers.

Deleting a Package Through the Package Browser

While we are on the subject of the package browser, we'd like to point out that you can also delete the package from the emulator using the following steps in the package browser mentioned in the previous section:

1. Highlight the package.
2. Click Menu.
3. Click Delete Package to delete a package.

On a real device, or through the emulator you can also do this as follows:

1. Choose Settings ➤ Applications ➤ Manage Applications.
2. Click the desired application package.
3. Choose Uninstall.

Revisiting the Package Signing Process

Because a process is tied to a package name, and a package name is tied to its signature, signatures play a role in securing the data belonging to a package. To fully understand the implications of this, let's investigate the nature of the package signing process.

In Chapter 14, we introduced the mechanics of signing an application prior to installing on the device. However, we haven't explored the need for and implications of the package signing process.

For example, when we download an application and install it on Windows or another operating systems, we don't need to sign it. Why is signing mandated on an Android device? What does the signing process really mean? What does it ensure? Are there any real-world parallels to the signing process that we can quickly relate to? We will explore these questions in this section.

As packages are installed onto a device, it is necessary that each installed package has a unique or distinct Java package name. If you try to install a new package with an existing name, the device will disallow the installation until the previous package is removed. To allow this type of package upgrading, you must ensure that the same application publisher is associated with that package. This is done with digital signatures. After going through the following sections of this chapter, you will see that signing an `.apk` file ensures that, as a developer, you reserve that package name for you through your digital signature.

Let's walk through a couple of scenarios so you fully understand digital signatures.

Understanding Digital Signatures: Scenario 1

Imagine you are a wine collector located in a very un-wine-like place, like the Sahara. Furthermore, say wine makers around the world are sending you wine casks to archive or sell.

As a wine collector, you notice that each cask and the wine inside it have a specific color that is distinct from others. On further investigation, you find out that if two casks or the wine inside them have the same hue, they *always* come from the same wine maker. Even if a fake wine maker wants to produce the same hue as another reputable (or otherwise) wine maker, they can't.

On digging further, you find out that each vintner has a secret hue recipe that is kept locked up in a cellar and never revealed. This explains why each wine is different and why two wines with the same hue *must* come from the same wine maker. Of course, this identification by no means reveals the identity of the wine maker—just that the vintner is distinct and unique.

The hue becomes a signature of the wine maker, like a family stamp, and the wine maker hides the means of producing the signature from everyone else.

An important distinction in this example is that there is no way for you, as a collector, to know *which* wine maker sent a particular shipment of wine—there is no name or address associated with that signature.

Understanding Digital Signatures: Scenario 2

Let's consider another scenario for naturally occurring signatures. When you visit a foreign land, you turn on the radio and hear many songs. You can tell there are different singers, and you can identify each separately but not know who they are or know their names. This is self-signing (in this case, with their vocal chords). When a friend of yours tells you about a singer and associates that singer with a voice you have heard, it is analogous to third-party signing.

One singer can imitate another's voice to confuse or trick the listener. However, it is far, far harder to emulate a digital signature because of the mathematical algorithms that are used to encode signatures.

A Pattern for Understanding Digital Signatures

When we talk about someone signing a JAR file, that JAR file is uniquely "colored" and can be distinguished from other set of JAR files. However, there is no way to identify the source developer or company with authority. Such JAR files are called *self-signed* JAR files.

To know the *source* in the wine-collector scenario, you need a third-party company that the wine collector trusts to tell you that the color red comes from Company 1. Now, every time you see "color-red", you know that the wine is from Company 1. These are called third-party-signed JAR files. These are useful in your browsers to tell you that you are downloading a file from Company 1 or installing an application manufactured by Company 1 (authoritatively).

So How Do You Digitally Sign?

Digital signatures, which follow semantics similar to those explained in the earlier scenarios, are technically implemented through what is called *public/private key encryption*. Mathematics can be applied to generate two numbers whereby if you encode with the first number (the private key), only the second number (the public key) can decrypt it. These keys are asymmetric. Even if everyone knows the public key, there is no way they can encrypt a message that the public key can decrypt. Only its matching private key can do that.

Let's consider the idea of public and private keys in the context of the wine example.

A wine maker who wants to distinguish wine through digital signatures, as opposed to hues, creates a code (hue) for their casket using the private key. Because the private key is used to generate the code (hue), only a corresponding public key can decrypt the code.

The wine maker then boldly writes down the public key name and the encrypted code (the one generated through the private key) on top of the cask, or transfers the public key once through a courier.

When you, the wine collector, take that public key and successfully unravel the encrypted code, you know that the public key is correct and the message is only encrypted by the wine maker who wrote the public key. In this scenario, even if another imposter wine maker copies the public key of the real wine maker and writes it on a casket, the imposter will not possess the ability to write a secret message that the public key will decrypt.

In essence, the public key becomes the signature detector of the wine maker. Even if someone else were to claim the public key, that person wouldn't be able to produce a message that could be decrypted with the public key.

With this comparison of digital signatures with real signatures, we have established a parallel to help you grasp digital signatures. We already covered, in Chapter 14, the mechanics of using the JDK-based `keytool` and `jarsigner` commands to accomplish the signing process. In that chapter, we also covered how the newer Eclipse ADTs makes this a quick process using an export wizard.

Implications of the Signing Process

We now can see that we cannot have two distinct signatures for the same package name. Signatures are sometimes referred to as public key infrastructure (PKI) certificates. More accurately stated, you would use a PKI certificate to sign a bundle, a JAR file, or a DLL or an application.

The PKI certificate is tied to the package name to ensure that two developers cannot install a package that carries the same package name. However, the same certificate can be used to sign any number of packages. In other words *one PKI certificate supports many packages*. This relationship is one-to-many. However *one package has one, and only one, signature through its PKI certificate*. A developer then protects the private key of a certificate with a password.

These facts are important not only for new releases of the same package but also to share data between packages when the packages are signed with the same signature.

Sharing Data Among Packages

In previous chapters, we established that each package runs in its own process. All assets that are installed or created through this package belong to the user whose ID is assigned to the package. You also know that Android allocates a unique Linux-based user ID to run that package. In Figure 16–1, you can see what this user ID look like. According to the Android SDK documentation

> *This user ID is assigned when the application is installed on the device, and remains constant for the duration of its life on that device. Any data stored by an application will be assigned that application's user ID, and not normally accessible to other packages. When creating a new file with `getSharedPreferences(String, int)`, `openFileOutput(String, int)`, or `openOrCreateDatabase(String, int, SQLiteDatabase.CursorFactory)`, you can use the `MODE_WORLD_READABLE` and/or `MODE_WORLD_WRITEABLE` flags to allow any other package to read/write the file. When setting these flags, the file is still owned by your application, but it's global read and/or write permissions have been set appropriately so any other application can see it.*

If your intention is to allow a set of cooperating applications that depend on a common set of data, you have an option to explicitly specify a user ID that is unique to you and common for your needs. This shared user ID is also defined in the manifest file, similar to the definition of a package name. Listing 16–1 shows an example.

Listing 16–1. *Shared User ID Declaration*

```
<manifest xmlns:android="http://schemas.android.com/apk/res/android"
      package="com.androidbook.somepackage"
      sharedUserId="com.androidbook.mysharedusrid"
      ...
>
...the rest of the xml nodes
</manifest>
```

The Nature of Shared User IDs

Multiple applications can specify the same shared user ID if they share the same signature (signed with the same PKI certificate). Having a shared user ID allows multiple applications to share data and even run in the same process. To avoid the duplication of a shared user ID, use a convention similar to naming a Java class. Here are some examples of shared user IDs found in the Android system:

```
"android.uid.system"
"android.uid.phone"
```

> **NOTE:** A shared ID must be specified as a raw string and not a string resource.

As a note of caution, if you are planning to use shared user IDs, the recommendation is to use them from the start. Otherwise, they don't work well when you upgrade your application from a nonshared user ID to one with a shared ID. One of the cited reasons is that Android will not run `chown` on the old resources because of the user ID change. Therefore, we strongly advised that you

- Use a shared user ID from the start if needed.
- Don't change a user ID once it's in use.

A Code Pattern for Sharing Data

This section explores the opportunities we have when two applications want to share resources and data. As you know, the resources and data of each package are owned and protected by that package's context during runtime. It is no surprise then that you need access to the context of the package from which you want to share the resources or data.

Android provides an API called `createPackageContext()` to help with this. You can use the `createPackageContext()` API on any existing context object (such as your activity) to get a reference to the target context that you want to interact with. Listing 16–2 provides an example (this is an example only to show you the usage and not intended to be compiled).

Listing 16–2. *Using the* `createPackageContext()` *API*

```
//Identify package you want to use
String targetPackageName="com.androidbook.samplepackage1";

//Decide on an appropriate context flag
int flag=Context.CONTEXT_RESTRICTED;

//Get the target context through one of your activities
Activity myContext = ......;
Context targetContext =
        myContext.createPackageContext(targetPackageName, flag);

//Use context to resolve file paths
Resources res = targetContext.getResources();
File path = targetContext.getFilesDir();
```

Notice how we are able to get a reference to the context of a given package name such as `com.androidbook.samplepackage1`. This `targetContext` in Listing 16–2 is identical to the context that is passed to the target application when that application is launched. As the name of the method indicates (in its "create" prefix), each call returns a new context object. However, the documentation assures us that this returned context object is designed to be lightweight.

This API is applicable regardless of whether you have a shared user ID. If you share the user ID, it is well and good. If you don't share a user ID, the target application would need to declare its resources accessible to the outside users.

`createPackageContext()` uses one of three flags:

- If the flag is `CONTEXT_INCLUDE_CODE`, Android allows you to load the target application code into the current process. That code will then run as yours. This will succeed only if both packages have the same signature and a shared user ID. If the shared user IDs don't match, using this flag will result in a security exception.
- If the flag is `CONTEXT_RESTRICTED`, we still should be able to access the resource paths without going to the extreme case of requesting a code load.
- If the flag is `CONTEXT_IGNORE_SECURITY`, the certificates are ignored and the code is loaded, but it will run under your user ID. The documentation as a consequence suggests severe caution if you use this flag.

Now we know how packages, signatures, and shared user IDs can be used in concert in controlling access to what applications own and create.

Library Projects

As we talk through sharing code and resources, one question worth asking is, will the idea of a "library" project help? To investigate this, we first need to understand what library projects are, how to create them, and how these projects are used.

What Is a Library Project?

Starting with the ADT 0.9.7 Eclipse plug-in, Android supports the idea of library projects. Since the last edition of the book, the approach to building libraries has changed a bit. In this edition, we are going to cover the latest for Android libraries as of SDK Tools 15.0 and Android SDK 4.0. We will draw attention to older approaches where applicable.

> **NOTE:** At the time of this writing, the ADT is at release 15.0. Don't be too alarmed by the increase in numbers for the ADT. At 0.9.9, the release number sequence was changed to 8.0 to match the SDK Tools release. ADT 15.0 is designed to be used with SDK Tools release 15.0. You can learn about these dependencies at `http://developer.android.com/sdk/eclipse-adt.html`.

Because the same SDK Tools 15.0 is used to build programs for previous SDKs as well, the library building approach doesn't change based on the Android SDK number; it only depends on the SDK Tools release and the ADT release numbers.

A library project is a collection of Java code and resources that looks like a regular project but never ends up in an `.apk` file by itself. Instead, the code and resources of a library project become part of another project and get compiled into that main project's `.apk` file.

Library Project Predicates

Here are some facts about these library projects:

- A library project can have its own package name.
- A library project will not be compiled into its own `.apk` file and instead gets absorbed into an `.apk` file of the project that uses it as a dependency.
- A library project can use other JAR files.
- A library project cannot be made into a complete JAR file by itself yet, although efforts are under way to make this happen in future releases.
- Eclipse ADT will compile the library Java source files into a JAR file that is then compiled with the application project. This is a difference from the previous approach where source files were brought in and recompiled.
- Except for the Java files, the rest of the files belonging to a library project (such as resources) are kept with the library project. The presence of the library project is required in order to compile the application project that includes that library as a dependency.
- Starting with SDK Tools 15.0, the resource IDs generated for library projects are not final. (This is explained later in the chapter.)
- Both the library project and the main project can access the resources from the library project through their respective `R.java` files.
- You can have duplicate resource IDs between the main project and a library project. Resource IDs from the main project will take precedence over those in the library project.
- If you would like to distinguish resource IDs between the two projects, you can use different resource prefixes, such as `lib_` for the library project resources.
- A main project can reference any number of library projects.
- You can set precedence for the library projects to see whose resources are more important.
- Components, such as an activity, of a library need to be defined in the target main project manifest file. When this is done, the component name from the library package must be fully qualified with the library package name.
- It is not necessary to define the components in a library manifest file, although it may be a good practice to know quickly what components it supports.

- Creating a library project starts with creating a regular Android project and then choosing the `Is Library` flag in its properties window.
- You can set the dependent library projects for a main project through the project properties screen as well.
- Clearly, being a library project, any number of main projects can include a library project.
- One library project cannot reference another library project as of this release, although there seems to be a desire to be able to do so in future releases.

Let's explore library projects by creating a library project and a main project. The goal of this sample project is to do the following:

1. Create a simple activity in a library project.
2. Create a menu for the activity in step 1 by defining some menu resources.
3. Create a main project activity that uses the library project as a dependency.
4. Create an activity in the main project from step 3.
5. Create a menu for the main activity in step 4.
6. Have a menu item from the main activity invoke the activity from the library project.

Once these is done, Figure 16–2 shows the activity from the main project (the activity from step 4).

Figure 16–2. *A sample activity with menus in a main project*

When you click the Invoke Lib menu item from the main project activity, you will see the activity shown in Figure 16–3 served from the library project.

Figure 16–3. *A sample activity from the Library project*

The menus in this library activity come from resources of the library project. Clicking these menus simply logs a message on the screen that a particular menu item is clicked. Let's start the exercise by creating a library project first.

Creating a Library Project

This sample library project will have the following files:

- `TestLibActivity.java` (Listing 16–3)
- `layout/lib_main.xml` (Listing 16–4)
- `menu/lib_main_menu.xml` (Listing 16–5)
- `AndroidManifest.xml` (Listing 16–6)

These files should be sufficient to create your own Android library project and are shown in the following listings.

> **NOTE:** We will give you a URL at the end of the chapter that you can use to download projects from this chapter. This will allow you to import these projects into Eclipse directly.

Listing 16–3. *Sample Library Project Activity:* `TestLibActivity.java`

```
package com.androidbook.library.testlibrary;

//...basic imports here
//use CTRL-SHIFT-O to have eclipse generate
//necessary imports. Keep an eye out for duplicates.

public class TestLibActivity extends Activity
{
    public static final String tag="TestLibActivity";
    @Override
    public void onCreate(Bundle savedInstanceState) {
        super.onCreate(savedInstanceState);
        setContentView(R.layout.lib_main);
    }
    @Override
    public boolean onCreateOptionsMenu(Menu menu) {
        super.onCreateOptionsMenu(menu);
       MenuInflater inflater = getMenuInflater(); //from activity
       inflater.inflate(R.menu.lib_main_menu, menu);
        return true;
    }
    @Override
    public boolean onOptionsItemSelected(MenuItem item) {
        appendMenuItemText(item);
        if (item.getItemId() == R.id.menu_clear){
            this.emptyText();
            return true;
        }
        return true;
    }
    private TextView getTextView(){
        return (TextView)this.findViewById(R.id.text1);
    }
    public void appendText(String abc){
        TextView tv = getTextView();
        tv.setText(tv.getText() + "\n" + abc);
    }
    private void appendMenuItemText(MenuItem menuItem){
       String title = menuItem.getTitle().toString();
       TextView tv = getTextView();
       tv.setText(tv.getText() + "\n" + title);
    }
    private void emptyText(){
          TextView tv = getTextView();
          tv.setText("");
    }
}
```

Listing 16–4 shows the supporting layout file for this activity: just a single text view that is used to write out the name of the menu item clicked.

Listing 16–4. *Sample Library Project Layout File:* `layout/lib_main.xml`

```
<?xml version="1.0" encoding="utf-8"?>
<LinearLayout xmlns:android="http://schemas.android.com/apk/res/android"
    android:orientation="vertical"
    android:layout_width="fill_parent"
    android:layout_height="fill_parent"
    >
<TextView
    android:id="@+id/text1"
    android:layout_width="fill_parent"
    android:layout_height="wrap_content"
    android:text="Your debug will appear here "
    />
</LinearLayout>
```

Listing 16–5 provides the menu file to support the menus shown in the library activity of Figure 16–3.

Listing 16–5. *Library Project Menu File:* `menu/lib_main_menu.xml`

```
<menu xmlns:android="http://schemas.android.com/apk/res/android">
    <!-- This group uses the default category. -->
    <group android:id="@+id/menuGroup_Main">
        <item android:id="@+id/menu_clear"
         android:title="clear" />
        <item android:id="@+id/menu_testlib_1"
            android:title="Lib Test Menu1" />
        <item android:id="@+id/menu_testlib_2"
            android:title="Lib Test Menu2" />
    </group>
</menu>
```

And the manifest file for the library project is contained in Listing 16–6.

Listing 16–6. *Library Project Manifest File:* `AndroidManifest.xml`

```
<?xml version="1.0" encoding="utf-8"?>
<manifest xmlns:android="http://schemas.android.com/apk/res/android"
      package="com.androidbook.library.testlibrary"
      android:versionCode="1"
      android:versionName="1.0.0">
    <uses-sdk android:minSdkVersion="3" />
    <application android:icon="@drawable/icon"
         android:label="Test Library Project">
        <activity android:name=".TestLibActivity"
                  android:label="Test Library Activity">
        </activity>
    </application>
</manifest>
```

As pointed out in the "Library Project Predicates" section, the activity definition in the library project manifest file is merely for documentation; executing it is optional.

When these files are assembled, you start by creating a regular Android project. Once the project is set up, right-click the project name, and click the properties context menu to show the properties dialog for the library project. This dialog is shown in Figure 16–4.

(The available build targets in this figure may vary with your version of the Android SDK.) Simply select Is Library from this dialog to set up this project as a library project.

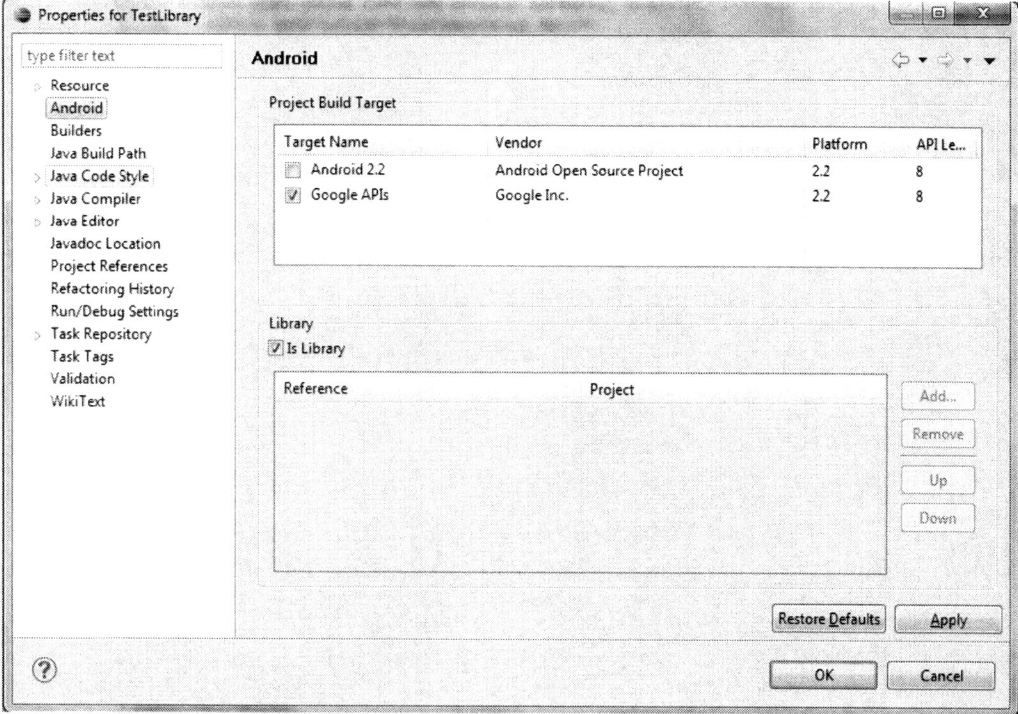

Figure 16–4. *Designating a project as a library project*

With that, we have completed creating a library project. Although Figure 16–4 indicates the build target as 2.3, this will work equally well for other SDK targets including 3.x and 4.x. Let's see now how to create an application project that can use this library project.

Creating an Android Project That Uses a Library

We will use a similar set of files to create an application project and then go on to use the library project from the previous section as a dependency. Here is the list of files we will be using to create the main project:

- `TestAppActivity.java` (Listing 16–7)
- `layout/main.xml` (Listing 16–8)
- `menu/main_menu.xml` (Listing 16–9)
- `AndroidManifest.xml` (Listing 16–10)

Listing 16–7 shows `TestAppActivity.java`.

Listing 16–7. *Main Project Activity Code:* *TestAppActivity.java*

```
package com.androidbook.library.testlibraryapp;
import com.androidbook.library.testlibrary.*;
//...other imports

public class TestAppActivity extends Activity
{
    public static final String tag="TestAppActivity";
    @Override
    public void onCreate(Bundle savedInstanceState) {
        super.onCreate(savedInstanceState);
        setContentView(R.layout.main);
    }
    @Override
    public boolean onCreateOptionsMenu(Menu menu) {
        super.onCreateOptionsMenu(menu);
       MenuInflater inflater = getMenuInflater(); //from activity
       inflater.inflate(R.menu.main_menu, menu);
        return true;
    }
    @Override
    public boolean onOptionsItemSelected(MenuItem item) {
        appendMenuItemText(item);
        if (item.getItemId() == R.id.menu_clear)
        {
            this.emptyText();
            return true;
        }
        if (item.getItemId() == R.id.menu_library_activity){
            this.invokeLibActivity(item.getItemId());
            return true;
        }
        return true;
    }
    private void invokeLibActivity(int mid)
    {
    Intent intent = new Intent(this,TestLibActivity.class);
    //Pass the menu id as an intent extra
    //incase if the lib activity wants it.
    intent.putExtra("com.androidbook.library.menuid", mid);
    startActivity(intent);
    }
    private TextView getTextView(){
        return (TextView)this.findViewById(R.id.text1);
    }
    public void appendText(String abc){
        TextView tv = getTextView();
        tv.setText(tv.getText() + "\n" + abc);
    }
    private void appendMenuItemText(MenuItem menuItem){
       String title = menuItem.getTitle().toString();
       TextView tv = getTextView();
       tv.setText(tv.getText() + "\n" + title);
    }
    private void emptyText(){
          TextView tv = getTextView();
```

```
            tv.setText("");
    }
}
```

Please note that, after creating this file, you may get a compile error on the reference to the activity class that is in the library project. This will not go away until you read a bit further and discover how to specify the previous library project as a dependency of the application project.

The corresponding layout file to support the activity is in Listing 16–8.

Listing 16–8. *Main Project layout file:* `layout/main.xml`

```
<?xml version="1.0" encoding="utf-8"?>
<LinearLayout xmlns:android="http://schemas.android.com/apk/res/android"
    android:orientation="vertical"
    android:layout_width="fill_parent"
    android:layout_height="fill_parent"
    >
<TextView
    android:id="@+id/text1"
    android:layout_width="fill_parent"
    android:layout_height="wrap_content"
    android:text="Debug Text Will Appear here"
    />
</LinearLayout>
```

The Java code in the main project activity (Listing 16–7) is using a menu item called `R.id.menu_library_activity` to invoke the `TestLibActivity`. Here is the code extracted from the Java file (Listing 16–7):

```
private void invokeLibActivity(int mid)
{
    Intent intent = new Intent(this,TestLibActivity.class);
    //Pass the menu id as an intent extra
    //incase if the lib activity wants it.
    intent.putExtra("com.androidbook.library.menuid", mid);
    startActivity(intent);
}
```

Notice how we have used `TestLibActivity.class` as if it is a local class, except that we have imported the Java classes from the library package:

```
import com.androidbook.library.testlibrary.*;
```

And the menu file is in Listing 16–9.

Listing 16–9. *Main Project Menu File:* `menu/main_menu.xml`

```
<menu xmlns:android="http://schemas.android.com/apk/res/android">
    <!-- This group uses the default category. -->
    <group android:id="@+id/menuGroup_Main">
        <item android:id="@+id/menu_clear"
         android:title="clear" />
        <item android:id="@+id/menu_library_activity"
         android:title="invoke lib" />
    </group>
</menu>
```

The manifest file to complete the project creation is shown in Listing 16–10.

Listing 16–10. *Main Project Manifest File:* `AndroidManifest.xml`

```
<?xml version="1.0" encoding="utf-8"?>
<manifest xmlns:android="http://schemas.android.com/apk/res/android"
      package="com.androidbook.library.testlibraryapp"
      android:versionCode="1"
      android:versionName="1.0.0">
    <application android:icon="@drawable/icon" android:label="Test Library App">
        <activity android:name=".TestAppActivity"
                  android:label="Test Library App">
            <intent-filter>
                <action android:name="android.intent.action.MAIN" />
                <category android:name="android.intent.category.LAUNCHER" />
            </intent-filter>
        </activity>
        <activity android:name=
"com.androidbook.library.testlibrary.TestLibActivity"
                  android:label="Test Library Activity"/>
    </application>
    <uses-sdk android:minSdkVersion="3" />
</manifest>
```

In this main application manifest file, notice how we have defined the activity `TestLibActivity` from the library project. We have also used the fully qualified package name for the activity definition. Also notice that the package names for the library project could be different from the main application project.

Associating the Library Project with the Main Application Project

Once you have set up an Android project with these files, you can use the following project properties dialog (see Figure 16–5) to indicate that this main project depends on the library project that was created earlier.

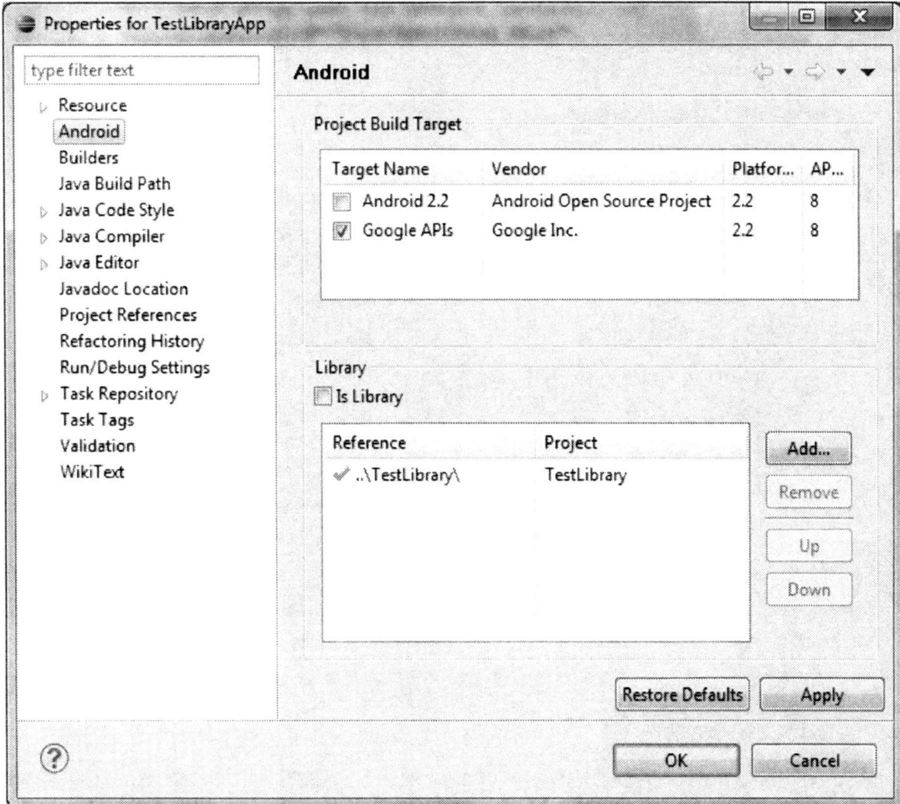

Figure 16–5. *Declaring a library project dependency*

Notice the Add button in the dialog. You can use this to add the library in Figure 16–5 as a reference. You don't need to do anything else.

Structure of the Application Project with a Library Dependency

Once the library project is set up as a dependency for the main application project, the library project appears as a compiled JAR file in the application project under the node `Library Projects` (see Figure 16–6).

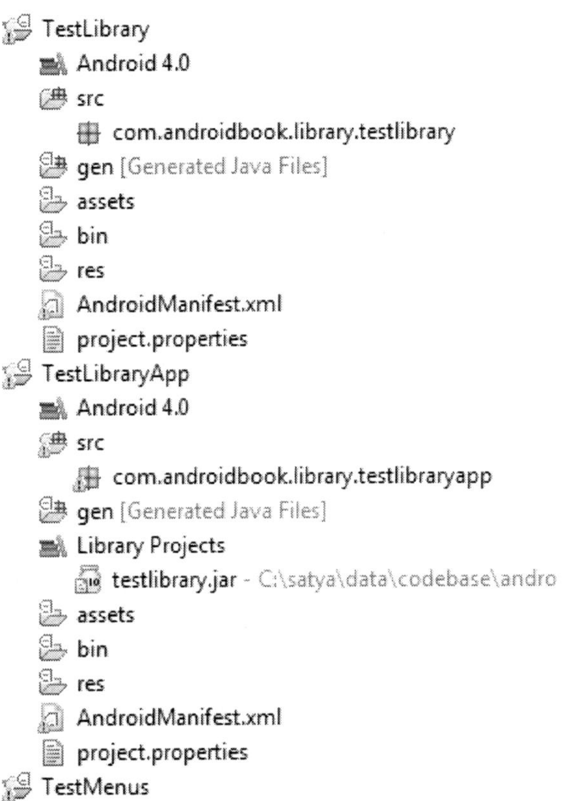

Figure 16–6. *Absorbed library project in the main project view as a JAR file*

Notice the node where the ADT has included the compiled library JAR (`testlibrary.jar`) into the main application. These precompiled library JAR files save time in compiling the main application.

Nature of Resources in Application and Library Projects

In previous releases, this inclusion of libraries was source-code based. This meant that every time the main project was compiled, all the library source files had to be compiled as well. In the new scheme, the libraries are precompiled and included in the target application as JAR files.

This compile-time assimilation of libraries poses a couple of unexpected challenges. Previously, the `R.java` file for the application project was regenerated when things were compiled in the target application, including the library source files. This meant that if you had ten different libraries, `R.java` was generated once for all of them, and the generated IDs could be guaranteed to be unique. However, when you precompile, these IDs belonging to the library projects can get frozen in the compiled JAR files. This will lead to duplicated IDs.

To fix this, Android has temporarily used a local `R.java` file for a library to generate the necessary Java classes, but it doesn't package the corresponding `R.class` in the generated JAR file for that library project. Instead, it relies on the `R.java` file that is re-created and made available in the application project. As long as the application project has an `R.java` file under the same Java package as the library project, the scheme will work. Figure 16–7 shows how Android has multiple `R.java` classes/files in the `gen` subdirectory of the application project. You see an `R.java` file for the application and one `R.java` file for each of the library projects.

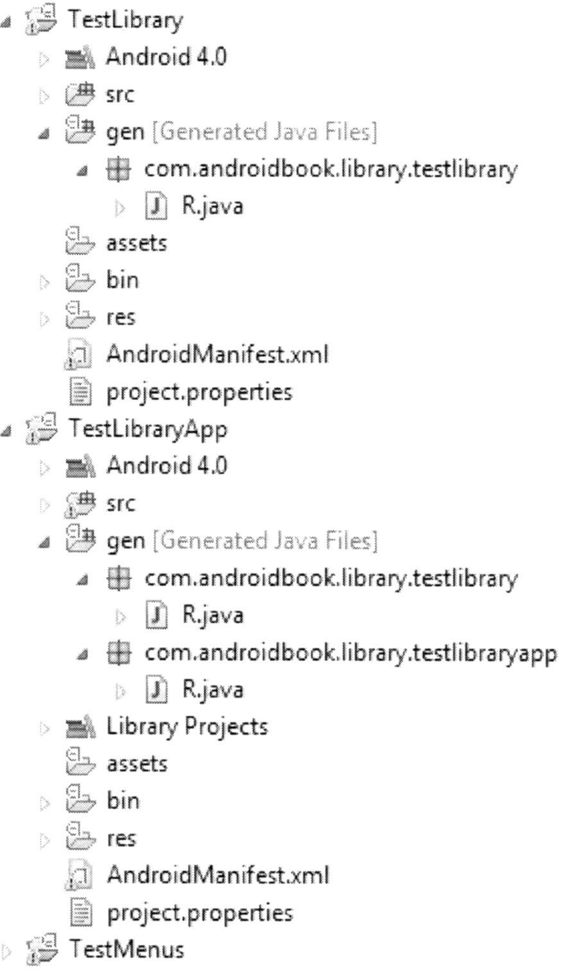

Figure 16–7. *Multiple `R.java` files in the application project*

There is one more challenge that needs solving when compiled library JARs are included in the target application project. When Android compiles the Java classes in the library project, those Java classes reference the local library `R.java` constants, because at the time of compiling the library project, the library project is all that is available. The application project that uses the library is a future prospect.

If the library's R.java constants were to be declared static final, then the compiler would hard-code the constant numbers (such as 0x7778989) in the compiled code. This needs to be stopped if you wish to avoid such duplicate numbers from multiple library JAR files. The solution Android has adapted is to declare the constants in R.java files as non-final.

Listing 16–11 shows the R.java file created for our TestLibrary in the library project. Notice that the IDs are not declared final. They are merely static variables in a Java class. Typically, these would have been final as well.

Listing 16–11. *Non-Final Resource IDs in the Library Project R.java File*

```
package com.androidbook.library.testlibrary;
public final class R {
    public static final class attr {
    }
    public static final class drawable {
        public static int icon=0x7f020000;
        public static int robot=0x7f020001;
    }
    public static final class id {
        public static int menuGroup_Main=0x7f050001;
        public static int menu_clear=0x7f050002;
        public static int menu_testlib_1=0x7f050003;
        public static int menu_testlib_2=0x7f050004;
        public static int text1=0x7f050000;
    }
    public static final class layout {
        public static int lib_main=0x7f030000;
    }
    public static final class menu {
        public static int lib_main_menu=0x7f040000;
    }
}
```

These IDs in the library version of the R.java file will help with compiling the Java source files in the library project. By keeping the IDs as non-final variables, Android prevents the values for these IDs from getting (hard-coded) into the compiled java class files for the library.

Now, as you saw in Figure 16–6, these Java JAR files from the library are included into the application project. Figure 16–7 also showed that the R.java file from the library (Listing 16–11) is reproduced in the application project. Listing 6-12 shows this re-created library's R.java file in the application project.

Listing 16–12. *Re-created R.java File for the Library's Resources in the Application Project*

```
package com.androidbook.library.testlibrary;
public final class R {
    public static final class attr {
    }
    public static final class drawable {
        public static final int icon=0x7f020000;
        public static final int robot=0x7f020001;
    }
    public static final class id {
```

```
        public static final int menuGroup_Main=0x7f060001;
        public static final int menu_clear=0x7f060002;
        public static final int menu_library_activity=0x7f060005;
        public static final int menu_testlib_1=0x7f060003;
        public static final int menu_testlib_2=0x7f060004;
        public static final int text1=0x7f060000;
    }
    public static final class layout {
        public static final int lib_main=0x7f030000;
        public static final int main=0x7f030001;
    }
    public static final class menu {
        public static final int lib_main_menu=0x7f050000;
        public static final int main_menu=0x7f050001;
    }
    public static final class string {
        public static final int app_name=0x7f040001;
        public static final int hello=0x7f040000;
    }
}
```

All the IDs from the library's `R.java` file are re-created in the duplicated `R.java` file of the application project. This file also contains IDs from the main application. It is not entirely clear why there is a need to place app constants in the `R.java` file belonging to the library's Java package.

Also, curiously, the `R.java` file belonging to the application's Java project is identical except for the Java package name at the top. Listing 6-13 contains that file just to show you how identical this file is to the one in Listing 16–12. The only difference, as you can see, is the Java package name.

Perhaps this answers the previous question that we have pondered: if there is no harm in mixing the main resources and the library's resources, why create two files? Just create one, and copy it with different Java package names at the top.

Listing 16–13. *Main Application's `R.java` File Containing Combined Resources*

```
package com.androidbook.library.testlibraryapp;
public final class R {
    public static final class attr {
    }
    public static final class drawable {
        public static final int icon=0x7f020000;
        public static final int robot=0x7f020001;
    }
    public static final class id {
        public static final int menuGroup_Main=0x7f060001;
        public static final int menu_clear=0x7f060002;
        public static final int menu_library_activity=0x7f060005;
        public static final int menu_testlib_1=0x7f060003;
        public static final int menu_testlib_2=0x7f060004;
        public static final int text1=0x7f060000;
    }
    public static final class layout {
        public static final int lib_main=0x7f030000;
        public static final int main=0x7f030001;
    }
```

```
    public static final class menu {
        public static final int lib_main_menu=0x7f050000;
        public static final int main_menu=0x7f050001;
    }
    public static final class string {
        public static final int app_name=0x7f040001;
        public static final int hello=0x7f040000;
    }
}
```

Implication of Runtime Library Dependency

There is an implication tied to the fact that IDs in the library's `R.java` file are not final. It is common to use a `switch` statement to respond to menu items based on a menu item ID. This language construct will fail at compile time when done in the library code if the IDs are not final. This is because the `case` statement in a `switch` clause has to be a real, constant number, like a `#define` in C.

So the `switch` statement in Listing 16–14 will not compile unless the IDs (such as `R.id.menu_item_1`) are actual literal numbers or static finals.

Listing 16–14. *Sample `switch` Statement to Demonstrate Non-Final Variables*

```
switch(menuItem.getItemId())
{
   case R.id.menu_item_1:
        Statment1;
        break
   case 0x7778888: // as an example for R.id.menu_item_2:
       statement;
       statement;
       break;
   default:
       statement;
       statement;
}
```

Because the IDs are defined as non-final for library projects, we are forced to use `if/else` statements instead of `switch/case` clauses.

The interesting thing is, though, that the same constants re-created from the library's `R.java` files are final in the application project (see `R.java` in Listing 16–12). Because these are final in the application project, you can use freely the `switch` clause.

Caveats to Using Library Projects

In their Java code, both projects (the library and the application) can refer to a resource by using the `R.some-id`. The value of the constant may be the same, but you will have that resource ID available in both Java namespaces: the library package namespace and the main project package namespace.

Also, pay attention to the menu names: `lib_main_menu` and `main_menu`. We would be in a real pickle if we name these two menu resource files the same but had different menu items inside them. The bottom line is that the resources are aggregated and available in one place for the main application. Pay special attention to the resources that are at the file level, such as menus and layouts, and the IDs that are generated for internal items of those resource files.

With so many caveats, we believe the library support in Android is a work in progress and may take a few more releases to stabilize.

Now that you understand library projects, are we any closer to answering the shared-data questions posed earlier?

As you can see, library projects are compile-time constructs. Clearly, any resources that belong to the library get absorbed and merged into the main project. There is not a question of sharing at runtime, because there is just one package file with the name of the main package. One suggestion often mentioned is that you can potentially develop free versions and paid versions of an application by having both versions share a library.

References

Here are some useful links to further strengthen your understanding of this chapter:

- `http://developer.android.com/guide/publishing/app-signing.html`: A good read for information about signing `.apk` files.
- `http://java.sun.com/j2se/1.3/docs/tooldocs/win32/keytool.html`: Excellent documentation on `keytool`, `jarsigner`, and the signing process itself.
- `www.androidbook.com/item/3493`: Author's notes, including a conceptual model, on understanding what it means to sign a JAR file.
- `www.androidbook.com/item/3279`: Our research on understanding Android packages. You will see how to sign .apk files, further links to how to share data between packages, more on shared user IDs, and instructions to install and uninstall packages.
- `www.androidbook.com/item/3908`: Our research notes on all aspects of Android library support, including older screen shots, newer screen shots, useful URLs, sample code, and more.
- `http://developer.android.com/guide/developing/projects/projects-eclipse.html`: An article that will help you understand how to create and use library projects. This is the primary SDK reference on the subject of libraries
- `http://android-developers.blogspot.com/2011/10/changes-to-library-projects-in-android.html`: What has changed in libraries at the time of Android 4.0 and the reasons for the changes. This blog also talks about future directions for working with libraries

- http://tools.android.com/tips/non-constant-fields: Insightful discussion of the role of non-final variables and how they affect switch statements.
- http://tools.android.com/knownissues: Android documentation of known issues in the SDK Tools and the ADT releases. Also note the domain name of this URL; this site is dedicated to all aspects of Android tooling.
- www.androidbook.com/item/3826: How to use the Eclipse ADT wizard to create a signed .apk file to deploy to the market.
- www.androidbook.com/proandroid4/projects: A list of downloadable projects related to this book. For this chapter, look for a file called ProAndroid4_Ch16_TestAndroidLibraries.zip. This ZIP file contains both projects in this chapter in separate root directories, so you can import them into Eclipse ADT.

Summary

In this chapter, we covered the following:

- Working with packages and processes
- Sharing code and data among packages
- Creating Android library projects

Interview Questions

The following set of questions should consolidate your understanding of this chapter:

1. How do you see what apps or packages are installed on your device?
2. How do you force-stop or uninstall an application?
3. How is PKI used to ensure the ownership of packages?
4. What does it mean to say that the public key and private key are asymmetrical keys?
5. Can a package have more than one certificate/key?
6. Can you use the same certificate to sign multiple packages?
7. When should you sign multiple packages with the same key?
8. What is the relationship between a Linux process and a package file?
9. Under what user ID does a .apk process run?
10. What is a shared user ID?

11. What is the prerequisite for sharing a user ID?
12. What are the advantages of sharing a user ID?
13. What API do you use to read files owned by another package?
14. What is a library project, and when do you use it?
15. Can a library project contain UI components such as activities?
16. If you don't know which app will use a lib in the future, how do you name a library package?
17. Can a library have a package name different from an app?
18. Can a library project use other JAR files?
19. Do you need to define an activity from a library in the manifest file of the application?
20. How do you indicate that a project is a library project?
21. Can you edit the source files belonging to the library project directly under the project that uses the library?
22. Starting in 4.0, why are the IDs in the library's `R.java` file non-final?
23. Why can't you use `switch` statements in library Java source files?
24. Does the entire library project, along with its resources, become a JAR file to be included in the main application project that uses the library?

Chapter 17

Exploring Handlers

We showed in Chapter 16 that each package runs in its own process. In this chapter, we will explain the organization of threads within this process. This will lead us to why we need handlers.

Most code in an Android application runs in the context of a component such as an activity or a service. We will show how these components of an application interact with threads. Most of the time there is only one thread running in an Android process, called the main thread. We will talk about the implications of sharing this main thread among various components. Primarily, this can lead to Application Not Responding (ANR) messages (the "A" stands for "application" and not "Annoying"). We will show you how you can use handlers, messages, and threads to break the dependency on the main thread when long-running operations are needed.

We will start this chapter by looking at the components of an Android application and the thread context they run under.

Android Components and Threading

As you have gathered by now from many of the previous chapters, an Android process has four primary components:

- `Activity`
- `Service`
- `ContentProvider` (often referred as just a provider)
- `BroadcastReceiver` (often referred as just a receiver)

Most code you write in an Android application is part of one of these components or called by one of these components. Each of these components gets its own XML node under an application node specification in the Android project manifest file. To recall, here are these nodes:

```
<application>
    <activity/>
    <service/>
    <receiver/>
    <provider/>
</application>
```

With some exceptions (such as external process calls to content providers), Android uses the same thread to process (or run through) code in these components. This thread is called the *main thread* of the application. When these components are called, the call can be either a synchronous call, such as when you call a content provider for data, or a deferred one through a message queue, such as when you invoke functionality by calling a start service.

Figure 17–1 describes the relationship between threads and these four components. The goal of this diagram is to show how threads weave through the Android framework and its components. The diagram does not indicate the order in which a thread might weave through the various components. The diagram is merely showing that the processing continues from one component to another in a sequential fashion. We explore aspects of this diagram in the next few subsections.

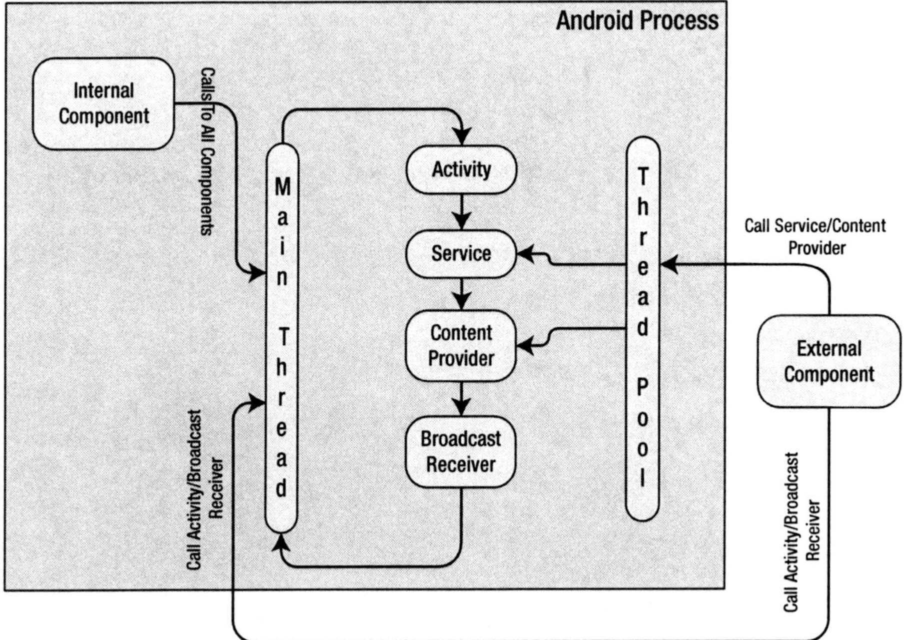

Figure 17–1. *Android components and threading framework*

Activities Run on the Main Thread

As indicated in Figure 17–1, the main thread does the heavy lifting. It runs through all the components. Moreover, it does this through a message queue. For example, as you select menus or buttons on the device screen, the device will translate these actions as messages and drop them on to the main queue of the process that is in focus. The main thread sits in a loop and processes each message. If any message takes more than five seconds or so, Android throws an ANR message.

Broadcast Receivers Run on the Main Thread

Similarly, in response to a menu item, if you were to invoke a broadcast message, Android again drops a message on the main queue of the package process in which the registered receiver is to be invoked from. The main thread will come around to that message at a later time to invoke the receiver. The main thread does the work for a broadcast receiver as well. If the main thread is busy responding to a menu action, the broadcast receiver will have to wait until the main thread gets freed up.

Services Run on the Main Thread

The same is true with a service. When you start a local service with `startService` from a menu item, a message is dropped on to the main queue, and the main thread will come around to process it via the service code.

Content Provider Runs on the Main Thread

Calls to a local content provider are slightly different. A content provider still runs on the main thread, but a call to it is synchronous and does not use message queues.

There are many content providers on an Android device. For example, all your contacts are maintained in a contact database exposed as a content provider. See Chapter 4 for more details on the architecture of content providers.

Implications of a Singular Main Thread

You may ask, "Why is it important whether most code in an Android application runs on the main thread or otherwise?" This is important because the main thread has the responsibility to get back to its queue so that UI events are responded to. As a consequence, you should not hold up the main thread. If there is something that is going to take longer than five seconds, you should get that done in a separate thread or defer it by asking the main thread to come back to it when it is freed up from other processing.

As it turns out, doing work in a separate thread is not as simple as it initially appears. We will return to that later in this chapter and also the next chapter, but for now let's talk about the thread pool that is identified in Figure 17–1.

Thread Pools, Content Providers, and External Service Components

When external clients or components outside of the process make a call to the content provider for data, then that call is allocated a thread from a thread pool. The same is true with external clients connecting to services.

Thread Utilities: Discovering Your Threads

After much talk about main threads and worker threads, it is quite instructive to use the utility class in Listing 17–1 to figure out which thread is running your part of the code. You can then verify what we have covered so far by monitoring LogCat and seeing which thread ID is being printed.

Remember that we have a URL for the downloadable project at the end of the chapter. You can use that project to further examine the various source files pointed out in this chapter. You can import this project and run the application to test all aspects we touch on.

Listing 17–1. *Thread Utilities*

```
//utils.java
public class Utils
{
   public static long getThreadId() {
      Thread t = Thread.currentThread();
      return t.getId();
   }

   public static String getThreadSignature(){
      Thread t = Thread.currentThread();
      long l = t.getId();
      String name = t.getName();
      long p = t.getPriority();
      String gname = t.getThreadGroup().getName();
      return (name
            + ":(id)" + l
            + ":(priority)" + p
            + ":(group)" + gname);
   }

   public static void logThreadSignature(){
      Log.d("ThreadUtils", getThreadSignature());
   }

   public static void sleepForInSecs(int secs){
      try{
         Thread.sleep(secs * 1000);
      } catch(InterruptedException x){
         throw new RuntimeException("interrupted",x);
      }
   }
```

```
    //The following two methods are used by worker threads
    //that we will introduce later.
    public static Bundle getStringAsABundle(String message){
        Bundle b = new Bundle();
        b.putString("message", message);
        return b;
     }
     public static String getStringFromABundle(Bundle b){
         return b.getString("message");
     }
}
```

If you use the `logThreadSignature()`, you can see which thread is executing the code. You can also use the Java library's `sleep()` method to see what happens if you pause the main thread and thereby disallow it to process the message queue. You can see this `sleep()` function wrapped in the utility method `sleepForInSecs()`.

We have briefly referred to the idea of deferring work on a main thread if needed. This is done through handlers. Handlers are extensively used throughout Android so that the main UI thread is not held up. They also play a role in communicating with the main thread from other spawned worker threads. Let's look at what handlers are and how they function in the next section.

Handlers

A *handler* is a mechanism to drop a message on the main queue (more precisely, the queue attached to the thread on which the handler is instantiated) so that the message can be processed at a *later point* in time by the main thread. The message that is dropped has an internal reference pointing to the handler that dropped it.

When the main thread gets around to processing that message, it invokes the handler that dropped the message through a callback method on the handler object. This callback method is called `handleMessage`. Figure 17–2 presents this relationship between handlers, messages, and the main thread.

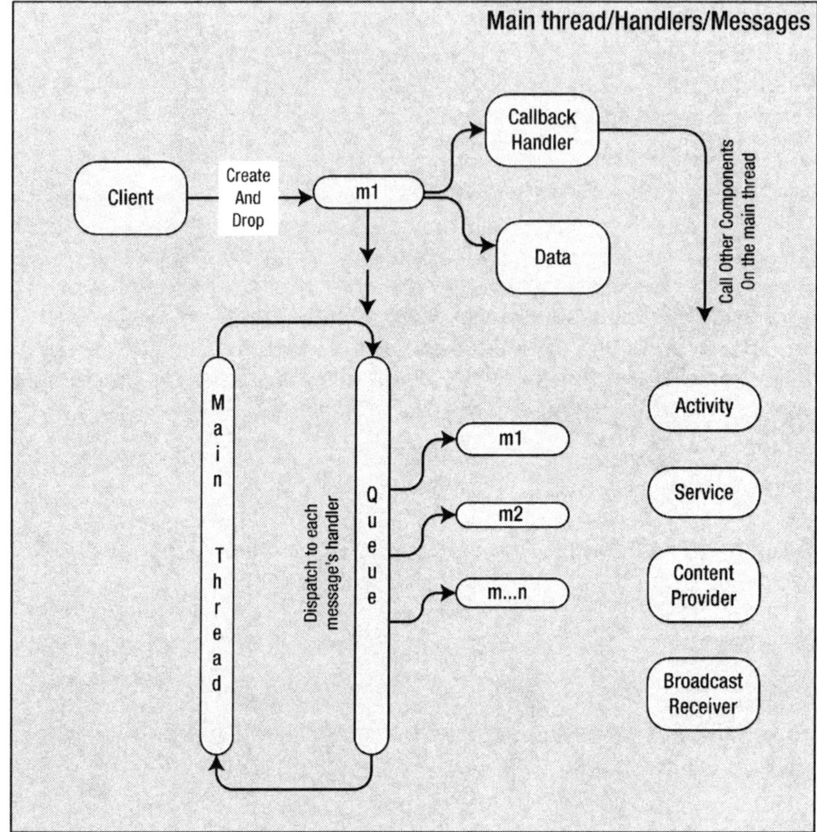

Figure 17–2. *Handler, message, message queue relationship*

Figure 17–2 illustrates the key players that work together when we talk about handlers:

- Main thread
- Main thread queue
- Handler
- Message

Out of these four, we are not exposed to the main thread or the queue directly. We primarily deal with the `Handler` object and the `Message` object. Even between these two, the `Handler` object coordinates most of the work.

Despite the importance of a `Handler` in this interaction, you should also note that, although a handler allows us to drop a message on to the queue, it is the message that actually holds a reference back to the handler. The `Message` object also holds a data structure that can be passed back to the handler. In Figure 17–2, the `Message` object is depicting this relationship by showing reference to a `Data` object.

Because of this seemingly inverted relationship between a handler and a message, and also the fact that the main thread and its queue are hidden from the programmer, a handler is best understood by an example.

For the example, we will have a menu item that invokes a function, and that function, in turn, performs an action five times at one-second intervals and reports back to the invoking activity each time.

Implications of Holding the Main Thread

If we didn't mind holding up the main thread, we could have coded the preceding scenario like the pseudo code in Listing 17–2

Listing 17–2. *Holding Up the Main Thread with a Sleep Method*

```
public class SomeActivity
{
    ....other methods

    void respondToMenuItem()
    {
        //Prove that we are on the main thread
        Utils.logThreadSignature();

        for (int i=0;i<5;i++)
        {
            sleepFor(1000);// put main thread to sleep for 1 sec
            dosomething();
            SomeTextView.setText("did something");
         }
    }
}
```

This will satisfy the requirement of the use case. However, if we do this, we *are* holding up the main thread, and we are guaranteed to have an ANR.

Using a Handler to Defer Work on the Main Thread

We can use a handler to avoid the ANR in the previous example. Pseudo code to do this via a handler will look like Listing 17–3.

Listing 17–3. *Instantiating a Handler from the Main Thread*

```
void respondToMenuItem()
{
    SomeHandlerDerivedFromHandler myHandler =
                 new SomeHandlerDerivedFromHandler();
    myHandler.doDeferredWork(); //invoke a function in 1 sec intervals
}
```

Now, the call `respondToMenuItem()` will allow the main thread to go back to its loop. The instantiated handler knows that it is invoked on the main thread and hooks itself up to the queue. The method `doDeferredWork()` will schedule work so that the main thread can get back to this work once it is free. So how does it do that? Here are the steps for implementing this function:

1. Construct a message object so that it can be dropped off on the queue.

2. Send the message object to the queue so that it can invoke a callback in 1 second.

3. Respond to the handleMessage() callback from the main thread.

To investigate this protocol, let's see the actual source code for a proper handler. The code in Listing 17–4 in the next section demonstrates this handler, which is called DeferWorkHandler.

In the pseudo code in Listing 17–3, the indicated handler SomeHandlerDerivedFromHandler is equivalent to DeferWorkHandler. Similarly, the indicated method doDeferredWork() is implemented on the DeferWorkHandler in Listing 17–4.

Sample Handler Source Code That Defers Work

Before we explain each of the steps in the previous section, the code for DeferWorkHandler is presented in Listing 17–4. Keep in mind that the source code for the main driver activity that invokes this handler is available in the downloadable project. This driver activity is very similar to what is presented in Listing 17–3.

This parent driver activity is indicated as the variable parentActivity in Listing 17–4. This variable is not critical for understanding this code, and it is primarily used to report the status of the work (via log messages) that is occurring in the handler. If you would like to examine the class TestHandlersDriverActivity corresponding to parentActivity, you can refer to the downloadable project.

Listing 17–4. *DeferWorkHandler Source Code*

```
public class DeferWorkHandler extends Handler
{
    public static final String tag = "DeferWorkHandler";

    //Keep track of how many times we sent the message
    private int count = 0;

    //A parent driver activity we can use
    //to inform of status.
    private TestHandlersDriverActivity parentActivity = null;

    //During construction we take in the parent
    //driver activity.
    public DeferWorkHandler(TestHandlersDriverActivity inParentActivity){
        parentActivity = inParentActivity;
    }
    @Override
    public void handleMessage(Message msg)
```

```
    {
        String pm = new String(
                "message called:" + count + ":" +
                msg.getData().getString("message"));

        Log.d(tag,pm);
        this.printMessage(pm);

        if (count > 5)
        {
            return;
        }
        count++;
        sendTestMessage(1);
    }
    public void sendTestMessage(long interval)
    {
        Message m = this.obtainMessage();
        prepareMessage(m);
        this.sendMessageDelayed(m, interval * 1000);
    }
    public void doDeferredWork()
    {
        count = 0;
        sendTestMessage(1);
    }
    public void prepareMessage(Message m)
    {
        Bundle b = new Bundle();
        b.putString("message", "Hello World");
        m.setData(b);
        return ;
    }
    //This method just prints a message
    //in a text box in the parent activity.
    private void printMessage(String xyz)
    {
        parentActivity.appendText(xyz);
    }
}
```

Let's look at the primary aspects of this source code.

Constructing a Suitable Message Object

As we have indicated before, when the `DeferWorkHandler` is constructed, it already knows how to hook itself up to the main queue, because it inherited that property from the base `Handler` class. The base handler offers a series of methods to send messages to the queue to be responded to later.

`sendMessage()` and `sendMessageDelayed()` are two examples of these send methods. `sendMessageDelayed()`, which we used in the example, allows us to drop a message on the main queue with a given amount of time delay. `sendMessage()`, in contrast, drops the message to be processed immediately.

When you call sendMessage() or sendMessageDelayed(), you will need an instance of the Message object. It is best that you ask the handler to give it to you, because when the handler returns the Message object, it hides itself in the belly of the Message. That way, when the main thread comes along, it knows which handler to call based solely on the message.

In Listing 17–4, the message is obtained using the following code:

```
Message m = this.obtainMessage();
```

The variable this refers to the handler object instance. As the name indicates, the method does not create a new message but instead gets one from a global message pool. At a later point, once this message is processed, it will be recycled. The method obtainMessage() has the variations listed in Listing 17–5

Listing 17–5. *Constructing a Message Through a Handler*

```
obtainMessage();
obtainMessage(int what);
obtainMessage(int what, Object object);
obtainMessage(int what, int arg1, int arg2)
obtainMessage(int what, int arg1, int arg2, Object object);
```

Each method variation sets the corresponding fields on the message object. There are some restrictions on the Object object argument when the message crosses process boundaries. In such cases, it needs to be parcelable. It is much safer and compatible in such cases to use the setData() method explicitly on the message object, which takes a bundle. In Listing 17–4, we have used setData(). You are encouraged to use arg1 or arg2 instead if what you are intending to pass are simple indicators that can be accommodated with integer values.

The argument what allows you to dequeue message or enquire if there are messages of this type in the queue. See the operations on the Handler class for more details. This chapter's "References" section has a URL for the API documentation of the Handler class.

Sending Message Objects to the Queue

Once we obtain a message from the handler, we can optionally modify the data contents of that message. In our example, we have used the setData() function by passing it a bundle object. After we have categorized or identified the data of the message, we can send the message to the queue through sendMessage() or sendMessageDelayed(). When these methods are called, the main thread will return to attending the queue.

Responding to the handleMessage Callback

The class DeferWorkHandler is derived from Handler. Once the messages are delivered to the queue, the handler sits and waits (figuratively speaking) until the main thread retrieves those messages and calls the handler's handleMessage().

If you want to see this handler and main thread interaction more clearly, you can write a `logcat` message when you are sending the message and in the `handleMessage()` callback. You will notice the time stamps differ as the main thread would have taken a few more milliseconds to come back to the `handleMessage()` method.

This is also a good way to know that both the `sendMessage()` and the `handleMessage()` run on the main thread. You can use the `Utils.logThreadSignature()` method (see Listing 17–1) to illustrate this.

In our example, each `handleMessage()`, after processing one message, sends another message to the queue so that it can be called again. It does this five times, and when the counter reaches five, it quits sending messages to the queue.

In our example of a handler, `DeferWorkHandler` (as indicated earlier) also takes the parent activity as an input so that it can report back any information using the methods provided by that activity.

Using Worker Threads

When we use a handler like the one in the previous section, the code is still executed on the main thread. Each call to `handleMessage()` still should return within the time stipulations of the main thread (in other words, each message invocation should complete in less than five seconds to avoid Android Not Responding). If your goal is to extend that time of execution further, you will need to start a separate thread, keep the thread running until it finishes the work, and allow for that subthread to report back to the main activity, which is running on the main thread. This type of a subthread is often called a *worker thread*.

It is a no-brainer to start a separate thread while responding to a menu item. However, the clever trick is to allow the worker thread to post a message to the queue of the main thread that something is happening and that the main thread should look at it when it gets to that message.

A reasonable solution that involves a worker thread is as follows:

1. Create a handler in the main thread while responding to the menu item. Keep it aside.

2. Create a separate thread (a worker thread) that does the actual work. Pass the handler from step 1 to the worker thread. This handler allows the worker thread to communicate with the main thread.

3. The worker thread code can now do the actual work for longer than five seconds and, while doing it, can call the handler to send status messages to communicate with the main thread

4. These status messages now get processed by the main thread, because the handler belonged to the main thread. The main thread can process these messages while the worker thread is doing its work.

Let's show you some sample code for a menu item that starts the process for a worker thread

Invoking a Worker Thread from a Menu

The code in Listing 17–6 illustrates a function called `testThread()` that can be invoked in response to a menu item on the main thread.

Listing 17–6. *Instantiating a Subthread from a Main Thread*

```
//Keep a couple of local variables
//so that they are not re-created with every menu click
//in your activity

//Holds a pointer to the handler
Handler statusBackHandler = null;

//An instance of the thread
Thread workerThread = null;

//this method will be invoked by a menu
private void testThread()
{
    if (statusBackHandler == null)
    {
        //Menu item was never clicked before
        //The classes referred here are listed later in the chapter
        statusBackHandler = new ReportStatusHandler(this);
        workerThread = new Thread(new WorkerThreadRunnable(statusBackHandler));
        workerThread.start();
        return;
    }

    //Thread is already there
    if (workerThread.getState() != Thread.State.TERMINATED)
    {
        Log.d(tag, "thread is new or alive, but not terminated");
    }
    else
    {
        Log.d(tag, "thread is likely dead. starting now");
        //you have to create a new thread.
        //no way to resurrect a dead thread.
        workerThread = new Thread(new WorkerThreadRunnable(statusBackHandler));
        workerThread.start();
    }
}
```

The code looks a bit roundabout, but the crux of the code is

```
statusBackHandler = new ReportStatusHandler(this);
workerThread = new Thread(new WorkerThreadRunnable(statusBackHandler));
workerThread.start();
```

Basically, we have created a handler (one that is responsible for reporting the status), passed it to the worker thread, and started the worker thread. The code outside of this block in Listing 17–6 is there so that, if we were to press the menu item twice or thrice while the thread is doing its work and not terminated, we won't create another thread and handler.

Communicating Between the Worker and the Main Threads

We will now cover the classes `ResportStatusHandler` and `WorkerThreadRunnable`. We didn't present them earlier because we wanted to drive your understanding using a top-down approach, where we plan and tell you what is required at a high level and then go into the details of how each concept is executed.

WorkerThreadRunnable Implementation

Let's see now what the worker thread is doing through the `WorkerThreadRunnable` class. The source code for the `WorkerThreadRunnable` class is in Listing 17–7. Take a quick look at this listing, especially the comments in the code, to get a feel for what it might be doing. Following the listing, we will explain the key concepts.

Listing 17–7. *Worker Thread Implementation*

```
//Primary Responsibilities
//1. Do the work
//2. Inform the parent activity
public class WorkerThreadRunnable implements Runnable
{
    //the handler to communicate with the main thread
    //Set this in the constructor
    Handler statusBackMainThreadHandler = null;

    public WorkerThreadRunnable(Handler h)
    {
        statusBackMainThreadHandler = h;
    }

    //usual debug tag
    public static String tag = "WorkerThreadRunnable";
    public void run()
    {
        Log.d(tag,"start execution");
        //see which thread is running this code
        //The following method is from Listing 17-1
        //It prints out the thread id and name
        Utils.logThreadSignature();

        //Tell parent that the worker thread has
        //started working
        informStart();
        for(int i=1;i <= 5;i++)
        {
            //In the real world instead of sleeping
            //work will be done here.
```

```
            Utils.sleepForInSecs(1);
            //Report back the work is progressing
            informMiddle(i);
        }
        informFinish();
    }

    public void informMiddle(int count)
    {
        Message m = this.statusBackMainThreadHandler.obtainMessage();
        m.setData(Utils.getStringAsABundle("done:" + count));
        this.statusBackMainThreadHandler.sendMessage(m);
    }

    public void informStart()
    {
        Message m = this.statusBackMainThreadHandler.obtainMessage();
        m.setData(Utils.getStringAsABundle("starting run"));
        this.statusBackmainThreadHandler.sendMessage(m);
    }
    public void informFinish()
    {
        Message m = this.statusBackMainThreadHandler.obtainMessage();
        m.setData(Utils.getStringAsABundle("Finishing run"));
        this.statusBackMainThreadHandler.sendMessage(m);
    }
}
```

There are two important things in Listing 17–7. In the `run()` method, we put the thread to sleep for one second and call the inform methods to tell the main thread whether the worker thread is at the beginning, middle, or end of the processing.

We have also included a call to the `Utils.logThreadSignature()` to identify the thread.

However, in the real world, instead of the `sleep()` method, this code will be calling a useful function for as longs as necessary. You can think of `sleep()` as simulating a work item that takes that many seconds.

ReportStatusHandler Implementation

All of the inform methods in Listing 17–7 create an appropriate string message and send it to the main thread through `ReportStatusHandler`, which is shown in Listing 17–8.

Listing 17–8. *Sending Status to the Main Thread*

```
public class ReportStatusHandler extends Handler
{
    public static final String tag = "ReportStatusHandler";

    //Remember the parent activity so that
    //so that we can inform it of the progress
    private TestHandlersDriverActivity
                parentTestHandlersDriverActivity = null;

    public ReportStatusHandler(
               TestHandlersDriverActivity inParentActivity){
```

```
        parentTestHandlersDriverActivity = inParentActivity;
    }

    @Override
    public void handleMessage(Message msg)
    {
        //Get string data from the message
        String pm = Utils.getStringFromABundle(msg.getData());
        Log.d(tag,pm);
        //Tell the parent activity that something happened
        this.printMessage(pm);
        //Assert that this runs on the main thread
        Utils.logThreadSignature();
    }

    private void printMessage(String xyz){
        parentTestHandlersDriverActivity.appendText(xyz);
    }
}
```

The code in this class is straightforward. When this handler receives the handleMessage(), it tells the parent driver activity that the worker thread has sent a status string through the appendText() method. The parent activity can choose whatever is necessary for that message. In the test downloadable project, we just log it to the activity screen.

In summary, we have demonstrated the following with handler examples:

- Through DeferWorkHandler, we have shown how the main thread can schedule a message (or messages) to be processed at a later time (or deferred). This technique can also be used to do repetitive processing without using a timer or alarm manager.
- Through ReportStatusHandler and a WorkerThread, we have shown how you can start a separate worker thread and have that worker thread communicate back to the main UI through a handler.

Component and Process Lifetimes

As you have seen so far, as messages get dropped on the queue, the main thread comes around to process those messages. What would happen to the messages that are pending in the main queue if you were to hit the Back button or the Home button? What would happen to the worker thread that is executing? We will explain what happens by considering the life cycle of each of the Android components.

Although we discuss the component life cycles here, please note that this is not a full discussion of those life cycles. The activity life cycle is already described with the help of a diagram in Chapter 2. Similarly, the service life cycle is elaborated on in Chapter 15. The discussion here is limited to addressing only those aspects that affect message processing and worker threads.

Activity Life Cycle

We will start with the `Activity` component. Figure 17–3 shows the activity life cycle with respect to its visibility and lifetime (the state transitions of an activity between its life cycle methods are described in Chapter 2).

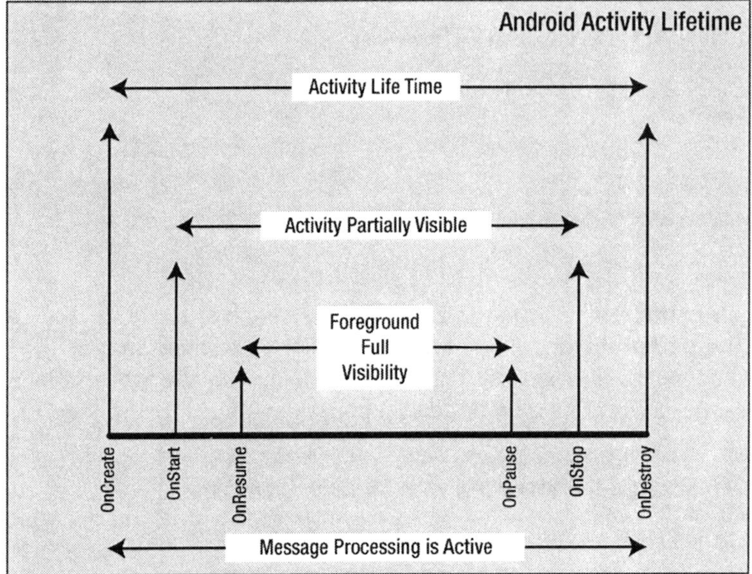

Figure 17–3. *Activity life cycle*

Once an activity comes to life (due to a start), it is fully visible, partially visible, or completely hidden. You can detect each boundary through callback methods.

An activity calls `onPause` when it is moving into partially visible state. It then may call the `onStop` method when it goes into the completely hidden state. Finally, when the process is taken out, its `onDestroy` method is called. When the `onDestroy` method is called, the view state is destroyed right after the call. Prior to that, the view state is still intact.

When an activity is moving into a full visibility state, its `onResume` is called. When it is moving out of the invisible state, it first calls `onStart` and then `onResume` (or it may call `onStop` if the activity gets hidden again). Between `onResume` and `onPause`, the activity is in a fully visible state.

Although an application may be partially or fully invisible, the message queue will still be active and so will your worker thread. You can see this by monitoring the activity life cycle methods. The downloadable project for this chapter contains these methods. You will see in that project that the messages from the worker thread and the handler are still active when `onPause` and `onStop` are called.

You can test this hypothesis by clicking the Home button when you are on an activity. Doing so will send this activity to the background and invokes `onPause` , `onStop`, and maybe even onDestroy. You will see the messages all the way until onDestroy is called (assuming you have sent that many messages).

If the process is not active when an activity is requested, it will be started and brought to life. Under low-memory conditions or when the application is completely hidden and nothing else is going on in that process, Android will remove the process.

> **NOTE:** The key thing to know is that if an activity is stopped for any of these needs, it will not automatically be brought back to life. A user has to explicitly invoke the activity either by clicking it or through other indirect means, such as starting another activity that results in invoking this activity. The only time an activity is stopped and started automatically is when the device configuration changes (such as going from portrait to landscape). As you can imagine, this can happen quite often as a phone is moved from vertical to horizontal and back.

Service Life Cycle

A service component acts differently from an activity in one primary respect—a service component is fundamentally sticky. Android makes every effort to keep a service running. Even if the service process is reclaimed due to memory conditions, it will be restarted if there are pending messages. We will go into a lot more detail of this interaction in the next chapter when we discuss broadcast receivers and long-running services.

However, a thing common to a service component and an activity component is that they both can be taken down under low-memory conditions. Android will try its best to keep a service running, but even still, there are no guarantees it will run to completion.

> **NOTE:** Services and activities should be coded such that they can be gracefully stopped through onDestroy when they have worker threads running and doing work for them. You can do this by having the thread monitor a common variable frequently to see if it is being asked to stop.

Receiver Life Cycle

Broadcast receivers use a call-and-be-gone model. The process hosting the broadcast receiver will be around only for the lifetime of the receiver and no longer. Also, the broadcast receiver runs on the main thread, and it has a hard ten-second timeframe to finish its work. You have to follow a pretty roundabout protocol to accomplish more complicated and time-consuming work in a broadcast receiver. This, indeed, is the topic of Chapter 19. But briefly, if you have a broadcast receiver that takes longer than ten seconds, you will need to follow a protocol such as the following:

1. Get hold of a wakelock in the receiver code (no later) so that the device is at least partially awake.
2. Issue a `startService()` call so that the process is tagged as sticky and restartable, if needed, and hangs around. Note that you cannot do the work in the service directly, because it would take more than ten seconds and that would hold up the main thread. This is because the service also runs on the main thread.
3. Start a worker thread from the service.
4. Have the worker thread post a message through a handler to the service or issue a `stopService()` call on the service.

As promised, we will go through this protocol in more detail in Chapter 19.

Provider Life Cycle

Content providers are another story. Clients, both internal and external, interact with a content provider synchronously. For external clients, content providers use a thread pool to satisfy this requirement. Like broadcast receivers, content providers do not have a particular life cycle. They get started when needed and stay around as long as the process stays around. Even though they are synchronous for external clients, they will run not on the main thread but on a thread pool of the process that they reside in, similar to a web client and a web server. The client thread will wait until the call comes back. When there are on clients around, the process gets reclaimed as per the reclamation rules of a process, depending on what other components are defined and active in that process.

References

As you learn about the topics in this chapter, you may want to keep the following reference URLs handy. We have indicated what you will learn from each URL:

- `http://developer.android.com/reference/android/os/Handler.html`: A reference to the Handler API. You will see here method signatures for how to construct a handler, obtain a message, override `handleMessage()` and `sendMessage()`, and so on.
- `http://developer.android.com/reference/android/os/Message.html`: A reference to the Message API. Although you use this API less because equivalent functions are available on the Handler API, it is good to know the underpinnings of a message. We recommend taking a look at this API reference.

- http://developer.android.com/guide/topics/fundamentals.html: More detail about omponent life cycles. This primarily explains activity and service life cycles and a bit about broadcast receivers. This resource is silent about content providers.
- www.science.uva.nl/ict/ossdocs/java/tutorial/java/threads/states.html: A very zippy, and necessary, introduction to threads.
- www.netmite.com/android/mydroid/1.6/frameworks/base/core/java/android/app/IntentService.java: The source code listing of IntentService.java. It shows an excellent use of handlers by core Android code in implementing the IntentService class. With the background information provided in this chapter, we strongly urge you to go over this source code as an exercise to solidify your understanding of threads in Android.
- www.androidbook.com/item/3514: One of the author's research on long-running services.
- www.androidbook.com/proandroid4/projects: A list of downloadable projects from this books. For this chapter, look for a ZIP file named ProAndroid4_Ch17_TestHandlers.zip.

Summary

In this chapter, we explored

- How the main thread coordinates between various components of an Android process.
- How handlers and threads can be used to extend the reach of a main thread, as well as how a main thread must return in five seconds to avoid ANR messages. This rule applies to broadcast receivers as well, except that the limit for a broadcast receiver is ten seconds.
- How component life cycles impact both main threads and subthreads.

Interview Questions

The following questions will help you to consolidate what is covered in this chapter:

1. What are the four components of an Android process?
2. How many threads are typically active in an Android process?
3. What thread is used to run a broadcast receiver?

4. Why do you get an ANR if you spend too much time in a broadcast receiver?
5. Can you avoid an ANR by calling a service?
6. What is the best way to break up a task into smaller chunks?
7. If you want to carry on a long-running task, what do you need to do?
8. How can you communicate between a worker thread and a main thread?
9. What is the best way to stop a worker thread?

Chapter 18

Exploring the AsyncTask

In Chapter 17, you saw the need for handlers and worker threads. These handlers helped you to run long-running tasks on a worker thread while the main thread kept the house in order.

Android SDK has recognized this as a pattern and abstracted all of the handler and thread details into a utility class called AsyncTask. You can use AsyncTask to run tasks that take longer than five seconds in the context of UI. (We will cover how to run really long-running tasks, ranging from minutes to even hours, through services in Chapter 19.)

Listing 18–1 shows a high-level pseudo code of using an AsyncTask from a menu handler.

Listing 18–1. *Usage Pattern for an* AsyncTask *by an Activity*

```
public class MyActivity
{
    //menu handler
    void respondToMenuItem()
    {
       performALongTask();
    }

    //Use an async task
    void performLongTask()
    {
        //Derive from an async task
        //Instantiate this AsyncTask
        MyLongTask myLongTask = new MyLongTask(...CallBackObjects...);

        //start the work on a worker thread
        myLongTask.execute(...someargs...);
        //have the main thread get back to its business
    }

    //Hear back from the async task
    void someCallBackFromAsyncTask(SomeParameterizedType x)
    {
        //Although invoked by the AsyncTask this
        //code runs on the main thread
        //report back to the user of the progress
```

```
    }
}
```

The use of an async task starts with inheriting from AsyncTask first. This allows you to specialize not only the execute() method but also some callback back methods that get called before, during, and after execution.

The execute() method is called on the main thread. As an implementer, you will use a method called doInBackground() to do your actual work by overriding it. This method runs on the worker thread.

To hear back from the worker thread of doInBackground(), you will override a method called onProgressUpdate(), which in turn could call someCallBackFromAsyncTask() (see Listing 18–1). To facilitate this callback into the client activity you will pass a reference to the activity (for example) through the AsyncTask constructor.

The pointer to the async task object that a client has created will also allow the client to cancel the task if needed.

Implementing a simple AsyncTask

Let's get into the details now. We will show you the following through source code:

- How to extend AsyncTask
- How to pass constructor arguments
- What methods to override
- How to set up a progress dialog through the preexecute() method
- Where to do the actual work: the doInBackground() method
- How to trigger progress callbacks
- How to override the progress method to report progress
- How to detect the end of work through the postExecute() method

We will start with how to extend AsyncTask as it uses generics and is worth spending a few minutes on.

GettingPpast the Generics in AsyncTask

The AsyncTask class employs generics to provide type safety to its methods. AsyncTask (through generics) wants you to specify the following types when you extend it:

- *The type of parameters to the* execute() *method:* When extending AsyncTask, you will need to indicate the type of parameters that you will pass to the execute() method. If you say that your type is String, then the execute() method will expect any number of strings separated by commas in its invocation.

- *Parameter types to the progress callback method:* This type indicates the array of values passed back to the caller while reporting progress through the method callback `onProgressUpdate(Progress... progressValuesArray)`. The ability to pass an array of progress values allows situations where multiple aspects of a task can be monitored and reported on. This could be useful if an async task is working on multiple subtasks.
- *Return type from the* `execute()` *method:* This type indicates the return value that is sent back as the final result from the execution through the callback `onPostExecute(Result finalResult)`.

To understand how you specify these types, you have to look at the definition (partially specified in Listing 18–2) of the `AsyncTask` class.

Listing 18–2. *A Quick Look at the* `AsyncTask` *Class Definition*

```
public class
AsyncTask<Params, Progress, Result>
{
    //A client will call this method
    AsyncTask<Params, Progress, Result>
    execute(Params.... params);

    //Do your work here
    //Frequently trigger onProgressUpdate()
    Result doInBackGround(Params... params);

    //Callback: After the work is complete
    void onPostExecute(Result result);

    //Callback: As the work is progressing
    void onProgressUpdate(Progress.... progressValuesArray);
}
```

Notice how the class definition of `AsyncTask` puts the burden on the derived classes to specify the type names for the following:

- `Params`
- `Result`
- `Progress`

Subclassing an Async Task

For example, suppose we decide on the following for a specific async task:

- `Params`: A `String` array
- `Result`: An `int`
- `Progress`: An `Integer` array

We declare the class as shown in Listing 18–3.

Listing 18–3. *Extending the Generic* `AsyncTask`

```
public class MyLongTask
extends AsyncTask<String,Integer,Integer>
{
...other constructors stuff

    //We just need to call execute() and
    //hence no overriding is needed
    Integer doInBackground(String... params);
    void onPostExecute(Integer result);
    void onProgressUpdate(Integer.... progressValuesArray);

....other methods

}
```

Notice how this concrete class `MyLongTask` has disambiguated the type names and arrived at function signatures that are type safe.

Implementing Your First Async Task

Let's look at a simple, but complete, implementation of `MyLongTask`. We have amply commented the code inline to indicate which methods run on which thread. Also pay attention to the constructor of `MyLongTask` where it receives object references of the calling context (usually an activity) and also a specific simple interface such as `IReportBack` to log progress messages.

The `IReportBack` interface is not critical to your understanding because it is merely a wrapper to a log. You can also see these additional classes in the downloadable project for this chapter. This URL is specified in the "References" section at the end of this chapter.

Listing 18–4 lays out the complete code for `MyLongTask`.

Listing 18–4. *Complete Source Code for Implementing an Async Task*

```
//AsyncTask comes from android.os package
import android.os.AsyncTask;
//Use Ctrl-shift-o to fill in the imports

//Start by specializing it
//The generics of AsyncTask are used define
//type safe methods for the class.

public class MyLongTask
extends AsyncTask<String,Integer,Integer>
{
    IReportBack r;
    Context ctx;
    public String tag = null;
    ProgressDialog pd = null;
    MyLongTask(IReportBack inr, Context inCtx, String inTag)
    {
        r = inr;
```

```
        ctx = inCtx;
        tag = inTag;
    }
    protected void onPreExecute()
    {
        //Runs on the main ui thread
        Utils.logThreadSignature(this.tag);
        pd = ProgressDialog.show(ctx, "title", "In Progress...",true);
    }
    protected void onProgressUpdate(Integer... progress)
    {
        //Runs on the main ui thread
        Utils.logThreadSignature(this.tag);
        this.reportThreadSignature();

        //will be called multiple times
        //triggered by onPostExecute
        Integer i = progress[0];
        r.reportBack(tag, "Progress:" + i.toString());
    }
    protected void onPostExecute(Integer result)
    {
        //Runs on the main ui thread
        Utils.logThreadSignature(this.tag);
        r.reportBack(tag, "onPostExecute result:" + result);
        pd.cancel();
    }
    protected Integer doInBackground(String...strings)
    {
        //Runs on a worker thread
        //May even be a pool if there are
        //more tasks.
        Utils.logThreadSignature(this.tag);

        for(String s :strings)
        {
            Log.d(tag, "Processing:" + s);
            //r.reportTransient(tag, "Processing:" + s);
        }
        for (int i=0;i<3;i++)
        {
            Utils.sleepForInSecs(2);
            publishProgress(i);
        }
        return 1;
    }
    protected void reportThreadSignature()
    {
        String s = Utils.getThreadSignature();
        r.reportBack(tag,s);
    }
}
```

We will go into the details of each of the methods here after covering briefly how a client would make use of (or call) `MyLongTask`.

Calling an Async Task

Once we have the class MyLongTask implemented, a client will then utilize this class as shown in Listing 18–5.

Listing 18–5. *Calling an Async Task*

```
Void respondToMenuItem()
{
   //An interface to log some messages back to the activity
   //See downloadable project if you need the details.
   IReportBack reportBackObject = this; //activity
   Context ctx = this; //activity
   String tag = "Task1";

   //Instantiate and execute the long task
   MyLongTask mlt = new MyLongTask(reportBackObject,ctx,tag);
   mlt.execute("String1","String2","String3");
}
```

Notice how the execute() method is called. Because we have indicated one of the generic types as a String and that the execute() methods takes a variable number of arguments for this type, we are able to pass any number of strings to the execute() method. In the example in Listing 18–5, we have passed three string arguments. You can pass more if you wish to.

Once we call the execute() method on the async task, this will result in a call to the onPreExecute() method followed by a call to the doInBackground() method. The system will also call the onPostExecute() callback once the doInBackground() method finishes.

onPreExecute() Callback and Progress Dialog

In the onPreExecute() method, we will do something important: we will start a progress dialog to indicate to the user that something is going on.

Figure 18–1 shows an image of that dialog.

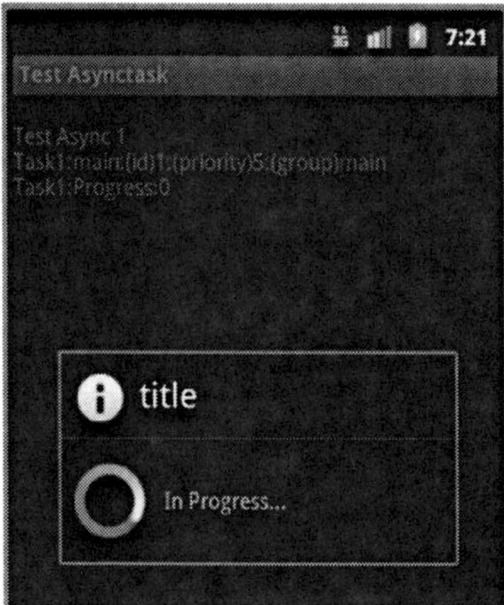

Figure 18–1. *A simple progress dialog interacting with an async task*

The code segment (taken from Listing 18–4) that shows the progress dialog is reproduced in Listing 18–6.

Listing 18–6. *Showing a (Indeterministic) Progress Dialog*

```
pd = ProgressDialog.show(context, "title", "In Progress...",true);
```

The variable pd was already declared in the constructor (see Listing 18–4). This call in Listing 18–6 will create a progress dialog and display it as shown in Figure 18–1. The last argument to the show() method in Listing 18–6 indicates if the dialog is *indeterministic*" (whether the dialog can estimate beforehand how much work there is). We will cover the deterministic case in a later section.

doInBackground() method

All the background work carried out by the async task is done in the doInBackground() method. This method is orchestrated by the async task to run on a worker thread. As a result, this work can take more than five seconds to complete. This is is the main reason for abstracting this async task.

In our example from Listing 18–4, in the doInBackground() method we simply retrieve each of the strings as if they are an array. In this method definition we haven't defined an explicit string array. However, the single argument to this function is defined as a variable-length argument, as shown in Listing 18–7.

Listing 18–7. *doInBackground() Method Signature*

```
protected Integer doInBackground(String...strings)
```

Java then treats the argument as if it is an array inside the function.

So in our code in the doInBackground() method, we read each of the strings and log them to indicate that we know what they are.

We then wait long enough to simulate a long-running operation. Because this method is running in a worker thread, we have no access to the UI functionality of Android. For instance, you won't be able to update any Views directly that you may have access to. You cannot even send a Toast from here.

Triggering onProgressUpdate()

We know that the `doInBackground()` method could take a short while to complete. It is reasonable then to expect this method to tell us frequently how much of the task it has accomplished. So `doInBackground()` is also responsible for triggering `onProgressUpdate()` by calling the `publishProgress()` method.

The triggered `onProgressUpdate()` method then runs on the main thread. This allows the `onProgressUpdate()` method to update UI elements such as Views appropriately. You can also send a Toast from here. In Listing 18–4, we simply log a message.

Once all the work is done, we return from the `doInBackground()` method with a result code.

onPostExecute() Method

This result code from the `doInBackground()` method is then passed to the `onPostExecute()` callback method. This callback is also executed on the main thread. In this method, we tell the progress dialog to close. Being on the main thread, you can access any UI elements in this method without restriction.

This is all there is to using `AsyncTask`.

Upgrading to the Deterministic Progress Dialog

In the previous example in Listing 18–4, we used a very basic progress dialog (Figure 18–1) that doesn't tell us what portion of the work is complete. This progress dialog is called an *indeterministic* progress dialog.

If you set the `indeterministic` property to `false` on this progress dialog, you will see a progress dialog that tracks progress. This is shown in Figure 18–2.

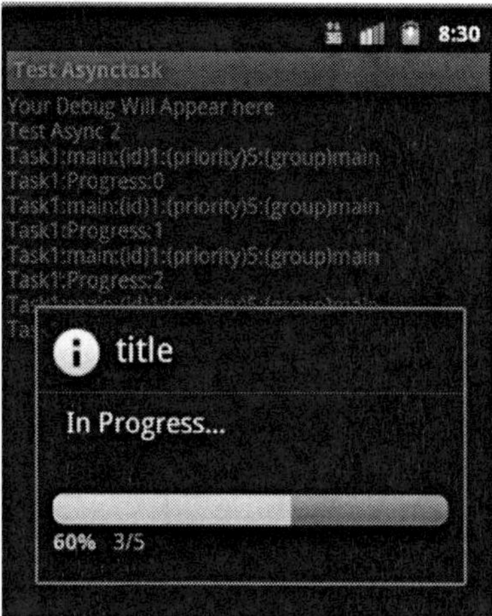

Figure 18–2. *A Progress dialog showing explicit progress, interacting with an async task*

Listing 18–8 shows the previous task from Listing 18–4 rewritten to change the behavior of the progress dialog to a deterministic progress dialog. Key portions of the code are highlighted.

Listing 18–8. *A Long Task Utilizing a Deterministic Progress Dialog*

```
public class MyLongTask1
extends AsyncTask<String,Integer,Integer>
implements OnCancelListener
{
    IReportBack r;
    Context ctx;
    public String tag = null;
    ProgressDialog pd = null;
    MyLongTask1(IReportBack inr, Context inCtx, String inTag)
    {
        r = inr;
        ctx = inCtx;
        tag = inTag;
    }
    protected void onPreExecute()
    {
        //Runs on the main ui thread
        Utils.logThreadSignature(this.tag);
        //pd = ProgressDialog.show(ctx, "title", "In Progress...",false);
        pd = new ProgressDialog(ctx);
        pd.setTitle("title");
        pd.setMessage("In Progress...");
        pd.setCancelable(true);
        pd.setOnCancelListener(this);
        pd.setIndeterminate(false);
        pd.setProgressStyle(ProgressDialog.STYLE_HORIZONTAL);
        pd.setMax(5);
        pd.show();
    }
    protected void onProgressUpdate(Integer... progress)
```

```
    {
        //Runs on the main ui thread
        Utils.logThreadSignature(this.tag);
        this.reportThreadSignature();

        //will be called multiple times
        //triggered by onPostExecute
        Integer i = progress[0];
        r.reportBack(tag, "Progress:" + i.toString());
        pd.setProgress(i);
    }
    protected void onPostExecute(Integer result)
    {
        //Runs on the main ui thread
        Utils.logThreadSignature(this.tag);
        r.reportBack(tag, "onPostExecute result:" + result);
        pd.cancel();
    }
    protected Integer doInBackground(String...strings)
    {
        //Runs on a worker thread
        //May even be a pool if there are
        //more tasks.
        Utils.logThreadSignature(this.tag);

        for(String s :strings)
        {
            Log.d(tag, "Processing:" + s);
            //r.reportTransient(tag, "Processing:" + s);
        }
        //break work into manageable units (say 5 units)
        for (int i=0;i<5;i++)
        {
            //simulate work
            Utils.sleepForInSecs(2);
            //frequently publish progress
            publishProgress(i);
        }
        //alternatively you could have broken down for each string as well

        return 1;
    }
    protected void reportThreadSignature()
    {
        String s = Utils.getThreadSignature();
        r.reportBack(tag,s);
    }
    public void onCancel(DialogInterface d)
    {
        r.reportBack(tag,"Cancel Called");
        this.cancel(true);
    }
}
```

Notice how we have prepared the progress dialog. The code is reproduced in Listing 18–9.

Listing 18–9. *Creating and Showing a Deterministic Progress Dialog*

```
//pd = ProgressDialog.show(ctx, "title", "In Progress...",false);
pd = new ProgressDialog(ctx);
pd.setTitle("title");
pd.setMessage("In Progress...");
pd.setCancelable(true);
pd.setOnCancelListener(this);
pd.setIndeterminate(false);
pd.setProgressStyle(ProgressDialog.STYLE_HORIZONTAL);
pd.setMax(5);
pd.show();
```

In this case we haven't used the static method `show()` (in contrast to what we did in Listing 18–4) on the progress dialog. Instead, we have explicitly instantiated the progress dialog. The variable `ctx` stands for the context (or activity) in which this UI progress dialog operates.

Then we individually set the various properties on the dialog, including its deterministic or indeterministic character.

The method `setMax()` indicates how many steps the progress dialog has. We have also passed ourselves as a listener when `cancel` is triggered. On this cancel callback, we explicitly issue a `cancel` on the async task.

You can see SDK docs about this function and its behavior at

`http://developer.android.com/reference/android/os/AsyncTask.html#cancel(boolean)`

Primarily, this `cancel` will try to stop the worker thread. This method can take a boolean to force-stop the worker thread.

Nature of an Async Task

Consider the code in Listing 18–10, where a menu item is invoking two async tasks one after the other.

Listing 18–10. *Invoking Two Long-Running Tasks*

```
void respondToMenuItem()
{
    MyLongTask mlt = new MyLongTask(this.mReportTo,this.mContext,"Task1");
    mlt.execute("String1","String2","String3");

    MyLongTask mlt1 = new MyLongTask(this.mReportTo,this.mContext,"Task2");
    mlt1.execute("String1","String2","String3");
}
```

Here we are executing two tasks on the main thread. You would expect both the tasks to get started close to each other. However, the preferred default behavior is to run them sequentially using a single thread drawn out of a pool of threads.

Here is what the SDK documentation tells us about this behavior:

> *The execute() method executes the task with the specified parameters. The task returns itself (this) so that the caller can keep a reference to it. Note: this function schedules the task on a queue for a single background thread or pool of threads depending on the platform version. When first introduced, AsyncTasks were executed serially on a single background thread. Starting with DONUT, this was changed to a pool of threads allowing multiple tasks to operate in parallel. After HONEYCOMB, it is planned to change this back to a single thread to avoid common application errors caused by parallel execution. If you truly want parallel execution, you can use the executeOnExecutor(Executor, Params...) version of this method with THREAD_POOL_EXECUTOR; however, see commentary there for warnings on its use.*

So according to this quote from the SDK, you can plan for multiple threads, but the default behavior is a serial execution of these tasks.

Also per the documentation, it is not valid to call the `execute()` method more than once on a single async task. If you want to do that, you have to instantiate a new task and call the `execute()` method again.

Device Rotation and AsyncTask

There is a fundamental flaw in both the `AsyncTask` examples we have shown in Listing 18–4 and Listing 18–8. We are keeping a pointer to the parent activity through a local variable in the `AsyncTask` implementation. However, this local variable to the currently running activity will no longer be valid when the device is rotated. You will end up with an orphaned pointer, and you won't be able to perform any UI operations.

This is because when the device rotates, the `AsyncTask` continues to be valid and continues to run on the worker thread. However, the activity is destroyed and re-created and will have a new pointer.

To fix this, the async task and the activity need to be loosely bound. So when the activity comes back up again, it needs to inform the async task of its new pointer or the async task needs to realize that the activity could have gone away and it needs to reacquire the pointer.

One approach is to use a weak reference to the activity instead of a hard reference. This approach has two benefits. On one side, when the activity is dropped and re-created, a weak reference will allow the old activity to get garbage collected. The weak reference also allows the async task to know whether the activity pointer/reference is `null`. Knowing that the activity is `null`, the async task can stop calling UI methods at each progress step. When the activity comes alive, it can locate the async tasks it started in a registry and reestablish its pointer. We have a reference on how to use weak references at the end of this chapter.

Life Cycle Methods and AsyncTask

What will happen if a user clicks the Back button when a progress dialog initiated by an async task is visible? This will cancel the dialog as per UI guidelines. However, if you don't take the precaution of also cancelling the async task, it will continue to run. So a good practice is to capture the `oncancel` of the dialog and explicitly cancel the async task.

What happens if you are not using a progress dialog but are the progress through some other means on the activity? What happens if a user navigates away from the activity either through a Back button or a press on the Home key? In both these cases, there is no expectation that the user will come back any time soon. Many times, the right thing to do is recognize this life cycle state of the activity and then accordingly cancel the async task.

In short, an async task needs to be fully aware of the life cycle states of the activity. This may behoove you to actually implement life cycle methods on your task and have the activity call these life cycle methods so that an async task behaves as if it is a part and parcel of the activity.

References

The following references will help you learn more about the topics discussed in this chapter:

- `http://developer.android.com/reference/android/os/AsyncTask.html`: The main API reference. You will see here a comprehensive list of methods available on the `AsyncTask` object. This is also the key resource that definitively documents the behavior of `AsyncTask`.
- `www.androidbook.com/item/3536`: Our research notes on `AsyncTask` that we gathered in preparing this chapter.
- `www.androidbook.com/item/3537`: Our research notes on Java generics. Android is increasingly using Java generics in its API. A good understanding of Java generics will help demystify some of the APIs. We have collected our notes at this URL as we ourselves brushed up on Java generics.
- `www.androidbook.com/item/3528`: Our research notes on Java weak references. Java weak references are very useful when part of your program gets restarted. An activity is an example of this. The notes here will help you understand what weak references are.
- `www.androidbook.com/proandroid4/projects`: A list of downloadable projects from this book referenced. For this chapter, look for a ZIP file named `ProAndroid4_Ch18_TestAsyncTask.zip`.

Summary

At a high level, this chapter has covered the following:

- Using an async task instead of handlers to do work in a background worker thread
- Using a progress dialog to monitor the progress of background work
- Dealing with device rotation
- Why you should care about an activity's life cycle methods
- Best practices when writing an async task

Interview Questions

The following questions will help you to consolidate what is covered in this chapter:

1. When do you use an async task?
2. How does an async task improve upon using a handler to do the same thing?
3. What are the three generic types used by an async task?
4. What method of `AsyncTask` runs on the worker thread?
5. Can you issue a Toast from the `doInBackground()` method?
6. From what method can you report progress back to the main thread?
7. How do you start work on an async task? What method do you call?
8. Can you call `execute()` twice on an async task?
9. Do multiple async tasks issued by the main thread run simultaneously?
10. What is the role of thread pools on async tasks?
11. What is an orphan pointer?
12. How do you reestablish a link to the activity when the device rotates?
13. What should you do when the user backs away from a progress dialog?
14. What is a good practice for an activity when the user navigates away from the activity?
15. What is a weak reference? (Use the resource listed in "References" to read up on this topic. This is a useful thing to know, but we couldn't cover it due to space limitations.)
16. How does a weak reference help an async task?

Chapter 19

Broadcast Receivers and Long-Running Services

Through previous chapters, you have been exposed to activities, content providers, and services. We haven't talked much about broadcast receivers, so we will do that in this chapter.

We'll show you how to invoke single and multiple broadcast receivers. We will explore how broadcast receivers can reside in processes outside of the client processes. And we will demonstrate how a broadcast receiver issues notifications.

We will talk about the ten-second limit on a broadcast receiver to respond before the system throws Application Not Responding (ANR) messages and suggest known mechanisms to work around this. We will develop a framework where you can start viewing a long-running service as a special abstraction of a broadcast service, and finally, we'll talk about wake locks in the context of long running services.

Broadcast Receivers

A broadcast receiver is another component of an Android process, along with activities, content providers, and services. As the name indicates, a broadcast receiver is a component that can respond to a broadcast message sent by a client. The message itself is an Android broadcast intent. A broadcast intent (message) can invoke (or be responded to by) more than one receiver.

A component such as an activity or a service uses the `sendBroadcast()` method available on the `Context` class to send a broadcast event. The argument to this method is an intent.

Receiving components of the broadcast intent will need to inherit from a `Receiver` class available in the Android SDK. These receiving components (broadcast receivers) then need to be registered in the manifest file through a `receiver` tag to indicate that the class is interested in responding to a certain type of broadcast intent.

NOTE: You can register receivers at runtime as well without mentioning them in the manifest file. We are not covering that aspect in this chapter. We have included the relevant API documentation URL in the "References" section of this chapter for further information.

Sending a Broadcast

Listing 19–1 shows sample code, taken from an activity class, that sends a broadcast event. This code creates an intent with a unique, specific action, puts an extra message on it, and calls the `sendBroadcast()` method. Putting an extra message on the intent is optional; many times, receiving an intent is sufficient for a receiver, and an extra is not needed.

Listing 19–1. *Broadcasting an Intent*

```
private void testSendBroadcast(Activity activty)
{
    //Create an intent with an action
    String uniqueActionString = "com.androidbook.intents.testbc";
    Intent broadcastIntent = new Intent(uniqueActionString);
    broadcastIntent.putExtra("message", "Hello world");
    activity.sendBroadcast(broadcastIntent);
}
```

In the code in Listing 19–1, the action is an arbitrary identifier that is suitable for your needs. To make this action string unique, you may want to use a namespace similar to a Java class. Although an intent is created exactly the same way, whether to invoke a service or to invoke an activity or to invoke a broadcast receiver, the usage of that intent is unique to each target. For example, broadcast intents are not kept in the same pool as activity intents.

Now, let's look at how we can respond to this broadcast intent.

Coding a Simple Receiver: Sample Code

Listing 19–2 shows a receiver that can respond to the broadcasted intent from Listing 19–1.

Listing 19–2. *Sample Receiver Code*

```
public class TestReceiver extends BroadcastReceiver
{
    private static final String tag = "TestReceiver";
    @Override
    public void onReceive(Context context, Intent intent)
    {
        Utils.logThreadSignature(tag);
        Log.d("TestReceiver", "intent=" + intent);
        String message = intent.getStringExtra("message");
        Log.d(tag, message);
    }
}
```

Creating a broadcast receiver is quite simple. Just extend the BroadcastReceiver class and override the onReceive() method. We are able to see the intent in the receiver and extract the message from it.

In Listing 19–2, if the broadcast intent doesn't have an extra that is called "message," it will return a null. In our example, because we know that we are setting that extra, we haven't checked for a null value. Once we retrieve the extra, we are just logging the retrieved message. Also notice that in Listing 19–2 we have used a method called Utils.logThreadSignature() to log the thread that is running the broadcast receiver code. This Utils class was introduced in Chapter 16; the name of the method speaks for itself.

Once we have the receiver code available, from Listing 19–2, we need to register it in the manifest file as a receiver.

Registering a Receiver in the Manifest File

Listing 19–3 shows how you can declare a receiver as the recipient of the intent whose action is com.androidbook.intents.testbc.

Listing 19–3. *A Receiver Definition in the Manifest File*

```
<manifest>
<application>
...
<activity .....>
...
<receiver android:name=".TestReceiver">
    <intent-filter>
        <action android:name="com.androidbook.intents.testbc"/>
    </intent-filter>
</receiver>
...
</application>
</manifest>
```

The receiver element is a child node of the application element like the other component nodes. This is all you need to test your receiver.

With the receiver (Listing 19–2) and its registration in the manifest file (Listing 19–3) available, you can invoke the receiver using the client code in Listing 19–1. We have included a complete downloadable project that exercises these concepts. The URL for the project is included in the "References" section at the end of this chapter. Import and run the application, and you will see a "broadcast" menu item, as shown in Figure 19–1.

Figure 19–1. *A sample activity with a menu to test a broadcast*

When you click the "broadcast" menu item, the `TestReceiver` in Listing 19–2 will be invoked, and LogCat will show the `helloworld` message that was loaded into the broadcast intent by the activity.

Accommodating Multiple Receivers

The idea of a broadcast is that there is a possibility for more than one receiver. So let's replicate `TestReceiver` (see Listing 19–2) as `TestReceiver2` and see if both get invoked. The code for `TestReceiver2` is presented in Listing 19–4.

Listing 19–4. *TestReceiver2*

```
public class TestReceiver2 extends BroadcastReceiver
{
    private static final String tag = "TestReceiver2";
    @Override
    public void onReceive(Context context, Intent intent)
    {
        Utils.logThreadSignature(tag);
        Log.d(tag, "intent=" + intent);
        String message = intent.getStringExtra("message");
        Log.d(tag, message);
    }
}
```

Once you have this code, you can add this receiver to a manifest file as shown in Listing 19–5.

Listing 19–5. *TestReceiver2 Definition in the Manifest File*

```
<receiver android:name=".TestReceiver2">
    <intent-filter>
        <action android:name="com.androidbook.intents.testbc"/>
    </intent-filter>
</receiver>
```

Now, if you fire off the event as in Listing 19–1 by invoking the "broadcast" menu item again (from Figure 19–1), both receivers will be called.

We have shown you how to use the method `Utils.logThreadSignature(tag)` (in Chapter 17, Section: Thread Utilities: Discover Your Threads), so you can use this method to see which thread the broadcast receivers are running. From the LogCat, you will realize that this is indeed the main thread.

We can modify the code in Listing 19–1 slightly to further examine the main thread's behavior. We can introduce a LogCat message before and after the `SendBroadcast()` method and see if the log messages from the receivers come after both the "before" and "after" messages. Listing 19–6 shows the modified code that you can use to test this.

Listing 19–6. *Monitoring Main Thread Behavior*

```
private void testSendBroadcast(Activity activty)
{
    //Create an intent with an action
    String uniqueActionString = "com.androidbook.intents.testbc";
    Intent broadcastIntent = new Intent(uniqueActionString);
    broadcastIntent.putExtra("message", "Hello world");

    Log.d("tag","before");
    activity.sendBroadcast(broadcastIntent);
    Log.d("tag","before");
}
```

You will see that the log messages that were placed before and after `sendBroadcast()` in Listing 19–6 were both printed before the receiver messages from Listing 19–2 and Listing 19–4 and with the same main thread signature.

This proves that the main thread is going around in a round-robin fashion and finally attending to the broadcast receivers at a later time from the message queue. So, the `SendBroadcast()` is clearly an asynchronous message that lets the main thread get back to its queue.

To see further proof, you can hold up the main thread a bit longer so that the time stamps are clearly demarcated. Let's write another receiver that delays the main thread by sleeping a little while. The source code for such a time delay receiver is presented in Listing 19–7.

Listing 19–7. *A Receiver with a Time Delay*

```
/*
 * This receiver is introduced to see
 * how the main thread schedules broadcast receivers
 *
 * it helps answer such questions as
 * 1. Do they get invoked in the order they are specified?
```

```
 * 2. Do they get invoked one after the other? or do they get invoked parallel
 *
 * The time delay here shows that the main thread
 * gets halted for those many secs. You can see this
 * in the Log.d output
 */
public class TestTimeDelayReceiver extends BroadcastReceiver
{
    private static final String tag = "TestTimeDelayReceiver";
    @Override
    public void onReceive(Context context, Intent intent)
    {
        Utils.logThreadSignature(tag);
        Log.d(tag, "intent=" + intent);
        Log.d(tag, "going to sleep for 2 secs");
        Utils.sleepForInSecs(2);
        Log.d(tag, "wake up");
        String message = intent.getStringExtra("message");
        Log.d(tag, message);
    }
}
```

Now, if you insert this receiver as the second receiver in the manifest file, you can see the traversal of the main thread through the main logic and the broadcast receiver's logic. In LogCat, you will see that the first receiver is executed first. Then, the second receiver is invoked, and the main thread waits there for two seconds and proceeds with the third receiver. Moreover, all receivers are invoked only after the `SendBroadcast()` call returns.

Out-of-Process Receivers

The intention of a broadcast is more likely that the process responding to it is an unknown one and separate from the client process. You can easily prove this by replicating one of your receivers and creating a separate `.apk` file from it. Then when you fire off the event from Listing 19–1, you will see that both the in-process receivers (those that are in the same project or `.apk` file) and out-of-process receivers (those that are in a separate `.apk` file) are invoked. You will also see through the LogCat messages that the in-process and out-of-process receivers run in their respective main threads.

We have included an additional separate stand-alone project in the chapter's downloadable ZIP file to test this concept. To try it, you have to deploy both the invoking project and the stand-alone receiver's project on the emulator.

Using Notifications from a Receiver

Broadcast receivers often need to communicate to the user about something that happened or a status. This is usually done by alerting the user through a notification icon in the systemwide notification bar. We will show you, in this section, how to create a notification from a broadcast receiver, send it, and view it through the notification manager.

Monitoring Notifications Through the Notification Manager

Android shows icons of notifications as alerts in the notification area. The notification area is located at the top of device in a strip that looks like Figure 19–2. The look and placement of the notification area may change based on whether the device is a tablet or a phone and may at times also change based on Android release.

Figure 19–2. *Android notification icon status bar*

The notification area shown in Figure 19–2 is called the *status bar*. This status bar is the staple of phone form factor. It contains system indicators such as battery strength, signal strength, and so on.

In 3.0, for tablet form factors, Android introduced a new *system bar* that sits at the bottom of the tablet, taking the place of the status bar for tablets. This system bar also includes navigation icons such as home, back, and search.

When the phone and tablet APIs merged in 4.0, the *navigation bar* was introduced for phone form factors. A navigation bar takes the place of the system bar for phones. However, for phones, the status bar is still in play and continues to show notifications and system indicators. The navigation bar for phones primarily shows the navigation icons: home, back, and search.

Of course, we devote a full chapter to *action bars* (Chapter 10), which are solely owned by the activities in your application.

When we deliver a notification, the notification will appear as an icon in the area shown Figure 19–2. The notification icon is illustrated in Figure 19–3.

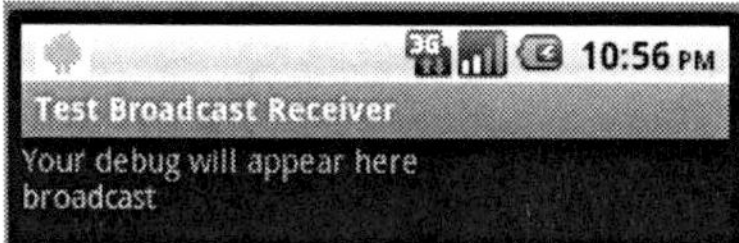

Figure 19–3. *Status bar showing a notification icon*

Figure 19–3 shows both the notification area and an activity, in addition to the notification icon. For an activity, we just happened to be sitting in an application that is issuing the broadcast. It can be any activity or even the home page.

The notification icon is an indicator to the user that something needs to be observed. To see the full notification, you have to hold a finger on the icon and drag the title strip shown in Figure 19–2 down like a curtain. This will expand the notification area, as shown in Figure 19–4.

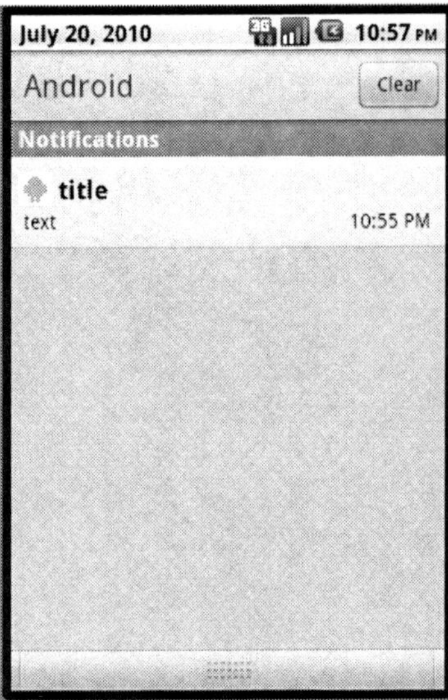

Figure 19–4. *Expanded notification view*

In the expanded view of the notification in Figure 19–4, you get to see the details supplied to the notification. You can also click a notification detail to fire off the intent to bring up the full application to which the notification belongs.

As you can also see from Figure 19–4, you can use this view to clear notifications.

You can also reach the notification detail view shown in Figure 19–4 from the menu on the home page. Figure 19–5 shows the available menu on the home page of the emulator. Depending on the device and the Android release, this homepage menu may differ.

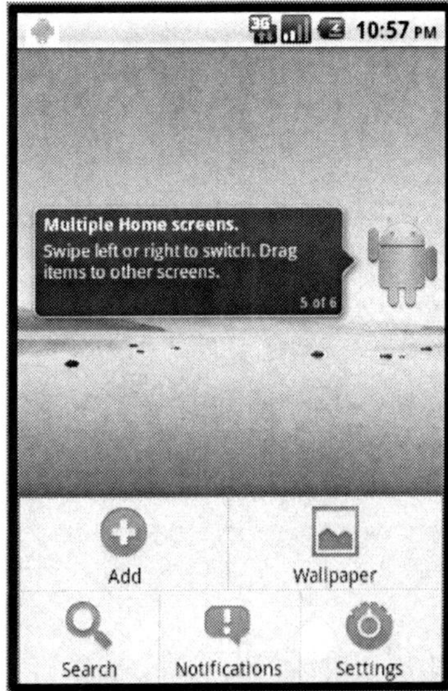

Figure 19–5. *The Notifications menu item from home menu*

Clicking the Notifications icon in Figure 19–5 will bring up the notification screen in Figure 19–4.

Let's see now how to generate a notification icon like the one shown in Figures 19–3 and 19–4.

Sending a Notification

Let's get started. The process of sending a notification has the following three steps:

1. Create a suitable notification.
2. Get access to the notification manager.
3. Send the notification to the notification manager.

When you create a notification, you'll need to ensure that it had the following basic parts:

- An icon to display
- Ticker text like "hello world"
- The time when it is delivered

Once you have a notification object constructed with these details, you get the notification manager by asking the context to give you a system service named `Context.NOTIFICATION_SERVICE`. Once you have the notification manager reference, call the notify method on the notification manager reference to send the notification.

Listing 19–8 presents the source code for a broadcast receiver that sends the notification shown in Figures 19–3 and 19–4.

Listing 19–8. *A Receiver That Sends a Notification*

```
public class NotificationReceiver extends BroadcastReceiver
{
    private static final String tag = "Notification Receiver";
    @Override
    public void onReceive(Context context, Intent intent)
    {
        Utils.logThreadSignature(tag);
        Log.d(tag, "intent=" + intent);
        String message = intent.getStringExtra("message");
        Log.d(tag, message);
        this.sendNotification(context, message);
    }
    private void sendNotification(Context ctx, String message)
    {
        //Get the notification manager
        String ns = Context.NOTIFICATION_SERVICE;
        NotificationManager nm =
            (NotificationManager)ctx.getSystemService(ns);

        //Create Notification Object
        int icon = R.drawable.robot;
```

```
        CharSequence tickerText = "Hello";
        long when = System.currentTimeMillis();

        Notification notification =
            new Notification(icon, tickerText, when);

        //Set ContentView using setLatestEvenInfo
        Intent intent = new Intent(Intent.ACTION_VIEW);
        intent.setData(Uri.parse("http://www.google.com"));
        PendingIntent pi = PendingIntent.getActivity(ctx, 0, intent, 0);
        notification.setLatestEventInfo(ctx, "title", "text", pi);

        //Send notification
        //The first argument is a unique id for this notification.
        //This id allows you to cancel the notification later
        //This id also allows you to update your notification
        //by creating a new notification and resending it against that id
        //This id is unique with in this application
        nm.notify(1, notification);
    }
}
```

In the source code in Listing 19–8, we have referenced an alert icon called `R.drawable.robot`. You can create your own alert icon and drop it into the `res/drawable` subdirectory and name it `robot` with a proper image extension. Or you can refer to the downloadable ZIP file for this project (a URL is included in the "References" section).

When you create a notification with the basic parameters (icon, text, and time) and send it to the notification manager, it looks like it is not sufficient (the first part of creating the notification in Listing 19–8). You will also have to set up something called a *content view* for that notification using this (inexplicably named) method:

```
setLatestEventInfo(...)
```

The content view of a notification is displayed when the notification is expanded. This is what you see in Figure 19–4. Typically, the content view needs to be a `RemoteViews` object. However, we don't pass a content view directly to the `setLatestEventInfo` method. This `setLatestEventInfo()` method is a shortcut for setting the standard predefined content view using a title and the text to display.

This method `setLatestEventInfo()` also takes a pending intent, called a *content intent*, that gets fired when this expanded view is clicked. Look back at Listing 19–8 to see what parameters we have used to pass to this method.

You also have an option to create a remote view yourself and set it as the content view, without using `setLatestEventInfo()`.

The steps for using remote views for a content view of a notification follow:

1. Create a layout file.
2. Create a `RemoteViews` object using the package name and the layout file ID.
3. Call set methods on the `RemoteViews` to set text, icons, and so on.

4. Call `setContentView()` on the notification object before sending it to the notification manager.

Keep in mind that only the following limited set of controls may participate in a remote view as of Android release 2.2:

- FrameLayout
- LinearLayout
- RelativeLayout
- AnalogClock
- Button
- Chronometer
- ImageButton
- ImageView
- ProgressBar
- TextView

Refer to Chapter 25 (Home Screen Widgets) to learn more about constructing these remote views as widget views on the homepage are essentially remote views. In that chapter on home screen widgets you will also see an updated list of possible RemoteViews in releases 2.3 and 3.0.

The code in Listing 19–8 creates a notification and uses `setLatestEventInfo()` to set the implicit content view (through title and text) and the intent to fire (in our case, this intent is the browser intent).

This method `setLatestEventInfo()` is named as such because this method lets you create or adjust a new notification based on the status. Once the notification is created with the new information, it can be resent through the notification manager and the unique ID of the notification. The ID of the notification, which is set to 1 in Listing 19–8, is unique within this application context. This uniqueness allows us to continuously update what is happening to that notification and also cancel it if needed.

You may also want to look at the various flags available while creating a notification, such as FLAG_NO_CLEAR and FLAG_ONGOING_EVENT to control the persistence of these notifications. The URL to check these flags is

http://developer.android.com/reference/android/app/Notification.html

Starting an Activity in a Broadcast Receiver

Although you're well advised to use the notification manager when a user needs to be informed, Android does allow you to spawn an activity explicitly. You can do this by using the usual `startActivity()` method but with the following flags:

- Intent.FLAG_ACTIVITY_NEW_TASK
- Intent.FLAG_FROM_BACKGROUND
- Intent.FLAG_ACTIVITY_SINGLETOP

Long-Running Receivers and Services

So far, we have covered the happy path of broadcast receivers where the execution of a broadcast receiver is unlikely to take more than ten seconds. As it turns out, the problem space becomes a bit complicated if we want to perform tasks that take longer than ten seconds.

To understand why, let's quickly review a few facts about broadcast receivers:

- A broadcast receiver, like other components of an Android process, runs on the main thread.
- Holding up the code in a broadcast receiver will hold up the main thread and will result in ANR.
- The time limit on a broadcast receiver is ten seconds compared to five seconds for an activity. It is a touch of a reprieve, but the limit is still there.
- The process hosting the broadcast receiver will start and terminate along with the broadcast receiver execution. Hence the process will not stick around after the broadcast receiver's `onReceive()` method returns. Of course, this is assuming that the process contains only the broadcast receiver. If the process contains other components, such as activities or services, that are already running, then the lifetime of the process takes these component life cycles into account as well.
- Unlike a service process, a broadcast receiver process will not get restarted.
- If a broadcast receiver were to start a separate thread and return to the main thread, Android will assume that the work is complete and will shut down the process even if there are threads running, bringing those threads to abrupt stop.
- Android acquires a partial wake lock when invoking a broadcast service and releases it when it returns from the service in the main thread. A wake lock is a mechanism and an API class available in the SDK to keep the device from going to sleep or wake it up if it is already asleep.

Given these predicates, how can we execute longer-running code in response to a broadcast event?

Long-Running Broadcast Receiver Protocol

The answer lies in resolving the following:

- We will clearly need a separate thread so that the main thread can get back and avoid ANR messages.
- To stop Android from killing the process and hence the worker thread, we need to tell Android that this process contains a component, such as a service, with a life cycle. So we need to create or start that service. The service itself cannot directly do the work for more than five seconds because that happens on the main thread, so the service needs to start a worker thread and let the main thread go.
- For the duration of the worker thread's execution, we need to hold on to the partial wake lock so that the device won't go to sleep. A partial wake lock will allow the device to run code without turning on the screen and so on, which allows for longer battery life.
- The partial wake lock must be obtained in the main line code of the receiver; otherwise, it will be too late. For example, you cannot do this in the service, because it may be too late between the `startService()` being issued by the broadcast receiver and the `onStartCommand()` of a service that begins execution.
- Because we are creating a service, the service itself can be brought down and brought back up because of low-memory conditions. If this happens, we need to acquire the wake lock again.
- When the worker thread started by the `onStartCommand()` method of the service completes its work, it needs to tell the service to stop so that it can be put to bed and not brought back to life by Android.
- It is also possible that more than one broadcast event can occur. Given that, we need to be cautious about how many worker threads we need to spawn.

Given these facts, the recommended protocol for extending the life of a broadcast receiver is as follows:

1. Get a (static) partial wake lock in the `onReceive()` method of the broadcast receiver. The partial wake lock needs to be static to allow communication between the broadcast receiver and the service. There is no other way of passing a reference of the wake lock to the service, as the service is invoked through a default constructor that takes no parameters.
2. Start a local service so that the process won't be killed.

3. In the service, start a worker thread to do the work. Do not do the work in the `onStart()` method of the service. If you do, you are basically holding up the main thread again.
4. When the worker thread is done, tell the service to stop itself either directly or through a handler.
5. Have the service turn off the static wake lock. To repeat, a static wake lock is the only way to communicate between a service and its invoker, in this case the broadcast service, because there is no way to pass a wake lock reference to the service.

IntentService

Recognizing the need for a service to not hold up the main thread, Android has provided a utility local service implementation called `IntentService` to offload work to a worker thread so that the main thread can be released after scheduling the work to the subthread. Under this scheme, when you call `startService()` on an `IntentService`, the `IntentService` will queue that request to a subthread using a looper and a handler so that a derived method of the `IntentService` is called to do the actual work on a single worker thread.

Here is what the API documentation for `IntentService` says:

> *IntentService is a base class for Services that handle asynchronous requests (expressed as Intents) on demand. Clients send requests through startService(Intent) calls; the service is started as needed, handles each Intent in turn using a worker thread, and stops itself when it runs out of work. This "work queue processor" pattern is commonly used to offload tasks from an application's main thread. The IntentService class exists to simplify this pattern and take care of the mechanics. To use it, extend IntentService and implement onHandleIntent(Intent). IntentService will receive the Intents, launch a worker thread, and stop the service as appropriate. All requests are handled on a single worker thread -- they may take as long as necessary (and will not block the application's main loop), but only one request will be processed at a time.*

This idea can be clearly demonstrated using a simple example, as in Listing 19–9. You extend the `IntentService` and provide what you want to do in the `onHandleIntent()` method.

Listing 19–9. *Using* `IntentService`

```
public class MyService extends IntentService
{
    protected abstract void onHandleIntent(Intent intent)
```

```
    {
        Utils.logThreadSignature("MyService");
        //do the work in this subthread
        //and return
    }
}
```

Once you have a service like this, you can register this service in the manifest file and use client code to invoke this service as `context.startService(new Intent(MyService.class))`. This invocation will result in a call to `onHandleIntent()` in Listing 19–9.

You will notice that the `logThreadSignature()` method will print the ID of the worker thread and not the main thread.

IntentService Source Code

In Chapter 17, we covered the main thread and the role of handlers. In that context, it is very instructive to study the source code of the `IntentService` as the SDK has implemented it. This will consolidate how handlers and the main thread are used to accomplish a long-running service using a worker thread. Analyzing the `IntentService` code is also helpful for a future abstraction we are planning later in this chapter (Section: Long Running Receiver), which will improve upon the `IntentService`.

Let's consider the source code of `IntentService` (taken from the source code distribution of Android) in Listing 19–10.

Listing 19–10. *`IntentService`* ***Source Code***

```
public abstract class IntentService extends Service {
    private volatile Looper mServiceLooper;
    private volatile ServiceHandler mServiceHandler;
    private String mName;

    private final class ServiceHandler extends Handler {
        public ServiceHandler(Looper looper) {
            super(looper);
        }
        @Override
        public void handleMessage(Message msg) {
            onHandleIntent((Intent)msg.obj);
            stopSelf(msg.arg1);
        }
    }

    public IntentService(String name) {
        super();
        mName = name;
    }
    @Override
    public void onCreate() {
        super.onCreate();
        HandlerThread thread =
          new HandlerThread("IntentService[" + mName + "]");
        thread.start();
```

```
        mServiceLooper = thread.getLooper();
        mServiceHandler = new ServiceHandler(mServiceLooper);
    }

    @Override
    public void onStart(Intent intent, int startId) {
        super.onStart(intent, startId);
        Message msg = mServiceHandler.obtainMessage();
        msg.arg1 = startId;
        msg.obj = intent;
        mServiceHandler.sendMessage(msg);
    }
    @Override
    public void onDestroy() {
        mServiceLooper.quit();
    }
    @Override
    public IBinder onBind(Intent intent) {
        return null;
    }
    protected abstract void onHandleIntent(Intent intent);
}
```

Let's step through an explanation of this code:

1. Create a separate worker thread in the `onCreate()` method of the service. Typically, you will have started worker threads in the `onStartCommand` method of a service. However, that would have resulted in multiple worker threads, one for each `startService`. `IntentService` wants to do this by having a single worker thread that services all of the `startService` invocations, so we set up the worker thread in the `onCreate()` method, which is invoked only once: when the service is brought into memory (not exactly but similar to a singleton).

2. Set up a looper (and thereby a queue to receive and dispatch messages) on that worker thread. This allows the same worker thread to respond to many messages one by one instead of creating a new worker thread for each request

3. Establish a handle on the worker thread so that the main thread of the service can drop a message via the handler. We need this worker thread, because every time a client uses `startService()`, that call goes to the main thread of the `IntentService`, and we don't want to hold up the main thread of the `IntentService`. We need a mechanism to queue this request so that the worker thread can process it when it becomes available. This is feasible by having the main thread hold a handler for the worker thread. Notice the `onStart()` method that runs on the main thread. If you want to prove this, just override this method and call its parent while you log the thread signature. You will see that `onStart()`

runs on the main thread and onHandleMessage() runs on the secondary worker thread.

4. Finally, when onHandleIntent() returns, the handler will call the stopSelf() method of the service. This stopSelf() will succeed in stopping the service if there are no pending messages. The stopSelf() method is reference counted. This means even if you call it multiple times, there must be an equal number of startService invocations. This is why we are able to call stopSelf() after handling every startService invocation.

Extending IntentService for a Broadcast Receiver

From the perspective of a broadcast receiver, an IntentService is a wonderful thing. It lets us execute long-running code with out blocking the main thread. So can we use the IntentService for the needs of a long-running operation? Yes and no.

Yes, because the IntentService does two things: first, it keeps the process running because it is a service. And second, it lets the main thread go and avoids related ANR messages.

To understand the "no" answer, you need to understand wake locks a bit more. When a broadcast receiver is invoked, especially through an alarm manager, the device may not be on. So the alarm manager partially turns on the device (just enough to run the code without any UI) by making a call to the power manager and requesting a wake lock. And this wake lock gets released as soon as the broadcast receiver returns.

This leaves the IntentService invocation without a wake lock, so the device may go to sleep before the actual code runs. However, IntentService, being a general-purpose extension to a service, it does not acquire a wake lock.

So we need further props on top of an IntentService. We need an abstraction.

Mark Murphy has created a variant of the IntentService called WakefulIntentService that keeps the semantics of using an IntentService but also acquires the wake lock and releases it properly under a variety of conditions. You can look at his implementation at http://github.com/commonsguy/cwac-wakeful.

Long-Running Broadcast Service Abstraction

WakefulIntentService is a fine abstraction. However, we want to go a step further so that our abstraction parallels the method of extending IntentService as in Listing 19–10 and does everything that an IntentService does but also provides a few more benefits:

- Acquire and release wake locks (similar to WakefulIntentService).

- Pass the original intent that was passed to the broadcast receiver to the overridden method onHandleIntent. This allows us to largely hide the broadcast receiver.
- Deal with a service being restarted.
- Allow a uniform way to deal with the wake lock for multiple receivers and multiple services in the same process.

We will call this abstract class ALongRunningNonStickyBroadcastService. As the name suggests, we want this service to allow for long-running work. It will also be specifically built for a broadcast receiver. This service will also be nonsticky (we will explain this concept later in the chapter, but briefly, this indicates that Android will not start the service if there are no messages in the queue). To allow for the behavior of an IntentService, it will extend the IntentService and override the onHandleIntent method.

Combining these ideas, the abstract ALongRunningNonStickyBroadcastService service will have a signature that looks like Listing 19–11.

Listing 19–11. *Long-Running Service Abstract Idea*

```
public abstract class ALongRunningNonStickyBroadcastService
extends IntentService
{
...other implementation detials
   protected abstract void
   handleBroadcastIntent(Intent broadcastIntent);
...other implementation details

}
```

The implementation details for this ALongRunningNonStickyBroadcastService are quite involved, and we will cover them later, as soon as we explain why we are going after this type of service. We want to demonstrate first the utility and simplicity of having it.

Once we have this abstract class, the MyService example in Listing 19–9 can be rewritten as in Listing 19–12.

Listing 19–12. *Long-Running Service Sample Usage*

```
public class MyService extends ALongRunningNonStickyBroadcastService
{
    protected abstract void handleBroadcastIntent(Intent broadcastIntent)
    {
        Utils.logThreadSignature("MyService");
        //do the work here
        //and return
    }
}
```

Isn't that simple? Especially the fact that you are receiving directly, unmodified, the same intent that invoked the broadcast receiver. It is as if the broadcast receiver has disappeared from the solution.

As you can see, you can extend this new long-running service class (just like IntentService and WakefulIntentService) and override a single method and do very little to nothing in the broadcast receiver. Your work will be done in a worker thread (thanks to IntentService) without blocking the main thread.

Listing 19–12 is a simple example demonstrating the concept. Let's now turn to a more complete implementation that implements a long-running service that can run for 60 seconds in response to a broadcast event (proving that we can run for more than 10 seconds and avoid an ANR message). We will call this service appropriately Test60SecBCRService (BCR stands for broadcast receiver), and its implementation is shown in Listing 19–13.

Listing 19–13. *Test60SecBCRService*

```
public class Test60SecBCRService
extends ALongRunningNonStickyBroadcastService
{
   public static String tag = "Test60SecBCRService";
   //Required by IntentService to pass the classname
   public Test60SecBCRService(){
      super("com.androidbook.service.Test60SecBCRService");
   }

   /*
    * Perform long running operations in this method.
    * This is executed in a separate thread.
    */
   @Override
   protected void handleBroadcastIntent(Intent broadcastIntent)
   {
      Utils.logThreadSignature(tag);
      Log.d(tag,"Sleeping for 60 secs");
      Utils.sleepForInSecs(60);
      String message =
         broadcastIntent.getStringExtra("message");
      Log.d(tag,"Job completed");
      Log.d(tag,message);
   }
}
```

As you can see, this code successfully simulates doing work for 60 seconds and still avoids the ANR message.

A Long-Running Receiver

Once we have the long-running service in Listing 19–13, we need to be able to invoke the service from a broadcast receiver. Again we are going after an abstraction to hide the broadcast receiver as much as possible.

The first goal of a long-running broadcast receiver is to delegate the work to the long-running service. To do this, the long-running receiver will need the class name of the long-running service to invoke it.

The second goal for this long-running receiver is to acquire a wake lock if we want to ensure the code will continue to run when the receiver returns.

The third goal for the long-running receiver is to transfer the original intent that the broadcast receiver is invoked on to the service. We will do this by sticking the original intent as a `Parcelable` in the intent extras. We will use `original_intent` as the name for this extra. The long-running service then extracts `original_intent` and passes it to the overridden method of the long-running service (you will see this later in the implementation of the long-running service). This facility thus gives the impression that the long-running service is indeed an extension of the broadcast receiver.

Although we could instruct every long-running receiver to do these two things every time, it is better that we abstract out these and provide a base class. The long-running receiver abstraction will then use the derived class to supply the name of the long-running service (LRS) class through an abstract method called `getLRSClass()`.

Before we let you go on to the implementation of this abstraction, we have to talk a little bit about the direction we took on wake locks. Wake locks need to be coordinated between the broadcast receiver and the corresponding service they invoke. Although the idea is simple, in the implementation, we need to worry about many places and conditions where this needs to happen. So we have conceptually abstracted out the wake lock using a concept called `LightedGreenRoom`. We will present this class later, but for now, treat this as just a wake lock that you turn on and off.

Putting these needs together, the source code for the implementation of the abstract class `ALongRunningReceiver` is in Listing 19–14.

Listing 19–14. *ALongRunningReceiver*

```
public abstract class  ALongRunningReceiver
extends BroadcastReceiver
{
    private static final String tag = "ALongRunningReceiver";
    @Override
    public void onReceive(Context context, Intent intent)
    {
       Log.d(tag,"Receiver started");
       //LightedGreenRoom abstracts the Android WakeLock
       //to keep the device partially on.
       //In short this is equivalent to turning on
       //or acquiring the wakelock.
       LightedGreenRoom.setup(context);
       startService(context,intent);
       Log.d(tag,"Receiver finished");
    }
    private void startService(Context context, Intent intent)
    {
       Intent serviceIntent = new Intent(context,getLRSClass());
       serviceIntent.putExtra("original_intent", intent);
       context.startService(serviceIntent);
    }
    /*
     * Override this method to return the
     * "class" object belonging to the
```

```
     * nonsticky service class.
     */
    public abstract Class getLRSClass();
}
```

Once this abstraction is available, you'll need a receiver that works hand in hand with the 60-second long-running service in Listing 19–13. Such a receiver is provided in Listing 19–15.

Listing 19–15. *A Sample Long-Running Broadcast Receiver, Test60SecBCR*

```
public class Test60SecBCR
extends ALongRunningReceiver
{
   @Override
   public Class getLRSClass()
   {
      Utils.logThreadSignature("Test60SecBCR");
      return Test60SecBCRService.class;
   }
}
```

Just like the service abstraction in Listings 19–12 and 19–13, the code in Listing 19–15 uses an abstraction for the broadcast receiver. The receiver abstraction starts the service indicated by the service class returned by the `getLRSClass()` method.

Thus far, we have demonstrated why we needed the two important abstractions to implement long-running services invoked by broadcast receivers:

- `ALongRunningNonStickyBroadcastService` (Listing 19–11)
- `ALongRunningReceiver` (Listing 19–14)

However, we have postponed showing the implementation for one of these classes due to the level of detail involved. We also have not presented the implementation of a common class, `LightedGreenRoom`, that both these abstractions use. We are now at a point to explain and present the code for these two remaining classes. We will start with the common class `LightedGreenRoom`.

Abstracting a Wake Lock with LightedGreenRoom

As mentioned earlier, the primary purpose of the `LightedGreenRoom` abstraction is to simplify the interaction with the wake lock, and a wake lock is used to keep the device on during background processing. Listing 19–16 shows how a wake lock is used typically as stated in the SDK.

Listing 19–16. *WakeLock API*

```
//Get access to the power manager service
PowerManager pm =
   (PowerManager)inCtx.getSystemService(Context.POWER_SERVICE);

//Get hold of a wake lock
PowerManager.WakeLock wl =
   pm.newWakeLock(PowerManager.PARTIAL_WAKE_LOCK, tag);
```

```
//Acquire the wake lock
wl.acquire();

//do some work
//while this work is being done the device will be on partially

//release the wakelock
wl.release();
```

Given this interaction, the broadcast receiver is supposed acquire the lock, and when the long-running service is finished, it needs to release the lock. However, there is no good way to pass the wake lock variable to the service from the broadcast receiver. The only way the service knows about this wake lock is to use a static or application-level variable.

Another difficulty in acquiring and releasing a wake lock is the reference count. So as a broadcast receiver is invoked multiple times, if the invocations overlap, there are going to be multiple calls to acquire the wake lock. Similarly, there are going to be multiple calls to release. If the number of acquire and release calls don't match, we will end up with a wake lock that at worst keeps the device on for far longer than needed. Also, when the service is no longer needed and the garbage collection runs, if the wake lock counts are mismatched, there will be a runtime exception in the LogCat.

These issues have prompted us to do our best to abstract the wake lock to ensure proper usage.

> **NOTE:** Now that you are aware of the issues and the need for wake locks, you are encouraged to tinker with `LightedGreenRoom` and replace it with another class if you find that to be simpler. This disclaimer is to reassure you that there is no magic about `LightedGreenRoom` and that it is quite simple at its heart.

We will now explain the conceptual thought that went into seeing the wake lock as `LightedGreenRoom`.

A Lighted Green Room

Let's start with a *green room*, which is a room that allows visitors. The room starts out dark, and the first one to enter turns on the lights. Subsequent visitors have no effect if the lights are already on. The last visitor to leave will turn off the lights. It is called a green room because it uses energy efficiently. The enter and leave methods need to be synchronized to keep their states, as they could happen between multiple threads.

So what, then, is a lighted green room? Unlike a green room that starts with the lights off, a lighted green room starts with the lights on, even before the first visitor arrives. We can assume that, with the lights off, a visitor cannot find the way to the green room. This relates to the fact that if a device is off, no service can run. Still, the last one to leave will turn off the lights. This is useful for a broadcast receiver, because it needs to turn on the lights first and then transfer to the service.

Starting a service is considered equivalent to a visitor coming in. Stopping a service equates to a visitor leaving the room. Please note that you need to distinguish between the *creation* of a service and *starting* a service. Creation and destruction happen only once per service, whereas starting and stopping can happen many times.

There could be, and typically is, a time delay between setting up the wake lock (the lighted green room) in the receiver and starting the service, essentially a call to `onStartCommand` (having the first visitor enter the room).

Because a wake lock is reference counted, if a service is to be taken down because of low-memory conditions, we would like to explicitly release the locks. If you were to use the same lighted green room to serve multiple services, you might want to track the last service to be destroyed and release the locks only once that service was finished.

To allow for this pattern, we will create a client. Each service will register with the lighted green room as a client so that its destroy method will work.

On top of that, we need to keep track of the `enter` and `leave` of each `startService`.

Lighted Green Room Implementation

Combining all the concepts from the last section, the implementation of a lighted green room looks like Listing 19–17. Please note that this seemed to work well with our limited testing. Tinker with it and adjust it to your needs, as it is difficult for us to consider each possibility that might exist in your development environment. (In other words, think of this example as experimental.)

Listing 19–17. *Lighted Green Room Implementation*

```
public class LightedGreenRoom
{
   //debug tag
   private static String tag="LightedGreenRoom";

   //Keep count of visitors to know the last visitor.
   //On destroy set the count to zero to clear the room.
   private int count;

   //Needed to create the wake lock
   private Context ctx = null;

   //Our switch
   PowerManager.WakeLock wl = null;

   //Multi-client support
   private int clientCount = 0;

   /*
    * This is expected to be a singleton.
    * One could potentially make the constructor
    * private.
```

```
 */
public LightedGreenRoom(Context inCtx)
{
   ctx = inCtx;
   wl = this.createWakeLock(inCtx);
}

/*
 * Setting up the green room using a static method.
 * This has to be called before calling any other methods.
 * what it does:
 *       1. Instantiate the object
 *       2. acquire the lock to turn on lights
 * Assumption:
 *       It is not required to be synchronized
 *       because it will be called from the main thread.
 *       (Could be wrong. need to validate this!!)
 */
private static LightedGreenRoom s_self = null;

public static void setup(Context inCtx)
{
   if (s_self == null)
   {
      Log.d(LightedGreenRoom.tag,"Creating green room and lighting it");
      s_self = new LightedGreenRoom(inCtx);
      s_self.turnOnLights();
   }
     }
     public static boolean isSetup()
{
   return (s_self != null) ? true: false;
}

/*
 * The methods "enter" and "leave" are
 * expected to be called in tandem.
 *
 * On "enter" increment the count.
 *
 * Do not turn the lights or off
 * as they are already turned on.
 *
 * Just increment the count to know
 * when the last visitor leaves.
 *
 * This is a synchronized method as
 * multiple threads will be entering and leaving.
 *
 */
synchronized public int enter()
{
   count++;
   Log.d(tag,"A new visitor: count:" + count);
   return count;
}
/*
```

```
     * The methods "enter" and "leave" are
     * expected to be called in tandem.
     *
     * On "leave" decrement the count.
     *
     * If the count reaches zero turn off the lights.
     *
     * This is a synchronized method as
     * multiple threads will be entering and leaving.
     *
     */
    synchronized public int leave()
    {
       Log.d(tag,"Leaving room:count at the call:" + count);
       //if the count is already zero
       //just leave.
       if (count == 0)
       {
          Log.w(tag,"Count is zero.");
          return count;
       }
       count--;
       if (count == 0)
       {
          //Last visitor
          //turn off lights
          turnOffLights();
       }
       return count;
    }
  synchronized public int getCount()
    {
       return count;
    }

    /*
     * acquire the wake lock to turn the lights on
     * it is up to other synchronized methods to call
     * this at the appropriate time.
     */
    private void turnOnLights()
    {
       Log.d(tag, "Turning on lights. Count:" + count);
       this.wl.acquire();
    }

    /*
     * Release the wake lock to turn the lights off.
     * it is up to other synchronized methods to call
     * this at the appropriate time.
     */
    private void turnOffLights()
    {
       if (this.wl.isHeld())
       {
          Log.d(tag,"Releasing wake lock. No more visitors");
          this.wl.release();
```

```
        }
    }
    /*
     * Standard code to create a partial wake lock
     */
    private PowerManager.WakeLock createWakeLock(Context inCtx)
    {
        PowerManager pm =
            (PowerManager)inCtx.getSystemService(Context.POWER_SERVICE);

        PowerManager.WakeLock wl = pm.newWakeLock
            (PowerManager.PARTIAL_WAKE_LOCK, tag);
        return wl;
    }
    private int registerClient()
    {
        Utils.logThreadSignature(tag);
        this.clientCount++;
        Log.d(tag,"registering a new client:count:" + clientCount);
        return clientCount;
    }
    private int unRegisterClient()
    {
        Utils.logThreadSignature(tag);
        Log.d(tag,"un registering a new client:count:" + clientCount);
        if (clientCount == 0){
            Log.w(tag,"There are no clients to unregister.");
            return 0;
        }
        //clientCount is not zero
        clientCount--;
        if (clientCount == 0){
            emptyTheRoom();
        }
        return clientCount;
    }
    synchronized public void emptyTheRoom()
    {
        Log.d(tag, "Call to empty the room");
        count = 0;
        this.turnOffLights();
    }
    //*************************************************
    //*  static members: Purely helper methods
    //*   Delegates to the underlying singleton object
    //*************************************************
    public static int s_enter(){
        assertSetup();
        return s_self.enter();
    }
    public static int s_leave(){
        assertSetup();
        return s_self.leave();
    }
    //Don't directly call this method
    //probably will be deprecated.
    //Call register and unregister client methods instead
```

```
    public static void ds_emptyTheRoom(){
        assertSetup();
        s_self.emptyTheRoom();
        return;
    }
    public static void s_registerClient(){
        assertSetup();
        s_self.registerClient();
        return;
    }
    public static void s_unRegisterClient(){
        assertSetup();
        s_self.unRegisterClient();
        return;
    }
    private static void assertSetup(){
        if (LightedGreenRoom.s_self == null){
            Log.w(LightedGreenRoom.tag,"You need to call setup first");
            throw new RuntimeException("You need to setup GreenRoom first");
        }
    }
}
```

A reasonable approach for the broadcast receiver and the service to communicate with each other is through a static variable. Instead of making wakelock static, we have made the entire LightedGreenRoom a static instance. However, every other variable inside LightedGreenRoom stays local and nonstatic.

Every public method of LightedGreenRoom is also exposed as a static method for convenience. You can choose, instead, to get rid of the static methods and directly call the single object instance of LightedGreenRoom.

Long-Running Service Implementation

Now that the LightedGreenRoom implementation is finished, we are almost ready to present the long-running service abstraction. However, we have to take one more detour to explain the lifetime of a service and how it relates to the implementation of onStartCommand. This is the method that is ultimately responsible for starting the worker thread and the semantics of a service.

You know that the broadcast receiver invokes the service using a startService call and that this call will result in calling the onStartCommand method of the service. The lifetime of the service is controlled by what this method returns.

To understand what happens in this method, you need detailed background on the nature of local services. We covered the basics of local services in Chapter 15, and now we need to dig a bit deeper.

When a service is started, it gets created first, and its onStartCommand method is called. Android has enough provisions to keep this process in memory so that the service can serve multiple incoming client requests.

There is a difference between a service process being in memory and running. A service runs only in response to `startService`, which calls its `onStartCommand` method. Just because this method is not executing doesn't mean the service process is not in memory. Sometimes, people refer to this service as running even though it is just sitting there and claiming some resources but not actually executing anything. This is what it typically means when Android claims that it keeps the service running.

In fact, if a `startService` call, resulting in an `onStartCommand`, takes more than five to ten seconds, this will result in ANR message and could kill the process hosting the service. Without a worker thread, a service cannot run for longer than ten seconds. So you should distinguish between a service that's available and one that's running.

Android does its best to keep a service available in memory. However, under demanding memory conditions, Android may choose to reclaim the process and call the `onDestroy()` method of the service.

> **NOTE:** Android tries to call the `onDestroy()` method for a service to reclaim its resources when the service is not executing its `onCreate()`, `onStart()`, or `onDestroy()` method.

However, unlike an activity that is shut down, a service is scheduled to restart again when resources are available if there are pending `startService` intents in the queue. The service will be woken up and the next intent delivered to it via `onStartCommand()`. Of course, `onCreate()` will be called when the service is brought back.

> **NOTE:** Because services are automatically restarted if they are not explicitly stopped, it is reasonable to think that, unlike activities and other components, a service component is fundamentally a sticky component.

Details of a Nonsticky Service

What is a *nonsticky* service then?

Let's talk about a situation when a service is not automatically restarted. After a client calls `startService`, the service is created and `OnStartCommand is` called to do its work. This service will not be automatically restarted if a client explicitly calls `stopService`.

This `stopService` method depending on how many clients are still connected, can move the service into a stopped state, at which time the service's `onDestroy` method is called and the service life cycle is complete. Once a service has been stopped like this by its last client, the service will not be brought back.

This protocol works well when everything happens as per design, where start and stop methods are called and executed in sequence and without a miss.

Prior to Android 2.0, devices have seen a lot of services hanging around and claiming resources even though there was no work to be done, meaning Android brought the services back into memory even though there were no messages in the queue. This would have happened when stopService was not invoked either because of an exception or because the process was taken out between onStartCommand and stopService.

Android 2.0 introduced a solution so that we can indicate to the system, if there are no pending intents, that it shouldn't bother restarting the service. This is OK because whoever started the service to do the work will call it again, such as the alarm manager. This is done by returning the nonsticky flag (Service.START_NOT_STICKY) from onStartCommand.

However, nonsticky is not really that nonsticky. Remember, even if we mark the service as nonsticky, if there are pending intents, Android will bring the service back to life. This setting applies only when there are no pending intents.

Details of a Sticky Service

What does it mean for a service to be really sticky then?

The sticky flag (Service.START_STICKY) means that Android should restart the service even if there are no pending intents. When the service is restarted, call onCreate and onStartCommand with a null intent. This will give the service an opportunity, if need be, to call stopSelf if that is appropriate. The implication is that a service that is sticky needs to deal with null intents on restarts.

A Variation of Nonsticky: Redeliver Intents

Local services in particular follow a pattern where onStart and stopSelf are called in pairs. A client calls onStart. The service, when it finishes that work, calls stopSelf. You can see this clearly in the implementation of the IntentService utility class in Listing 19–10.

If a service takes, say, 30 minutes to complete a task, it will not call stopSelf for 30 minutes. Meanwhile, the service is reclaimed. If we use the nonsticky flag, the service will not wake up, and we would never have called stopSelf.

Many times, this is OK. However, if you want to make sure whether these two calls happen for sure, you can tell Android to not to unqueue the start event until stopSelf is called. This ensures that, when the service is reclaimed, there is always a pending event unless the stopSelf is called. This is called redeliver mode, and it can be indicated in reply to the onStartCommand method by returning the Service.START_REDELIVER flag.

Specifying Service Flags in OnStartCommand

Interestingly, stickiness is tied to onStartCommand, and not to onCreate, for a service. This is a bit odd, because so far, we have been talking about a service being in sticky, nonsticky, or redeliver mode as if these were service-level attributes. However, this determination for the nature of a service is made based on the return value from `OnStartCommand`. Wonder what the goal here is? Because for the same service instance, `OnStartCommand` is called many times, once for each `startService`. What if the method returns different flags indicating different service behaviors? Perhaps the best guess is that the last returned value is what determines the service behavior.

Picking Suitable Stickiness

Given the combination of possible service behaviors, what type of service is suitable for a long-running broadcast receiver? We believe a simple, nonsticky service, which just assumes the service will stop if there are no pending messages in the queue, will do. We are finding it hard to think that there is a use case for sticky long-running broadcast receivers, especially if we want to use `IntentService`, which expects the service to stop if there are no pending intents.

You will see this conclusion in the implementation of our long-running service abstraction in the upcoming Listing 19–18, where we have returned the nonsticky flag.

Controlling the Wake Lock from Two Places

Before presenting the source code for the long-running service, let's talk about the responsibilities of the service regarding keeping the device on.

When the service code is running, we should have the partial wake lock in effect. To do this, when the service is created, we need to turn on the wake lock by creating the lighted green room. You might say that this is done by the broadcast receiver, which is true. However, the service may be woken up by itself, in which case we would have missed the setup of the lighted room. So we need to control the wake lock from both places.

The long-running broadcast receiver code in Listing 19–14 initializes the wake lock using `LightedGreenRoom.setup()`. We will do the same in the service creation callback.

In addition to setting up the lighted green room, our service needs to register itself as a client to the lighted green room. This allows for cleanup when the service component gets destroyed through `onDestroy()`.

Long-Running Service Implementation

Now that you have the background on `IntentService`, service-start flags, and the lighted green room, we're ready to take a look at the long-running service in Listing 19–18.

Listing 19–18. *A Long-Running Service*

```
public abstract class ALongRunningNonStickyBroadcastService
extends IntentService
{
    public static String tag = "ALongRunningBroadcastService";
    protected abstract void
```

```
handleBroadcastIntent(Intent broadcastIntent);

public ALongRunningNonStickyBroadcastService(String name){
    super(name);
}
/*
 * This method can be invoked under two circumstances
 * 1. When a broadcast receiver issues a "startService"
 * 2. when android restarts it due to pending "startService" intents.
 *
 * In case 1, the broadcast receiver has already
 * setup the "lightedgreenroom".
 *
 * In case 2, we need to do the same.
 */
@Override
public void onCreate()
{
    super.onCreate();

    //Set up the green room
    //The setup is capable of getting called multiple times.
    LightedGreenRoom.setup(this.getApplicationContext());

    //It is possible that more than one service
    //of this type is running.
    //Knowing the number will allow us to clean up
    //the locks in ondestroy.
    LightedGreenRoom.s_registerClient();
}
@Override
public int onStartCommand(Intent intent, int flag, int startId)
{
    //Call the IntentService "onstart"
    super.onStart(intent, startId);

    //Tell the green room there is a visitor
    LightedGreenRoom.s_enter();

    //mark this as non sticky
    //Means: Don't restart the service if there are no
    //pending intents.
    return Service.START_NOT_STICKY;
}
/*
 * Note that this method call runs
 * in a secondary thread setup by the IntentService.
 *
 * Override this method from IntentService.
 * Retrieve the original broadcast intent.
 * Call the derived class to handle the broadcast intent.
 * finally tell the lighted room that you are leaving.
 * if this is the last visitor then the lock
 * will be released.
 */
@Override
final protected void onHandleIntent(Intent intent)
```

```
    {
        try {
            Intent broadcastIntent
            = intent.getParcelableExtra("original_intent");
            handleBroadcastIntent(broadcastIntent);
        }
        finally {
            LightedGreenRoom.s_leave();
        }
    }
    /*
     * If Android reclaims this process,
     * this method will release the lock
     * irrespective of how many visitors there are.
     */
    @Override
    public void onDestroy() {
        super.onDestroy();
        LightedGreenRoom.s_unRegisterClient();
    }
}
```

Clearly, this class extends `IntentService` and gets all the benefits of a worker thread as set up by `IntentService`. In addition, it specializes the `IntentService` further so that it is set up as a nonsticky service. From a developer's perspective, the primary method to focus on is the abstract `handleBroadcastIntent()` method.

Testing Long-Running Services

So far, we have presented all the code necessary to quickly run code that lasts longer than ten seconds in response to a broadcast event. The abstractions we have developed allow a clear path from the broadcast intent to the code that processes that intent. The abstraction takes care of everything in the middle.

Your Responsibilities

To summarize, as an implementer of a long-running service, your responsibilities are as follows:

1. Inherit from the abstracted long-running service and implement the single method: five lines of code (see, for example, Listing 19–13). Then register the service in the manifest file (see Listing 19–5 or 19–19).
2. Inherit from the long-running broadcast receiver: five lines of code (for example, Listing 19–15). Then register the broadcast receiver in the manifest file (see Listing 19–5 or 19–19).
3. Register a user permission to work with the wake lock. You are finished.

Listing 19–19 shows an example of the manifest file where a long-running service and the corresponding broadcast receiver are registered.

Listing 19–19. *The Long-Running Receiver and Service Definition*

```
<manifest…>
……
<application….>
<receiver android:name=".Test60SecBCR">
    <intent-filter>
       <action android:name="com.androidbook.intents.testbc"/>
    </intent-filter>
</receiver>
<service android:name=".Test60SecBCRService"/>
</application>
…..
<uses-permission android:name="android.permission.WAKE_LOCK"/>
</manifest>
```

Also notice that you will need the wake lock permission to run this long-running receiver abstraction.

Framework Responsiblities

The framework will do the rest:

- It will manage the wake locks.
- It will transfer the intent transparently from your receiver to the service method.
- It will manage the restart of services and the associated wake locks.

The framework in essence gives you the impression that your code is invoked directly from the caller that raised the broadcast event. For you, as a programmer, it is as if everything in the middle has disappeared and allowed the following client programming abstraction:

1. Raise event `event1`.
2. Process event `event1` in method `method1()` for 15 minutes if needed.

Again, in summary, the classes that orchestrates this framework are

- `LongRunningService` (Listing 19–18)
- `LightedGreenRoom` (Listing 19–17)
- `LongRunningBroadcastReceiver` (Listing 19–14)

The complete working source code for all the examples and the framework is available in the downloadable ZIP file.

A Few Notes about the Project Download File

This chapter has two projects: one to test the broadcast receiver (called `TestBCR`) and one to test the stand-alone receivers, including the long-running receiver and service (called StandaloneBCR). Both these projects are zipped and available in the download file; the URL for this is listed in the "References" section. We strongly suggest that you download the ZIP file and unzip it to see these projects individually.

Once you compile these projects in Eclipse, we suggest that you deploy StandaloneBCR first on the device. This is a headless .apk file. Then you install the menu-driven TestBCR application. This TestBCR application will present menus to kick off broadcast events.

These events will be responded to by regular and long-running receivers in the StandaloneBCR project. You will see the results in the LogCat. In the LogCat you can also see the thread ID that is used to run the code of the receivers.

References

Here are some helpful references to topics you may wish to explore further:

- http://developer.android.com/reference/android/content/BroadcastReceiver.html: The BroadcastReceiver API. In this chapter, we have covered the most basic version of a broadcast receiver. You will find at this link more about ordered broadcasts and a little about the BroadcastReceiver life cycle.
- http://developer.android.com/reference/android/app/Service.html: The Service API. This reference is especially good to have while working with long-running services.
- http://developer.android.com/reference/android/app/NotificationManager.html: The NotificationManager API.
- http://developer.android.com/reference/android/app/Notification.html: The Notification API. You will see here the various options available for working with a notification, such as content views and sound effects.
- http://developer.android.com/reference/android/widget/RemoteViews.html: The RemoteViews API. RemoteViews are used to construct custom detailed views of notifications.
- www.androidbook.com/item/3514: The authors' research on long-running services.
- www.androidbook.com/item/3482: The authors' research on broadcast receivers. This note also explains how to start an activity from a receiver.
- www.androidbook.com/proandroid4/projects: A list of downloadable projects from this book. For this chapter, look for a ZIP file named ProAndroid4_Ch19_TestReceivers.zip.

Summary

We have covered the following in this chapter:

- Sending and responding to broadcast events through broadcast receivers
- Creating and sending notifications from a broadcast receiver by utilizing the notification manager
- The role of wake locks while responding to a broadcast receiver
- The need for and use of long-running services when responding to a broadcast receiver

Interview Questions

The following list of questions should further consolidate what is covered in this chapter:

1. What method is used to send a broadcast event?
2. What class do you have to inherit from to be a broadcast receiver?
3. What are the inputs to the `onReceive()` method of a broadcast receiver?
4. What tag is used to register a broadcast receiver?
5. How do you indicate that you are interested in a particular broadcast intent?
6. Can broadcasts be received by receivers in an external process?
7. How do you get access to a notification manager?
8. What class is used to construct a notification?
9. What are the essential components of a notification?
10. What method is used to create the expanded notification view?
11. What is the use of `setContentView()` for a notification object?
12. What are the limitations of `RemoteViews`?
13. What is the goal of `IntentService`?
14. If many clients call an `IntentService`, how many threads are running on behalf of the clients?
15. What is the use of `WakefulIntentService`?
16. How do you get access to a wake lock?

17. What is relevance of a reference count on a wake lock?
18. How can you transfer a broadcast intent to a service execution method?
19. What method do you override on an `IntentService`?
20. What is the use of a looper?
21. What is the use of `stopSelf()` on a service?
22. What permission do you need to ask to use a wake lock in your app?
23. How do you start an activity from a broadcast receiver?

Chapter 20

Exploring the Alarm Manager

In Android, you can use the alarm manager to trigger events. These events can be at a specific time or at regular intervals. We will start the chapter with the basics of the alarm manager where we set a simple alarm. We will then cover setting an alarm that repeats, cancelling an alarm, the role of pending intents (specifically the role their uniqueness plays), and setting multiple alarms. By the end of the chapter, you will have learned both the basics and the practical nitty-gritty of the Android alarm manager.

Alarm Manager Basics: Setting Up a Simple Alarm

We will start the chapter with setting an alarm at a particular time and having it call a broadcast receiver. Once the broadcast receiver is invoked, we can use the information from Chapter 19 to perform both simple and long-running operations in that broadcast receiver.

The steps are as follows:

1. Get access to the alarm manager.
2. Come up with a time to set the alarm.
3. Create a receiver to be invoked.
4. Create a pending intent that can be passed to the alarm manager to invoke the receiver at the appointed time.
5. Use the time instance from step 2 and the pending intent from step 4 to set the alarm.
6. See the receiver from step 3 invoked.

Getting Access to the Alarm Manager

Getting access to the alarm manager is simple as in Listing 20–1.

Listing 20–1. *Getting an Alarm Manager*

```
AlarmManager am =
    (AlarmManager)
         mContext.getSystemService(Context.ALARM_SERVICE);
```

The variable mContext refers to a context object. For example, if you are invoking this code from an activity menu, the context variable will be the activity.

Setting Up the Time for the Alarm

To set the alarm for a particular date and time, you will need an instance in time identified by a Java Calendar object. Listing 20–2 has a few utility functions to work with the Calendar object.

Listing 20–2. *A Few Useful Calendar Utilities*

```
public class Utils {
    public static Calendar getTimeAfterInSecs(int secs) {
        Calendar cal = Calendar.getInstance();
        cal.add(Calendar.SECOND,secs);
        return cal;
    }
    public static Calendar getCurrentTime(){
        Calendar cal = Calendar.getInstance();
        return cal;
    }
    public static Calendar getTodayAt(int hours){
        Calendar today = Calendar.getInstance();
        Calendar cal = Calendar.getInstance();
        cal.clear();

        int year = today.get(Calendar.YEAR);
        int month = today.get(Calendar.MONTH);
        //represents the day of the month
        int day = today.get(Calendar.DATE);
        cal.set(year,month,day,hours,0,0);
        return cal;
    }
    public static String getDateTimeString(Calendar cal){
        SimpleDateFormat df = new SimpleDateFormat("MM/dd/yyyy hh:mm:ss");
        df.setLenient(false);
        String s = df.format(cal.getTime());
        return s;
    }
}
```

From this list of utilities, we will use the function getTimeAfterInSecs(), as shown in Listing 20–3 , to look for a time instance that is 30 seconds from now.

Listing 20–3. *Obtaining a Time Instance*

```
        Calendar cal = Utils.getTimeAfterInSecs(30);
```

Creating a Receiver for the Alarm

Now, we need a receiver to set against the alarm. A simple receiver is shown in Listing 20–4.

Listing 20–4. *TestReceiver to Test Alarm Broadcasts*

```
public class TestReceiver extends BroadcastReceiver
{
    private static final String tag = "TestReceiver";
    @Override
    public void onReceive(Context context, Intent intent)
    {
        Log.d tag, "intent=" + intent);
        String message = intent.getStringExtra("message");
        Log.d(tag, message);
    }
}
```

You will need to register this receiver in the manifest file using the <receiver> tag, as shown in Listing 20–5. This process is covered in detail in Chapter 19.

Listing 20–5. *Registering a Broadcast Receiver*

```
  <receiver android:name=".TestReceiver"/>
```

Creating a PendingIntent Suitable for an Alarm

Once we have a receiver, we can set up a PendingIntent, which is needed to set the alarm. However, we need an intent to create a pending intent. So, we start by creating a regular intent that can invoke the TestReceiver from Listing 20–4. This intent creation is shown in Listing 20–6.

Listing 20–6. *Creating an Intent Pointing to TestReceiver*

```
Intent intent =
    new Intent(mContext, TestReceiver.class);
intent.putExtra("message", "Single Shot Alarm");
```

The variable mContext is an activity. We have used the TestReceiver class name directly (instead of using an intent filter against an intent action as we did in Chapter 19 for receivers). We also have an opportunity to load the intent with "extras" while creating this intent.

Once we have this regular intent pointing to a receiver, we need to create a pending intent that is necessary to pass to an alarm manager. Listing 20–7 shows how to create a PendingIntent from the intent in Listing 20–6.

Listing 20–7. *Creating a Pending Intent*

```
PendingIntent pi =
    PendingIntent.getBroadcast(
```

```
        mContext,     //context, or activity, or service
        1,            //request id, used for disambiguating this intent
        intent,       //intent to be delivered
        0);           //pending intent flags
```

Notice that we have asked the PendingIntent class to construct a pending intent that is suitable for a broadcast explicitly. The other variations of creating a pending intent are as follows:

```
PendingIntent.getActivity() //useful to start an activity
PendingIntent.getService() //useful to start a service
```

We will discuss the request id argument, which we set to 1, in greater detail later in the chapter. Briefly, it is used to separate two intent objects that are similar in all other respects.

The pending intent flags have little to no influence on the alarm manager. Our recommendation is to use no flags at all and use 0 for their values. The intent flags are typically useful in controlling the lifetime of the pending intent. However, in this case, the lifetime is maintained by the alarm manager. For example, to cancel a pending intent, you ask the alarm manager to cancel it.

Setting the Alarm

Once we have the time instance in milliseconds as a Calendar object and the pending intent pointing to the receiver, we can set up an alarm by calling the set() method of the alarm manager, as illustrated in Listing 20–8.

Listing 20–8. *Alarm Manager* set() *Method*

```
am.set(AlarmManager.RTC_WAKEUP,
        cal.getTimeInMillis(),
        pi);
```

If you use AlarmManager.RTC_WAKEUP, the alarm will wake up the device. Or you can use AlarmManager.RTC in its place to deliver the intent when the device wakes up.

The time specified by the second argument is the instance in time specified by the calendarObject that we created earlier (see Listing 20–3). This time is in milliseconds since 1970. This also coincides with the Java Calendar object default.

When this method is called, the alarm manager will invoke the TestReceiver in Listing 20–4 30 seconds after the time when the method was called.

Test Project

We have a test project for this chapter that exercises the code listed so far. Investigating this test project is entirely optional. The previous sections have listed all the code you need to understand basic alarm features. The test project merely provides a driver activity, a menu, and a manifest file to test the code that is introduced.

The URL for the downloadable project is given in the "References" section of this chapter. If you compile and deploy this project, you will see an activity similar to the one shown in Figure 20–1.

> **NOTE:** To work with an alarm manager, there are no specific entries needed in the manifest file other than the receiver.

The code covered in the previous section will be invoked if you click the Alarm Once menu item.

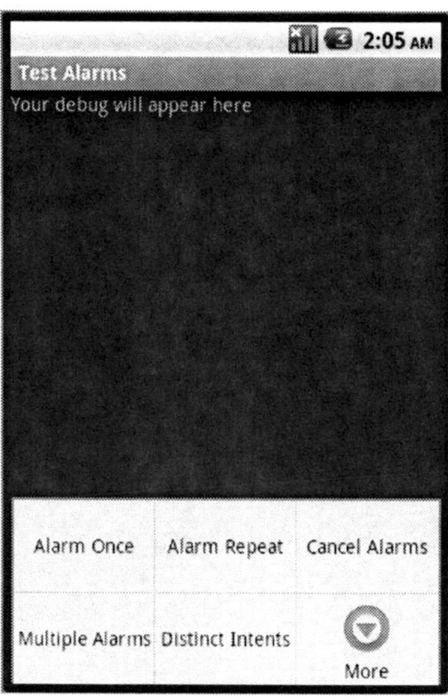

Figure 20–1. *A sample activity to test the alarm manager*

Figure 20–1 is showing only a portion of the available menu items from the test program. To see the other menu items, click the More icon. This view is shown in Figure 20–2.

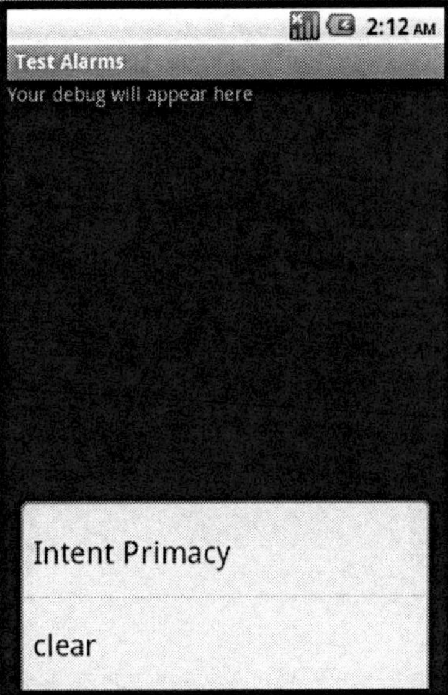

Figure 20–2. *Expanded menus for our sample activity*

We will discuss the use of the other menu items as we continue the chapter and present the relevant code.

Exploring Alarm Manager Alternate Scenarios

Now that we have explained the basics of setting an alarm, we will cover a few additional scenarios, such as setting off an alarm repeatedly and cancelling alarms. We will also show you exception conditions that you may run into while using the alarm manager.

Setting Off an Alarm Repeatedly

We have already covered how to set a simple one-time alarm, so let's now consider how we can set an alarm that goes of repeatedly; see Listing 20–9.

Listing 20–9. *Setting a Repeating Alarm*

```
public void sendRepeatingAlarm()
{
    Calendar cal = Utils.getTimeAfterInSecs(30);
    String s = Utils.getDateTimeString(cal);

    //Get an intent to invoke the receiver
    Intent intent =
        new Intent(this.mContext, TestReceiver.class);
    intent.putExtra("message", "Repeating Alarm");

    PendingIntent pi = this.getDistinctPendingIntent(intent, 2);
    // Schedule the alarm!
    AlarmManager am =
```

```
        (AlarmManager)
            this.mContext.getSystemService(Context.ALARM_SERVICE);

    am.setRepeating(AlarmManager.RTC_WAKEUP,
            cal.getTimeInMillis(),
            5*1000, //5 secs repeat
            pi);
}

protected PendingIntent getDistinctPendingIntent
                (Intent intent, int requestId)
{
    PendingIntent pi =
        PendingIntent.getBroadcast(
          mContext,     //context, or activity
          requestId,     //request id
          intent,         //intent to be delivered
          0);
    return pi;
}
```

Key elements of the code in Listing 20–9 are highlighted. The repeating alarm is set by invoking the `setRepeating()` method on the alarm manager object. One of the inputs to this method is a pending intent pointing to a receiver. We have used the same intent that was created in Listing 20–6, pointing to `TestReceiver`.

However, when we make a pending intent out of it, we use a unique request code, such as 2. If we don't do this, we will see a bit of odd behavior. Say you click a menu item for the repeating alarm first. This schedules the alarm to go off repeatedly and calls `TestReceiver`. Say this repeating alarm starts in 30 seconds. Now, you go ahead and click a menu item such as Alarm Once. This will schedule the alarm to go off just one time in 30 seconds and call the same `TestReceiver`.

If both these menu items worked, we would see both types of alarms go off. However, you will notice that the alarm goes off only one time. To make this work right, you have to use a different `requestcode` on the pending intent. We will go into the reasoning behind `requestcode` in the "Intent Primacy in Setting Off Alarms" section.

Again, if you make use of our downloadable test program, you can choose the Alarm Repeat menu item shown in Figure 20–1 to invoke the code in Listing 20–9 (setting the alarm repeatedly).

Cancelling an Alarm

To help you understand how to cancel an alarm, we'll use the code from Listing 20–10.

Listing 20–10. *Cancelling a Repeating Alarm*

```
public void cancelRepeatingAlarm()
{
    //Get an intent that was originally
    //used to invoke TestReceiver class
    Intent intent =
        new Intent(this.mContext, TestReceiver.class);
```

```
    //To cancel, extra is not necessary to be filled in
    //intent.putExtra("message", "Repeating Alarm");

    PendingIntent pi = this.getDistinctPendingIntent(intent, 2);

    // Cancel the alarm!
    AlarmManager am =
        (AlarmManager)
           this.mContext.getSystemService(Context.ALARM_SERVICE);
    am.cancel(pi);
}
```

To cancel an alarm, we have to construct a pending intent first and then pass it to the alarm manager as an argument to the `cancel()` method.

However, you must pay attention to make sure that the `PendingIntent` is constructed the exact same way when setting the alarm, including the request code and targeted receiver. In Listing 20–10, we have used the method `getDistinctPendingIntent()` again. Originally, this method was provided in Listing 20–9.

In constructing the cancel intent, you can ignore the intent extras from the original intent (Listing 20–10) because intent extras don't play a role in cancelling that intent. (This is because intent extras don't play a role in deciding the uniqueness of intents. This topic is discussed later in the chapter.)

If you are using the downloadable test project, you can test the cancel functionality by first selecting the Alarm Repeat menu item (see Figure 20–1). This will start updating the LogCat every five seconds. Now, if you click the Cancel Alarms menu item, the messages will stop.

Working with Multiple Alarms

When it comes to setting multiple alarms pointing to the same receiver, in our opinion, there is a bit of unintuitive behavior —if you invoke an alarm pointing to a particular receiver multiple times, only the last invocation takes effect.

To explain this behavior, first examine the Listing 20–11. There are two methods in this listing. The first one, `scheduleSameIntentMultipleTimes()`, schedules the same intent multiple times. The second function, `scheduleDistinctIntents()`, does the same but distinguishes the intents with the aid of a request ID.

Listing 20–11. *Working with Multiple Alarms*

```
/*
 * Same intent cannot be scheduled multiple times.
 * If you do, only the last one will take affect.
 *
 * Notice you are using the same request id.
 */
public void scheduleSameIntentMultipleTimes()
{
   //Get multiple time instances
```

```
    Calendar cal = Utils.getTimeAfterInSecs(30);
    Calendar cal2 = Utils.getTimeAfterInSecs(35);
    Calendar cal3 = Utils.getTimeAfterInSecs(40);
    Calendar cal4 = Utils.getTimeAfterInSecs(45);

    //Print to the debug view that we are
    //scheduling at a specific time
    //mReportTo.reportBack() is just a method to log
    //See the downloadable project for full details
    //Or you can delete the following two lines or use Log.d()
    String s = Utils.getDateTimeString(cal);
    mReportTo.reportBack(tag, "Scheduling alarm at: " + s);


    //Get an intent to invoke a receiver
    Intent intent =
        new Intent(mContext, TestReceiver.class);
    intent.putExtra("message", "Same intent multiple times");

    PendingIntent pi = this.getDistinctPendingIntent(intent, 1);

    // Schedule this same intent multiple times
    AlarmManager am =
        (AlarmManager)
            mContext.getSystemService(Context.ALARM_SERVICE);

    am.set(AlarmManager.RTC_WAKEUP,
            cal.getTimeInMillis(),
            pi);

    am.set(AlarmManager.RTC_WAKEUP,
            cal2.getTimeInMillis(),
            pi);
    am.set(AlarmManager.RTC_WAKEUP,
            cal3.getTimeInMillis(),
            pi);
    am.set(AlarmManager.RTC_WAKEUP,
            cal4.getTimeInMillis(),
            pi);
}
/*
 * Same intent can be scheduled multiple times
 * if you change the request id on the pending intent.
 * Request id identifies an intent as a unique intent.
 */
public void scheduleDistinctIntents()
{
    //Get the instance in time that is
    //30 secs from now.
    Calendar cal = Utils.getTimeAfterInSecs(30);
    Calendar cal2 = Utils.getTimeAfterInSecs(35);
    Calendar cal3 = Utils.getTimeAfterInSecs(40);
    Calendar cal4 = Utils.getTimeAfterInSecs(45);

    //If you want to point to 11:00 hours today.
    //Calendar cal = Utils.getTodayAt(11);
```

```
        //Print to the debug view that we are
        //scheduling at a specific time
        String s = Utils.getDateTimeString(cal);

        //Get an intent to invoke
        //TestReceiver class
        Intent intent =
            new Intent(mContext, TestReceiver.class);
        intent.putExtra("message", "Schedule distinct alarms");

        //Schedule the same intent but with different req ids.
        AlarmManager am =
            (AlarmManager)
                 mContext.getSystemService(Context.ALARM_SERVICE);

        am.set(AlarmManager.RTC_WAKEUP,
                cal.getTimeInMillis(),
                getDistinctPendingIntent(intent,1));

        am.set(AlarmManager.RTC_WAKEUP,
                cal2.getTimeInMillis(),
                getDistinctPendingIntent(intent,2));
        am.set(AlarmManager.RTC_WAKEUP,
                cal3.getTimeInMillis(),
                getDistinctPendingIntent(intent,3));
        am.set(AlarmManager.RTC_WAKEUP,
                cal4.getTimeInMillis(),
                getDistinctPendingIntent(intent,4));
    }
```

In the method `scheduleSameIntentMultipleTimes()`, we have started with an intent and scheduled the same intent four different times. You will see that when you test this by selecting the Multiple Alarms menu item from Figure 20–1, only the last alarm is fired, and all the previous ones are ignored.

The recommended way to fix this is to change each pending intent (when their underlying intents don't differ) to use a different request ID. This is why we have a function `getDistinctPendingIntent()`, which quickly creates pending intents based on request ID. Listing 20–9 shows the source code for this function.

You can fix the duplicate intent problem by looking at the `scheduleDistinctIntents()` method from Listing 20–11. Here, we have varied the request ID, so the `TestReceiver` will get called multiple times; you will see the evidence of this in LogCat.

Intent Primacy in Setting Off Alarms

We have mentioned a number of times so far that if you set alarms on the same type of intent, only the last alarm will take effect. Let's explore the reason behind this. Throughout the code examples, you might think that we are setting an alarm on the alarm manager. At least, that is the impression the API is giving us by exposing the following method:

```
alarmManager.set(time, intent);
```

However, assume we do the following:

```
alarmManager.set(time1, intent1);
alarmManager.setRepeated(time2, interval,  intent1);
```

You might have expected that the `intent1` object would just be a passive receiver and get invoked by both the alarms. However, in practice only the last `set` method counts. This is as if we are doing a set on the intent as in the following example:

```
intent1.set(...)
intent1.setRepeated(..)
```

In this case, it probably makes sense that you have just one intent object and one alarm against it and that if you set it multiple times you are resetting the previous alarm, just like an alarm clock on your desk.

This idea can be tested using Listing 20–12.

Listing 20–12. *Code to Test Intent Primacy*

```
/*
 * It is not the alarm that matters but the pending intent.
 * Even with a repeating alarm for an intent,
 * if you schedule the same intent again for one time,
 * the later one takes affect.
 *
 * It is as if you are setting the alarm on an existing intent multiple
 * times and not the other way around.
 */
public void alarmIntentPrimacy()
{
    Calendar cal = Utils.getTimeAfterInSecs(30);
    String s = Utils.getDateTimeString(cal);

    //Get an intent to invoke
    //TestReceiver class
    Intent intent =
        new Intent(this.mContext, TestReceiver.class);
    intent.putExtra("message", "Repeating Alarm");

    PendingIntent pi = getDistinctPendingIntent(intent,0);
    AlarmManager am =
        (AlarmManager)
            this.mContext.getSystemService(Context.ALARM_SERVICE);

    am.setRepeating(AlarmManager.RTC_WAKEUP,
            cal.getTimeInMillis(),
            5*1000, //5 secs
            pi);

    am.set(AlarmManager.RTC_WAKEUP,
            cal.getTimeInMillis(),
            pi);
}
```

Why does the later alarm replace the prior one if set on the same intent?

Many folks in the Android developer group pointed out that two intents are really the same and will result in the same PendingIntent object if their attributes are the same. Setting those intents as targets for multiple alarms is like setting multiple alarm times on the same intent.

However, what is really happening becomes obvious when we look at the source code of the AlarmManagerService (this is an implementation of the IAlarmManager interface) from the Android SDK. Listing 20–13 contains the code segment that is used to set an alarm (all alarm sets ultimately flow through this SDK code).

Listing 20–13. *AlarmManagerService Implementation Extract from Android Source*

```
    public void setRepeating(int type, long triggerAtTime, long interval,
            PendingIntent operation) {
        if (operation == null) {
            Slog.w(TAG, "set/setRepeating ignored because there is no intent");
            return;
        }
        synchronized (mLock) {
            Alarm alarm = new Alarm();
            alarm.type = type;
            alarm.when = triggerAtTime;
            alarm.repeatInterval = interval;
            alarm.operation = operation;

            // Remove this alarm if already scheduled.
            removeLocked(operation);

            if (localLOGV) Slog.v(TAG, "set: " + alarm);

            int index = addAlarmLocked(alarm);
            if (index == 0) {
                setLocked(alarm);
            }
        }
    }
```

Notice that, in the middle of a set method, the code is calling removeLocked(operation), where the operation argument is the PendingIntent. This essentially removes the previous alarm. In fact, when we call cancel(pendingIntent), it ends up calling the same removeLocked(pendingIntent).

In essence, the SDK chose to cancel the previous alarms and keep only the latest for that particular pending intent. If you want to do otherwise, you will need to qualify the pending intent with a request ID.

This also becomes clear when we take a closer look at the cancel() API, which just takes the PendingIntent object. If the relationship between an alarm and a PendingIntent is not unique, what would be the meaning of cancelling an alarm based on a PendingIntent and nothing else?

Of course, you can also use this feature to your advantage if your goal is to cancel any previous alarms and set a new one for that particular receiver.

Persistence of Alarms

Another note on alarms is the fact that they are not persisted across device reboots. This means you will need to persist the alarm settings and pending intents in a persistent store and reregister them based on device reboot broadcast messages, and possibly time-change messages (e.g., `android.intent.action.BOOT_COMPLETED`, `ACTION_TIME_CHANGED`, `ACTION_TIMEZONE_CHANGED`).

Alarm Manager Predicates

Let's conclude the chapter by providing a quick summary of the facts surrounding alarms, pending intents, and the alarm manager:

- Pending intents are intents that are kept in a pool and reused. You cannot new a pending intent. You really locate a pending intent with an option to reuse, update, and so on.
- An intent is uniquely distinguished by its action, data URI, and category. The details of this uniqueness are specified in the `filterEquals()` API of the intent class.
- A pending intent is further qualified (in addition to the base intent it depended on) by the request code.
- Alarms and pending intents (even intents, for that matter) are not independent. A given pending intent cannot be used for multiple alarms. The last alarm will override the previous alarms.
- Alarms are not persistent across boots. Whatever alarms you have set through the alarm manager will be lost when the device reboots.
- You will need to persist alarm parameters yourself if you would like to retain them beyond device reboots. You will need to listen to broadcast boot events and time-change events to reset these alarms as needed.
- The implication of the intent-based cancel API is that, when you use or persist alarms, you will also need to persist intents so that those alarms can be cancelled at a later time when needed.

References

The following references are useful to support the material in this chapter. Especially note the last URL listed, which allows you to download projects developed for this chapter:

- `http://developer.android.com/reference/android/app/AlarmManager.html`: The alarm manager API. You will see here signatures for methods like `set`, `setRepeating`, and `cancel`.
- `http://developer.android.com/reference/android/app/PendingIntent.html`: How to construct a pending intent. Don't pay too much attention to the pending intent flags; they are not that critical to the alarm manager.
- `www.androidbook.com/item/1040`: Quick examples and some references for working with date and time classes.
- `www.androidbook.com/item/3503`: Our basic research on alarm managers.
- `http://download.oracle.com/docs/cd/E17476_01/javase/1.4.2/docs/api/java/util/Calendar.html`: How to work with the `Calendar` object.
- `www.androidbook.com/proandroid4/projects`: A list of downloadable projects from this book. For this chapter, look for a ZIP file named `ProAndroid4_Ch20_TestAlarmManager.zip`.

Summary

In this chapter, we have covered the following:

- Setting a single alarm
- Setting a repeating alarm
- Canceling an alarm
- Making pending intents unique using the request ID

Interview Questions

The following questions will further consolidate what is covered in this chapter about Android alarms:

1. How do you get access to an alarm manager?
2. What objects do you need to set an alarm?
3. What is the starting time/year for the Java `Calendar` object?

4. Do you need any special permissions in the manifest file to access an alarm manager?
5. What method of the alarm manager is used to set an alarm repeatedly?
6. How do you make a pending intent out of an intent for broadcast?
7. What role does extra data in an intent play on alarms?
8. What is the role of the request ID on a pending intent?
9. What is the role of the request ID on the uniqueness of pending intents?
10. How do you cancel an alarm? What is the input?
11. What flags are available on an alarm manager for wakeup?
12. Why does the last intent take precedence when multiple alarms are set on the same intent?

Chapter 21

Exploring 2D Animation

Animation allows an object on a screen to change its color, position, size, or orientation over time. Animation capabilities in Android are practical, fun, and simple. They are used frequently.

Android 2.3 and prior releases support three types of animation: frame-by-frame animation, which occurs when a series of frames is drawn one after the other at regular intervals; layout animation, where you animate views inside a container view such as lists and tables; and view animation, in which you animate any view. The latter two fall into the category of tweening animation, which involves the drawings in between the key drawings.

> **NOTE:** Android 3.0 enhanced animation by introducing the ability to animate the properties of UI elements. Some of these features, especially as they apply to the concept of fragments, are covered in Chapter 8. We will cover this topic in much greater depth later in this chapter, because this new method is becoming the preferred approach to animations.

Another way of explaining tweening animation is to say that it is *not* frame-by-frame animation. If you are able to accomplish animating a figure without repeating frames, you are primarily doing tweening animation. For example, if a figure is at location A now and will be at location B in four seconds, we can change the location every second and redraw the same figure. This will make the figure look like it is moving from A to B.

The idea is that knowing the beginning and ending states of a drawing allows an artist to vary certain aspects of the drawing in time. The varying aspect could be color, position, size, or some other element. With computers, you accomplish this kind of animation by changing the intermediate values at regular intervals and redrawing the surface.

In this chapter, we will cover frame-by-frame, layout, and view animation using working examples and in-depth analysis.

NOTE: We have given a URL at the end of the chapter that you can use to download projects from this chapter and import these projects into Eclipse directly.

Frame-by-Frame Animation

Frame-by-frame animation is the simple process of showing a series of images in succession at quick intervals so that the final effect is that of an object moving or changing. This is how movie projectors work. We'll explore an example in which we'll design an image and save that image as a number of distinct images, where each one differs slightly from the others. Then, we will take the collection of those images and run them through the sample code to simulate animation.

Planning for Frame-by-Frame Animation

Before you start writing code, you first need to plan the animation sequence using a series of drawings. As an example of this planning exercise, Figure 21–1 shows a set of same-sized circles with a colored ball on each of the circles placed at a different position. You can create a series of these pictures showing the circle at the same size and position with the colored ball at different points along the circle's border. Once you save seven or eight of these frames, you can use animation to suggest that the colored ball is moving around the circle.

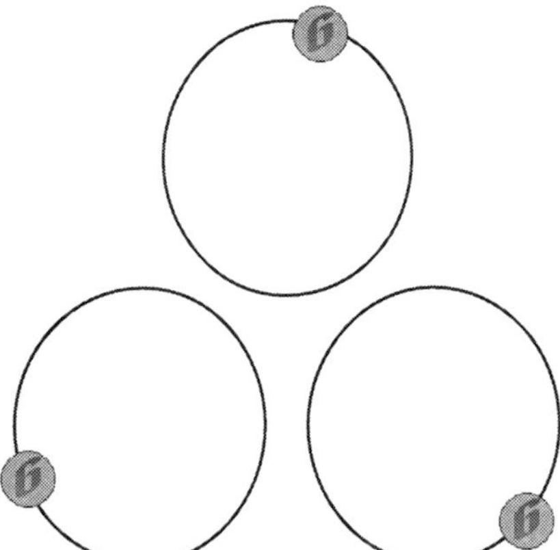

Figure 21–1. *Designing your animation before coding it*

Give the image a base name of `colored-ball`, and store eight of these images in the `/res/drawable` subdirectory so that you can access them using their resource IDs. The name of each image will have the pattern `colored-ballN`, where `N` is the digit representing the image number. When you have finished with the animation, you want it to look like Figure 21–2.

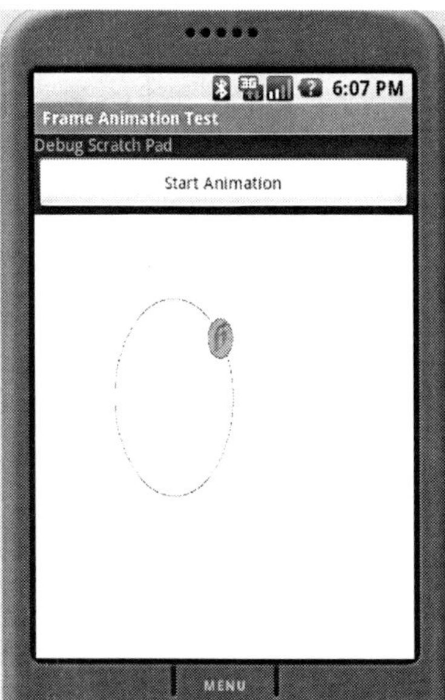

Figure 21–2. *A frame-by-frame animation test harness*

The primary area in this activity is used by the animation view. We have included a button to start and stop the animation to observe its behavior. We have also included a debug scratch pad at the top, so you can write any significant events to it as you experiment with this program. Let's see now how we could create the layout for such an activity.

Creating the Activity

Start by creating the basic XML layout file in the `/res/layout` subdirectory (see Listing 21–1) in a file named `frame_animations_layout.xml`.

Listing 21–1. *XML Layout File for the Frame Animation Example*

```
<?xml version="1.0" encoding="utf-8"?>
<!--filename: /res/layout/frame_animations_layout.xml -->
<LinearLayout xmlns:android="http://schemas.android.com/apk/res/android"
    android:orientation="vertical"
    android:layout_width="fill_parent"
    android:layout_height="fill_parent"
    >
<TextView android:id="@+id/textViewId1"
    android:layout_width="fill_parent"
    android:layout_height="wrap_content"
    android:text="Debug Scratch Pad"
    />
<Button
   android:id="@+id/startFAButtonId"
    android:layout_width="fill_parent"
    android:layout_height="wrap_content"
    android:text="Start Animation"
/>
```

```
<ImageView
        android:id="@+id/animationImage"
        android:layout_width="fill_parent"
       android:layout_height="wrap_content"
        />
</LinearLayout>
```

The first control is the debug-scratch text control, which is a simple `TextView`. You then add a button to start and stop the animation. The last view is the `ImageView`, where you will play the animation. Once you have the layout, create an activity to load this view (see Listing 21–2).

Listing 21–2. ***Activity to Load the*** *`ImageView`*

```
public class FrameAnimationActivity extends Activity
{
    @Override
    public void onCreate(Bundle savedInstanceState)
    {
        super.onCreate(savedInstanceState);
        setContentView(R.layout.frame_animations_layout);
    }
}
```

You will be able to run this activity from any menu item you might have in your current application by executing the following code:

```
Intent intent = new Intent(inActivity,FrameAnimationActivity.class);
inActivity.startActivity(intent);
```

At this point, you will see an activity that looks like the one in Figure 21–3.

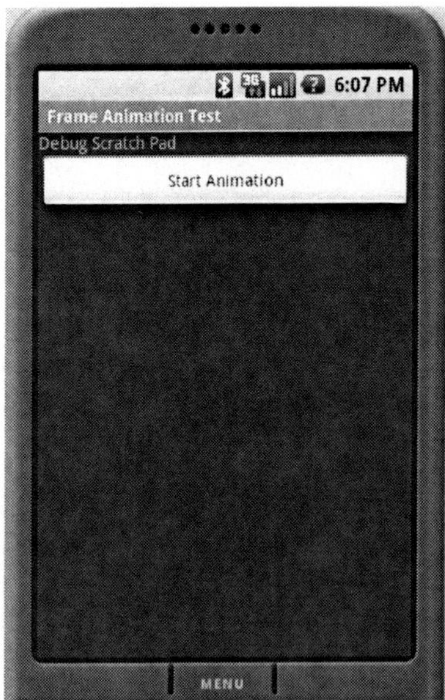

Figure 21–3. *Frame-by-frame animation activity*

Adding Animation to the Activity

Now that you have the activity and layout in place, we'll show you how to add animation to this sample. In Android, you accomplish frame-by-frame animation through a class in the graphics package called `AnimationDrawable`.

You can tell from its name that it is like any other drawable that can work as a background for any view (for example, the background bitmaps are represented as `Drawables`). This class, `AnimationDrawable`, in addition to being a `Drawable`, can take a list of other `Drawable` resources (like images) and render them at specified intervals. This class is really a thin wrapper around the animation support provided by the basic `Drawable` class.

> **TIP:** The `Drawable` class enables animation by asking its container or view to invoke a `Runnable` class that essentially redraws the `Drawable` using a different set of parameters. Note that you don't need to know these internal implementation details to use the `AnimationDrawable` class. But if your needs are more complex, you can look at the `AnimationDrawable` source code for guidance in writing your own animation protocols.

To make use of the `AnimationDrawable` class, start with a set of `Drawable` resources (for example, a set of images) placed in the `/res/drawable` subdirectory. In our case, these will be the eight similar, but slightly different, images that we talked about in the "Planning for Frame-by-Frame Animation" section. You will then construct an XML file that defines the list of frames (see Listing 21–3). This XML file will need to be placed in the `/res/drawable` subdirectory as well.

Listing 21–3. *XML File Defining the List of Frames to Be Animated*

```
<animation-list xmlns:android="http://schemas.android.com/apk/res/android"
android:oneshot="false">
   <item android:drawable="@drawable/colored_ball1" android:duration="50" />
   <item android:drawable="@drawable/colored_ball2" android:duration="50" />
   <item android:drawable="@drawable/colored_ball3" android:duration="50" />
   <item android:drawable="@drawable/colored_ball4" android:duration="50" />
   <item android:drawable="@drawable/colored_ball5" android:duration="50" />
   <item android:drawable="@drawable/colored_ball6" android:duration="50" />
   <item android:drawable="@drawable/colored_ball7" android:duration="50" />
   <item android:drawable="@drawable/colored_ball8" android:duration="50" />
</animation-list>
```

NOTE: As you prepare this list of images, we have to draw attention to some limitations of the AnimationDrawable class. This class loads all images into memory before it starts animation. When we tested this in the Android 2.3 emulator, a set of images greater than six exceeds the memory limitations allocated for each application. Depending on your test bed, you may need to restrict how many frames you have. To overcome this limitation, you will need to directly use the animation capabilities of Drawable and roll your own solution. Unfortunately, we haven't covered the Drawable class in detail in this edition of this book. Please check www.androidbook.com, as we intend to post an update soon.

Each frame points to one of the colored-ball images you have assembled through their resource IDs. The animation-list tag essentially gets converted into an AnimationDrawable object representing the collection of images. You then need to set this Drawable as a background resource for our ImageView in the sample. Assuming that the file name for this XML file is frame_animation.xml and that it resides in the /res/drawable subdirectory, you can use the following code to set the AnimationDrawable as the background of the ImageView:

```
view.setBackGroundResource(R.drawable.frame_animation);
```

With this code, Android realizes that the resource ID R.drawable.frame_animation is an XML resource and accordingly constructs a suitable AnimationDrawable Java object for it before setting it as the background. Once this is set, you can access this AnimationDrawable object by doing a get on the view object like this:

```
Object  backgroundObject = view.getBackground();
AnimationDrawable ad = (AnimationDrawable)backgroundObject;
```

Once you have the AnimationDrawable object, you can use its start() and stop() methods to start and stop the animation. Here are two other important methods on this object:

```
setOneShot();
addFrame(drawable, duration);
```

The setOneShot() method runs the animation once and then stops. The addFrame() method adds a new frame using a Drawable object and sets its display duration. The functionality of the addFrame() method resembles that of the XML tag android:drawable.

Put this all together to get the complete code for our frame-by-frame animation test harness (see Listing 21–4).

Listing 21–4. *Complete Code for the Frame-by-Frame Animation Test Harness*

```
public class FrameAnimationActivity extends Activity {
    @Override
    public void onCreate(Bundle savedInstanceState)
    {
        super.onCreate(savedInstanceState);
        setContentView(R.layout.frame_animations_layout);
        this.setupButton();
    }
```

```
    private void setupButton()
    {
       Button b = (Button)this.findViewById(R.id.startFAButtonId);
       b.setOnClickListener(
             new Button.OnClickListener(){
                public void onClick(View v)
                {
                   parentButtonClicked(v);
                }
             });
    }
    private void parentButtonClicked(View v)
    {
       animate();
    }
    private void animate()
    {
        ImageView imgView =
          (ImageView)findViewById(R.id.animationImage);
        imgView.setVisibility(ImageView.VISIBLE);
        imgView.setBackgroundResource(R.drawable.frame_animation);

        AnimationDrawable frameAnimation =
           (AnimationDrawable) imgView.getBackground();

        if (frameAnimation.isRunning())
        {
            frameAnimation.stop();
        }
        else
        {
            frameAnimation.stop();
            frameAnimation.start();
        }
    }
}//eof-class
```

The `animate()` method locates the `ImageView` in the current activity and sets its background to the `AnimationDrawable` identified by the resource `R.drawable.frame_animation`. The code then retrieves this object and performs the animation. The Start/Stop button is set up such that if the animation is running, clicking the button will stop it; if the animation is in a stopped state, clicking the button will start it.

Note that, if you set the `OneShot` parameter of the animation list to `true`, the animation will stop after executing once. However, there is no clear-cut way to know when that happens. Although the animation ends when it plays the last picture, you have no callback telling you when it finishes. Because of this, there isn't a direct way to invoke another action in response to the completed animation.

That drawback aside, you can bring great visual effects to bear by drawing a number of images in succession through the simple process of frame-by-frame animation.

Layout Animation

Like frame-by-frame animation, layout animation is pretty simple. As the name suggests, layout animation is dedicated to certain types of views laid out in a particular manner. For instance, you'll use layout animation with `ListView` and `GridView`, which are two commonly used layout controls in Android. Specifically, you'll use layout animation to add visual effects to the way each item in a `ListView` or `GridView` is displayed. In fact, you can use this type of animation on all controls derived from a `ViewGroup`.

Unlike frame-by-frame animation, layout animation is not achieved through repeating frames. Instead, it is achieved by changing the various properties of a view over time. Every view in Android has a transformation matrix that maps the view to the screen. By changing this matrix in a number of ways, you can accomplish scaling, rotation, and movement (translation) of the view. By changing the transparency of the view from `0` to `1`, for example, you can accomplish what is called an alpha animation.

Basic Tweening Animation Types

These are the basic tweening animation types in a bit more detail:

- *Scale animation*: You use this type of animation to make a view smaller or larger either along the x axis or on the y axis. You can also specify the pivot point around which you want the animation to take place.
- *Rotate animation*: You use this to rotate a view around a pivot point by a certain number of degrees.
- *Translate animation*: You use this to move a view along the x axis or the y axis.
- *Alpha animation*: You use this to change the transparency of a view.

You can define these animations as XML files in the `/res/anim` subdirectory. Listing 21–5 shows a quick sample of how one of these animations can be declared in an XML file

Listing 21–5. ***A Scale Animation Defined in an XML File at*** `/res/anim/scale.xml`

```
<set xmlns:android="http://schemas.android.com/apk/res/android"
android:interpolator="@android:anim/accelerate_interpolator">
   <scale
         android:fromXScale="1"
         android:toXScale="1"
         android:fromYScale="0.1"
         android:toYScale="1.0"
         android:duration="500"
         android:pivotX="50%"
         android:pivotY="50%"
         android:startOffset="100" />
</set>
```

All of the parameter values associated with these animation XML definitions have a "from" and a "to" flavor because you must specify the starting and ending values for the animation.

Each animation also allows duration and a time interpolator as arguments. We'll cover interpolators at the end of the section on layout animation, but for now, know that interpolators determine the rate of change of the animated argument during animation.

Once you have this declarative animation file, you can associate this animation with a layout to animate the layout's constituent views.

> **NOTE:** This is a good place to point out that each of these animations is represented as a Java class in the `android.view.animation` package. The Java documentation for each of these classes describes not only its Java methods but also the allowed XML arguments for each type of animation.

Now that you have enough background on animation types and understand a little bit about layout animation, let's design an example.

Planning the Layout Animation Test Harness

You can test all the layout-animation concepts we've covered using a simple `ListView` set in an activity. Once you have a `ListView`, you can attach an animation to it so that each list item will go through that animation.

Assume you have a scale animation in mind that makes a view grow from a tenth to its original size on the y axis. Visually this is equivalent to a line of text starting as a thin horizontal line and become fatter to grow to its actual font size.

You can attach such an animation to a `ListView`. When this happens, the `ListView` will animate each item in that list using this animation.

You can set some additional parameters that extend the basic animation, such as animating the list items from top to bottom or from bottom to top. You specify these parameters through an intermediate class that acts as a mediator between the individual animation XML file and the list view.

You can define both the individual animation and the mediator in XML files in the `/res/anim` subdirectory. Once you have the mediator XML file, you can use that file as an input to the `ListView` in its XML layout definition. When you have this basic setup working, you can start altering the individual animations to see how they impact the `ListView` display.

Before we embark on this exercise, let us show you what the `ListView` will look like after the animation completes (see Figure 21–4).

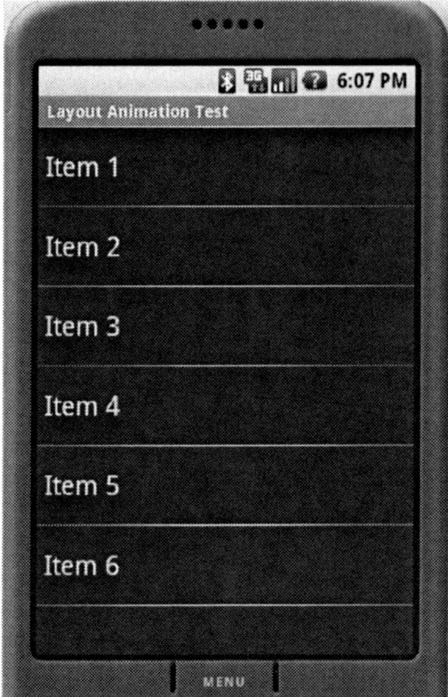

Figure 21–4. *The `ListView` we will animate*

Creating the Activity and the ListView

Start by creating an XML layout for the `ListView` in Figure 21–4 so you can load that layout in a basic activity. Listing 21–6 contains a simple layout with a `ListView` in it. You will need to place this file in the `/res/layout` subdirectory. Assuming the file name is `list_layout.xml`, your complete file will reside in `/res/layout/list_layout.xml`.

Listing 21–6. *XML Layout File Defining the ListView*

```
<?xml version="1.0" encoding="utf-8"?>
<!-- filename: /res/layout/list_layout.xml -->
<LinearLayout xmlns:android="http://schemas.android.com/apk/res/android"
    android:orientation="vertical"
    android:layout_width="fill_parent"
    android:layout_height="fill_parent"
    >

    <ListView
        android:id="@+id/list_view_id"
        android:layout_width="fill_parent"
        android:layout_height="fill_parent"
        />
</LinearLayout>
```

Listing 21–6 shows a simple `LinearLayout` with a single `ListView` in it. However, we should take this opportunity to mention one point about the `ListView` definition that is somewhat tangentially related to this chapter. You often see the ID for a `ListView` specified as `@android:id/list`. The resource reference `@android:id/list` points to an ID that is predefined in the `android` namespace. The question is, when do we use `android:id` versus our own ID such as `@+id/list_view_id`?

You will need to use @android:id/list only if the activity is a ListActivity. A ListActivity assumes that a ListView identified by this predetermined ID is available for loading. In the example here, you're using a general-purpose activity rather than a ListActivity, and you are going to explicitly populate the ListView yourself. As a result, there are no restrictions on the kind of ID you can allocate to represent this ListView. However, you do have the option of also using @android:id/list because it doesn't conflict with anything as there is no ListActivity in sight.

This surely is a digression, but it's worth noting as you create your own ListViews outside a ListActivity. Now that you have the layout needed for the activity, you can write the code for the activity to load this layout file so you can generate your UI (see Listing 21–7).

Listing 21–7. *Code for the Layout-Animation Activity*

```
public class LayoutAnimationActivity extends Activity
{
    @Override
    public void onCreate(Bundle savedInstanceState)
    {
        super.onCreate(savedInstanceState);
        setContentView(R.layout.list_layout);
        setupListView();
    }
    private void setupListView()
    {
          String[] listItems = new String[] {
                "Item 1", "Item 2", "Item 3",
                "Item 4", "Item 5", "Item 6",
          };

          ArrayAdapter<String> listItemAdapter =
               new ArrayAdapter<String>(this
                       ,android.R.layout.simple_list_item_1
                       ,listItems);
          ListView lv = (ListView)this.findViewById(R.id.list_view_id);
          lv.setAdapter(listItemAdapter);
    }
}
```

Refer to Chapter 6 on how to work with list views and populate a list view. You can now invoke the activity LayoutAnimationActivity from any menu item in your application using the following code:

```
Intent intent = new Intent(inActivity,LayoutAnimationActivity.class);
inActivity.startActivity(intent);
```

However, as with any other activity invocation, you will need to register the LayoutAnimationActivity in the AndroidManifest.xml file for the preceding intent invocation to work. Here is the code for it:

```
<activity android:name=".LayoutAnimationActivity"
        android:label="View Animation Test Activity"/>
```

Animating the ListView

Now that you have the test harness ready (see Listings 21–6 and 21–7), you'll learn how to apply scale animation to this `ListView`. Listing 21–8 shows how this scale animation is defined in an XML file.

Listing 21–8. *Defining Scale Animation in an XML File*

```
<?xml version="1.0" encoding="utf-8"?>
<!-- filename: /res/anim/scale.xml -->
<set xmlns:android="http://schemas.android.com/apk/res/android"
android:interpolator="@android:anim/accelerate_interpolator">
   <scale
         android:fromXScale="1"
         android:toXScale="1"
         android:fromYScale="0.1"
         android:toYScale="1.0"
         android:duration="500"
         android:pivotX="50%"
         android:pivotY="50%"
         android:startOffset="100" />
</set>
```

As indicated earlier, these animation-definition files reside in the `/res/anim` subdirectory.

Let's break down these XML attributes into plain English:

- The `from` and `to` scales point to the starting and ending magnification factors. The magnification starts at `1` and stays at `1` on the x axis. This means the list items will not grow or shrink on the x axis.
- On the y axis, however, the magnification starts at `0.1` and grows to `1.0`. In other words, the object being animated starts at one-tenth of its normal size and then grows to reach its normal size.
- The scaling operation will take `500` milliseconds to complete.
- The center of action is halfway (`50%`) in both x and y directions.
- The `startOffset` value refers to the number of milliseconds to wait before starting the animation.
- The parent node of the scale animation points to an animation set that could allow more than one animation to be in effect. We will cover one of those examples as well, but for now, there is only one animation in this set.

Name this file `scale.xml`, and place it in the `/res/anim` subdirectory. You are not yet ready to set this animation XML as an argument to the `ListView`; the `ListView` first requires another XML file that acts as a mediator between itself and the animation set. The XML file that describes that mediation is shown in Listing 21–9.

Listing 21–9. *Definition for a Layout-Controller XML File*

```
<?xml version="1.0" encoding="utf-8"?>
<!-- filename: /res/anim/list_layout_controller.xml -->
<layoutAnimation xmlns:android="http://schemas.android.com/apk/res/android"
        android:delay="30%"
        android:animationOrder="reverse"
        android:animation="@anim/scale" />
```

You will also need to place this XML file in the /res/anim subdirectory. For our example, assume that the file name is list_layout_controller. Once you look at this definition, you can see why this intermediate file is necessary.

This XML file specifies that the animation in the list should proceed in reverse, and that the animation for each item should start with a 30 percent delay with respect to the total animation duration. This XML file also refers to the individual animation file, scale.xml. Also notice that instead of the file name, the code uses the resource reference @anim/scale.

Now that you have the necessary XML input files, we'll show you how to update the ListView XML definition to include this animation XML as an argument. First, review the XML files you have so far:

```
// individual scale animation
/res/anim/scale.xml

// the animation mediator file
/res/anim/list_layout_controller.xml

// the activity view layout file
/res/layout/list_layout.xml
```

With these files in place, you need to modify the XML layout file list_layout.xml to have the ListView point to the list_layout_controller.xml file (see Listing 21–10).

Listing 21–10. *The Updated Code for the list_layout.xml File*

```
<?xml version="1.0" encoding="utf-8"?>
<LinearLayout xmlns:android="http://schemas.android.com/apk/res/android"
    android:orientation="vertical"
    android:layout_width="fill_parent"
    android:layout_height="fill_parent"
    >
    <ListView
        android:id="@+id/list_view_id"
        android:persistentDrawingCache="animation|scrolling"
        android:layout_width="fill_parent"
        android:layout_height="fill_parent"
        android:layoutAnimation="@anim/list_layout_controller" />
        />
</LinearLayout>
```

The changed lines are highlighted in bold. android:layoutAnimation is the key tag, which points to the mediating XML file that defines the layout controller using the XML tag layoutAnimation (see Listing 21–9). The layoutAnimation tag, in turn, points to the individual animation, which in this case is the scale animation defined in scale.xml.

Android also recommends setting the persistentDrawingCache tag to optimize for animation and scrolling. Refer to the Android SDK documentation for more details on this tag.

When you update the list_layout.xml file, as shown in Listing 21–10, Eclipse's ADT plug-in will automatically recompile the package taking this change into account. If you were to run the application now, you would see the scale animation take effect on the individual items. We have set the duration to 500 milliseconds so that you can observe the scale change clearly as each item is drawn.

Now you're in a position to experiment with different animation types. You'll try alpha animation next. To do this, create a file called /res/anim/alpha.xml, and populate it with the content from Listing 21–11.

Listing 21–11. *The alpha.xml File to Test Alpha Animation*

```
<alpha xmlns:android="http://schemas.android.com/apk/res/android"
       android:interpolator="@android:anim/accelerate_interpolator"
       android:fromAlpha="0.0" android:toAlpha="1.0" android:duration="1000" />
```

Alpha animation is responsible for controlling the fading of color. In this example, you are asking the alpha animation to go from invisible to full color in 1,000 milliseconds, or 1 second. Make sure the duration is one second or longer; otherwise, the color change is hard to notice.

Every time you want to change the animation of an individual item like this, you will need to change the mediator XML file (see Listing 21–9) to point to this new animation file. Here is how to change the animation from scale animation to alpha animation:

```
<layoutAnimation xmlns:android="http://schemas.android.com/apk/res/android"
        android:delay="30%"
        android:animationOrder="reverse"
        android:animation="@anim/alpha" />
```

The changed line in the layoutAnimation XML file is highlighted. Let's now try an animation that combines a change in position with a change in color gradient. Listing 21–12 shows the sample XML for this animation.

Listing 21–12. *Combining Translate and Alpha Animations Through an Animation Set*

```
<set xmlns:android="http://schemas.android.com/apk/res/android"
android:interpolator="@android:anim/accelerate_interpolator">
    <translate android:fromYDelta="-100%" android:toYDelta="0"
android:duration="500" />
    <alpha android:fromAlpha="0.0" android:toAlpha="1.0"
android:duration="500" />
</set>
```

Notice how we have specified two animations in the animation set. The translate animation will move the text from top to bottom in its currently allocated display space. The alpha animation will change the color gradient from invisible to visible as the text item descends into its slot. The duration setting of 500 will allow the user to perceive the change in a comfortable fashion. Of course, you will have to change the layoutAnimation mediator XML file again with a reference to this file name. Assuming the file name for this combined animation is /res/anim/translate_alpha.xml, your layoutAnimation XML file will look like this:

```
<layoutAnimation xmlns:android="http://schemas.android.com/apk/res/android"
        android:delay="30%"
        android:animationOrder="reverse"
        android:animation="@anim/translate_alpha" />
```

Let's now look at how to use rotate animation (see Listing 21–13).

Listing 21–13. *Rotate Animation XML File*

```
<rotate xmlns:android="http://schemas.android.com/apk/res/android"
       android:interpolator="@android:anim/accelerate_interpolator"
       android:fromDegrees="0.0"
      android:toDegrees="360"
      android:pivotX="50%"
      android:pivotY="50%"
      android:duration="500" />
```

The code in Listing 21–13 will spin each text item in the list one full circle around the midpoint of the text item. The duration of 500 milliseconds is a good amount of time for the user to perceive the rotation. As before, to see this effect you must change the layout controller XML file and the ListView XML layout file and then rerun the application.

Now we've covered the basic concepts in layout animation, where we start with a simple animation file and associate it with a ListView through an intermediate layoutAnimation XML file. That's all you need to do to see the animated effects. However, we need to talk about one more thing with regard to layout animation: interpolators.

Using Interpolators

Interpolators tell an animation how a certain property, such as a color gradient, changes over time. Will it change in a linear or exponential fashion? Will it start quickly but slow down toward the end? Consider the alpha animation that we introduced in Listing 21–11:

```
<alpha xmlns:android="http://schemas.android.com/apk/res/android"
       android:interpolator="@android:anim/accelerate_interpolator"
       android:fromAlpha="0.0" android:toAlpha="1.0" android:duration="1000" />
```

The animation identifies the interpolator it wants to use—accelerate_interpolator, in this case. There is a corresponding Java object that defines this interpolator. Also, note that we've specified this interpolator as a resource reference. This means there must be a file corresponding to the anim/accelerate_interpolator that describes what this Java object looks like and what additional parameters it might take. That indeed is the case. Look at the XML file definition for @android:anim/accelerate_interpolator:

```
<accelerateInterpolator
  xmlns:android="http://schemas.android.com/apk/res/android"
  factor="1" />
```

You can see this XML file in the following subdirectory within the Android package:

/res/anim/accelerate_interpolator.xml

The `accelerateInterpolator` XML tag corresponds to a Java object with this name:

```
android.view.animation.AccelerateInterpolator
```

You can look up the Java documentation for this class to see what XML tags are available. This interpolator's goal is to provide a multiplication factor given a time interval based on a hyperbolic curve. The source code for the interpolator illustrates this:

```
public float getInterpolation(float input)
{
   if (mFactor == 1.0f)
   {
      return (float)(input * input);
   }
   else
   {
      return (float)Math.pow(input, 2 * mFactor);
   }
}
```

Every interpolator implements this `getInterpolation` method differently. In this case, if the interpolator is set up so that the factor is `1.0`, it will return the square of the factor. Otherwise, it will return a power of the input that is further scaled by the factor. So if the factor is `1.5`, you will see a cubic function instead of a square function.

The supported interpolators include

- `AccelerateDecelerateInterpolator`
- `AccelerateInterpolator`
- `CycleInterpolator`
- `DecelerateInterpolator`
- `LinearInterpolator`
- `AnticipateInterpolator`
- `AnticipateOvershootInterpolator`
- `BounceInterpolator`
- `OvershootInterpolator`

To see how flexible these interpolators can be, take a quick look at the `BounceInterpolator` which bounces the object (that is, moves it back and forth) toward the end of the following animation:

```
public class BounceInterpolator implements Interpolator {
    private static float bounce(float t) {
        return t * t * 8.0f;
    }

    public float getInterpolation(float t) {
        t *= 1.1226f;
        if (t < 0.3535f) return bounce(t);
        else if (t < 0.7408f) return bounce(t - 0.54719f) + 0.7f;
        else if (t < 0.9644f) return bounce(t - 0.8526f) + 0.9f;
```

```
        else return bounce(t - 1.0435f) + 0.95f;
    }
}
```

You can find the behavior of these interpolators described at the following URL:

http://developer.android.com/reference/android/view/animation/package-summary.html

The Java documentation for each of these classes also points out the XML tags available to control them. However, the description of what each interpolator does is hard to figure out from the documentation. The best approach is to try it out in an example and see the effect produced.

This concludes our section on layout animation. We will now move to the third section on view animation, in which we'll discuss animating a view programmatically.

View Animation

Now that you're familiar with frame-by-frame animation and layout animation, you're ready to tackle view animation—the more complex of the three animation types. View animation allows you to animate any arbitrary view by manipulating the transformation matrix that is in place for displaying the view.

Understanding View Animation

When a view is displayed on a presentation surface in Android, it goes through a transformation matrix. In graphics applications, you use transformation matrices to transform a view in some way. The process involves taking the input set of pixel coordinates and color combinations and translating them into a new set of pixel coordinates and color combinations. At the end of a transformation, you will see an altered picture in terms of size, position, orientation, or color.

You can achieve all of these transformations mathematically by taking the input set of coordinates and multiplying them in some manner using a transformation matrix to arrive at a new set of coordinates. By changing the transformation matrix, you can impact how a view will look.

A matrix that *doesn't* change the view when you multiply with it is called an *identity matrix*. You typically start with an identity matrix and apply a series of transformations involving size, position, and orientation. You then take the final matrix and use that matrix to draw the view.

Android exposes the transformation matrix for a view by allowing you to register an animation object with that view. The animation object will have a callback that lets it obtain the current matrix for a view and change it in some manner to arrive at a new view. We will go through this process now.

Let's start by planning an example for animating a view. You'll begin with an activity where you'll place a `ListView` with a few items, similar to the way you began the example in the "Layout Animation" section. You will then create a button at the top of the screen to start the ListView animation when clicked (see Figure 21–5). Both the button and the ListView appear, but nothing has been animated yet. You'll use the button to trigger the animation.

When you click the Start Animation button in this example, you want the view to start small in the middle of the screen and gradually become bigger until it consumes all the space that is allocated for it. We'll show you how to write the code to make this happen. Listing 21–14 shows the XML layout file that you can use for the activity.

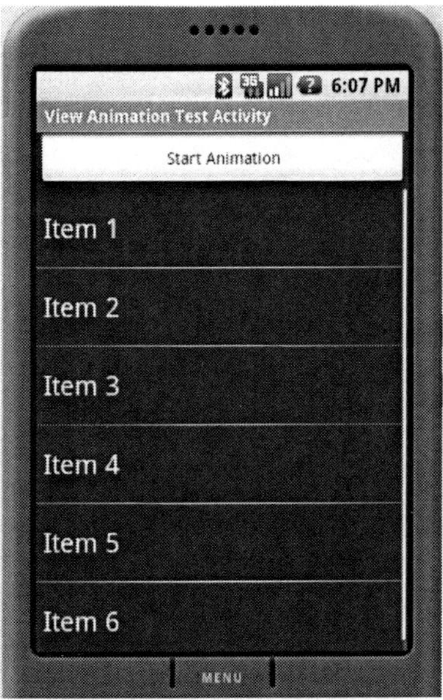

Figure 21–5. *The view-animation activity*

Listing 21–14. *XML Layout File for the View-Animation Activity*

```
<?xml version="1.0" encoding="utf-8"?>
<!-- This file is at /res/layout/list_layout.xml -->
<LinearLayout xmlns:android="http://schemas.android.com/apk/res/android"
    android:orientation="vertical"
    android:layout_width="fill_parent"
    android:layout_height="fill_parent"
    >
<Button
   android:id="@+id/btn_animate"
    android:layout_width="fill_parent"
    android:layout_height="wrap_content"
    android:text="Start Animation"
/>
<ListView
     android:id="@+id/list_view_id"
     android:persistentDrawingCache="animation|scrolling"
     android:layout_width="fill_parent"
     android:layout_height="fill_parent"
 />
</LinearLayout>
```

Notice that the file location and the file name are embedded at the top of the XML file for your reference. This layout has two parts: the first is the button named `btn_animate` to animate a view, and the second is the `ListView`, which is named `list_view_id`.

Now that you have the layout for the activity, you can create the activity to show the view and set up the Start Animation button (see Listing 21–15).

Listing 21–15. *Code for the View-Animation Activity, Before Animation*

```
public class ViewAnimationActivity extends Activity {

    @Override
    public void onCreate(Bundle savedInstanceState)
    {
        super.onCreate(savedInstanceState);
        setContentView(R.layout.list_layout);
        setupListView();
        this.setupButton();
    }
    private void setupListView()
    {
         String[] listItems = new String[] {
               "Item 1", "Item 2", "Item 3",
               "Item 4", "Item 5", "Item 6",
         };

         ArrayAdapter<String> listItemAdapter =
             new ArrayAdapter<String>(this
                     ,android.R.layout.simple_list_item_1
                     ,listItems);
         ListView lv = (ListView)this.findViewById(R.id.list_view_id);
         lv.setAdapter(listItemAdapter);
    }
    private void setupButton()
    {
       Button b = (Button)this.findViewById(R.id.btn_animate);
       b.setOnClickListener(
           new Button.OnClickListener(){
             public void onClick(View v)
             {
                //animateListView();
             }
           });
    }
}
```

The code for the view-animation activity in Listing 21–15 closely resembles the code for the layout-animation activity in Listing 21–7. We have similarly loaded the view and set up the `ListView` to contain six text items. We've set up the button in such a way that it would call `animateListView()` when clicked. But for now, comment out that part until you get this basic example running.

You can invoke this activity as soon as you register it in the `AndroidManifest.xml` file:

```
<activity android:name=".ViewAnimationActivity"
        android:label="View Animation Test Activity">
```

Once this registration is in place, you can invoke this view-animation activity from any menu item in your application by executing the following code:

```
Intent intent = new Intent(this, ViewAnimationActivity.class);
startActivity(intent);
```

When you run this program, you will see the UI as laid out in Figure 21–5.

Adding Animation

Our aim in this example is to add animation to the ListView shown in Figure 21–5. To do that, you need a class that derives from android.view.animation.Animation. You then need to override the applyTransformation method to modify the transformation matrix. Call this derived class ViewAnimation. Once you have the ViewAnimation class, you can do something like this on the ListView class:

```
ListView lv = (ListView)this.findViewById(R.id.list_view_id);
lv.startAnimation(new ViewAnimation());
```

Let us go ahead and show you the source code for ViewAnimation and discuss the kind of animation we want to accomplish (see Listing 21–16).

Listing 21–16. *Code for the ViewAnimation Class*

```
public class ViewAnimation extends Animation
{
  @Override
  public void initialize(int width, int height,
                         int parentWidth,
                         int parentHeight)
    {
        super.initialize(width, height, parentWidth, parentHeight);
        setDuration(2500);
        setFillAfter(true);
        setInterpolator(new LinearInterpolator());
    }
  @Override
  protected void
  applyTransformation(float interpolatedTime, Transformation t)
    {
        final Matrix matrix = t.getMatrix();
        matrix.setScale(interpolatedTime, interpolatedTime);
     }
}
```

The initialize method is a callback method that tells us about the dimensions of the view. This is also a place to initialize any animation parameters you might have. In this example, we have set the duration to be 2,500 milliseconds (2.5 seconds). We have also specified that we want the animation effect to remain intact after the animation completes by setting FillAfter to true. Plus, we've indicated that the interpolator is a linear interpolator, meaning that the animation changes in a gradual manner from start to finish. All of these properties come from the base android.view.animation.Animation class.

The main part of the animation occurs in the applyTransformation method. The Android framework will call this method again and again to simulate animation. Every time Android calls the method, interpolatedTime has a different value. This parameter changes from 0 to 1 depending on where you are in the 2.5-second duration that you set during initialization. When interpolatedTime is 1, you are at the end of the animation.

Our goal, then, is to change the transformation matrix that is available through the transformation object called t in the applyTransformation method. You will first get the matrix and change something about it. When the view gets painted, the new matrix will take effect. You can find the kinds of methods available on the Matrix object by looking up the API documentation for android.graphics.Matrix:

http://developer.android.com/reference/android/graphics/Matrix.html

In Listing 21–16, here is the code that changes the matrix:

```
matrix.setScale(interpolatedTime, interpolatedTime);
```

The setScale method takes two parameters: the scaling factor in the x direction and the scaling factor in the y direction. Because interpolatedTime goes between 0 and 1, you can use that value directly as the scaling factor.

So when you start the animation, the scaling factor is 0 in both the x and y directions. Halfway through the animation, this value will be 0.5 in both the x and y directions. At the end of the animation, the view will be at its full size because the scaling factor will be 1 in both the x and y directions. The end result of this animation is that the ListView starts out tiny and grows to full size.

Listing 21–17 shows the complete source code for the ViewAnimationActivity that includes the animation.

Listing 21–17. *Code for the View-Animation Activity, Including Animation*

```
public class ViewAnimationActivity extends Activity {

    @Override
    public void onCreate(Bundle savedInstanceState)
    {
        super.onCreate(savedInstanceState);
        setContentView(R.layout.list_layout);
        setupListView();
        this.setupButton();
    }
    private void setupListView()
    {
         String[] listItems = new String[] {
               "Item 1", "Item 2", "Item 3",
               "Item 4", "Item 5", "Item 6",
         };

         ArrayAdapter<String> listItemAdapter =
             new ArrayAdapter<String>(this
                     ,android.R.layout.simple_list_item_1
                     ,listItems);
         ListView lv = (ListView)this.findViewById(R.id.list_view_id);
         lv.setAdapter(listItemAdapter);
```

```
    }
    private void setupButton()
    {
       Button b = (Button)this.findViewById(R.id.btn_animate);
       b.setOnClickListener(
           new Button.OnClickListener(){
             public void onClick(View v)
             {
                animateListView();
             }
           });
    }
    private void animateListView()
    {
       ListView lv = (ListView)this.findViewById(R.id.list_view_id);
       lv.startAnimation(new ViewAnimation());
    }
}
```

When you run the code in Listing 21–17, you will notice something odd. Instead of uniformly growing larger from the middle of the screen, the `ListView` grows larger from the top-left corner. The reason is that the origin for the matrix operations is at the top-left corner. To get the desired effect, you first have to move the whole view so that the view's center matches the animation center (top-left). Then, you apply the matrix and move the view back to the previous center.

The rewritten code from Listing 21–16 for doing this is shown in Listing 21–18, with key elements highlighted.

Listing 21–18. *View Animation using `preTranslate` and `postTranslate`*

```
public class ViewAnimation extends Animation {
    float centerX, centerY;
    public ViewAnimation(){}

    @Override
    public void initialize(int width, int height, int parentWidth, int parentHeight) {
        super.initialize(width, height, parentWidth, parentHeight);
        centerX = width/2.0f;
        centerY = height/2.0f;
        setDuration(2500);
        setFillAfter(true);
        setInterpolator(new LinearInterpolator());
    }
    @Override
    protected void applyTransformation(float interpolatedTime, Transformation t) {
        final Matrix matrix = t.getMatrix();
        matrix.setScale(interpolatedTime, interpolatedTime);
        matrix.preTranslate(-centerX, -centerY);
        matrix.postTranslate(centerX, centerY);
    }
}
```

The `preTranslate` and `postTranslate` methods set up a matrix before the scale operation and after the scale operation. This is equivalent to making three matrix transformations in tandem. The following code

```
        matrix.setScale(interpolatedTime, interpolatedTime);
        matrix.preTranslate(-centerX, -centerY);
        matrix.postTranslate(centerX, centerY);
```

is equivalent to

```
move to a different center
scale it
move to the original center
```

You will see this pattern of pre and post applied again and again. You can also accomplish this result using other methods on the Matrix class, but this technique is the most common—plus, it's succinct. We will, however, cover these other methods toward the end of this section.

More important, the Matrix class allows you not only to scale a view but also to move it around through translate methods and change its orientation through rotate methods. You can experiment with these methods and see what the resulting animation looks like. In fact, the animations presented in the preceding "Layout Animation" section are all implemented internally using the methods on this Matrix class.

Using Camera to Provide Depth Perception in 2D

The graphics package in Android provides another animation-related—or more accurately, transformation-related—class called Camera. You can use this class to provide depth perception by projecting onto a 2D surface a 2D image moving in 3D space. For example, you can take our ListView and move it back from the screen by 10 pixels along the z axis and rotate it by 30 degrees around the y axis. Listing 21–19 is an example of manipulating the matrix using Camera.

Listing 21–19. *Using Camera*

```
...
public class ViewAnimation extends Animation {
    float centerX, centerY;
    Camera camera = new Camera();
    public ViewAnimation(float cx, float cy){
        centerX = cx;
        centerY = cy;
    }
    @Override
    public void initialize(int width, int height, int parentWidth, int parentHeight) {
        super.initialize(width, height, parentWidth, parentHeight);
        setDuration(2500);
        setFillAfter(true);
        setInterpolator(new LinearInterpolator());
    }
    @Override
    protected void applyTransformation(float interpolatedTime, Transformation t) {
        applyTransformationNew(interpolatedTime,t);
    }
    protected void applyTransformationNew(float interpolatedTime, Transformation t)
    {
        final Matrix matrix = t.getMatrix();
        camera.save();
```

```
        camera.translate(0.0f, 0.0f, (1300 - 1300.0f * interpolatedTime));
        camera.rotateY(360 * interpolatedTime);
        camera.getMatrix(matrix);

        matrix.preTranslate(-centerX, -centerY);
        matrix.postTranslate(centerX, centerY);
        camera.restore();
    }
}
```

This code animates the ListView by first placing the view 1,300 pixels back on the z axis and then bringing it back to the plane where the z coordinate is 0. While doing this, the code also rotates the view from 0 to 360 degrees around the y axis. Let's see how the code relates to this behavior by looking at the following method:

```
camera.translate(0.0f, 0.0f, (1300 - 1300.0f * interpolatedTime));
```

This method tells the camera object to translate the view such that when interpolatedTime is 0 (at the beginning of the animation), the z value will be 1300. As the animation progresses, the z value will get smaller and smaller until the end, when the interpolatedTime becomes 1 and the z value becomes 0.

The method camera.rotateY(360 * interpolatedTime) takes advantage of 3D rotation around an axis by the camera. At the beginning of the animation, this value will be 0. At the end of the animation, it will be 360.

The method camera.getMatrix(matrix) takes the operations performed on the Camera so far and imposes those operations on the matrix that is passed in. Once the code does that, the matrix has the translations it needs to get the end effect of having a Camera. Now the Camera is out of the picture (no pun intended) because the matrix has all the operations embedded in it. Then, you do the pre and post on the matrix to shift the center and bring it back. At the end, you set the Camera to its original state that was saved earlier.

When you plug this code into our example, you will see the ListView arriving from the center of the view in a spinning manner toward the front of the screen, as we intended when we planned our animation.

Here is some example code to invoke the AnimationView:

```
ListView lv = (ListView)this.findViewById(R.id.list_view_id);
float cx = (float)(lv.getWidth()/2.0);
float cy = (float)(lv.getHeight()/2.0);
lv.startAnimation(new ViewAnimation(cx, cy));
```

As part of our discussion about view animation, we showed you how to animate any view by extending an Animation class and then applying it to a view. In addition to letting you manipulate matrices (both directly and through a Camera class), the Animation class lets you detect various stages in an animation. We will cover this next.

Exploring the AnimationListener Class

Android uses a listener interface called AnimationListener to monitor animation events (see Listing 21–20). You can listen to these animation events by implementing the AnimationListener interface and setting that implementation against the Animation class implementation.

Listing 21–20. *An Implementation of the AnimationListener Interface*

```
public class ViewAnimationListener
implements Animation.AnimationListener {

    public ViewAnimationListener(){}

    public void onAnimationStart(Animation animation)
    {
        Log.d("Animation Example", "onAnimationStart");
    }
    public void onAnimationEnd(Animation animation)
    {
        Log.d("Animation Example", "onAnimationEnd");
    }
    public void onAnimationRepeat(Animation animation)
    {
        Log.d("Animation Example", "onAnimationRepeat");
    }
}
```

The ViewAnimationListener class just logs messages. You can update the animateListView method in the view-animation activity example (see Listing 21–17) to take the animation listener into account:

```
private void animateListView()
{
   ListView lv = (ListView)this.findViewById(R.id.list_view_id);
   ViewAnimation animation = new ViewAnimation();
   animation.setAnimationListener(new ViewAnimationListener());
   lv.startAnimation(animation);
}
```

Notes on Transformation Matrices

As you have seen in this chapter, matrices are key to transforming views and animations. We will now briefly explore some key methods of the Matrix class. These are the primary operations on a matrix:

- *matrix.reset():* Resets a matrix to an identity matrix, which causes no change to the view when applied
- *matrix.setScale():* Changes size
- *matrix.setTranslate():* Changes position to simulate movement
- *matrix.setRotate():* Changes orientation
- *matrix.setSkew():* Distorts a view

The last four methods have input parameters.

You can concatenate matrices or multiply them together to compound the effect of individual transformations. Consider the following example, where m1, m2, and m3 are identity matrices:

```
m1.setScale();
m2.setTranlate()
m3.setConcat(m1,m2)
```

Transforming a view by m1 and then transforming the resulting view with m2 is equivalent to transforming the same view by m3. Note that m3.setConcat(m1,m2) is different from m3.setConcat(m2,m1).

You have already seen the pattern used by the preTranslate and postTranslate methods to affect matrix transformation. In fact, pre and post methods are not unique to translate, and you have versions of pre and post for every one of the set transformation methods. Ultimately, a preTranslate such as m1.preTranslate(m2) is equivalent to

```
m1.setConcat(m2,m1)
```

In a similar manner, the method m1.postTranslate(m2) is equivalent to

```
m1.setConcat(m1,m2)
```

By extension, the following code

```
matrix.setScale(interpolatedTime, interpolatedTime);
matrix.preTranslate(-centerX, -centerY);
matrix.postTranslate(centerX, centerY);
```

is equivalent to

```
Matrix matrixPreTranslate = new Matrix();
matrixPreTranslate.setTranslate(-centerX, -centerY);

Matrix matrixPostTranslate = new Matrix();
matrixPostTranslate.setTranslate(cetnerX, centerY);

matrix.concat(matrixPreTranslate,matrix);
matrix.postTranslate(matrix,matrixpostTranslate);
```

Property Animations: The New Animation API

Now that you are feeling comfortable with the numerous animation options you have seen so far, we need to tell you that the animation API is entirely overhauled in 3.0 and 4.0. The new approach to animations is called *property animation*. If you have read Chapter 8 on fragments, you have seen how fragments are transitioned into their positions through animators. These animators are part of the new property animation API. We covered only a small portion of the full property animation API in that chapter.

The Property Animation API is extensive and different enough to refer to the previous animation API (prior to 3.x) as the legacy API.

NOTE: The old animation API is in the package `android.view.animation`. The new animation API is in the package `android.animation`.

Key concepts in the new animation API are

- Animators
- Value animators
- Object animators
- Animator sets
- Animator builders
- Animation listeners
- Property value holders
- Type evaluators
- View property animators
- Layout transitions
- Animators defined in XML files

We will cover each of these concepts in the rest of the chapter.

Property Animation

The new animation is called property animation because, at the core, it changes the value of a property over time. This property can be anything, such as a stand-alone integer, a float, or a specific property of an object.

NOTE: Most, if not all, of the interfaces and classes you see in the subsequent code are in the `android.animation` package.

For example, you can define a time sequence to change an `int` value from 10 to 200 over a time period of 5,000 milliseconds by using an animator class called `ValueAnimator` (see Listing 21–21).

Listing 21–21. *A Simple Value Animator*

```
//Define an animator to change an int value from 10 to 200
ValueAnimator anim = ValueAnimator.ofInt(10f, 200f);

//set the duration for the animation
anim.setDuration(5000); //5 seconds, default 300 ms

//Provide a callback to monitor the changing value
anim.addUpdateListener(
```

```
        new ValueAnimator.AnimatorUpdateListener()
        {
            public void onAnimationUpdate(ValueAnimator animation)
            {
                Int value = (Int) animation.getAnimatedValue();
                // this code gets called many many times for 5 seconds.
                // The value will range from 10 to 200
            }
        }
);
anim.start();
```

The idea is easy to grasp. A `ValueAnimator` is giving us a mechanism to do something every 10 ms (this is the default). In the corresponding callback that is called every ten milliseconds, you can choose to update a view or any other aspect to affect animation.

Listing 21–21 shows one animation callback in action. It's reproduced in Listing 21–22 for clarity.

Listing 21–22. *A Simple Value Animator*

```
public static interface
ValueAnimator.AnimatorUpdateListener
{
   abstract void  onAnimationUpdate(ValueAnimator animation);
}
```

This callback is available on the `ValueAnimator` class. The other callbacks for animators are defined in the interface tied to the base `Animator` class as shown in Listing 21–23.

Listing 21–23. *`Animator` Callback Interface*

```
public static interface
Animator.AnimatorListener
{
abstract void onAnimationStart(Animator animation);
abstract void onAnimationRepeat(Animator animation);
abstract void onAnimationCancel(Animator animation);
abstract void onAnimationEnd(Animator animation);
}
```

You can use these callbacks to further act on objects of interest during or after an animation.

Planning a Test Bed for Property Animation

Starting with the basic idea of value animators, Android provides a number of derived ways to animate any arbitrary object, especially, views. To demonstrate these mechanisms, we will take a simple text view in a linear layout and animate its `alpha` property (simulating transparency animation) and also the x and y positions (simulating movement).

We will use Figure 21–6 as an anchor to explain property animation concepts.

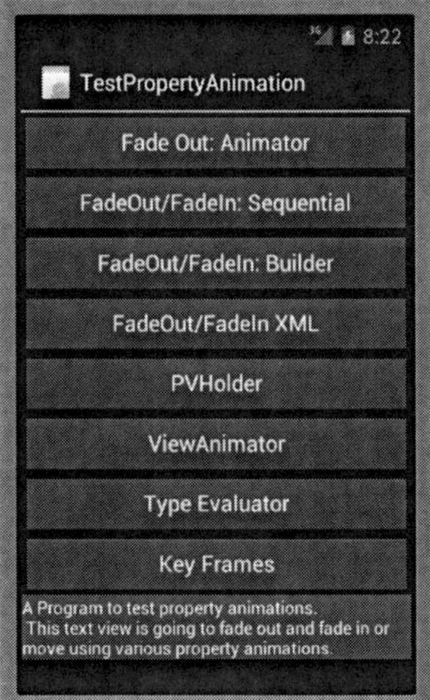

Figure 21–6. *Property animations*

Each button in Figure 21–6 uses a separate mechanism to animate the text view at the bottom of the figure. The mechanisms we will demonstrate are as follows:

- *Button 1:* Using object animators, fade out and fade in a view alternatively at the click of a button
- *Button 2:* Using an `AnimatorSet`, run a fade out animation followed by fade in animation in a sequential manner.
- *Button 3:* Use an `AnimationSetBuilder` object to tie multiple animations together in a "before," "after," or "with" relationship. Use this approach to run the same animation as button 2.
- *Button 4:* Define an XML file for button 2's sequence animation, and attach it to the text view for the same animation affect.
- *Button 5:* Using a `PropertyValuesHolder` object, animate multiple properties of the text view in the same animation. We will change the x and y values to move the text view from bottom-right to top-left.
- *Button 6:* Use `ViewPropertyAnimation` to move the text view from bottom-right to top-left (same animation as button 5).
- *Button 7:* Use a `TypeEvaluator` on custom point objects to move the text view from bottom-right to top-left (same animation as button 5).
- *Button 8:* Use key frames to affect movement and also alpha changes on the text view (same animation as button 5, but staggered).

Before we show you the code for each of the animations, Listing 21–24 shows the layout file for the activity in Figure 21–6.

Listing 21–24. *Layout File for Figure 21–6*

```
<?xml version="1.0" encoding="utf-8"?>
<LinearLayout xmlns:android="http://schemas.android.com/apk/res/android"
    ..... other stuff
    android:orientation="vertical" >
    <Button ...other attribs
        android:onClick="toggleAnimation"
        android:text="Fade Out: Animator" />
    <Button ...other attribs
        android:onClick="sequentialAnimation"
        android:text="FadeOut/FadeIn: Sequential" />
    <Button ...other attribs
        android:onClick="testAnimationBuilder"
        android:text="FadeOut/FadeIn: Builder" />
    <Button ...other attribs
        android:onClick="sequentialAnimationXML"
        android:text="FadeOut/FadeIn XML" />
    <Button ...other attribs
        android:onClick="testPropertiesHolder"
        android:text="PVHolder" />
    <Button ...other attribs
        android:onClick="testViewAnimator"
        android:text="ViewAnimator" />
    <Button ...other attribs
        android:onClick="testTypeEvaluator"
        android:text="Type Evaluator" />
    <Button ...other attribs
        android:onClick="testKeyFrames"
        android:text="Key Frames" />
    <LinearLayout
        android:layout_width="fill_parent"
        android:layout_height="wrap_content">
        <TextView
            android:id="@+id/tv_id"
            ...other attribs
            android:text="text you see there in the figure" />
    </LinearLayout>
</LinearLayout>
```

We have shortened the layout file to highlight only the key elements; you can download the project if you would like to build this project (The URL is at the end of this chapter.) The layout in Listing 21–24 is pretty straightforward: a series of buttons followed by a text view. The text view is embedded in a second linear layout so that the text view has its own boundaries that are different from the buttons'. This allows us to animate the text view with in the boundaries of its parent linear layout.

You can also see that each button is calling a specific method to affect the animation. By exploring each method, we will cover all aspects of property animations. Let's start with the first button.

Basic View Animation with Object Animators

As you can see from the layout file in Listing 21–24, the first button invokes the toggleAnimation(View) method. This method is shown in Listing 21–25.

Listing 21–25. *Basic View Animation wit h Object Animators*

```
public void toggleAnimation(View btnView)
{
    //The button we have pressed
    Button tButton = (Button)btnView;

    //m_tv: is the pointer to the text view
    if (m_tv.getAlpha() != 0)
    {
        //Animate the alpha from current value to 0
        //this will make it invisible
        ObjectAnimator fadeOut =
            ObjectAnimator.ofFloat(m_tv, "alpha", 0f);

        fadeOut.setDuration(5000);
        fadeOut.start();

        tButton.setText("Fade In");
    }
    else
    {
        //Animate the alpha from current value to 1
        //this will make it visible
        ObjectAnimator fadeIn =
            ObjectAnimator.ofFloat(m_tv, "alpha", 1f);

        fadeIn.setDuration(5000);
        fadeIn.start();

        tButton.setText("Fade out");
    }
}
```

This code in Listing 21–25 first examines the alpha value of the text view. If this value is greater than 0, then the code assumes that the text view is visible and runs a fade-out animation. At the end of the fade-out animation, the text view will be invisible.

If the alpha value of the text view is 0, then the code assumes the text view is invisible and runs a fade-in animation to make the text view visible again.

Let's examine the use of ObjectAnimator by looking at some key lines from Listing 21–25. This extracted ObjectAnimator code is presented in Listing 21–26.

Listing 21–26. *Object Animator in Action*

```
ObjectAnimator fadeOut =
       ObjectAnimator.ofFloat(m_tv, "alpha", 0f);
fadeOut.setDuration(5000);
fadeOut.start();
```

The object animator's static method offFloat() takes three arguments. The first argument is an object. In our case, this target object is the text view: m_tv. (Note that in a real example, you will need to save this reference to the text view m_tv as a local variable in your activity.)

The second argument is the property name as a string. In our case, this property name is alpha. The convention is that the target object needs to have a public method to match this name. For example, for a property named alpha, the corresponding view object needs to have the following set method:

```
view.setAlpha(float f);
```

The third argument is the value of the property at the end of the animation. If you specify a fourth argument, then the third argument is the starting value and the fourth is the target value. You can pass more arguments as long as they are all floats. The animation will use those values as intermediate values in the animation process.

If you specify just the "to" value, then the "from" value is taken from the current value by using

```
view.getAlpha();
```

The methods setDuration() and start() in Listing 21–26 are self explanatory. The duration is specified in milliseconds.

When you play this animation, the text view will gradually disappear first. The code in Listing 21–25 then renames the button to "Fade in". Now if you click the button again, the second animation is run, and the text view will appear gradually over five seconds.

Sequential Animation with AnimatorSet

Button 2 in Figure 21–6 runs two animations one after the other: a fade out followed by a fade in. We could use animation listener callbacks to wait for the first animation to finish and then start the second animation. Instead, there is an automated way to run animations in tandem through the class AnimatorSet to get the same affect.

Button 2 proves this point by invoking the method in Listing 21–27.

Listing 21–27. ***Sequential Animation through an AnimatorSet***

```
public void sequentialAnimation(View bView)
{
    ObjectAnimator fadeOut =
        ObjectAnimator.ofFloat(m_tv, "alpha", 0f);

    ObjectAnimator fadeIn =
        ObjectAnimator.ofFloat(m_tv, "alpha", 1f);

    AnimatorSet as = new AnimatorSet();
    as.playSequentially(fadeOut,fadeIn);

    as.setDuration(5000); //5 secs
    as.start();
}
```

In Listing 21–27, we have created two animators: a fade-out animator and a fade-in animator. Then we create an animator set and tell it to play both animations one after the other.

You can also choose to play animations together using an animator set by calling the method `playTogether()`. Both of these methods, `playSequentially()` and `playTogether()`, can take a variable number of `Animator` objects.

> **NOTE:** Many methods in the new animation API take a variable number of arguments. Keep an eye on the SDK reference to see which ones.

When you play this animation, the text view will gradually disappear and then reappear, much like the animation you saw earlier.

Setting Animation Relationships with AnimationSetBuilder

You have seen that the class `AnimatorSet` allows you to play animations sequentially or in parallel. This same class provides a bit more elaborate way to link animations through a utility class called `AnimatorSetBuilder`.

The code in Listing 21–28 demonstrates this

Listing 21–28. *Using an `AnimatorSetBuilder`*

```
public void testAnimationBuilder(View v)
{
    ObjectAnimator fadeOut =
        ObjectAnimator.ofFloat(m_tv, "alpha", 0f);
    ObjectAnimator fadeIn =
        ObjectAnimator.ofFloat(m_tv, "alpha", 1f);

    AnimatorSet as = new AnimatorSet();

    //play() returns the nested class: AnimatorSetBuilder
    as.play(fadeOut).before(fadeIn);

    as.setDuration(5000); //5 secs
    as.start();
}
```

The `play` method on an `AnimatorSet` returns a class called `AnimatorSetBuilder`. This is purely a utility class. The methods on this class are

- `after(animator)`
- `before(animator)`
- `with(animator)`

This class is initialized with the first animator you supply through the `play` method. Every other call on this object is with respect to this original animator.

So when we say

```
AnimatorSetBuilder builder = someSet.play(main_animator);
builder.before(animator1)
```

then animator1 will play before main_animator. But when we say

```
builder.after(animator2)
```

the animation of animator2 will play after main_animator.

The method with(animator) means to play them together.

The key point is that the relationship established via before(), after(), and with() is not chained, but only tied to the original animator that was obtained from play() method. Also, the animation start() method is not on the builder object but on the original animator set.

Button 3 invokes this function in Listing 21–28. When you play this animation, the text view will gradually disappear and then reappear, much as in the previous animation.

Using XML to Load Animators

It is expected that the Android SDK allows animators to be described in XML resource files. The Android SDK has a new resource type called R.animator to distinguish animator resources files. Although they don't have to be, these XML files are typically stored in the /res/animator subdirectory.

Listing 21–29 is an example of an animator set defined in an XML file.

Listing 21–29. *An Animator XML resource file*

```
<?xml version="1.0" encoding="utf-8" ?>
<set xmlns:android="http://schemas.android.com/apk/res/android"
    android:ordering="sequentially">
<objectAnimator
    android:interpolator="@android:interpolator/accelerate_cubic"
    android:valueFrom="1"
    android:valueTo="0"
    android:valueType="floatType"
    android:propertyName="alpha"
    android:duration="5000" />
<objectAnimator
    android:interpolator="@android:interpolator/accelerate_cubic"
    android:valueFrom="0"
    android:valueTo="1"
    android:valueType="floatType"
    android:propertyName="alpha"
    android:duration="5000" />
</set>
```

You will naturally wonder what XML nodes are available for you to define these animations. As of 4.0 the allowed XML tags are as follows:

- *animator:* Binds to ValueAnimator
- *objectAnimator:* Binds to ObjectAnimator
- *animatorSet:* Binds to AnimatorSet

You can see a basic discussion of these tags at the following Android SDK URL:

http://developer.android.com/guide/topics/graphics/prop-animation.html#declaring-xml

The complete XML reference for the animation tags can be found at the following URL:

http://developer.android.com/guide/topics/resources/animation-resource.html#Property

Once you have this XML file, you can play this animation using the method shown in Listing 21–30.

Listing 21–30. *Loading an Animator XML Resource File*

```
public void sequentialAnimationXML(View bView)
{
    AnimatorSet set = (AnimatorSet)
        AnimatorInflater.loadAnimator(this, R.animator.fadein);
    set.setTarget(m_tv);
    set.start();
}
```

Notice how it is necessary to load the animation XML file first followed by explicitly setting the object to animate. In our case, the object to animate is the text view represented by `m_tv`.

The method is Listing 21–30 is called by button 4 in Figure 21–6. When this animation runs, the text view will fade out first and then reappear by fading in, just as in the previous alpha animations.

Using PropertyValuesHolder

So far, we have seen how to animate a single value in a single animation. The class `PropertyValueHolder` will allow us to animate multiple values during the same animation cycle.

The code in Listing 21–31 demonstrates the use of the `PropertyValuesHolder` class.

Listing 21–31. *Using the `PropertyValueHolder` class*

```
public void testPropertiesHolder(View v)
{
    //Get the current coordinates of the text view
    //this will allow us to know starting and ending
    //positions to animate
    float h = m_tv.getHeight();
    float w = m_tv.getWidth();
    float x = m_tv.getX();
    float y = m_tv.getY();

    //Set the view to the bottom right
    //as a starting point
    m_tv.setX(w);
    m_tv.setY(h);

    //from the right bottom animate "x" to its
    //original position which is top left
    PropertyValuesHolder pvhX =
```

```
            PropertyValuesHolder.ofFloat("x", x);

        //from the right bottom animate "y" to its
        //original position which is left top
        PropertyValuesHolder pvhY =
            PropertyValuesHolder.ofFloat("y", y);

        //when you dont specify the from position
        //the animation will take the current position
        //as the from position.

        //Tell the object animator to consider both
        //"x" and "y" properties to animate to their respective
        //target values.
        ObjectAnimator oa
        = ObjectAnimator.ofPropertyValuesHolder(m_tv, pvhX, pvhY);

        //set the duration
        oa.setDuration(5000); //5 secs

        //here is a way to set an interpolator
        //on any animator
        oa.setInterpolator(
                new  AccelerateDecelerateInterpolator());
        oa.start();
    }
```

Key portions of Listing 21–31 are highlighted. As you can see, a `PropertyValuesHolder` class allows you to hold a property name and its target value. Then you can define many of these `PropertyValueHolders` with their own property to animate.

You can supply this set of `PropertyValueHolders` to the object animator. The object animator will then set these properties to their respective values on the target object. With each refresh of the animation, all the values from each `PropertyValueHolder` will be applied all at once. This is more efficient than applying multiple animations in parallel.

Button 5 in Figure 21–6 invokes the function in Listing 21–31. When this animation runs, the text view will emerge from bottom-right and migrate toward the top-left in five seconds.

View Properties Animation

If you are primarily animating views, the Android SDK has an optimized approach to animate the views' various properties. This is done through a class called `ViewPropertyAnimator`.

The code in Listing 21–32 uses this class to move the text view from bottom-right to top-left.

Listing 21–32. *Using a `ViewPropertyAnimator`*

```
public void testViewAnimator(View v)
{
    //Remember current boundaries
    float h = m_tv.getHeight();
    float w = m_tv.getWidth();
    float x = m_tv.getX();
    float y = m_tv.getY();

    //Position the view at bottom right
    m_tv.setX(w);
    m_tv.setY(h);

    //Get a ViewPropertyAnimator from the text view
    ViewPropertyAnimator vpa = m_tv.animate();

    //Just set as many target values you want to set
    vpa.x(x);
    vpa.y(y);

    //Set duration and interpolators
    vpa.setDuration(5000); //2 secs
    vpa.setInterpolator(
            new  AccelerateDecelerateInterpolator());

    //The animation automatically starts when the UI thread
    //gets to it.
    //No need to explicitly call the start method.
    //vpa.start();
}
```

As per the documentation, `ViewPropertyAnimator` is expected to be efficient by applying multiple value changes in a single animation cycle. From Listing 21–32, you can see that the steps to use `ViewPropertyAnimator` are as follows:

1. Get a `ViewPropertyAnimator` by calling the `animate()` method on a view.
2. Use the `ViewPropertyAnimator` object to set various properties of that view, such as `x`, `y`, `scale`, `alpha`, and so on. You may want to set the initial properties for the view explicitly, as in Listing 21–32.
3. Let the UI thread proceed by returning from the function. The animation will automatically start.

This animation is invoked by button 6. When this animation runs, the text view will migrate from bottom right to top left.

Type Evaluators

As we have seen, an object animator directly sets a particular value on a target object with each animation cycle. These values so far have been single point values such as `floats`, `ints`, and so on. What happens if your target object has a property that is an object itself? This is where type evaluators come into play.

TypeEvaluator is a helper object that knows how to proportionately set a composite value such as a two-dimensional or three-dimensional point. In this scenario, an ObjectAnimator will take the starting composite value (which itself is a value object), an ending composite value, and a TypeEvaluator helper object. When the animation cycle comes around, the ObjectAnimator will call the TypeEvaluator to supply the new composite value. This composite value is then set on the target object.

Let's first see how an ObjectAnimator uses a TypeEvaluator by examining the code in Listing 21–33.

Listing 21–33. *Using a TypeEvaluator*

```
public void testTypeEvaluator(View v)
{
    float h = m_tv.getHeight();
    float w = m_tv.getWidth();
    float x = m_tv.getX();
    float y = m_tv.getY();

    PointF startingPoint = new PointF(w,h);
    PointF endingPoint = new PointF(x,y);

    //m_atv: You will need this code in your activity
    //earlier as a local variable:
    //MyAnimatableView m_atv = new MyAnimatableView(m_tv);


    ObjectAnimator tea =
        ObjectAnimator.ofObject(m_atv
            ,"point"
            ,new MyPointEvaluator()
            ,startingPoint
            ,endingPoint);

    tea.setDuration(5000);
    tea.start();
}
```

Notice in Listing 21–33 that the ObjectAnimator is using the method ofObject() as opposed to ofFloat() or ofInt(). Also notice that the starting value and ending value for the animation are composite values represented by the class PointF. The goal of the object animator is now to come up with an intermediate value for PointF and then pass it to the method setPoint(PointF) on the custom class MyAnimatableView. The class MyAnimatableView can accordingly set the respective individual properties on the contained text view.

Now that you have the idea, let's see how to use a TypeEvaluator to come up with an intermediate composite value for a class of type PointF. This is shown in Listing 21–34.

Listing 21–34. *Coding a* `TypeEvaluator`

```
public class MyPointEvaluator
implements TypeEvaluator<PointF>
{
    public PointF evaluate(float fraction,
            PointF startValue,
            PointF endValue)
    {
        PointF startPoint = (PointF) startValue;
        PointF endPoint = (PointF) endValue;
        return new PointF(
            startPoint.x + fraction * (endPoint.x - startPoint.x),
            startPoint.y + fraction * (endPoint.y - startPoint.y));
    }
}
```

From Listing 21–34 you can see that you need to inherit from a `TypeEvaluator` class and override the `evaluate()` method. In this method, you will be passed the fraction of the animation's total progress. You can use that fraction to adjust your intermediate composite value and return it as a typed value.

Listing 21–35 shows how we encapsulate a regular view for which we know how to change x and y. The encapsulation will allow the animation to call once for both x and y through the `PointF` abstraction. We will provide a `setPoint(PointF)` method and then, inside that method, parse out x and y and set them on the view. Because we set them once every animation cycle, we end up animating the contained view.

Listing 21–35. *Animating a View Through a* `TypeEvaluator`

```
public class MyAnimatableView
{
    PointF curPoint = null;
    View m_v = null;
    public MyAnimatableView(View v)
    {
        curPoint = new PointF(v.getX(),v.getY());
        m_v = v;
    }

    public PointF getCurPointF()
    {
        return curPoint;
    }
    public void setPoint(PointF p)
    {
        curPoint = p;
        m_v.setX(p.x);
        m_v.setY(p.y);
    }
}
```

This animation in Listing 21–33 is invoked by button 7. When this animation runs, the view will migrate from bottom right to top left.

Key Frames

Key frames are useful places to put key time markers (significant instances in time) during the animation time cycle. Listing 21–36 demonstrates key-frame animation.

Listing 21–36. *Animating a View Using Key Frames*

```
public void testKeyFrames(View v)
{
    float h = m_tv.getHeight();
    float w = m_tv.getWidth();
    float x = m_tv.getX();
    float y = m_tv.getY();

    //Start frame : 0.2
    //alpha: 0.8
    Keyframe kf0 = Keyframe.ofFloat(0.2f, 0.8f);

    //Middle frame: 0.5
    //alpha: 0.2
    Keyframe kf1 = Keyframe.ofFloat(.5f, 0.2f);

    //end frame: 0.8
    //alpha: 0.8
    Keyframe kf2 = Keyframe.ofFloat(0.8f, 0.8f);

    PropertyValuesHolder pvhAlpha =
        PropertyValuesHolder.ofKeyframe("alpha", kf0, kf1, kf2);

    PropertyValuesHolder pvhX =
        PropertyValuesHolder.ofFloat("x", w, x);

    //end frame
    ObjectAnimator anim =
        ObjectAnimator.ofPropertyValuesHolder(m_tv, pvhAlpha,pvhX);
    anim.setDuration(5000);
    anim.start();
}
```

A key frame specifies a particular value for a property at a given moment in time. The time is between 0 (beginning of animation) and 1 (end of animation). Once you gather these key-frame values, you set them against a particular property such as `alpha`, `x`, or `y`. This association of key frames to their respective properties is done through the `PropertyValuesHolder` class. You then tell the `ObjectAnimator` to animate the resulting `PropertyValuesHolder` objects.

The animation in Listing 21–36 is invoked by button 8. When this animation runs, you will see the text view start its journey from bottom right to top left. When 20 percent of the time has passed, `alpha` will change to 80 percent. The `alpha` value will reach 20 percent at half way and change back to 80 percent at the eightieth percentile of the animation time.

Layout Transitions

The Property Animation API also provides layout-based animations through the `LayoutTransition` class. This class is well documented as part of the standard API Java doc at the following URL.

`http://developer.android.com/reference/android/animation/LayoutTransition.html`

We will summarize here only the key points of layout transitions. To enable layout transitions on a view group (most layouts are view groups), you will need to use the code shown in Listing 21–37.

Listing 21–37. *Setting a Layout Transition*

```
viewgroup.setLayoutTransition(
  new LayoutTransition()
);
```

A layout transition object comes with its own set of four default animators: one for each of the four types of layout transition. The four types of layout transitions are as follows:

- Add a view (appearing)
- Change appearing (rest of the items in the layout)
- Remove a view (disappearing)
- Change disappearing (rest of the items in the layout)

Constants are defined for each of these layout transition types. Let's examine a couple of useful methods on this class by looking at Listing 21–38.

Listing 21–38. *Layout Transition Methods*

```
//Here is how you get a new layout transition
LayoutTransition lt
= new LayoutTransition();

//You can set this layout transition on a layout
someLayout.setLayoutTransition(lt);

//obtain a default animator if you
//need to remember
Animator defaultAppearAnimator
= lt.getAnimator(APPEARING);

//create a new animator
ObjectAnimator someNewObjectAnimator;

//set it as your custom animator for appearing transition
lt.setAnimator(APPEARING, someNewObjectAnimator);
```

Because the animator you supply to a layout transition applies to each view, the animators are internally cloned before being applied to each view.

There are some limitations to layout transition animation, as documented in the API Java doc mentioned earlier. A key limitation is that if views are in motion, clicks on those views may not be predictable.

Resources

Here are some useful links to further strengthen your understanding of this chapter:

- `www.androidbook.com/item/3901`: Author research notes on Android property animations.
- `http://android-developers.blogspot.com/2011/02/animation-in-honeycomb.html`: A key blog resource from Chet Hasse that will help you understand property animations in 3.0 and beyond.
- `http://android-developers.blogspot.com/2011/05/introducing-viewpropertyanimator.html`: A key blog resource from Chet Hasse that will help you understand view property animations.
- `http://developer.android.com/guide/topics/graphics/prop-animation.html`: Primary documentation on property animations from the Android SDK.
- `http://developer.android.com/guide/topics/graphics/animation.html`: Android documentation links to all animation types, including property animations and old-style animations.
- `http://developer.android.com/reference/android/view/animation/package-summary.html`: The Java doc API for the older animation package `android.view.animation`.
- `http://developer.android.com/guide/topics/resources/animation-resource.html`: XML tags for various animation types.
- `www.androidbook.com/item/3550`: Our cumulative research notes on the earlier animation API defined in `android.view.animation`.
- `www.androidbook.com/proandroid4/projects`: Downloadable test projects for this chapter. The names of the zip files are `ProAndroid4_ch21_SampleFrameAnimation.zip`, `ProAndroid4_ch21_SampleLayoutAnimation.zip`, `ProAndroid4_ch21_SampleViewAnimation.zip`, and `ProAndroid4_ch21_SamplePropertyAnimation.zip`.

Summary

In this chapter, we covered the following topics:

- A fun way to enhance UI elements by extending them with animation capabilities

- All the major types of animation supported by Android: frame-by-frame animation, layout animation, and view animation
- Animation concepts such as interpolators and transformation matrices
- Depth perception for 2D views through the `Camera` object
- Most aspects of the new Property Animation API

Now that you have this background, we encourage you to go through the API samples that come with the Android SDK to examine the sample XML definitions for a variety of animations.

Interview Questions

The following interview questions should further consolidate what we have covered in this chapter:

1. How does frame-by-frame animation differ from tweening animation?
2. What Java class encapsulates frame animation in Android?
3. How do you initialize an `AnimationDrawable` in an XML file?
4. What methods are available on `AnimationDrawable`?
5. What is layout animation?
6. How do you define an animation in an XML file?
7. What is an interpolator?
8. How do you specify an interpolator in an animation?
9. What are the four types of tweening animation?
10. What is a list layout controller animation XML file?
11. How is the `android:layoutAnimation` tag used for a `ListView`?
12. How many built-in interpolators are available?
13. What kinds of files can be found in the `anim` subdirectory?
14. How do you animate any view?
15. How do you use the `ViewAnimation` class?
16. How do you manipulate a transformation matrix?
17. How do you provide depth perception through animation for a 2D view?
18. How do you use a `Camera` object to affect the translation matrix?
19. How do you use the `AnimationListener` class?

20. Will the operations on a matrix change the matrix or return the changed matrix?
21. How do you use `pre` and `post` translations?
22. How do the new property animations differ from the old API?
23. How do you use `ObjectAnimator`?
24. How do you use `AnimatorSet`?
25. What is the `R.animator` resource type?
26. How do you load animators from XML files?
27. How do you use `PropertyValuesHolder`?
28. How do you use `TypeEvaluator`?
29. How do you use `ViewPropertyAnimator`?
30. How do you use key frames in property animation?
31. How do you use the `LayoutTransition` class?
32. In which Android Java package is the set of new animation classes defined?
33. Where can you find out what XML tags are allowed in the animator XML files?

Chapter 22

Exploring Maps and Location-based Services

In this chapter, we are going to talk about maps and location-based services. Location-based services form one of the more exciting pieces of the Android SDK. This portion of the SDK provides APIs to let application developers display and manipulate maps, obtain real-time device-location information, and take advantage of other exciting features.

The location-based services facility in Android sits on two pillars: the mapping and location-based APIs. Each of these APIs is isolated with respect to its own package. For example, the mapping package is `com.google.android.maps`, and the location package is `android.location`. The mapping APIs in Android provide facilities for you to display a map and manipulate it. For example, you can zoom and pan; you can change the map mode (from satellite view to traffic view, for example); you can add custom data to the map, and so on. The other end of the spectrum is Global Positioning System (GPS) data and real-time location data, both of which are handled by the location package.

These APIs often reach across the Internet to invoke services from Google servers. Therefore, you will usually need to have Internet connectivity for these to work. In addition, Google has Terms of Service that you must agree to before you can develop applications with these Android Maps API services. Read the terms carefully; Google places some restrictions on what you can do with the service data. For example, you can use location information for users' personal use, but certain commercial uses are restricted, as are applications involving automated control of vehicles. The terms will be presented to you when you sign up for a Maps API key.

In this chapter, we'll go through each of these packages. We'll start with the mapping APIs and show you how to use maps with your applications. As you'll see, mapping in Android boils down to using the `MapView` UI control and the `MapActivity` class in addition to the mapping APIs, which integrate with Google Maps. We will also show you how to place custom data onto the maps that you display and how to show the current location of the device on a map. After talking about maps, we'll delve into location-based services, which extend the mapping concepts. We will show you how to use the Android

Geocoder class and the LocationManager service. We will also touch on threading issues that surface when you use these APIs.

> **NOTE:** As of this writing, there is still no support for a MapFragment. Although there is a known workaround for this issue, we won't be covering it here. We hope that by the time you read this, Google will have released an update so you can do maps the right way.

Understanding the Mapping Package

As we mentioned, the mapping APIs are one of the components of Android's location-based services. The mapping package contains everything you'll need to display a map on the screen, handle user interaction with the map (such as zooming), display custom data on top of the map, and so on.

You may have noticed that your Android SDK Manager shows Google API packages in addition to the Android SDK platforms. The Google API packages contain the mapping API jar file, so to build any applications with maps requires one of these. You will need to install one of these packages and then specify it as your Android build target when creating a new application.

The first step to working with this package is to display a map. To do that, you'll use the MapView view class. Using this class, however, requires some preparation work. Specifically, before you can use the MapView, you'll need to get a Maps API key from Google. The *Maps API key* enables Android to interact with Google Maps services to obtain map data. The next section explains how to obtain a Maps API key.

Obtaining a Maps API Key from Google

The first thing to understand about the Maps API key is that you'll actually need two keys: one for development with the emulator and another for production (on devices). The reason for this is that the certificate used to obtain the Maps API key will differ between development and production (as we discussed in Chapter 14).

For example, during development, the ADT plug-in generates the .apk file and deploys it to the emulator. Because the .apk file must be signed with a certificate, the ADT plug-in uses the debug certificate during development. For production deployment, you'll likely use a self-signed certificate to sign your .apk file. The good news is that you can obtain one Maps API key for development and another for production, and it's then easy to swap the keys before exporting the production build.

To obtain a Maps API key, you need the certificate that you'll use to sign your application (in the case of the emulator, the debug certificate). You'll get the MD5 fingerprint of your certificate, and then you'll enter it on Google's web site to generate an associated Maps API key.

First, you must locate your debug certificate, which is generated and maintained by Eclipse. You can find the exact location using the Eclipse IDE. From Eclipse's Preferences menu, go to Android ➤ Build. The debug certificate's location will be displayed in the Default Debug Keystore field, as shown in Figure 22–1. (See Chapter 2 if you have trouble finding the Preferences menu.)

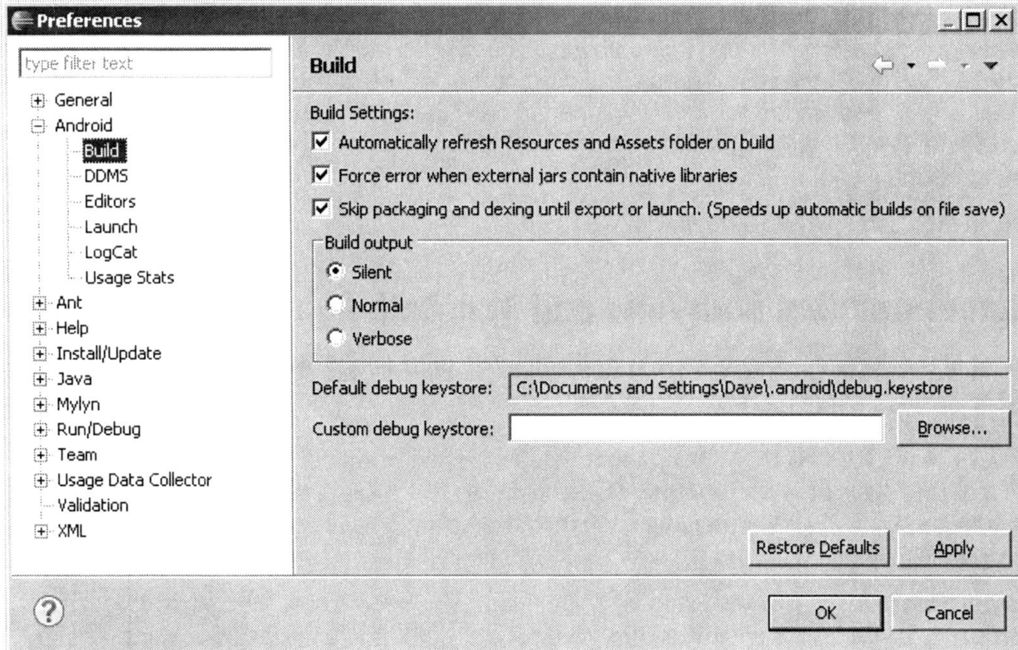

Figure 22–1. *The debug certificate's location*

To extract the MD5 fingerprint, you can run the keytool with the -list option, as shown here:

```
keytool -list -alias androiddebugkey -keystore
"FULL PATH OF YOUR debug.keystore FILE" -storepass android -keypass android
```

Note that the alias you want from the debug store is androiddebugkey. Similarly, the keystore password is android, and the private key password is also android. When you run this command, the keytool provides the fingerprint (see Figure 22–2).

```
C:\WINDOWS\system32\cmd.exe
C:\Documents and Settings\Dave\.android>keytool -list -alias androiddebugkey -keystore debug.keystore
Enter keystore password:
androiddebugkey, Sep 6, 2009, PrivateKeyEntry,
Certificate fingerprint (MD5):

C:\Documents and Settings\Dave\.android>
```

Figure 22–2. *The keytool output for the* `list` *option (actual fingerprint smudged on purpose)*

Now, paste your certificate's MD5 fingerprint in the appropriate field on this Google site:

http://code.google.com/android/maps-api-signup.html

Read through the Terms of Service. If you agree to the terms, click the Generate API Key button to get a corresponding Maps API key from the Google Maps service. The Maps API key is active immediately, so you can start using it to obtain map data from Google. Note that you will need a Google account to obtain a Maps API key; when you try to generate the Maps API key, you will be prompted to log in to your Google account.

Remember from Chapter 14 that when your debug certificate expires, so too will your development Maps API key. If you change your debug certificate, you'll need to repeat these steps, with the new debug certificate, to get a new development Maps API key. This is good motivation for creating a debug certificate that lasts longer than the default one year. See Chapter 14 for more details on creating a debug certificate that lasts a long time.

Now, let's start playing with maps.

Understanding MapView and MapActivity

A lot of the mapping technology in Android relies on the `MapView` UI control and an extension of `android.app.Activity` called `MapActivity`. The `MapView` and `MapActivity` classes take care of the heavy lifting when it comes to displaying and manipulating a map in Android. One of the things that you'll have to remember about these two classes is that they have to work together. Specifically, to use a `MapView`, you need to instantiate it within a `MapActivity`. In addition, when instantiating a `MapView`, you need to supply the Maps API key.

If you instantiate a `MapView` using an XML layout, you need to set the `android:apiKey` property. If you create a `MapView` programmatically, you have to pass the Maps API key to the `MapView` constructor. Finally, because the underlying data for the map comes from Google Maps, your application will need permission to access the Internet. This means you need at least the following permission request in your `AndroidManifest.xml` file:

```
<uses-permission android:name="android.permission.INTERNET" />
```

Listing 22–1 shows in bold the entries required in `AndroidManifest.xml` to make a map application work.

Listing 22–1. *Tags Needed in `AndroidManifest.xml` for a Map Application*

```
<?xml version="1.0" encoding="utf-8"?>
<manifest xmlns:android="http://schemas.android.com/apk/res/android"
      package="com.androidbook"
      android:versionCode="1"
      android:versionName="1.0">
    <application android:icon="@drawable/icon"
                 android:label="@string/app_name">
        <uses-library android:name="com.google.android.maps" />
        <activity android:name=".MapViewDemoActivity"
                  android:label="@string/app_name">
          <intent-filter>
             <action android:name="android.intent.action.MAIN" />
             <category android:name="android.intent.category.LAUNCHER" />
          </intent-filter>
```

```
        </activity>
    </application>
    <uses-permission android:name="android.permission.INTERNET"/>
    <uses-sdk android:minSdkVersion="4" />
</manifest>
```

There's another modification you need to make to the AndroidManifest.xml file. The definition of your map application needs to reference a mapping library (this line was also included in Listing 22–1). With the prerequisites out of the way, have a look at Figure 22–3.

Figure 22–3. *A MapView control*

Figure 22–3 shows an application that displays a map. The application also demonstrates how you can zoom in, zoom out, and change the map's view mode. The XML layout is shown in Listing 22–2.

NOTE: We will give you a URL at the end of the chapter which you can use to download projects from this chapter. This will allow you to import these projects into Eclipse directly.

Listing 22–2. ***XML Layout of the* `MapView` *Demonstration***

```
<?xml version="1.0" encoding="utf-8"?>
<!-- This file is /res/layout/mapview.xml -->
<LinearLayout xmlns:android="http://schemas.android.com/apk/res/android"
    android:orientation="vertical" android:layout_width="fill_parent"
    android:layout_height="fill_parent">

    <LinearLayout
        xmlns:android="http://schemas.android.com/apk/res/android"
        android:orientation="horizontal"
        android:layout_width="fill_parent"
        android:layout_height="wrap_content">

        <Button android:id="@+id/zoomin"
            android:layout_width="wrap_content"
            android:layout_height="wrap_content" android:text="+"
            android:onClick="myClickHandler" android:padding="12px" />

        <Button android:id="@+id/zoomout"
            android:layout_width="wrap_content"
            android:layout_height="wrap_content" android:text="-"
            android:onClick="myClickHandler" android:padding="12px" />

        <Button android:id="@+id/sat"
            android:layout_width="wrap_content"
            android:layout_height="wrap_content" android:text="Satellite"
            android:onClick="myClickHandler" android:padding="8px" />

        <Button android:id="@+id/traffic"
            android:layout_width="wrap_content"
            android:layout_height="wrap_content" android:text="Traffic"
            android:onClick="myClickHandler" android:padding="8px" />

        <Button android:id="@+id/normal"
            android:layout_width="wrap_content"
            android:layout_height="wrap_content" android:text="Normal"
            android:onClick="myClickHandler" android:padding="8px" />

    </LinearLayout>

    <com.google.android.maps.MapView
        android:id="@+id/mapview" android:layout_width="fill_parent"
        android:layout_height="wrap_content" android:clickable="true"
        android:apiKey="YOUR MAPS API KEY GOES HERE" />

</LinearLayout>
```

As shown in Listing 22–2, a parent `LinearLayout` contains a child `LinearLayout` and a `MapView`. The child `LinearLayout` contains the buttons shown at the top of Figure 22–3.

Also note that you need to update the `MapView` control's `android:apiKey` value with the value of your own Maps API key.

The code for our sample mapping application is shown in Listing 22–3.

Listing 22–3. *The `MapActivity` Extension Class That Loads the XML Layout*

```
public class MapViewDemoActivity extends MapActivity
{
    private MapView mapView;

    @Override
    protected void onCreate(Bundle savedInstanceState) {
        super.onCreate(savedInstanceState);
        setContentView(R.layout.mapview);

        mapView = (MapView)findViewById(R.id.mapview);
    }

    public void myClickHandler(View target) {
        switch(target.getId()) {
        case R.id.zoomin:
            mapView.getController().zoomIn();
            break;
        case R.id.zoomout:
            mapView.getController().zoomOut();
            break;
        case R.id.sat:
            mapView.setSatellite(true);
            break;
        case R.id.traffic:
            mapView.setTraffic(true);
            break;
        case R.id.normal:
            mapView.setSatellite(false);

            mapView.setTraffic(false);
            break;
        }
        // The following line should not be required but it is,
        // up through Froyo (Android 2.2)
        mapView.postInvalidateDelayed(2000);
    }

    @Override
    protected boolean isLocationDisplayed() {
        return false;
    }

    @Override
    protected boolean isRouteDisplayed() {
        return false;
    }
}
```

As shown in Listing 22–3, displaying the `MapView` using `onCreate()` is no different from displaying any other control. That is, you set the content view of the UI to a layout file that contains the MapView, and that takes care of it. Surprisingly, supporting zoom features is also fairly easy. To zoom in or out, you use the MapController class of the MapView. Do this by calling mapView.getController() and then calling the approproiate zoomIn() or zoomOut() method. Zooming this way produces a one-level zoom; users need to repeat the action to increase the amount of magnification or reduction.

You'll also find it straightforward to offer the ability to change view modes. The MapView supports modes:

- Map is the default mode.
- Satellite mode shows aerial photographs of the map, so you can see the actual tops of buildings, trees, roads, and so on.
- Traffic mode shows traffic information on the map with colored lines to represent traffic that is moving well as opposed to traffic that is backed up. Note that traffic mode is supported on a limited number of major highways and roads.

To change modes, you must call the appropriate setter method with true. To turn off a mode, set that mode to false. We'll be talking about Overlays in just a bit, but for now, know that the traffic mode does *not* use Overlays.

> **NOTE:** The statement mapView.postInvalidateDelayed(2000) was used to work around an issue with traffic mode. The issue is with the way threads are used internally to fetch the data for displaying the traffic lines. See Android Issue 10317 for more information at http://code.google.com/p/android/issues/detail?id=10317.

To make the map move sideways, set the attribute android:clickable="true" for the MapView in XML; otherwise, users will only be able to zoom in and out, not move laterally. You can also set this in code using the setClickable(true) method call on your mapView.

The final things to mention from this example are the two methods isLocationDisplayed() and isRouteDisplayed(). The documentation for these methods says their use is required by the Google Terms of Service, although when requesting a Maps API key, there is no mention of these methods in those Terms of Service. We're not lawyers, but we'd recommend implementing these methods. Your application is obligated to respond with true or false to indicate to the map server whether or not the current device location is being displayed or if any route information is being displayed, such as driving directions.

You'll probably agree that the amount of code required to display a map and to implement zoom and mode changes is minimal with Android (see Listing 22–3). However, there's an even easier way to implement zoom controls. Take a look at the XML layout and code shown in Listing 22–4.

Listing 22–4. *Zooming Made Easier*

```
<?xml version="1.0" encoding="utf-8"?>
<!-- This file is /res/layout/mapview.xml -->
<RelativeLayout xmlns:android="http://schemas.android.com/apk/res/android"
        android:orientation="vertical" android:layout_width="fill_parent"
        android:layout_height="fill_parent">

     <com.google.android.maps.MapView android:id="@+id/mapview"
             android:layout_width="fill_parent"
             android:layout_height="wrap_content"
             android:clickable="true"
             android:apiKey="YOUR MAPS API KEY GOES HERE"
             />
</RelativeLayout>

public class MapViewDemoActivity extends MapActivity
{
    private MapView mapView;
    @Override
    protected void onCreate(Bundle savedInstanceState) {
        super.onCreate(savedInstanceState);

        setContentView(R.layout.mapview);
        mapView = (MapView)findViewById(R.id.mapview);

        mapView.setBuiltInZoomControls(true);
    }

    @Override
    protected boolean isLocationDisplayed() {
        return false;
    }

    @Override
    protected boolean isRouteDisplayed() {
        return false;
    }
}
```

The difference between Listing 22–4 and Listing 22–3 is that we changed the XML layout for our view to use `RelativeLayout`. We removed all the zoom controls and view mode controls. The magic in this example is in the code and not the layout. The `MapView` already has controls that allow you to zoom in and out. All you have to do is turn them on using the `setBuiltInZoomControls()` method. Figure 22–4 shows the `MapView`'s default zoom controls.

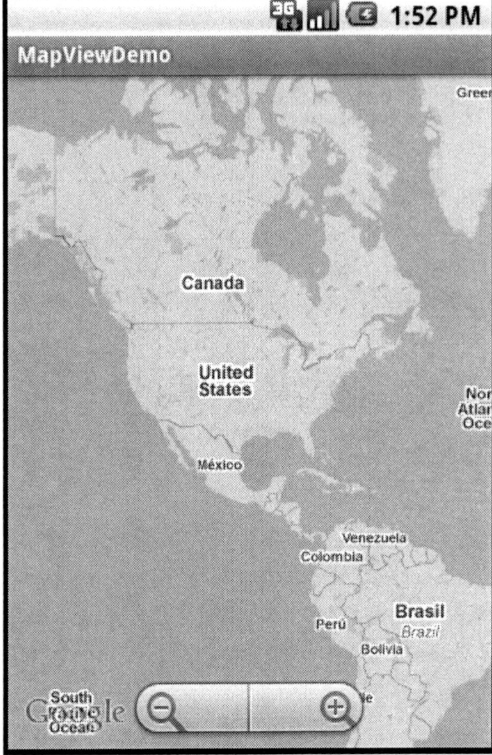

Figure 22–4. *The `MapView`'s built-in zoom controls*

Now, let's discuss how to add custom data to the map.

Adding Markers Using Overlays

Google Maps provides a facility that allows you to place custom data on top of the map. You can see an example of this if you search for pizza restaurants in your area: Google Maps places pushpins, or balloon markers, to indicate each location. Google Maps provides this facility by allowing you to add a layer on top of the map. Android provides several classes that help you to add layers to a map. The key class for this type of functionality is `Overlay`, but you can use an extension of this class called `ItemizedOverlay`. Listing 22–5 shows an example of the Java code. The layout XML file from Listing 22–4 can be used for this project as well.

Listing 22–5. *Marking Up a Map Using `ItemizedOverlay`*

```
public class MappingOverlayActivity extends MapActivity {
    private MapView mapView;

    @Override
    protected void onCreate(Bundle savedInstanceState) {
        super.onCreate(savedInstanceState);
```

```
        setContentView(R.layout.mapview);

        mapView = (MapView) findViewById(R.id.mapview);
        mapView.setBuiltInZoomControls(true);

        Drawable marker=getResources().getDrawable(R.drawable.mapmarker);
        marker.setBounds( (int) (-marker.getIntrinsicWidth()/2),
                                -marker.getIntrinsicHeight(),
                                (int) (marker.getIntrinsicWidth()/2),
                                0);

        InterestingLocations funPlaces =
                    new InterestingLocations(marker);
        mapView.getOverlays().add(funPlaces);

        GeoPoint pt = funPlaces.getCenterPt();
        int latSpan = funPlaces.getLatSpanE6();
        int lonSpan = funPlaces.getLonSpanE6();
        Log.v("Overlays", "Lat span is " + latSpan);
        Log.v("Overlays", "Lon span is " + lonSpan);

        MapController mc = mapView.getController();
        mc.setCenter(pt);
        mc.zoomToSpan((int)(latSpan*1.5), (int)(lonSpan*1.5));
    }

    @Override
    protected boolean isLocationDisplayed() {
        return false;
    }

    @Override
    protected boolean isRouteDisplayed() {
        return false;
    }

    class InterestingLocations extends ItemizedOverlay {
        private ArrayList<OverlayItem> locations =
                new ArrayList<OverlayItem>();
        private GeoPoint center = null;

        public InterestingLocations(Drawable marker)
        {
            super(marker);

            // create locations of interest
            GeoPoint disneyMagicKingdom =
                new GeoPoint((int)(28.418971*1000000),
                             (int)(-81.581436*1000000));
            GeoPoint disneySevenLagoon =
                new GeoPoint((int)(28.410067*1000000),
                             (int)(-81.583699*1000000));

            locations.add(new OverlayItem(disneyMagicKingdom ,
                          "Magic Kingdom", "Magic Kingdom"));
            locations.add(new OverlayItem(disneySevenLagoon ,
                          "Seven Seas Lagoon", "Seven Seas Lagoon"));

            populate();
```

```
        }

        //  We added this method to find the middle point of the cluster
        //  Start each edge on its opposite side and move across with
        //  each point. The top of the world is +90, the bottom -90,
        //  the west edge is -180, the east +180
        public GeoPoint getCenterPt() {
            if(center == null) {
                int northEdge = -90000000;    // i.e., -90E6 microdegrees
                int southEdge = 90000000;
                int eastEdge = -180000000;
                int westEdge = 180000000;
                Iterator<OverlayItem> iter = locations.iterator();
                while(iter.hasNext()) {
                    GeoPoint pt = iter.next().getPoint();
                    if(pt.getLatitudeE6() > northEdge)
                        northEdge = pt.getLatitudeE6();
                    if(pt.getLatitudeE6() < southEdge)
                        southEdge = pt.getLatitudeE6();
                    if(pt.getLongitudeE6() > eastEdge)
                        eastEdge = pt.getLongitudeE6();
                    if(pt.getLongitudeE6() < westEdge)
                        westEdge = pt.getLongitudeE6();
                }
                center = new GeoPoint((int)((northEdge +southEdge)/2),
                        (int)((westEdge + eastEdge)/2));
            }
            return center;
        }

        @Override
        public void draw(Canvas canvas, MapView mapView, boolean shadow)
        {
            // Hide the shadow by setting shadow to false
            shadow = false;
            super.draw(canvas, mapView, shadow);
        }

        @Override
        protected OverlayItem createItem(int i) {
            return locations.get(i);
        }

        @Override
        public int size() {
            return locations.size();
        }

    }
}
```

Listing 22–5 demonstrates how you can overlay markers onto a map. The example places two markers: one at Disney's Magic Kingdom and another at Disney's Seven Seas Lagoon, both near Orlando, Florida (see Figure 22–5).

NOTE: To run this demonstration, you'll need to get a drawable to serve as your map marker. This image file must be saved into your `/res/drawable` folder so that the resource ID reference in the `getDrawable()` call matches the file name you choose for your image file. If possible, make the area surrounding your marker transparent. Some sample markers are provided with the source code for this chapter.

In order for you to add markers onto a map, you have to create and add an extension of `com.google.android.maps.Overlay` to the map. The `Overlay` class itself cannot be instantiated, so you'll have to extend it or use one of the extensions. In our example, we have implemented `InterestingLocations`, which extends `ItemizedOverlay`, which in turn extends `Overlay`. The `Overlay` class defines the contract for an overlay, and `ItemizedOverlay` is a handy implementation that makes it easy for you to create a list of locations that can be marked on a map.

The general usage pattern is to extend the `ItemizedOverlay` class and add your items—interesting locations—in the constructor. After you instantiate your points of interest, you call the `populate()` method of `ItemizedOverlay`. The `populate()` method is a utility that caches any `OverlayItems`. Internally, the class calls the `size()` method to determine the number of overlay items and then enters a loop, calling `createItem(i)` for each item. In the `createItem` method, you return the already-created item given the index in the array.

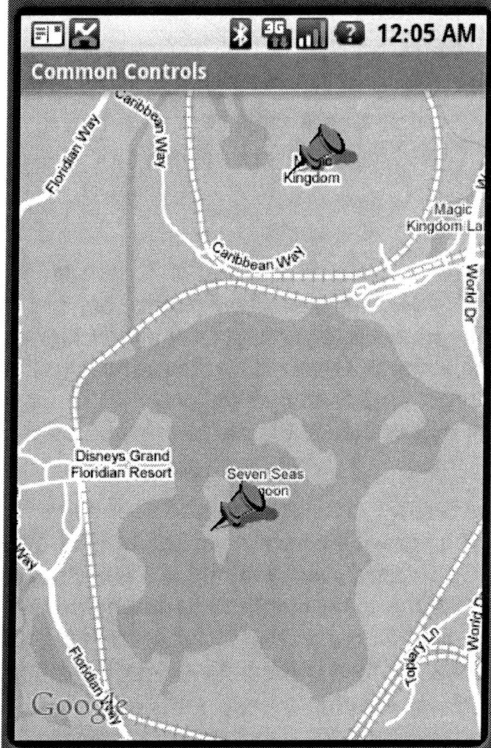

Figure 22–5. *`MapView` with markers*

As Listing 22–5 shows, you simply create the points and call `populate()` to show markers on a map. The `Overlay` contract manages the rest. To make it all work, the `onCreate()` method of the activity creates the `InterestingLocations` instance, passing in the `Drawable` that's used as a default for the markers. Then, `onCreate()` adds the `InterestingLocations` instance to the overlay collection (`mapView.getOverlays().add()`).

The `Drawable` you choose needs to be prepared for use with an `ItemizedOverlay`. The Maps API needs to know where the (0, 0) point is on the `Drawable`. This point will be used to mark the exact spot on the map that the marker is supposed to represent. You can do this yourself using the `setBounds()` method of the `Drawable` class as shown in our example. The arguments represent the left, top, right, and bottom coordinates, and we can use the `getIntrinsicHeight()` and `getIntrinsicWidth()` methods to figure out how tall and wide our `Drawable` is.

In our example, the (0, 0) coordinate would be halfway across the bottom edge. Remember that the coordinate system starts from the left and increases as you go right, and from the top increasing as you go down. Therefore, our top coordinate must be less than the 0 at the bottom, and so is negative.

Android provides a couple of convenience methods in the `ItemizedOverlay` class to set bounds on `Drawables`. They are `boundCenterBottom()` and `boundCenter()`. The first method acts on our Drawable in the exact same way that we did, resulting in (0, 0) being halfway across the bottom edge of the Drawable. The second method would put (0, 0) in the very center of the Drawable. A common practice is to use one of these methods as the first call in your constructor. We could have done the following instead of using setBounds() earlier:

```
public InterestingLocations(Drawable marker)
{
    super(boundCenterBottom(marker));
    [ … ]
```

You'll also notice that we can use any size or shape Drawable we want. One thing that makes our markers look good is to use the transparent color around the shape we want. The bubbles you're used to seeing on Google Maps are not square, and because they use a transparent color around them, you can see the map where there is no marker. This is also good because the Maps API will paint a shadow of your marker onto the map, and you want your shadow to be your shape and not a rectangle (OK, really a parallelogram).

But what if you don't want a shadow? No problem. Simply override the draw() method of your ItemizedOverlay extension class and set shadow to false when calling the parent's draw() method. Check out the draw() method in our example. We mentioned that the Drawable used to create the ItemizedOverlay is the default marker. Each OverlayItem can instead have a unique marker by using its setMarker() method with some other Drawable. You could set the unique markers when instantiating the OverlayItems, or you could set them later. We'll revisit markers in Chapter 25 when we cover touch screens and show you how to have even more fun with markers.

Now that the overlay is associated to our map, we still need to move into the right position to actually see the markers in the display. To do this, we need to set the center of the displayed map to a point. The getCenter() method of the ItemizedOverlay class returns the first ranked point, not the center point, as you might expect. An ItemizedOverlay will sort the points it contains, and it will choose one to be first. Therefore, to find the center of the points, we implemented our own getCenterPt() method to iterate through the points and find the center. The setCenter() method of the MapvVew's controller sets the center of what's displayed, and we pass it our calculated center point.

MapController's setZoom() method sets how high we are above the map. It takes a value from 1 to 21, where 21 is zoomed in as close as we can go, and 1 is as far away as we can go. But because we're not exactly sure what value to use here to see all our points at once, we use the zoomToSpan() method of the MapController. We need to pass in the height and width of the rectangle that contains all our points. Fortunately, ItemizedOverlay has two methods to tell us the height and width of that rectangle, to give us our latitude span, getLatSpanE6(), and our longitude span, getLonSpanE6(), respectively; we can then use these values with zoomToSpan(). Notice that we chose to expand our rectangle by a factor of 1.5, so our points are not right at the edges of the map when displayed.

Another interesting aspect of Listing 22–5 is the creation of the OverlayItem(s). To create an OverlayItem, you need an object of type GeoPoint. The GeoPoint class represents a location by its latitude and longitude, in micro degrees. In our example, we obtained the latitude and longitude of Magic Kingdom and Seven Seas Lagoon using geocoding sites on the Web. (As you'll see shortly, you can use geocoding to convert an address to a latitude/longitude pair, for example.) We then converted the latitude and longitude to micro degrees—because the APIs operate on micro degrees—by multiplying by 1,000,000 and performing a cast to an integer.

So far, we've shown you how to place markers on a map. But overlays are not restricted to showing pushpins or balloons. They can be used to do other things. For example, we could show animations of products moving across maps, or we could show symbols such as weather fronts or thunderstorms.

All in all, you'll agree that placing markers on a map couldn't be easier. Or could it? We don't have a database of latitude/longitude pairs, but we're guessing that we'll need to somehow create one or more GeoPoints using a real address. That's when you can use the Geocoder class, which is part of the location package that we'll discuss next.

Understanding the Location Package

The android.location package provides facilities for location-based services. In this section, we are going to discuss two important pieces of this package: the Geocoder class and the LocationManager service. We'll start with Geocoder.

Geocoding with Android

If you are going to do anything practical with maps, you'll likely have to convert an address (or location) to a latitude/longitude pair. This concept is known as *geocoding*, and the android.location.Geocoder class provides this facility. In fact, the Geocoder class provides both forward and backward conversion—it can take an address and return a latitude/longitude pair, and it can translate a latitude/longitude pair into a list of addresses. The class provides the following methods:

- List<Address> getFromLocation(double latitude, double longitude, int maxResults)
- List<Address> getFromLocationName(String locationName, int maxResults, double lowerLeftLatitude, double lowerLeftLongitude, double upperRightLatitude, double upperRightLongitude)
- List<Address> getFromLocationName(String locationName, int maxResults)

It turns out that computing an address is not an exact science because of the various ways a location can be described. For example, the getFromLocationName() methods can take the name of a place, the physical address, an airport code, or simply a well-known name for the location. Thus, the methods return a list of addresses and not a single address. Because the methods return a list, which could be quite long (and take a long time to return), you are encouraged to limit the result set by providing a value for `maxResults` that ranges between `1` and `5`. Now, let's consider an example.

Listing 22–6 shows the XML layout and corresponding code for the user interface shown in Figure 22–6. To run the example, you'll need to update the listing with your own Maps API key.

Listing 22–6. ***Working with the Android*** `Geocoder` ***Class***

```
<?xml version="1.0" encoding="utf-8"?>
<!-- This file is /res/layout/geocode.xml -->
<RelativeLayout xmlns:android="http://schemas.android.com/apk/res/android"
        android:layout_width="fill_parent"
        android:layout_height="fill_parent">

        <LinearLayout android:layout_width="fill_parent"
            android:layout_alignParentBottom="true"
            android:layout_height="wrap_content"
            android:orientation="vertical" >

            <EditText android:layout_width="fill_parent"
                android:id="@+id/location"
                android:layout_height="wrap_content"
                android:text="White House"/>

            <Button android:id="@+id/geocodeBtn"
                android:layout_width="wrap_content"
                android:layout_height="wrap_content"
                android:onClick=”doClick” android:text="Find Location"/>
        </LinearLayout>

        <com.google.android.maps.MapView
                android:id="@+id/geoMap" android:clickable="true"
                android:layout_width="fill_parent"
                android:layout_height="320px"
                android:apiKey="YOUR MAPS API KEY GOES HERE"
                />

</RelativeLayout>


public class GeocodingDemoActivity extends MapActivity
{
    Geocoder geocoder = null;
    MapView mapView = null;

    @Override
    protected boolean isLocationDisplayed() {
        return false;
    }

    @Override
    protected boolean isRouteDisplayed() {
        return false;
    }
```

```
    @Override
    protected void onCreate(Bundle savedInstanceState)
    {
        super.onCreate(savedInstanceState);

        setContentView(R.layout.geocode);
        mapView = (MapView)findViewById(R.id.geoMap);
        mapView.setBuiltInZoomControls(true);

        // lat/long of Jacksonville, FL
        int lat = (int)(30.334954*1000000);
        int lng = (int)(-81.5625*1000000);
        GeoPoint pt = new GeoPoint(lat,lng);
        mapView.getController().setZoom(10);
        mapView.getController().setCenter(pt);

        geocoder = new Geocoder(this);
    }

    public void doClick(View arg0) {
        try {
            EditText loc = (EditText)findViewById(R.id.location);
            String locationName = loc.getText().toString();

            List<Address> addressList =
                    geocoder.getFromLocationName(locationName, 5);
            if(addressList!=null && addressList.size()>0)
            {
                int lat =
                    (int)(addressList.get(0).getLatitude()*1000000);
                int lng =
                    (int)(addressList.get(0).getLongitude()*1000000);

                GeoPoint pt = new GeoPoint(lat,lng);
                mapView.getController().setZoom(15);
                mapView.getController().setCenter(pt);
            }
        } catch (IOException e) {
            e.printStackTrace();
        }
    }
}
```

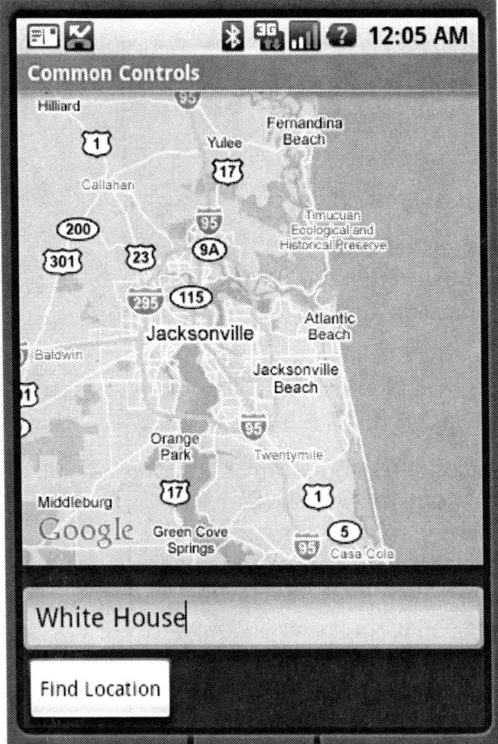

Figure 22–6. *Geocoding to a point given the location name*

To demonstrate the uses of geocoding in Android, type the name or address of a location in the `EditText` field, and then click the Find Location button. To find the address of a location, we call the `getFromLocationName()` method of `Geocoder`. The location can be an address or a well-known name such as "White House." Geocoding can be a prolonged operation, so we recommend that you limit the results to five, as the Android documentation suggests.

The call to `getFromLocationName()` returns a list of addresses. The sample application takes the list of addresses and processes the first one if any were found. Every address has a latitude and longitude, which you use to create a `GeoPoint`. You then get the map controller and navigate to the point. The zoom level can be set to an integer between 1 and 21, inclusive. As you move from 1 toward 21, the zoom level increases by a factor of 2. We could have presented a dialog to display multiple found locations if we wanted to, but for now, we'll just display the first location returned to us.

In our example application, we only read the latitude and longitude of our returned `Address`. In fact, there can be a ton of data about `Addresses` returned to us, including the location's common name, street, city, state, postal/zip code, country, and even phone number and web site URL.

NOTE: Location-based services do not use micro degrees like the Maps API does. Forgetting to convert from one to the other is a common cause of errors. To pass a `Location`'s latitude and longitude to a Maps API method, you must multiply by 1,000,000 first.

You should understand a few points with respect to geocoding:

- First, a returned address is not always an exact address. Obviously, because the returned list of addresses depends on the accuracy of the input, you need to make every effort to provide an accurate location name to the `Geocoder`.
- Second, whenever possible, set the `maxResults` parameter to a value between `1` and `5`.
- Finally, you should seriously consider doing the geocoding operation in a different thread from the UI thread. There are two reasons for this. The first is obvious: the operation is time-consuming, and you don't want the UI to hang while you do the geocoding, causing Android to kill your activity. The second reason is that, with a mobile device, you always need to assume that the network connection can be lost and that the connection is weak. Therefore, you need to handle input/output (I/O) exceptions and timeouts appropriately. Once you have computed the addresses, you can post the results to the UI thread. Let's investigate this a bit more.

Geocoding with Background Threads

Using background threads to handle time-consuming operations is very common. You don't want the user getting Application Not Responding (ANR) pop-up dialogs because an operation on the main thread takes too long. The general pattern is to handle a UI event (such as a button click) to initiate a long-running operation in the background. From the event handler, you create a new thread to execute the work, and then you start the new thread. The UI thread returns to the user interface to handle the interaction with the user while the background thread works. After the background thread completes, a part of the UI might have to be updated, or the user might have to be notified. The background thread does not update the UI directly; instead, the background thread notifies the UI thread to update itself. Listing 22–7 demonstrates this idea using geocoding. We'll use the same `geocode.xml` file as before. We can also use the same `AndroidManifest.xml` file as before.

Listing 22–7. *Geocoding in a Separate Thread*

```
public class GeocodingDemoActivity extends MapActivity
{
    Geocoder geocoder = null;
    MapView mapView = null;
    ProgressDialog progDialog=null;
    List<Address> addressList=null;
```

```
@Override
protected boolean isRouteDisplayed() {
    return false;
}

@Override
protected void onCreate(Bundle icicle) {
    super.onCreate(icicle);

    setContentView(R.layout.geocode);
    mapView = (MapView)findViewById(R.id.geoMap);
    mapView.setBuiltInZoomControls(true);

    // lat/long of Jacksonville, FL
    int lat = (int)(30.334954*1000000);
    int lng = (int)(-81.5625*1000000);
    GeoPoint pt = new GeoPoint(lat,lng);
    mapView.getController().setZoom(10);
    mapView.getController().animateTo(pt);

    geocoder = new Geocoder(this);
}

public void doClick(View view) {
    EditText loc = (EditText)findViewById(R.id.location);
    String locationName = loc.getText().toString();

    progDialog = ProgressDialog.show(GeocodingDemoActivity.this,
                "Processing...", "Finding Location...", true, false);

    findLocation(locationName);
}

private void findLocation(final String locationName)
{
    Thread thrd = new Thread()
    {
        public void run()
        {
            try {
                // do background work
                addressList =
                    geocoder.getFromLocationName(locationName, 5);
                //send message to handler to process results
                uiCallback.sendEmptyMessage(0);

            } catch (IOException e) {
                e.printStackTrace();
            }
        }
    };
    thrd.start();
}

// ui thread callback handler
private Handler uiCallback = new Handler()
{
    @Override
```

```
            public void handleMessage(Message msg)
            {
                // tear down dialog
                progDialog.dismiss();

                if(addressList!=null && addressList.size()>0)
                {
                    int lat =
                       (int)(addressList.get(0).getLatitude()*1000000);
                    int lng =
                       (int)(addressList.get(0).getLongitude()*1000000);
                    GeoPoint pt = new GeoPoint(lat,lng);
                    mapView.getController().setZoom(15);
                    mapView.getController().animateTo(pt);

                }
                else
                {
                    Dialog foundNothingDlg = new
                        AlertDialog.Builder(GeocodingDemoActivity.this)
                               .setIcon(0)
                               .setTitle("Failed to Find Location")
                               .setPositiveButton("Ok", null)
                               .setMessage("Location Not Found...")
                               .create();
                    foundNothingDlg.show();
                }
            }
        };
}
```

Listing 22–7 is a modified version of the example in Listing 22–6. The difference is that, now, in the `doClick()` method, you display a progress dialog and call `findLocation()` (see Figure 22–7). `findLocation()` then creates a new thread and calls the `start()` method, which ultimately results in a call to the thread's `run()` method. In the `run()` method, you use the `Geocoder` class to search for the location. When the search is done, you must post the message to something that knows how to interact with the UI thread, because you need to update the map. Android provides the `android.os.Handler` class for this purpose. From the background thread, call the `uiCallback.sendEmptyMessage(0)` method to have the UI thread process the results from the search. In our case, we don't need to actually send any content in the message, because the data is being shared through the `addressList`. The code calls the handler's callback, which dismisses the dialog and then looks at the `addressList` returned by the `Geocoder`. After that, the callback updates the map with the result or displays an alert dialog to indicate that the search returned nothing. The UI for this example is shown in Figure 22–7.

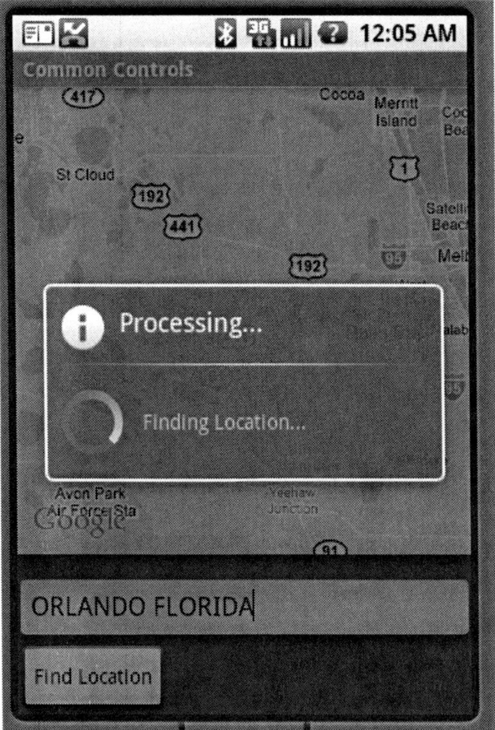

Figure 22–7. *Showing a progress window during long operations*

Understanding the LocationManager Service

The `LocationManager` service is one of the key services offered by the `android.location` package. This service provides two things: a mechanism for you to obtain the device's geographical location and a facility for you to be notified (via an intent) when the device enters a specified geographical location.

In this section, you are going to learn how the `LocationManager` service works. To use the service, you must first obtain a reference to it. Listing 22–8 shows a simple usage of the `LocationManager` service.

Listing 22–8. *Using the `LocationManager` Service*

```
public class LocationManagerDemoActivity extends Activity
{

    @Override
    protected void onCreate(Bundle savedInstanceState)
    {
        super.onCreate(savedInstanceState);

        LocationManager locMgr = (LocationManager)
            this.getSystemService(Context.LOCATION_SERVICE);
```

```
        Location loc =
            locMgr.getLastKnownLocation(LocationManager.GPS_PROVIDER);

        List<String> providerList = locMgr.getAllProviders();
    }
}
```

The LocationManager service is a system-level service. *System-level services* are services that you obtain from the context using the service name; you don't instantiate them directly. The android.app.Activity class provides a utility method called getSystemService() that you can use to obtain a system-level service. As shown in Listing 22–8, you call getSystemService() and pass in the name of the service you want, in this case, Context.LOCATION_SERVICE.

The LocationManager service provides geographical location details by using location providers. Currently, there are three types of location providers:

- *GPS* providers use a Global Positioning System to obtain location information.
- *Network* providers use cell-phone towers or Wi-Fi networks to obtain location information.
- The *passive* provider is like a location update sniffer, and it passes to your application location updates that are requested by other applications, without your application having to specifically request any location updates. Of course, if no one else is requesting location updates, you won't get any either.

The LocationManager class can provide the device's last known location via the getLastKnownLocation() method. Location information is obtained from a provider, so the method takes as a parameter the name of the provider you want to use. Valid values for provider names are LocationManager.GPS_PROVIDER, LocationManager.NETWORK_PROVIDER, and LocationManager.PASSIVE_PROVIDER. In order for your application to successfully get location information, it must have the appropriate permissions in the AndroidManifest.xml file. android.permission.ACCESS_FINE_LOCATION is required for GPS and for passive providers, whereas android.permission.ACCESS_COARSE_LOCATION or android.permission.ACCESS_FINE_LOCATION can be used for network providers, depending on what you need. For instance, assume your application will use GPS or network data for location updates. Because you need ACCESS_FINE_LOCATION for GPS, you've also satisfied permissions for network access, so you do not need to also specify ACCESS_COARSE_LOCATION. If you're only going to use the network provider, you could get by with only ACCESS_COARSE_LOCATION in the manifest file.

Calling getLastKnownLocation() returns an android.location.Location instance, or null if no location is available. The Location class provides the location's latitude and longitude, the time the location was computed, and possibly the device's altitude, speed, and bearing. A Location object can also tell you which provider it came from using getProvider(), which will be either GPS_PROVIDER or NETWORK_PROVIDER. If you're getting location updates via the PASSIVE_PROVIDER, remember that you're only really sniffing location updates, so all updates are ultimately from either GPS or the network.

Because the LocationManager operates on providers, the class provides APIs to obtain providers. For example, you can get all of the known providers by calling getAllProviders(). You can obtain a specific provider by calling getProvider(), passing the name of the provider as an argument (such as LocationManager.GPS_PROVIDER). One thing to watch out for is that getAllProviders() will return providers that you may not have access to or that are currently disabled. Fortunately, you are able to determine the status of providers using other methods, such as isProviderEnabled(String providerName) or getProviders(boolean enabledOnly), which you could call with a value of true to get only providers you are able to use immediately.

There's another way to get a suitable provider, and that is to use the getProviders(Criteria criteria, boolean enabledOnly) method of LocationManager. By specifying criteria for location updates, and by setting enabledOnly to true so you get providers that are enabled and ready to go, you can get a list of provider names returned to you without having to know the specifics of which provider you got. This could be more portable, because a device may have a custom LocationProvider that meets your needs without you having to know about it in advance. The Criteria object can be set with parameters that include accuracy level and the need for information about speed, bearing, altitude, cost, and power requirements. If no providers meet your criteria, a null list will be returned, allowing you to either bail out or relax the criteria and try again.

How to Enable Location Providers

You might think there's a simple API to enable a location provider (such as GPS) if it's not turned on when your application runs. Unfortunately, this is not the case. To get a location service turned on, the user must do that from within the Settings screens of their device. Your application can make this a lot simpler for the user by launching that particular Settings screen. The location settings source screen is really just an activity, and this activity is set up to respond to an intent. So all your application needs to do is request an activity using the correct intent. The code you might use looks like this:

```
startActivityForResult(new Intent(
    android.provider.Settings.ACTION_LOCATION_SOURCE_SETTINGS), 0);
```

Remember that to handle a response, you must implement the onActivityResult() callback in your activity (covered in Chapter 5). And also keep in mind that although you hope the user turns on a location provider such as GPS, they may not. You will need to check again to see if the user has enabled a location provider, and take appropriate action based on the result.

What Can You Do With a Location?

As mentioned before, Locations can tell you the latitude and longitude, when the Location was computed, the provider that computed this Location, and optionally the altitude, speed, bearing, and accuracy level. Depending on the provider where the Location came from, there could be extra information as well. For example, if the Location came from a GPS provider, there is an extras Bundle that will tell you how many satellites were used to compute the Location. The optional values may or may not be present, depending on the provider. To know if a Location has one of these values, the Location class provides a set of has...() methods that return a boolean value, for example hasAccuracy(). Before relying on the return value of getAccuracy(), it would be wise to call hasAccuracy() first.

The Location class has some other useful methods, including a static method distanceBetween(), which will return the shortest distance between two Locations. Another distance-related method is distanceTo(), which will return the shortest distance between the current Location object and the Location object passed to the method. Note that distances are in meters and that the distance calculations take into account the curvature of the Earth. But also be aware that the distances are not provided in terms of the distance you would have to go by car, for example.

If you want to get driving directions or driving distances, you will need to have your beginning and ending Locations, but to do the calculations, you will likely need to use the Google Maps JavaScript API services. For example, there is a Google Directions API. The Directions API would allow your application to show how to get from your beginning to your ending location.

Sending Location Updates to Your Application During Development

When doing development testing, LocationManager needs location information, and the emulator doesn't have access to GPS or cell towers. In order for you to test your LocationManager service application in the emulator, you manually send location updates from Eclipse. Listing 22–9 shows a simple example to illustrate how to do this.

Listing 22–9. *Registering for Location Updates*

```
public class LocationUpdateDemoActivity extends Activity
{
    LocationManager locMgr = null;
    LocationListener locListener = null;

    @Override
    public void onCreate(Bundle savedInstanceState)
    {
        super.onCreate(savedInstanceState);

        locMgr = (LocationManager)
            getSystemService(Context.LOCATION_SERVICE);

        locListener = new LocationListener()
```

```
        {
            public void  onLocationChanged(Location location)
            {
                if (location != null)
                {
                    Toast.makeText(getBaseContext(),
                        "New location latitude [" +
                        location.getLatitude() +
                        "] longitude [" +
                        location.getLongitude()+"]",
                        Toast.LENGTH_SHORT).show();
                }
            }

            public void  onProviderDisabled(String provider)
            {
            }

            public void  onProviderEnabled(String provider)
            {
            }

            public void  onStatusChanged(String provider,
                            int status, Bundle extras)
            {
            }        };
    }

    @Override
    public void onResume() {
        super.onResume();

        locMgr.requestLocationUpdates(
            LocationManager.GPS_PROVIDER,
            0,                // minTime in ms
            0,                // minDistance in meters
            locListener);
    }

    @Override
    public void onPause() {
        super.onPause();
        locMgr.removeUpdates(locListener);
    }
}
```

We're not displaying a user interface for this example, so the standard initial layout XML file will do. This is also why we don't need to extend a `MapActivity` for this application, because we're not displaying any maps.

One of the primary uses of the `LocationManager` service is to receive notifications of the device's location. Listing 22–9 demonstrates how you can register a listener to receive location-update events. To register a listener, you call the `requestLocationUpdates()` method, passing the provider type as one of the parameters. When the location changes, the `LocationManager` calls the `onLocationChanged()` method of the listener with the new `Location`. It is very important that you remove any registrations for location updates at the appropriate time. In our example, we do registration in `onResume()`, and we remove that registration in `onPause()`. If we aren't going to be around to do anything with location updates, we should tell the provider not to send them. There's also the possibility that our activity could be destroyed (for example, if the user rotates their device and our activity is restarted), in which case our old activity could still exist, be receiving updates, displaying them with Toast, and taking up memory.

In our example, we set the `minTime` and `minDistance` to zero. This tells the `LocationManager` to send us updates as often as possible. These are not desired settings in real life, but we use them here to make the demonstrations run better. (In real life, you would not want the hardware trying to figure out our current position so often, as this drains the battery.) Set these values appropriately for the situation, trying to minimize how often you truly need to be notified of a change in position.

To test this in the emulator, you can use the Dalvik Debug Monitor Service (DDMS) perspective that ships with the ADT plug-in for Eclipse. The DDMS UI provides a screen for you to send the emulator a new location (see Figure 22–8).

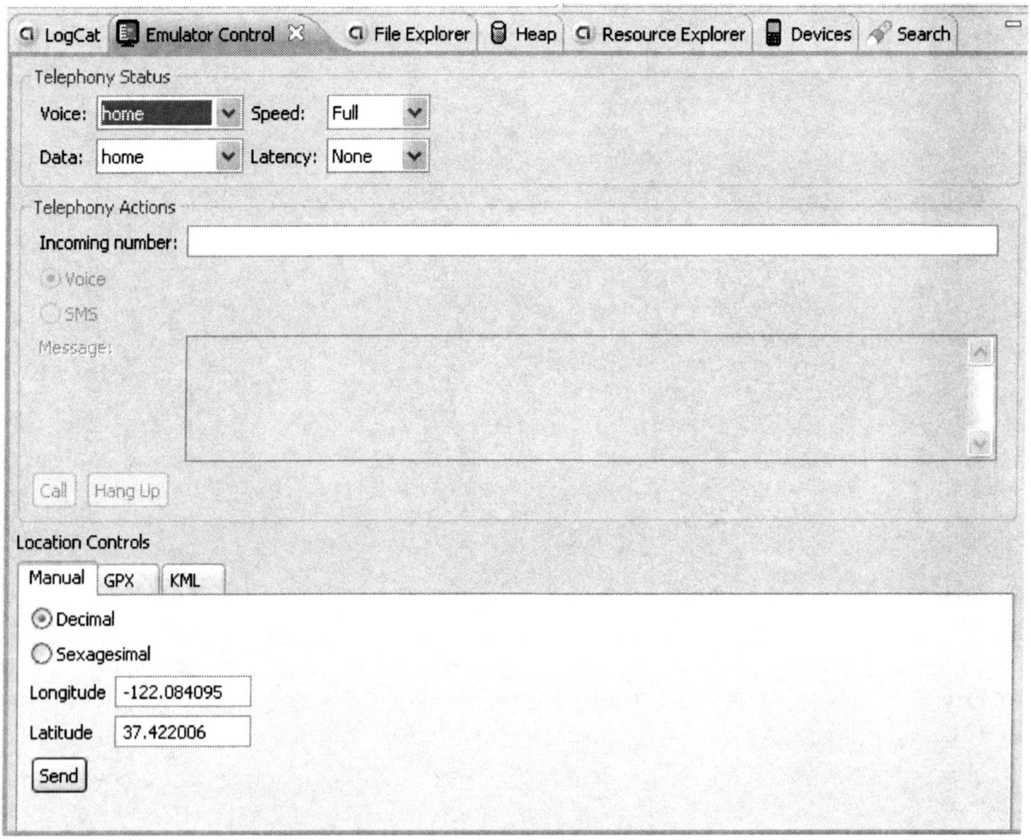

Figure 22–8. *Using the DDMS UI in Eclipse to send location data to the emulator*

To get to the DDMS in Eclipse, use Window ➤ Open Perspective ➤ DDMS. The Emulator Control view should already be there for you, but if not, use Window ➤ Show View ➤ Other ➤ Android ➤ Emulator Control to make it visible in this perspective. You may need to scroll down in the emulator control to find the location controls. As shown in Figure 22–8, the Manual tab in the DDMS user interface allows you to send a new GPS location (latitude/longitude pair) to the emulator. Sending a new location will fire the `onLocationChanged()` method on the listener, which will result in a message to the user conveying the new location.

You can send location data to the emulator using several other techniques, as shown in the DDMS user interface (see Figure 22–8). For example, the DDMS interface allows you to submit a GPS Exchange Format (GPX) file or a Keyhole Markup Language (KML) file. You can obtain sample GPX files from these sites:

- `www.topografix.com/gpx_resources.asp`
- `http://tramper.co.nz/?view=gpxFiles`
- `www.gpsxchange.com/`

Similarly, you can use the following KML resources to obtain or create KML files:

- `http://bbs.keyhole.com/`
- `http://code.google.com/apis/kml/documentation/kml_tut.html`

NOTE: Some sites provide KMZ files. These are zipped KML files, so simply unzip them to get to the KML file. Some KML files need to have their XML namespace values altered in order to play properly in DDMS. If you have trouble with a particular KML file, make sure it has this:

`<kml xmlns="http://earth.google.com/kml/2.x">`.

You can upload a GPX or KML file to the emulator and set the speed at which the emulator will play back the file (see Figure 22–9). The emulator will then send location updates to your application based on the configured speed. As Figure 22–9 shows, a GPX file contains points, shown in the top part, and paths, shown in the bottom part. You can't play a point, but when you click a point, it will be sent to the emulator. You click a path, and then the Play button will be enabled so you can play the points.

NOTE: There have been reports that not all GPX files are understandable by the emulator control. If you attempt to load a GPX file and nothing happens, try a different file from a different source.

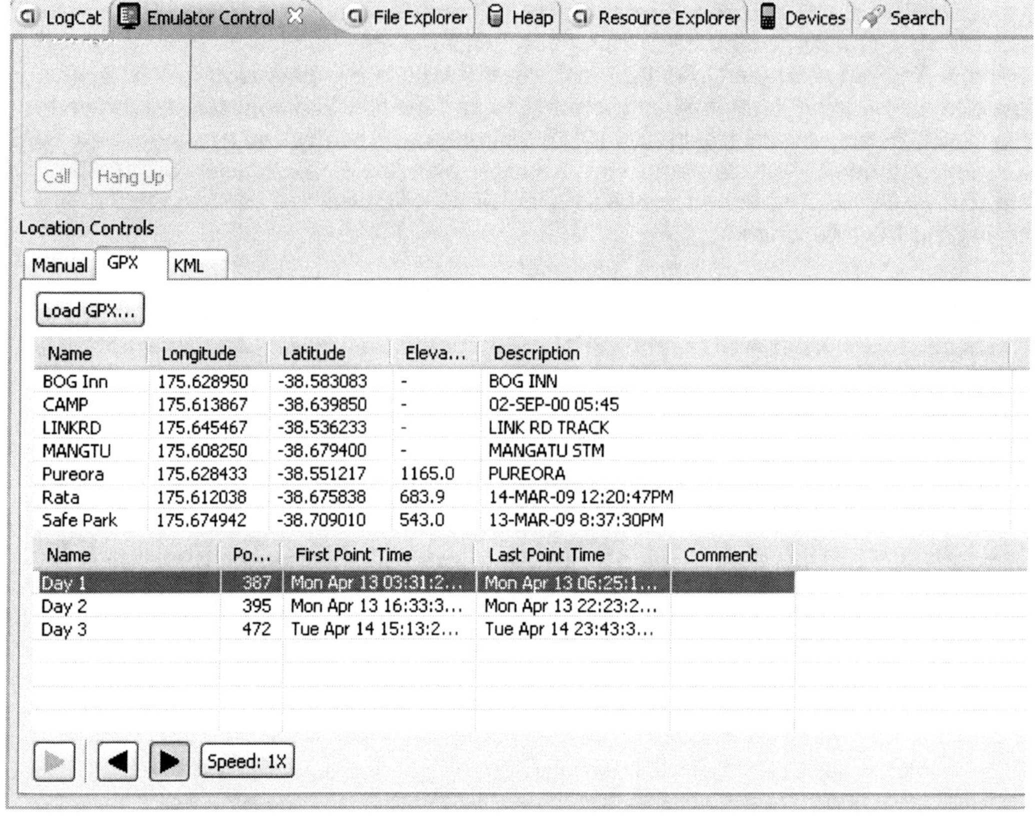

Figure 22–9. *Uploading GPX and KML files to the emulator for playback*

Listing 22–9 includes some additional methods for `LocationListener` we haven't mentioned yet. They are the callbacks `onProviderDisabled()`, `onProviderEnabled()`, and `onStatusChanged()`. For our sample, we did not do anything with these, but in your application, you could be notified when a location provider, such as `gps`, is disabled or enabled by the user, or when a status changes with one of the location providers. Statuses include `OUT_OF_SERVICE`, `TEMPORARILY_UNAVAILABLE`, and `AVAILABLE`. Even if a provider is enabled, it does not mean that it will be sending any location updates, and you can tell that using statuses. Note that `onProviderDisabled()` will be invoked immediately if a `requestLocationUpdates()` is called for a disabled provider.

Sending Location Updates from the Emulator Console

Eclipse has some easy-to use-tools for sending location updates to your application, but there's another way to do it. Remember from Chapter 2 that to launch the emulator console, you use the following command from a tools window

```
telnet localhost emulator_port_number
```

where emulator_port_number is the number associated to the instance of the AVD that's already running, displayed in the title bar of the emulator window. Once you're connected, you can use the geo fix command to send in location updates. To send in latitude/longitude coordinates with altitude (altitude is optional), use this form of the command:

```
geo fix lon lat [ altitude ]
```

For example, the following command will send the location of Jacksonville, Florida to your application with an altitude of 120 meters.

```
geo fix  -81.5625  30.334954  120
```

Please pay careful attention to the order of the arguments to the geo fix command. Longitude is the *first* argument, and latitude is the second.

Alternate Ways of Getting Location Updates

Earlier we showed you how to get location updates sent to your activity using the requestLocationUpdates() method of the LocationManager. There are actually several different signatures of this method, including ones that use a PendingIntent. This gives you the ability to direct location updates to services or broadcast receivers. You can also direct location updates to other Looper threads instead of the main thread, giving you lots of flexibility for your application, although some of these methods are only available since Android 2.3.

Showing Your Location Using MyLocationOverlay

A common use for GPS and maps is to show users where they are. Fortunately, Android makes this easy to do by supplying a special overlay called MyLocationOverlay. By adding this overlay to your MapView, you can quite easily add a blinking blue dot to your map showing where the LocationManager service says the device is.

For this example, we're going to combine a bunch of concepts together into one application. Using Listing 22–10, we can modify our previous example by updating the main.xml and MyLocationDemoActivity.java files. Or simply create a new project from the existing source of Chapter 22. Don't forget to put your Maps API key into your layout file.

Listing 22–10. *Using MyLocationOverlay*

```
<?xml version="1.0" encoding="utf-8"?>
<!-- This file is /res/layout/main.xml -->
<RelativeLayout
        xmlns:android="http://schemas.android.com/apk/res/android"
        android:layout_width="fill_parent"
        android:layout_height="fill_parent">

    <com.google.android.maps.MapView
        android:id="@+id/geoMap" android:clickable="true"
        android:layout_width="fill_parent"
        android:layout_height="fill_parent"
        android:apiKey="YOUR MAPS API KEY GOES HERE"
```

```
        />

</RelativeLayout>
```

```
public class MyLocationDemoActivity extends MapActivity {

    MapView mapView = null;
    MapController mapController = null;
    MyLocationOverlay whereAmI = null;

    @Override
    protected boolean isLocationDisplayed() {
        return whereAmI.isMyLocationEnabled();
    }

    @Override
    protected boolean isRouteDisplayed() {
        return false;
    }

    /** Called when the activity is first created. */
    @Override
    public void onCreate(Bundle savedInstanceState) {
        super.onCreate(savedInstanceState);
        setContentView(R.layout.main);

        mapView = (MapView)findViewById(R.id.geoMap);
        mapView.setBuiltInZoomControls(true);

        mapController = mapView.getController();
        mapController.setZoom(15);

        whereAmI = new MyLocationOverlay(this, mapView);
        mapView.getOverlays().add(whereAmI);
        mapView.postInvalidate();
    }

    @Override
    public void onResume()
    {
        super.onResume();
        whereAmI.enableMyLocation();
        whereAmI.runOnFirstFix(new Runnable() {
            public void run() {
                mapController.setCenter(whereAmI.getMyLocation());
            }
        });
    }

    @Override
    public void onPause()
    {
        super.onPause();
        whereAmI.disableMyLocation();
    }
}
```

Notice that, in this example, `isLocationDisplayed()` will return true if we are now showing the current location of the device on a map.

Once you launch this application in the emulator, you need to start sending it location updates before it gets very interesting. To do this, go to the DDMS Emulator Control view in Eclipse as described earlier in this section:

1. You need to find a sample GPX file from somewhere on the Internet. The sites listed earlier for GPX files have lots of them. Just pick one, and download it to your workstation.
2. Load this file into the emulator control using the Load GPX button on the GPX tab under Location Controls.
3. Select a path from the bottom list, and click the play button (the green arrow). Notice the Speed button also. This should start sending a stream of location updates to the emulator, which will be picked up by your application.
4. Click the Speed button to make the updates happen more often.

Figure 22–10 shows what your screen might look like.

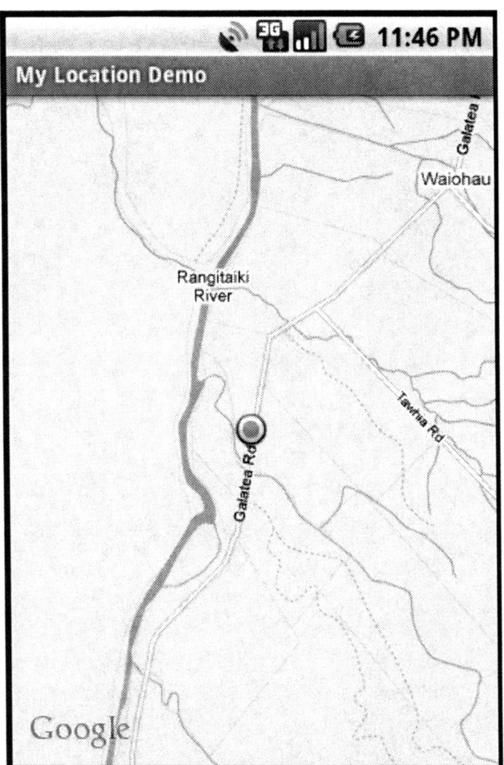

Figure 22–10. *Displaying our current location with* `MyLocationOverlay`

The preceding code is very straightforward. After setting up the basics of a MapView, turning on the zoom controls and zooming in close, we create the MyLocationOverlay overlay. We add the new overlay to the MapView and call postInvalidate() on the MapView, so the new overlay will appear on the screen. Without this last call, the overlay will be created, but it will not show up.

Remember that our application will call onResume() even when it's just starting up, as well as after waking up. Therefore, we want to enable location tracking in onResume() and disable it in onPause(). There's no sense in draining the battery with location requests if we're not going to be around to consume them. In addition to enabling location requests in onResume(), we also want to jump to where we're at right now. The MyLocationOverlay class has a helpful method for this: runOnFirstFix(). This method allows us to set up code that will run as soon as we have a location at all. This could be immediately, because we've got a last location, or it could be later when we get something from GPS_PROVIDER, NETWORK_PROVIDER, or PASSIVE_PROVIDER. When we have a fix, we center on it. After that, we don't need to do anything ourselves, because the MyLocationOverlay is getting location updates and putting the blinking blue dot in that location. If the blue dot gets close to the edge of the map, the map will recenter itself so the blue dot is back in the middle of the screen.

Customizing MyLocationOverlay

You might have noticed that you are able to zoom in and out while the location updates are occurring, and you can even pan away from the current location. This could be a good thing or a bad thing depending on your point of view. If you pan away and don't remember where you are, it will be difficult to find yourself again unless you zoom way out and look for the blue dot. The recentering trick only works if the blue dot gradually approaches the edge of the map on its own. Once you've panned away so the blue dot is no longer visible, it won't put itself back into view. This situation can also occur if the blue dot jumps off the map without coming close to the edge first.

If you want the current location to always be displayed near the center of the screen, we need to make sure we keep animating to the current location, and we can do that relatively easily. For the next version of this exercise we'll reuse everything in our MyLocationDemo project except for a very small change to our Activity, and we're going to add a new class to our package, an extension of MyLocationOverlay, so we can tweak its behavior just a bit. The new extension of MyLocationOverlay is shown in Listing 22–11.

Listing 22–11. *Extending MyLocationOverlay and Keeping Our Location in View*

```
public class MyCustomLocationOverlay extends MyLocationOverlay {
    MapView mMapView = null;

    public MyCustomLocationOverlay(Context ctx, MapView mapView) {
        super(ctx, mapView);
        mMapView = mapView;
    }

    public void onLocationChanged(Location loc) {
```

```
        super.onLocationChanged(loc);
        GeoPoint newPt = new GeoPoint((int) (loc.getLatitude()*1E6),
                (int) (loc.getLongitude()*1E6));
        mMapView.getController().animateTo(newPt);
    }
}
```

The only thing we need to change from Listing 22–10 is to use MyCustomLocationOverlay instead of MyLocationOverlay in our activity's onCreate() method, like so:

```
whereAmI = new MyCustomLocationOverlay(this, mapView);
```

Go ahead and run this in the emulator, and then send it new locations through the emulator control. If you're sending in a stream of location updates using a GPX file, you'll notice that the blue dot is always moved to the center of the map. Even if you pan completely away from the blue dot, the map returns to show it in the center.

Using Proximity Alerts

We mentioned earlier that the LocationManager can notify you when the device enters a specified geographical location. The method to set this up is addProximityAlert() from the LocationManager class. Basically, you tell the LocationManager that you want an Intent to be fired when the location of the device goes into, or leaves, a circle of a certain radius with a center at a latitude/longitude position. The Intent can trigger a BroadcastReceiver or a Service to be called, or an Activity to be started. There is also an optional time limit placed on the alert, so it could time out before the Intent fires.

Internally, the code for this method registers listeners for both the GPS and network providers and sets up location updates for once per second and a minDistance of 1 meter. You don't have any way to override this behavior or set parameters. Therefore, if you leave this running for a long time, you could end up draining the battery very quickly. If the screen goes to sleep, proximity alerts will only be checked once every four minutes, but again, you have no control over the time duration here.

It could be much better to do your own thing to decide if the device is within a certain distance of a latitude/longitude position using the techniques we've shown you in this chapter. For example, if you maintain a list of locations that you want to check against, you could measure the distance from the current location to each location in the list. Depending on how far away you are, you could decide to wait quite a while before checking the current location again. For example, if the nearest location is 100 miles away and we want to know when we're within 300 meters, clearly, we don't need to check in 1 second from now.

If you do wish to use this method, though, we'll show you how. Listing 22–12 shows the Java code for our main Activity, as well as the BroadcastReceiver that will receive the broadcasts.

Listing 22–12. *Setting Up a Proximity Alert with a* `BroadcastReceiver`

```
// This file is ProximityActivity.java
public class ProximityActivity extends Activity {
    private final String PROX_ALERT =
        "com.androidbook.intent.action.PROXIMITY_ALERT";
    private ProximityReceiver proxReceiver = null;
    private LocationManager locMgr = null;
    PendingIntent pIntent1 = null;
    PendingIntent pIntent2 = null;

    /** Called when the activity is first created. */
    @Override
    public void onCreate(Bundle savedInstanceState) {
        super.onCreate(savedInstanceState);

        locMgr = (LocationManager)
                 this.getSystemService(LOCATION_SERVICE);

        double lat = 30.334954;      // Coordinates for Jacksonville, FL
        double lon = -81.5625;
        float radius = 5.0f * 1609.0f; // 5 miles x 1609 meters per mile

        String geo = "geo:"+lat+","+lon;

        Intent intent = new Intent(PROX_ALERT, Uri.parse(geo));
        intent.putExtra("message", "Jacksonville, FL");

        pIntent1 = PendingIntent.getBroadcast(getApplicationContext(), 0,
                intent, PendingIntent.FLAG_CANCEL_CURRENT);

        locMgr.addProximityAlert(lat, lon, radius, -1L, pIntent1);

        lat = 28.54;         // Coordinates for Orlando, FL
        lon = -81.38;
        geo = "geo:"+lat+","+lon;

        intent = new Intent(PROX_ALERT, Uri.parse(geo));
        intent.putExtra("message", "Orlando, FL");

        pIntent2 = PendingIntent.getBroadcast(getApplicationContext(), 0,
                intent, PendingIntent.FLAG_CANCEL_CURRENT);

        locMgr.addProximityAlert(lat, lon, radius, -1L, pIntent2);

        proxReceiver = new ProximityReceiver();

        IntentFilter iFilter = new IntentFilter(PROX_ALERT);
        iFilter.addDataScheme("geo");

        registerReceiver(proxReceiver, iFilter);
    }

    protected void onDestroy() {
        super.onDestroy();
        unregisterReceiver(proxReceiver);
        locMgr.removeProximityAlert(pIntent1);
        locMgr.removeProximityAlert(pIntent2);
```

```
    }
}

// This file is ProximityReceiver.java
public class ProximityReceiver extends BroadcastReceiver {

    private static final String TAG = "ProximityReceiver";

    @Override
    public void onReceive(Context arg0, Intent intent) {
        Log.v(TAG, "Got intent");
        if(intent.getData() != null)
            Log.v(TAG, intent.getData().toString());
        Bundle extras = intent.getExtras();
        if(extras != null) {
            Log.v(TAG, "Message: " + extras.getString("message"));
            Log.v(TAG, "Entering? " +
              extras.getBoolean(LocationManager.KEY_PROXIMITY_ENTERING));
        }
    }
}
```

Because we're not actually displaying any positions on a map, we do not need to use a `MapActivity`, the Google Map APIs library, or a target. However, we do need to add a permission to our manifest file for `android.permission.ACCESS_FINE_LOCATION`, because the `LocationManager` will be attempting to use the GPS provider. It also attempts to use the network provider, but because we already require `ACCESS_FINE_LOCATION`, we're covered permissionwise. We register our `BroadcastReceiver` in code in the `onCreate()` method, so we do not need to set up a receiver in the manifest file. If you put the receiver into a separate application, then you *would* need to add an entry in that manifest file for the receiver. For our sample in Listing 22–12, it could look like the manifest snippet in Listing 22–13.

Listing 22–13. *`AndroidManifest.xml` Snippet for a `BroadcastReceiver` for a Proximity Alert*

```
<application … >

    <receiver android:name=".ProximityReceiver">
        <intent-filter>
            <action android:name="com.androidbook.android.intent.PROXIMITY_ALERT" />
            <data android:scheme="geo" />
        </intent-filter>
    </receiver>
</application>
```

The proximity alert capability in Android works by receiving a `PendingIntent` object, the coordinates of our latitude/longitude point of interest, the radius (in meters) around that point that we want to check, and a time duration for how long to check. These arguments are all passed in using the `addProximityAlert()` method of `LocationManager`. The `PendingIntent` contains an `Intent` that will be the thing fired if the device either enters, or leaves, the circle we've defined. For our sample, we've chosen to use a broadcast intent, so we called the `getBroadcast()` method of the `PendingIntent` class, passing in our application's context plus our `Intent` that contains the alert action and the Uri of our Location point. If the device enters or leaves our circle of interest, our Intent will be broadcast to any receivers registered to receive it.

We chose not to set a timeout for our alerts, using a value of -1L for the duration. If you want to set a timeout, this value would be the number of milliseconds LocationManager waits before giving up and deleting your PendingIntent. You will not be notified if LocationManager deletes it before it fires.

For our sample, we get a reference to the LocationManager, create our first Intent and PendingIntent, and then call addProximityAlert() to set up our first alert. Later, when our Intent fires, the only thing the LocationManager will add to it (in extras) is a boolean that says whether we're entering or leaving the circle. It does not add the current latitude/longitude position of the device, or the latitude/longitude that we used in the call to addProximityAlert(). Therefore, in order for us to know which Location we're near in our BroadcastReceiver, we've added some data to our Intent, which is the latitude/longitude of our Location of interest. For fun, we've also added a message, in extras, with the description of this Location. We could have added doubles for the latitude and longitude if it would help on the receiving end.

After adding our first alert, we set up a second alert in the same fashion as before. Finally, we register a BroadcastReceiver to receive our Intents when they are broadcast by the LocationManager. We use an IntentFilter with both the alert as the action and geo as the scheme. We need both things so we catch the broadcasts, because the broadcasts contain data; we could catch broadcasts without specifying a scheme if the broadcasts did not contain any data. The last thing we need to do is make sure we clean up after ourselves in the onDestroy() method, by unregistering our receiver and removing our proximity alerts from LocationManager using our saved PendingIntents. This is why we keep references to our PendingIntents, so we can remove the alerts later.

Our ProximityReceiver class is very simple. Upon receiving a broadcast message, it looks for information to print out in LogCat. Here is where you can see the extra data that LocationManager inserts for us, to tell us if we're entering or leaving the circle.

When you start up this sample application in the emulator, you'll see a blank screen with our application title. Now, you can send in location updates, using either the DDMS Emulator Control screen, or the emulator console with the geo fix command. When you send in locations such that you've transitioned across the edge of one of our circles (either the five-mile circle around Jacksonville or the five-mile circle around Orlando), you should see messages in LogCat from our BroadcastReceiver. Figure 22–11 shows what your LogCat window might look like once you've sent in some location updates that trigger the broadcasts.

Log

tag	Message
ActivityManager	Start proc com.androidbook.location.proximity for activity com.androidbook.location.proximity/.Pro
dalvikvm	LinearAlloc 0x0 used 638596 of 5242880 (12%)
ddm-heap	Got feature list request
LocationManager	Constructor: service = android.location.ILocationManager$Stub$Proxy@43d07da8
LocationManager	addProximityAlert: latitude = 30.334954, longitude = -81.5625, radius = 8045.0, expiration = 2000
GpsLocationProvider	setMinTime 1000
GpsLocationProvider	startNavigating
LocationManager	addProximityAlert: latitude = 28.54, longitude = -81.38, radius = 8045.0, expiration = 60000, inte
ActivityManager	Displayed activity com.androidbook.location.proximity/.ProximityActivity: 2164 ms (total 2164 ms)
dalvikvm	GC freed 540 objects / 46120 bytes in 126ms
dalvikvm	GC freed 1860 objects / 108208 bytes in 147ms
GpsLocationProvider	TTFF: 17381
ProximityReceiver	Got intent
ProximityReceiver	geo:28.54,-81.38
ProximityReceiver	Message: Orlando, FL
ProximityReceiver	Entering? true

Figure 22–11. *LogCat window with messages from our `BroadcastReceiver`*

Because these are broadcasts, we cannot rely on the order in which they are received. For example, if we're inside the Orlando circle and we jump inside the Jacksonville circle, we could receive the broadcast that says we're inside the Jacksonville circle *before* we get the broadcast that says we left the Orlando circle.

Because we're dealing with `Locations`, we're using the `geo` scheme for the URI, which is one of the known schemes and quite perfect for passing latitude and longitude information. You should note that the structure of the `geo` URI puts latitude before longitude, but when we use the `geo fix` command in our emulator console, we put longitude before latitude. This can trip you up if you're not paying attention, and you could end up spending a lot of time trying to debug your application when the problem is simply the order in which you're sending in location updates. You could always use a GPX or KML file to send in locations and preselect locations for testing where your circle will overlap with the path from that file.

Our sample application is very simple. In a real application, the `BroadcastReceiver` could do notifications or start a service. Instead of a broadcast, `PendingIntent` could be for an activity or a service, even in some other application. Our application could be a mentioned service instead.

References

Here is a helpful reference you may wish to explore further.

- `http://www.androidbook.com/proandroid4/projects`. A list of downloadable projects related to this book. For this chapter, look for a zip file called `ProAndroid4_Ch22_Maps.zip`. This zip file contains all projects from this chapter, listed in separate root directories. There is also a `README.TXT` file that describes exactly how to import projects into Eclipse from one of these zip files.
- `http://developer.android.com/guide/topics/location/index.html`. The Android developer's guide for Location and Maps.

- `http://code.google.com/android/add-ons/google-apis/maps-overview.html`. The Maps API documentation which is separate from the rest of the online Android documentation. The Maps API documentation includes the API Reference.

Summary

Let's conclude this chapter by quickly enumerating what you have learned about maps so far:

- How to get your own Maps API key from Google.
- `MapView` and `MapActivity`.
- The modifications you need to make to your `AndroidManifest.xml` file to get a maps application to work.
- Using a `MapView` tag in a layout, and where the Maps API key goes.
- The map controller.
- Zooming in and out, and the different ways to accomplish this.
- Including different modes such as satellite and traffic.
- Using overlays to add markers to your maps.
- Using the `zoomToSpan()` method to set a zoom level that accommodates a specific set of markers.
- The Geocoder, and how it converts from address to latitude/longitude, or from latitude/longitude to addresses and places of interest.
- Putting the `Geocoder` into a background thread to avoid nasty Application Not Responding (ANR) pop-ups.
- The `LocationManager` service, which uses GPS and/or network towers to pinpoint the location of the device.
- Selecting a location provider, and what to do if the desired location provider is not enabled.
- Using methods of the `Location` class to, for example, calculate distances between points
- Using the emulator's features to send location events to your application for testing. This includes using special files that record entire series of location events.
- Using `LocationOverlays`, a special type of map overlay, to easily show the device's current location on the map.
- Customizing the `LocationOverlay`.

- Alerting on proximity—that is, setting up a proximity and being alerted when the device enters or leaves that proximity.

Interview Questions

You can use the following questions as a guide to consolidate your understanding of this chapter:

1. Can a `MapView` be used without a `MapActivity`? Hint: this is why you can't currently use a `MapView` inside a fragment.
2. How is a Maps API key related to your keystore certificate?
3. What are the two changes you need to make to a maps application's AndroidManifest.xml file to make the app work? Why?
4. What are different ways to do zooming with maps?
5. What do you set to allow a map to be moved sideways?
6. What is an ItemizedOverlay used for?
7. How do you get rid of a marker's shadow?
8. Why do you call `setBounds()` on a marker `Drawable`?
9. Why would you want to restrict how many results come back from `Geocoder`?
10. When would it make sense to use the passive location provider?
11. Do you need to specify `ACCESS_COARSE_LOCATION` if you've already specified `ACCESS_FINE_LOCATION` in the `AndroidManifest.xml` file?
12. Why might `getLastKnownLocation()` not provide an accurate value for the device's current position?
13. Name some of the methods that you can call on a `Location` object.
14. What's the proper way to set the values for `minTime` and `minDistance` when invoking `requestLocationUpdates()`?
15. Where do you go in Eclipse to simulate GPS events for an emulated app?
16. How can you simulate GPS events from a command line?
17. Why do you need to return a good value from `isLocationDisplayed()`?
18. What are the parameters for setting up a proximity alert, and what do they mean?

Chapter 23

Using the Telephony APIs

Many Android devices are smartphones, but so far, we haven't talked about how to program applications that use phone features. In this chapter, we will show you how to send and receive Short Message Service (SMS) messages. We will also touch on several other interesting aspects of the telephony APIs in Android, including the Session Initiation Protocol (SIP) functionality. SIP is an Internet Engineering Task Force (IETF) standard for implementing Voice over Internet Protocol (VoIP) where the user can make telephone-like calls over the Internet. SIP can also handle video.

Working with SMS

SMS stands for Short Message Service, but it's commonly called *text messaging*. The Android SDK supports sending and receiving text messages. We'll start by discussing various ways to send SMS messages with the SDK.

Sending SMS Messages

To send a text message from your application, you need to add the `android.permission.SEND_SMS` permission to your manifest file and then use the `android.telephony.SmsManager` class. See Listing 23–1 for the layout XML file and the Java code for this example. If you need to see where the permission goes in the manifest XML file, you can sneak ahead to Listing 23–2.

> **NOTE:** We will give you a URL at the end of the chapter that you can use to download projects of this chapter. This will allow you to import these projects into Eclipse directly.

Listing 23–1. *Sending SMS (Text) Messages*

```
<?xml version="1.0" encoding="utf-8"?>
<!-- This file is /res/layout/main.xml -->
<LinearLayout xmlns:android="http://schemas.android.com/apk/res/android"
    android:orientation="vertical"
    android:layout_width="fill_parent"
```

```
        android:layout_height="fill_parent">

    <LinearLayout
        xmlns:android="http://schemas.android.com/apk/res/android"
        android:orientation="horizontal"
        android:layout_width="fill_parent"
        android:layout_height="wrap_content">

        <TextView android:layout_width="wrap_content"
            android:layout_height="wrap_content"
            android:text="Destination Address:" />

        <EditText android:id="@+id/addrEditText"
            android:layout_width="fill_parent"
            android:layout_height="wrap_content"
            android:phoneNumber="true"
            android:text="9045551212" />

    </LinearLayout>

    <LinearLayout
        xmlns:android="http://schemas.android.com/apk/res/android"
        android:orientation="vertical"
        android:layout_width="fill_parent"
        android:layout_height="wrap_content">

        <TextView android:layout_width="wrap_content"
            android:layout_height="wrap_content"
            android:text="Text Message:" />

        <EditText android:id="@+id/msgEditText"
            android:layout_width="fill_parent"
            android:layout_height="wrap_content"
            android:text="hello sms" />

    </LinearLayout>

    <Button android:id="@+id/sendSmsBtn"
        android:layout_width="wrap_content"
        android:layout_height="wrap_content"
        android:text="Send Text Message"
        android:onClick="doSend" />

</LinearLayout>


// This file is TelephonyDemo.java
import android.app.Activity;
import android.os.Bundle;
import android.telephony.SmsManager;
import android.view.View;
import android.widget.EditText;
import android.widget.Toast;

public class TelephonyDemo extends Activity
{
    @Override
```

```
    protected void onCreate(Bundle savedInstanceState) {
        super.onCreate(savedInstanceState);
        setContentView(R.layout.main);
    }

    public void doSend(View view) {
        EditText addrTxt =
              (EditText) findViewById(R.id.addrEditText);

        EditText msgTxt =
              (EditText) findViewById(R.id.msgEditText);

        try {
            sendSmsMessage(
                addrTxt.getText().toString(),
                msgTxt.getText().toString());
            Toast.makeText(this, "SMS Sent",
                    Toast.LENGTH_LONG).show();
        } catch (Exception e) {
            Toast.makeText(this, "Failed to send SMS",
                    Toast.LENGTH_LONG).show();
            e.printStackTrace();
        }
    }

    @Override
    protected void onDestroy() {
        super.onDestroy();
    }

    private void sendSmsMessage(String address,String message)throws Exception
    {
        SmsManager smsMgr = SmsManager.getDefault();
        smsMgr.sendTextMessage(address, null, message, null, null);
    }
}
```

The example in Listing 23–1 demonstrates sending SMS text messages using the Android SDK. Looking at the layout snippet first, you can see that the user interface has two `EditText` fields: one to capture the SMS recipient's destination address (the phone number) and another to hold the text message. The user interface also has a button to send the SMS message, as shown in Figure 23–1.

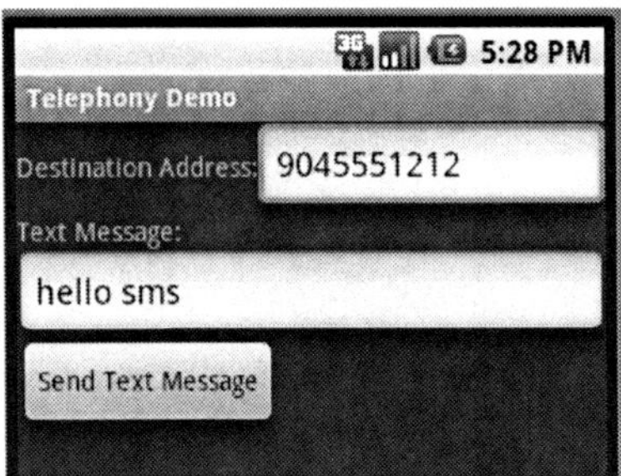

Figure 23–1. *The UI for the SMS example*

The interesting part of the sample is the sendSmsMessage() method. The method uses the SmsManager class's sendTextMessage() method to send the SMS message. Here's the signature of SmsManager.sendTextMessage():

```
sendTextMessage(String destinationAddress, String smscAddress,
    String textMsg, PendingIntent sentIntent,
    PendingIntent deliveryIntent);
```

In this example, you populate only the destination address and the text-message parameters. You can, however, customize the method so it doesn't use the default SMS center (the address of the server on the cellular network that will dispatch the SMS message). You can also implement a customization in which pending intents are broadcast when the message is sent (or failed) and when a delivery notification has been received.

There are two main steps to sending an SMS message: sending and delivering. As each step is reached, if provided by your application, a pending intent is broadcast. You can put whatever you want into the pending intent, such as the action, but the result code passed to your BroadcastReceiver will be specific to SMS sending or delivery. Also, you may get extra data related to radio errors or status reports depending on the implementation of the SMS system.

Without pending intents, your code can't tell if the text message was sent successfully or not. While testing, though, you can. If you launch this sample application in an emulator and launch *another* instance of an emulator (either from the command line or from the Eclipse Window ➤ Android SDK and AVD Manager screen), you can use the port number of the other emulator as the destination address. The port number is the number that appears in the emulator window title bar; it's usually something like 5554. After clicking the Send Text Message button, you should see a notification appear in the other emulator, indicating that your text message has been received on the other side.

The SMSManager class provides two other ways to send SMS messages:

- sendDataMessage() takes an additional argument to specify a port number and, instead of a String message, takes a byte array.
- sendMultipartTextMessage() allows for sending text messages when the whole message is larger than is allowed in the SMS specification. The sendMultipartTextMessage() method takes an array of Strings, but note that it also then takes an optional array of pending intents for both sending and delivery. The SMSManager class provides a divideMessage() method to help split up big messages into multiple parts.

All in all, sending an SMS message is about as simple as it gets with Android. Realize that, with the emulator, your SMS messages are not actually sent to their destinations. You can, however, assume success if the sendTextMessage() method returns without an exception. As shown in Listing 23–1, you can use the Toast class to display a message in the UI to indicate whether the SMS message was sent successfully.

Sending SMS messages is only half the story. Now, we'll show you how to monitor incoming SMS messages.

Monitoring Incoming SMS Messages

We're going to use the same application that you just created to send SMS messages, and we're going to add a BroadcastReceiver to listen for the action android.provider.Telephony.SMS_RECEIVED. This action is broadcast by Android when an SMS message is received by the device. When we register our receiver, our application will be notified whenever an SMS message is received. The first step in monitoring incoming SMS messages is to request permission to receive them. To do that, we must add the android.permission.RECEIVE_SMS permission to the manifest file. To implement the receiver, we must write a class that extends android.content.BroadcastReceiver and then register the receiver in the manifest file. Listing 23–2 includes both the AndroidManifest.xml file and our receiver class. Notice that both permissions are present in the manifest file because we still need the send permission for the activity we created earlier.

Listing 23–2. *Monitoring SMS Messages*

```
<?xml version="1.0" encoding="utf-8"?>
<!-- This file is AndroidManifest.xml -->
<manifest xmlns:android="http://schemas.android.com/apk/res/android"
    package="com.androidbook.telephony" android:versionCode="1"
    android:versionName="1.0">
    <application android:icon="@drawable/icon"
            android:label="@string/app_name">
        <activity android:name=".TelephonyDemo"
                android:label="@string/app_name">
          <intent-filter>
            <action android:name="android.intent.action.MAIN" />
            <category android:name="android.intent.category.LAUNCHER" />
          </intent-filter>
        </activity>
        <receiver android:name="MySMSMonitor">
          <intent-filter>
            <action
                android:name="android.provider.Telephony.SMS_RECEIVED"/>
          </intent-filter>
        </receiver>

    </application>
    <uses-sdk android:minSdkVersion="4" />

    <uses-permission android:name="android.permission.SEND_SMS"/>
    <uses-permission android:name="android.permission.RECEIVE_SMS"/>

</manifest>


// This file is MySMSMonitor.java
import android.content.BroadcastReceiver;
import android.content.Context;
import android.content.Intent;
```

```
import android.telephony.SmsMessage;
import android.util.Log;

public class MySMSMonitor extends BroadcastReceiver
{
    private static final String ACTION =
                "android.provider.Telephony.SMS_RECEIVED";
    @Override
    public void onReceive(Context context, Intent intent)
    {
        if(intent!=null && intent.getAction()!=null &&
                ACTION.compareToIgnoreCase(intent.getAction())==0)
        {
            Object[] pduArray= (Object[]) intent.getExtras().get("pdus");
            SmsMessage[] messages = new SmsMessage[pduArray.length];
            for (int i = 0; i<pduArray.length; i++) {
                messages[i] = SmsMessage.createFromPdu(
                                    (byte[])pduArray [i]);
                Log.d("MySMSMonitor", "From: " +
                        messages[i].getOriginatingAddress());
                Log.d("MySMSMonitor", "Msg: " +
                        messages[i].getMessageBody());
            }
            Log.d("MySMSMonitor","SMS Message Received.");
        }
    }
}
```

The top portion of Listing 23–2 is the manifest definition for the BroadcastReceiver to intercept SMS messages. The SMS monitor class is MySMSMonitor. The class implements the abstract onReceive() method, which is called by the system when an SMS message arrives. One way to test the application is to use the Emulator Control view in Eclipse. Run the application in the emulator, and choose Window ➤ Show View ➤ Other ➤ Android ➤ Emulator Control. The user interface allows you to send data to the emulator to emulate receiving an SMS message or phone call. As shown in Figure 23–2, you can send an SMS message to the emulator by populating the Incoming Number field and selecting the SMS radio button. Next, type some text in the Message field, and click the Send button. Doing this sends an SMS message to the emulator and invokes your BroadcastReceiver's onReceive() method.

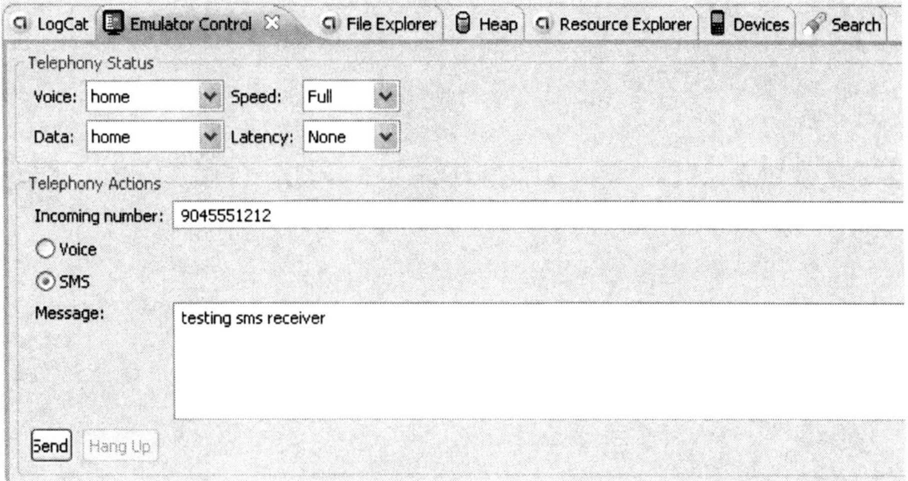

Figure 23–2. *Using the Emulator Control UI to send SMS messages to the emulator*

The onReceive() method will have the broadcast intent, which will contain the SmsMessage in the bundle property. You can extract the SmsMessage by calling intent.getExtras().get("pdus"). This call returns an array of objects defined in Protocol Description Unit (PDU) mode—an industry-standard way of representing an SMS message. You can then convert the PDUs to Android SmsMessage objects, as shown in Listing 23–2. As you can see, you get the PDUs as an object array from the intent. You then construct an array of SmsMessage objects, equal to the size of the PDU array. Finally, you iterate over the PDU array and create SmsMessage objects from the PDUs by calling SmsMessage.createFromPdu(). What you do after reading the incoming message must be quick. A broadcast receiver gets high priority in the system, but its task must be finished quickly, and it does not get put into the foreground for the user to see. Therefore, your options are limited. You should not do any direct UI work. Issuing a notification is fine, as is starting a service to continue work there. Once the onReceive() method completes, the hosting process of the onReceive() method could get killed at any time. Starting a service is OK, but binding to one is not, because that would require your process to exist for a while, which might not happen. For more information on BroadcastReceivers, see Chapter 19.

Now, let's continue our discussion about SMS by looking at how you can work with various SMS folders.

Working with SMS Folders

Accessing the SMS inbox is another common requirement. To get started, you need to add read SMS permission (android.permission.READ_SMS) to the manifest file. Adding this permission gives you the ability to read from the SMS inbox.

To read SMS messages, you need to execute a query on the SMS inbox, as shown in Listing 23–3.

Listing 23–3. *Displaying the Messages from the SMS Inbox*

```
<?xml version="1.0" encoding="utf-8"?>
<!-- This file is /res/layout/sms_inbox.xml -->
<LinearLayout xmlns:android="http://schemas.android.com/apk/res/android"
    android:orientation="vertical"
    android:layout_width="fill_parent"
    android:layout_height="fill_parent" >

  <TextView android:id="@+id/row"
    android:layout_width="fill_parent"
    android:layout_height="fill_parent"/>

</LinearLayout>


// This file is SMSInboxDemo.java
import android.app.ListActivity;
import android.database.Cursor;
import android.net.Uri;
import android.os.Bundle;
import android.widget.ListAdapter;
import android.widget.SimpleCursorAdapter;

public class SMSInboxDemo extends ListActivity {

    private ListAdapter adapter;
    private static final Uri SMS_INBOX =
            Uri.parse("content://sms/inbox");

    @Override
    public void onCreate(Bundle bundle) {
        super.onCreate(bundle);
        Cursor c = getContentResolver()
                .query(SMS_INBOX, null, null, null, null);
        startManagingCursor(c);
        String[] columns = new String[] { "body" };
        int[]      names = new int[]    { R.id.row };
        adapter = new SimpleCursorAdapter(this, R.layout.sms_inbox,
                c, columns, names);

        setListAdapter(adapter);
    }
}
```

Listing 23–3 opens the SMS inbox and creates a list in which each item contains the body portion of an SMS message. The layout portion of Listing 23–3 contains a simple `TextView` that will hold the body of each message in a list item. To get the list of SMS messages, you create a URI pointing to the SMS inbox (`content://sms/inbox`) and then execute a simple query. You then filter on the body of the SMS message and set the list adapter of the `ListActivity`. After executing the code from Listing 23–3, you'll see a list of SMS messages in the inbox. Make sure you generate a few SMS messages using the Emulator Control before running the code on the emulator.

Because you can access the SMS inbox, you would expect to be able to access other SMS-related folders, such as the Sent and Draft folders. The only difference between accessing the inbox and accessing the other folders is the URI you specify. For example, you can access the Sent folder by executing a query against `content://sms/sent`. Following is the complete list of SMS folders and the URI for each folder:

- *All*: `content://sms/all`
- *Inbox*: `content://sms/inbox`
- *Sent*: `content://sms/sent`
- *Draft*: `content://sms/draft`
- *Outbox*: `content://sms/outbox`
- *Failed*: `content://sms/failed`
- *Queued*: `content://sms/queued`
- *Undelivered*: `content://sms/undelivered`
- *Conversations*: `content://sms/conversations`

Android combines MMS and SMS and allows you to access content providers for both at the same time, using an AUTHORITY of `mms-sms`. Therefore, you can access a URI such as this:

```
content://mms-sms/conversations
```

Sending E-mail

Now that you've seen how to send SMS messages in Android, you might assume that you can access similar APIs to send e-mail. Unfortunately, Android does not provide APIs for you to send e-mail. The general consensus is that users don't want an application to start sending e-mail on their behalf without them knowing about it. Instead, to send e-mail, you have to go through a registered e-mail application. For example, you could use `ACTION_SEND` to launch the e-mail application, as shown in Listing 23–4.

Listing 23–4. *Launching the E-mail Application via an Intent*

```
Intent emailIntent=new Intent(Intent.ACTION_SEND);

String subject = "Hi!";
String body = "hello from android....";

String[] recipients = new String[]{"aaa@bbb.com"};
emailIntent.putExtra(Intent.EXTRA_EMAIL, recipients);

emailIntent.putExtra(Intent.EXTRA_SUBJECT, subject);
emailIntent.putExtra(Intent.EXTRA_TEXT, body);
emailIntent.setType("message/rfc822");

startActivity(emailIntent);
```

This code launches the default e-mail application and allows the user to decide whether to send the e-mail. Other extras you can add to an email intent include EXTRA_CC and EXTRA_BCC.

Let's assume you want to send an e-mail attachment with your message. To do this, you can use something like the following, where the Uri is a reference to the file you want as the attachment:

```
emailIntent.putExtra(Intent.EXTRA_STREAM,
    Uri.fromFile(new File(myFileName)));
```

Next, we'll talk about the telephony manager.

Working with the Telephony Manager

The telephony APIs also include the telephony manager (android.telephony.TelephonyManager), which you can use to obtain information about the telephony services on the device, get subscriber information, and register for telephony state changes. A common telephony use case requires that an application execute business logic on incoming phone calls. For example, a music player might pause itself for an incoming call and resume when the call has been completed. The easiest way to listen for phone state changes is to implement a broadcast receiver on android.intent.action.PHONE_STATE. You could do this in the same way we listened for incoming SMS messages earlier. The other way is to use TelephonyManager.

In this section, we are going to show you how to register for telephony state changes and how to detect incoming phone calls. Listing 23–5 shows the details.

Listing 23–5. *Using the Telephony Manager*

```
<?xml version="1.0" encoding="utf-8"?>
<!-- This file is res/layout/main.xml -->
<LinearLayout xmlns:android="http://schemas.android.com/apk/res/android"
    android:orientation="vertical"
    android:layout_width="fill_parent"
    android:layout_height="fill_parent"
    >
<Button
    android:id="@+id/callBtn"
    android:layout_width="wrap_content"
    android:layout_height="wrap_content"
    android:text="Place Call"
    android:onClick="doClick"
    />
<TextView
    android:id="@+id/textView"
    android:layout_width="fill_parent"
    android:layout_height="fill_parent"
    />
</LinearLayout>


// This file is PhoneCallActivity.java
package com.androidbook.phonecall.demo;
```

```
import android.app.Activity;
import android.content.Context;
import android.content.Intent;
import android.net.Uri;
import android.os.Bundle;
import android.telephony.PhoneStateListener;
import android.telephony.TelephonyManager;
import android.view.View;
import android.widget.TextView;

public class PhoneCallActivity extends Activity {
    private TelephonyManager teleMgr = null;
    private MyPhoneStateListener myListener = null;
    private String logText = "";
    private TextView tv;

    @Override
    protected void onCreate(Bundle savedInstanceState)
    {
        super.onCreate(savedInstanceState);
        setContentView(R.layout.main);

        tv = (TextView)findViewById(R.id.textView);

        teleMgr =
                (TelephonyManager)getSystemService(Context.TELEPHONY_SERVICE);
        myListener = new MyPhoneStateListener();
    }

    protected void onResume() {
        super.onResume();
        teleMgr.listen(myListener, PhoneStateListener.LISTEN_CALL_STATE);
    }

    protected void onPause() {
        super.onPause();
        teleMgr.listen(myListener, PhoneStateListener.LISTEN_NONE);
    }

    public void doClick(View target) {
        Intent intent = new Intent(Intent.ACTION_VIEW,
                Uri.parse("tel:5551212"));
        startActivity(intent);
    }

    class MyPhoneStateListener extends PhoneStateListener
    {
        @Override
        public void onCallStateChanged(int state, String incomingNumber)
        {
            super.onCallStateChanged(state, incomingNumber);

            switch(state)
            {
                case TelephonyManager.CALL_STATE_IDLE:
                    logText = "call state idle...incoming number is["+
```

```
                        incomingNumber + "]\n" + logText;
                break;
            case TelephonyManager.CALL_STATE_RINGING:
                logText = "call state ringing...incoming number is["+
                        incomingNumber + "]\n" + logText;
                break;
            case TelephonyManager.CALL_STATE_OFFHOOK:
                logText = "call state Offhook...incoming number is["+
                        incomingNumber + "]\n" + logText;
                break;
            default:
                logText = "call state [" + state +
                        "]incoming number is[" +
                        incomingNumber + "]\n" + logText;
                break;
            }
            tv.setText(logText);
        }
    }
}
```

When working with the telephony manager, be sure to add the android.permission.READ_PHONE_STATE permission to your manifest file, so you can access phone state information. As shown in Listing 23–5, you get notified about phone state changes by implementing a PhoneStateListener and calling the listen() method of the TelephonyManager. When a phone call arrives or the phone state changes, the system will call your PhoneStateListener's onCallStateChanged() method with the new state. As you will see when you try this out, the incoming phone number is only available when the state is CALL_STATE_RINGING. You write a message to the screen in this example, but your application could implement custom business logic in its place, such as pausing the playback of audio or video. To emulate incoming phone calls, you can use Eclipse's Emulator Control UI—the same one you used to send SMS messages (see Figure 23–2), but choose Voice instead of SMS.

Notice that we tell the TelephonyManager to stop sending us updates in onPause(). It is always important to turn off messages when our activity is being paused. Otherwise, the telephony manager could keep a reference to our object and prevent it from being cleaned up later.

This example deals with only one of the phone states that are available for listening. Check out the documentation on PhoneStateListener for others, including for example LISTEN_MESSAGE_WAITING_INDICATOR. When dealing with phone state changes, you might also need to get the subscriber's (user's) phone number; TelephonyManager.getLine1Number() will return that for you.

You may be wondering if it's possible to answer a phone via code. Unfortunately, at this time, the Android SDK does not provide a way to do this, even though the documentation implies that you can fire off an intent with an action of ACTION_ANSWER. In practice, this approach does not yet work, although you may want to check to see if this has been fixed since the time of this writing.

Similarly, you may want to place an outbound phone call via code. Here, you will find things easier. The simplest way to make an outbound call is to invoke the Dialer application via an intent with code such as the following:

```
Intent intent = new Intent(Intent.ACTION_CALL, Uri.parse("tel:5551212"));
startActivity(intent);
```

Note that for this to actually dial, your application will need the `android.permission.CALL_PHONE` permission. Otherwise, when your application attempts to invoke the Dialer application, you will get a `SecurityException`. To do dialing without this permission, change the action of the intent to `Intent.ACTION_VIEW`, which will cause the Dialer application to appear with your desired number to dial, but the user will need to click the Send button to initiate the call.

One other thing to keep in mind when dealing with phone features in your application is that other applications could very well respond to incoming phone calls and cause your activity to pause. In that case, you'll stop receiving notifications, although you will get an immediate notification when your `onResume()` method is called again and you reregister with `TelephonyManager`. Be prepared for that when deciding what to do in your handler for phone state notifications.

Your other option for detecting changes in the phone's state is to register a broadcast receiver for phone state changes (`android.intent.action.PHONE_STATE`). This can be done in code, or you can specify a `<receiver>` tag in your manifest file. See Chapter 19 for more information about broadcast receivers.

Session Initiation Protocol (SIP)

Android 2.3 (Gingerbread) introduced new features to support SIP, in the `android.net.sip` package. SIP is an IETF standard for orchestrating the sending of voice and video over a network connection to link people together in calls. This technology is sometimes called Voice over IP (VoIP), but note that there is more than one way to do VoIP. Skype for instance, uses a proprietary protocol to do VoIP and is incompatible with SIP. SIP also is not the same as Google Voice. Google Voice does not (as of this writing) support SIP directly, although there are ways to integrate Google Voice with a SIP provider to tie things together. Google Voice sets up a new telephone number for you that you can then connect with other phones such as your home, work, or mobile phone. Some SIP providers will generate a telephone number that can be used with Google Voice, but in this case Google Voice does not really know that the number is for a SIP account. A search of the Internet will reveal quite a few SIP providers, many with reasonable calling rates, and some that are free.

It is important to note that the SIP standard does not address passing audio and video data over a network. SIP is only involved in setting up and tearing down the direct connections between devices to allow audio and video data to flow. Client computer programs use SIP, as well as audio and video codecs and other libraries, to set up calls between users. Other standards often involved with SIP calls include Real-time Transport Protocol (RTP), Real-time Streaming Protocol (RTSP) and Session Description Protocol (SDP). Android 3.1 added direct support for RTP in the `android.net.rtp` package. RTSP has been supported by the MediaPlayer for some time, although not all RTSP servers are compatible with Android's MediaPlayer. SDP is an application-level protocol for describing multimedia sessions, so you see message content in SDP format.

Users can make SIP calls from desktop computers without incurring long-distance charges. The computer program can just as easily be running on a mobile device such as an Android smartphone or tablet. SIP computer programs are often called *soft phones*. The real advantage of a soft phone on a mobile device occurs when the device is connecting to the Internet using Wi-Fi, so that the user is not using any wireless minutes but is still able to make or receive a call. On the receiving end, a soft phone must have registered its location and capabilities with a SIP provider so the provider's SIP server can respond to invite requests to set up the direct connection. If the receiver's soft phone is not available, the SIP server can direct the inbound request to a voicemail account, for example.

Experimenting with SipDemo

Google provides a demonstration application for SIP called SipDemo. We'd like to explore that application with you now and help you understand how it works. Certain aspects are not obvious if you are new to SIP. If you'd like to experiment with SipDemo, you're probably going to need a physical Android device that supports SIP. This is because the Android emulators, as of this writing, do not support SIP (or Wi-Fi for that matter). There are some attempts on the Internet to make SIP work in the emulator, and by the time you read this, some may be easy to implement and robust.

To play with SipDemo, you will also need to get a SIP account from a SIP provider. You will need to have your SIP ID, SIP domain name (or proxy), and your SIP password. These will be plugged into the SipDemo application's preferences screen to be used by the application. Finally, you will need a Wi-Fi connection from your device to the Internet. If you don't want to actually experiment with SipDemo on a device, you should still be able to understand the rest of this section. The SipDemo looks like Figure 23–3.

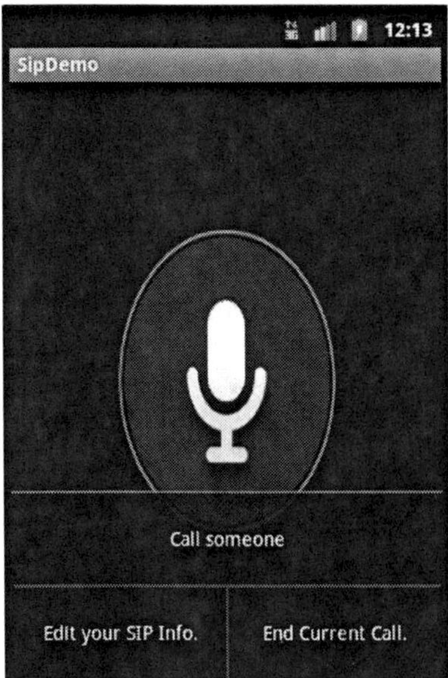

Figure 23–3. *The SipDemo application with the menu showing*

To load SipDemo as a new project into Eclipse, use the New Android Project wizard, but click the Create Project from Existing Sample option, choose Android 2.3 or higher in the Build Target section, and use the drop-down Samples menu to choose SipDemo. Click Finish, and Eclipse will create the new project for you. You can run this project with no changes to it, but as mentioned before, it won't do anything unless the device supports SIP, Wi-Fi is enabled, you've got a SIP account somewhere, you've used the Menu button to edit your SIP info, and you use the Menu button to initiate a call. You will need some other SIP account to call to test out the application. Pressing the big microphone image on the screen allows you to talk to the other side. This demo application can also receive an incoming call. Now, let's talk about the inner workings of the `android.net.sip` package.

The android.net.sip package

The package has four basic classes: `SipManager`, `SipProfile`, `SipSession`, and `SipAudioCall`. `SipManager` is at the core of this package and provides access to the rest of the SIP functionality. You invoke the static `newInstance()` method of `SipManager` to get a `SipManager` object. With a `SipManager` object, you can then get a `SipSession` for most SIP activity, or you can get a `SipAudioCall` for an audio-only call. This means Google has provided features in the `android.net.sip` package beyond what standard SIP provides, namely the ability to setup an audio call.

`SipProfile` is used to define the SIP accounts that will be talking to each other. This does not point directly to an end user's device, but rather the SIP account at a SIP provider. The servers will assist in the rest of the details to set up actual connections.

A SipSession is where the magic happens. Setting up a session includes your SipProfile so your application can make itself known to your SIP provider's server. You also pass a SipSession.Listener instance, which is going to be notified when things are happening. Once you've set up a SipSession object, your application is ready to make calls to another SipProfile or to receive incoming calls. The listener has a bunch of callbacks so your application can properly deal with the changing states of the session.

As of Honeycomb, the easiest thing to do is use SipAudioCall. The logic is all there to hook up the microphone and the speaker to the data streams so that you can carry on a conversation with the other side. There are lots of methods on SipAudioCall for managing mute, hold, and so on. All of the audio pieces are also handled for you. For anything more than that, you have work to do.

The SipSession class has the makeCall() method for placing an outbound call. The main parameter is the session description (as a String). This is where things require more work. Building a session description requires formatting according to SDP. Understanding a received session description means parsing it according to SDP. The standards documentation for SDP is here: http://tools.ietf.org/html/rfc4566; unfortunately, the Android SDK does not provide any support for SDP. Thanks to some very kind people, there are a couple of free SIP applications for Android that have built this capability. They are sipdroid (http://code.google.com/p/sipdroid/) and CSipSimple (http://code.google.com/p/csipsimple/).

We haven't even started talking about the codecs for managing video streams between SIP clients, although Sipdroid has this capability. Other aspects of SIP that are very appealing are the ability to set up conference calls among more than two people. These topics are beyond the scope of this book, but we hope you can appreciate what SIP can do for you.

Note that SIP applications will need at a minimum the android.permission.USE_SIP and android.permission.INTERNET permissions in order to function properly. If you use SipAudioCall, you also need the android.permission.RECORD_AUDIO permission. And assuming you're using Wi-Fi, you should add android.permission.ACCESS_WIFI_STATE and android.permission.WAKE_LOCK. It is also a good idea to add the following tag to your AndroidManifest.xml file, as a child of <manifest>, so that your application will only be installable on devices that have hardware support for SIP:

```
<uses-feature android:name="android.hardware.sip.voip" />
```

References

Here are some helpful links to topics you may wish to explore further:

- http://androidbook.com/proandroid4/projects: A list of downloadable projects related to this book. For this chapter, look for the file ProAndroid4_Ch23_Telephony.zip. This zip file contains all projects from this chapter, listed in separate root directories. There is also a README.TXT file that describes exactly how to import projects into Eclipse from one of these zip files.

- `http://en.wikipedia.org/wiki/Session_Initiation_Protocol`: The Wikipedia page for SIP.
- `http://tools.ietf.org/html/rfc3261`: The official IETF standard for SIP.
- `http://tools.ietf.org/html/rfc4566`: The official IETF standard for SDP.
- `www.ietf.org/rfc/rfc3551.txt`: The official IETF standard for RTP. See also the `android.net.rtp` package.
- `http://code.google.com/p/sipdroid/`, `http://code.google.com/p/csipsimple/`: Two open source applications for Android that implement SIP clients.

Summary

This chapter talked about the Android telephony APIs:

- Sending and receiving an SMS message
- SMS folders and reading SMS messages
- Sending e-mail from an application
- `TelephonyManager` and how to detect an incoming call
- Using SIP to create a VoIP client program

Interview Questions

You can use the following questions as a guide to consolidate your understanding of this chapter:

1. Can an SMS message contain more than 140 characters?
2. True or false: You get an `SmsManager` instance by calling `Context.getSystemService(MESSAGE_SERVICE)`.
3. Where is the ADT feature that allows you to send a test SMS message to an emulator?
4. Can an application send an e-mail without the user knowing?
5. Can an application send an SMS message without the user knowing?
6. Can an application make a phone call without the user knowing?
7. Is SIP the same as Skype?
8. What are the four main classes of the `android.net.sip` package?
9. Which SIP class defines the SIP accounts that will be talking to each other?

10. What tag do you put into the `AndroidManifest.xml` file to ensure that a SIP app is only seen by devices that support SIP?
11. What permissions are needed to make SIP work properly?

Chapter 24

Understanding the Media Frameworks

Now we are going to explore a very interesting part of the Android SDK: the media frameworks. We will show you how to play and record audio and video, from a variety of sources. We'll also cover how to take photos with the camera. Any discussion of media would be incomplete without explaining Secure Digital (SD) cards and how to work with them, because you'll use SD cards often to read and write media files.

Using the Media APIs

Android supports playing audio and video content under the `android.media` package. In this chapter, we are going to explore the media APIs from this package.

At the heart of the `android.media` package is the `android.media.MediaPlayer` class. The `MediaPlayer` class is responsible for playing both audio and video content. The content for this class can come from the following sources:

- *Web*: You can play content from the Web via a URL.
- `.apk` *file*: You can play content that is packaged as part of your `.apk` file. You can package the media content as a resource or as an asset (within the `assets` folder).
- *SD card*: You can play content that resides on the device's SD card.

The MediaPlayer is capable of decoding quite a few different content formats, including 3rd Generation Partnership Project (3GPP, `.3gp`), MP3 (`.mp3`), MIDI (`.mid` and others), Ogg Vorbis (`.ogg`), PCM/WAVE (`.wav`), and MPEG-4 (`.mp4`). RTSP, HTTP/HTTPS live streaming, and M3U playlists are also supported, although playlists that include URLs are not, at least as of this writing. For a complete list of supported media formats, go to `http://developer.android.com/guide/appendix/media-formats.html`.

Using SD Cards

Before we get into creating and using our different types of media, let's look at how to work with SD cards. SD cards are used in Android phones for storing lots of user data, usually media content such as pictures, audio, and video. They are basically pluggable memory chips that keep their data even when they lose power. On a real phone, the SD card plugs into a memory slot and is accessible to the device. Many devices have one slot, and it's not expected that you will replace the SD card. On some devices, you can have multiple cards, switching among them with your device, and you can use them across different devices. Fortunately for us, the Android emulator can simulate SD cards, using space on your workstation's hard drive as if it were a plug-in SD card.

When you created your first Android Virtual Device (AVD) in Chapter 2, you specified a size for an SD card, which made it available to your application when you ran it in the emulator. If you look inside the AVD directory that was created, you will see a file called `sdcard.img` with the file size you specified. We didn't use the SD card then, but we'll be using it in this chapter.

As a developer, once you have an SD card, you can use the Android tools within Eclipse to push media files (or any other files) to the SD card. You can also use the Android Debug Bridge (`adb`) utility to push or pull files to and from an SD card. The `adb` utility is located in the `tools` subdirectory of the Android SDK; it is easy to get to from a tools window, as described in Chapter 2.

You already know how to get an SD card by creating an AVD. And, of course, you could create lots of AVDs that are the same except for the size of the SD card. Here's the other way to go: the Android SDK tools bundle contains a utility called `mksdcard` that can create an SD card image. Actually, the utility creates a formatted file that is used as an SD card. To use this utility, first find or create a folder for the image file, at `c:\Android\sdcard\`, for example. Then open a tools window and run a command like the following, using an appropriate path to the SD card image file:

```
mksdcard 256M c:\Android\sdcard\sdcard.img
```

This example command creates an SD card image at `c:\Android\sdcard\` with the file name `sdcard.img`. The size of the SD card will be 256MB. To specify other sizes, you can use K for kilobytes; G doesn't work yet for gigabytes, so you'll need to specify multiples of 1024MB to get gigabyte sizes. You can also simply specify an integer value representing the total number of bytes. Also note that the Android emulator won't work with SD card sizes below 8MB.

The Android Development Tools (ADT) in Eclipse offer a way to specify extra command-line arguments when launching the emulator. To find the field for the emulator options, go to the Preferences window of Eclipse, and then choose Android ➤ Launch. In theory, you could add `-sdcard "PATH_TO_YOUR_SD_CARD_IMAGE_FILE"` here, and it would override the SD card file path for your AVD. But this hasn't worked for a few Android releases now, and you always get the SD card image file that was created along with the AVD. The most reliable way to use a separate SD card with your AVD is to launch the emulator from the command line and specify the SD card image to use there. From within a tools window (to see how to get a tools window, refer to Chapter 2), the following command launches a named AVD but uses the specified SD card image file instead of the SD card image file that was created with the AVD:

```
emulator -avd AVDName -sdcard "PATH_TO_YOUR_SD_CARD_IMAGE_FILE"
```

When your SD card is first created, there are no files on it. You can add files by using the File Explorer tool in Eclipse. Start the emulator, and wait until it initializes. Then go to the Java, Debug, or Dalvik Debug Monitor Service (DDMS) perspective in Eclipse, and look for the File Explorer tab, as shown in Figure 24–1.

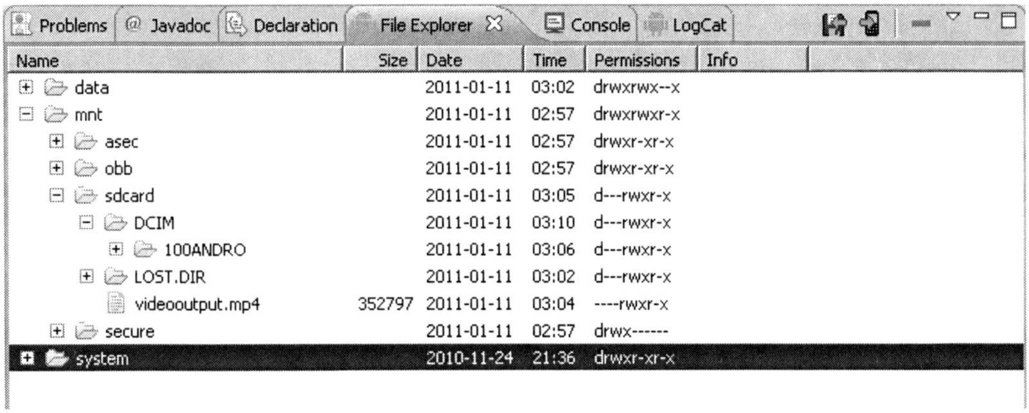
Problems
Javadoc
Declaration
File Explorer
Console
LogCat
Name
Size
Date
Time
Permissions
Info
data 2011-01-11 03:02 drwxrwx--x
mnt 2011-01-11 02:57 drwxrwxr-x
asec 2011-01-11 02:57 drwxr-xr-x
obb 2011-01-11 02:57 drwxr-xr-x
sdcard 2011-01-11 03:05 d---rwxr-x
DCIM 2011-01-11 03:10 d---rwxr-x
100ANDRO 2011-01-11 03:06 d---rwxr-x
LOST.DIR 2011-01-11 03:02 d---rwxr-x
videooutput.mp4 352797 2011-01-11 03:04 ----rwxr-x
secure 2011-01-11 02:57 drwx------
system 2010-11-24 21:36 drwxr-xr-x

Figure 24–1. *The File Explorer view*

If the File Explorer is not shown, you can bring it up by going to Window ➤ Show View ➤ Other ➤ Android and selecting File Explorer. Or, you can show the DDMS perspective by going to Window ➤ Open Perspective ➤ Other ➤ DDMS. The File Explorer view is by default on the DDMS perspective. The list of available views in Eclipse for Android is shown in Figure 24–2.

Figure 24–2. *Enabling Android views*

To push a file onto the SD card, select the `sdcard` folder in the File Explorer and click the button with the right-facing arrow (in the upper-right corner) pointing into what looks like a phone. This launches a dialog box that lets you select a file. Select the file that you want to upload to the SD card. The button next to it looks like a left arrow pointing into a floppy disk. Click this button for pulling a file from the device onto your workstation, after selecting the file you want to pull from within the File Explorer.

If the File Explorer displays an empty view, you either don't have the emulator running, Eclipse has disconnected from the emulator, or the AVD that you are running in the emulator is not selected under the Devices tab. To get a Devices tab, follow the same procedure described for the File Explorer. Devices should also be available by default on the DDMS perspective.

The other way to move files onto and off of the SD card is to use the `adb` utility. To try this, open a tools window, and then type a command such as

```
adb push c:\path_to_my_file\filename /mnt/sdcard/newfile
```

This will push a file from your workstation to the SD card. Note that the device always uses forward slashes to separate directories. Use whatever directory separator character is appropriate for your workstation for the file that's being pushed, and use an appropriate path for the file on your workstation. Conversely, the following command will pull a file from the SD card to your workstation:

```
adb pull /mnt/sdcard/devicefile c:\path_to_where_its_going\filename
```

One of the nice features of this command is that it will create directories as needed, in either direction (push or pull), to get the file to the desired destination. Unfortunately, you cannot use adb to copy multiple files at the same time. You must do each file separately.

> **NOTE:** Until Android 2.2, the SD card was most likely at /sdcard. Since Android 2.2, the SD card is most likely at /mnt/sdcard, however, there is a symbolic link called /sdcard that points to `/mnt/sdcard` for backward compatibility.

You may have noticed a directory on the SD card called `DCIM`. This is the Digital Camera Images directory. It is an industry standard to put a `DCIM` directory within the root directory of an SD card that's used for digital images. It's also an industry standard to put a directory underneath `DCIM` that represents a camera, in the format *123ABCDE*—three digits followed by five letters. The emulator creates a directory called `100ANDRO` under `DCIM`, but makers of digital cameras, and Android phone makers, can call this directory whatever they want. The emulator—and some Android phones—has a directory called `Camera` under the `DCIM` directory, but this isn't compliant with the standard. Nevertheless, you may find image files under `Camera` and you may find them under `100ANDRO`, or you may find some other directory under `DCIM` where image files are stored.

Unfortunately, there is not a method call to tell you which directory might be used underneath the `DCIM` directory for camera pictures. There are a couple of methods though to tell you where the top of the SD card is. The first is `Environment.getExternalStorageDirectory()` and it returns a `File` object for the top-level directory for the SD card. On pre-Android 2.2 devices, this was most likely `/sdcard`, but not on all devices. With Android 2.2, most devices will have `/mnt/sdcard`. It is much better to use this `Environment` method than to assume you know the name of the SD card's root directory. We will describe the other method next.

Since Android 2.2 (a.k.a. Froyo), there are some new constants available in the `Environment` class for locating directories, and there's also a new method in this class for locating directories. Previously, the SD card was a bit of a free-for-all, with no standardized directory names other than `DCIM`. With Froyo, there are several standardized directory names, as described in Table 24–1. The third column is the directory name used in the emulator, where the top of the SD card will most likely be `/mnt/sdcard` (this may vary by device). The variance in directories is why you should always use an `Environment` method to find the desired directory on the SD card.

Table 24–1. *The Standardized Directories of the SD card*

Directory Constant	Description	Directory in Emulator from Top of SD Card
`DIRECTORY_ALARMS`	When Android looks for audio files to use for alarms, it looks in this standard directory.	`Alarms`
`DIRECTORY_DCIM`	Industry-standard directory to look in for pictures and video taken using the camera.	`DCIM`
`DIRECTORY_DOWNLOADS`	Standard directory to hold files the user has downloaded.	`Download` (note: not plural)
`DIRECTORY_MOVIES`	When Android looks for movie files for the user, it looks in this standard directory.	`Movies`
`DIRECTORY_MUSIC`	When Android looks for audio files to use as regular music for the user to listen to, it looks in this standard directory.	`Music`
`DIRECTORY_NOTIFICATIONS`	When Android looks for audio files to use for notifications, it looks in this standard directory.	`Notifications`
`DIRECTORY_PICTURES`	When Android looks for image files not taken by the camera, it looks in this standard directory.	`Pictures`
`DIRECTORY_PODCASTS`	When Android looks for audio files to use as podcasts, it looks in this standard directory.	`Podcasts`
`DIRECTORY_RINGTONES`	When Android looks for audio files to use for ringtones, it looks in this standard directory.	`Ringtones`

The new method for locating directories is `Environment.getExternalStoragePublicDirectory(String type)`, where the `type` parameter is one of the constants from Table 24–1. This method returns a `File` object representing the requested directory. This method doesn't exist on older devices (older than Froyo), and even on newer devices you may find you need to accommodate differences. For example, Samsung has devices with two SD cards, so these methods are not sufficient to figure out all external storage on those.

And finally, a word about security. You need to add this permission to your manifest file in order for your application to be able to write to the SD card:

```
<uses-permission android:name="android.permission.WRITE_EXTERNAL_STORAGE" />
```

Now that you know the basics of SD cards, let's get into audio.

Playing Media

To get started, we'll show you how to build a simple application that plays an MP3 file located on the Web (see Figure 24–3). After that, we will talk about using the setDataSource() method of the MediaPlayer class to play content from the .apk file or the SD card. MediaPlayer isn't the only way to play audio, though, so we'll also cover the SoundPool class, as well as JetPlayer, AsyncPlayer, and, for the lowest level of working with audio, the AudioTrack class. After that, we will discuss some of the shortfalls of the MediaPlayer class. Finally, we'll see how to play video content.

Playing Audio Content

Figure 24–3 shows the user interface for our first example. This application will demonstrate some of the fundamental uses of the MediaPlayer class, such as starting, pausing, restarting, and stopping the media file. Look at the layout for the application's user interface.

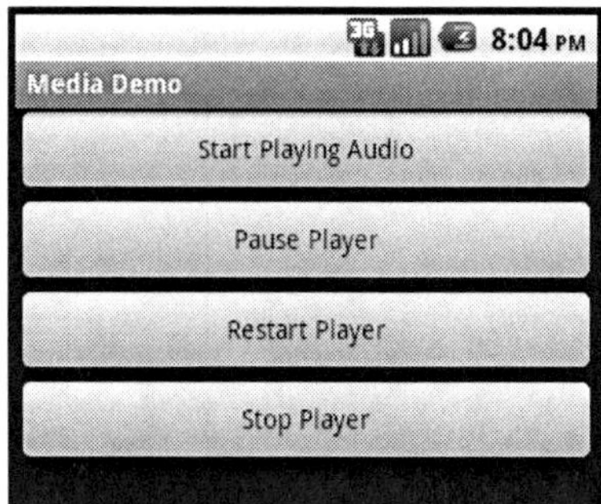

Figure 24–3. *The user interface for the media application*

The user interface consists of a LinearLayout with four buttons: one to start the player, one to pause the player, one to restart the player, and one to stop the player. The code and layout file for the application are shown in Listing 24–1. We're going to assume you're building against Android 2.2 or later for this example, because we're using the getExternalStoragePublicDirectory() method of Environment. If you want to build this against an older version of Android, simply use getExternalStorageDirectory() instead, and adjust where you put the media files so your application will find them.

NOTE: See the "References" section at the end of this chapter for the URL from which you can import these projects into Eclipse directly, instead of copying and pasting code.

Listing 24–1. *The Layout and Code for the Media Application*

```
<?xml version="1.0" encoding="utf-8"?>
<!-- This file is /res/layout/main.xml -->
<LinearLayout xmlns:android="http://schemas.android.com/apk/res/android"
    android:layout_width="fill_parent"
    android:layout_height="fill_parent"
    android:orientation="vertical" >

  <Button android:id="@+id/startPlayerBtn"
    android:layout_width="fill_parent"
    android:layout_height="wrap_content"
    android:text="Start Playing Audio" android:onClick="doClick" />

  <Button android:id="@+id/pausePlayerBtn"
    android:layout_width="fill_parent"
    android:layout_height="wrap_content"
    android:text="Pause Player" android:onClick="doClick" />

  <Button android:id="@+id/restartPlayerBtn"
    android:layout_width="fill_parent"
    android:layout_height="wrap_content"
    android:text="Restart Player" android:onClick="doClick" />

  <Button android:id="@+id/stopPlayerBtn"
    android:layout_width="fill_parent"
    android:layout_height="wrap_content"
    android:text="Stop Player" android:onClick="doClick" />
</LinearLayout>


// This file is MainActivity.java
import android.app.Activity;
import android.content.res.AssetFileDescriptor;
import android.media.MediaPlayer;
import android.media.MediaPlayer.OnPreparedListener;
import android.os.Bundle;
import android.os.Environment;
import android.util.Log;
import android.view.View;

public class MainActivity extends Activity implements OnPreparedListener
{
    static final String AUDIO_PATH =
     "http://www.androidbook.com/akc/filestorage/android/documentfiles/3389/play.mp3";
//   "http://listen.radionomy.com/Radio-Mozart";
//    Environment.getExternalStoragePublicDirectory(
//        Environment.DIRECTORY_MUSIC) +
//        "/music_file.mp3";
//    Environment.getExternalStoragePublicDirectory(
//        Environment.DIRECTORY_MOVIES) +
//        " /movie.mp4";
```

```
    private MediaPlayer mediaPlayer;
    private int playbackPosition=0;

    /** Called when the activity is first created. */
    @Override
    public void onCreate(Bundle savedInstanceState) {
        super.onCreate(savedInstanceState);
        setContentView(R.layout.main);
    }

    public void doClick(View view) {
        switch(view.getId()) {
        case R.id.startPlayerBtn:
            try {
            // Only have one of these play methods uncommented
                playAudio(AUDIO_PATH);
//              playLocalAudio();
//              playLocalAudio_UsingDescriptor();
            } catch (Exception e) {
                e.printStackTrace();
            }
            break;
        case R.id.pausePlayerBtn:
            if(mediaPlayer != null && mediaPlayer.isPlaying()) {
                playbackPosition = mediaPlayer.getCurrentPosition();
                mediaPlayer.pause();
            }
            break;
        case R.id.restartPlayerBtn:
            if(mediaPlayer != null && !mediaPlayer.isPlaying()) {
                mediaPlayer.seekTo(playbackPosition);
                mediaPlayer.start();
            }
            break;
        case R.id.stopPlayerBtn:
            if(mediaPlayer != null) {
                mediaPlayer.stop();
                playbackPosition = 0;
            }
            break;
        }
    }

    private void playAudio(String url) throws Exception
    {
        killMediaPlayer();

        mediaPlayer = new MediaPlayer();
        mediaPlayer.setAudioStreamType(AudioManager.STREAM_MUSIC);
        mediaPlayer.setDataSource(url);
        mediaPlayer.setOnPreparedListener(this);
        mediaPlayer.prepareAsync();
    }

    private void playLocalAudio() throws Exception
    {
```

```
        mediaPlayer = MediaPlayer.create(this, R.raw.music_file);
        mediaPlayer.setAudioStreamType(AudioManager.STREAM_MUSIC);
        // calling prepare() is not required in this case
        mediaPlayer.start();
    }

    private void playLocalAudio_UsingDescriptor() throws Exception {

        AssetFileDescriptor fileDesc = getResources().openRawResourceFd(
                    R.raw.music_file);
        if (fileDesc != null) {

            mediaPlayer = new MediaPlayer();
            mediaPlayer.setAudioStreamType(AudioManager.STREAM_MUSIC);
            mediaPlayer.setDataSource(fileDesc.getFileDescriptor(),
                    fileDesc.getStartOffset(), fileDesc.getLength());

            fileDesc.close();

            mediaPlayer.prepare();
            mediaPlayer.start();
        }
    }

    // This is called when the MediaPlayer is ready to start
    public void onPrepared(MediaPlayer mp) {
        mp.start();
    }

    @Override
    protected void onDestroy() {
        super.onDestroy();
        killMediaPlayer();
    }

    private void killMediaPlayer() {
        if(mediaPlayer!=null) {
            try {
                mediaPlayer.release();
            }
            catch(Exception e) {
                e.printStackTrace();
            }
        }
    }
}
```

In this first scenario, you are playing an MP3 file from a web address. Therefore, you will need to add `android.permission.INTERNET` to your manifest file. Listing 24–1 shows that the `MainActivity` class contains three members: a `final` string that points to the URL of the MP3 file, a `MediaPlayer` instance, and an integer member called `playbackPosition`. Our `onCreate()` method just sets up the user interface from our layout XML file. In the button-click handler, when the Start Playing Audio button is pressed, the `playAudio()` method is called. In the `playAudio()` method, a new instance of the `MediaPlayer` is created, and the data source of the player is set to the URL of the MP3 file.

The prepareAsync() method of the player is then called to prepare the MediaPlayer for playback. We're in the main UI thread of our activity, so we don't want to take too long to prepare the MediaPlayer. There is a prepare() method on MediaPlayer, but it blocks until the prepare is complete. If this takes a long time, the user could think the application is stuck or, worse, get an error message. The prepareAsync() method returns immediately but sets up a background thread to handle the prepare() method of the MediaPlayer. When the preparation is complete, our activity's onPrepared() callback is called. This is where we ultimately start the MediaPlayer playing. We have to tell the MediaPlayer who the listener is for the onPrepared() callback, which is why we call setOnPreparedListener() just before the call to prepareAsync(). You don't have to use the current activity as the listener; we do here because it's simpler for this demonstration.

Now look at the code for the Pause Player and Restart Player buttons. You can see that when the Pause Player button is selected, you get the current position of the player by calling getCurrentPosition(). You then pause the player by calling pause(). When the player has to be restarted, you call seekTo(), passing in the position obtained earlier from getCurrentPosition(), and then call start().

The MediaPlayer class also contains a stop() method. Note that if you stop the player by calling stop(), you need to prepare the MediaPlayer again before calling start() again. Conversely, if you call pause(), you can call start() again without having to prepare the player. Also, be sure to call the release() method of the media player once you are done using it. In this example, you do this as part of the killMediaPlayer() method.

There is a second URL in the sample application for an audio source, but it is not an MP3 file, it's a streaming audio feed (Radio-Mozart). This also works with the MediaPlayer and shows again why you need to call prepareAsync() instead of prepare(). Preparing an audio stream for playback can take a while, depending on the server, network traffic, and so on.

Listing 24–1 shows you how to play an audio file located on the Web. The MediaPlayer class also supports playing media local to your .apk file. Listing 24–2 shows how to reference and play back a file from the /res/raw folder of your .apk file. Go ahead and add the raw folder under /res if it's not already there in the Eclipse project. Then, copy the MP3 file of your choice into /res/raw with the file name music_file.mp3.

Listing 24–2. *Using the MediaPlayer to Play Back a File Local to the Application*

```
    private void playLocalAudio()throws Exception
    {
        mediaPlayer = MediaPlayer.create(this, R.raw.music_file);
        mediaPlayer.setAudioStreamType(AudioManager.STREAM_MUSIC);        // calling
prepare() is not required in this case
        mediaPlayer.start();
    }
```

If you need to include an audio or video file with your application, you should place the file in the /res/raw folder. You can then get a MediaPlayer instance for the resource by passing in the resource ID of the media file. You do this by calling the static create()

method, as shown in Listing 24–2. Note that the MediaPlayer class provides a few other static create() methods that you can use to get a MediaPlayer rather than instantiating one yourself. In Listing 24–2, the create() method is equivalent to calling the constructor MediaPlayer(Context context,int resourceId) followed by a call to prepare(). You should only use the create() method when the media source is local to the device, because it always uses prepare() and not prepareAsync().

Understanding the setDataSource Method

In Listing 24–2, we called the create() method to load the audio file from a raw resource. With this approach, you don't need to call setDataSource(). Alternatively, if you instantiate the MediaPlayer yourself using the default constructor, or if your media content is not accessible through a resource ID or a Uri, you'll need to call setDataSource().

The setDataSource() method has overloaded versions that you can use to customize the data source for your specific needs. For example, Listing 24–3 shows how you can load an audio file from a raw resource using a FileDescriptor.

Listing 24–3. *Setting the MediaPlayer's data source using a FileDescriptor*

```
private void playLocalAudio_UsingDescriptor() throws Exception {

    AssetFileDescriptor fileDesc = getResources().openRawResourceFd(
            R.raw.music_file);
    if (fileDesc != null) {

        mediaPlayer = new MediaPlayer();
        mediaPlayer.setAudioStreamType(AudioManager.STREAM_MUSIC);
        mediaPlayer.setDataSource(fileDesc.getFileDescriptor(),
                fileDesc.getStartOffset(), fileDesc.getLength());

        fileDesc.close();

        mediaPlayer.prepare();
        mediaPlayer.start();
    }
}
```

Listing 24–3 assumes that it's within the context of an activity. As shown, you call the getResources() method to get the application's resources and then use the openRawResourceFd() method to get a file descriptor for an audio file within the /res/raw folder. You then call the setDataSource() method using the AssetFileDescriptor, the starting position to begin playback, and the ending position. You can also use this version of setDataSource() if you want to play back a specific portion of an audio file. If you always want to play the entire file, you can call the simpler version of setDataSource(FileDescriptor desc), which does not require the initial offset and length.

In this case, we chose to use prepare() followed by start(), only to show you what it might look like. We should be able to get away with it because the audio resource is local, but it couldn't hurt to use prepareAsync() as before.

We have one more source for audio content to talk about: the SD card. Earlier we showed you how to put content onto the SD card. Using it with MediaPlayer is pretty easy. In our example, we used setDataSource() to access content on the Internet by passing in a URL for an MP3 file. If you've got an audio file on your SD card, you can use the same setDataSource() method but instead pass it the path to your audio file on the SD card. For example, if you put an MP3 file in the standard Music directory and called the file music_file.mp3, you could modify the AUDIO_PATH variable and it would play, like so:

```
static final String AUDIO_PATH =
Environment.getExternalStoragePublicDirectory(
    Environment.DIRECTORY_MUSIC) +
    "/music_file.mp3";
```

You may have noticed that we did not implement onResume() and onPause() in our example. This means that when our activity goes into the background, it continues to play audio—at least, until the activity is killed, or until access to the audio source is turned off. For example, if we do not hold a wake lock, the CPU could be shut down, thus ending the playing of music. In addition, if MediaPlayer is playing an audio stream over Wi-Fi, and if our activity does not obtain a lock on Wi-Fi, Wi-Fi could be turned off, and we'll lose our connection to the stream. MediaPlayer has a method called setWakeMode() that allows us to set a PARTIAL_WAKE_LOCK to keep the CPU alive while playing. However, in order to lock Wi-Fi, we need to do that separately through WifiManager and WifiManager.WifiLock.

The other aspect of continuing to play audio in the background is that we need to know when not to do so, perhaps because there's an incoming phone call, or because an alarm is going off. Android has an AudioManager to help with this. The methods to call include requestAudioFocus() and abandonAudioFocus(), and there's a callback method called onAudioFocusChange() in the interface AudioManager.OnAudioFocusChangeListener. For more information, see the Media page in the Android Developer's Guide.

Using SoundPool for Simultaneous Track Playing

The MediaPlayer is an essential tool in our media toolbox, but it only handles one audio or video file at a time. What if we want to play more than one audio track simultaneously? One way is to create multiple MediaPlayers and work with them at the same time. If you only have a small amount of audio to play, and you want snappy performance, Android has the SoundPool class to help you. Behind the scenes, SoundPool uses MediaPlayer, but we don't get access to the MediaPlayer API, just the SoundPool API.

One of the other differences between MediaPlayer and SoundPool is that SoundPool is designed to work with local media files only. That is, you can load audio from resource files, files elsewhere using file descriptors, or files using a pathname. There are several other nice features that SoundPool provides, such as the ability to loop an audio track, pause and resume individual audio tracks, or pause and resume all audio tracks.

There are some downsides to SoundPool, though. There is an overall audio buffer size in memory for all the tracks that SoundPool will manage, and it's not very large: 1MB. This might seem large when you look at MP3 files that are only a few kilobytes in size. But SoundPool expands the audio in memory to make the playback fast and easy. The size of an audio file in memory depends on the bit rate, number of channels (stereo versus mono), sample rate, and length of the audio. If you have trouble getting your sounds loaded into SoundPool, you could try playing with these parameters of your source audio file to make the audio smaller in memory.

We're going to show you an example application that loads and plays animal sounds. One of the sounds is of crickets and it plays constantly in the background. The other sounds play at different intervals of time. Sometimes all you hear are crickets; other times you will hear several animals all at the same time. We'll also put a button in the user interface to allow for pausing and resuming. Listing 24–4 shows our layout XML file and the Java code of our activity. Your best bet is to download this from our web site, in order to get the sound files as well as the code. See the "References" section at the end of this chapter for information on how to locate the downloadable source code.

Listing 24–4. *Playing Audio with SoundPool*

```
<?xml version="1.0" encoding="utf-8"?>
<LinearLayout xmlns:android="http://schemas.android.com/apk/res/android"
    android:orientation="vertical"
    android:layout_width="fill_parent"  android:layout_height="fill_parent"
    >
<ToggleButton android:id="@+id/button"
    android:textOn="Pause"  android:textOff="Resume"
    android:layout_width="wrap_content"  android:layout_height="wrap_content"
    android:onClick="doClick" android:checked="true" />
</LinearLayout>


// This file is MainActivity.java
import java.io.IOException;
import android.app.Activity;
import android.content.Context;
import android.content.res.AssetFileDescriptor;
import android.media.AudioManager;
import android.media.SoundPool;
import android.os.Bundle;
import android.os.Handler;
import android.util.Log;
import android.view.View;
import android.widget.ToggleButton;

public class MainActivity extends Activity implements SoundPool.OnLoadCompleteListener {
    private static final int SRC_QUALITY = 0;
    private static final int PRIORITY = 1;
    private SoundPool soundPool = null;
    private AudioManager aMgr;

    private int sid_background;
    private int sid_roar;
    private int sid_bark;
    private int sid_chimp;
```

```
    private int sid_rooster;

    @Override
    public void onCreate(Bundle savedInstanceState) {
        super.onCreate(savedInstanceState);
        setContentView(R.layout.main);
    }

    @Override
    protected void onResume() {
        soundPool = new SoundPool(5, AudioManager.STREAM_MUSIC,
                SRC_QUALITY);
        soundPool.setOnLoadCompleteListener(this);

        aMgr =
            (AudioManager)this.getSystemService(Context.AUDIO_SERVICE);

        sid_background = soundPool.load(this, R.raw.crickets, PRIORITY);

        sid_chimp = soundPool.load(this, R.raw.chimp, PRIORITY);
        sid_rooster = soundPool.load(this, R.raw.rooster, PRIORITY);
        sid_roar = soundPool.load(this, R.raw.roar, PRIORITY);

        try {
            AssetFileDescriptor afd =
                    this.getAssets().openFd("dogbark.mp3");
            sid_bark = soundPool.load(afd.getFileDescriptor(),
                         0, afd.getLength(), PRIORITY);
            afd.close();
        } catch (IOException e) {
            e.printStackTrace();
        }
        //sid_bark = soundPool.load("/mnt/sdcard/dogbark.mp3", PRIORITY);

        super.onResume();
    }

    public void doClick(View view) {
        switch(view.getId()) {
        case R.id.button:
            if(((ToggleButton)view).isChecked()) {
                soundPool.autoResume();
            }
            else {
                soundPool.autoPause();
            }
            break;
        }
    }

    @Override
    protected void onPause() {
        soundPool.release();
        soundPool = null;
        super.onPause();
    }

    @Override
```

```
    public void onLoadComplete(SoundPool sPool, int sid, int status) {
        Log.v("soundPool", "sid " + sid + " loaded with status " +
                status);

        final float currentVolume =
            ((float)aMgr.getStreamVolume(AudioManager.STREAM_MUSIC)) /
            ((float)aMgr.getStreamMaxVolume(AudioManager.STREAM_MUSIC));

        if(status != 0)
            return;
        if(sid == sid_background) {
            if(sPool.play(sid, currentVolume, currentVolume,
                    PRIORITY, -1, 1.0f) == 0)
                Log.v("soundPool", "Failed to start sound");
        } else if(sid == sid_chimp) {
            queueSound(sid, 5000, currentVolume);
        } else if(sid == sid_rooster) {
            queueSound(sid, 6000, currentVolume);
        } else if(sid == sid_roar) {
            queueSound(sid, 12000, currentVolume);
        } else if(sid == sid_bark) {
            queueSound(sid, 7000, currentVolume);
        }
    }

    private void queueSound(final int sid, final long delay,
        final float volume)
    {
        new Handler().postDelayed(new Runnable() {
            @Override
            public void run() {
                if(soundPool == null) return;
                if(soundPool.play(sid, volume, volume,
                        PRIORITY, 0, 1.0f) == 0)
                    Log.v("soundPool", "Failed to start sound (" + sid +
                        ")");
                queueSound(sid, delay, volume);
            }}, delay);
    }
}
```

The structure of this example is fairly straightforward. We have a user interface with a single ToggleButton on it. We'll use this to pause and resume the active audio streams. When our app starts, we create our SoundPool and load it up with audio samples. When the samples are properly loaded, we start playing them. The crickets sound plays in a neverending loop; the other samples play after a delay and then set themselves up to play again after the same delay. By choosing different delays, we get a somewhat random effect of sounds on top of sounds.

Creating a SoundPool requires three parameters:

- The first is the maximum number of samples that the SoundPool will play simultaneously. This is not how many samples the SoundPool can hold.

- The second parameter is which audio stream the samples will play on. The typical value is AudioManager.STREAM_MUSIC, but SoundPool can be used for alarms or ringtones. See the AudioManager reference page for the complete list of audio streams.
- The SRC_QUALITY value should just be set to 0 when creating the SoundPool.

The code demonstrates several different load() methods of SoundPool. The most basic is to load an audio file from /res/raw as a resource. We use this method for the first four audio files. Then we show how you could load an audio file from the /assets directory of the application. This load() method also takes parameters that specify the offset and the length of the audio to load. This would allow us to use a single file with multiple audio samples in it, pulling out just what we want to use. Finally, we show in comments how you might access an audio file from the SD card. Up through Android 4.0, the PRIORITY parameter should just be 1.

For our example, we chose to use some of the features introduced in Android 2.2, specifically the onLoadCompleteListener interface for our Activity, and the autoPause() and autoResume() methods in our button callback.

When loading sound samples into a SoundPool, we must wait until they are properly loaded before we can start playing them. Within our onLoadComplete() callback, we check the status of the load, and, depending on which sound it is, we then set it up to play. If the sound is the crickets, we play with looping turned on (a value of -1 for the fifth parameter). For the others, we queue the sound up to play after a short period of time. The time values are in milliseconds. Note the setting of the volume. Android provides the AudioManager to let us know the current volume setting. We also get the maximum volume setting from AudioManager so we can calculate a volume value for play() that is between 0 and 1 (as a float). The play() method actually takes a separate volume value for the left and right channels, but we just set both to the current volume. Again, PRIORITY should just be set to 1. The last parameter on the play() method is for setting the playback rate. This value should be between 0.5 and 2.0, with 1.0 being normal.

Our queueSound() method uses a Handler to basically set up an event into the future. Our Runnable will run after the delay period has elapsed. We check to be sure we still have a SoundPool to play from, then we play the sound once, and schedule the same sound to play again after the same interval as before. Because we call queueSound() with different sound IDs and different delays, the effect is a somewhat random playing of animal sounds.

When you run this example, you'll hear crickets, a chimp, a rooster, a dog and a roar (a bear, we think). The crickets are constantly chirping while the other animals come and go. One nice thing about SoundPool is that it lets us play multiple sounds at the same time with no real work on our part. Also, we're not taxing the device too badly, because the sounds were decoded at load time, and we simply need to feed the sound bits to the hardware.

If you click the button, the crickets will stop, as will any other animal sound currently being played. However, the autoPause() method does not prevent new sounds from being played. You'll hear the animal sounds again within seconds (except for the crickets). Because we've been queuing up sounds into the future, we will still hear those sounds. In fact, SoundPool does not have a way to stop all sounds now and in the future. You'll need to handle stopping on your own. The crickets will only come back if we click the button again to resume the sounds. But even then, we might have lost the crickets because SoundPool will throw out the oldest sound to make room for newer sounds if the maximum number of simultaneously playing samples is reached.

Playing Sounds with JetPlayer

SoundPool is not too bad a player, but the memory limitations can make it difficult to get the job done. An alternative when you need to play simultaneous sounds is JetPlayer. Tailored for games, JetPlayer is a very flexible tool for playing lots of sounds and for coordinating those sounds with user actions. The sounds are defined using Musical Instrument Digital Interface (MIDI).

JetPlayer sounds are created using a special JETCreator tool. This tool is provided under the Android SDK tools directory, although you'll also need to install Python in order to use it. The resulting JET file can be read into your application, and the sounds set up for playback. The whole process is somewhat involved and beyond the scope of this book, so we'll just point you to more information in the "References" section at the end of this chapter.

Playing Background Sounds with AsyncPlayer

If all you want is some audio played, and you don't want to tie up the current thread, the AsyncPlayer may be what you're looking for. The audio source is passed as a URI to this class, so the audio file could be local or remote over the network. This class automatically creates a background thread to handle getting the audio and starting the playback. Because it is asynchronous, you won't know exactly when the audio will start. Nor will you know when it ends, or even if it's still playing. You can, however, call stop() to get the audio to stop playing. If you call play() again before the previous audio has finished playing, the previous audio will immediately stop and the new audio will begin at some time in the future when everything has been set up and fetched. This is a very simple class that provides an automatic background thread. Listing 24–5 shows how your code should look to implement this.

Listing 24–5. *Playing Audio with AsyncPlayer*

```
private static final String TAG = "AsyncPlayerDemo";
private AsyncPlayer mAsync = null;

[ ... ]

    mAsync = new AsyncPlayer(TAG);
    mAsync.play(this, Uri.parse("file://" + "/perry_ringtone.mp3"),
            false, AudioManager.STREAM_MUSIC);
```

```
[ ... ]

@Override
protected void onPause() {
    mAsync.stop();
    super.onPause();
}
```

Low-level Audio Playback Using AudioTrack

So far, we've been dealing with audio from files, be they local files or remote files. If you want to get down to a lower level, perhaps playing audio from a stream, you need to investigate the AudioTrack class. Besides the usual methods like play() and pause(), AudioTrack provides methods for writing bytes to the audio hardware. This class gives you the most control over audio playback, but it is much more complicated than the audio classes discussed so far in this chapter. We'll be showing a sample application a little later in this chapter that uses the AudioRecord class. The AudioRecord class is very much like the AudioTrack class, so to get a better understanding of the AudioTrack class, refer to the AudioRecord sample later on.

More About MediaPlayer

In general, the MediaPlayer is very systematic, so you need to call operations in a specific order to initialize a MediaPlayer properly and prepare it for playback. The following list summarizes some of the other details you should know of using the media APIs:

- Once you set the data source of a MediaPlayer, you cannot easily change it to another one—you'll have to create a new MediaPlayer or call the reset() method to reinitialize the state of the player.
- After you call prepare(), you can call getCurrentPosition(), getDuration(), and isPlaying() to get the current state of the player. You can also call the setLooping() and setVolume() methods after the call to prepare(). If you used prepareAsync(), you should wait until onPrepared() is called before using any of these other methods.
- After you call start(), you can call pause(), stop(), and seekTo().
- Every MediaPlayer you create uses a lot of resources, so be sure to call the release() method when you are done with the media player. The VideoView takes care of this in the case of video playback, but you'll have to do it manually if you decide to use MediaPlayer instead of VideoView. More about VideoView in the next sections.

- MediaPlayer works with several listeners you can use for additional control over the user experience, including OnCompletionListener, OnErrorListener, and OnInfoListener. For example, if you're managing a playlist of audio, OnCompletionListener will be called when a piece is finished so you can queue up the next piece.

This concludes our discussion about playing audio content. Now we'll turn our attention to playing video. As you will see, referencing video content is similar to referencing audio content.

Playing Video Content

In this section, we are going to discuss video playback using the Android SDK. Specifically, we will discuss playing a video from a web server and playing one from an SD card. As you can imagine, video playback is a bit more involved than audio playback. Fortunately, the Android SDK provides some additional abstractions that do most of the heavy lifting.

> **NOTE:** Playing back video in the emulator is not very reliable. If it works, great. But if it doesn't, try running on a device instead. Because the emulator must use only software to run video, it can have a very hard time keeping up with video, and you will likely get unexpected results.

Playing video requires more effort than playing audio, because there's a visual component to take care of in addition to the audio. To take some of the pain away, Android provides a specialized view control called android.widget.VideoView that encapsulates creating and initializing the MediaPlayer. To play video, you create a VideoView widget in your user interface. You then set the path or URI of the video and fire the start() method. Listing 24–6 demonstrates video playback in Android.

Listing 24–6. *Playing Video Using the Media APIs*

```
<?xml version="1.0" encoding="utf-8"?>
<!-- This file is /res/layout/main.xml -->
<LinearLayout
 android:layout_width="fill_parent" android:layout_height="fill_parent"
 xmlns:android="http://schemas.android.com/apk/res/android">

    <VideoView  android:id="@+id/videoView"
        android:layout_width="200px"  android:layout_height="200px" />

</LinearLayout>


// This file is MainActivity.java
import android.app.Activity;
import android.net.Uri;
import android.os.Bundle;
import android.widget.MediaController;
import android.widget.VideoView;
```

```
public class MainActivity extends Activity {
    /** Called when the activity is first created. */
    @Override
    protected void onCreate(Bundle savedInstanceState) {
        super.onCreate(savedInstanceState);
        this.setContentView(R.layout.main);

        VideoView videoView =
               (VideoView)this.findViewById(R.id.videoView);
        MediaController mc = new MediaController(this);
        videoView.setMediaController(mc);
        videoView.setVideoURI(Uri.parse(
              "http://www.androidbook.com/akc/filestorage/android/" +
               "documentfiles/3389/movie.mp4"));
 /* videoView.setVideoPath(
    Environment.getExternalStoragePublicDirectory(
    Environment.DIRECTORY_MOVIES) +
    "/movie.mp4");
 */
        videoView.requestFocus();
        videoView.start();
    }
}
```

Listing 24–6 demonstrates video playback of a file located on the Web at `www.androidbook.com/akc/filestorage/android/documentfiles/3389/movie.mp4`, which means the application running the code will need to request the `android.permission.INTERNET` permission. All of the playback functionality is hidden behind the `VideoView` class. In fact, all you have to do is feed the video content to the video player. The user interface of the application is shown in Figure 24–4.

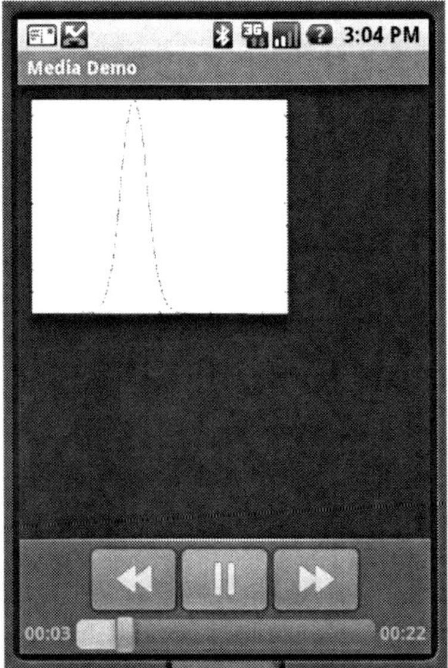

Figure 24–4. *The video playback UI with media controls enabled*

When this application runs, you will see the button controls along the bottom of the screen for about three seconds, and then they disappear. You get them back by clicking anywhere within the video frame. When we were doing playback of audio content, we only needed to display the button controls to start, pause, and restart the audio. We did not need a view component for the audio itself. With video, of course, we need button controls as well as something to view the video in. For this example, we're using a VideoView component to display the video content. But instead of creating our own button controls (which we could still do if we chose to), we create a MediaController that provides the buttons for us. As shown in Figure 24–4 and Listing 24–6, you set the VideoView's media controller by calling setMediaController() to enable the play, pause, and seek-to controls. If you want to manipulate the video programmatically with your own buttons, you can call the start(), pause(), stopPlayback(), and seekTo() methods.

Keep in mind that we're still using a MediaPlayer in this example—we just don't see it. You can in fact "play" videos directly in MediaPlayer. If you go back to the example from Listing 24–1, put a movie file on your SD card, and plug in the movie's file path in AUDIO_PATH, you will find that it plays the audio quite nicely even though you can't see the video.

Whereas MediaPlayer has a setDataSource() method, VideoView does not. VideoView instead uses the setVideoPath() or setVideoURI() methods. Assuming you put a movie file onto your SD card, you change the code from Listing 24–6 to comment out the setVideoURI() call and uncomment the setVideoPath() call, adjusting the path to the movie file as necessary. When you run the application again, you will now hear *and see* the video in the VideoView. Technically, we could have called setVideoURI() with the following to get the same effect as setVideoPath():

```
videoView.setVideoURI(Uri.parse("file://" +
    Environment.getExternalStoragePublicDirectory(
    Environment.DIRECTORY_MOVIES) + "/movie.mp4"));
```

You might have noticed that VideoView does not have a method to read data from a file descriptor as MediaPlayer did. You may also have noticed that MediaPlayer has a couple of methods for adding a SurfaceHolder to a MediaPlayer (a SurfaceHolder is like a view port for images or video). One of the MediaPlayer methods is create(Context context, Uri uri, SurfaceHolder holder), and the other is setDisplay(SurfaceHolder holder).

Now let's explore recording media.

Recording Media

As we've shown, there are many ways to play media from within Android. For recording, there are fewer options. The main workhorse of recording is the MediaRecorder class, which is used for both audio and video. In this section, we'll show you how to use MediaRecorder for both types of media. The other class for recording audio is AudioRecord, and we'll demonstrate this with another sample application. Sometimes you don't want to write code to accomplish something when an existing application can do it for you. So we'll also show you how to fire off an intent to record audio, as well as to capture still camera images using the Camera application.

Exploring Audio Recording with MediaRecorder

The Android media framework supports recording audio. One way you record audio is through the `android.media.MediaRecorder` class. In this section, we'll show you how to build an application that records audio content and then plays the content back. The user interface of the application is shown in Figure 24–5.

Figure 24–5. *The user interface of the audio-recorder example*

As shown in Figure 24–5, the application contains four buttons: two to control recording, and two to start and stop playback of the recorded content. Listing 24–7 shows the layout file and activity class for the UI.

Listing 24–7. *Media Recording and Playback in Android*

```
<?xml version="1.0" encoding="utf-8"?>
<!-- This file is /res/layout/record.xml -->
<LinearLayout xmlns:android="http://schemas.android.com/apk/res/android"
    android:orientation="vertical"
    android:layout_width="fill_parent"
    android:layout_height="fill_parent">

  <Button android:id="@+id/beginBtn"  android:text="Begin Recording"
    android:layout_width="fill_parent"
    android:layout_height="wrap_content"
    android:onClick="doClick" />

  <Button android:id="@+id/stopBtn"  android:text="Stop Recording"
    android:layout_width="fill_parent"
    android:layout_height="wrap_content"
    android:onClick="doClick" />

  <Button android:id="@+id/playRecordingBtn"
    android:text="Play Recording"
    android:layout_width="fill_parent"
    android:layout_height="wrap_content"
    android:onClick="doClick" />

  <Button android:id="@+id/stopPlayingRecordingBtn"
    android:text="Stop Playing Recording"
```

```
    android:layout_width="fill_parent"
    android:layout_height="wrap_content"
    android:onClick="doClick" />

</LinearLayout>


// RecorderActivity.java
import java.io.File;
import android.app.Activity;
import android.media.MediaPlayer;
import android.media.MediaRecorder;
import android.os.Bundle;
import android.os.Environment;
import android.view.View;

public class RecorderActivity extends Activity {
    private MediaPlayer mediaPlayer;
    private MediaRecorder recorder;
    private String OUTPUT_FILE;

    @Override
    protected void onCreate(Bundle savedInstanceState) {
        super.onCreate(savedInstanceState);
        setContentView(R.layout.record);

        OUTPUT_FILE = Environment.getExternalStorageDirectory() +
                        "/recordaudio3.3gpp";
    }

    public void doClick(View view) {
        switch(view.getId()) {
        case R.id.beginBtn:
            try {
                beginRecording();
            } catch (Exception e) {
                e.printStackTrace();
            }
            break;
        case R.id.stopBtn:
            try {
                stopRecording();
            } catch (Exception e) {
                e.printStackTrace();
            }
            break;
        case R.id.playRecordingBtn:
            try {
                playRecording();
            } catch (Exception e) {
                e.printStackTrace();
            }
            break;
        case R.id.stopPlayingRecordingBtn:
            try {
                stopPlayingRecording();
            } catch (Exception e) {
```

```
                e.printStackTrace();
            }
            break;
        }
    }

    private void beginRecording() throws Exception {
        killMediaRecorder();

        File outFile = new File(OUTPUT_FILE);

        if(outFile.exists()) {
            outFile.delete();
        }
        recorder = new MediaRecorder();
        recorder.setAudioSource(MediaRecorder.AudioSource.MIC);
        recorder.setOutputFormat(MediaRecorder.OutputFormat.THREE_GPP);
        recorder.setAudioEncoder(MediaRecorder.AudioEncoder.AMR_NB);
        recorder.setOutputFile(OUTPUT_FILE);
        recorder.prepare();
        recorder.start();
    }

    private void stopRecording() throws Exception {
        if (recorder != null) {
            recorder.stop();
        }
    }

    private void killMediaRecorder() {
        if (recorder != null) {
            recorder.release();
        }
    }

    private void killMediaPlayer() {
        if (mediaPlayer != null) {
            try {
                mediaPlayer.release();
            } catch (Exception e) {
                e.printStackTrace();
            }
        }
    }

    private void playRecording() throws Exception {
        killMediaPlayer();

        mediaPlayer = new MediaPlayer();
        mediaPlayer.setDataSource(OUTPUT_FILE);

        mediaPlayer.prepare();
        mediaPlayer.start();
    }

    private void stopPlayingRecording() throws Exception {
        if(mediaPlayer != null) {
```

```
                mediaPlayer.stop();
            }
        }

        @Override
        protected void onDestroy() {
            super.onDestroy();

            killMediaRecorder();
            killMediaPlayer();
        }
}
```

Before we jump into Listing 24–7, you'll need to add the following permission to your manifest file in order to record audio:

```
<uses-permission android:name="android.permission.RECORD_AUDIO" />
```

As discussed earlier in the section on SD cards, you will also need to add a uses-permission tag for "android.permission.WRITE_EXTERNAL_STORAGE". Finally, if you are going to try this out with the emulator, you'll need to provide a microphone input on your workstation.

If you look at the onCreate() method in Listing 24–7, you'll see that the only thing we need to do there is create the file pathname for our output audio file. Our doClick() method uses the standard pattern of switching on the button that was pressed, and we invoke the appropriate function call to perform each desired action. The beginRecording() method handles recording. To record audio, you must create an instance of MediaRecorder and set the audio source, output format, audio encoder, and output file.

For audio sources, there is usually the microphone. There are also three audio sources related to phone calls. You can record the entire call (MediaRecorder.AudioSource.VOICE_CALL), the uplink side only (MediaRecorder.AudioSource.VOICE_UPLINK), or the downlink side only (MediaRecorder.AudioSource.VOICE_DOWNLINK). The uplink side of a call would be the voice of the phone's user. The downlink side of the call would be sounds coming from the other end of the call.

With Android SDK 2.1, two more audio sources were added: CAMCORDER and VOICE_RECOGNITION. The CAMCORDER audio source would be a camera-related microphone; otherwise, this option will use the default main microphone of the device. The VOICE_RECOGNITION microphone is one tuned to doing voice recognition; otherwise, this option also will use the default main microphone of the device. The phrase "tuned to doing voice recognition" means that the audio stream will be as raw as possible, with no extra audio modifications in between the microphone and your application. For example, some HTC devices have Auto Gain Control (AGC) on the microphone, so using that audio source for voice recognition is problematic. The VOICE_RECOGNITION audio source bypasses this extra processing for better results doing voice recognition.

The most common output format for audio is 3GPP. Prior to Android 2.3.3 (Gingerbread), you must set the encoder to AMR_NB, which signifies the Adaptive Multi-

Rate (AMR) narrowband audio codec, because this is the only supported audio encoder. As of Android 2.3.3, you can also use AMR_WB (wideband) and Advanced Audio Coding (AAC) as audio encoders. The recorded audio in our example is written to the SD card as a file named `recordoutput.3gpp`. Note that Listing 24–7 assumes that you've created an SD card image and that you've pointed the emulator to the SD card. If you have not done this, refer to the section "Using SD Cards" for details on setting this up. Or you can use a real device, which is recommended when developing for audio and video.

There are some additional methods to the `MediaRecorder` that you might find useful. In order to limit the length and size of audio recordings, the methods `setMaxDuration(int length_in_ms)` and `setMaxFileSize(long length_in_bytes)` can be used. You set the maximum length of the recording in milliseconds or the maximum length of the recording file in bytes, to stop recording when these limits are reached.

Recording Audio with AudioRecord

So far, you've seen how to record audio directly to a file. But what if you want to do some processing on the audio data before it goes to a file? Or what if you don't even want to send the audio to a file? Android provides a class called `AudioRecord` for just these purposes. When you set up an `AudioRecord` object, Android will ensure that audio data is written to the internal buffer of the `AudioRecord`, and then your application can do whatever it wants with the audio data. Listing 24–8 shows an activity for reading and processing audio using an `AudioRecord`. There is no user interface for this activity, because we'll just be writing log messages to LogCat. The `AndroidManifest.xml` file is not shown, but you will need to add an Android permission for `android.permission.RECORD_AUDIO` for this to work.

Listing 24–8. *Recording Raw Audio with `AudioRecord`*

```
import android.app.Activity;
import android.media.AudioFormat;
import android.media.AudioRecord;
import android.media.MediaRecorder;
import android.os.Bundle;
import android.util.Log;

public class MainActivity extends Activity {
    protected static final String TAG = "AudioRecord";
    private int mAudioBufferSize;
    private int mAudioBufferSampleSize;
    private AudioRecord mAudioRecord;
    private boolean inRecordMode = false;

    public void onCreate(Bundle savedInstanceState) {
        super.onCreate(savedInstanceState);

        initAudioRecord();
    }

    @Override
    public void onResume() {
        super.onResume();
```

```
        Log.v(TAG, "Resuming...");
        inRecordMode = true;
        Thread t = new Thread(new Runnable() {

            @Override
            public void run() {
                getSamples();
            }
        });
        t.start();
    }

    protected void onPause() {
        Log.v(TAG, "Pausing...");
        inRecordMode = false;
        super.onPause();
    }

    @Override
    protected void onDestroy() {
        Log.v(TAG, "Destroying...");
        if(mAudioRecord != null) {
            mAudioRecord.release();
            Log.v(TAG, "Released AudioRecord");
        }
        super.onDestroy();
    }

    private void initAudioRecord() {
        try {
            int sampleRate = 8000;
            int channelConfig = AudioFormat.CHANNEL_IN_MONO;
            int audioFormat = AudioFormat.ENCODING_PCM_16BIT;
            mAudioBufferSize =
                    2 * AudioRecord.getMinBufferSize(sampleRate,
                    channelConfig, audioFormat);
            mAudioBufferSampleSize = mAudioBufferSize / 2;
            mAudioRecord = new AudioRecord(
                    MediaRecorder.AudioSource.MIC,
                    sampleRate,
                    channelConfig,
                    audioFormat,
                    mAudioBufferSize);
            Log.v(TAG, "Setup of AudioRecord okay. Buffer size = " +
                    mAudioBufferSize);
            Log.v(TAG, "   Sample buffer size = " +
                    mAudioBufferSampleSize);
        } catch (IllegalArgumentException e) {
            e.printStackTrace();
        }

        int audioRecordState = mAudioRecord.getState();
        if(audioRecordState != AudioRecord.STATE_INITIALIZED) {
            Log.e(TAG, "AudioRecord is not properly initialized");
            finish();
        }
        else {
```

```
            Log.v(TAG, "AudioRecord is initialized");
        }
    }

    private void getSamples() {
        if(mAudioRecord == null) return;

        short[] audioBuffer = new short[mAudioBufferSampleSize];

        mAudioRecord.startRecording();

        int audioRecordingState = mAudioRecord.getRecordingState();
        if(audioRecordingState != AudioRecord.RECORDSTATE_RECORDING) {
            Log.e(TAG, "AudioRecord is not recording");
            finish();
        }
        else {
            Log.v(TAG, "AudioRecord has started recording...");
        }

        while(inRecordMode) {
            int samplesRead = mAudioRecord.read(
                            audioBuffer, 0, mAudioBufferSampleSize);
            Log.v(TAG, "Got samples: " + samplesRead);
            Log.v(TAG, "First few sample values: " +
                    audioBuffer[0] + ", " +
                    audioBuffer[1] + ", " +
                    audioBuffer[2] + ", " +
                    audioBuffer[3] + ", " +
                    audioBuffer[4] + ", " +
                    audioBuffer[5] + ", " +
                    audioBuffer[6] + ", " +
                    audioBuffer[7] + ", " +
                    audioBuffer[8] + ", " +
                    audioBuffer[9] + ", "
                    );
        }

        mAudioRecord.stop();
        Log.v(TAG, "AudioRecord has stopped recording");
    }
}
```

Our sample application is fairly straightforward. We start by initializing our AudioRecord. This requires choosing the audio source, the frequency of sampling, the channel configuration (mono, stereo, left, right, and the like), the audio encoding format, and the internal buffer size. For the audio source, you'll choose from the set of options as defined in MediaRecorder.AudioSource. One word of caution here: not all devices have implemented VOICE_CALL because that acts like two inputs instead of one. For the sample frequency, you should choose one of the standard values, such as 8000, 16000, 44100, 22050, or 11025Hz. The channel configuration should be chosen from the CHANNEL* values described in AudioFormat. The encoding format will be either ENCODING_PCM_8BIT or ENCODING_PCM_16BIT. Note that your choice here will affect the kind of values you'll get back as raw audio data. If you don't need the precision of 16-bit, go with 8-bit—you'll use less memory and go faster. The documentation says that only the sample frequency of 44100 is guaranteed to work on all devices, but ironically the emulator only supports 8000Hz, CHANNEL_IN_MONO, and ENCODING_PCM_8BIT.

The AudioRecord class has a static helper method called getMinBufferSize(), which will take your desired parameter settings and return to you the minimum-sized buffer that you should specify to properly initialize your AudioRecord. This buffer is not directly accessible to you, but AudioRecord needs to have enough room internally to store audio data while you're processing the audio data you've retrieved previously. You can certainly go with the minimum size value, or you could bump it up a little. You definitely should not attempt to set a buffer size less than what is recommended by this helper method. In our sample, we chose a buffer size twice what the minimum is. You'll get an IllegalArgumentException if your parameters are not acceptable to the AudioRecord. For example, if you try a sample frequency value that is not supported on this hardware, you'll get this exception. Unfortunately, there is no convenient method to get a list of supported sample frequencies, so your only recourse is to try a desired sample frequency; if you get the exception, try another sample frequency until you find one that works.

As a last check within our initialize method, we make sure that our AudioRecord is properly initialized. Now we're ready to read audio samples.

We've chosen to turn on our sampling in the onResume() method of our Activity, and turn off sampling in onPause(). We do not want to tie up our main UI thread with sampling, so we create a separate thread to do the audio sampling in. We also set a boolean (inRecordMode) so we can tell our thread to stop sampling. Within the getSamples() method, we create our own buffer for the audio data. As mentioned before, we cannot directly access the internal audio data buffer of our AudioRecord, so we read into our sample buffer. Note that the size of our buffer is audioBufferSampleSize, not audioBufferSize. We're only reading the sample size so that's all we need in our buffer. We tell the AudioRecord to start recording, we check that the state has changed to RECORDING, and then we start looping on reads. These are blocking reads, but we're in a separate thread, so it's okay. As the AudioRecord gets to our sample size of data, our read returns so we can process that audio sample.

Meanwhile, the AudioRecord will be collecting additional audio data for us for the next time we call read. We only have a certain amount of time to do our processing before the AudioRecord's internal buffer fills up, so we definitely want to be careful not to do too much. Depending on what you want to do with the data, you could simply stop recording and start again later. In our example, we simply report in LogCat that we got samples, and we display the first ten values. As you run this sample application, make different sounds into the microphone to see the values change in LogCat.

Our looping continues until the boolean inRecordMode changes to false, which happens when the application is being hidden or is being killed.

While perusing the documentation on AudioRecord, you may notice some callback interfaces. These allow you to set up listeners on either reaching a marker within the audio stream or triggering a periodic callback every so often. We modified the previous example by adding the statements in Listing 24–9. For the complete source code of this project, see the book's web site.

Listing 24–9. ***Recording Raw Audio with* `AudioRecord` *and Callbacks***

```
// This code goes inside of our Activity class
public OnRecordPositionUpdateListener mListener =
        new OnRecordPositionUpdateListener() {

    public void onPeriodicNotification(AudioRecord recorder) {
        Log.v(TAG, "in onPeriodicNotification");
    }

    public void onMarkerReached(AudioRecord recorder) {
        Log.v(TAG, "in onMarkerReached");
        inRecordMode = false;
    }
};


// These statements go inside of initAudioRecord() after the
// creation of mAudioRecord and before the check of the state
// of mAudioRecord.
    mAudioRecord.setNotificationMarkerPosition(10000);
    mAudioRecord.setPositionNotificationPeriod(1000);
    mAudioRecord.setRecordPositionUpdateListener(mListener);
```

Notice how the listener has two separate callback methods. The first one is called every time we read 1,000 frames of audio, which we set up in our initialization method. This frame count is independent of our sample size buffer. Although we may be reading 1,600 frames at a time, the first callback is invoked every 1,000 frames. We could therefore see our callback invoked twice within one read loop. The second callback is called when our absolute frame count is reached. In our example application, we set this to 10,000 frames, and when this count is reached, we turn off recording by setting the boolean to `false`. If we had only logged a message and not turned off recording, we would not have seen this callback invoked again no matter how many frames were read in the future. The marker is relative to when `startRecording()` is called on the `AudioRecord`.

Exploring Video Recording

Since the introduction of Android SDK 1.5, you can capture video using the media framework. This works in a similar way to recording audio and, in fact, recorded video usually includes an audio track. There is one big exception with video, however. Beginning with Android SDK 1.6, recording video requires that you preview the camera images onto a `Surface` object. In basic applications, this is not much of an issue, because the user probably wants to be viewing what the camera sees. For more sophisticated applications, however, this could be a problem. If your application doesn't need to show the video feed to the user as it happens, you still need to provide a `Surface` object so the camera can preview the video. We expect this requirement will be relaxed in future versions of the Android SDK, so that applications could work directly with the video buffers without having to copy to a UI component as well. For now, though, we'll have to work with a Surface.

The next example shows how to do this. This sample application is a bit long, so we've broken it down into pieces so we can describe what the pieces do as we go along. You'll most likely want to import this project into Eclipse after downloading from our web site. See the "References" section at the end of this chapter for instructions on how to do that. We start with the layout for our application in Listing 24–10.

Listing 24–10. *Record Video's XML layout*

```
<?xml version="1.0" encoding="utf-8"?>
<!-- This file is /res/layout-land/main.xml -->
<LinearLayout xmlns:android="http://schemas.android.com/apk/res/android"
    android:layout_width="fill_parent"
    android:layout_height="fill_parent"
    android:orientation="horizontal" >
  <LinearLayout
    android:orientation="vertical" android:layout_width="wrap_content"
    android:layout_height="wrap_content">

    <Button android:id="@+id/initBtn"
        android:layout_width="wrap_content"
        android:layout_height="wrap_content"
        android:text="Initialize Recorder"  android:onClick="doClick"
        android:enabled="false" />

    <Button android:id="@+id/beginBtn"
        android:layout_width="wrap_content"
        android:layout_height="wrap_content"
        android:text="Begin Recording"  android:onClick="doClick"
        android:enabled="false" />

    <Button android:id="@+id/stopBtn"
        android:layout_width="wrap_content"
        android:layout_height="wrap_content"
        android:text="Stop Recording"  android:onClick="doClick" />

    <Button android:id="@+id/playRecordingBtn"
        android:layout_width="wrap_content"
        android:layout_height="wrap_content"
        android:text="Play Recording"  android:onClick="doClick" />

    <Button android:id="@+id/stopPlayingRecordingBtn"
        android:layout_width="wrap_content"
        android:layout_height="wrap_content"
        android:text="Stop Playing"  android:onClick="doClick" />
  </LinearLayout>
  <LinearLayout android:orientation="vertical"
        android:layout_width="fill_parent"
        android:layout_height="fill_parent" >
    <TextView android:id="@+id/recording" android:text=" "
        android:textColor="#FF0000"
        android:layout_width="wrap_content"
        android:layout_height="wrap_content" />
    <VideoView android:id="@+id/videoView"
        android:layout_width="250dip"  android:layout_height="200dip" />
```

```
    </LinearLayout>
</LinearLayout>
```

The result of this layout will look like Figure 24–6. This image was snapped during a recording of video on a real device, looking at Eclipse on the workstation.

Figure 24–6. *The Record Video UI*

The layout is composed of two `LinearLayouts` side by side in a containing `LinearLayout`. On the left are five buttons that our application will enable and disable as the demonstration progresses. On the right is the main `VideoView`, and above it is the "RECORDING" message, which turns on when the application is actually recording video. As you've probably figured out, we've forced this application to be in landscape mode by setting the `android:screenOrientation="landscape"` attribute in the `<activity>` tag in AndroidManifest.xml. Let's start exploring this application with `MainActivity`, as shown in Listing 24–11.

Listing 24–11. *Record Video's* `MainActivity`

```
public class MainActivity extends Activity implements
        SurfaceHolder.Callback, OnInfoListener, OnErrorListener {

    private static final String TAG = "RecordVideo";
    private MediaRecorder mRecorder = null;
    private String mOutputFileName;
    private VideoView mVideoView = null;
    private SurfaceHolder mHolder = null;
    private Button mInitBtn = null;
    private Button mStartBtn = null;
    private Button mStopBtn = null;
    private Button mPlayBtn = null;
    private Button mStopPlayBtn = null;
    private Camera mCamera = null;
    private TextView mRecordingMsg = null;

    /** Called when the activity is first created. */
    @Override
    public void onCreate(Bundle savedInstanceState) {
        super.onCreate(savedInstanceState);
        Log.v(TAG, "in onCreate");
        setContentView(R.layout.main);
```

```
        mInitBtn = (Button) findViewById(R.id.initBtn);
        mStartBtn = (Button) findViewById(R.id.beginBtn);
        mStopBtn = (Button) findViewById(R.id.stopBtn);
        mPlayBtn = (Button) findViewById(R.id.playRecordingBtn);
        mStopPlayBtn = (Button)
             findViewById(R.id.stopPlayingRecordingBtn);
        mRecordingMsg = (TextView) findViewById(R.id.recording);

        mVideoView = (VideoView)this.findViewById(R.id.videoView);
    }
        // The rest of this class is in the listings that will follow.
}
```

We're using a standard activity for this application, but we're also implementing three interfaces. The first interface, `SurfaceHolder.Callback`, is used to receive an indication of when the `Surface` is ready for displaying a video image. The `Surface` in our case comes from the `VideoView`. We also want to be told if there are any messages coming from our `MediaRecorder`, which is why we implement both `OnInfoListener` and `OnErrorListener`. The methods of these interfaces will be coming up shortly.

There are several member fields for our activity that we'll need later, and we initialize several of them in the `onCreate()` method. For now, we're only showing a comment where the rest of the `MainActivity` class goes. Those class methods will be covered in the subsequent listings, starting with Listing 24–12, where we show our standard `onResume()` and `onPause()` methods.

Listing 24–12. ***Record Video's Resume and Pause Code***

```
@Override
protected void onResume() {
    Log.v(TAG, "in onResume");
    super.onResume();
    mInitBtn.setEnabled(false);
    mStartBtn.setEnabled(false);
    mStopBtn.setEnabled(false);
    mPlayBtn.setEnabled(false);
    mStopPlayBtn.setEnabled(false);
    if(!initCamera())
        finish();
}

@Override
protected void onPause() {
    Log.v(TAG, "in onPause");
    super.onPause();
    releaseRecorder();
    releaseCamera();
}
```

> **NOTE:** Listing 24–12 contains methods of our MainActivity class; we've only separated them into different listings to make it easier to follow along. The same is true of the rest of the listings for the Record Video application.

These are pretty standard methods. In onResume(), we simply set our buttons to their initialized state, and then we initialize the camera (that method is coming up next). In onPause() we need to release both our MediaRecorder and Camera. This way, anytime our application goes out of view, recording will stop and the camera is released so another application can use it. If the user comes back to our application, things will restart, and the user will be able to record video again. Next up, in Listing 24–13, is the initialization code for the camera, the Surface.Callback callbacks, plus the release methods for both Camera and MediaRecorder.

Listing 24–13. *Record Video's initCamera() and Release Methods*

```
private boolean initCamera() {
    try {
        mCamera  = Camera.open();
        Camera.Parameters camParams = mCamera.getParameters();
        mCamera.lock();
        //mCamera.setDisplayOrientation(90);
        // Could also set other parameters here and apply using:
        //mCamera.setParameters(camParams);

        mHolder = mVideoView.getHolder();
        mHolder.addCallback(this);
        mHolder.setType(SurfaceHolder.SURFACE_TYPE_PUSH_BUFFERS);
    }
    catch(RuntimeException re) {
        Log.v(TAG, "Could not initialize the Camera");
        re.printStackTrace();
        return false;
    }
    return true;
}

@Override
public void surfaceCreated(SurfaceHolder holder) {
    Log.v(TAG, "in surfaceCreated");

    try {
        mCamera.setPreviewDisplay(mHolder);
        mCamera.startPreview();
    } catch (IOException e) {
        Log.v(TAG, "Could not start the preview");
        e.printStackTrace();
    }
    mInitBtn.setEnabled(true);
}

@Override
public void surfaceDestroyed(SurfaceHolder holder) {
    Log.v(TAG, "in surfaceDestroyed");
}

@Override
public void surfaceChanged(SurfaceHolder holder, int format,
        int width, int height) {
    Log.v(TAG, "surfaceChanged: Width x Height = " +
              width + "x" + height);
```

```
    }

    private void releaseRecorder() {
        if(mRecorder != null) {
            mRecorder.release();
            mRecorder = null;
        }
    }

    private void releaseCamera() {
        if(mCamera != null) {
            try {
                mCamera.reconnect();
            } catch (IOException e) {
                e.printStackTrace();
            }
            mCamera.release();
            mCamera = null;
        }
    }
```

The `initCamera()` method is called to set up our access to the device's camera. It's the beginning of everything. For this sample application, we are using the default parameters of `Camera`, but we could easily get the current parameter values, update them, and write them back. The commented code shows where you could change the camera's behavior and appearance. Once the camera is set, we grab the `SurfaceHolder` where the video images will appear.

The `surfaceCreated()` callback is where we give the `Camera` object a place to show the current view: in other words, the camera preview. Once the preview has been started, we can enable the button to initialize the `MediaRecorder`. The camera preview is a very useful feature that allows the user to see what the camera sees before it starts recording. Whether you're doing video recording or still photography, you would most likely do a preview, and it would be done this way for either case.

For completeness, we've shown the `releaseRecorder()` and `releaseCamera()` methods. These get called in `onPause()` as was shown in Listing 24–12.

At this point in our application, we've set up the camera, initialized our buttons and are showing a preview of what the camera sees. Now the user can start clicking buttons, although the only one that is enabled at the start is the Initialize Recorder button. When a button is clicked, the code in Listing 24–14 executes. Each of the five actions corresponding to each button is provided in this listing. As each action executes, the buttons will enable and disable appropriately for the next action. For example, once the recorder has been initialized, the Initialize Recorder button is disabled and the Begin Recording button is enabled.

Listing 24–14. *Record Video's Button-Processing Code*

```
public void doClick(View view) {
    switch(view.getId()) {
    case R.id.initBtn:
        initRecorder();
        break;
```

```
        case R.id.beginBtn:
            beginRecording();
            break;
        case R.id.stopBtn:
            stopRecording();
            break;
        case R.id.playRecordingBtn:
            playRecording();
            break;
        case R.id.stopPlayingRecordingBtn:
            stopPlayingRecording();
            break;
        }
    }

    private void initRecorder() {
        if(mRecorder != null) return;

        mOutputFileName = Environment.getExternalStorageDirectory() +
                                 "/videooutput.mp4";

        File outFile = new File(mOutputFileName);
        if(outFile.exists()) {
            outFile.delete();
        }

        try {
            mCamera.stopPreview();
            mCamera.unlock();
            mRecorder = new MediaRecorder();
            mRecorder.setCamera(mCamera);

            mRecorder.setAudioSource(MediaRecorder.AudioSource.CAMCORDER);
            mRecorder.setVideoSource(MediaRecorder.VideoSource.CAMERA);
            mRecorder.setOutputFormat(MediaRecorder.OutputFormat.MPEG_4);
            mRecorder.setVideoSize(176, 144);
            mRecorder.setVideoFrameRate(15);
            mRecorder.setVideoEncoder(MediaRecorder.VideoEncoder.MPEG_4_SP);
            mRecorder.setAudioEncoder(MediaRecorder.AudioEncoder.AMR_NB);
            mRecorder.setMaxDuration(7000); // limit to 7 seconds
            mRecorder.setPreviewDisplay(mHolder.getSurface());
            mRecorder.setOutputFile(mOutputFileName);

            mRecorder.prepare();
            Log.v(TAG, "MediaRecorder initialized");
            mInitBtn.setEnabled(false);
            mStartBtn.setEnabled(true);
        }
        catch(Exception e) {
            Log.v(TAG, "MediaRecorder failed to initialize");
            e.printStackTrace();
        }
    }

    private void beginRecording() {
        mRecorder.setOnInfoListener(this);
        mRecorder.setOnErrorListener(this);
```

```
        mRecorder.start();
        mRecordingMsg.setText("RECORDING");
        mStartBtn.setEnabled(false);
        mStopBtn.setEnabled(true);
    }

    private void stopRecording() {
        if (mRecorder != null) {
            mRecorder.setOnErrorListener(null);
            mRecorder.setOnInfoListener(null);
            try {
                mRecorder.stop();
            }
            catch(IllegalStateException e) {
                // This can happen if the recorder has already stopped.
                Log.e(TAG, "Got IllegalStateException in stopRecording");
            }
            releaseRecorder();
            mRecordingMsg.setText("");
            releaseCamera();
            mStartBtn.setEnabled(false);
            mStopBtn.setEnabled(false);
            mPlayBtn.setEnabled(true);
        }
    }

    private void playRecording() {
        MediaController mc = new MediaController(this);
        mVideoView.setMediaController(mc);
        mVideoView.setVideoPath(mOutputFileName);
        mVideoView.start();
        mStopPlayBtn.setEnabled(true);
    }

    private void stopPlayingRecording() {
        mVideoView.stopPlayback();
    }
```

The initRecorder() method is where a lot of our setup happens. The recorder needs to know where to record to, so we provide a file path name. We delete the file if it already exists. Notice how we then stop the preview of the camera, unlock it, then we turn around and connect it to the MediaRecorder? The camera is somewhat sensitive to locking and unlocking, and sometimes you need to lock the camera to prevent others from getting to it, and other times you need to unlock it so you can do what you want to with it. This is one of those times when you need to unlock it to connect it to the MediaRecorder.

Once the camera is connected, which we do first, we proceed to set the rest of the MediaRecorder attributes, including audio source and video source. But wait, didn't we just connect the camera to the recorder? Well, yes we did. But we still need to set the video source explicitly. By setting the camera in the recorder, we avoid having to destroy the Camera object only to have the recorder object build a new one. We also set the audio and video encoders and a path to the output file on the SD card before calling the prepare() method. The prepare() method comes at the end and gets us ready to actually record something. Unlike MediaPlayer, there is no prepareAsync() method on MediaRecorder. We end this method by enabling the Begin Recording button.

The beginRecording() method is fairly straightforward by comparison. It adds the listeners, calls start(), and then sets the recording message string and changes the buttons. When this method reaches the end, our application should be recording video and the red "RECORDING" message should be displayed, as it was in Figure 24–6.

The stopRecording() method is a little more complicated, in part because it could be called from more than one place. We'll get to the second place in a bit, but for now assume that the Stop Recording button has triggered this method. If we still have a valid recorder, we disable the callbacks and then call stop(). Because it is possible that stop() could be called on a recorder that is already stopped, we handle the exception that says we tried to stop a recorder that was already stopped. Then we release the recorder and the camera and set the "RECORDING" message to blank. Finally, the buttons change to switch from recording to playback.

The playRecording() method is also straightforward. We grab a MediaController for our VideoView, point it to our new file, then call start(). Our stopPlayingRecording() method is even simpler; we just stop the playback of the video. When we're in playback mode, it's harmless to click the Play button when the video is already playing or to click Stop when the video is stopped.

We mentioned before that the recording action can be stopped from more than one place. One of the settings on the recorder was a maximum duration of seven seconds. This means that recording will stop after seven seconds, and our info callback will get called. Let's take a look at these now in Listing 24–15.

Listing 24–15. *Record Video's Info Callbacks*

```
    @Override
    public void onInfo(MediaRecorder mr, int what, int extra) {
        Log.i(TAG, "got a recording event");
        if(what ==
           MediaRecorder.MEDIA_RECORDER_INFO_MAX_DURATION_REACHED) {
            Log.i(TAG, "...max duration reached");
            stopRecording();
            Toast.makeText(this,
              "Recording limit has been reached. Stopping the recording",
                    Toast.LENGTH_SHORT).show();
        }
    }

    @Override
    public void onError(MediaRecorder mr, int what, int extra) {
        Log.e(TAG, "got a recording error");
        stopRecording();
        Toast.makeText(this,
            "Recording error has occurred. Stopping the recording",
                Toast.LENGTH_SHORT).show();
    }
}
```

These two callbacks are very similar. The only difference between them is the circumstances under which they are called. In the onInfo() method, the messages are not considered errors. onInfo() could be called because we reached the maximum recording time, or the maximum file size, if we set either of these options on the recorder. For onError(), the documentation doesn't say specifically why this might be called, but it could be because the recorder runs out of space where the video file is being written. If onInfo() was called because we hit our time limit, or if we got some sort of recording error, we will stop recording.

As before when recording audio, we need to set the same permissions for audio (android.permission.RECORD_AUDIO) and the SD card (android.permission.WRITE_EXTERNAL_STORAGE), and now we need to add permission to access the camera (android.permission.CAMERA). For completeness, the AndroidManifest.xml file is shown in Listing 24–16. You'll notice that we force the orientation of our application to be landscape, which is why our layout file is in /res/layout-land/main.xml.

Listing 24–16. Record Video's *AndroidManifest.xml* File

```
<?xml version="1.0" encoding="utf-8"?>
<manifest xmlns:android="http://schemas.android.com/apk/res/android"
      package="com.androidbook.record.video"
      android:versionCode="1"
      android:versionName="1.0">
    <application android:icon="@drawable/icon"
            android:label="@string/app_name">
        <activity android:name=".MainActivity"
                  android:label="@string/app_name"
                  android:screenOrientation="landscape">
            <intent-filter>
              <action android:name="android.intent.action.MAIN" />
              <category android:name="android.intent.category.LAUNCHER" />
            </intent-filter>
        </activity>
    </application>
    <uses-sdk android:minSdkVersion="4" />

<uses-permission
       android:name="android.permission.WRITE_EXTERNAL_STORAGE"/>
<uses-permission android:name="android.permission.RECORD_AUDIO"/>
<uses-permission android:name="android.permission.CAMERA"/>
</manifest>
```

Camera and Camcorder Profiles

In Listing 24–14, you saw in the initRecorder() method a series of very specific settings for the video recorder. The question is, how can you know what the capabilities of the device are that your application is running on? Prior to Android 2.2, there really wasn't a good answer to this question. The stock Camera application that comes with Android uses an undocumented SystemProperties class. Therefore, prior to Android 2.2 you had to choose values that would work across all the devices you wanted to target. This was less than satisfying, especially as better cameras became available on the newer devices. To rectify this situation, Android 2.2 introduced a couple of new classes: CameraProfile and CamcorderProfile. These classes are simply containers for the camera attributes that you care about. Although CameraProfile has only one value (JPEG Encoding Quality Parameter), CamcorderProfile tells you about the frame rate, frame size (height and width), and other video and audio parameters. Not only that, but the MediaRecorder class can accept a CamcorderProfile to set the various video-recording values that a CamcorderProfile contains. You just have to be careful to call the setProfile() method after setting the video and audio sources, and before setting the output file.

With the introduction of Android 2.3, methods that deal with cameras now may have an alternate method that will accept a camera identifier. Before Android 2.3, most devices had only one camera, and it usually faced the back of the device. With the newer devices with front-facing cameras in addition to back-facing cameras, code needs a way to specify which camera it wants to be dealing with. For example, in the Camera class, the open() method will return a Camera object for the back-facing camera, if one exists. There is an open(int cameraid) method that returns a specific camera, allowing your application to use the front-facing camera if one exists. To determine how many cameras are available and which one is which, the Camera.getNumberOfCameras() method will return the camera count, and Camera.getCameraInfo() will return information about a specific camera, including in which direction it faces.

Exploring the MediaStore Class

So far, we've dealt with media by directly instantiating classes to play and record media within our own application. One of the great things about Android is that you can access other applications to do work for you. The MediaStore class provides an interface to the media that is stored on the device, both internally and externally.

MediaStore also provides APIs for you to act on the media. These include mechanisms for you to search the device for specific types of media, intents for you to record audio and video to the store, ways for you to establish playlists, and more.

Because the MediaStore class supports intents for you to record audio and video, and the MediaRecorder class does recording also, an obvious question is: when do you use MediaStore versus MediaRecorder? As you saw with the preceding video-capture example and the audio-recording examples, MediaRecorder enables you to set various options on the source of the recording. These options include the audio/video input source, video frame rate, video frame size, output formats, and so on. MediaStore does not provide this level of granularity, but if you don't need it, you may find it easier to go through the MediaStore's intents. More important, content created with MediaRecorder is not automatically available to other applications that are looking at the media store. If you use MediaRecorder, you might want to add the recording to the media store using the MediaStore APIs, so it might be simpler just to use MediaStore in the first place.

Another significant difference is that calling MediaStore through an intent does not require your application to request permissions to record audio, or access the camera, or to write to the SD card. Your application is invoking a separate activity, and that other activity must have permission to record audio, access the camera, and write to the SD card. `MediaStore` activities already have these permissions. Therefore, your application doesn't have to. So, let's see how we can leverage the `MediaStore` APIs.

Recording Audio Using an Intent

As we've seen, recording audio was easy, but it gets much easier if you use an intent from `MediaStore`. Listing 24–17 demonstrates how to use an intent to record audio.

Listing 24–17. *Using an Intent to Record Audio*

```
<?xml version="1.0" encoding="utf-8"?>
<!-- This file is /res/layout/main.xml -->
<LinearLayout xmlns:android="http://schemas.android.com/apk/res/android"
    android:orientation="vertical"
    android:layout_width="fill_parent"
    android:layout_height="fill_parent" >
 <Button android:id="@+id/recordBtn"
    android:text="Record Audio"
    android:layout_width="wrap_content"
    android:layout_height="wrap_content" />
</LinearLayout>


import android.app.Activity;
import android.content.Intent;
import android.net.Uri;
import android.os.Bundle;
import android.util.Log;
import android.view.View;
import android.view.View.OnClickListener;
import android.widget.Button;

public class UsingMediaStoreActivity extends Activity {
    @Override
    protected void onCreate(Bundle savedInstanceState) {
        super.onCreate(savedInstanceState);

        setContentView(R.layout.main);

        Button btn = (Button)findViewById(R.id.recordBtn);
        btn.setOnClickListener(new OnClickListener(){

            @Override
            public void onClick(View view) {

                startRecording();

            }});
    }

    public void startRecording() {
        Intent intt =
            new Intent("android.provider.MediaStore.RECORD_SOUND");
```

```
        startActivityForResult(intt, 0);
    }

    @Override
    protected void onActivityResult(int requestCode, int resultCode, Intent data) {

        switch (requestCode) {
        case 0:
            if (resultCode == RESULT_OK) {
                Uri recordedAudioPath = data.getData();
                Log.v("Demo", "Uri is " + recordedAudioPath.toString());
            }
        }
    }
}
```

Listing 24–17 creates an intent requesting the system to begin recording audio. The code launches the intent against an activity by calling `startActivityForResult()`, passing the intent and the `requestCode`. When the requested activity completes, `onActivityResult()` is called with the `requestCode`. As shown in `onActivityResult()`, we look for a `requestCode` that matches the code that was passed to `startActivityForResult()` and then retrieve the URI of the saved media by calling `data.getData()`. You could then feed the URI to an intent to listen to the recording if you wanted to. The UI for Listing 24–17 is shown in Figure 24–7.

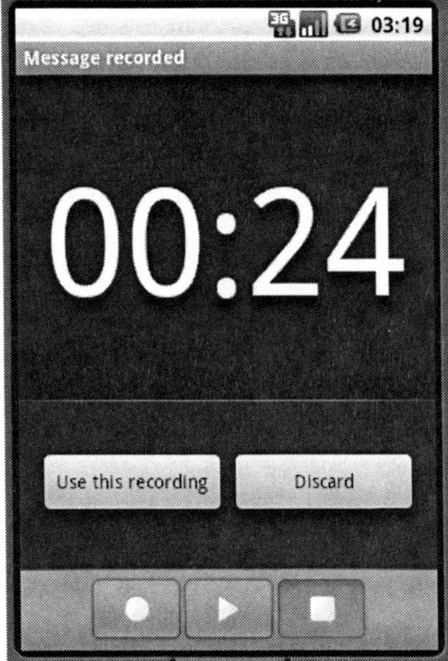

Figure 24–7. *Built-in audio recorder before and after a recording*

Figure 24–7 shows two screenshots. The image on the left displays the audio recorder during recording, and the image on the right shows the activity UI after the recording has been stopped.

Similar to the way it provides an intent for audio recording, MediaStore also provides an intent for you to take a picture. Listing 24–18 demonstrates this.

Listing 24–18. *Launching an Intent to Take a Picture*

```
<?xml version="1.0" encoding="utf-8"?>
<!-- This file is /res/layout/main.xml -->
<LinearLayout xmlns:android="http://schemas.android.com/apk/res/android"
    android:orientation="vertical"
    android:layout_width="fill_parent"
    android:layout_height="fill_parent" >
  <Button android:id="@+id/btn"  android:text="Take Picture"
      android:layout_width="wrap_content"
      android:layout_height="wrap_content"
      android:onClick="captureImage" />

</LinearLayout>


import android.app.Activity;
import android.content.ContentValues;
import android.content.Intent;
import android.net.Uri;
import android.os.Bundle;
import android.provider.MediaStore;
import android.provider.MediaStore.Images.Media;
import android.view.View;
import android.view.View.OnClickListener;
import android.widget.Button;

public class MainActivity extends Activity {

    Uri myPicture = null;

    @Override
    public void onCreate(Bundle savedInstanceState) {
        super.onCreate(savedInstanceState);
        setContentView(R.layout.main);

        setRequestedOrientation(ActivityInfo.SCREEN_ORIENTATION_LANDSCAPE);
    }

    public void captureImage(View view)
    {
        ContentValues values = new ContentValues();
        values.put(Media.TITLE, "My demo image");
        values.put(Media.DESCRIPTION, "Image Captured by Camera via an Intent");

        myPicture = getContentResolver().insert(Media.EXTERNAL_CONTENT_URI, values);

        Intent i = new Intent(MediaStore.ACTION_IMAGE_CAPTURE);
        i.putExtra(MediaStore.EXTRA_OUTPUT, myPicture);

        startActivityForResult(i, 0);
    }

    @Override
    protected void onActivityResult(int requestCode, int resultCode, Intent data) {
```

```
        if(requestCode==0 && resultCode==Activity.RESULT_OK)
        {
            // Now we know that our myPicture Uri
            // refers to the image just taken
        }
    }
}
```

The activity class shown in Listing 24–18 defines the `captureImage()` method. In this method, an intent is created where the action name of the intent is set to `MediaStore.ACTION_IMAGE_CAPTURE`. When this intent is launched, the camera application is brought to the foreground and the user takes a picture. Because we created the URI in advance, we can add additional details about the picture before the camera takes it. This is what the `ContentValues` class does for us. Additional attributes can be added to `values` besides `TITLE` and `DESCRIPTION`. Look up `MediaStore.Images.ImageColumns` in the Android reference for a complete list. After the picture is taken, our `onActivityResult()` callback is called. In our example, we've used the media content provider to create a new file. We could also have created a new URI from a new file on the SD card, as shown here:

```
myPicture = Uri.fromFile(new
          File(Environment.getExternalStoragePublicDirectory(DIRECTORY_DCIM) +
          "/100ANDRO/imageCaptureIntent.jpg"));
```

However, creating a URI this way does not so easily allow us to set attributes about the image, such as `TITLE` and `DESCRIPTION`. There is another way to invoke the camera intent in order to take a picture. If we do not pass any URI at all with the intent, we will get a bitmap object returned to us in the intent argument for `onActivityResult()`. The problem with this approach is that by default, the bitmap will be scaled down from the original size, apparently because the Android team does not want you to receive a large amount of data from the camera activity back to your activity. The bitmap will have a size of 50KB To get the `Bitmap` object, you'd do something like this inside of `onActivityResult()`:

```
Bitmap myBitmap = (Bitmap) data.getExtras().get("data");
```

`MediaStore` also has a video-capture intent that behaves similarly. You can use `MediaStore.ACTION_VIDEO_CAPTURE` to capture video.

Adding Media Content to the Media Store

One of the other features provided by Android's media framework is the ability to add information about content to the media store via the `MediaScannerConnection` class. In other words, if the media store doesn't know about some new content, we use a `MediaScannerConnection` to tell the media store about the new content. Then that content can be served up to others. Let's see how this works (see Listing 24–19).

Listing 24–19. *Adding a File to the Media Store*

```
<?xml version="1.0" encoding="utf-8"?>
<!-- This file is /res/layout/main.xml -->
<LinearLayout
```

```
    xmlns:android="http://schemas.android.com/apk/res/android"
    android:orientation="vertical"
    android:layout_width="fill_parent"
    android:layout_height="wrap_content">

    <EditText android:id="@+id/fileName"  android:hint="Enter new filename"
      android:layout_width="fill_parent"
      android:layout_height="wrap_content" />

    <Button android:id="@+id/scanBtn"  android:text="Add file"
      android:layout_width="wrap_content"
      android:layout_height="wrap_content"
      android:onClick=”startScan” />

</LinearLayout>
```

```
import java.io.File;
import android.app.Activity;
import android.content.Intent;
import android.media.MediaScannerConnection;
import android.media.MediaScannerConnection.MediaScannerConnectionClient;
import android.net.Uri;
import android.os.Bundle;
import android.util.Log;
import android.view.View;
import android.widget.EditText;
import android.widget.Toast;

public class MediaScannerActivity extends Activity implements
MediaScannerConnectionClient
{
    private EditText editText = null;
    private String filename = null;
    private MediaScannerConnection conn;

    @Override
    protected void onCreate(Bundle savedInstanceState) {
        super.onCreate(savedInstanceState);
        setContentView(R.layout.main);

        editText = (EditText)findViewById(R.id.fileName);
    }

    public void startScan(View view)
    {
        if(conn!=null) {
            conn.disconnect();
        }

        filename = editText.getText().toString();

        File fileCheck = new File(filename);
        if(fileCheck.isFile()) {
            conn = new MediaScannerConnection(this, this);
            conn.connect();
        }
```

```
        else {
            Toast.makeText(this,
                "That file does not exist",
                Toast.LENGTH_SHORT).show();
        }
    }

    @Override
    public void onMediaScannerConnected() {
        conn.scanFile(filename, null);
    }

    @Override
    public void onScanCompleted(String path, Uri uri) {
        try {
            if (uri != null) {
                Intent intent = new Intent(Intent.ACTION_VIEW);
                intent.setData(uri);
                startActivity(intent);
            }
            else {
                Log.e("MediaScannerDemo", "That file is no good");
            }
        } finally {
            conn.disconnect();
            conn = null;
        }
    }
}
```

Listing 24–19 shows an activity class that adds a file to the media store. If the add is successful, the added file is displayed to the user via an intent. What happens behind the scenes, is that the file is inspected by the MediaScanner to determine what type of file it is and other relevant details about it. We could have given the MediaScanner the MIME type of our file as the second argument to scanFile(). If MediaScanner can't determine what the type of the file is by the extension, it won't get added. If the file belongs in the MediaStore, a database entry is made into the media provider database. The file itself doesn't move. But now the media provider knows about this file. If you added an image file, you can now open the Gallery application and see it. If you added a music file, it will now show up in the Music application.

If you want to see inside the media provider's database, open a tools window, launch adb shell, and then navigate on the device to /data/data/com.android.providers.media/databases. There you will find databases, one of which is internal.db. There could be external database files there also, corresponding to one or more SD cards. Because you can use multiple SD cards with an Android phone, there could also be multiple external database files there. You can use the sqlite3 utility to inspect the tables in these databases. There are tables for audio, images, and video. See Chapter 4 for more information on using sqlite3.

Triggering MediaScanner for the Entire SD Card

In the previous example, we used the MediaScanner to look at a single, specific file. This is fine if you want to add a single file. But what if you want to rename a file, or delete a file, and you want the MediaStore to be updated? Fortunately, there's a very simple way to trigger this to happen. If you execute the following inside your application, it will be picked up by MediaScanner, which will rescan the entire SD card:

```
sendBroadcast(new Intent(Intent.ACTION_MEDIA_MOUNTED,
    Uri.parse("file://" +
    Environment.getExternalStorageDirectory())));
```

> **NOTE:** If MediaScanner encounters an empty file called .nomedia in a directory, that directory and all of its subdirectories are skipped during media scanning. You can use the .nomedia file to hide media files from Gallery or Music. Look for a .nomedia file if files that you think should be available in Gallery or Music are not displayed.

As an exercise, go ahead and create a simple application that just does this command in the onCreate().

This concludes our discussion of the media APIs. We hope you'll agree that playing and recording media content is not too difficult with Android.

References

Here are some helpful references to topics you may wish to explore further:

- www.androidbook.com/proandroid4/projects: A list of downloadable projects related to this book. For the projects in this chapter, look for a zip file called ProAndroid4_Ch24_Media.zip. This zip file contains all the projects from this chapter, listed in separate root directories. There is also a README.TXT file that describes exactly how to import projects into Eclipse from one of these zip files.
- http://developer.android.com/guide/topics/media/jet/jetcreator_manual.html: The user manual for the JETCreator tool. You can use this to create a JET sound file to be played using the JetPlayer. JETCreator is only available for Windows and Mac OS. To see JetPlayer in action, load the JetBoy sample project from the Android SDK into Eclipse, build it, and run it. Note that the Fire button is the center directional pad key.

Summary

Here is a summary the topics covered in this media chapter on audio and video:

- SD cards, and how to create SD card images
- Manually putting files onto and getting files from SD cards
- The standard DCIM directory for picture and video files
- Android constants for specifying other types of directories on SD cards, and the method calls to use them with
- Playing audio through a MediaPlayer
- Several ways to source audio for `MediaPlayer`, from local application resources, to files, to streaming over the network
- Steps to take with a MediaPlayer to get the audio to come out properly
- `SoundPool` and its ability to play several sounds simultaneously
- `SoundPool`'s limitations in terms of the amount of audio it can handle
- `AsyncPlayer`, which is useful because sounds generally need to be managed in the background
- `AudioTrack`, which provides low-level access to audio; and an example application demonstrating `AudioRecord`
- Playing video using `VideoView`
- Recording audio using `MediaRecorder`
- Specifying the audio source, the output format, the audio encoder, and the output destination
- Using callbacks with `AudioRecord` to receive audio data at its most raw
- Recording video and using the `Camera` preview
- Setup and locking required to do video recording
- Using callbacks to detect when some predetermined limit has been reached, such as time or size of the recording
- Using `Camera` and `Camcorder` profiles
- `MediaStore` and its ability to provide easier intent methods of recording audio and video
- `MediaStore`'s database of metadata about media, and adding to that database with a `MediaScanner`
- The secret file you can use to hide media files from the scanner

Interview Questions

Ask yourself the following questions to solidify your understanding of this chapter:

1. If you ran the mksdcard utility with a size of 2G, would you get an SD card image file that's 2GB in size?
2. How many SD cards can an Android device have at any one time?
3. Which Android constant do you use to locate the directory on the SD card where ringtones are stored?
4. What permission is needed to write new media files to an SD card?
5. What permission is needed to receive streaming media over the network?
6. What other permissions might you want to request if the application receives streamed media?
7. What's the difference between using the default constructor of MediaPlayer and using the static method MediaPlayer.create()?
8. Why should you always use prepareAsync() with MediaPlayer?
9. How much audio data can a SoundPool manage?
10. What is a major advantage of using AsyncPlayer?
11. What could be a potential disadvantage of using AsyncPlayer?
12. Why is the emulator so difficult to use to view video?
13. What are the two ways to specify a video source to a VideoView?
14. What are the five things you must do before calling start() on a MediaRecorder?
15. What permission is required in order to record audio?
16. What is the difference in the AudioRecord.OnRecordPositionUpdateListener class between the callbacks onPeriodicNotification() and onMarkerReached()?
17. Is it possible to record video without having to show it on the screen?
18. Does the MediaRecorder have a prepareAsync() method?
19. Which callbacks give you information about the status of video recording?
20. How can an application figure out what cameras are available on a device and what their parameters are?
21. What Intent action would you use to request an audio recording be made via MediaStore?
22. How many files does MediaScannerConnection work on at a time?

Chapter 25

Home Screen Widgets

In this chapter, we will cover Android's home screen widgets in detail. Home screen widgets present frequently changing information on the home screen of Android. From a high-level perspective, home screen widgets are disconnected views (albeit populated with data) that are displayed on the home screen. The data content of these views is updated at regular intervals by background processes.

For example, an e-mail home screen widget might alert you to the number of outstanding e-mails to be read. The widget may just show you the number of e-mails and not the messages themselves. Clicking the e-mail count may then take you to the activity that displays actual e-mails. These could even be external e-mail sources such as Yahoo, Gmail, and Hotmail, as long as the device has a way to access the counts through HTTP or other connectivity mechanisms.

We will divide this chapter into three sections. In the first section, we will introduce home screen widgets and their architecture. We will describe how Android uses `RemoteViews` for showing widgets, and co-opts broadcast receivers to update those `RemoteViews`. You will learn how to create activities to configure widgets on the home screen and discover the relationship between services and widgets. At the end of this section, you will have a clear understanding of the architecture and life cycle of home screen widgets.

In the second section, we will show you how to design and develop a home screen widget with annotated code. You will learn how to define widgets to Android and how to write broadcast receivers to update these widgets. We will show you how to manage widget state through shared preferences and how to write an activity to configure widgets.

In the third section, we will talk about suitability, limitations, and broader guidelines for working with widgets.We will also offer design suggestions to write widgets that require far more frequent updates. We will cover list-based widgets in Chapter 26.

Architecture of Home Screen Widgets

Let's start our discussion of home screen widgets architecture by considering what home screen widgets are in greater detail.

What Are Home Screen Widgets?

Home screen widgets are views that can be displayed on a home page and updated frequently. As a view, a widget's look and feel is defined through a layout XML file. For a widget, in addition to the layout of the view, you will need to define how much space the view of the widget will need on the home screen.

A widget definition also includes a couple of Java classes that are responsible for initializing the view and updating it frequently. These Java classes are responsible for managing the life cycle of the widget on the home screen. These classes respond when the widget is dragged onto the home page, when the widget needs to be updated, and when the widget is uninstalled by dragging it to the trash can.

> **NOTE:** The view and the corresponding Java classes are architected in such a way that they are disconnected from each other. For example, any Android service or activity can retrieve the view using its layout ID, populate that view with data (just like populating a template), and send it to the home screen. Once the view is sent to the home screen, it is dislodged from the underlying Java code.

At a minimum, a widget definition contains the following:

- A view layout to be displayed on the home screen, along with how big it should be (at a minimum) to fit on a home page. Keep in mind that this is just the view without any data. It will be the responsibility of a Java class to update the view.
- A timer that specifies the frequency of updates.
- A broadcast receiver Java class called a *widget provider* that can respond to timer updates in order to alter the view in some fashion by populating with data.

Once a widget is defined and the Java classes are provided, the widget will be available for the user to drag onto a home page .

Before we show you how to implement a widget from scratch, we'll first give you an overview of how a widget is used by an end user.

User Experience with Home Screen Widgets

Home screen widget functionality in Android allows you to choose a preprogrammed widget to be placed on the home screen. When placed, the widget will allow you to configure it using an activity (defined as part of the widget package), if necessary. It is important to understand this interaction before actually going into the details how a widget is implemented.

We are going to walk you through a widget called Birthday Widget that we have created for this chapter. We will present the source code for it later in the chapter. First, we are going to use this widget as an example for our walkthrough. As a consequence of source code coming later, we need your consideration to read along and follow the pictures and not look for this widget on your screen. If you follow the provided figures and explanation, you will know the nature and behavior of the Birthday Widget, which will make things clear when we code it subsequently.

Let's start this tour by locating the widget we want and creating an instance of it on the home screen.

Creating a Widget Instance on the Home Screen

To access the available widget list, you need to long-click the home page. This will bring up the home screen context menu, as shown in Figure 25–1. Note that depending on the Android SDK release, these screens may look a bit different. The images in the chapter were captured using 2.3 SDK. However, we have tested the program on the latest releases available.

> **NOTE:** In Android 4.0, the long-click is not the mechanism that is used to get to the list of widgets. Instead, you have to navigate to the list of applications. When you see the list of applications, a tab called Widgets appears at the top.

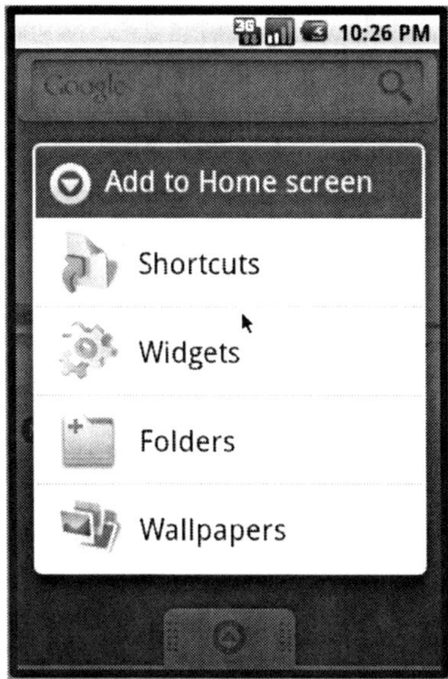

Figure 25–1. *Home screen context menu*

If you choose Widgets from this list, you will be shown another screen that is a pick list of available widgets, as shown in Figure 25–2.

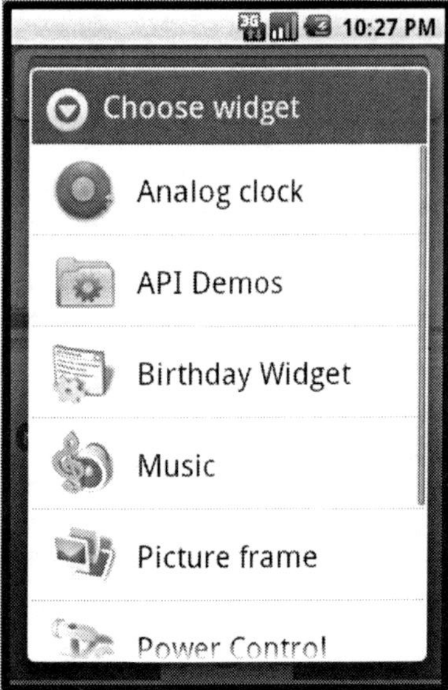

Figure 25–2. *Home screen widget pick list*

Most of these widgets come as part of Android. Depending on the release of Android you are looking at, these may vary.

> **NOTE:** In Android 4.0 instead of a list, you will see a panel full of widgets laid out in a grid. From that point on, picking a widget is the same in all releases.

In this list, the widget named Birthday Widget is the widget that we designed for this exercise.

If you choose that widget, it will create a corresponding widget instance on the home screen that looks like the example Birthday Widget shown in Figure 25–3.

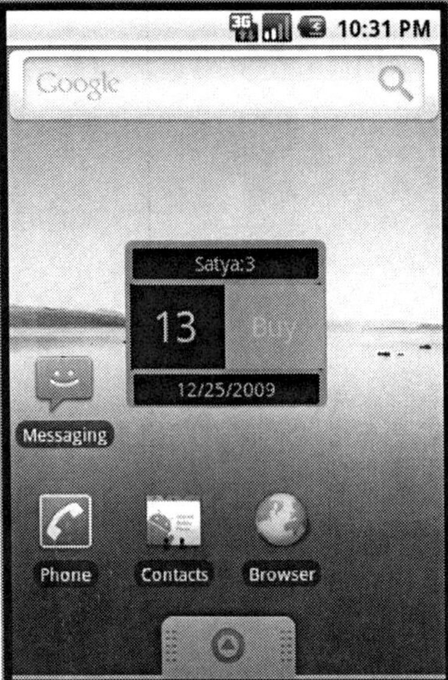

Figure 25–3. *An example Birthday Widget*

This Birthday Widget will indicate in its header the name of a person, how many days away this person's birthday is, the date of the birthday, and a link to buy gifts.

You may be wondering how the name of the person and the date of birth were configured. What if you want two of instances of this widget, each with the name and date of birth for a different person? This is where the widget configurator activity comes into play and is the topic we are covering next.

> **NOTE:** The view that is created on the home page for this widget definition is called a *widget instance*. The implication is that you can create more than one instance of this widget definition.

Understanding Widget Configurator

A widget definition optionally includes a specification of an activity called a widget configurator activity. When you choose a widget from the home page widget pick list to create the widget instance, Android invokes the corresponding widget configuration activity. This activity is something you need to write, which is then responsible for configuring the widget instance.

In the case of our Birthday Widget, this configuration activity will prompt you for the name of the person and the upcoming birth date, as shown in Figure 25–4. It is the responsibility of the configurator to save this information in a persistent place so that when an update is called on the widget provider, the widget provider will be able to locate this information and update the view with proper values that are set by the configurator.

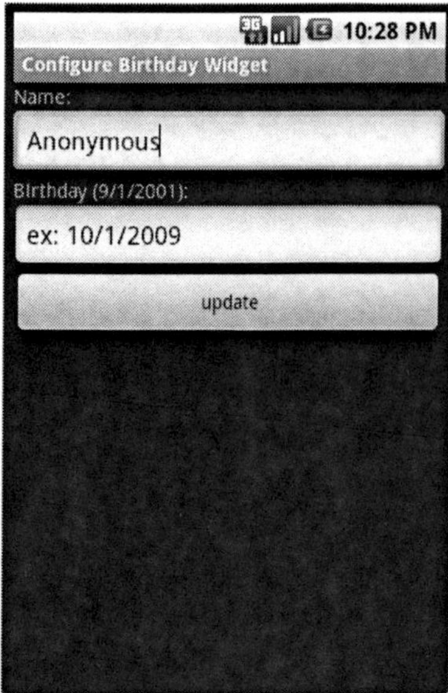

Figure 25–4. *Birthday widget configurator activity*

> **NOTE:** When a user chooses to create two Birthday Widget instances on the home screen, the configurator activity will be called twice (once for each widget instance).

Internally, Android keeps track of the widget instances by allocating them an ID. This ID is passed to the Java callbacks and to the configurator Java class so that initial configuration and updates can be directed to the right instance. In Figure 25–3, in the later part of the string `satya:3`, the `3` is the widget ID—or, more accurately, the widget instance ID. The widget itself is identified by its java component name (which is itself the class name and the package that the widget class is in); "Widget ID" and "widget instance ID" are interchangeably used in this chapter and refer to the widget instance ID. We have included the widget instance ID in Figure 25–3 to illustrate the point.

With this overview of a widget, we will now examine the life cycle of a widget in greater detail.

Life Cycle of a Widget

We have mentioned the widget definition a few times so far. We have also briefly talked about the role of Java classes. In this section, we will lay out both these ideas in a lot more detail and examine the life cycle of a widget.

The life cycle of a widget has the following phases:

1. Widget definition
2. Widget instance creation
3. `onUpdate()` (when the time interval expires)
4. Responses to clicks (on the widget view on the home screen)
5. Widget deletion (from the home screen)
6. Uninstallation

We will go through each of these phases in detail now.

Widget Definition Phase

The life cycle of a widget starts with the definition of the widget view. This definition tells Android to show the widget name in the widget pick list (Figure 25–2) invoked from the home page. You will need two things to complete this definition: a Java class that implements the `AppWidgetProvider` and a layout view for the widget.

You start off this widget definition with the following entry in the Android manifest file where you specify the `AppWidgetProvider` (Listing 25–1).

Listing 25–1. *Widget Definition in Android Manifest File*

```
<manifest..>
<application>
....
   <receiver android:name=".BDayWidgetProvider">
      <meta-data android:name="android.appwidget.provider"
             android:resource="@xml/bday_appwidget_provider" />
      <intent-filter>
           <action android:name="android.appwidget.action.APPWIDGET_UPDATE" />
      </intent-filter>
   </receiver>
   ...
   <activity>
      .....
   </activity>
<application>
</manifest>
```

This definition indicates that there is a broadcast receiver Java class called `BDayWidgetProvider` (as you will see, this inherits from the Android core class `AppWidgetProvider` from the widget package) that receives broadcast messages intended for application widget updates.

NOTE: Android delivers the update messages as broadcast messages based on the frequency of the time interval.

The widget definition in Listing 25–1 also points to an XML file in the /res/xml directory that, in turn, specifies the widget view and the update frequency, as shown in Listing 25–2.

Listing 25–2. *Widget View Definition in Widget Provider Information XML File*

```
<appwidget-provider xmlns:android="http://schemas.android.com/apk/res/android"
    android:minWidth="150dp"
    android:minHeight="120dp"
    android:updatePeriodMillis="43200000"
    android:initialLayout="@layout/bday_widget"
    android:configure="com.androidbook.BDayWidget.ConfigureBDayWidgetActivity"
    android:previewImage="@drawable/some_preview_image_icon"
    android:resizeMode="horizontal|vertical"
    android:previewImage="@drawable/some_preview_image_icon"
    >
</appwidget-provider>
```

This XML file is called the App widget provider information file. Internally, this gets translated to the AppWidgetProviderInfo Java class. This file identifies the width and height of the layout to be 150dp and 120dp, respectively. This definition file also indicates the update frequency to be 12 hours translated to milliseconds. The definition also points to a layout file (Listing 25–7) that describes what the widget view looks like (see Figure 25–5).

Starting with SDK 3.1, users have the ability to resize a widget that is placed on one of their images. The user sees resize handles when they long-click the widget and can then use these handles to resize. This resize can be horizontal, vertical, or none. You can combine horizontal and vertical to resize the widget in both dimensions, as shown in Listing 25–2. However, to take advantage of this, your widget controls should be laid out in such a way that they can expand and contract using their layout parameters. There is no callback to tell you what size your widget is. The update is not triggered—at least, the documentation doesn't mention such a fact.

The preview image attribute in Listing 25–2 indicates what image or icon is used to show your widget in the list of available widgets. This attribute is included starting with SDK 3.0 (API 11). If you omit it, the default behavior is to show the main icon for your application package, which is indicated in the manifest file. The emulator caches this preview image even if you uninstall and reinstall the package. You may want to restart your emulator to see this effect.

The layout for widget views is restricted to contain only certain types of view elements. The views allowed in a widget layout fall under a class of views called RemoteViews, and only certain types of child views are allowed for these remote views. Some of the allowed subview elements are shown in Listing 25–3. Note that their subclasses are not supported, only those that are included in Listing 25–3.

Listing 25–3. *Allowed View Controls in* `RemoteViews`

```
FrameLayout
LinearLayout
RelativeLayout

AnalogClock
Button
Chronometer
ImageButton
ImageView
ProgressBar
TextView
ViewFlipper
ListView
GridView
StackView
AdapterViewFlipper
```

This list may grow with each release. The primary reason for restricting what is allowed in a remote view is that these views are disconnected from the processes that actually control them. These widget views are hosted by an application like the Home application. The controllers for these views are background processes that get invoked by timers. For this reason, these views are called remote views. There is a corresponding Java class called `RemoteViews` that allows access to these views. In other words, programmers do not have direct access to these views to call methods on them. You have access to these views only through the `RemoteViews` (like a gatekeeper).

We will cover the relevant methods of a `RemoteViews` class when we explore the example in the next main section. For now, remember that only a limited set of views is allowed in the widget layout file (see Listing 25–3).

The widget definition (Listing 25–2) also includes a specification of the configuration activity that needs to be invoked when the user creates a widget instance. This configuration activity in Listing 25–2 is the `ConfigureBDayWidgetActivity`. This activity is like any other Android activity with a number of form fields. The form fields are used to collect the information needed by a widget instance.

Widget Instance Creation Phase

Once all the XML pieces needed by a widget definition are in place and all the widget Java classes are available, let's see what happens when a user chooses the widget name in the widget pick list (see Figure 25–2) to create a widget instance. Android invokes the configurator activity (see Figure 25–3) and expects that configurator activity to do the following:

1. Receive the widget instance ID from the invoking intent that started the configurator.

2. Prompt the user through a set of form fields to collect the widget-instance-specific information.

3. Persist the widget instance information so that subsequent calls to widget `update` have access to this information.

4. Prepare to display the widget view for the first time by retrieving the widget view layout and create a `RemoteViews` object with it.

5. Call methods on the `RemoteViews` object to set values on individual view objects, such as text and images.

6. Also use the `RemoteViews` object to register any `onClick` events on any of the subviews of the widget.

7. Tell the `AppWidgetManager` to paint the `RemoteViews` on the home screen using the instance ID of that widget.

8. Return the widget ID, and close.

Notice that the first painting of the widget in this case is done by the configurator and not `AppWidgetProvider`'s `onUpdate()` method.

> **NOTE:** The configurator activity is optional. If the configurator activity is not specified, the call goes directly to the `onUpdate()` method of the `AppWidgetProvider`. It is up to `onUpdate()` to update the view.

Android will repeat this process for each widget instance that the user creates. Also note that there is no direct documented support for restricting the user to a single widget instance.

Besides invoking the configurator activity, Android also invokes the `onEnabled` callback of the `AppWidgetProvider`. Let's briefly consider the callbacks on an `AppWidgetProvider` class by taking a look at the shell of our `BDayWidgetProvider` (see Listing 25–4). We will examine the complete listing of this file later in Listing 25–9.

Listing 25–4. *A Widget Provider Shell*

```
public class BDayWidgetProvider extends AppWidgetProvider
{
    public void onUpdate(Context context,
                         AppWidgetManager appWidgetManager,
                         int[] appWidgetIds){}

    public void onDeleted(Context context, int[] appWidgetIds){}
    public void onEnabled(Context context){}
    public void onDisabled(Context context) {}
}
```

The `onEnabled()` callback method indicates that there is at least one instance of the widget up and running on the home screen. This means a user must have dropped the widget on the home page at least once. In this call, you will need to enable receiving messages for this broadcast receiver component (you will see this in Listing 25–9). The base class AppWidgetProvider has the functionality to enable or disable receiving broadcast messages.

The onDeleted() callback method is called when a user drags the widget instance view to the trash can. This is where you will need to delete any persistent values you are holding for that widget instance.

The onDisabled() callback method is called after the last widget instance is removed from the home screen. This happens when a user drags the last instance of a widget to the trash. You should use this method to unregister your interest in receiving any broadcast messages intended for this component (you will see this in Listing 25–9).

The onUpdate() callback method is called every time the timer specified in Listing 25–2 expires. This method is also called the very first time the widget instance is created if there is no configurator activity. If there is a configurator activity, this method is not called at the creation of a widget instance. This method will subsequently be called when the timer expires at the frequency indicated.

onUpdate Phase

Once the widget instance is on the home screen, the next significant event is the expiration of the timer. Android will call onUpdate() in response to that timer. Because onUpdate() is called is through a broadcast receiver, the corresponding Java process will be loaded and will remain alive until the end of that call. Once the call returns, the process will be ready to be taken down.

It is recommended that you use a mechanism such as a long-running broadcast receiver as documented in Chapter 19 if your response is going to take more than 10 seconds to work. Once you have the necessary data available to update the widget in the onUpdate() method, you can invoke the AppWidgetManager to paint the remote view. If you were to invoke a long-running service to do the update instead, you would need to pass the widget ID as extra data to the intent that starts the service.

This goes to show that the AppWidgetProvider class is stateless and may even be incapable of maintaining static variables between invocations. This is because the Java process containing this broadcast receiver class could be taken down and reconstructed between two invocations, resulting in re-initialization of static variables.

As a result, you will need to come up with a scheme to remember state if that is required. When the updates are not too frequent, such as every few seconds, it is reasonable to save the state of the widget instance in a persistent store such as a file, shared preferences, or a SQLite database. In the examples in this chapter, we will use shared preferences as the persistence API.

WARNING: To save power, Google recommends that the duration of the updates be more than an hour, so the device won't wake up too often. Starting with the 2.0 API, there is a restriction of 30 minutes or more for the update timeout.

For durations that are shorter, such as only seconds, you need to call this `onUpdate()` method yourself by using the facilities in the `AlarmManager` class. When you use `AlarmManager`, you also have the option not to call `onUpdate()` but, instead, do the work of `onUpdate()` in alarm callbacks. Refer to Chapter 20 for working with the alarm manager.

This is what you typically need to do in an `onUpdate()` method:

1. Make sure the configurator has finished its work; otherwise, just return. This should not be problem in releases 2.0 and above, where the duration is expected to be longer. Otherwise, it is possible that `onUpdate()` will be called before the user has finished configuring the widget in the configurator.
2. Retrieve the persisted data for that widget instance.
3. Retrieve the widget view layout, and create a `RemoteViews` object with it.
4. Call methods on the `RemoteViews` to set values on individual view objects such as text and images.
5. Register any `onClick` events on any of the views by using pending intents.
6. Tell the `AppWidgetManager` to paint the `RemoteViews` using the instance ID.

As you can see, there is a lot of overlap between what a configurator does initially and what the `onUpdate()` method does. You may want to reuse this functionality between the two places.

Widget View Mouse Click Event Callbacks

As stated, the `onUpdate()` method keeps the widget views up to date. The widget view and subelements in that view could have callbacks registered for a mouse click. Typically, the `onUpdate()` method uses a pending intent to register an action for an event like a mouse click. This action could then start a service or start an activity such as opening up a browser.

This invoked service or activity can then communicate back with the view, if needed, using the widget instance ID and the `AppWidgetManager`. Hence, it is important that the pending intent carries with it the widget instance ID.

Deleting a Widget Instance

Another distinct event that can happen to a widget instance is that it can get deleted. To do this, a user has to long-press the widget on the home screen. This will enable the trash can to show at the bottom of the home screen. The user can then drag the widget instance to the trash can to delete the widget instance from the screen.

Doing so calls the `onDelete()` method of the widget provider. If you have saved any state information for this widget instance, you will need to delete that data in this `onDelete` method.

Android also calls `onDisable()` if the widget instance that is just deleted is the last of the widget instances of this type. You will use this callback to clean up any persistence attributes that are stored for all widget instances and also unregister for callbacks from the widget `onUpdate()` broadcasts (see Listing 22-9).

Uninstalling Widget Packages

There is a need to clean up the widgets if you are planning to uninstall and install a new release of your `.apk` file containing these widgets.

It is recommended that you remove or delete all widget instances before trying to uninstall the package. Follow the directions in the "Deleting a Widget Instance" section to delete each widget instance until none remains.

Then, you can uninstall and install the new release. This is especially important if you are using the Eclipse ADT to develop your widgets, because during the development time, ADT tries to do this every time you run the application. So, between runs, make sure you remove the widget instances.

A Sample Widget Application

So far, we have covered the theory and approach behind widgets. Let's create the sample widget whose behavior has been used as the example to explain widget architecture. We will develop, test, and deploy this Birthday Widget.

Each Birthday Widget instance will show a name, the date of the next birthday, and how many days from today until the birthday. It will also create an `onClick` area where you can click to buy gifts. This click will open a browser and take you to `www.google.com`.

The layout of the finished widget should look like Figure 25–5.

Figure 25–5. *Birthday Widget look and feel*

The implementation of this widget consists of the following widget-related files. The entire project is also available for download at the URL mentioned in the "References" section of this chapter.

The basic files are

- `AndroidManifest.xml`: Where the `AppWidgetProvider` is defined (see Listing 25–5)
- `res/xml/bday_appwidget_provider.xml`: Widget dimensions and layout (see Listing 25–6)
- `res/layout/bday_widget.xml`: The widget layout (see Listing 25–7)
- `res/drawable/box1.xml`: Provides boxes for sections of the widget layout (see Listing 25–8)
- `src/.../BdayWidgetProvider.java`: Implementation of the AppWidgetProvider class (see Listing 25–9)

These files manage the state of a widget:

- `src/.../IWidgetModelSaveContract.java`: Contract for saving a widget model (see Listing 25–10)
- `src/.../APrefWidgetModel.java`: Abstract preference-based widget model (see Listing 25–11)
- `src/.../BDayWidgetModel.java`: Widget model holding the data for a widget view (see Listing 25–12)
- `src/.../Utils.java`: A few utility classes (see Listing 25–13)

These files implement the widget configuration activity:

- src/.../ConfigureBDayWidgetActivity.java: Configuration activity (see Listing 25–14)
- layout/edit_bday_widget.xml: Layout for taking the name and birthday (see Listing 25–15)

We will walk through each file and explain any additional concepts that bear further consideration.

Defining the Widget Provider

Definition of a widget starts in the Android application manifest file. This is where you specify the widget provider (because it is a receiver), the widget configuration activity (because it is an activity), and a pointer to a widget definition XML file that further defines the widget layout.

For the Birthday Widget, you can see all of these bolded in the following Android manifest file (see Listing 25–5). Notice the definition of BDayAppWidgetProvider as a broadcast receiver and also the definition for the configuration activity ConfigureBDayWidgetActivity.

Listing 25–5. *Android Manifest File for BDayWidget Sample Application*

```
<?xml version="1.0" encoding="utf-8"?>
<manifest xmlns:android="http://schemas.android.com/apk/res/android"
      package="com.androidbook.BDayWidget"
      android:versionCode="1"
      android:versionName="1.0.0">
<application android:icon="@drawable/icon"
             android:label="Birthday Widget">
<!--
**********************************************************************
*  Birthday Widget Provider Receiver
**********************************************************************
 -->
   <receiver android:name=".BDayWidgetProvider">
      <meta-data android:name="android.appwidget.provider"
             android:resource="@xml/bday_appwidget_provider"/>
      <intent-filter>
           <action android:name="android.appwidget.action.APPWIDGET_UPDATE"/>
      </intent-filter>
   </receiver>
<!--
**********************************************************************
*  Birthday Provider Configurator Activity
**********************************************************************
 -->
   <activity android:name=".ConfigureBDayWidgetActivity"
             android:label="Configure Birthday Widget">
        <intent-filter>
            <action android:name="android.appwidget.action.APPWIDGET_CONFIGURE"/>
        </intent-filter>
   </activity>
```

```
    </application>
    <uses-sdk android:minSdkVersion="3"/>
</manifest>
```

The application label identified by "Birthday Widget" in the following line

```
<application android:icon="@drawable/icon" android:label="Birthday Widget">
```

is what shows up in the widget pick list (see Figure 25–2) of the home page. You can also indicate in the widget definition XML file (Listing 25–2) an alternate icon to be shown when the widget is listed (also called a *preview*).

If you are creating a widget definition for the first time, make sure the following line is replicated exactly:

```
<meta-data android:name="android.appwidget.provider"
```

The specification "android.appwidget.provider" is an Android-specific key to identify the widget XML definition file and should be mentioned as such; the same is true of the broadcast receiver intent filter lines shown here:

```
<intent-filter>
    <action android:name="android.appwidget.action.APPWIDGET_UPDATE"/>
</intent-filter>
```

These lines indicate that the widget provider is invoked by receiving the appwidget update broadcast messages.

Finally, the configuration activity definition is like any other normal activity, except that it needs to declare itself as capable of responding to APPWIDGET_CONFIGURE actions.

Defining Widget Size

Listing 25–6 shows the widget provider information file (/res/xml/bday_appwidget_provider.xml).

Listing 25–6. *Widget View Definition for* BDayWidget

```
<!-- res/xml/bday_appwidget_provider.xml -->
<appwidget-provider xmlns:android="http://schemas.android.com/apk/res/android"
    android:minWidth="150dp"
    android:minHeight="120dp"
    android:updatePeriodMillis="4320000"
    android:initialLayout="@layout/bday_widget"
    android:configure="com.androidbook.BDayWidget.ConfigureBDayWidgetActivity"
    >
</appwidget-provider>
```

This file indicates to Android the width and height that you want in pixels. However, Android will round them to the nearest cell. Android organizes its home screen area into a matrix of cells; each cell carries 74 density-independent pixels (dp) in width and height. Android recommends that you specify your width and height in multiples of these cells minus 2 pixels (to adjust for rounding.).

This file also indicates how often the `onUpdate()` method needs to be called. Android highly recommends that this value be no more than a few times a day. You can put a value of 0 to indicate never to call the update. This is useful when you want to control your own updates through the `AlarmManager` class.

The initial layout attribute points to the actual layout of the widget (see Listing 25–7). Finally, the `configure` attribute points to the configuration activity class. This class needs to be fully qualified in its definition.

Let's examine the actual layout for the widget now.

Widget Layout-Related Files

From the previous section and Listing 25–6, you can see that the layout of a widget is defined in a layout file. This layout file is just like any other layout file for a view in Android.

However, to guide standardization around widgets, Android published a set of widget design guidelines. You can access these guidelines at

http://developer.android.com/guide/practices/ui_guidelines/widget_design.html

In addition to the guidelines, this link has a set of view backgrounds that you can use to improve the look and feel of your widgets. In this example, we took a different route and used the traditional approach of view layouts with background shapes instead.

Widget Layout File

Listing 25–7 shows the layout file we used to produce the widget layout shown in Figure 25–5.

Listing 25–7. *Widget View Layout Definition for* `BDayWidget`

```
<?xml version="1.0" encoding="utf-8"?>
<!-- res/layout/bday_widget.xml -->
<LinearLayout xmlns:android="http://schemas.android.com/apk/res/android"
    android:orientation="vertical"
    android:layout_width="150dp"
    android:layout_height="120dp"
    android:background="@drawable/box1"
    >
<TextView
    android:id="@+id/bdw_w_name"
    android:layout_width="fill_parent"
    android:layout_height="30dp"
    android:text="Anonymous"
    android:background="@drawable/box1"
    android:gravity="center"
    />
<LinearLayout
    android:orientation="horizontal"
    android:layout_width="fill_parent"
    android:layout_height="60dp"
```

```
        >
        <TextView
            android:id="@+id/bdw_w_days"
            android:layout_width="wrap_content"
            android:layout_height="fill_parent"
            android:text="0"
            android:gravity="center"
            android:textSize="30sp"
            android:layout_weight="50"
            />
        <TextView
          android:id="@+id/bdw_w_button_buy"
            android:layout_width="wrap_content"
            android:layout_height="fill_parent"
            android:textSize="20sp"
            android:text="Buy"
            android:layout_weight="50"
            android:background="#FF6633"
            android:gravity="center"
        />
    </LinearLayout>
    <TextView
        android:id="@+id/bdw_w_date"
        android:layout_width="fill_parent"
        android:layout_height="30dp"
        android:text="1/1/2000"
        android:background="@drawable/box1"
        android:gravity="center"
        />
    </LinearLayout>
```

This layout uses nested LinearLayout nodes to get the desired effect. Some of the controls also use a shape definition file called box1.xml to define the borders.

Widget Background Shape File

The code for the shape definition file is shown in Listing 25–8 (this file should be in the /res/drawable subdirectory).

Listing 25–8. *A Boundary Box Shape Definition*

```
<!-- res/drawable/box1.xml -->
<shape xmlns:android="http://schemas.android.com/apk/res/android">
    <stroke android:width="4dp" android:color="#888888"/>
    <padding android:left="2dp" android:top="2dp"
            android:right="2dp" android:bottom="2dp"/>
    <corners android:radius="4dp"/>
</shape>
```

We have used this layout approach because it is handy not only for widgets but also for other non-widget layouts.

You may want to build an activity and test these layouts separately before actually testing them with your widget (at least, that is what we did). It took us a number of trials to get the look and feel right. It can be tedious to attempt to experiment directly with widgets; every time you run the application, you have to delete the widgets, uninstall, install, and then drag them back to the home page.

The files discussed so far complete the XML definitions needed by a typical widget. Let's see now how we will respond to the life cycle events of widgets by examining the widget provider class.

Implementing a Widget Provider

The Java code in Listing 25–9 demonstrates the implementation of the widget provider class.

Listing 25–9. *Sample Widget Provider: BDayWidgetProvider*

```
///src/<your-package>/BDayWidgetProvider.java
public class BDayWidgetProvider extends AppWidgetProvider
{
    private static final String tag = "BDayWidgetProvider";
    public void onUpdate(Context context,
                         AppWidgetManager appWidgetManager,
                         int[] appWidgetIds)  {
        final int N = appWidgetIds.length;
        for (int i=0; i<N; i++)
        {
            int appWidgetId = appWidgetIds[i];
            updateAppWidget(context, appWidgetManager, appWidgetId);
        }
    }

public void onDeleted(Context context, int[] appWidgetIds)
{
     final int N = appWidgetIds.length;
     for (int i=0; i<N; i++)
     {
        BDayWidgetModel bwm =
           BDayWidgetModel.retrieveModel(context, appWidgetIds[i]);
        bwm.removePrefs(context);
     }
}
    public void onEnabled(Context context) {
       BDayWidgetModel.clearAllPreferences(context);
        PackageManager pm = context.getPackageManager();
        pm.setComponentEnabledSetting(
                new ComponentName("com.androidbook.BDayWidget",
                       ".BDayWidgetProvider"),
                PackageManager.COMPONENT_ENABLED_STATE_ENABLED,
                PackageManager.DONT_KILL_APP);
    }

    public void onDisabled(Context context) {
        BDayWidgetModel.clearAllPreferences(context);
        PackageManager pm = context.getPackageManager();
        pm.setComponentEnabledSetting(
                new ComponentName("com.androidbook.BDayWidget",
                       ".BDayWidgetProvider"),
                PackageManager.COMPONENT_ENABLED_STATE_DISABLED,
```

```
                PackageManager.DONT_KILL_APP);
    }

    private void updateAppWidget(Context context,
                        AppWidgetManager appWidgetManager,
                        int appWidgetId) {
        BDayWidgetModel bwm = BDayWidgetModel.retrieveModel(context, appWidgetId);
        if (bwm == null) {
            return;
        }
        ConfigureBDayWidgetActivity
                .updateAppWidget(context, appWidgetManager, bwm);
    }
}
```

Refer to the “Architecture of Home Screen Widgets” section to see what needs to happen in each of these methods. For the Birthday Widget, all these methods in turn make use of methods from the `BDayWidgetModel` class. Some of these methods are `removePrefs()`, `retrievePrefs()`, and `clearAllPreferences()`.

The `BDayWidgetModel` class is used to encapsulate the state of each Birthday Widget instance (we will cover this class in the next section). To understand this widget provider class, all you need to know is that we are using a model class to retrieve data needed for this widget instance. This data is kept in preferences, which is why the methods are named `removePrefs()`, `retrievePrefs()`, and `clearAllPreferences()`.

As indicated, the update method is called for all the widget instances. This method must update all the widget instances. The widget instances are passed in as an array of IDs. For each `id`, the `onUpdate()` method will locate the corresponding widget instance model and call the same method that is used by the configurator activity (see Listing 25–14) to display the retrieved widget model.

In the `onDeleted()` method, we have instantiated a `BDayWidgetModel` and then asked it to remove itself from the preferences persistence store.

In the `onEnabled()` method, because it is called only once when the first instance comes into play, we have cleared all persistence of the widget models so that we start with a clean slate. We do the same in the `onDisabled()` method so that no memory of widget instances exists.

In the `onEnabled()` method, we enable the widget provider component so that it can receive broadcast messages. In the `onDisabled()` method, we disable the component so that it won't look for any broadcast messages.

By employing the idea of widget models, the code stays clean. We'll explore the widget models and their implementation next.

Implementing Widget Models

What is a widget model? The widget model is not an Android concept. If you are familiar with traditional UI programming, you will recall the concept of Model, View, Controller (MVC) architecture, where the model holds data needed by a view; the view is responsible for display; and the controller is responsible for mediating between the view and the model.

Although the Android SDK does not mandate a specific approach, we have used the MVC idea to simplify widget programming. In this approach, for every widget instance view, you will have an equivalent Java class that is a widget model. This model will have all the methods that can supply the needed data for the widget instances.

In addition to supplying the data, we have created some base classes for these models so that they know how to save and retrieve themselves from a persistent store such as shared preferences. We will go through the model class hierarchy and show you how we use shared preferences to store and retrieve data. You can refer to Chapter 13 to read more about preferences.

Interface for a Widget Model

We will start this discussion with an interface that acts as a contract for a widget model so that the widget model can declare the fields to be saved in a persistent data base. The contract also defines how to set a field when that field is retrieved from a database. The interface, in addition, provides an `init()` callback so that it is called when a model is newly retrieved from the database and before being passed on to a requesting client.

Listing 25–10 shows the source code for the widget contract interface.

Listing 25–10. *Saving Widget State: The Contract*

```
//filename: src/…/IWidgetModelSaveContract.java
public interface IWidgetModelSaveContract
{
    //Name of preferences file
    public String getPrefname();

    //facilitate model population
    public void setValueForPref(String key, String value);

    //return key value pairs you want to be saved
    public Map<String,String> getPrefsToSave();

    //gets called after restore
    public void init();
}
```

This interface is designed in such a way that a derived abstract class will provide an implementation using a specific persistence store while relying on this contract of the leaf-level Java class . Then the leaf-level Java class will need to implement this contract to provide what needs to be saved and retrieved.

We will use the shared preferences facility of Android as the persistence store. As the name of this interface indicates, it is purely a save contract. The clients such as the `BDayWidgetProvider` will still rely on the most-often derived class of this interface for specific methods.

The implementer of this interface will need to provide the name of a preference file in response to the method getPrefname(). This preference file is then used to save the key/value pairs obtained from getPrefsToSave(). In an inverse operation (setValueForPref()), the derived class is asked to set its internal value given a key and value restored from the preferences store.

Finally, the method init() is called on the derived class to indicate that the values have been restored from the persistent store or any other initializations that could happen.

> **NOTE:** Please remember that, in a real-world application, you would structure this inheritance a bit differently; you would probably use a delegation mechanism for reuse instead of inheritance. However, this inheritance hierarchy will work well for our test case to demonstrate widget models.

Let's consider now the abstract implementation that stores the data fields of a widget as shared preferences.

Abstract Implementation of a Widget Model

All the code that is responsible for interacting with a persistent store is implemented in the APrefWidgetModel class (see Listing 25–11). The Pref in this class stands for "preference," because this class uses the SharedPreferences facility of Android to store the widget model data.

In addition, this class represents the idea of a basic widget. The field id represents the instance ID of the widget. This class always needs a constructor that takes the widget instance ID as an argument to accommodate the instance ID requirement.

Let's take a look at the source code of this class in Listing 25–11. Key methods of this class are highlighted.

Listing 25–11. *Implementing Widget Saves Through Shared Preferences*

```
//filename: /src/…/APrefWidgetModel.java
public abstract class APrefWidgetModel
implements IWidgetModelSaveContract
{
   private static String tag = "AWidgetModel";

   public int iid;
   public APrefWidgetModel(int instanceId) {
      iid = instanceId;
   }
  //abstract methods
   public abstract String getPrefname();
   public abstract void init();
   public Map<String,String> getPrefsToSave(){   return null;}

    public void savePreferences(Context context){
       Map<String,String> keyValuePairs = getPrefsToSave();
```

```
    if (keyValuePairs == null){
        return;
    }
    //going to save some values
    SharedPreferences.Editor prefs =
        context.getSharedPreferences(getPrefname(), 0).edit();

    for(String key: keyValuePairs.keySet()){
        String value = keyValuePairs.get(key);
        savePref(prefs,key,value);
    }
    //finally commit the values
    prefs.commit();
}

private void savePref(SharedPreferences.Editor prefs,
                      String key, String value) {
    String newkey = getStoredKeyForFieldName(key);
    prefs.putString(newkey, value);
}
private void removePref(SharedPreferences.Editor prefs, String key) {
    String newkey = getStoredKeyForFieldName(key);
    prefs.remove(newkey);
}
protected String getStoredKeyForFieldName(String fieldName){
    return fieldName + "_" + iid;
}
public static void clearAllPreferences(Context context, String prefname) {
    SharedPreferences prefs=context.getSharedPreferences(prefname, 0);
    SharedPreferences.Editor prefsEdit = prefs.edit();
    prefsEdit.clear();
    prefsEdit.commit();
}

public boolean retrievePrefs(Context ctx) {
    SharedPreferences prefs = ctx.getSharedPreferences(getPrefname(), 0);
    Map<String,?> keyValuePairs = prefs.getAll();
    boolean prefFound = false;
    for (String key: keyValuePairs.keySet()){
        if (isItMyPref(key) == true){
            String value = (String)keyValuePairs.get(key);
            setValueForPref(key,value);
            prefFound = true;
        }
    }
    return prefFound;
}
public void removePrefs(Context context) {
    Map<String,String> keyValuePairs = getPrefsToSave();
    if (keyValuePairs == null){
        return;
    }
    //going to save some values
     SharedPreferences.Editor prefs =
        context.getSharedPreferences(getPrefname(), 0).edit();

     for(String key: keyValuePairs.keySet()){
```

```
            removePref(prefs,key);
        }
        //finally commit the values
        prefs.commit();
    }
    private boolean isItMyPref(String keyname) {
        if (keyname.indexOf("_" + iid) > 0){
            return true;
        }
        return false;
    }
    public void setValueForPref(String key, String value) {
        return;
    }
}
```

Let's see how the key methods of this class are implemented. We'll start by saving the widget model attributes in a shared preferences file:

```
public void savePreferences(Context context)
{
    Map<String,String> keyValuePairs = getPrefsToSave();
    if (keyValuePairs == null){ return; }

    //going to save some values
    SharedPreferences.Editor prefs =
    context.getSharedPreferences(getPrefname(), 0).edit();

     for(String key: keyValuePairs.keySet()){
         String value = keyValuePairs.get(key);
         savePref(prefs,key,value);
     }
     //finally commit the values
     prefs.commit();
}
```

This method starts off by asking the derived classes to return a map of key/value pairs, where the keys are the attributes of the model, and values are string representations of those attribute values. It will then ask the Android `context` to get hold of a `SharedPreferences` file through `context.getSharedPreferences()`. This API needs a unique name for this package. The derived model is responsible for supplying this.

Once we get the shared preferences, by following the Android documentation, we will ask to get an editable version of the shared preferences. Then, we update the preferences one by one. Once that is complete, we run the `commit()` method, so the preferences are persisted.

Read the API references and Chapter 13 for more information about the `SharedPreferences` and the `SharedPreferences.Editor` classes; the "Resources" section of this chapter has URLs pointing out where this information is. It is also worth noting that these shared preference files are XML files and can be found in the data directory of the package.

Because we have used a single file to store data for all widget instances, we need a way to distinguish field names among multiple widget instances. For example, if we have two widget instances named 1 and 2, we will need two keys to store the Name attribute so that there is a name_1 and a name_2. We do this translation in the following method:

```
protected String getStoredKeyForFieldName(String fieldName) {
    return fieldName + "_" + iid;
}
```

The derived class also uses this method to examine which field to update when it is called with a setValue() method.

Implementation of a Widget Model for Birthday Widget

Ultimately the most-often derived class in this hierarchy of widget models is responsible for actually maintaining all the fields needed by the view. It relies on its base classes to store and retrieve. We have designed this most-often derived class in such a way that the clients that are dealing with these models directly deal with the most-often derived class, as this is the class that is most pertinent to them.

For example, when a widget instance is first created by the configurator activity, the configurator activity instantiates one of these classes and fills up its values and asks to save itself.

This class, because of the needs of the view, maintains three fields:

- name: Name of the person
- bday: The date the next birthday falls on
- url: The URL to go to for buying gifts

The class then has a calculated attribute called howManyDays, which represents the number of days from today to the date of the next birthday.

You will also notice that this class is responsible for fulfilling the save contract. These methods are as follows:

```
public void setValueForPref(String key, String value);
public String getPrefname();
public Map<String,String> getPrefsToSave();
```

Listing 25–12 lays out the code that orchestrates all of this.

Listing 25–12. *BDayWidgetModel: Implementing a State Model*

```
//filename: /src/…/BDayWidgetModel.java
public class BDayWidgetModel extends APrefWidgetModel
{
   private static String tag="BDayWidgetModel";

   // Provide a unique name to store date
   private static String BDAY_WIDGET_PROVIDER_NAME=
      "com.androidbook.BDayWidget.BDayWidgetProvider";

   // Variables to paint the widget view
   private String name = "anon";
```

```
    private static String F_NAME = "name";

    private String bday = "1/1/2001";
    private static String F_BDAY = "bday";



    // Constructor/gets/sets
    public BDayWidgetModel(int instanceId){
       super(instanceId);
    }
    public BDayWidgetModel(int instanceId, String inName, String inBday){
       super(instanceId);
       name=inName;
       bday=inBday;
    }
     public void init(){}
     public void setName(String inname){name=inname;}
     public void setBday(String inbday){bday=inbday;}

     public String getName(){return name;}
     public String getBday(){return bday;}

     public long howManyDays(){
        try      {
           return Utils.howfarInDays(Utils.getDate(this.bday));
        }
        catch(ParseException x){
           return 20000;
        }
     }

    //Implement save contract

     public void setValueForPref(String key, String value){
        if (key.equals(getStoredKeyForFieldName(BDayWidgetModel.F_NAME))){
           this.name = value;
           return;
        }
        if (key.equals(getStoredKeyForFieldName(BDayWidgetModel.F_BDAY))){
           this.bday = value;
           return;
        }
     }
    public String getPrefname()   {
       return BDayWidgetModel.BDAY_WIDGET_PROVIDER_NAME;
    }

    //return key value pairs you want to be saved
    public Map<String, String> getPrefsToSave()   {
       Map<String, String> map
       = new HashMap<String, String>();
       map.put(BDayWidgetModel.F_NAME, this.name);
       map.put(BDayWidgetModel.F_BDAY, this.bday);
       return map;
    }
    public String toString()   {
```

```
        StringBuffer sbuf = new StringBuffer();
        sbuf.append("iid:" + iid);
        sbuf.append("name:" + name);
        sbuf.append("bday:" + bday);
        return sbuf.toString();
    }
    public static void clearAllPreferences(Context ctx){
        APrefWidgetModel.clearAllPreferences(ctx,
                BDayWidgetModel.BDAY_WIDGET_PROVIDER_NAME);
    }

    public static BDayWidgetModel retrieveModel(Context ctx, int widgetId){
        BDayWidgetModel m = new BDayWidgetModel(widgetId);
        boolean found = m.retrievePrefs(ctx);
        return found ? m:null;
    }
}
```

As you can see, this class uses a couple of date-related utilities. We will show you the source code for these utilities before moving on to explaining the widget configuration activity implementation.

A Few Date-Related Utilities

Listing 25–13 contains a utility class that is used to work with dates. It takes a date string and validates if it is a valid date. It also calculates how far a date is from today. The code is self-explanatory. We have included it here for completeness.

Listing 25–13. *Date Utilities*

```
public class Utils
{
    private static String tag = "Utils";
    public static Date getDate(String dateString)
    throws ParseException {
        DateFormat a = getDateFormat();
        Date date = a.parse(dateString);
        return date;
    }
    public static String test(String sdate){
        try {
            Date d = getDate(sdate);
            DateFormat a = getDateFormat();
            String s = a.format(d);
            return s;
        }
        catch(Exception x){
            return "problem with date:" + sdate;
        }
    }
    public static DateFormat getDateFormat(){
        SimpleDateFormat df = new SimpleDateFormat("MM/dd/yyyy");
        //DateFormat df = DateFormat.getDateInstance(DateFormat.SHORT);
        df.setLenient(false);
        return df;
    }
```

```
    //valid dates: 1/1/2009, 11/11/2009,
    //invalid dates: 13/1/2009, 1/32/2009
    public static boolean validateDate(String dateString){
       try {
          SimpleDateFormat df = new SimpleDateFormat("MM/dd/yyyy");
          df.setLenient(false);
          Date date = df.parse(dateString);
          return true;
       }
       catch(ParseException x) {
          return false;
       }
    }
    public static long howfarInDays(Date date){
       Calendar cal = Calendar.getInstance();
       Date today = cal.getTime();
       long today_ms = today.getTime();
       long target_ms = date.getTime();
       return (target_ms - today_ms)/(1000 * 60 * 60 * 24);
    }
}
```

Now, let's look at the implementation of the configuration activity that we have talked about already.

Implementing Widget Configuration Activity

In the "Architecture of Home Screen Widgets" section, we explained the role of configuration activity and its responsibilities. For the Birthday Widget example, these responsibilities are implemented in an activity class called ConfigureBDayWidgetActivity. You can see the source code for this class in Listing 25–14.

This class collects the name of the person and the next birthday. It then creates a BDayWidgetModel and stores it in shared preferences.

Listing 25–14. *Implementing a Configurator Activity*

```
public class ConfigureBDayWidgetActivity extends Activity
{
   private static String tag = "ConfigureBDayWidgetActivity";
   private int mAppWidgetId = AppWidgetManager.INVALID_APPWIDGET_ID;

    /** Called when the activity is first created. */
    @Override
    public void onCreate(Bundle savedInstanceState) {
        super.onCreate(savedInstanceState);
        setContentView(R.layout.edit_bday_widget);
        setupButton();

        Intent intent = getIntent();
        Bundle extras = intent.getExtras();
        if (extras != null) {
            mAppWidgetId = extras.getInt(
                    AppWidgetManager.EXTRA_APPWIDGET_ID,
```

```
                AppWidgetManager.INVALID_APPWIDGET_ID);
    }

}

private void setupButton(){
   Button b = (Button)this.findViewById(R.id.bdw_button_update_bday_widget);
   b.setOnClickListener(
         new Button.OnClickListener(){
            public void onClick(View v)
            {
               parentButtonClicked(v);
            }
         });

}
private void parentButtonClicked(View v){
   String name = this.getName();
   String date = this.getDate();
   if (Utils.validateDate(date) == false){
      this.setDate("wrong date:" + date);
      return;
   }
   if (this.mAppWidgetId == AppWidgetManager.INVALID_APPWIDGET_ID){
      return;
   }
   updateAppWidgetLocal(name,date);
   Intent resultValue = new Intent();
   resultValue.putExtra(AppWidgetManager.EXTRA_APPWIDGET_ID, mAppWidgetId);
   setResult(RESULT_OK, resultValue);
   finish();
}
private String getName(){
    EditText nameEdit = (EditText)this.findViewById(R.id.bdw_bday_name_id);
    String name = nameEdit.getText().toString();
    return name;
}
private String getDate(){
    EditText dateEdit = (EditText)this.findViewById(R.id.bdw_bday_date_id);
    String dateString = dateEdit.getText().toString();
    return dateString;
}
private void setDate(String errorDate){
    EditText dateEdit = (EditText)this.findViewById(R.id.bdw_bday_date_id);
    dateEdit.setText("error");
    dateEdit.requestFocus();
}
private void updateAppWidgetLocal(String name, String dob){
   BDayWidgetModel m = new BDayWidgetModel(mAppWidgetId,name,dob);
   updateAppWidget(this,AppWidgetManager.getInstance(this),m);
   m.savePreferences(this);
}

public static void updateAppWidget(Context context,
        AppWidgetManager appWidgetManager,
        BDayWidgetModel widgetModel)
{
```

```
        RemoteViews views = new RemoteViews(context.getPackageName(),
                      R.layout.bday_widget);

        views.setTextViewText(R.id.bdw_w_name
           , widgetModel.getName() + ":" + widgetModel.iid);

        views.setTextViewText(R.id.bdw_w_date
              , widgetModel.getBday());

        //update the name
        views.setTextViewText(R.id.bdw_w_days,
                             Long.toString(widgetModel.howManyDays()));

          Intent defineIntent = new Intent(Intent.ACTION_VIEW,
                Uri.parse("http://www.google.com"));
          PendingIntent pendingIntent =
             PendingIntent.getActivity(context,
                      0 /* no requestCode */,
                      defineIntent,
                      0 /* no flags */);
          views.setOnClickPendingIntent(R.id.bdw_w_button_buy, pendingIntent);

        // Tell the widget manager
        appWidgetManager.updateAppWidget(widgetModel.iid, views);
    }
}
```

If you look at the code for the function `updateAppWidgetLocal()`, you will notice that it is the function that creates and stores the model. It then uses the function `updateAppWidget()` to display it. It is worth noting how this function `updateAppWidget()` uses a pending intent to register a callback. The pending intent takes a primary intent such as

```
Intent defineIntent = new Intent(Intent.ACTION_VIEW,
      Uri.parse("http://www.google.com"));
```

and creates a pending intent to start an activity. In contrast, a pending intent can be used to start a service as well. It is also noteworthy that this function works with `RemoteViews` and `AppWidgetManager`. Notice that this function accomplishes the following tasks:

- Obtaining `RemoteViews` from the layout
- Setting text values on the `RemoteViews`
- Registering a pending intent through `RemoteViews`
- Invoking the `AppWidgetManager` to send the `RemoteViews` to the widget
- Returning at the end with a result

> **NOTE:** The static function `udpateAppWidget` can be called from anywhere as long as you know the widget ID. This suggests that you can update a widget from anywhere on your device and from any process, both visual and nonvisual.

It is also important that you use the following code to end the widget configuration activity:

```
Intent resultValue = new Intent();
resultValue.putExtra(AppWidgetManager.EXTRA_APPWIDGET_ID, mAppWidgetId);
setResult(RESULT_OK, resultValue);
finish();
```

Notice how we are passing the widget ID back to the caller. This is how AppWidgetManager knows that the configurator activity is completed for that widget instance.

Let's conclude this discussion of widget configuration by presenting the layout for the widget configuration activity in Listing 25–15. This view is pretty straightforward: it has a couple of text boxes and edit controls with an update button. You can also see this visually in Figure 25–4.

Listing 25–15. *Layout Definition for Configurator Activity*

```
<?xml version="1.0" encoding="utf-8"?>
<!-- res/layout/edit_bday_widget.xml -->
<LinearLayout xmlns:android="http://schemas.android.com/apk/res/android"
   android:id="@+id/root_layout_id"
    android:orientation="vertical"
    android:layout_width="fill_parent"
    android:layout_height="fill_parent"
    >
<TextView
   android:id="@+id/bdw_text1"
    android:layout_width="fill_parent"
    android:layout_height="wrap_content"
    android:text="Name:"
    />
<EditText
    android:id="@+id/bdw_bday_name_id"
    android:layout_width="fill_parent"
    android:layout_height="wrap_content"
    android:text="Anonymous"
   />
<TextView
    android:id="@+id/bdw_text2"
    android:layout_width="fill_parent"
    android:layout_height="wrap_content"
    android:text="Birthday (9/1/2001):"
    />
<EditText
    android:id="@+id/bdw_bday_date_id"
    android:layout_width="fill_parent"
    android:layout_height="wrap_content"
    android:text="ex: 10/1/2009"
   />
<Button
    android:id="@+id/bdw_button_update_bday_widget"
    android:layout_width="fill_parent"
    android:layout_height="wrap_content"
    android:text="update"
/>
</LinearLayout>
```

This concludes our discussion on implementing a sample widget.

Widget Preview Tool

We have briefly talked about how you can specify a preview image for a widget in the widget definition XML file (see Listing 25–2).

The emulator provides a helper application that will capture a screen shot of your widget, which can be used later used as a preview image. As a developer, you can invoke this application by looking at the application list in the emulator and locating the widget preview application. This application allows you to choose the widget that you want to preview. Once you select the widget, the corresponding widget configurator (if available) kicks in. At the end, you see an instance of your widget. The widget preview application will then let you take a screen shot and will send you an e-mail containing the file.

Or you can provide any iconified preview image, instead. That is all there is to a preview: it's just an icon. How you get it is up to you.

Widget Limitations and Extensions

Android home widgets appear simple when you look at them. However, they have nuances when you start writing widgets.

If your widget doesn't require any state management and doesn't need to be invoked more than a few times a day, you have a widget that is very simple to write.

The next level of widget is one where you will need to manage the state, but the widget is invoked infrequently, like the one we have shown here. These types of widget can benefit from a state management framework. We have shown in this chapter a bare-bones state management framework. We assume that more sophisticated ones will be available or that you could write one that is more robust and flexible.

The next level of widgets must be invoked at the levels of seconds and milliseconds. For these widgets, you will need to rig your own update calls using the `AlarmManager`. You will also likely need a service to manage state frequently, rather than relying on a persistent framework. For example, if you were to write a widget for a `StopWatch`, you would need to have a timer that counts at least every second, and you would also need to keep track of your counters, which implies state. AlarmManager and long-running services are covered in earlier chapters of this book.

Another factor to consider is that the `RemoteViews` on which the widget view framework relies have no mechanism to edit directly on a widget (at least none that is documented). `RemoteViews` also put restrictions on what kinds of views and layouts can be used. You don't have direct control of the views, only control through the methods supplied by the `RemoteViews` class.

Based on the current design and intentions of widgets, Google seems to expect that the widgets mostly fall under category 1 or 2. There is lot of opportunity to expand the widget framework in coming releases.

Collection-Based Widgets

Starting with SDK 3.0, Android has expanded the widgets to include widgets based on collections. We will cover these collection-based widgets in Chapter 26.

Resources

As we prepared material for this chapter, we found the following resources to be useful, and we present them here in the order of their utility:

- `http://developer.android.com/guide/topics/appwidgets/index.html`: The official Android SDK documentation on app widgets.
- `http://developer.android.com/reference/android/content/SharedPreferences.html`: The `SharedPreferences` API for managing state.
- `http://developer.android.com/reference/android/content/SharedPreferences.Editor.html`: The `SharedPreferences.Editor` API, which is related to shared preferences.
- `http://developer.android.com/guide/practices/ui_guidelines/widget_design.html`: Design pleasing widget layouts.
- `http://developer.android.com/reference/android/widget/RemoteViews.html`: The `RemoteViews` API, used to paint and manipulate widget views.
- `http://developer.android.com/reference/android/appwidget/AppWidgetManager.html`: Widgets themselves are managed by a widget manager class.
- `www.androidbook.com/item/3938`: The research notes used in writing this chapter, including a summary, research, code snippets, and useful URLs.
- `www.androidbook.com/proandroid4/projects`: Downloadable test projects for this chapter. The name of the ZIP file for this chapter is `ProAndroid4_ch25_TestWidgets.zip`.

Summary

We have covered in this chapter:

- The theory behind widgets
- Working examples to illustrate widget nuances.
- The need for widget models and widget state management
- State management implementation

- Design issues and limitations of widgets

Interview Questions

The following questions will further consolidate what is covered in this chapter:

1. What is the connection between widgets and remote views?
2. How do you create a widget on your home screen?
3. How do you delete a widget from your home screen?
4. Can you create a widget multiple times on the home screen?
5. What is a widget configuration activity?
6. What is a widget provider class?
7. Why can't you configure a widget for an update frequency less than 30 minutes?
8. What is the starting point for a widget definition?
9. In what directory is the widget view definition kept?
10. What can you specify in a widget definition XML file?
11. What Java class do you use to update content on a remote view?
12. What paints the widget remote views for the first time: the widget provider or the configurator?
13. Is the widget configurator mandatory?
14. If a configurator does not exist for a widget, what does the update for the first time?
15. What is the difficulty of maintaining state between two app widget updates?
16. What are the four callbacks of a widget provider class?
17. How do you enable and disable a widget provider to accept widget provider update messages?
18. Is there a race condition between `onEnabled()` and the widget configurator invocation?
19. Why should you reuse code between `onUpdate()` and the widget configurator?
20. What metadata key is used to indicate the XML widget configuration file?

21. What intent filter action is necessary to define a widget configuration activity?
22. What are the four methods that a widget provider should implement?
23. How are the widget instance IDs passed to the widget provider callbacks?
24. What API do you use to get access to a shared preferences file?
25. How do you specify a preferences file name?
26. What API do you use to save a key/value pair in a preferences file?
27. Where are the shared preference files stored on the device?
28. How do you attach a pending intent to start an activity to a widget view?

Chapter 26

Exploring List Widgets

As you learned in Chapter 25, remote views form the core of home screen widgets. A home screen widget is essentially a remote view that is painted on the home screen. A remote view is a view that is entirely disconnected from the underlying data, much like a web page is disconnected from its server.

Collection views such as lists and grids were not part of allowed widgets in the 2.3 release. In release 3.0 they are, allowing for a richer experience on the home screen. Release 3.0 also offers a mini-framework around these collection-based widgets to load and present data asynchronously. There are new classes and methods in 3.0 to support these aspects. As you may know, the 3.0 SDK is optimized and unique for tablets and not available for phones. In 4.0, the APIs of 2.3 and 3.0 are merged to form the same API going forward for both tablets and phones. Although we use the tablet UI in this chapter to demonstrate the concepts, thanks to 4.0, the API is equally applicable to the phone form factor.

A Quick Note on Remote Views

The class `RemoteViews` can't be constructed by passing explicit view objects. Nor can view objects be directly added to `RemoteViews`. Instead, you construct a `RemoteViews` object by passing a layout file to its constructor. As of 4.0, only the following views are allowed to be in these layout files

- `FrameLayout`
- `LinearLayout`
- `RelativeLayout`
- `AnalogClock`
- `Button`
- `Chronometer`
- `ImageButton`
- `ProgressBar`

- ListView
- GridView
- StackView
- TextView
- DateTimeView
- ImageView
- AdapterViewFlipper
- ViewFlipper

More remote views may be added in future releases. The key to finding out which of the current UI objects are enabled for RemoteViews is the fact that these classes are annotated with an interface called RemoteViews.RemoteView.

Armed with this information, you can use Eclipse to figure out which classes in a project use this annotation. Here's how you do it:

1. In your source code, put an import statement for the RemoteView interface.
2. Highlight that interface name.
3. Right-click, and go to the References tab.
4. Choose to look for references of this interface in this project.

This will present a list of classes that are annotated with the RemoteView interface.

Working with Lists in Remote Views

In Chapter 25, we covered the existing set of classes in the SDK that support home screen widgets. The primary ones are AppWidgetProvider, AppWidgetManager, RemoteViews, and an activity that can be used to configure an AppWidgetProvider with initialization parameters.

Briefly, here is the core idea of how home screen widgets work (knowing this should make the rest of this section a bit easier to follow). An AppWidgetProvider is a broadcast receiver that gets invoked every once in a while based on a timer interval that you specify in a configuration file. This AppWidgetProvider then loads a RemoteViews instance based on a layout file. This RemoteViews object is then passed to the AppWidgetManager to be displayed on the home screen.

Optionally, you can tell Android that you have an activity that needs to be invoked before placing the widget for the first time on the home screen. This allows the configuration activity to set initialization parameters for the widget.

You also can set up onClick events on the remote views of the widget so that intents can get fired based on those events. These intents then can invoke whatever components necessary, including sending messages to the `AppWidgetProvider` broadcast receiver.

At a high level, this is all there is to home screen widgets. The rest is the mechanics and variations on each of these basic ideas. To support list-based remote views, Android 3.0/4.0 has added the following new classes:

- *RemoteViewsFactory:* This class allows you to populate a list-based remote view much like list adapters populate regular list views. (See Chapter 6.) This class is a thin wrapper around a list view adapter to supply individual remote views to the list remote view in an asynchronous manner. So, the main job of this class is to provide a remote view for each item in the list. This factory loads the layout for each item and returns that layout after populating it with data. Clearly this layout can only allow remote views as part of its children.
- *RemoteViewsService:* This class is a service that is responsible for returning a `RemoteViewsFactory` given a list-based view ID in a layout file. It is the responsibility of the `AppWidgetProvider` to tie a remote views service to a list remote view. This is done by attaching an intent that knows how to invoke this service to the list remote view ID. This service allows you to extend the life of the process containing the `AppWidgetProvider`. Otherwise, when the broadcast receiver returns, the process can be reclaimed. Chapter 19 explains the symbiotic relationship between broadcast receivers and services that need to run beyond the lifetime of a broadcast receiver.

The following new API methods have been added to support list-based remote views:

- *RemoteViews.setPendingIntentTemplate():* This method allows you to set a pending intent template on a list remote view in order to respond to click events on the list items. We will talk about templates when we cover the details later.
- *RemoteViews.setOnClickFillIntent():* This is set on the individual list items of the list remote view and works closely with the previous method.

These additional two methods in concert will let you respond to clicks on list-based remote views. These two methods are designed so that as few pending intents are created as possible.

We will cover these classes and methods in detail as we go through this chapter. Given these new features, here are the general steps to work with a list view on a home screen widget. Reread the brief overview of home screen widgets (from earlier in this section) to help you understand these steps:

1. *Prepare a remote layout for your widget:* Create a suitable remote layout with a list view in it. A remote layout is a regular layout with only allowed remotable views. This is no different than what you have to do for any home screen widget (and is clearly shown in Chapter 25).

2. *Load the remote layout:* In the `onUpdate()` method of the widget provider, load the compound remote layout view from the previous step as a remote view. Here, also, there is no difference.

3. *Set up the `RemoteViewsService`:* Locate the list view in your layout by its ID, and set an intent on that list remote view ID so that the intent invokes a list remote view service.

4. *Set up `RemoteViewsFactory`:* The list remote view service from step 3 will need to return a list `RemoteViewsFactory` that knows how to populate the list remote view.

5. *Set up click events:* As part of setting up the list remote view in the `AppWidgetProvider`, also set the `onClick` pending intent template so that you can respond to that intent. However, you will also need to correspondingly set up the individual clicks using the `RemoteViewsFactory` for each view in the list. This is because the items in the remote list view are populated from the list view factory.

6. *Respond to click events:* Someone needs to respond to the `onClick` events set on the remote list views. You can choose your `AppWidgetProvider` to be the receiver for these events. You need to prepare the broadcast receiver to receive and respond to `onClick` events from remote views.

Let's look at each of these steps with annotated sample code.

Preparing a Remote Layout

As described in the previous section, the layout for a remote view that can be displayed as a home widget can now include a list view. Listing 26–1 shows an example remotable layout with a list view in it.

Listing 26–1. *A Remote Layout File with a List View*

```
<?xml version="1.0" encoding="utf-8"?>
<!-- /res/layout/test_list_widget_layout.xml -->
<LinearLayout xmlns:android="http://schemas.android.com/apk/res/android"
    android:orientation="vertical"
    android:layout_width="150dp"
    android:layout_height="match_parent"
    android:background="@drawable/box1">
<TextView
    android:id="@+id/listwidget_header_textview_id"
    android:layout_width="fill_parent"
```

```
    android:layout_height="30dp"
    android:text="Header View"
    android:background="@drawable/box1"
    android:gravity="center"
    android:layout_weight="0"/>
<FrameLayout
    android:layout_width="match_parent"
    android:layout_height="match_parent"
    android:layout_weight="1"
    android:layout_gravity="center">
        <ListView android:id="@+id/listwidget_list_view_id"
           android:layout_width="match_parent"
           android:layout_height="match_parent"/>
        <TextView
        android:id="@+id/listwidget_empty_view_id"
        android:layout_width="match_parent"
        android:layout_height="match_parent"
        android:gravity="center"
        android:visibility="gone"
        android:textColor="#ffffff"
        android:text="Empty Records View"
        android:textSize="20sp" />
</FrameLayout>
<TextView
    android:id="@+id/listwidget_footer_textview_id"
    android:layout_width="fill_parent"
    android:layout_height="30dp"
    android:text="Footer View"
    android:background="@drawable/box1"
    android:gravity="center"
    android:layout_weight="0"/>
</LinearLayout>
```

In Listing 26–1, every XML node represents a valid remote view. This layout is presented in such a way that when shown as a home screen widget, the layout would look like that in Figure 26–1.

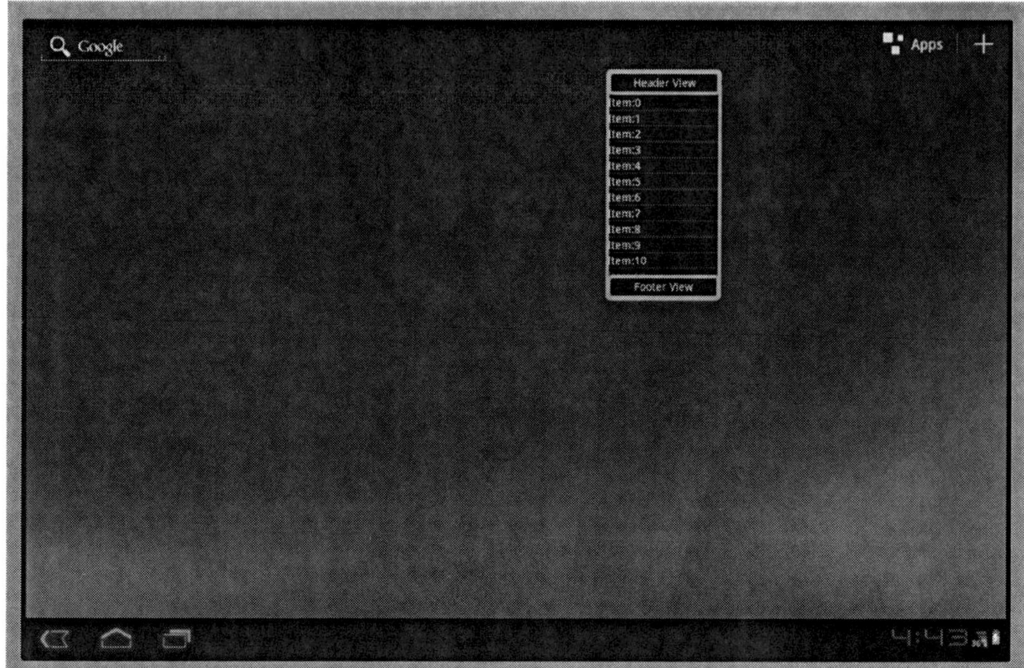

Figure 26–1. *Home screen populated with a list view widget*

The layout pattern in Listing 26–1 follows a simple header, body, footer format. The header and footer are both set at a fixed height; in this example, these heights are set to 30dp. However, you want the body height to be stretchable to take the rest of the vertical height. The way to accomplish this is to set `android:layout_weight` to `0` on the header and footer. On the body, you set `android:layout_weight` to `1` and `android:layout_height` to `match_parent`.

The `FrameLayout` that is taking the position of the body of this widget needs a bit of explanation. A `FrameLayout` chooses one of its children as the view exclusively. In this case, when you have data in the list, you will use the `ListView`. When the list is empty, you will use the empty text view. You can set this up using the `RemoteViewsFactory`.

Also in this layout file is a custom drawable identified by `@drawable/box1` to make the corners round. Listing 26–2 is the `box1.xml` file that needs to be placed in the `/res/drawable` subdirectory.

Listing 26–2. *res/drawable/box1.xml*

```
<shape xmlns:android="http://schemas.android.com/apk/res/android">
    <stroke android:width="4dp" android:color="#888888" />
    <padding android:left="2dp" android:top="2dp"
            android:right="2dp" android:bottom="2dp" />
    <corners android:radius="4dp" />
</shape>
```

Now that you have a sample layout for a home screen widget, let's discuss how you would go about loading this layout into a remote view.

Loading a Remote Layout

For a home screen widget, a remote view is loaded and displayed in the onUpdate() callback of the AppWidgetProvider. Listing 26–3 shows an example of how this is done.

Listing 26–3. *Loading a Remote Layout in onUpdate()*

```
public void onUpdate(Context context,
                     AppWidgetManager appWidgetManager,
                     int[] appWidgetIds)
{
    int N = appWidgetIds.length;
    for (int i=0; i<N; i++)
    {
        int appWidgetId = appWidgetIds[i];

        RemoteViews rv =
        new RemoteViews(context.getPackageName(),
                R.layout.test_list_widget_layout);

        rv.setEmptyView(R.id.listwidget_list_view_id,
                       R.id.listwidget_empty_view_id);

        //update this instance of the app widget
        appWidgetManager.updateAppWidget(appWidgetId, rv);
    }
    super.onUpdate(context,appWidgetManager, appWidgetIds);
}
```

Notice that a RemoteViews object is constructed using the ID of the layout file describing the entire widget. This layout file is the same one that is in Listing 26–1. You then take the resulting RemoteViews object and set an empty view for the specific list view resource (located by its ID) inside that layout file.

In the example in Listing 26–3, the layout file is identified by

```
R.layout.test_list_widget_layout
```

The list view resource within this file is identified by

```
R.id.listwidget_list_view_id
```

The empty view for this list view resource is identified by

```
R.id.listwidget_empty_view_id
```

With these IDs, the code in Listing 26–4 demonstrates how to construct a remote view and set an empty view for one of its list views.

Listing 26–4. *Loading Remote Views*

```
RemoteViews rv =
new RemoteViews(context.getPackageName(),
   R.layout.test_list_widget_layout);

rv.setEmptyView(R.id.bdw_list_view_id,
   R.id.empty_view_id);
```

Setting Up RemoteViewsService

So far, you have successfully loaded the remote views in the `onUpdate()` method of the `AppWidgetProvider`. Now you need to hook up the list remote view with a remote view service so that the remote view service can return the remote view adapter that can populate the list remote view.

Why a service? Why not directly hook up the remote view factory to the remote list view?

Because an `AppWidgetProvider` is a broadcast receiver, the `onUpdate()` method of the widget provider runs under the time constraints of a broadcast receiver. To avoid the time criticality, Android 3.0 delegated the job of populating the list view to a separate service that is inherited from `android.widget.RemoteViewsService`. This `RemoteViewsService` is then responsible for returning a list adapter that can populate the list. This adapter needs to be of type `RemoteViewsService.RemoteViewsFactory`. In a way, this is a rote procedure of ultimately getting the remote list view with the remote list view factory.

Listing 26–5 shows an example of how a remote view service is coded and how it returns the remote view factory.

Listing 26–5. *`RemoteViewsService` Example*

```
public class TestRemoteViewsService
extends android.widget.RemoteViewsService
{
    @Override
    public RemoteViewsFactory onGetViewFactory(Intent intent)
    {
        return new TestRemoteViewsFactory(
                this.getApplicationContext(), intent);
    }
}
```

Notice the following in Listing 26–5:

- You will need to inherit from `RemoteViewsService`.
- You will need to specialize a `RemoteViewsFactory` and return that factory. We will cover this factory soon.

Being a service, the inherited `RemoteViewsService` (`TestRemoteViewsService`, in this case) needs to be declared in the manifest file as well. Listing 26–6 shows an example.

Listing 26–6. *Declaring `RemoteViewsService` in the Manifest File*

```
<!-- The service serving the RemoteViews to the collection widget -->
<service android:name=".TestRemoteViewsService"
  android:permission="android.permission.BIND_REMOTEVIEWS"
  android:exported="false" />
```

Once you have this RemoteViewsService coded, you can attach this service to the list remote view object using the code in Listing 26–7. (Recall that this code runs in the onUpdate() method of the AppWidgetProvider.)

Listing 26–7. *Associating `RemoteViewsService` with a `RemoteViewList`*

```
final Intent intent =
    new Intent(context, TestRemoteViewsService.class);

rv.setRemoteAdapter(appWidgetId,
        R.id.listwidget_list_view_id, intent);
```

In Listing 26–7, you first create an explicit intent by identifying the RemoteViewsService class to this intent. You can then attach this intent to the remote list view by calling setRemoteAdapter() and passing the list view ID. The intent you are passing here is the same intent that is delivered to onGetViewFactory() method of the RemoteViewsService in Listing 26–5. Moreover Android uses this intent to cache the factory returned by onGetViewFactory(). The onGetViewFactory() method has the option to examine the nature of the intent and return different factories depending on the intent. This could be useful if you were to specify the same service as a target for multiple list-based views in a widget. We cannot think of a reason, but if you don't want your factory to be not cached, then you need to construct these intents so that each intent is unique. You can use the widget ID as an extra to enforce this uniqueness.

Setting Up RemoteViewsFactory

Although you have specified a RemoteViewsService to delegate the list population, ultimately a RemoteViewsFactory is responsible for populating the list view. So, to populate the list view, you will start by implementing the adapter-like interface RemoteViewsFactory. (See Chapter 6 to understand list controls and list adapters.)

Listing 26–8 shows the key method signatures of a class that implements this factory interface.

Listing 26–8. *A `RemoteViewsFactory` Contract*

```
class TestRemoteViewsFactory
implements RemoteViewsService.RemoteViewsFactory
{
    public TestRemoteViewsFactory(Context context, Intent intent);
    public void onCreate();
    public void onDestroy();
    public int getCount();
    public RemoteViews getViewAt(int position);
    public RemoteViews getLoadingView();
    public int getViewTypeCount();
    public long getItemId(int position);
```

```
    public boolean hasStableIds();
    public void onDataSetChanged();
}
```

Let's talk about each of these methods and what needs to be done in each of them, starting with the constructor.

RemoteViewsFactory Constructor

The signature of the RemoteViewsFactory constructor is

```
public TestRemoteViewsFactory(Context context, Intent intent);
```

The constructor takes two arguments. The first argument is a context. Because this factory is constructed by your own implementation of RemoteViewsService (as shown in Listing 26–5), you can use the getApplicationContext() method to get the context.

The second argument to the constructor is an intent. This intent is the same intent that is used to invoke the remote views service.

In the constructor, both these values (the context and the intent) could be maintained as local variables so that subsequent methods could make use of these variables. We would like to continue to impress that these factories are cached based on intents.

onCreate() Callback

The signature of onCreate() is

```
Public void onCreate()
```

Following the pattern of a number of components in Android, a RemoteViewsFactory provides onCreate() and onDestroy() methods.

> **NOTE:** To understand when and how the onCreate() and onDestroy() methods are called, you will need to look at two source files in the Android source code: RemoteViewsService.java and AppWidgetService.java.

As said, when the RemoteViewsService makes a call to create your instance of the RemoteViewsFactory (see Listing 26–5), it caches this factory. The cache is based on the uniqueness of the intent that is used to invoke the service.

When you create a widget on the home screen, repeated updates to that widget data will not necessarily create multiple instances of RemoteViewsFactory objects. Instead, the cached factory object is used. The first time the RemoteViewsFactory object is created, its onCreate() method is called.

The RemoteViewsFactory is also reused if you drop multiple instances of the same widget on the home page. To make this so, you shouldn't place the extra widget ID in the intent. Even if you do include the extra widget ID, you shouldn't make the intent unique by resetting the data URI for that intent.

Because you are the one implementing the constructor and also the one writing the onCreate() method, you can initialize your class in either place!

onDestroy() Callback

The signature of onDestroy() is

```
Public void onDestroy()
```

This is the complement of the onCreate() method. The mechanics of onDestroy() are a bit involved to understand.

Start with the fact that factories are cached based on intents. If there are ten widgets of the same type on a screen, the same intent (and hence the same service and the same factory) can serve all ten widgets. So, you want to keep the factory around.

Android applies the following logic to call destroy on the factory and remove it from the cache:

1. Determine the widget ID that is being removed, because we dragged the widget to the trash can.
2. Walk through all the intents (an intent may be serving multiple widget IDs).
3. For each intent, remove the widget ID from its list.
4. If the intent has no widget IDs in the process, then call destroy on that factory.

By this logic, for performance reasons it is better not to pass the widget ID along with the service intent and make that service intent unique. Otherwise, we will be unnecessarily creating more factories than we need.

When the last widget of a certain type is removed from the home page (uninstalled), that will trigger onDestroy().

getCount() Callback

The signature of the getCount() method is

```
public int getCount()
```

You will need to return the total number of items in this list view. This method is very much like the corresponding method for list adapters in Chapter 6.

getViewAt() Callback

The signature of the getViewAt() method is

```
public RemoteViews getViewAt(int position)
```

The responsibility of this method is to return a remote view appropriate for this position in the list view. Typically, in this method you will load a layout that is specific to this type of remote view at this position and then set the values in that remote view using the position as an indicator to load the corresponding data. Listing 26–9 is an example of loading an individual layout for a list view item.

Listing 26–9. *Loading an Individual List View Item Layout*

```
RemoteViews rv =
    new RemoteViews(
        this.mContext.getPackageName(),
        R.layout.list_item_layout);
```

The layout that is referred to in Listing 26–9 could look like the layout in Listing 26–10.

Listing 26–10. *An Individual List View Item Layout*

```
<?xml version="1.0" encoding="utf-8"?>
<TextView  xmlns:android="http://schemas.android.com/apk/res/android"
    android:id="@+id/textview_widget_list_item_id"
    android:layout_width="fill_parent"
    android:layout_height="wrap_content"
    android:text="Temporary text"
/>
```

Once you load the remote view (Listing 26–9), you can return that remote view to the calling list remote view to be painted. This is also the place where you can set onClick behavior for this particular list item view.

getLoadingView() Callback

The signature of the getLoadingView() method is

```
public RemoteViews getLoadingView()
```

This method returns a custom loading view that appears between the time getViewAt(position) is called and returns. You can return null if you want to use the default loading view.

getViewTypeCount() Callback

The signature of the getViewTypeCount() method is

```
public int getViewTypeCount()
```

If the remote list view contains only one type of view as a child, this method will return 1. If there is more than one type of view, this method will need to return as many types of child views as are present.

getItemId() Callback

The signature of the getItemId() method is

```
public long getItemId(int position)
```

This method returns the appropriate ID of the underlying item for this position in the list view. This method is very much like the corresponding method for list adapters documented in Chapter 6.

hasStableIds() Callback

The signature of the hasStableIds() method is

```
public boolean hasStableIds()
```

This method should return true if the same item ID from getItemId() points to the same object. This method is very much like the corresponding method for list adapters in Chapter 6.

onDataSetChanged() Callback

The signature of onDataSetChanged() is

```
public void onDataSetChanged()
```

This method is called when someone tells the AppWidgetManager that the widget containing this remote list view has changed. This call to the widget manager will eventually trickle down to the remote view factory as onDataSetChanged(). In response, you will need to set up the underlying data so that other callbacks such as getViewAt()and getCount() can respond with new data. The documentation assures that long-running operations are permitted in this method to set up the data.

This completes the discussion of how to make a remote list view visible in a widget. Let's now tackle how to attach click events to a list view and even to its child views.

Setting Up onClick Events

Setting up click events for a list remote view is a two-step process. First, you register an onClick on the list view in the onUpdate() method of the widget provider. Then you register onClick events for each of the individual child views of that list view in the remote view factory's getViewAt() method.

First you'll learn how to register for click events on the main list view. When you set up a click event on a remote view, you need an intent to fire when that list remote view is clicked. Because an AppWidgetProvider is a broadcast receiver, you can set up this underlying AppWidgetProvider as a target for this intent. You then need to make provisions in the AppWidgetProvider to specialize the onReceive() callback so that you can handle this intent.

The code snippet in Listing 26–11 shows how you can set up an onClick intent with a widget provider as its target.

Listing 26–11. *Creating an Intent to Self-Invoke the* `AppWidgetProvider`

```
Intent onListClickIntent =
    new Intent(context,TestListWidgetProvider.class);
```

Notice how you set up the class name of a widget provider as the target component for this intent. This intent will be delivered to the widget provider. However, a widget provider is already responding to intents coming in with other widget-related actions. To distinguish this intent from other intents, you need to set up an explicit action for it. Listing 26–12 shows an example.

Listing 26–12. *Defining a Unique Action for an* `onClick` *in the Widget Provider*

```
onListClickIntent.setAction(
        TestListWidgetProvider.ACTION_LIST_CLICK);
```

Of course, the action `TestListWidgetProvider.ACTION_LIST_CLICK` is custom and is best defined as part of the widget provider `TestListWidgetProvider`.

Because the clicks could happen on multiple instances of this widget, you need to load the widget ID as an extra on the invoking intent. Listing 26–13 shows how to do this.

Listing 26–13. *Loading Widget ID into the* `onClick` *Intent*

```
onListClickIntent.putExtra(
        AppWidgetManager.EXTRA_APPWIDGET_ID, appWidgetId);
```

Now this intent is almost ready to be set on the remote list view as an `onClick` intent. You need to do one more thing to this intent. When intents are set to invoke at a later point of time, they are set as pending intents. See Chapter 5 and Chapter 20 for more detail on pending intents.

A pending intent does not take into account any subsequent extras you set on the underlying intent unless that intent is unique after taking into account the extras. However, intents don't take into account their extras when considering if they are unique. To circumvent this issue you need to use a method called `toUri()` on an intent.

This `toUri()` method takes all the extras of an intent and then makes a long string representing this intent with extras at the end. When you take this long string and set it as the data portion of the same intent, you essentially made this intent unique. This is because an intent will take its data portion under consideration for uniqueness. Listing 26–14 is an example of making an intent unique by using its `toUri()` method.

Listing 26–14. *Use of* `toUri()` *Method*

```
onListClickIntent.setData(
    Uri.parse(
        onListClickIntent.toUri(Intent.URI_INTENT_SCHEME)));
```

Once you made the intent unique, you can get the necessary broadcast pending intent, as shown in Listing 26–15.

Listing 26–15. *Getting a Broadcast Pending Intent from an Intent*

```
PendingIntent onListClickPendingIntent =
    PendingIntent.getBroadcast(context, 0,
        onListClickIntent,
        PendingIntent.FLAG_UPDATE_CURRENT);
```

In Listing 26–15, the FLAG_UPDATE_CURRENT flag means that if you find a similar underlying intent, just update its extras. You'll understand why this may be necessary when we discuss how this pending intent is utilized by the remote views.

Once you have the necessary pending intent, such as the one from Listing 26–15, you can set the click behavior for the list view. Use a method called setPendingIntentTemplate() to do this association between a pending intent and a list view. Listing 26–16 shows an example of how to use the setPendingIntentTemplate() method.

Listing 26–16. *Using setPendingIntentTemplate()*

```
RemoteViews rv;
rv.setPendingIntentTemplate(R.id.listwidget_list_view_id,
        onListClickPendingIntent);
```

In Listing 26–16, the first argument is the list view ID for the list view in the main layout (see Listing 26–1). The second argument is the pending intent you have created and prepared in Listings 26–11 through 26–14. Note in Listing 26–16 that you are calling the pending intent a *pending intent template*. What's up with the word *template*?

As per the SDK docs, Android doesn't want to create a pending intent for each of the rows in a list. It wants to create one pending intent for the whole list and then just override its extras as users click the individual items of that list. The way Android has facilitated this is to create one pending intent at the list level and then reissue that intent with different extras. This is why the pending intent in Listing 26–15 is set with a flag of update for its extras.

Let's now see how the extras are supplied from the individual list item RemoteViews. As you might expect, this is done in the same place where the list remote view items are constructed. This is in the getViewAt() method of the remote view factory (see Listing 26–9). Listing 26–17 shows how to attach intents with extras to a list view item when it is clicked.

Listing 26–17. *Attaching Intents with Extras to a List Item View when Clicked*

```
//Load your list item remote view
RemoteViews listItemRv;

//Get a fresh new intent
Intent ei = new Intent();

//Load it with whatever extra you want
ei.putExtra("com.androidbook.widgets.some_unique_extra_string_key",
       "Position of the item Clicked:" + position);

//Set it on the list remote view
listItemRv.setOnClickFillInIntent(R.id.textview_widget_list_item_id, ei);
```

In Listing 26–17, the key method is setOnClickFillIntent(). This method allows you to supply a fresh intent loaded with whatever extras what you want to load. Internally, the framework will take these extras and superimpose them on the pending intent template that you set up as part of the view onClick.

In Listing 26–17, you just took the text from the current row and embellished it a little and then set it as the extra. With this code, if you were to click the list item on the widget, it would raise an intent that would be sent to the broadcast receiver with the extras. Let's see how to prepare the broadcast receiver and retrieve this extra that is specific for each list view item.

Responding to onClick Events

In the list view pending intent template (Listing 26–16), you see the following two things:

- The component to invoke is the widget provider itself.
- The action is set to a specific action that is unique to this widget provider.

In response, the widget provider needs to do the following:

1. Declare a string action that it can recognize.
2. Override the `onReceive()` method, and deal with the action in step 1.

Listing 26–18 shows how to define the unique action in the provider as a string constant.

Listing 26–18. *Custom Action Definition*

```
public static final String ACTION_LIST_CLICK =
    "com.androidbook.homewidgets.listclick";
```

Listing 26–19 shows how to override `onReceive()`. It shows how to test for the action of the intent and call the `dealWithThisAction()` method. At the end of this method, you must call the base class's `onReceive()` method for all other actions. If you don't do so, the widget itself will not receive widget-based actions.

Listing 26–19. *Overriding* `onReceive()`

```
@Override
public void onReceive(Context context, Intent intent)
{
    if (intent.getAction()
            .equals(TestListWidgetProvider.ACTION_LIST_CLICK))
    {
        //this action is not one of the widget's usual actions
        //this is a specific action that is directed here.
        dealwithListAction(context,intent);
        return;
    }

    //make sure you call this
    super.onReceive(context, intent);
}
```

Listing 26–20 shows the `dealWithThisAction()` method where you retrieve the extra that you have loaded with the intent in Listing 26–17.

Listing 26–20. *Responding to the List View Item* `onClick`

```
public void dealwithListAction(Context context, Intent  intent)
{
    String clickedItemText =
        intent.getStringExtra(
                TestListWidgetProvider.EXTRA_LIST_ITEM_TEXT);
    if (clickedItemText == null)
    {
        clickedItemText = "Error";
    }
    clickedItemText =
        clickedItemText
        + "You have clicked on item:"
        + clickedItemText;

    Toast t =
        Toast.makeText(context,clickedItemText,Toast.LENGTH_LONG);
    t.show();
}
```

In Listing 26–20, you retrieved the extra through a predefined constant and provided a toast. This method runs on the main thread, so you need to make sure you don't run long-running operations on it. (See Chapter 19 to understand this aspect in greater depth.)

This completes the conceptual understanding of all the new features provided around list widgets. Let's now look at a working example to test and demonstrate these features in action. Much of the code presented so far has been taken from this working sample, so the working sample should be easy to follow.

Working Sample: Test Home Screen List Widget

This home screen list widget sample will demonstrate the ideas covered thus far about list-based home screen widgets. At the end of this sample, you will see a list-based widget that you can drag onto the home screen. When you drag it, you will see a widget displaying 20 rows of list items filled with sample text. When you click one of these list item rows, you will see a Toast on the home screen containing text from that specific row of the list.

Here is the list of files you will need:

- `TestListWidgetProvider.java` is the primary class; it's the test widget provider that implements a widget with a list view as one of its views (Listing 26–21).
- `TestRemoteViewsFactory.java` is the class that provides a list of items to show for the list view loaded by the widget provider (Listing 26–22).
- `TestRemoteViewsService.java` is the remote views service that instantiates the `TestRemoteViewsFactory` (Listing 26–23).

- layout\test_list_widget_layout.xml is the primary layout for the whole widget loaded by the widget provider (Listing 26–1).
- layout\list_item_layout.xml is the layout file for the individual list item view. This layout is loaded by the remote view factory (Listing 26–10).
- drawable\box1.xml is a simple layout helper class to provide rounded corners for the main widget layout (Listing 26–2).
- xml\test_list_appwidget_provider.xml is the metadata file for defining the widget to Android (Listing 26–24).
- AndroidManifest.xml is the configurations file for the application where you define the widget provider and the remote view service (Listing 26–25).

Creating the Test Widget Provider

The process of creating a home screen widget starts with creating a widget provider inheriting from AppWidgetProvider and overloading its onUpdate() method to provide a view for the widget. This process is explained in great detail in Chapter 22. In this example, the example provider is called TestListWidgetProvider. Listing 26–21 provides the source code with comments for this class.

Listing 26–21. *TestListWidgetProvider.java*

```
package com.androidbook.homewidgets.listwidget;

/*
 * Use CTRL-SHIFT-O in Eclipse to fill in imports
 */
public class TestListWidgetProvider extends AppWidgetProvider
{
    private static final String tag = "TestListWidgetProvider";

    public static final String ACTION_LIST_CLICK =
        "com.androidbook.homewidgets.listclick";

    public static final String EXTRA_LIST_ITEM_TEXT =
        "com.androidbook.homewidgets.list_item_text";

    public void onUpdate(Context context,
                        AppWidgetManager appWidgetManager,
                        int[] appWidgetIds)
   {
        Log.d(tag, "onUpdate called");
        final int N = appWidgetIds.length;
        Log.d(tag, "Number of widgets:" + N);
        for (int i=0; i<N; i++)
        {
            int appWidgetId = appWidgetIds[i];
            updateAppWidget(context, appWidgetManager, appWidgetId);
        }
```

```
        super.onUpdate(context,appWidgetManager, appWidgetIds);
    }

    public void onDeleted(Context context, int[] appWidgetIds)
    {
        Log.d(tag, "onDelete called");
        super.onDeleted(context,appWidgetIds);
    }

    public void onEnabled(Context context)
    {
        Log.d(tag, "onEnabled called");
        super.onEnabled(context);
    }

    public void onDisabled(Context context)
    {
        Log.d(tag, "onDisabled called");
        super. onDisabled (context);
    }

    private void updateAppWidget(Context context,
                        AppWidgetManager appWidgetManager,
                         int appWidgetId)
   {
        Log.d(tag, "onUpdate called for widget:" + appWidgetId);

        final RemoteViews rv =
        new RemoteViews(context.getPackageName(),
                R.layout.test_list_widget_layout);

        rv.setEmptyView(R.id.listwidget_list_view_id,
                R.id.listwidget_empty_view_id);

        // Specify the service to provide data for the
        // collection widget.
        final Intent intent =
            new Intent(context, TestRemoteViewsService.class);

        //This is purely for debugging. Unnecessary otherwise
        intent.putExtra(AppWidgetManager.EXTRA_APPWIDGET_ID,
                                            appWidgetId);

        rv.setRemoteAdapter(appWidgetId,
                R.id.listwidget_list_view_id, intent);

        //setup a list view callback.
        //you need a pending intent that is unique
        //for this widget id. Send a message to
        //ourselves which you will catch in OnReceive.
        Intent onListClickIntent =
            new Intent(context,TestListWidgetProvider.class);

        //set an action so that this receiver can distinguish it
        //from other widget related actions
        onListClickIntent.setAction(
                TestListWidgetProvider.ACTION_LIST_CLICK);
```

```
        //because this receiver serves all instances
        //of this app widget. You need to know which
        //specific instance this message is targeted for.
        onListClickIntent.putExtra(
                AppWidgetManager.EXTRA_APPWIDGET_ID, appWidgetId);

        //Make this intent unique as you are getting ready
        //to create a pending intent with it.
        //The toUri method loads the extras as
        //part of the uri string.
        //The data of this intent is not used at all except
        //to establish this intent as a unique pending intent.
        //See intent.filterEquals() method to see
        //how intents are compared to see if they are unique.
        onListClickIntent.setData(
            Uri.parse(
                onListClickIntent.toUri(Intent.URI_INTENT_SCHEME)));

        //you need to deliver this intent later when
        //the remote view is clicked as a broadcast intent
        //to this same receiver.
        final PendingIntent onListClickPendingIntent =
            PendingIntent.getBroadcast(context, 0,
                onListClickIntent,
                PendingIntent.FLAG_UPDATE_CURRENT);

        //Set this pending intent as a template for
        //the list item view.
        //Each view in the list will then need to specify
        //a set of additional extras to be appended
        //to this template and then broadcast the
        //final template.
        //See how the remoteviewsfactory() sets up
        //the each item in the list remoteview.
        //See also docs for RemoteViews.setFillIntent()
        rv.setPendingIntentTemplate(R.id.listwidget_list_view_id,
                onListClickPendingIntent);

        //update the widget
        appWidgetManager.updateAppWidget(appWidgetId, rv);
    }

    @Override
    public void onReceive(Context context, Intent intent)
    {
        if (intent.getAction()
                .equals(TestListWidgetProvider.ACTION_LIST_CLICK))
        {
            //this action is not one of usual widget actions
            //such as onDeleted, onEnabled etc.
            //Instead this is a specific action that is directed here
            //by the intents loaded into the list view items
            dealWithListAction(context,intent);
            return;
        }
```

```
        //make sure you call this
        super.onReceive(context, intent);
    }
    public void dealWithListAction(Context context, Intent  intent)
    {
        String clickedItemText =
            intent.getStringExtra(
                    TestListWidgetProvider.EXTRA_LIST_ITEM_TEXT);
        if (clickedItemText == null)
        {
            clickedItemText = "Error";
        }
        clickedItemText =
            clickedItemText
            + "Clicked on item text:"
            + clickedItemText;

        Toast t =
            Toast.makeText(context,clickedItemText,Toast.LENGTH_LONG);
        t.show();
    }

}//eof-class
```

With the background information provided, much of what this class needs to do is already explained. The source code is amply peppered with comments to restate much that was discussed; however, here's a quick overview of the functionality:

1. In `onUpdate()`, load the remote view.
2. Locate the list remote view and hook it up with a remote view factory via a remote view service.
3. Set the remote view with a pending intent template for the `onClick` behavior.
4. Override the `onReceive()` method, and deal with the specialized `onClick` action.

Creating the Remote Views Factory

Listing 26–22 provides the source code for the remote view factory that is responsible for populating the list view.

Listing 26–22. *TestRemoteViewsFactory.java*

```
package com.androidbook.homewidgets.listwidget;
/*
 * Use CTRL-SHIFT-O in Eclipse to fill in imports
 */
class TestRemoteViewsFactory
implements RemoteViewsService.RemoteViewsFactory
{
    private Context mContext;
    private int mAppWidgetId;
```

```
private static String tag="TRVF";
public TestRemoteViewsFactory(Context context, Intent intent)
{
    mContext = context;
    //Purely for debugging. Unnecessary otherwise.
    mAppWidgetId =
        intent.getIntExtra(
            AppWidgetManager.EXTRA_APPWIDGET_ID,
            AppWidgetManager.INVALID_APPWIDGET_ID);

    Log.d(tag,"factory created");
}

//Called when your factory is first constructed.
//The same factory may be shared across multiple
//RemoteViewAdapters depending on the intent passed.
public void onCreate()
{
    Log.d(tag,"onCreate called for widget id:" + mAppWidgetId);
}

//Called when the last RemoteViewsAdapter that is
//associated with this factory is unbound.
public void onDestroy()
{
    Log.d(tag,"destroy called for widget id:" + mAppWidgetId);
}

//The total number of items
//in this list
public int getCount()
{
   return 20;
}

public RemoteViews getViewAt(int position)
{
     Log.d(tag,"getview called:" + position);
     RemoteViews rv =
         new RemoteViews(
             this.mContext.getPackageName(),
             R.layout.list_item_layout);
     String itemText  = "Item:" + position;
     rv.setTextViewText(
         R.id.textview_widget_list_item_id, itemText);

     this.loadItemOnClickExtras(rv, position);
     return rv;
}
private void loadItemOnClickExtras(RemoteViews rv, int position)
{
    Intent ei = new Intent();
    ei.putExtra(TestListWidgetProvider.EXTRA_LIST_ITEM_TEXT,
            "Position of the item Clicked:" + position);
    rv.setOnClickFillInIntent(R.id.textview_widget_list_item_id, ei);
       }
```

```
    //This allows for the use of a custom loading view
    //which appears between the time that getViewAt(int)
    //is called and returns. If null is returned,
    //a default loading view will be used.
    public RemoteViews getLoadingView()
    {
        return null;
    }

//How many different types of views
    //are there in this list.
    public int getViewTypeCount()
    {
        return 1;
    }

    //The internal id of the item
    //at this position
    public long getItemId(int position)
    {
        return position;
    }

    //True if the same id
    //always refers to the same object.
    public boolean hasStableIds()
    {
        return true;
    }

    //Called when notifyDataSetChanged() is triggered
    //on the remote adapter. This allows a RemoteViewsFactory
    //to respond to data changes by updating
    //any internal references.
    //Note: expensive tasks can be safely performed
    //synchronously within this method.
    //In the interim, the old data will be displayed
    //within the widget.
    public void onDataSetChanged()
    {
         Log.d(tag,"onDataSetChanged");
    }
}
```

Much of this code has been explained already. At a high level, this class assumes there are 20 rows. Each row's layout is loaded from a layout file and its text set to the corresponding position. It then loads the text from each position into the `onClick` intent. This is the text that you would see as a Toast.

Coding Remote Views Service

Listing 26–23 shows the source code for the class that returns the remote view factory.

Listing 26–23. *TestRemoteViewsService.java*

```
package com.androidbook.homewidgets.listwidget;
import android.content.Intent;

public class TestRemoteViewsService
extends android.widget.RemoteViewsService
{
    @Override
    public RemoteViewsFactory onGetViewFactory(Intent intent)
    {
        return new TestRemoteViewsFactory(
                this.getApplicationContext(), intent);
    }
}
```

You can get quite creative with this class implementation. You can use a single service to create multiple factories depending on the incoming intent so that you don't have to define multiple services if you have multiple list views in a single widget or have multiple widgets in your package. However, to invoke `onGetViewFactory()`, the intents have to be unique. So if you are using extras to identify which factory to return, make sure you set the data URI that includes the extras.

Main Widget Layout File

The main layout file that corresponds to how the widget looks on the home page needs to be at `\res\layout\test_list_widget_layout.xml` (note that this layout file was presented in Listing 26–1). This main layout file also requires rounded corners, which are provided a box drawable located at `\res\drawable\box1.xml`, which was presented in Listing 26–2.

Layout for the Individual List Items

This layout file corresponds to the layout of the individual list item inside the list. This layout file needs to be at `layout\list_item_layout.xml`. This layout file was presented in Listing 26–10.

Widget Provider Metadata

A widget provider needs to specify a metadata XML file when that widget provider is declared in the Android manifest file. This file needs to be at `\res\xml\test_list_appwidget_provider.xml`. Listing 26–24 shows this widget metadata information file.

Listing 26–24. *Widget Information File*

```
<!-- xml/test_list_appwidget_provider.xml -->
<appwidget-provider xmlns:android="http://schemas.android.com/apk/res/android"
    android:minWidth="222dp"
    android:minHeight="222dp"
    android:updatePeriodMillis="1000000"
```

```
    android:initialLayout="@layout/test_list_widget_layout"
    android:label="Test List Widget"
    >
</appwidget-provider>
```

This provider metadata file specifies the size for the widget and how often to fire the `onUpdate` callback on the widget, specified in milliseconds. Note that this file is discussed in greater detail in Chapter 22.

AndroidManifest.xml

Listing 26–25 shows the configuration file for the application. The widget provider definition and the remote view service definition are highlighted.

Listing 26–25. *Android Manifest File*

```
<?xml version="1.0" encoding="utf-8"?>
<manifest xmlns:android="http://schemas.android.com/apk/res/android"
      package="com.androidbook.homewidgets.listwidget"
      android:versionCode="1"
      android:versionName="1.0.0">
    <application android:icon="@drawable/icon"
       android:label="Test List Widget Application">
<!--
**********************************************************************
*  Test List Widget Provider
**********************************************************************
 -->
    <receiver android:name=".TestListWidgetProvider">
       <meta-data android:name="android.appwidget.provider"
         android:resource="@xml/test_list_appwidget_provider" />
       <intent-filter>
         <action
           android:name="android.appwidget.action.APPWIDGET_UPDATE" />
       </intent-filter>
    </receiver>

   <!-- The service serving the RemoteViews to the collection widget -->
   <service android:name=".TestRemoteViewsService"
     android:permission="android.permission.BIND_REMOTEVIEWS"
     android:exported="false" />

    </application>
    <uses-sdk android:minSdkVersion="11" />
</manifest>
```

Testing the Test List Widget

Once you build and deploy this project, you will see in Eclipse that the project is successfully deployed. Because this project doesn't contain an activity that is identified to run at startup, you won't see anything on the emulator by default.

To install the widget created in this sample, you need to see a list of available widgets first. Clicking the home screen will bring up a list of available widgets, as shown in Figure 26–2.

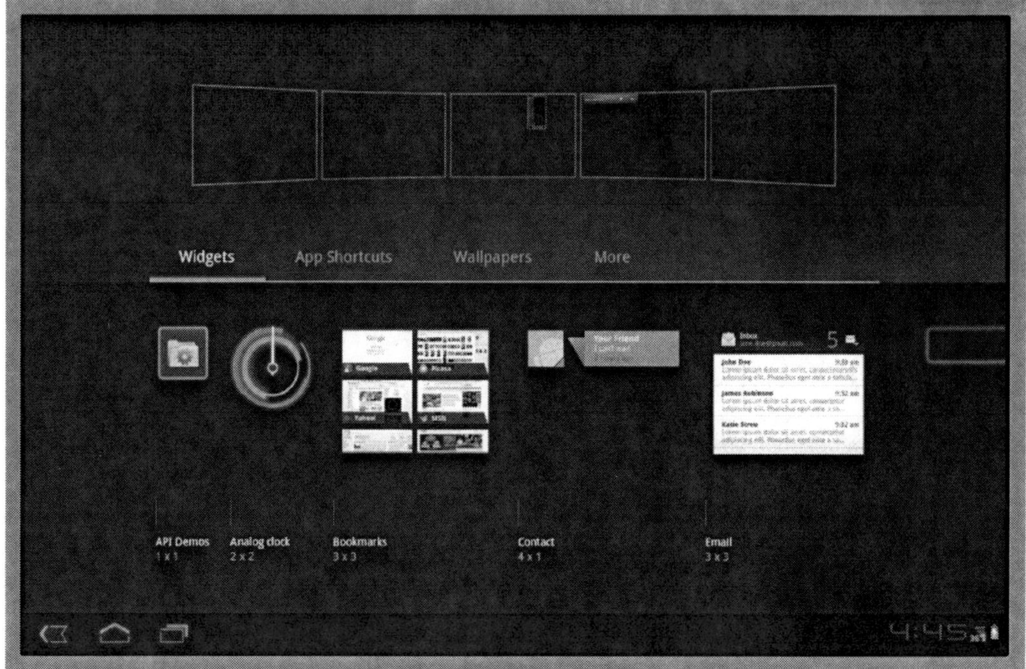

Figure 26–2. *List of widgets*

The name of your widget is Test List Widget Application, so it may be the farthest to the right; you may have to scroll to the right to see it, as shown in Figure 26–3.

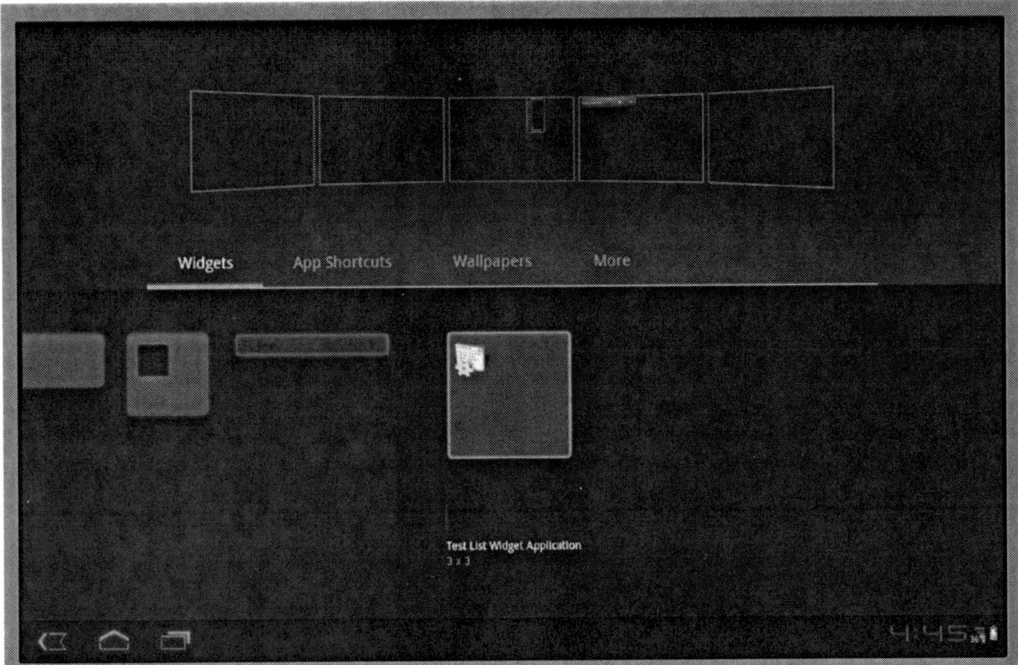

Figure 26–3. *Scrolling to the right to find the Test List Widget Application*

Now you can drag the Test List Widget Application to the home screen of your choice. Once your drag is recognized, you can select the Home button at the bottom to go to the home screen. At that time, you will see the widget in its main form, as shown previously in Figure 26–1. If you click one of the list items, a Toast message appropriate to the line item you have clicked will appear (see Figure 26–4).

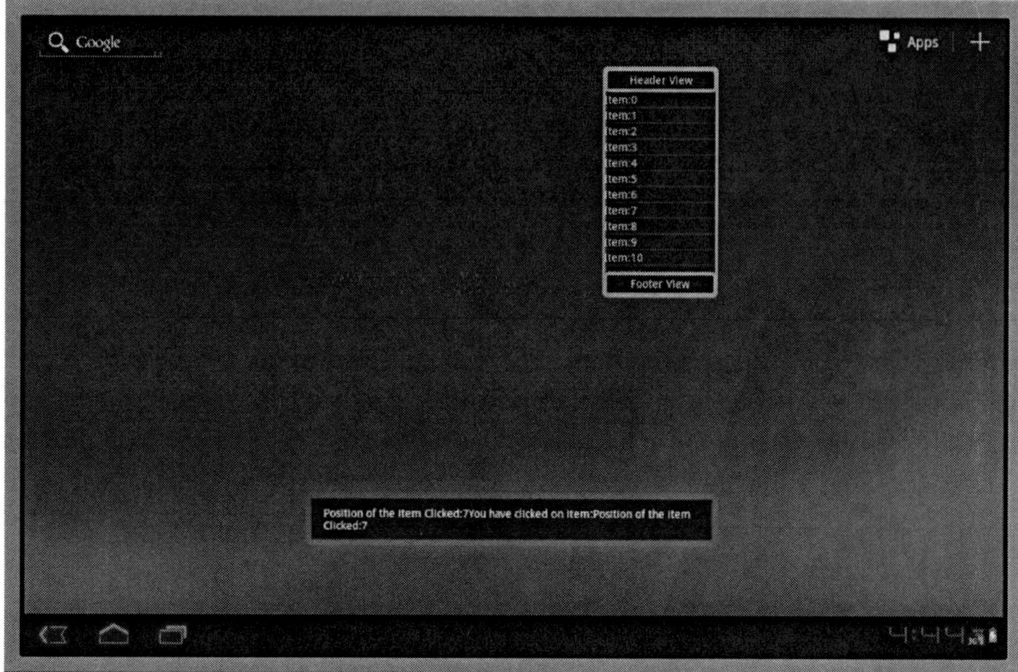

Figure 26–4. *Toast in response to a list view item click*

You can also use this program to test the concepts related to `onCreate()` and `onDestroy()`. As a first test, when you create a single widget on the home page, you will see the `onCreate()` method log a message to LogCat. Now, if you remove the widget, `onDestroy()` is called.

As a second test, create two widgets on the home page by dragging the widget twice. You will notice that `onCreate()` on the factory gets called only once. Now remove one widget by dragging it to the trash can. `onDestroy()` is not called. This is because the factory is still required to support the last widget, which is still on the home screen. If you drag this last widget to the trash can, `onDestroy()` is called, and a log message will appear in LogCat.

References

Here are some useful links to further strengthen your understanding of this chapter:

- `www.androidbook.com/item/3938`: Our work notes for preparing the material for home screen widgets. You'll find links to APIs, code snippets, open questions, and more research.
- `www.androidbook.com/item/3637`: Our notes on `RemoteViews`, updated with 4.0 material including code samples, pondered questions, and internal and external references.

- http://developer.android.com/guide/topics/appwidgets/index.html: The main document on app widgets from the previous releases.
- http://developer.android.com/reference/android/appwidget/AppWidgetManager.html: The reference page for the important AppWidgetManager API.
- http://developer.android.com/reference/android/widget/RemoteViewsService.RemoteViewsFactory.html: The RemoteViewsFactory API reference.
- http://developer.android.com/reference/android/widget/RemoteViews.html: The RemoteViews API reference.
- http://developer.android.com/reference/android/widget/RemoteViewsService.html: The RemoteViewsService API reference.
- www.androidbook.com/proandroid4/projects: The URL to download the test projects for this book. The name of the ZIP file for this chapter is ProAndroid4_ch26_TestListWidget.zip.

Summary

In this chapter, we have covered these topics:

- Loading and populating list-based remote views through the remote views service and a remote views factory.
- The life cycle of RemoteViewsFactory
- Setting up onClick events for list-based widgets
- Using the AppWidgetProvider itself to respond to onClick events from list widgets

Interview Questions

The following questions should further consolidate what is covered in this chapter:

1. What are the new key classes/methods introduced to support list widgets?
2. What is a RemoteViewsService?
3. What is a RemoteViewsFactory?
4. How is a RemoteViewsFactory similar to a list adapter?
5. Why is a RemoteViewsFactory cached?
6. When is the onDestroy() method of a RemoteViewsFactory called?
7. What is the disadvantage of passing an extra widget ID to the remote views service intent?

8. Can a single `RemoteViewsService` instantiate multiple `RemoteViewsFactorys`?
9. What is `RemoteViews.setPendingIntentTemplate()`?
10. Why, when, and how do you make a pending intent unique based on its extras?
11. What is `RemoteViews.setOnClickFillIntent()`?
12. How can you set an empty view for a `ListView` in a widget?

Chapter 27

Touch Screens

Many Android devices incorporate touch screens. When a device does not have a physical keyboard, much of the user input *must* come through the touch screen. Therefore your applications will often need to be able to deal with touch input from the user. You've most likely already seen the virtual keyboard that displays on the screen when text input is required from the user. We used touch with mapping applications in Chapter 22 to pan the maps sideways. The implementations of the touch screen interface have been hidden from you so far, but now we'll show you how to take advantage of the touch screen.

This chapter is made up of four major parts. The first section will deal with `MotionEvent` objects, which is how Android tells an application that the user is touching a touch screen. We'll also cover the `VelocityTracker`. The second section will deal with multitouch, where a user can have more than one finger at a time on the touch screen. The third section covers touches with maps, because there are some special classes and methods to help us with maps and touch screens. Finally, we will include a section on gestures, a specialized type of capability in which touch sequences can be interpreted as commands.

Understanding MotionEvents

In this section, we're going to cover how Android tells applications about touch events from the user. For now, we will only be concerned with touching the screen one finger at a time (we'll cover multitouch in a later section).

At the hardware level, a touch screen is made up of special materials that can pick up pressure and convert that to screen coordinates. The information about the touch is turned into data, and that data is passed to the software to deal with it.

The MotionEvent Object

When a user touches the touch screen of an Android device, a `MotionEvent` object is created. The `MotionEvent` contains information about where and when the touch took place, as well as other details of the touch event. The MotionEvent object gets passed to an appropriate method in your application. This could be the onTouchEvent() method of a View object. Remember that the View class is the parent of quite a few classes in Android, including Layouts, Buttons, Lists, Surfaces, Clocks, and more. This means we can interact with all of these different types of View objects using touch events. When the method is called, it can inspect the MotionEvent object to decide what to do. For example, a MapView could use touch events to move the map sideways to allow the user to pan the map to other points of interest. Or a virtual keyboard object could receive touch events to activate the virtual keys to provide text input to some other part of the user interface (UI).

Receiving MotionEvent Objects

A MotionEvent object is one of a sequence of events related to a touch by the user. The sequence starts when the user first touches the touch screen, continues through any movements of the finger across the surface of the touch screen, and ends when the finger is lifted from the touch screen. The initial touch (an ACTION_DOWN action), the movements sideways (ACTION_MOVE actions) and the up event (an ACTION_UP action) of the finger all create MotionEvent objects. You could receive quite a few ACTION_MOVE events as the finger moves across the surface before you receive the final ACTION_UP event. Each MotionEvent object contains information about what action is being performed, where the touch is taking place, how much pressure was applied, how big the touch was, when the action occurred, and when the initial ACTION_DOWN occurred. There is a fourth possible action, which is ACTION_CANCEL. This action is used to indicate that a touch sequence is ending without actually doing anything. Finally, there is ACTION_OUTSIDE, which is set in a special case where a touch occurs outside of our window but we still get to find out about it.

There is another way to receive touch events, and that is to register a callback handler for touch events on a View object. The class to receive the events must implement the View.OnTouchListener interface, and the View object's setOnTouchListener() method must be called to set up the handler for that View. The implementing class of the View.OnTouchListener must implement the onTouch() method. Whereas the onTouchEvent() method takes just a MotionEvent object as a parameter, onTouch() takes both a View and a MotionEvent object as parameters. This is because the OnTouchListener could receive MotionEvent objects for multiple views. This will become clearer with our next example application.

If a MotionEvent handler (either through the onTouchEvent() or onTouch() method) consumes the event and no one else needs to know about it, the method should return true. This tells Android that the event does not need to be passed to any other views. If the View object is not interested in this event *or any future events related to this touch sequence*, it returns false. The onTouchEvent() method of the base class View doesn't do anything and returns false. Subclasses of View may or may not do the same. For example, a Button object will consume a touch event, because a touch is equivalent to a click, and therefore returns true from the onTouchEvent() method. Upon receiving an ACTION_DOWN event, the Button will change its color to indicate that it is in the process of being clicked. The Button also wants to receive the ACTION_UP event to know when the user has let go, so it can initiate the logic of clicking the button. If a Button object returned false from onTouchEvent(), it would not receive any more MotionEvent objects to tell it when the user lifted a finger from the touch screen.

When we want touch events to do something new with a particular View object, we can extend the class, override the onTouchEvent() method, and put our logic there. We can also implement the View.OnTouchListener interface and set up a callback handler on the View object. By setting up a callback handler with onTouch(), MotionEvents will be delivered there first before they go to the View's onTouchEvent() method. Only if the onTouch() method returned false would our View's onTouchEvent() method get called. Let's get to our example application where this should be easier to see.

> **NOTE:** We will give you a URL at the end of the chapter which you can use to download projects of this chapter. This will allow you to import these projects into Eclipse directly.

Setting Up an Example Application

Listing 27–1 shows the XML of a layout file. Create a new Android project in Eclipse starting with this layout.

Listing 27–1. *XML Layout File for TouchDemo1*

```
<?xml version="1.0" encoding="utf-8"?>
<!-- This file is res/layout/main.xml -->
<LinearLayout xmlns:android="http://schemas.android.com/apk/res/android"
    android:layout_width="fill_parent"
    android:layout_height="fill_parent"
    android:orientation="vertical" >

  <RelativeLayout  android:id="@+id/layout1"
    android:tag="trueLayoutTop"  android:orientation="vertical"
    android:layout_width="fill_parent"
    android:layout_height="wrap_content"
    android:layout_weight="1" >

    <com.androidbook.touch.demo1.TrueButton android:text="returns true"
    android:id="@+id/trueBtn1"  android:tag="trueBtnTop"
    android:layout_width="wrap_content"
    android:layout_height="wrap_content" />

    <com.androidbook.touch.demo1.FalseButton android:text="returns false"
    android:id="@+id/falseBtn1"  android:tag="falseBtnTop"
    android:layout_width="wrap_content"
    android:layout_height="wrap_content"
    android:layout_below="@id/trueBtn1" />

  </RelativeLayout>
  <RelativeLayout  android:id="@+id/layout2"
    android:tag="falseLayoutBottom"  android:orientation="vertical"
    android:layout_width="fill_parent"
    android:layout_height="wrap_content"
```

```
    android:layout_weight="1"  android:background="#FF00FF" >

    <com.androidbook.touch.demo1.TrueButton android:text="returns true"
    android:id="@+id/trueBtn2"  android:tag="trueBtnBottom"
    android:layout_width="wrap_content"
    android:layout_height="wrap_content" />

    <com.androidbook.touch.demo1.FalseButton android:text="returns false"
    android:id="@+id/falseBtn2"  android:tag="falseBtnBottom"
    android:layout_width="wrap_content"
    android:layout_height="wrap_content"
    android:layout_below="@id/trueBtn2" />

  </RelativeLayout>
</LinearLayout>
```

There are a couple of things to point out about this layout. We've incorporated tags on our UI objects, and we'll be able to refer to these tags in our code as events occur on them. We've used custom objects (TrueButton and FalseButton). You'll see in the Java code that these are classes extended from the Button class. Because these are Buttons, we can use all of the same XML attributes we would use on other buttons. Figure 27–1 shows what this layout looks like, and Listing 27–2 shows our button Java code.

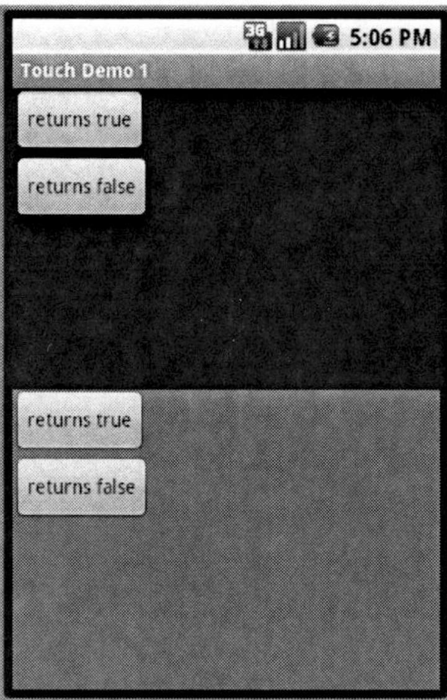

Figure 27–1. *The UI of our TouchDemo1 application*

Listing 27–2. *Java Code for the Button Classes for TouchDemo1*

```
// This file is BooleanButton.java
import android.content.Context;
import android.util.AttributeSet;
import android.util.Log;
import android.view.MotionEvent;
import android.widget.Button;
```

```
public abstract class BooleanButton extends Button {
    protected boolean myValue() {
        return false;
    }

    public BooleanButton(Context context, AttributeSet attrs) {
        super(context, attrs);
    }

    @Override
    public boolean onTouchEvent(MotionEvent event) {
        String myTag = this.getTag().toString();
        Log.v(myTag, "-----------------------------------");
        Log.v(myTag, MainActivity.describeEvent(this, event));
        Log.v(myTag, "super onTouchEvent() returns " +
                super.onTouchEvent(event));
        Log.v(myTag, "and I'm returning " + myValue());
        return(myValue());
    }
}


// This file is TrueButton.java
import android.content.Context;
import android.util.AttributeSet;

public class TrueButton extends BooleanButton {
    protected boolean myValue() {
        return true;
    }

    public TrueButton(Context context, AttributeSet attrs) {
        super(context, attrs);
    }
}


// This file is FalseButton.java
import android.content.Context;
import android.util.AttributeSet;

public class FalseButton extends BooleanButton {

    public FalseButton(Context context, AttributeSet attrs) {
        super(context, attrs);
    }
}
```

The `BooleanButton` class was built so we can reuse the `onTouchEvent()` method, which we've customized by adding the logging. Then, we created `TrueButton` and `FalseButton`, which will respond differently to the `MotionEvents` passed to them. This will be made clearer when you look at the main activity code, which is shown in Listing 27–3.

Listing 27–3. *Java Code for Our Main Activity*

```
// This file is MainActivity.java
import android.app.Activity;
import android.os.Bundle;
```

```
import android.util.Log;
import android.view.MotionEvent;
import android.view.View;
import android.view.View.OnTouchListener;
import android.widget.Button;
import android.widget.RelativeLayout;

public class MainActivity extends Activity implements OnTouchListener {
    /** Called when the activity is first created. */
    @Override
    public void onCreate(Bundle savedInstanceState) {
        super.onCreate(savedInstanceState);
        setContentView(R.layout.main);

        RelativeLayout layout1 =
                (RelativeLayout) findViewById(R.id.layout1);
        layout1.setOnTouchListener(this);
        Button trueBtn1 = (Button)findViewById(R.id.trueBtn1);
        trueBtn1.setOnTouchListener(this);
        Button falseBtn1 = (Button)findViewById(R.id.falseBtn1);
        falseBtn1.setOnTouchListener(this);

        RelativeLayout layout2 =
                (RelativeLayout) findViewById(R.id.layout2);
        layout2.setOnTouchListener(this);
        Button trueBtn2 = (Button)findViewById(R.id.trueBtn2);
        trueBtn2.setOnTouchListener(this);
        Button falseBtn2 = (Button)findViewById(R.id.falseBtn2);
        falseBtn2.setOnTouchListener(this);
    }

    @Override
    public boolean onTouch(View v, MotionEvent event) {
        String myTag = v.getTag().toString();
        Log.v(myTag, "-----------------------------");
        Log.v(myTag, "Got view " + myTag + " in onTouch");
        Log.v(myTag, describeEvent(v, event));
        if( "true".equals(myTag.substring(0, 4))) {
        /*  Log.v(myTag, "*** calling my onTouchEvent() method ***");
            v.onTouchEvent(event);
            Log.v(myTag, "*** back from onTouchEvent() method ***"); */
            Log.v(myTag, "and I'm returning true");
            return true;
        }
        else {
            Log.v(myTag, "and I'm returning false");
            return false;
        }
    }

    protected static String describeEvent(View view, MotionEvent event) {
        StringBuilder result = new StringBuilder(300);
        result.append("Action: ").append(event.getAction()).append("\n");
        result.append("Location: ").append(event.getX()).append(" x ")
                 .append(event.getY()).append("\n");
        if(    event.getX() < 0 || event.getX() > view.getWidth() ||
               event.getY() < 0 || event.getY() > view.getHeight()) {
```

```
            result.append(">>> Touch has left the view <<<\n");
        }
        result.append("Edge flags: ").append(event.getEdgeFlags());
        result.append("\n");
        result.append("Pressure: ").append(event.getPressure());
        result.append("   ").append("Size: ").append(event.getSize());
        result.append("\n").append("Down time: ");
        result.append(event.getDownTime()).append("ms\n");
        result.append("Event time: ").append(event.getEventTime());
        result.append("ms").append("  Elapsed: ");
        result.append(event.getEventTime()-event.getDownTime());
        result.append(" ms\n");
        return result.toString();
    }
}
```

Our main activity code sets up callbacks on our buttons and the layouts so we can process the touch events (the `MotionEvent` objects) for everything in our UI. We've added lots of logging, so you'll be able to tell exactly what's going on as touch events occur. One other good idea is to add the following tag to your manifest file so Android Market will know your application requires a touch screen to work: <uses-configuration android:reqTouchScreen="finger" />. For example, Google TVs don't have touch screens, so it wouldn't make sense to try to run this app there. When you compile and run this application, you should see a screen that looks like Figure 27–1.

Running the Example Application

To get the most out of this application, you need to open LogCat in Eclipse to watch the messages fly by as you touch the touch screen. This works in the emulator as well as on a real device. We also advise you to maximize the LogCat window, so you can more easily scroll up and down to see all of the generated events from this application. To maximize the window, just double-click the LogCat tab. Now, go to the application UI, and touch and release on the topmost button marked Returns True (if you're using the emulator, use your mouse to click and release the button). You should see at least two events logged in LogCat. The messages are tagged as coming from `trueBtnTop` and were logged from the `onTouch()` method in `MainActivity`. See `MainActivity.java` for the `onTouch()` method's code. As you view the LogCat output, see which method calls are producing the values. For example, the value displayed after Action comes from the `getAction()` method. Listing 27–4 shows a sample of what you might see in LogCat from the emulator, and Listing 27–5 shows a sample of what you might see from a real device.

Listing 27–4. *Sample LogCat Messages from TouchDemo1 from the Emulator*

```
trueBtnTop        -----------------------------
trueBtnTop        Got view trueBtnTop in onTouch
trueBtnTop        Action: 0
trueBtnTop        Location: 52.0 x 20.0
trueBtnTop        Edge flags: 0
trueBtnTop        Pressure: 0.0   Size: 0.0
trueBtnTop        Down time: 163669ms
trueBtnTop        Event time: 163669ms  Elapsed: 0 ms
```

```
trueBtnTop         and I'm returning true
trueBtnTop         ----------------------------
trueBtnTop         Got view trueBtnTop in onTouch
trueBtnTop         Action: 1
trueBtnTop         Location: 52.0 x 20.0
trueBtnTop         Edge flags: 0
trueBtnTop         Pressure: 0.0   Size: 0.0
trueBtnTop         Down time: 163669ms
trueBtnTop         Event time: 163831ms  Elapsed: 162 ms
trueBtnTop         and I'm returning true
```

Listing 27–5. *Sample LogCat Messages from TouchDemo1 from a Real Device*

```
trueBtnTop         ----------------------------
trueBtnTop         Got view trueBtnTop in onTouch
trueBtnTop         Action: 0
trueBtnTop         Location: 42.8374 x 25.293747
trueBtnTop         Edge flags: 0
trueBtnTop         Pressure: 0.05490196   Size: 0.2
trueBtnTop         Down time: 24959412ms
trueBtnTop         Event time: 24959412ms  Elapsed: 0 ms
trueBtnTop         and I'm returning true
trueBtnTop         ----------------------------
trueBtnTop         Got view trueBtnTop in onTouch
trueBtnTop         Action: 2
trueBtnTop         Location: 42.8374 x 25.293747
trueBtnTop         Edge flags: 0
trueBtnTop         Pressure: 0.05490196   Size: 0.2
trueBtnTop         Down time: 24959412ms
trueBtnTop         Event time: 24959530ms  Elapsed: 118 ms
trueBtnTop         and I'm returning true
trueBtnTop         ----------------------------
trueBtnTop         Got view trueBtnTop in onTouch
trueBtnTop         Action: 1
trueBtnTop         Location: 42.8374 x 25.293747
trueBtnTop         Edge flags: 0
trueBtnTop         Pressure: 0.05490196   Size: 0.2
trueBtnTop         Down time: 24959412ms
trueBtnTop         Event time: 24959567ms  Elapsed: 155 ms
trueBtnTop         and I'm returning true
```

Understanding MotionEvent Contents

The first event has an action of 0, which is ACTION_DOWN. The last event has an action of 1, which is ACTION_UP. If you used a real device, you might see more than two events. Any events in between ACTION_DOWN and ACTION_UP will most likely have an action of 2, which is ACTION_MOVE. The other possibilities are an action of 3, which is ACTION_CANCEL, or 4, which is ACTION_OUTSIDE. When using real fingers on a real touch screen, you can't always touch and release without a slight movement on the surface, so some ACTION_MOVE events are not unexpected.

There are some other differences between the emulator and a real device. Notice that the precision of the location within the emulator is in whole numbers (52 by 20), whereas on a real device you see fractions (42.8374 by 25.293747). The location for a MotionEvent has an X and Y component, where X represents the distance from the left-hand side of the View object to the point touched and Y represents the distance from the top of the View object to the point touched.

You should also notice that the pressure in the emulator is 0, as is the size. For a real device, the pressure represents how hard the finger pressed down, and size represents how large the touch is. If you touch lightly with the tip of your pinky finger, the values for pressure and size will be small. If you press hard with your thumb, both pressure and size will be larger. The documentation says that the values of pressure and size will be between 0 and 1. However, due to differences in hardware, it may be very difficult to use any absolute numbers in your application for making decisions about pressure and size. It would be fine to compare pressure and size between MotionEvents as they occur in your application, but you may run into trouble if you decide that pressure must exceed a value such as 0.8 to be considered a hard press. On that particular device, you might never get a value above 0.8. You might not even get a value above 0.2.

The down time and event time values operate in the same way between the emulator and a real device, the only difference being that the real device has much larger values. The elapsed times work the same.

The edge flags are for detecting when a touch has reached the edge of the physical screen. The Android SDK documentation says that the flags are set to indicate that a touch has intersected with an edge of the display (top, bottom, left, or right). However, the getEdgeFlags() method may always return zero, depending on what device or emulator it is used on. With some hardware, it is too difficult to actually detect a touch at the edge of the display, so Android is supposed to pin the location to the edge and set the appropriate edge flag for you. This doesn't always happen, so you should not rely on the edge flags being set properly. The MotionEvent class provides a setEdgeFlags() method so you can set the flags yourself if you want to.

The last thing to notice is that our onTouch() method returns true, because our TrueButton is coded to return true. Returning true tells Android that the MotionEvent object has been consumed and there is no reason to give it to someone else. It also tells Android to keep sending touch events from this touch sequence to this method. That's why we got the ACTION_UP event, as well as the ACTION_MOVE event in the case of the real device.

Now touch the Returns False button near the top of the screen. For the remainder of this section, we will show only sample LogCat output from a real device. The differences have been explained, so if you are working with the emulator, you should understand why you are seeing what you are seeing. Listing 27–6 shows a sample LogCat output for your Returns False touch.

Listing 27–6. *Sample LogCat from Touching the Top Returns False Button*

```
falseBtnTop        -----------------------------
falseBtnTop        Got view falseBtnTop in onTouch
falseBtnTop        Action: 0
falseBtnTop        Location: 61.309372 x 44.281494
falseBtnTop        Edge flags: 0
falseBtnTop        Pressure: 0.0627451   Size: 0.26666668
falseBtnTop        Downtime: 28612178ms
```

```
falseBtnTop         Event time: 28612178ms  Elapsed: 0 ms
falseBtnTop         and I'm returning false
falseBtnTop         ----------------------------------
falseBtnTop         Action: 0
falseBtnTop         Location: 61.309372 x 44.281494
falseBtnTop         Edge flags: 0
falseBtnTop         Pressure: 0.0627451   Size: 0.26666668
falseBtnTop         Downtime: 28612178ms
falseBtnTop         Event time: 28612178ms  Elapsed: 0 ms
falseBtnTop         super onTouchEvent() returns true
falseBtnTop         and I'm returning false
trueLayoutTop         ----------------------------
trueLayoutTop         Got view trueLayoutTop in onTouch
trueLayoutTop         Action: 0
trueLayoutTop         Location: 61.309372 x 116.281494
trueLayoutTop         Edge flags: 0
trueLayoutTop         Pressure: 0.0627451   Size: 0.26666668
trueLayoutTop         Downtime: 28612178ms
trueLayoutTop         Event time: 28612178ms  Elapsed: 0 ms
trueLayoutTop         and I'm returning true
trueLayoutTop         ----------------------------
trueLayoutTop         Got view trueLayoutTop in onTouch
trueLayoutTop         Action: 2
trueLayoutTop         Location: 61.309372 x 111.90039
trueLayoutTop         Edge flags: 0
trueLayoutTop         Pressure: 0.0627451   Size: 0.26666668
trueLayoutTop         Downtime: 28612178ms
trueLayoutTop         Event time: 28612217ms  Elapsed: 39 ms
trueLayoutTop         and I'm returning true
trueLayoutTop         ----------------------------
trueLayoutTop         Got view trueLayoutTop in onTouch
trueLayoutTop         Action: 1
trueLayoutTop         Location: 55.08958 x 115.30792
trueLayoutTop         Edge flags: 0
trueLayoutTop         Pressure: 0.0627451   Size: 0.26666668
trueLayoutTop         Downtime: 28612178ms
trueLayoutTop         Event time: 28612361ms  Elapsed: 183 ms
trueLayoutTop         and I'm returning true
```

Now you're seeing very different behavior, so we'll explain what happened. Android receives the ACTION_DOWN event in a MotionEvent object and passes it to our onTouch() method in the MainActivity class. Our onTouch() method records the information in LogCat and returns false. This tells Android that our onTouch() method did not consume the event, so Android looks to the next method to call, which in our case is the overridden onTouchEvent() method of our FalseButton class. Because FalseButton is an extension of the BooleanButton class, refer to the onTouchEvent() method in BooleanButton.java to see the code. In the onTouchEvent() method, we again write information to LogCat, we call the parent class's onTouchEvent() method, and then we also return false. Notice that the location information in LogCat is exactly the same as before. This should be expected because we're still in the same View object, the FalseButton. We see that our parent class wants to return true from onTouchEvent(), and we can see why. If you look at the button in the UI, it should be a different color from the Returns True button. Our Returns False button now looks like it's partway through being pressed. That is, it looks like a button looks when it has been pressed but has not been released. Our custom method returned false instead of true. Because we again told Android that we did not consume this event, by returning false, Android never sends the ACTION_UP event to our button, so our button doesn't know that the finger ever lifted from the touch screen. Therefore, our button is still in the pressed state. If we had returned true like our parent wanted to, we would eventually have received the ACTION_UP event, so we could change the color back to the normal button color. To recap, every time we return false from a UI object for a received MotionEvent object, Android stops sending MotionEvent objects to that UI object, and Android keeps looking for another UI object to consume our MotionEvent object.

You might have realized that when we touched our Returns True button, we didn't get a color change in the button. Why is that? Well, our onTouch() method was called before any actual button methods got called, and onTouch() returned true, so Android never bothered to call the Returns True button's onTouchEvent() method. If you add a v.onTouchEvent(event); line to the onTouch() method just before returning true, you will see the button change color. You will also see more log lines in LogCat, because our onTouchEvent() method is also writing information to LogCat.

Let's keep going through the LogCat output. Now that Android has tried twice to find a consumer for the ACTION_DOWN event and failed, it goes to the next View in the application that could possibly receive the event, which in our case is the layout underneath the button. We called our top layout trueLayoutTop, and we can see that it received the ACTION_DOWN event.

Notice that our onTouch() method got called again, although now with the layout view and not the button view. Everything about the MotionEvent object passed to onTouch() for trueLayoutTop is the same as before, including the times, except for the Y coordinate of the location. The Y coordinate changed from 44.281494 for the button to 116.281494 for the layout. This makes sense because the button is not in the upper-left corner of the layout, it's below the Returns True button. Therefore the Y coordinate of the touch relative to the layout is larger than the Y coordinate of the same touch relative to the button; the touch is further away from the top edge of the layout than it is from the top edge of the button. Because onTouch() for the trueLayoutTop returns true, Android sends the rest of the touch events to the layout, and we see the log records corresponding to the ACTION_MOVE and the ACTION_UP events. Go ahead and touch the top Returns False button again, and notice that the same set of log records occurs. That is, onTouch() is called for falseBtnTop, onTouchEvent() is called for falseBtnTop, and then onTouch() is called for trueLayoutTop for the rest of the events. Android only stops sending the events to the button for one touch sequence at a time. For a new sequence of touch events, Android will send to the button unless it gets another return of false from the called method, which it still does in our sample application.

Now touch your finger on the top layout but not on either button, and then drag your finger around a bit and lift it off the touch screen (if you're using the emulator, just use your mouse to make a similar motion). Notice a stream of log messages in LogCat, where the first record has an action of ACTION_DOWN, and then many ACTION_MOVE events are followed by an ACTION_UP event.

Now, touch the top Returns True button, and before lifting your finger from the button, drag your finger around the screen and then lift it off. Listing 27–7 shows some new information in LogCat.

Listing 27–7. *LogCat Records Showing a Touch Outside of Our View*

```
[ … log messages of an ACTION_DOWN event followed by some ACTION_MOVE events … ]

trueBtnTop          Got view trueBtnTop in onTouch
trueBtnTop          Action: 2
trueBtnTop          Location: 150.41768 x 22.628128
trueBtnTop          >>> Touch has left the view <<<
trueBtnTop          Edge flags: 0
trueBtnTop          Pressure: 0.047058824   Size: 0.13333334
trueBtnTop          Downtime: 31690859ms
trueBtnTop          Event time: 31691344ms  Elapsed: 485 ms
trueBtnTop          and I'm returning true

[ … more ACTION_MOVE events logged … ]

trueBtnTop          Got view trueBtnTop in onTouch
trueBtnTop          Action: 1
trueBtnTop          Location: 291.5864 x 223.43854
trueBtnTop          >>> Touch has left the view <<<
trueBtnTop          Edge flags: 0
trueBtnTop          Pressure: 0.047058824   Size: 0.13333334
trueBtnTop          Downtime: 31690859ms
trueBtnTop          Event time: 31692493ms  Elapsed: 1634 ms
trueBtnTop          and I'm returning true
```

Even after your finger drags itself off of the button, we continue to get notified of touch events related to the button. The first record in Listing 27–7 shows an event record where we're no longer on the button. In this case, the X coordinate of the touch event is to the right of the edge of our button object. However, we keep getting called with `MotionEvent` objects until we get an `ACTION_UP` event, because we continue to return `true` from the `onTouch()` method. Even when you finally lift your finger off of the touch screen, and even if your finger isn't on the button, our `onTouch()` method still gets called to give us the `ACTION_UP` event because we keep returning `true`. This is something to keep in mind when dealing with `MotionEvents`. When the finger has moved off of the view, we could decide to cancel whatever operation might have been performed and return `false` from the `onTouch()` method, so we don't get notified of further events. Or we could choose to continue to receive events (by returning `true` from the `onTouch()` method) and only perform the logic if the finger returns to our view before lifting off.

The touch sequence of events got associated to our top Returns True button when we returned `true` from `onTouch()`. This told Android that it could stop looking for an object to receive the `MotionEvent` objects and just send all future `MotionEvent` objects for this touch sequence to us. Even if we encounter another view when dragging our finger, we're still tied to the original view for this sequence.

Exercising the Bottom Half of the Example Application

Let's see what happens with the lower half of our application. Go ahead and touch the Returns True button in the bottom half. We see the same thing as happened with the top Returns True button. Because onTouch() returns true, Android sends us the rest of the events in the touch sequence until the finger is lifted from the touch screen. Now, touch the bottom Returns False button. Once again, the onTouch() method and onTouchEvent() methods return false (both associated with the falseBtnBottom view object). But this time, the next view to receive the MotionEvent object is the falseLayoutBottom object, and it also returns false. Now, we're finished.

Because the onTouchEvent() method called the super's onTouchEvent() method, the button has changed color to indicate it's halfway through being pressed. Again, the button will stay this way, because we never get the ACTION_UP event in this touch sequence, because our methods return false all the time. Unlike before, even the layout is not interested in this event. If you were to touch the bottom Returns False button and hold it down and then drag your finger around the display, you would not see any extra records in LogCat, because no more MotionEvent objects are sent to us. We always returned false, so Android won't bother us with any more events for this touch sequence. Again, if we start a new touch sequence, we can see new LogCat records showing up. If you initiate a touch sequence in the bottom layout and not on a button, you will see a single event in LogCat for falseLayoutBottom that returns false and then nothing after that (until you start a new touch sequence).

So far, we've used buttons to show you the effects of MotionEvent events from touch screens. It's worth pointing out that, normally, you would implement logic on buttons using the onClick() method. We used buttons for this sample application, because they're easy to create and they are subclasses of View that can therefore receive touch events just like any other view. Remember that these techniques apply to any View object in your application, be it a standard or customized view class.

Recycling MotionEvents

You may have noticed the recycle() method of the MotionEvent class in the Android reference documentation. It is tempting to want to recycle the MotionEvents that you receive in onTouch() or onTouchEvent(), but don't do it. If your callback method is not consuming the MotionEvent object and you're returning false, the MotionEvent object is likely to be handed to some other method or view or our activity, so you don't want Android recycling it yet. Even if you consumed the event and returned true, the event object doesn't belong to you, so you should not recycle it.

If you look at MotionEvent, you will see a few variations of a method called obtain(). This is either creating a copy of a MotionEvent or a brand new MotionEvent. Your copy, or your brand-new event object, is the event object that you should recycle when you are done with it. For example, if you want to hang onto an event object that is passed to you via a callback, you should use obtain() to make a copy, because once you return from the callback, that event object will be recycled by Android, and you may get strange results if you continue to use it. When you are finished using *your copy*, you invoke recycle() on it.

Using VelocityTracker

Android provides a class to help handle touch screen sequences, and that class is VelocityTracker. When a finger is in motion on a touch screen, it might be nice to know how fast it is moving across the surface. For example, if the user is dragging an object across the screen and lets go, your application probably wants to show that object flying across the screen accordingly. Android provides VelocityTracker to help with the math involved.

To use VelocityTracker, you first get an instance of a VelocityTracker by calling the static method VelocityTracker.obtain(). You can then add MotionEvent objects to it with the addMovement(MotionEvent ev) method. You would call this method in your handler that receives MotionEvent objects, from a handler method such as onTouch(), or from a view's onTouchEvent(). The VelocityTracker uses the MotionEvent objects to figure out what is going on with the user's touch sequence. Once VelocityTracker has at least two MotionEvent objects in it, we can use the other methods to find out what's happening.

The two VelocityTracker methods—getXVelocity() and getYVelocity()—return the corresponding velocity of the finger in the X and Y directions, respectively. The value returned from these two methods will represent pixels per time period. This could be pixels per millisecond or per second or really anything you want. To tell the VelocityTracker what time period to use, and before you can call these two getter methods, you need to invoke the VelocityTracker's computeCurrentVelocity(int units) method. The value of units represents how many milliseconds are in the time period for measuring the velocity. If you want pixels per millisecond, use a units value of 1; if you want pixels per second, use a units value of 1000. The value returned by the getXVelocity() and getYVelocity() methods will be positive if the velocity is toward the right (for X) or down (for Y). The value returned will be negative if the velocity is toward the left (for X) or up (for Y).

When you are finished with the VelocityTracker object you got with the obtain() method, call the VelocityTracker object's recycle() method. Listing 27–8 shows a sample onTouchEvent() handler for an activity. It turns out that an activity has an onTouchEvent() callback, which is called whenever no views have handled the touch event. Because we're using a stock, empty layout, we have no views consuming our touch events.

Listing 27–8. *Sample Activity That Uses VelocityTracker*

```
import android.app.Activity;
import android.os.Bundle;
import android.util.Log;
import android.view.MotionEvent;
import android.view.VelocityTracker;

public class MainActivity extends Activity {
```

```
    private static final String TAG = "VelocityTracker";

    /** Called when the activity is first created. */
    @Override
    public void onCreate(Bundle savedInstanceState) {
        super.onCreate(savedInstanceState);
        setContentView(R.layout.main);
    }

    private VelocityTracker vTracker = null;

    public boolean onTouchEvent(MotionEvent event) {
        int action = event.getAction();
        switch(action) {
            case MotionEvent.ACTION_DOWN:
                if(vTracker == null) {
                    vTracker = VelocityTracker.obtain();
                }
                else {
                    vTracker.clear();
                }
                vTracker.addMovement(event);
                break;
            case MotionEvent.ACTION_MOVE:
                vTracker.addMovement(event);
                vTracker.computeCurrentVelocity(1000);
                Log.v(TAG, "X velocity is " + vTracker.getXVelocity() +
                        " pixels per second");
                Log.v(TAG, "Y velocity is " + vTracker.getYVelocity() +
                        " pixels per second");
                break;
            case MotionEvent.ACTION_UP:
            case MotionEvent.ACTION_CANCEL:
                vTracker.recycle();
                break;
        }
        return true;
    }
}
```

There are a few key things to note about VelocityTracker. Obviously, when you've only added one MotionEvent to a VelocityTracker (the ACTION_DOWN event), the velocities cannot be computed as anything other than zero. But we need to add the starting point so that the subsequent ACTION_MOVE events can calculate velocities then. It turns out that the velocities reported after ACTION_UP is added to our VelocityTracker are also zero. Therefore, do not read the X and Y velocities after adding ACTION_UP expecting to get motion. For example, if you're writing a gaming application in which the user is throwing an object on the screen, use the velocities after adding the last ACTION_MOVE event to calculate the object's trajectory across the game view.

VelocityTracker is somewhat costly in terms of performance, so use it sparingly. Also, make sure that you recycle it as soon as you are done with it in case someone else wants to use one. There can be more than one VelocityTracker in use in Android, but they can take up a lot of memory, so give yours back if you're not going to continue to use it. In Listing 27–8, we also use the clear() method if we're starting a new touch sequence (that is, if we get an `ACTION_DOWN` event and our `VelocityTracker` object already exists) instead of recycling this one and obtaining a new one.

Multitouch

Now that you've seen single touches in action, let's move on to multitouch. Multitouch has gained a lot of interest ever since the TED conference in 2006 at which Jeff Han demonstrated a multitouch surface for a computer user interface. Using multiple fingers on a screen opens up a lot of possibilities for manipulating what's on the screen. For example, putting two fingers on an image and moving them apart could zoom in on the image. By placing multiple fingers on an image and turning clockwise, you could rotate the image on the screen. These are standard touch operations in Google Maps, for instance.

Android introduced support for multitouch with Android SDK 2.0. In that release you were able to (technically) use up to three fingers on a screen at the same time to perform actions such as zoom, rotate, or whatever else you could imagine doing with multiple touches (we say "technically" because the first Android *devices* to support multitouch only supported two fingers). If you think about it, though, there is no magic to this. If the screen hardware can detect multiple touches as they initiate on the screen, notify your application as those touches move in time across the surface of the screen, and notify you when those touches lift off of the screen, your application can figure out what the user is trying to do with those touches. Although it's not magic, it isn't easy either. We're going to help you understand multitouch in this section.

> **NOTE:** Prior to Android 2.2, the `MotionEvent` class made it more difficult to determine the actions and indexes of a MotionEvent object. Some of the methods we're about to use did not exist prior to 2.2. Our website has a sample application for 2.1 for your reference.

The Basics of Multitouch

The basics of multitouch are exactly the same as for single touches. `MotionEvent` objects get created for touches, and these `MotionEvent` objects are passed to your methods just like before. Your code can read the data about the touches and decide what to do. At a basic level, the methods of `MotionEvent` are the same; that is, we call `getAction()`, `getDownTime()`, `getX()`, and so on. However, when more than one finger is touching the screen, the `MotionEvent` object must include information from all fingers, with some caveats. The action value from `getAction()` is for one finger, not all. The down time value is for the very first finger down and measures the time as long as at least one finger is down. The location values `getX()` and `getY()`, as well as `getPressure()` and `getSize()`, can take an argument for the finger; therefore, you need to use a pointer index value to request the information for the finger you're interested in. There are method calls that we used previously that did not take any argument to specify a finger (for example, `getX()`, `getY()`), so which finger would the values be for if we used those methods? You can figure it out, but it takes some work. Therefore, if you don't take into account multiple fingers all of the time, you might end up with some strange results. Let's dig into this to figure out what to do.

The first method of MotionEvent you need to know about for multitouch is getPointerCount(). This tells you how many fingers are represented in the MotionEvent object but doesn't necessarily tell you how many fingers are actually touching the screen; that depends on the hardware and on the implementation of Android on that hardware. You may find that, on certain devices, getPointerCount() does not report all fingers that are touching, just some. But let's press on. As soon as you've got more than one finger being reported in MotionEvent objects, you need to start dealing with the pointer indexes and the pointer IDs.

The MotionEvent object contains information for pointers starting at index 0 and going up to the number of fingers being reported in that object. The pointer index always starts at 0; if three fingers are being reported, pointer indexes will be 0, 1, and 2. Calls to methods such as getX() must include the pointer index for the finger you want information about. Pointer IDs are integer values representing which finger is being tracked. Pointer IDs start at 0 for the first finger down but don't always start at 0 once fingers are coming and going on the screen. Think of a pointer ID as the name of that finger while it is being tracked by Android. For example, imagine a pair of touch sequences for two fingers, starting with finger 1 down, and followed by finger 2 down, finger 1 up, and finger 2 up. The first finger down will get pointer ID 0. The second finger down will get pointer ID 1. Once the first finger goes up, the second finger will still be pointer ID 1. At that point, the pointer index for the second finger becomes 0, because the pointer index always starts at 0. In this example, the second finger (pointer ID 1) starts as pointer index 1 when it first touches down and then shifts to pointer index 0 once the first finger leaves the screen. Even when the second finger is the only finger on the screen, it remains as pointer ID 1. Your applications will use pointer IDs to link together the events associated to a particular finger even as other fingers are involved. Let's look at an example.

Listing 27–9 shows our new XML layout plus our Java code for a multitouch application. Create a new application using Listing 27–9, and run it. Figure 27–2 shows what it should look like.

Listing 27–9. *XML Layout and Java for a Multitouch Demonstration*

```
<?xml version="1.0" encoding="utf-8"?>
<!-- This file is /res/layout/main.xml -->
<RelativeLayout  xmlns:android="http://schemas.android.com/apk/res/android"
    android:id="@+id/layout1"
    android:tag="trueLayout"  android:orientation="vertical"
    android:layout_width="fill_parent"
    android:layout_height="wrap_content"
    android:layout_weight="1"
    >

    <TextView android:text="Touch fingers on the screen and look at LogCat"
    android:id="@+id/message"
    android:tag="trueText"
    android:layout_width="wrap_content"
```

```
    android:layout_height="wrap_content"
    android:layout_alignParentBottom="true" />

</RelativeLayout>


// This file is MainActivity.java
import android.app.Activity;
import android.os.Bundle;
import android.util.Log;
import android.view.MotionEvent;
import android.view.View;
import android.view.View.OnTouchListener;
import android.widget.RelativeLayout;

public class MainActivity extends Activity implements OnTouchListener {
    /** Called when the activity is first created. */
    @Override
    public void onCreate(Bundle savedInstanceState) {
        super.onCreate(savedInstanceState);
        setContentView(R.layout.main);

        RelativeLayout layout1 =
                (RelativeLayout) findViewById(R.id.layout1);
        layout1.setOnTouchListener(this);
    }

    public boolean onTouch(View v, MotionEvent event) {
        String myTag = v.getTag().toString();
        Log.v(myTag, "-----------------------------");
        Log.v(myTag, "Got view " + myTag + " in onTouch");
        Log.v(myTag, describeEvent(event));
        logAction(event);
        if( "true".equals(myTag.substring(0, 4))) {
            return true;
        }
        else {
            return false;
        }
    }

    protected static String describeEvent(MotionEvent event) {
        StringBuilder result = new StringBuilder(500);
        result.append("Action: ").append(event.getAction()).append("\n");
        int numPointers = event.getPointerCount();
        result.append("Number of pointers: ");
        result.append(numPointers).append("\n");
        int ptrIdx = 0;
        while (ptrIdx < numPointers) {
            int ptrId = event.getPointerId(ptrIdx);
            result.append("Pointer Index: ").append(ptrIdx);
            result.append(", Pointer Id: ").append(ptrId).append("\n");
            result.append("   Location: ").append(event.getX(ptrIdx));
            result.append(" x ").append(event.getY(ptrIdx)).append("\n");
            result.append("   Pressure: ");
            result.append(event.getPressure(ptrIdx));
            result.append("   Size: ").append(event.getSize(ptrIdx));
```

```
            result.append("\n");

            ptrIdx++;
        }
        result.append("Downtime: ").append(event.getDownTime());
        result.append("ms\n").append("Event time: ");
        result.append(event.getEventTime()).append("ms");
        result.append("  Elapsed: ");
        result.append(event.getEventTime()-event.getDownTime());
        result.append(" ms\n");
        return result.toString();
    }

    private void logAction(MotionEvent event) {
        int action = event.getActionMasked();
        int ptrIndex = event.getActionIndex();
        int ptrId = event.getPointerId(ptrIndex);

        if(action == 5 || action == 6)
            action = action - 5;

        Log.v("Action", "Pointer index: " + ptrIndex);
        Log.v("Action", "Pointer Id: " + ptrId);
        Log.v("Action", "True action value: " + action);
    }
}
```

Figure 27–2. *Our multitouch demonstration application*

If you only have the emulator, this application will still work, but you won't be able to get multiple fingers simultaneously on the screen. You'll see output similar to what we saw in the previous application. Listing 27–10 shows sample LogCat messages for a touch sequence like we described earlier. That is, the first finger presses on the screen, and then the second finger presses, the first finger leaves the screen, and the second finger leaves the screen.

Listing 27–10. *Sample LogCat Output for a Multitouch Application*

```
trueLayout      ----------------------------
trueLayout      Got view trueLayout in onTouch
trueLayout      Action: 0
trueLayout      Number of pointers: 1
trueLayout      Pointer Index: 0, Pointer Id: 0
trueLayout         Location: 114.88211 x 499.77502
trueLayout         Pressure: 0.047058824   Size: 0.13333334
trueLayout      Downtime: 33733650ms
trueLayout      Event time: 33733650ms  Elapsed: 0 ms
Action          Pointer index: 0
Action          Pointer Id: 0
Action          True Action value: 0
trueLayout      ----------------------------
trueLayout      Got view trueLayout in onTouch
trueLayout      Action: 2
trueLayout      Number of pointers: 1
trueLayout      Pointer Index: 0, Pointer Id: 0
trueLayout         Location: 114.88211 x 499.77502
trueLayout         Pressure: 0.05882353   Size: 0.13333334
trueLayout      Downtime: 33733650ms
trueLayout      Event time: 33733740ms  Elapsed: 90 ms
Action          Pointer index: 0
Action          Pointer Id: 0
Action          True Action value: 2
trueLayout      ----------------------------
trueLayout      Got view trueLayout in onTouch
trueLayout      Action: 261
trueLayout      Number of pointers: 2
trueLayout      Pointer Index: 0, Pointer Id: 0
trueLayout         Location: 114.88211 x 499.77502
trueLayout         Pressure: 0.05882353   Size: 0.13333334
trueLayout      Pointer Index: 1, Pointer Id: 1
trueLayout         Location: 320.30692 x 189.67395
trueLayout         Pressure: 0.050980393   Size: 0.13333334
trueLayout      Downtime: 33733650ms
trueLayout      Event time: 33733962ms  Elapsed: 312 ms
Action          Pointer index: 1
Action          Pointer Id: 1
Action          True Action value: 0
trueLayout      ----------------------------
trueLayout      Got view trueLayout in onTouch
trueLayout      Action: 2
trueLayout      Number of pointers: 2
trueLayout      Pointer Index: 0, Pointer Id: 0
trueLayout         Location: 111.474594 x 499.77502
trueLayout         Pressure: 0.05882353   Size: 0.13333334
trueLayout      Pointer Index: 1, Pointer Id: 1
trueLayout         Location: 320.30692 x 189.67395
trueLayout         Pressure: 0.050980393   Size: 0.13333334
trueLayout      Downtime: 33733650ms
trueLayout      Event time: 33734189ms  Elapsed: 539 ms
Action          Pointer index: 0
Action          Pointer Id: 0
```

```
Action          True Action value: 2
trueLayout      ----------------------------
trueLayout      Got view trueLayout in onTouch
trueLayout      Action: 6
trueLayout      Number of pointers: 2
trueLayout      Pointer Index: 0, Pointer Id: 0
trueLayout         Location: 111.474594 x 499.77502
trueLayout         Pressure: 0.05882353   Size: 0.13333334
trueLayout      Pointer Index: 1, Pointer Id: 1
trueLayout         Location: 320.30692 x 189.67395
trueLayout         Pressure: 0.050980393   Size: 0.13333334
trueLayout      Downtime: 33733650ms
trueLayout      Event time: 33734228ms  Elapsed: 578 ms
Action          Pointer index: 0
Action          Pointer Id: 0
Action          True Action value: 1
trueLayout      ----------------------------
trueLayout      Got view trueLayout in onTouch
trueLayout      Action: 2
trueLayout      Number of pointers: 1
trueLayout      Pointer Index: 0, Pointer Id: 1
trueLayout         Location: 318.84656 x 191.45105
trueLayout         Pressure: 0.050980393   Size: 0.13333334
trueLayout      Downtime: 33733650ms
trueLayout      Event time: 33734240ms  Elapsed: 590 ms
Action          Pointer index: 0
Action          Pointer Id: 1
Action          True Action value: 2
trueLayout      ----------------------------
trueLayout      Got view trueLayout in onTouch
trueLayout      Action: 1
trueLayout      Number of pointers: 1
trueLayout      Pointer Index: 0, Pointer Id: 1
trueLayout         Location: 314.95224 x 190.5625
trueLayout         Pressure: 0.050980393   Size: 0.13333334
trueLayout      Downtime: 33733650ms
trueLayout      Event time: 33734549ms  Elapsed: 899 ms
Action          Pointer index: 0
Action          Pointer Id: 1
Action          True Action value: 1
```

Understanding Multitouch Contents

We'll now discuss what is going on with this application. The first event we see is the ACTION_DOWN (action value of 0) of the first finger. We learn about this using the getAction() method. Please refer to the describeEvent() method in MainActivity.java to follow along with which methods produce which output. We get one pointer with index 0 and pointer ID 0. After that, you'll probably see several ACTION_MOVE events (action value of 2) for this first finger, even though we're only showing one of these in Listing 27–10. We still only have one pointer and the index and ID are still both 0.

A little later we get the second finger touching the screen. The action is now a decimal value of 261. What does this mean? The action value is actually made up of two parts: an indicator of which pointer the action is for and what action that pointer is doing.

Converting decimal 261 to hexadecimal, we get 0x00000105. The action is the smallest byte (5 in this case), and the pointer index is the next byte over (1 in this case). Note that this tells us the pointer index but not the pointer ID. If you pressed a third finger onto the screen, the action would be 0x00000205 (or decimal 517). A fourth finger would be 0x00000305 (or decimal 773) and so on. You haven't seen an action value of 5 yet, but it's known as `ACTION_POINTER_DOWN`. It's just like `ACTION_DOWN` except that it's used in multitouch situations.

Now, look at the next pair of records from LogCat in Listing 27–10. The first record is for an `ACTION_MOVE` event (action value of 2). Remember that it is difficult to keep fingers from moving on a real screen. We're only showing one `ACTION_MOVE` event, but you might see several when you try this for yourself. When the first finger is lifted off of the screen, we get an action value of 0x00000006 (or decimal 6). Like before, we have pointer index 0 and an action value that is `ACTION_POINTER_UP` (similar to `ACTION_UP` but for multitouch situations). If the second finger was lifted in a multitouch situation, we would get an action value of 0x00000106 (or decimal 262). Notice how we still have information for two fingers when we get the `ACTION_UP` for one of them.

The last pair of records in Listing 27–10 shows one more `ACTION_MOVE` event for the second finger, followed by an `ACTION_UP` for the second finger. This time, we see an action value of 1 (`ACTION_UP`). We didn't get an action value of 262, but we'll explain that next. Also, notice that once the first finger left the screen, the pointer index for the second finger has changed from 1 to 0, but the pointer ID has remained as 1.

`ACTION_MOVE` events do not tell you which finger moved. You will always get an action value of 2 for a move regardless of how many fingers are down or which finger is doing the moving. All down finger positions are available within the `MotionEvent` object, so you need to read the positions and then figure things out. If there's only one finger left on the screen, the pointer ID will tell you which finger it is that's still moving because it's the only finger left. In Listing 27–10, when the second finger was the only one left on the screen, the `ACTION_MOVE` event had a pointer index of 0 and a pointer ID of 1, so we knew it was the second finger that was moving.

Not only can a `MotionEvent` object contain move events for more than one finger, but it can also contain multiple move events per finger. It does this using historical values contained within the object. Android should report all history since the last `MotionEvent` object. See `getHistoricalSize()` and the other `getHistorical...()` methods.

Going back to the beginning of Listing 27–10, the first finger down is pointer index 0 and pointer ID 0, so why don't we get 0x00000005 (or decimal 5) for the action value when the first finger is pressed to the screen before any other fingers? Unfortunately, this question doesn't have a happy answer. We can get an action value of 5 in the following scenario: press the first finger to the screen and then the second finger, resulting in action values of 0 and 261 (ignoring the `ACTION_MOVE` events for the moment). Now, lift the first finger (action value of 6), and press it back down on the screen. The pointer ID of the second finger remained as 1. For the moment when the first finger was in the air, our application knew about pointer ID 1 only. Once the first finger touched the screen again, Android reassigned pointer ID 0 to the first finger and gave it pointer index 0 as well. Because now we know there are multiple fingers involved, we get an action value of 5 (pointer index of 0 and the action value of 5). The answer to the question, therefore, is backward compatibility, but it is not a happy answer. The action values of 0 and 1 are pre-multitouch, and applications written before multitouch will still work as long as only one finger is used.

When only one finger remains on the screen, Android treats it like a single-touch case. So we get the old `ACTION_UP` value of 1 instead of a multitouch `ACTION_UP` value of 6. Our code will need to consider these cases carefully. A pointer index of 0 could result in an `ACTION_DOWN` value of 0 or 5, depending on which pointers are in play. The last finger up will get an `ACTION_UP` value of 1 no matter which pointer ID it has.

There is another action we haven't mentioned so far: `ACTION_SCROLL` (value of 8), introduced in Android 3.1. This comes from an input device like a mouse, not a touch screen. In fact, as you can see from the methods in `MotionEvent`, these objects can be used for lots of things other than touch screen touches. We won't be covering these other input devices in this book.

Touches with Maps

Maps can receive touch events as well. You have already seen how touching a map can bring up a zoom control or allow us to pan the map sideways. These are built-in functions of maps. But what if we want to do something different? We're going to show you how to implement some interesting functionality with maps, including the ability to click a location and get its latitude and longitude. From there, we can do lots of very useful things.

One of the main classes for maps is `MapView`. This class has an `onTouchEvent()` method just like the `Views` we covered earlier and takes a `MotionEvent` object as its only argument. We can also use the `setOnTouchListener()` method to set up a callback handler for touch events on a `MapView`. Other main types of objects for maps are the set of `Overlays`, including `ItemizedOverlay` and `MyLocationOverlay`. These were all introduced in Chapter 22. These `Overlay` classes also have an `onTouchEvent()` method, although the signature is slightly different from the `onTouchEvent()` method on a regular `View`. For an `Overlay`, the method signature is

```
onTouchEvent(android.view.MotionEvent e, MapView mapView)
```

We can override this `onTouchEvent()` method if we want to do different things with maps. It is more common to override methods in an `Overlay` class than in `MapView`, so we will focus our attention there in this section. As before, the `onTouchEvent()` method for `Overlays` deals with `MotionEvent` objects. Even with maps, the `MotionEvent` object gives us X and Y coordinates of where the user has touched the touch screen. This is only marginally useful when dealing with maps, because we often want to know the actual location on the map where the user touched. Fortunately, there are ways to figure this out.

`MapView` provides an interface called `Projection`, and `Projection` has methods to convert from a pixel to a `GeoPoint` or from a `GeoPoint` to a pixel. To get a `Projection`, call the `MapView.getProjection()` method. Once you have the `Projection`, the methods `fromPixels()` and `toPixels()` can be used for the conversions. Keep in mind that the

Projection is only good while the map doesn't change in the view. Within your onTouchEvent() method, you can convert the X and Y location values to a GeoPoint using fromPixels().

An interesting and very useful method of Overlay is the onTap() method, which is similar to the onTouch() method you saw earlier in this chapter but different in a key way. Map Overlays do not have an onTouch() method. The signature of the onTap() method is

```
public boolean onTap(GeoPoint p, MapView mapView)
```

This means that when a user touches on our Overlay, our onTap() method gets called with the GeoPoint of where the user touched. This will save us a lot of time trying to figure out where on the map the user is touching. We no longer need to worry about converting from an X and Y coordinate location to a latitude and longitude coordinate; Android takes care of this for us.

We're now going to revisit the example from Chapter 22 in which we displayed a map with buttons for the different modes (Satellite, Traffic, and Normal). We're going to add the ability to determine the latitude/longitude of a location from the map. To do this we need to add an Overlay object to our MapView, and when the Overlay object receives a touch event, we'll convert that touch event to a location on the map. With the converted location, we'll launch a Toast to display the latitude/longitude of the point that was clicked. We'll start by making a copy in Eclipse of our MapViewDemo from Chapter 22 (see Listings 22-2 and 22-3). Then, we'll use Listing 27–11 to modify the onCreate() method of the main Activity, plus add a new class with the file ClickReceiver.java, also provided in this listing. The changes to the onCreate() method are shown in bold. The UI will still look just like it did in Figure 22-3.

Listing 27–11. *Adding Touch to Our Maps Demonstration*

```
    @Override
    protected void onCreate(Bundle savedInstanceState) {
        super.onCreate(savedInstanceState);
        setContentView(R.layout.mapview);

        mapView = (MapView)findViewById(R.id.mapview);

            ClickReceiver clickRecvr = new ClickReceiver(this);
            mapView.getOverlays().add(clickRecvr);
            mapView.invalidate();
    }


// This file is ClickReceiver.java
import android.content.Context;
import android.widget.Toast;
import com.google.android.maps.GeoPoint;
import com.google.android.maps.MapView;
import com.google.android.maps.Overlay;

public class ClickReceiver extends Overlay{
        private Context mContext;

        public ClickReceiver(Context context) {
```

```
        mContext = context;
        }

        @Override
        public boolean onTap(GeoPoint p, MapView mapView) {
        String msg = "Got a touch at lat,lon: " +
                (float)p.getLatitudeE6() / 1000000f +
                "," + (float)p.getLongitudeE6() / 1000000f;
        Toast.makeText(mContext, msg, Toast.LENGTH_SHORT).show();
        // Of course, now you could do a GeoCoder call to find
        // out what is at this location.
                return true;
        }
}
```

When you run your newly modified Maps Demo application, zoom in on a city so you can see the streets. Now, you can touch a street and the `onTap()` method of our `ClickReceiver` will be called, which in turn will pop a Toast message indicating the latitude and longitude of the map location that was touched. With the latitude/longitude of a location, we could use the `Geocoder` to find out what's around that location. We could use the location to navigate to it using turn-by-turn directions. We could measure how far away the location is from where we are. We can even store the location for later use.

Gestures

Gestures are a special type of a touch screen event. The term *gesture* is used for a variety of things in Android, from a simple touch sequence like a fling or a pinch to the formal `Gesture` class that we're going to talk about later in this section. Flings, pinches, long presses, and scrolls have expected behaviors with expected triggers. That is, it is pretty clear to most people that a fling is a gesture where a finger touches the screen, drags somewhat quickly off in a single direction, and then lifts up. For example, when someone use a fling in the Gallery application (the one that shows images in a left-to-right chain), the images will move sideways to show new images to the user.

The first gesture we want to cover is the pinch. The pinch gesture is not explicitly supported in Android prior to version 2.2, so to implement the pinch gesture in prior versions, you have to create code yourself to read event objects and take appropriate action. We provide a sample application on our website for version 2.1, but we're not going to talk about that here. From version 2.2 onward, we have some helpful new features to use with gestures such as the pinch.

Next, we're going to introduce some helpful classes for other gestures, such as flings and long presses. From there, we'll cover custom gestures: that is, gestures that you can prerecord to allow the user to initiate action in your application by dragging a finger in custom patterns. But first, let's get pinching!

The Pinch Gesture

One of the cool applications of multitouch is the pinch gesture, which is used for zooming. The idea is that if you place two fingers on the screen and spread them apart, the application should respond by zooming in. If your fingers come together, the application should zoom out. The application is usually showing images, which could be maps.

Before we get to the pinch gesture's native support, we first need to cover a class that's been around from the beginning—`GestureDetector`.

GestureDetector and OnGestureListeners

The first class to help us with gestures is `GestureDetector`, which has been around from the very beginning of Android. Its purpose in life is to receive `MotionEvent` objects and tell us when a sequence of events looks like a common gesture. We pass all of our event objects to the `GestureDetector` from our callback, and it calls other callbacks when it recognizes a gesture, such as a fling or long press. We need to register a listener for the callbacks from the `GestureDetector`, and this is where we put our logic that says what to do if the user has performed one of these common gestures. Unfortunately, this class does not tell us if a pinch gesture is taking place; for that, we need to use a new class, which we'll get to shortly.

There are a few ways to build the listener side. Your first option is to write a new class that implements the appropriate gesture listener interface: for example, the `GestureDetector.OnGestureListener` interface. There are several abstract methods that must be implemented for each of the possible callbacks.

Your second option is to pick one of the simple implementations of a listener and override the appropriate callback methods that you care about. For example, the `GestureDetector.SimpleOnGestureListener` class has implemented all of the abstract methods to do nothing and return `false`. All you have to do is extend that class and override the few methods you need to act on those few gestures you care about. The other methods have their default implementations. It's more future-proof to choose the second option even if you decide to override all of the callback methods, because if a future version of Android adds another abstract callback method to the interface, the simple implementation will provide a default callback method, so you're covered.

Android 2.2 introduced the `ScaleGestureDetector` class, and this is the one that figures out the pinch gesture for us. We're going to explore this, plus the corresponding listener class, to see how to use the pinch gesture to resize an image. In this example, we extend the simple implementation (`ScaleGestureDetector.SimpleOnScaleGestureListener`) for our listener. Listing 27–12 has the XML layout and the Java code for our `MainActivity`.

Listing 27–12. *Layout and Java Code for the Pinch Gesture Using* `ScaleGestureDetector`

```
<?xml version="1.0" encoding="utf-8"?>
<LinearLayout xmlns:android="http://schemas.android.com/apk/res/android"
    android:id="@+id/layout"  android:orientation="vertical"
    android:layout_width="fill_parent"
    android:layout_height="fill_parent" >

  <TextView  android:text=
        "Use the pinch gesture to change the image size"
    android:layout_width="fill_parent"
    android:layout_height="wrap_content" />

  <ImageView android:id="@+id/image"  android:src="@drawable/icon"
    android:layout_width="match_parent"
    android:layout_height="match_parent"
    android:scaleType="matrix" />

</LinearLayout>


// This file is MainActivity.java
public class MainActivity extends Activity {
    private static final String TAG = "ScaleDetector";
    private ImageView image;
    private ScaleGestureDetector mScaleDetector;
    private float mScaleFactor = 1f;
    private Matrix mMatrix = new Matrix();
    @Override
    public void onCreate(Bundle savedInstanceState) {
        super.onCreate(savedInstanceState);
        setContentView(R.layout.main);

        image = (ImageView)findViewById(R.id.image);
        mScaleDetector = new ScaleGestureDetector(this,
                new ScaleListener());
    }

    @Override
    public boolean onTouchEvent(MotionEvent ev) {
        Log.v(TAG, "in onTouchEvent");
        // Give all events to ScaleGestureDetector
        mScaleDetector.onTouchEvent(ev);

        return true;
    }

    private class ScaleListener extends
            ScaleGestureDetector.SimpleOnScaleGestureListener {
        @Override
        public boolean onScale(ScaleGestureDetector detector) {
            mScaleFactor *= detector.getScaleFactor();

            // Make sure we don't get too small or too big
            mScaleFactor = Math.max(0.1f, Math.min(mScaleFactor, 5.0f));

            Log.v(TAG, "in onScale, scale factor = " + mScaleFactor);
            mMatrix.setScale(mScaleFactor, mScaleFactor);
```

```
            image.setImageMatrix(mMatrix);
            image.invalidate();
            return true;
        }
    }
}
```

Our layout is straightforward. We have a simple TextView with our message to use the pinch gesture, and we have our ImageView with the standard Android icon. We're going to resize this icon image using a pinch gesture. Of course, feel free to substitute your own image file instead of the icon. Just copy your image file into a drawable folder, and be sure to change the android:src attribute in the layout file. Notice the android:scaleType attribute in the XML layout for our image. This tells Android that we'll be using a graphics matrix to do scaling operations on the image. Although a graphics matrix can also do movement of our image within the layout, we're only going to focus on scaling for now. Also notice that we set the ImageView size to as big as possible. As we scale the image, we don't want it clipped by the boundaries of the ImageView.

The code is also straightforward. Within onCreate(), we get a reference to our image and create our ScaleGestureDetector. Within our onTouchEvent() callback, all we do is pass every event object we get to the ScaleGestureDetector's onTouchEvent() method and return true so we keep getting new events. This allows the ScaleGestureDetector to see all events and decide when to notify us of gestures.

The ScaleListener is where the zooming happens. There are actually three callbacks within the listener class: onScaleBegin(), onScale(), and onScaleEnd(). We don't need to do anything special with the begin and end methods, so we didn't implement them here.

Within onScale(), the detector passed in can be used to find out lots of information about the scaling operation. The scale factor is a value that hovers around 1. That is, as the fingers pinch closer together, this value is slightly below 1; as the fingers move apart, this value is slightly larger than 1. Our mScaleFactor member starts at 1, so it gets progressively smaller or larger than 1 as the fingers move together or apart. If mScaleFactor equals 1, our image will be normal size. Otherwise, our image will be smaller or larger than normal as mScaleFactor moves below or above 1. We set some bounds on mScaleFactor with the elegant min/max function combination. This prevents our image from getting too small or too large. We then use mScaleFactor to scale the graphics matrix, and we apply the newly scaled matrix to our image. The invalidate() call forces a redraw of the image on the screen.

Prior to Android 2.2, to detect a pinch gesture, we had to deal with the event objects ourselves and figure out the gesture. Now we can concern ourselves with performing the appropriate application logic when the common gesture has been performed. To work with the OnGestureListener interface, you'd do something very similar to what we've done here with our ScaleListener, except that the callbacks will be for different common gestures.

Common gestures are one thing, but what if you want to have custom gestures with your application? For example, what if you wanted the user to be able to draw a checkmark on the screen and have your application perform some function? For that, we need custom gestures, which is where we turn next.

Custom Gestures

In the final section of this chapter, we'll cover the formal `Gesture` classes of Android. Formally, a gesture is a prerecorded touch screen motion that your application can expect from the user. If the user performs the same gesture as the prerecorded gesture when using your application, your application can invoke specific logic according to what that gesture means to your application. Gestures require an overlay that can detect a gesture by the user to pass it to the underlying activity. Using gestures can simplify a user interface by eliminating buttons or other controls in favor of finger swipes or drawing motions. They can also make for interesting game interfaces. In this section, we will explore how to record custom gestures and how to use them in your application. Note that the gesture-related classes we used earlier are not used in this example at all; this section explores a different set of gesture classes.

The Gestures Builder Application

Before we get into gesture code, let's play with the Gestures Builder application that comes with the Android SDK, which will help you understand what a gesture is. Gestures Builder creates and manages a gestures file that contains a library of gestures. Launch an emulator from Eclipse, unlock the emulator device, go to your applications, and choose Gestures Builder. Figure 27–3 shows the application icon.

Figure 27–3. *The Gestures Builder icon*

If you don't see Gestures Builder within your emulator, you'll have to create a new project in Eclipse. Gestures Builder is provided as a sample application under `samples` in the Android SDK directory. If you didn't download any samples using the Android SDK and AVD Manager, go do that now. You can create a new Android project in Eclipse using the Create project from existing sample option. Select the desired Android version as a Build Target to enable the Create project from existing sample drop-down menu, and then choose GestureBuilder from the drop-down menu. You can then deploy this application to your emulator.

The Gestures Builder application will open to a mostly blank screen. Click the Add gesture button. You will be prompted for a name; the name you give will be associated to the gesture you're about to record. This name will be used in your code to refer to the gesture and will serve as a sort of command name. When the user performs the gesture to your application, the name will be passed to your methods so your application can do what the user is expecting it to do. The name you give could be a noun like *spiral* or *checkmark*, or it could be like a command such as *fetch* or *stop*. For now, let's call our first gesture *checkmark*, so type **checkmark** for the Name. Now, draw a check mark in the big blank space underneath, either with your mouse if using the emulator, or with your finger if using a device. If you don't like your first attempt, simply redraw a new check mark; the old one will be erased as soon as you start drawing a new one. When you're happy with your check mark, click Done. You should see a screen like the one shown in Figure 27–4.

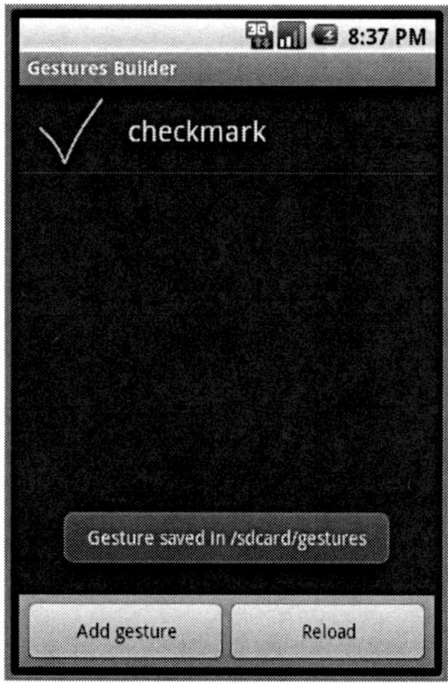

Figure 27–4. *Our check mark gesture saved to* `/sdcard`

Note that you could record different types of check marks and give them all the same name of *checkmark*. Record at least one more check mark-like gesture and name it *checkmark* too; it could be smaller or bigger or in some way different than your first check mark while still retaining the same basic shape. Add some different gestures with different names using the Add Gesture button. Each time you click Done, you add another gesture to your library. You might try to use a multitouch gesture, for example, drawing two fingers across the screen at the same time to make an equals sign. This doesn't work, and you get only one line. Maybe in the future, multitouch gestures—that is, gestures where two or more fingers are touching the screen at the same time—will be supported.

The Structure of a Gesture

Each gesture has a name and is made up of strokes. A *gesture stroke* is a touch sequence starting from when a finger touches down on the screen to when that finger lifts from the screen. As you learned earlier, a touch sequence is made up of `MotionEvent` objects. Similarly, a gesture stroke is made up of gesture points. Gestures get collected into a *gesture store*. A *gesture library* contains one gesture store. In Android, these are all classes that you can use in your code. See Figure 27–5 for a diagram that shows the classes' relationships.

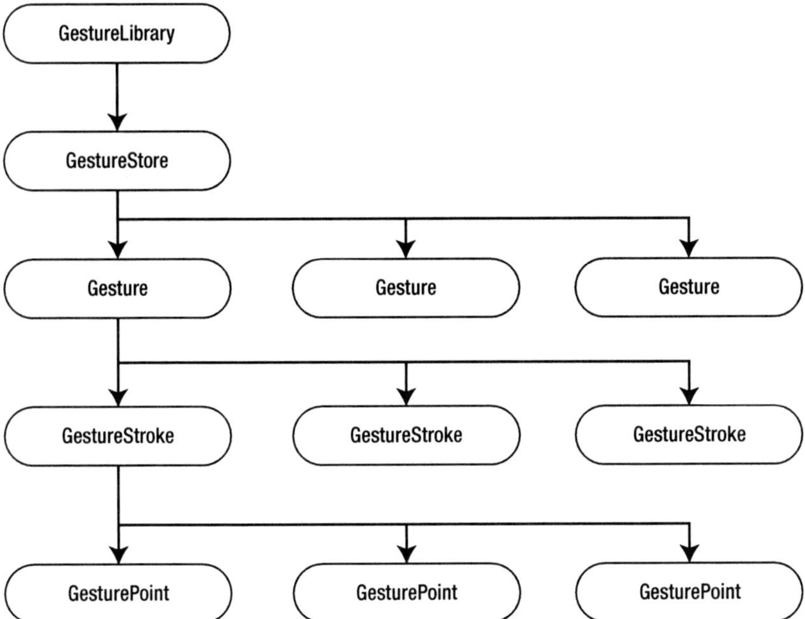

Figure 27–5. *The structure of gesture classes*

Although we can't use multitouch to create a custom gesture, we can have multiple gesture strokes in a single gesture. For example, to create a letter *E* gesture, you would need at least two gesture strokes: one gesture stroke could trace the top, back, and bottom sides of the *E*, and a second stroke could provide the center dash to complete the letter. You could also draw the back of the *E* with a vertical gesture stroke, followed by three separate horizontal gesture strokes to finish the letter. There are other ways you could draw an *E*, and fortunately, the gesture library allows you to give all of them the name *E* while recording different gestures. Go ahead and record *E* a few different ways, because your users might draw the letter in any one of those ways and you want your application to recognize an *E* however the user decides to draw it. Figure 27–6 shows different ways of recording an *E*.

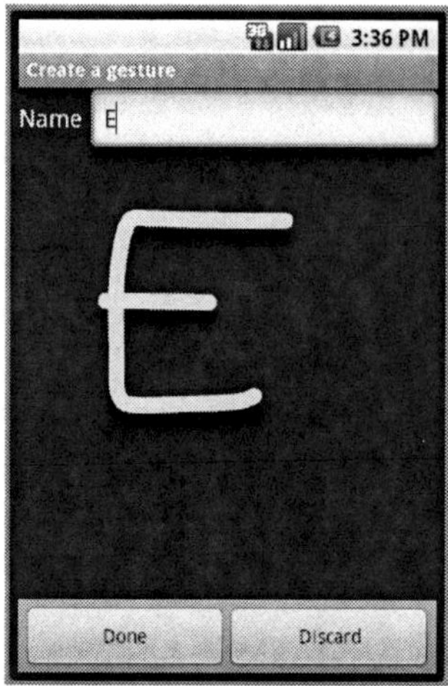

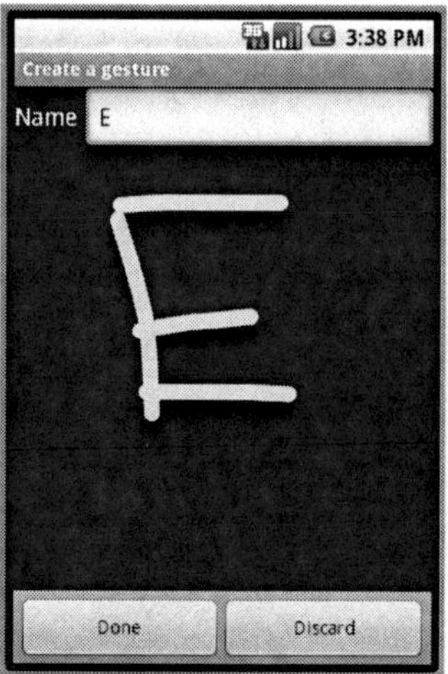

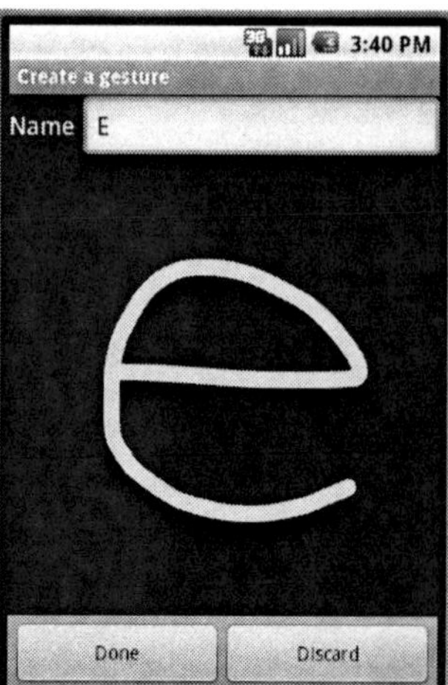

Figure 27–6. *Different ways to record an* E *gesture*

You may find it challenging to create a multistroke gesture in Gestures Builder in the emulator. As we noted earlier, you can simply redraw your gesture over the last one, and the preceding one will be erased. So how does Android know when you're starting over and when you're adding another gesture stroke to the current gesture? Android uses a value called the `FadeOffset`, which is a time value in milliseconds. If you wait longer than this time value to start the next gesture stroke, Android assumes you're starting over or starting a new gesture. By default, the time value is 420 milliseconds. This means that if you are drawing a gesture on the screen, and you lift your finger for longer than 420 milliseconds before drawing the next gesture stroke in your gesture, Android will assume you've already finished and will use what you've just drawn so far as the entirety of your gesture. On a real device, the default value might be long enough to start the next stroke of a gesture. On the emulator, though, it might not be. It depends on how fast your workstation is.

If you're having trouble getting Gestures Builder in the emulator to accept a multistroke gesture, you can create your own version of Gestures Builder and modify the default value of `FadeOffset`. We described earlier how to create a Gestures Builder project in Eclipse. Follow those instructions, and then go into the project's `/res/layout/create_gesture.xml` file to add the attribute `android:fadeOffset="1000"` to the `GestureOverlayView` element. This will extend `FadeOffset` to 1 second (1,000 milliseconds). You are free to choose a different value if you wish.

Let's investigate where these gestures went. The Toast message in Gestures Builder tells us the gestures are being saved to `/mnt/sdcard/gestures` (it could be a different directory depending on the device). Use File Explorer in Eclipse, or `adb`, to navigate to the `/mnt/sdcard` folder of the emulator. There, you will see a file called `gestures`. Notice that it is not very big. The `gestures` file is a binary file, so you will not be able to edit it by hand. To modify the contents, you will need to use the Gestures Builder application. When building your gesture-enabled application, you will need to copy the `gestures` file to your application's `/res/raw` directory. For this, you will need to use the File Copy feature of File Explorer, or use adb pull to get the gestures file onto your workstation so you can copy it into your project.

Besides adding new gestures in Gestures Builder, you can click and hold an existing gesture to bring up a menu. From the menu, you can change the gesture's name or delete it. You cannot record the gesture again, so if you don't like the gesture itself, you'll need to delete it and add it anew. As mentioned earlier, one thing you might want to do is record variations of gestures and give them the same name to account for user variation in inputting the gesture. The gesture name does not have to be unique, although gestures with the same name should be similar.

Example Application: Gesture Revealer

Now, we're going to create a sample application that uses our new gestures file. Using Eclipse, create a new Android Project. See Listing 27–13 for the XML of our layout file and for the code of our Activity class.

Listing 27–13. *Java Code for Our Gesture Revealer Application*

```
<?xml version="1.0" encoding="utf-8"?>
<!-- This file is /res/layout/main.xml -->
<LinearLayout xmlns:android="http://schemas.android.com/apk/res/android"
    android:orientation="vertical"
    android:layout_width="fill_parent"
    android:layout_height="fill_parent" >
  <TextView
    android:layout_width="fill_parent"
    android:layout_height="wrap_content"
    android:text="Draw gestures and I'll guess what they are" />

  <android.gesture.GestureOverlayView  android:id="@+id/gestureOverlay"
    android:layout_width="fill_parent"
    android:layout_height="fill_parent"
    android:gestureStrokeType="multiple"  android:fadeOffset="1000" />

</LinearLayout>


public class MainActivity extends Activity implements OnGesturePerformedListener {
    private static final String TAG = "Gesture Revealer";
    GestureLibrary gestureLib = null;

    @Override
    public void onCreate(Bundle savedInstanceState) {
        super.onCreate(savedInstanceState);
        setContentView(R.layout.main);

//        gestureLib = GestureLibraries.fromRawResource(this,
//                        R.raw.gestures);
        String filename =
            Environment.getExternalStorageDirectory().toString() +
                        "/gestures";
        gestureLib = GestureLibraries.fromFile(filename);
        if (!gestureLib.load()) {
```

```
            Toast.makeText(this, "Could not load " + filename,
                Toast.LENGTH_SHORT).show();
            finish();
        }

        // Let's take a look at the gesture library we have work with
        Log.v(TAG, "Library features:");
        Log.v(TAG, "  Orientation style: " +
                gestureLib.getOrientationStyle());
        Log.v(TAG, "  Sequence type: " + gestureLib.getSequenceType());
        for( String gestureName : gestureLib.getGestureEntries() ) {
            Log.v(TAG, "For gesture " + gestureName);
            int i = 1;
            for( Gesture gesture : gestureLib.getGestures(gestureName) )
            {
                Log.v(TAG, "    " + i + ": ID: " + gesture.getID());
                Log.v(TAG, "    " + i + ": Strokes count: " +
                        gesture.getStrokesCount());
                Log.v(TAG, "    " + i + ": Stroke length: " +
                        gesture.getLength());
                i++;
            }
        }

        GestureOverlayView gestureView =
            (GestureOverlayView) findViewById(R.id.gestureOverlay);
        gestureView.addOnGesturePerformedListener(this);
    }

    @Override
    public void onGesturePerformed(GestureOverlayView view,
            Gesture gesture)
    {
        ArrayList<Prediction> predictions =
                gestureLib.recognize(gesture);

        if (predictions.size() > 0) {
            Prediction prediction = (Prediction) predictions.get(0);
            if (prediction.score > 1.0) {
                Toast.makeText(this, prediction.name,
                        Toast.LENGTH_SHORT).show();
                for(int i=0;i<predictions.size();i++)
                    Log.v(TAG, "prediction " + predictions.get(i).name +
                            " - score = " + predictions.get(i).score);
            }
        }
    }
}
```

In this example, we're going to simply access the exact same file that the Gestures Builder application wrote to. In our `onCreate()` method, we use the `GestureLibraries.fromFile()` method to do this. But we also show in the comments how you would access a gestures file that is part of your application. If you were to use the `fromRawResource()` method, you'd use an argument like our regular resource IDs, and you'd put the gestures file into the `/res/raw` directory.

Our application doesn't do a whole lot, but running it will give you a better understanding of what is going on inside Android as it processes gestures. At startup, our application loads the gestures file and logs what it finds. It also logs the results of trying to match a sample gesture drawn into the input screen of our application. Go ahead and run the Gesture Revealer application, assuming, of course, that you've run Gestures Builder already and have some gestures in the `gestures` file. See how each gesture is logged with the ID, the number of strokes, and the length.

Use some gestures on the screen that you know exist in your gesture library. Then use some that you know do not exist. Watch the LogCat records to see what's happening. You may notice that sometimes what you draw is not recognized when you think it should be or that what Android recognized was not what you had in mind, but most of the time it correctly recognizes what you drew. You may also have noticed that when Android recognizes your input gesture, you get scores for all gestures in your library in predictions, but when Android doesn't recognize your input gesture, you don't get anything at all.

Also note what happens if you have a multistroke gesture, such as the letter *E*, and you take too long between strokes. The application will take what you've drawn so far and use that to compare to your gesture library, which is likely to result in an incorrect match or no match at all. This time delay is controlled by `FadeOffset`. Here is where it gets tricky. We want Android to begin matching gestures as soon as we're finished making our gesture, but we have no way to know if the user is finished unless we wait for some period of time and don't see the start of a new gesture stroke. Therefore, `FadeOffset` serves two purposes: one is to control how long to wait for a new gesture stroke as part of the current gesture, and the other is to control how long to wait to begin matching our gesture against the known gestures in our gesture library. Making `FadeOffset` very large means having to wait a long time before the matching process begins. Making `FadeOffset` too small means not being able to use a multistroke gesture because Android will think the gesture is finished before the user gets to the next gesture stroke. Whether 420 milliseconds is the right value to use is up to you. You might want to use a Preference value so users can adjust it for themselves.

While we're on the topic of multistroke gestures, note that the `GestureOverlayView` has a setting that controls whether or not multistroke gestures are expected. The attribute in XML is `android:gestureStrokeType`, and its value is either `single` (the default) or `multiple`. If you want to be able to draw multistroke gestures, this attribute must be set. You can also set it programmatically using `setGestureStrokeType(int type)`, using an argument of either `GestureOverlayView.GESTURE_STROKE_TYPE_SINGLE` or `GestureOverlayView.GESTURE_STROKE_TYPE_MULTIPLE`. `GestureOverlayView` also has XML attributes and methods for setting colors and line thicknesses.

To create your own gesture-aware application, you will need to decide what gestures your application will act on, create a library of those gestures, and implement the `onGesturePerformedListener` interface, probably in your `Activity`, to recognize the gestures and take appropriate action.

What if you want your users to be able to record their own gestures? For example, what about using a different gesture for an action in your application than the one that you provide? This is possible, but it means that you need to have a gesture library file that can be written to, and the logical place to put this is the SD card. It's fairly simple to create a new gesture library file, read out the default gestures from the gesture library file that comes with your application, and overwrite gestures that the user wants to replace. You can use the implementation of the Gestures Builder application as mentioned previously to see how to create a gesture recorder. Or maybe someone will write a Gestures Builder application that responds to intents, so you could simply invoke that activity to add a new gesture. Alternatively, you could record just the user's gestures into a new writable gesture library file and load two gesture libraries into your application, the user's and your original. Within the `onGesturePerformed()` method, you could first try `recognize()` on the user's library and then on your own. You could compare the top scores from any predictions from each library to decide which action to take.

Finally, you can change the orientation or sequence style of a `GestureLibrary` using the `setOrientationStyle()` and `setSequenceType()` methods. The arguments come from `GestureStore` constants. The critical thing to remember is to use these methods on your gesture library object before you call `load()` on the library. *Orientation invariance* means your application could better recognize rotated gestures. *Sequence invariance* means your application could better recognize gestures made up of gesture strokes in a different order than how the gesture was recorded.

References

Here are some helpful references to topics you may wish to explore further.

- `www.androidbook.com/proandroid4/projects`: Downloadable projects related to this book. For this chapter, look for a zip file called `ProAndroid4_Ch27_Touchscreens.zip`. This zip file contains all projects from this chapter, listed in separate root directories. There is also a `README.TXT` file that describes exactly how to import projects into Eclipse from one of these zip files.
- `www.ted.com/talks/jeff_han_demos_his_breakthrough_touchscreen.html`: Jeff Han demonstrates his multitouch computer user interface at TED in 2006—very cool.
- `http://android-developers.blogspot.com/2010/06/making-sense-of-multitouch.html`: An Android blog post about multitouch offers yet another way to implement a `GestureDetector` inside an extension of a view.

Summary

Let's conclude this chapter by quickly enumerating what you have learned about touch screens so far:

- MotionEvent as the foundation on which touch handling is done

- Different callbacks that handle touch events on a `View` object and through an `OnTouchListener`
- Different types of events that occur during a touch sequence
- How touch events travel through an entire view hierarchy, unless handled along the way
- Information that a `MotionEvent` object contains about touches, including for multiple fingers
- When to recycle a `MotionEvent` object and when not to
- Determining the speed at which a finger drags across a screen
- The wonderful world of multitouch, and the internal details of how it works
- Using the `onTap()` method with maps
- Implementing the pinch gesture, as well as other common gestures
- Recording and using custom gestures in the user interface
- Using internal details to tailor the user experience

Interview Questions

You can use the following questions as a guide to consolidate your understanding of this chapter:

1. What are some differences between `onTouchEvent()` and `onTouch()`?
2. What manifest tag should you use to indicate to Android Market that your application requires a touch screen?
3. If you want a view to continue to receive touch events in the current touch sequence, do you return `true` or `false` from the callback?
4. What is the first action you'll receive in a touch sequence?
5. What is the last action you'll receive?
6. True or false: a `MotionEvent` object is created for every single touch event.
7. What precautions should you take when using `getSize()` and `getPressure()` with `MotionEvent` objects?
8. When is it appropriate to recycle a `MotionEvent` object?
9. What happens if you ask a `VelocityTracker` to tell you the speed of the finger if you just added the `MotionEvent` object representing the last finger up event?

10. What method is used to determine how many fingers are being tracked by Android?
11. What is more likely to be the limiting factor in the number of fingers that can participate in a multitouch gesture: the hardware or the Android API?
12. What is the difference between a pointer index and a pointer ID?
13. If the raw action value is decimal 518, what does this mean?
14. What are the arguments to the `onTap()` method of `Overlay`?
15. What are the common gestures that a `GestureDetector` can detect?
16. What class specifically detects the pinch gesture? What callbacks does this class call during a pinch gesture?
17. Can you draw the organization of a `GestureLibrary`? What is the leaf node?
18. What property controls the maximum allowed time between strokes of the same gesture?

Chapter 28

Implementing Drag and Drop

In the last chapter, we covered touchscreens, the `MotionEvent` class, and gestures. You learned how to use touch to make things happen in your application. One area that we didn't cover was drag and drop. On the surface, drag and drop seems like it should be fairly simple: touch an object on the screen, drag it across the screen (usually over some other object), and let go, and the application should take the appropriate action. In many computer operating systems, this is a common way to delete a file from the desktop; you just drag the file's icon to the trash-bin icon, and the file gets deleted. In Android, you may have seen how to rearrange icons on the home screen by dragging them to new locations or to the trash.

This chapter is going to go in depth into drag and drop. We'll cover the drag-and-drop capabilities that were introduced with Android 3.0 (Honeycomb), and we'll give you a sample program. Prior to Android 3.0, developers were on their own when it came to drag and drop. But because there are still quite a few phones out there running Android 2.1 and 2.2, we'll show you how to do drag and drop on them as well. We'll show you the old way in the first section of this chapter, and then we'll show you the new way in the second part.

Exploring Drag and Drop

In this next example application, we're going to take a white dot and drag it to a new location in our user interface. We're also going to place three counters in our user interface, and if the user drags the white dot to one of the counters, that counter will increment and the dot will return back to its starting place. If the dot is dragged somewhere else on the screen, we'll just leave it there.

NOTE: See the "References" section at the end of this chapter for the URL from which you can import these projects into Eclipse directly. We'll only show code in the text to explain concepts. You'll need to download the code to create a working example application.

The first sample application for this chapter is called TouchDragDemo. There are two key files we want to talk about in this section:

- `/res/layout/main.xml`
- `/src/com/androidbook/touch/dragdemo/Dot.java`

The `main.xml` file contains our layout for the drag-and-drop demo. It is shown in Listing 28–1. Some of the key concepts we want you to notice are the use of a `FrameLayout` as the top-level layout, inside of which is a `LinearLayout` containing `TextView`s and a custom `View` class called `Dot`. Because the `LinearLayout` and `Dot` coexist within the `FrameLayout`, their positions and sizes don't really impact each other, but they will be sharing the screen real estate, one on top of the other. The UI for this application is shown in Figure 28–1.

Listing 28–1. *Example Layout XML for Our Drag Example*

```
<?xml version="1.0" encoding="utf-8"?>
<!-- This file is res/layout/main.xml -->
<FrameLayout xmlns:android="http://schemas.android.com/apk/res/android"
    android:layout_width="fill_parent"
    android:layout_height="fill_parent"
    android:background="#0000ff" >

  <LinearLayout android:id="@+id/counters"
    android:orientation="vertical"
    android:layout_width="fill_parent"
    android:layout_height="fill_parent" >

    <TextView android:id="@+id/top"
      android:text="0"
      android:background="#111111"
      android:layout_height="wrap_content"
      android:layout_width="60dp"
      android:layout_gravity="right"
      android:layout_marginTop="30dp"
      android:layout_marginBottom="30dp"
      android:padding="10dp" />

    <TextView android:id="@+id/middle"
      android:text="0"
      android:background="#111111"
      android:layout_height="wrap_content"
      android:layout_width="60dp"
      android:layout_gravity="right"
      android:layout_marginBottom="30dp"
      android:padding="10dp" />

    <TextView android:id="@+id/bottom"
```

```
        android:text="0"
        android:background="#111111"
        android:layout_height="wrap_content"
        android:layout_width="60dp"
        android:layout_gravity="right"
        android:padding="10dp" />
  </LinearLayout>

  <com.androidbook.touch.dragdemo.Dot
    android:id="@+id/dot"
    android:layout_width="fill_parent"
    android:layout_height="fill_parent" />

</FrameLayout>
```

Figure 28–1. *User Interface for TouchDragDemo*

Note that the package name in the layout XML file for the Dot element must match the package name you use for your application. As mentioned, the layout of Dot is separated from the LinearLayout. This is because we want the freedom to move the dot around the screen, which is why we chose the layout_width and layout_height of "fill_parent". When we draw the dot on the screen, we want it to be visible, and if we constrict the size of our dot's view to the diameter of the dot, we won't be able to see it when we drag it away from our starting place.

> **NOTE:** Technically, we could set android:clipChildren to true in the FrameLayout tag and set the layout width and height of the dot to wrap_content, but that doesn't feel as clean.

For each of the counters, we simply lay them out with a background, padding, margins and gravity to get them to show up along the right-hand side of the screen. We start them off at zero, but as you'll soon see, we'll be incrementing those values as dots are dragged over to them. Although we chose to use `TextViews` in this example, you could use just about any `View` object as a drop target. Now to look at the Java code for our `Dot` class, in Listing 28–2.

Listing 28–2. *Java Code for Our `Dot` Class*

```
public class Dot extends View {
    private static final String TAG = "TouchDrag";
    private float left = 0;
    private float top = 0;
    private float radius = 20;
    private float offsetX;
    private float offsetY;
    private Paint myPaint;
    private Context myContext;

    public Dot(Context context, AttributeSet attrs) {
        super(context, attrs);

        // Save the context (the activity)
        myContext = context;

        myPaint = new Paint();
        myPaint.setColor(Color.WHITE);
        myPaint.setAntiAlias(true);
    }

    public boolean onTouchEvent(MotionEvent event) {
        int action = event.getAction();
        float eventX = event.getX();
        float eventY = event.getY();
        switch(action) {
        case MotionEvent.ACTION_DOWN:
            // First make sure the touch is on our dot,
            // since the size of the dot's view is
            // technically the whole layout. If the
            // touch is *not* within, then return false
            // indicating we don't want any more events.
            if( !(left-20 < eventX && eventX < left+radius*2+20 &&
                top-20 < eventY && eventY < top+radius*2+20))
                return false;

            // Remember the offset of the touch as compared
            // to our left and top edges.
            offsetX = eventX - left;
            offsetY = eventY - top;
            break;
        case MotionEvent.ACTION_MOVE:
        case MotionEvent.ACTION_UP:
        case MotionEvent.ACTION_CANCEL:
            left = eventX - offsetX;
            top = eventY - offsetY;
            if(action == MotionEvent.ACTION_UP) {
                checkDrop(eventX, eventY);
            }
            break;
```

```
        }
        invalidate();
        return true;
    }

    private void checkDrop(float x, float y) {
        // See if the x,y of our drop location is near to
        // one of our counters. If so, increment it, and
        // reset the dot back to its starting position
        Log.v(TAG, "checking drop target for " + x + ", " + y);

        int viewCount = ((MainActivity)myContext).counterLayout
                         .getChildCount();

        for(int i = 0; i<viewCount; i++) {
            View view = ((MainActivity)myContext).counterLayout
                         .getChildAt(i);
            if(view.getClass() == TextView.class){
                Log.v(TAG, "Is the drop to the right of " +
                            (view.getLeft()-20));
                Log.v(TAG, "  and vertically between " +
                          (view.getTop()-20) +
                          " and " + (view.getBottom()+20) + "?");
                if(x > view.getLeft()-20 &&
                        view.getTop()-20 < y &&
                        y < view.getBottom()+20) {
                    Log.v(TAG, "    Yes. Yes it is.");

                    // Increase the count value in the TextView by one
                    int count =
                        Integer.parseInt(
                            ((TextView)view).getText().toString());
                    ((TextView)view).setText(String.valueOf( ++count ));

                    // Reset the dot back to starting position
                    left = top = 0;
                    break;
                }
            }
        }
    }

    public void draw(Canvas canvas) {
        canvas.drawCircle(left + radius, top + radius, radius, myPaint);
    }
}
```

When you run this application, you will see a white dot on a blue background. You can touch the dot and drag it around the screen. When you lift your finger, the dot stays where it is until you touch it again and drag it somewhere else. The `draw()` method puts the dot at its current location of left and top, adjusted by the dot's radius. By receiving `MotionEvent` objects in the `onTouchEvent()` method, we can modify the left and top values by the movement of our touch.

Because the user won't always touch the exact center of the object, the touch coordinates will not be the same as the location coordinates of the object. That is the purpose of the offset values: to get us back to the left and top edges of our dot from the position of the touch. But even before we start a drag operation, we want to be sure that the user's touch is considered close enough to the dot to be valid. If the user touches the screen far away from the dot, which is technically within the view layout of the dot, we don't want that to start a drag sequence. That is why we look to see if the touch is within the white dot itself; if it is not, we simply return `false`, which prevents receiving any more touch events in that touch sequence.

When your finger starts moving across the screen, we adjust the location of the object by the deltas in x and y based on the `MotionEvents` that we get. When you stop moving (`ACTION_UP`), we finalize our location using the last coordinates of your touch. We don't have to worry about scrollbars in this example, which could complicate the calculation of the object's position of our object on the screen. But the basic principle is still the same. By knowing the starting location of the object to be moved and keeping track of the delta values of a touch from `ACTION_DOWN` through to `ACTION_UP`, we can adjust the location of the object on the screen.

Dropping an object onto another object on the screen has much less to do with touch than it does with knowing where things are on the screen. As we drag an object around the screen, we are aware of its position relative to one or more reference points. We can also interrogate objects on the screen for their locations and sizes. We can then determine if our dragged object is "over" another object. The typical process of figuring out a drop target for a dragged object is to iterate through the available objects that can be dropped on and determine if our current position overlaps with that object. Each object's size and position (and sometimes shape) can be used to make this determination. If we get an `ACTION_UP` event, meaning that the user has let go of our dragged object, and the object is over something we can drop onto, we can fire the logic to process the drop action.

We used this approach in our sample application. When the `ACTION_UP` action is detected, we then look through the child views of the `LinearLayout`, and for each `TextView` that is found, we compare the location of the touch to the edges of the `TextView` (plus a little bit extra). If the touch is within that `TextView`, we grab the current numeric value of the `TextView`, increment it by one, and write it back. If this happens, the position of the dot is reset back to its starting place (left = 0, top = 0) for the next drag.

Our example shows you the basics of a way to do drag and drop in Android prior to 3.0. With this you could implement drag-and-drop features in your application. This might be the action of dragging something to the trash can, where the object being dragged should be deleted, or it could be dragging a file to a folder for the purposes of moving or copying it. To embellish your application, you could pre-identify which views are potential drop targets and cause them to visually change as a drag starts. If you wanted the dragged object to disappear from the screen when it is dropped, you could always programmatically remove it from the layout (see the various `removeView` methods in `ViewGroup`).

Now that you've seen the hard way to do drag and drop, we'd like to show you the drag-and-drop support that was added in Android 3.0.

Basics of Drag and Drop in 3.0+

Prior to Android 3.0, there was no direct support for drag and drop. You learned in the first section of this chapter how to drag a `View` around the screen; you also learned that it was possible to use the current location of the dragged object to determine if there was a drop target underneath. When the `MotionEvent` for the finger-up event was received, your code could figure out if that meant a drop had occurred. Although this was doable, it certainly wasn't as easy as having direct support in Android for the drag-and-drop operation. You now have that direct support.

At its most basic, the drag-and-drop operation starts with a view declaring that a drag has started; then all interested parties watch the drag take place until the drop event is fired. If a view catches the drop event and wants to receive it, then a drag and drop has just occurred. If there is no view to receive the drop, or if the view that receives it doesn't want it, then no drop takes place. Dragging is communicated through the use of a `DragEvent` object, which is passed to all of the drag listeners available.

Within the `DragEvent` object are descriptors for lots of information, depending on the initiator of the drag sequence. For example, the `DragEvent` can contain object references to the initiator itself, state information, textual data, URIs, or pretty much whatever you want to pass through the drag sequence.

Information could be passed that results in view-to-view dynamic communication; however, the originator data in a `DragEvent` object is set when the `DragEvent` is created, and it stays the same thereafter. In addition to this data, the `DragEvent` has an action value indicating what is going on with the drag sequence, and location information indicating where the drag is on the screen.

A `DragEvent` has six possible actions:

- `ACTION_DRAG_STARTED` indicates that a new drag sequence has begun.
- `ACTION_DRAG_ENTERED` indicates that the dragged object has been dragged into the boundaries of a specific view.
- `ACTION_DRAG_LOCATION` indicates that the dragged object has been dragged on the screen to a new location.
- `ACTION_DRAG_EXITED` indicates that the dragged object has been dragged outside the boundaries of a specific view.
- `ACTION_DROP` indicates that the user has let go of the dragged object. It is up to the receiver of this event to determine whether this truly means a drop has occurred.
- `ACTION_DRAG_ENDED` tells all drag listeners that the previous drag sequence has ended. The `DragEvent.getResult()` method indicates a successful drop or failure.

You might think that you need to set up a drag listener on each view in the system that could participate in a drag sequence; but, in fact, you can define a drag listener on just about anything in your application, and it will receive all of the drag events for all views in the system. This can make things a little confusing because the drag listener does not need to be associated with either the object being dragged or the drop target. The listener can manage all of the coordination of the drag and drop.

In fact, if you inspect the drag-and-drop example project that comes with the Android SDK, you will see that it sets up a listener on a `TextView` that has nothing to do with the actual dragging and dropping. The upcoming example project uses drag listeners that are tied to specific views. These drag listeners each receive a `DragEvent` object for the drag events that occur in the drag sequence. This means a view could receive a `DragEvent` object that can be ignored because it is really about a different view. This also means the drag listener must make that determination in code and that there must be enough information within the `DragEvent` object for the drag listener to figure out what to do.

If a drag listener got a `DragEvent` object that merely said there's an unknown object being dragged and it's at coordinates (15, 57), there isn't much the drag listener can do with it. It is much more helpful to get a `DragEvent` object that says a particular object is being dragged, it's at coordinates (15, 57), it's a copy operation, and the data is a specific URI. When that drops, there's enough information to be able to initiate a copy operation.

We're actually seeing two different kinds of dragging going on. In our first example application, we dragged a view across a frame layout, and we could let go and that view would stay where it was. We only got drag-and-drop behavior when we dropped our view on top of something else. The supported form of drag and drop works differently than this. Now, when you drag a view as part of a drag-and-drop sequence, the dragged view doesn't move at all. We get a shadow image of the dragged view which does travel across the screen, but if we let go of it, that shadow view goes away. What this means is that you might still have occasion to use the technique from the beginning of this chapter in an Android 3.0+ application, to move images around on the screen perhaps, without necessarily doing drag and drop.

Drag-and-Drop Example Application

For your next example application, you're going to employ a staple of 3.0, the fragments. This, among other things, will prove that drags can cross fragment boundaries. You'll create a palette of dots on the left and a square target on the right. When a dot is grabbed using a long click, you'll change the color of that dot in the palette and Android will show a shadow of the dot as you drag. When the dragged dot reaches the square target, the target will begin to glow. If you drop the dot on the square target, a message will indicate that you've just added one more drop to the drop count, the glowing will stop, and the original dot will go back to its original color.

List of Files

This application builds upon concepts we've covered throughout this book. We're only going to include the interesting files in the text. For the others, just look at them in your IDE at your leisure. Here are the ones that we've included in the text:

- `palette.xml` is the fragment layout for the dots on the left side (see Listing 28–3).
- `dropzone.xml` is the fragment layout for the square target on the right side, plus the drop-count message (see Listing 28–4).
- `DropZone.java` inflates the `dropzone.xml` fragment layout file and then implements the drag listener for the drop target (see Listing 28–5).
- `Dot.java` is your custom view class for the objects you're going to drag. It handles beginning the drag sequence, watching drag events, and drawing the dots (see Listing 28–6).

Laying Out the Example Drag-and-drop Application

Before we get into the code, Figure 28–2 shows what the application will look like.

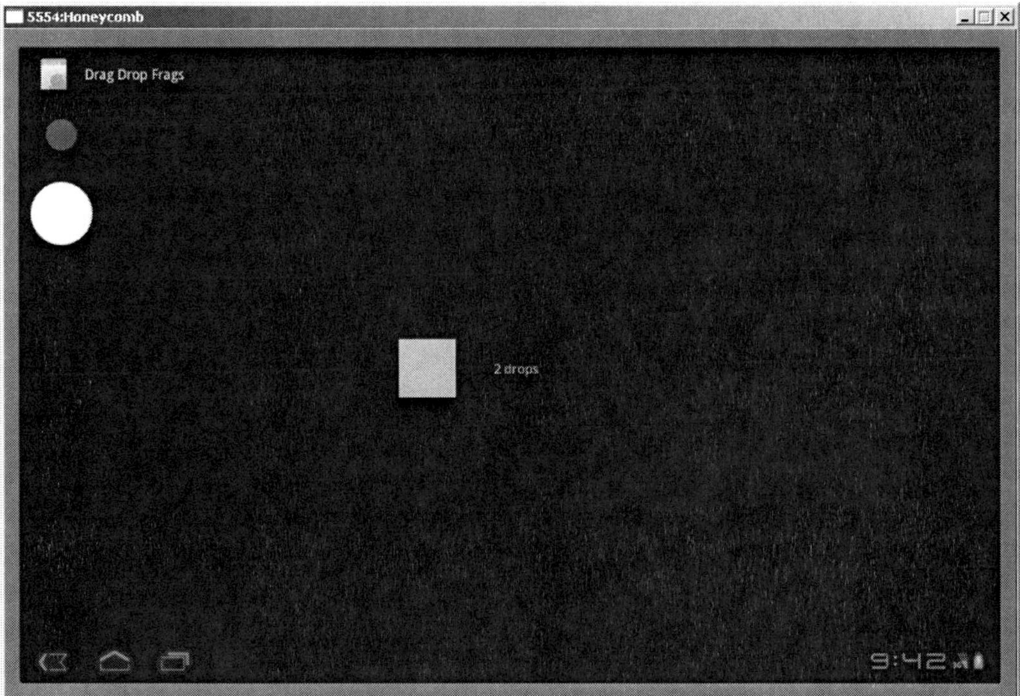

Figure 28–2. *Drag Drop Frags example application user interface*

The main layout file has a simple horizontal linear layout and two fragment specifications. The first fragment will be for the palette of dots and the second will be for the dropzone.

The palette fragment layout file (Listing 28–3) gets a bit more interesting. Although this layout represents a fragment, you don't need to include a fragment tag within this layout. This layout will be inflated to become the view hierarchy for your palette fragment. The dots are specified as custom dots, and there are two of them arranged vertically. Notice that there are a couple of custom XML attributes in the definition of your dots (`dot:color` and `dot:radius`). As you can see, these attributes specify the color and the radius of your dots. You might also have noticed that the layout width and height are `wrap_content`, not `fill_parent` as in the earlier example application in this chapter. The new drag-and-drop support makes things much easier.

Listing 28–3. ***The `palette.xml` Layout File for the `Dots`***

```
<?xml version="1.0" encoding="utf-8"?>
<!-- This file is res/layout/palette.xml -->
<LinearLayout
  xmlns:android="http://schemas.android.com/apk/res/android"
  xmlns:dot=
    "http://schemas.android.com/apk/res/com.androidbook.drag.drop.demo"
  android:layout_width="match_parent"
  android:layout_height="match_parent"
  android:orientation="vertical">

  <com.androidbook.drag.drop.demo.Dot android:id="@+id/dot1"
    android:layout_width="wrap_content"
    android:layout_height="wrap_content"
    android:padding="30dp"
    android:tag="Blue dot"
    dot:color="#ff1111ff"
    dot:radius="20dp"  />

  <com.androidbook.drag.drop.demo.Dot android:id="@+id/dot2"
    android:layout_width="wrap_content"
    android:layout_height="wrap_content"
    android:padding="10dp"
    android:tag="White dot"
    dot:color="#ffffffff"
    dot:radius="40dp"  />

</LinearLayout>
```

The dropzone fragment layout file in Listing 28–4 is also easy to understand. There's a green square and a text message arranged horizontally. This will be the drop zone for the dots you'll be dragging. The text message will be used to display a running count of the drops.

Listing 28–4. ***The `dropzone.xml` Layout File***

```
<?xml version="1.0" encoding="utf-8"?>
<!-- This file is res/layout/dropzone.xml -->
<LinearLayout
  xmlns:android="http://schemas.android.com/apk/res/android"
  android:layout_width="match_parent"
```

```
  android:layout_height="match_parent"
  android:orientation="horizontal" >

  <View android:id="@+id/droptarget"
    android:layout_width="75dp"
    android:layout_height="75dp"
    android:layout_gravity="center_vertical"
    android:background="#00ff00" />

  <TextView android:id="@+id/dropmessage"
    android:text="0 drops"
    android:layout_width="wrap_content"
    android:layout_height="wrap_content"
    android:layout_gravity="center_vertical"
    android:paddingLeft="50dp"
    android:textSize="17sp" />

</LinearLayout>
```

Responding to onDrag in the Dropzone

Now that you have the main application layout set, let's see how the drop target needs to be organized by examining Listing 28–5.

Listing 28–5. *The* `DropZone.java` *File*

```
public class DropZone extends Fragment {

    private View dropTarget;
    private TextView dropMessage;

    @Override
    public View onCreateView(LayoutInflater inflater,
            ViewGroup container, Bundle icicle)
    {
        View v = inflater.inflate(R.layout.dropzone, container, false);

        dropMessage = (TextView)v.findViewById(R.id.dropmessage);

        dropTarget = (View)v.findViewById(R.id.droptarget);
        dropTarget.setOnDragListener(new View.OnDragListener() {
            private static final String DROPTAG = "DropTarget";
            private int dropCount = 0;
            private ObjectAnimator anim;

            public boolean onDrag(View v, DragEvent event) {
                int action = event.getAction();
                boolean result = true;
                switch(action) {
                case DragEvent.ACTION_DRAG_STARTED:
                    Log.v(DROPTAG, "drag started in dropTarget");
                    break;
                case DragEvent.ACTION_DRAG_ENTERED:
                    Log.v(DROPTAG, "drag entered dropTarget");
                    anim = ObjectAnimator.ofFloat(
                                (Object)v, "alpha", 1f, 0.5f);
                    anim.setInterpolator(new CycleInterpolator(40));
```

```
                    anim.setDuration(30*1000); // 30 seconds
                    anim.start();
                    break;
                case DragEvent.ACTION_DRAG_EXITED:
                    Log.v(DROPTAG, "drag exited dropTarget");
                    if(anim != null) {
                        anim.end();
                        anim = null;
                    }
                    break;
                case DragEvent.ACTION_DRAG_LOCATION:
                    Log.v(DROPTAG, "drag proceeding in dropTarget: " +
                            event.getX() + ", " + event.getY());
                    break;
                case DragEvent.ACTION_DROP:
                    Log.v(DROPTAG, "drag drop in dropTarget");
                    if(anim != null) {
                        anim.end();
                        anim = null;
                    }

                    ClipData data = event.getClipData();
                    Log.v(DROPTAG, "Item data is " +
                           data.getItemAt(0).getText());

                    dropCount++;
                    String message = dropCount + " drop";
                    if(dropCount > 1)
                        message += "s";
                    dropMessage.setText(message);
                    break;
                case DragEvent.ACTION_DRAG_ENDED:
                    Log.v(DROPTAG, "drag ended in dropTarget");
                    if(anim != null) {
                        anim.end();
                        anim = null;
                    }
                    break;
                default:
                    Log.v(DROPTAG, "other action in dropzone: " +
                                   action);
                    result = false;
                }
                return result;
            }
        });
        return v;
    }
}
```

Now you're starting to get into interesting code. For the dropzone, you need to create the target upon which you want to drag the dots. As you saw earlier, the layout specifies a green square on the screen with a text message next to it. Because the dropzone is also a fragment, you're overriding the `onCreateView()` method of `DropZone`. The first thing to do is inflate the dropzone layout and then extract out the view reference for the square target (dropTarget) and for the text message (dropMessage). Then you need to set up a drag listener on the target so it will know when a drag is underway.

The drop-target drag listener has a single callback method in it: onDrag(). This callback will receive a view reference as well as a DragEvent object. The view reference relates to the view that the DragEvent is related to. As mentioned, the drag listener is not necessarily connected to the view that will be interacting with the drag event, so this callback must identify the view for which the drag event is taking place.

One of the first things you likely want to do in any onDrag() callback is read the action from the DragEvent object. This will tell you what's going on. For the most part, the only thing you want to do in this callback is log the fact that a drag event is taking place. You don't need to actually do anything for ACTION_DRAG_LOCATION, for example. But you do want to have some special logic for when the object is dragged within your boundaries (ACTION_DRAG_ENTERED) that will be turned off when the object is either dragged outside of your boundaries (ACTION_DRAG_EXITED) or when the object is dropped (ACTION_DROP).

You're using the ObjectAnimator class that was introduced in Chapter 8, only here you're using it in code to specify a cyclic interpolator that modifies the target's alpha. This will have the effect of pulsing the transparency of the green target square, which will be the visual indication that the target is willing to accept a drop of the object onto it. Because you turn on the animation, you must make sure to also turn it off when the object leaves or is dropped, or the drag and drop is ended. In theory, you shouldn't need to stop the animation on ACTION_DRAG_ENDED, but it's wise to do it anyway.

For this particular drag listener, you're only going to get ACTION_DRAG_ENTERED and ACTION_DRAG_EXITED if the dragged object interacts with the view with which you're associated. And as you'll see, the ACTION_DRAG_LOCATION events only happen if the dragged object is inside your target view.

The only other interesting condition is the ACTION_DROP itself (notice that DRAG_ is not part of the name of this action). If a drop has occurred on your view, it means the user has let go of the dot over the green square. Because you're expecting this object to be dropped on the green square, you can just go ahead and read the data from the first item and then log it to LogCat. In a production application, you might pay closer attention to the ClipData object that is contained in the drag event itself. By inspecting its properties, you could decide if you even want to accept the drop or not.

This is a good time to point out the result boolean in this onDrag() callback method. Depending on how things go, you want to let Android know either that you took care of the drag event (by returning true) or that you didn't (by returning false). If you don't see what you want to see inside of the drag event object, you could certainly return false from this callback, which would tell Android that this drop was not handled.

Once you log the information from the drag event in LogCat, you increment the count of the drops received; this is updated in the user interface, and that's about it for DropZone.

If you look this class over, it's really rather simple. You don't actually have any code in here that deals with MotionEvents, nor do you even need to make your own determination of whether there is a drag going on. You just get appropriate callback calls as a drag sequence unfolds.

Setting Up the Drag Source Views

Let's now consider how views corresponding to a drag source are organized, starting by looking at Listing 28–6.

Listing 28–6. *The Java for the Custom View:* `Dot`

```
public class Dot extends View
    implements View.OnDragListener
{
    private static final int DEFAULT_RADIUS = 20;
    private static final int DEFAULT_COLOR = Color.WHITE;
    private static final int SELECTED_COLOR = Color.MAGENTA;
    protected static final String DOTTAG = "DragDot";
    private Paint mNormalPaint;
    private Paint mDraggingPaint;
    private int mColor = DEFAULT_COLOR;
    private int mRadius = DEFAULT_RADIUS;
    private boolean inDrag;

    public Dot(Context context, AttributeSet attrs) {
        super(context, attrs);

        // Apply attribute settings from the layout file.
        // Note: these could change on a reconfiguration
        // such as a screen rotation.
        TypedArray myAttrs = context.obtainStyledAttributes(attrs,
                R.styleable.Dot);

        final int numAttrs = myAttrs.getIndexCount();
        for (int i = 0; i < numAttrs; i++) {
            int attr = myAttrs.getIndex(i);
            switch (attr) {
            case R.styleable.Dot_radius:
                mRadius = myAttrs.getDimensionPixelSize(attr,
                          DEFAULT_RADIUS);
                break;
            case R.styleable.Dot_color:
                mColor = myAttrs.getColor(attr, DEFAULT_COLOR);
                break;
            }
        }
        myAttrs.recycle();

        // Setup paint colors
        mNormalPaint = new Paint();
        mNormalPaint.setColor(mColor);
        mNormalPaint.setAntiAlias(true);

        mDraggingPaint = new Paint();
        mDraggingPaint.setColor(SELECTED_COLOR);
        mDraggingPaint.setAntiAlias(true);
```

```
        // Start a drag on a long click on the dot
        setOnLongClickListener(lcListener);
        setOnDragListener(this);
    }

    private static View.OnLongClickListener lcListener =
        new View.OnLongClickListener() {
        private boolean mDragInProgress;

        public boolean onLongClick(View v) {
            ClipData data =
            ClipData.newPlainText("DragData", (String)v.getTag());

            mDragInProgress =
            v.startDrag(data, new View.DragShadowBuilder(v),
                    (Object)v, 0);

            Log.v((String) v.getTag(),
              "starting drag? " + mDragInProgress);

            return true;
        }
    };

    @Override
    protected void onMeasure(int widthSpec, int heightSpec) {
        int size = 2*mRadius + getPaddingLeft() + getPaddingRight();
        setMeasuredDimension(size, size);
    }

    // The dragging functionality
    public boolean onDrag(View v, DragEvent event) {
        String dotTAG = (String) getTag();
        // Only worry about drag events if this is us being dragged
        if(event.getLocalState() != this) {
            Log.v(dotTAG, "This drag event is not for us");
            return false;
        }
        boolean result = true;

        // get event values to work with
        int action = event.getAction();
        float x = event.getX();
        float y = event.getY();

        switch(action) {
        case DragEvent.ACTION_DRAG_STARTED:
            Log.v(dotTAG, "drag started. X: " + x + ", Y: " + y);
            inDrag = true; // used in draw() below to change color
            break;
        case DragEvent.ACTION_DRAG_LOCATION:
            Log.v(dotTAG, "drag proceeding... At: " + x + ", " + y);
            break;
        case DragEvent.ACTION_DRAG_ENTERED:
            Log.v(dotTAG, "drag entered. At: " + x + ", " + y);
            break;
        case DragEvent.ACTION_DRAG_EXITED:
```

```
                Log.v(dotTAG, "drag exited. At: " + x + ", " + y);
                break;
            case DragEvent.ACTION_DROP:
                Log.v(dotTAG, "drag dropped. At: " + x + ", " + y);
                // Return false because we don't accept the drop in Dot.
                result = false;
                break;
            case DragEvent.ACTION_DRAG_ENDED:
                Log.v(dotTAG, "drag ended. Success? " + event.getResult());
                inDrag = false; // change color of original dot back
                break;
            default:
                Log.v(dotTAG, "some other drag action: " + action);
                result = false;
                break;
            }
            return result;
        }

        // Here is where you draw our dot, and where you change the color if
        // you're in the process of being dragged. Note: the color change
        // affects the original dot only, not the shadow.
        public void draw(Canvas canvas) {
            float cx = this.getWidth()/2 + getLeftPaddingOffset();
            float cy = this.getHeight()/2 + getTopPaddingOffset();
            Paint paint = mNormalPaint;
            if(inDrag)
                paint = mDraggingPaint;
            canvas.drawCircle(cx, cy, mRadius, paint);
            invalidate();
        }
    }
```

The Dot code looks somewhat similar to the code for DropZone. This is in part because you're also receiving drag events in this class. The constructor for a Dot figures out the attributes in order to set the correct radius and color, and then it sets up the two listeners: one for long clicks and another for the drag events.

The two paints are going to be used to draw your circle. You use the normal paint when the dot is just sitting there. But when the dot is being dragged, you want to indicate that by changing the color of the original to magenta.

The long-click listener is where you initiate a drag sequence. The only way you let the user start dragging a dot is if the user clicks and holds on a dot. When the long-click listener is firing, you create a new ClipData object using a string and the dot's tag. You happen to know that the tag is the name of the dot as specified in the XML layout file. There are several other ways to specify data into a ClipData object, so feel free to read the reference documentation on other ways to store data in a ClipData object.

The next statement is the critical one: startDrag(). This is where Android will take over and start the process of dragging. Note that the first argument is the ClipData object from before; then it's the drag-shadow object, then a local-state object, and finally the number zero.

The drag-shadow object is the image that will be displayed as the dragging is taking place. In your case, this does not replace the original dot image on the screen but shows a shadow of a dot as the dragging is taking place, in addition to the original dot on the screen. The default `DragShadowBuilder` behavior is to create a shadow that looks very much like the original, so for your purposes, you merely call it and pass in your view. You can get fancy here and create whatever sort of shadow view you want, but if you do override this class, you'll need to implement a few methods to make it work.

The `onMeasure()` method is here to supply dimension information to Android for the custom view you're using here. You have to tell Android how big your view is so it knows how to lay it out with everything else. This is standard practice for a custom view.

Finally, there's the `onDrag()` callback. As mentioned, each drag listener can receive drag events. They all get `ACTION_DRAG_STARTED` and `ACTION_DRAG_ENDED`, for example. So, when events happen, you must be careful what you do with the information. Because there are two dots in play in this example application, whenever you do something with the dots, you must be careful that you're affecting the correct one.

When both dots receive the `ACTION_DRAG_STARTED` action, only one should set the color of itself to magenta. To figure out which one is correct, compare the local state object passed in with yourself. If you look back where you set the local-state object, you passed the current view in. So now, when you've received the local-state object out, you compare it to yourself to see if you're the view that initiated the drag sequence.

If you aren't the same view, you write a log message to LogCat saying this is not for you, and you return `false` to say you're not handling this message.

If you are the view that should be receiving this drag event, you collect some values from the drag event, then you mostly just log the event to LogCat. The first exception to this is `ACTION_DRAG_STARTED`. If you got this action and it's for you, you then know that your dot has begun a drag sequence. Therefore, you set the `inDrag` boolean so the `draw()` method later on will do the right thing and display a different-colored dot. This different color only lasts until `ACTION_DRAG_ENDED` is received, at which time you restore the original color of the dot.

If a dot gets the `ACTION_DROP` action, this means the user tried to drop a dot on a dot—maybe even the original dot. This shouldn't do anything, so you just return `false` from this callback in this case.

Finally, the `draw()` method of your custom view figures out the location of the center point of your circle (dot) and then draws it with the appropriate paint. The `invalidate()` method is there to tell Android that you've modified the view and that Android should redraw the user interface. By calling `invalidate()`, you ensure that the user interface will be updated very shortly with whatever is new.

You now have all the files and the background necessary to compile and deploy this example drag-and-drop application.

Testing the Example Drag-and-Drop Application

Following is some example output from LogCat when we ran this example application. Notice how the log message used `Blue dot` to indicate messages from the blue dot, `White dot` for messages from the white dot, and `DropTarget` for the view where the drops are allowed to go.

```
White dot:  starting drag? true
Blue dot:   This drag event is not for us
White dot:  drag started. X: 53.0, Y: 206.0
DropTarget: drag started in dropTarget
DropTarget: drag entered dropTarget
DropTarget: drag proceeding in dropTarget: 29.0, 36.0
DropTarget: drag proceeding in dropTarget: 48.0, 39.0
DropTarget: drag proceeding in dropTarget: 45.0, 39.0
DropTarget: drag proceeding in dropTarget: 41.0, 39.0
DropTarget: drag proceeding in dropTarget: 40.0, 39.0
DropTarget: drag drop in dropTarget
DropTarget: Item data is White dot
ViewRoot:   Reporting drop result: true
White dot:  drag ended. Success? true
Blue dot:   This drag event is not for us
DropTarget: drag ended in dropTarget
```

In this particular case, the drag was started with the white dot. Once the long click has triggered the beginning of the drag sequence, we get the `starting drag?` message.

Notice how the next three lines all indicate that an `ACTION_DRAG_STARTED` action was received in three different views. Blue dot determined that the callback was not for it. It was also not for `DropTarget`.

Next, notice how the drag-proceeding messages show the drag happening through `DropTarget`, beginning with the `ACTION_DRAG_ENTERED` action. This means the dot was being dragged on top of the green square. The x and y coordinates reported in the drag event object are the coordinates of the drag point relative to the upper-left corner of the view. So, in the example app, the first record of the drag in the drop target is at (x, y) = (29, 36), and the drop occurred at (40, 39). See how the drop target was able to extract the tag name of the white dot from the event's `ClipData` to write it to LogCat.

Also see how once again, all drag listeners received the `ACTION_DRAG_ENDED` action. Only White dot determined that it's okay to display the results using `getResult()`.

Feel free to experiment with this example application. Drag a dot to the other dot, or even to itself. Go ahead and add another dot to `palette.xml`. Notice how when the dragged dot leaves the green square, there's a message saying that the drag exited. Note also that if you drop a dot somewhere other than the green square, the drop is considered failed.

References

Here are some helpful references to topics you may wish to explore further:

- `www.androidbook.com/proandroid4/projects`: A list of downloadable projects related to this book. For this chapter, look for a zip file called `ProAndroid4_Ch28_DragDrop.zip`. This zip file contains all the projects from this chapter, listed in separate root directories. There is also a `README.TXT file` that describes exactly how to import projects into Eclipse from one of these zip files.
- `http://developer.android.com/guide/topics/ui/drag-drop.html`: The Android developer's guide to Drag and Drop.

Summary

Let's summarize the topics covered in this chapter:

- Drag and drop support in Android 3.0, and implementing it prior to 3.0 using other methods
- Iterating through possible drop targets to see if a drop (that is, finger leaving the screen after dragging) occurred
- The difficulty of doing the math to keep track of where a dragged object is and whether it's over a drop target
- Drag-and-drop support in Android 3.0+, which is much nicer because it eliminates a lot of guesswork
- Drag listeners, which can be any objects and do not need to be draggables or drop-target views
- The fact that a drag can occur across fragments
- The `DragEvent` object, which can contain lots of great information about what is being dragged and why
- How Android takes care of the math to determine whether a drop is occurring on top of a view

Interview Questions

You can use the following questions as a guide to consolidate your understanding of this chapter:

1. What is a good layout to use for drag and drop pre-Android 3.0?
2. Post-Android 3.0, does it matter what layout you use for drag and drop?

3. What types of data can you add to a `DragEvent`?
4. Is it possible to watch a drag-and-drop sequence in Android 3.0+?
5. Can a drop target reject a drop operation? If so, how?
6. How can a dragged object determine whether a drop was successful?
7. What does `DragShadowBuilder` do, and when would you use it?

Chapter 29

Using Sensors

Android devices often come with hardware sensors built in, and Android provides a framework for working with those sensors. Working with sensors can be fun. Measuring the outside world and using that in software in a device is pretty cool. It is the kind of programming experience you just don't get on a regular computer that sits on a desk or in a server room. The possibilities for new applications that use sensors are huge, and we hope you are inspired to realize them.

In this chapter, we'll explore the Android sensor framework. We'll explain what sensors are and how we get sensor data, and then discuss some specifics of the kinds of data we can get from sensors and what we can do with it. While Android has defined several sensor types already, there are no doubt more sensors in Android's future, and we expect that future sensors will get incorporated into the sensor framework.

What Is a Sensor?

In Android, a *sensor* is a piece of hardware that has been wired into the device to feed data from the physical world to applications. Applications in turn use the sensor data to inform the user about the physical world, to control game play, to do augmented reality, or to provide useful tools for working in the real world. Sensors operate in one direction only; they're read-only (with one exception, the NFC sensor, which we'll cover). That makes using them fairly straightforward. You set up a listener to receive sensor data, and then you process the data as it comes in. GPS hardware is like the sensors we cover in this chapter. In Chapter 22, we set up listeners for GPS location updates, and we processed those location updates as they came in. But although GPS is similar to a sensor, it is not part of the sensor framework that is provided by Android.

Some of the sensor types that can appear in an Android device include

- Light sensor
- Proximity sensor
- Temperature sensor
- Pressure sensor

- Gyroscope sensor
- Accelerometer
- Magnetic field sensor
- Orientation sensor
- Gravity sensor (as of Android 2.3)
- Linear acceleration sensor (as of Android 2.3)
- Rotation vector sensor (as of Android 2.3)
- Relative humidity sensor (as of Android 4.0)
- Near Field Communication (NFC) sensor (as of Android 2.3)

The NFC sensor is not like the others in this list. We're going to cover the NFC sensor later on in this chapter, because it is accessed in a completely different way from the rest of these sensors.

Detecting Sensors

Please don't assume, however, that all Android devices have all of these sensors. In fact, many devices have just some of these sensors. The Android emulator, for example, has only an accelerometer. So how do you know which sensors are available on a device? There are two ways, one direct and one indirect.

The first way is that you ask the `SensorManager` for a list of the available sensors. It will respond with a list of sensor objects that you can then set up listeners for and get data from. We'll show you how a bit later in this chapter. This method assumes that the user has already installed your application onto a device, but what if the device doesn't have a sensor that your application needs?

That's where the second method comes in. Within the `AndroidManifest.xml` file, you can specify the features a device must have in order to properly support your application. If your application needs a proximity sensor, you specify that in your manifest file with a line such as the following:

```
<uses-feature android:name="android.hardware.sensor.proximity" />
```

Android Market will only install your app on a device that has a proximity sensor, so you know it's there when your application runs. The same cannot be said for all other Android app stores. That is, some Android app stores do not perform that kind of check to make sure your app can only be installed onto a device that supports the sensors you specify.

What Can We Know About a Sensor?

While using the `uses-feature` tags in the manifest file lets you know that a sensor your application requires exists on a device, it doesn't tell you everything you may want to know about the actual sensor. Let's build a simple application that queries the device for sensor information. Listing 29–1 shows the Java code of our `MainActivity`.

> **NOTE:** You can download this chapter's projects. We will give you the URL at the end of the chapter. This will allow you to import these projects into Eclipse directly.

Listing 29–1. *Java for a Sensor List App*

```
public class MainActivity extends Activity {
    @Override
    public void onCreate(Bundle savedInstanceState) {
        super.onCreate(savedInstanceState);
        setContentView(R.layout.main);

        TextView text = (TextView)findViewById(R.id.text);

        SensorManager mgr =
            (SensorManager) this.getSystemService(SENSOR_SERVICE);

        List<Sensor> sensors = mgr.getSensorList(Sensor.TYPE_ALL);

        StringBuilder message = new StringBuilder(2048);
        message.append("The sensors on this device are:\n");

        for(Sensor sensor : sensors) {
            message.append(sensor.getName() + "\n");
            message.append("  Type: " +
                    sensorTypes.get(sensor.getType()) + "\n");
            message.append("  Vendor: " +
                    sensor.getVendor() + "\n");
            message.append("  Version: " +
                    sensor.getVersion() + "\n");
            message.append("  Resolution: " +
                    sensor.getResolution() + "\n");
            message.append("  Max Range: " +
                    sensor.getMaximumRange() + "\n");
            message.append("  Power: " +
                    sensor.getPower() + " mA\n");
        }
        text.setText(message);
    }

    private HashMap<Integer, String> sensorTypes =
                  new HashMap<Integer, String>();

    {
      sensorTypes.put(Sensor.TYPE_ACCELEROMETER, "TYPE_ACCELEROMETER");
      sensorTypes.put(Sensor.TYPE_AMBIENT_TEMPERATURE,
                          "TYPE_AMBIENT_TEMPERATURE");
      sensorTypes.put(Sensor.TYPE_GRAVITY, "TYPE_GRAVITY");
      sensorTypes.put(Sensor.TYPE_GYROSCOPE, "TYPE_GYROSCOPE");
      sensorTypes.put(Sensor.TYPE_LIGHT, "TYPE_LIGHT");
      sensorTypes.put(Sensor.TYPE_LINEAR_ACCELERATION,
                          "TYPE_LINEAR_ACCELERATION");
      sensorTypes.put(Sensor.TYPE_MAGNETIC_FIELD, "TYPE_MAGNETIC_FIELD");
```

```
        sensorTypes.put(Sensor.TYPE_ORIENTATION,
                            "TYPE_ORIENTATION (deprecated)");
        sensorTypes.put(Sensor.TYPE_PRESSURE, "TYPE_PRESSURE");
        sensorTypes.put(Sensor.TYPE_PROXIMITY, "TYPE_PROXIMITY");
        sensorTypes.put(Sensor.TYPE_RELATIVE_HUMIDITY,
                            "TYPE_RELATIVE_HUMIDITY");
        sensorTypes.put(Sensor.TYPE_ROTATION_VECTOR,
                            "TYPE_ROTATION_VECTOR");
        sensorTypes.put(Sensor.TYPE_TEMPERATURE,
                            "TYPE_TEMPERATURE (deprecated)");
    }
}
```

Notice that we've used a ScrollView in this example, since we could easily get more rows than our screen can display at one time. Within our onCreate() method, we start by getting a reference to the SensorManager. There can be only one of these, so we retrieve it as a system service. We then call its getSensorList() method to get a list of sensors. For each sensor, we write out information about it. The output will look something like Figure 29–1.

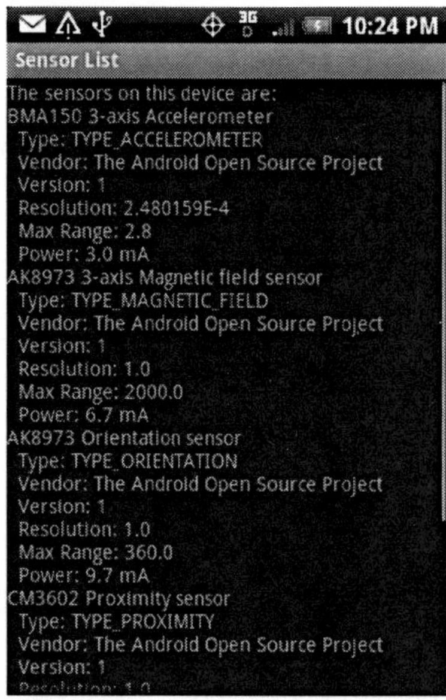

Figure 29–1. *Output from our Sensor List app*

There are a few things to know about this sensor information. The type value tells you the basic type of the sensor without getting specific. A light sensor is a light sensor, but you could get variations in light sensors from one device to another. For example, the resolution of a light sensor on one device could be different than on another device. When you specify that your app needs a light sensor in a <uses-feature> tag, you don't know in advance exactly what type of light sensor you're going to get. If it matters to your application, you'll need to query the device to find out, and adjust your code accordingly.

The values you get for resolution and maximum range will be in the appropriate units for that sensor. The power measurement is in milliamperes (mA) and represents the electrical current that the sensor draws from the device's battery; smaller is better.

Now that we know what sensors we have available to us, how do we go about getting data from them? As we explained earlier, we set up a listener in order to get sensor data sent to us. Let's explore that now.

Getting Sensor Events

Sensors provide data to our application once we register a listener to receive the data. When our listener is not listening, the sensor can be turned off, conserving battery life, so make sure you only listen when you really need to. Setting up a sensor listener is easy to do. Let's say that we want to measure the light levels from the light sensor. Listing 29–2 shows the Java code for a sample app that does this.

Listing 29–2. *Java Code for a Light Sensor Monitor App*

```
public class MainActivity extends Activity implements SensorEventListener {
    private SensorManager mgr;
    private Sensor light;
    private TextView text;
    private StringBuilder msg = new StringBuilder(2048);

    @Override
    public void onCreate(Bundle savedInstanceState) {
        super.onCreate(savedInstanceState);
        setContentView(R.layout.main);

        mgr = (SensorManager) this.getSystemService(SENSOR_SERVICE);

        light = mgr.getDefaultSensor(Sensor.TYPE_LIGHT);

        text = (TextView) findViewById(R.id.text);
    }

    @Override
    protected void onResume() {
        mgr.registerListener(this, light,
                SensorManager.SENSOR_DELAY_NORMAL);
        super.onResume();
    }

    @Override
    protected void onPause() {
        mgr.unregisterListener(this, light);
        super.onPause();
    }

    public void onAccuracyChanged(Sensor sensor, int accuracy) {
        msg.insert(0, sensor.getName() + " accuracy changed: " +
            accuracy + (accuracy==1?" (LOW)":(accuracy==2?" (MED)":
            " (HIGH)")) + "\n");
        text.setText(msg);
```

```
            text.invalidate();
        }

        public void onSensorChanged(SensorEvent event) {
            msg.insert(0, "Got a sensor event: " + event.values[0] +
                " SI lux units\n");
            text.setText(msg);
            text.invalidate();
        }
}
```

In this sample app, we again get a reference to the SensorManager, but instead of getting a list of sensors, we query specifically for the light sensor. We then set up a listener in the onResume() method of our activity, and we unregister the listener in the onPause() method. We don't want to be worrying about the light levels when our application is not in the foreground.

For the registerListener() method, we pass in a value representing how often we want to be notified of sensor value changes. This parameter could be

- SENSOR_DELAY_NORMAL
- SENSOR_DELAY_UI
- SENSOR_DELAY_GAME
- SENSOR_DELAY_FASTEST

It is important to select an appropriate value for this parameter. Some sensors are very sensitive and will generate a lot of events in a short amount of time. If you choose SENSOR_DELAY_FASTEST, you might even overrun your application's ability to keep up. Depending on what your application does with each sensor event, it is possible that you will be creating and destroying so many objects in memory that garbage collection will cause noticeable slowdowns and hiccups on the device. On the other hand, certain sensors pretty much demand to be read as often as possible; this is true of the rotation vector sensor in particular.

Because our Activity implements the SensorEventListener interface, we have two callbacks for sensor events: onAccuracyChanged() and onSensorChanged(). The first method will let us know if the accuracy changes on our sensor (or sensors, since it could be called for more than one). The value of the accuracy parameter will be 0, 1, 2, or 3 for unreliable, low, medium, or high accuracy, respectively. Unreliable accuracy does not mean that the device is broken; it normally means that the sensor needs to be calibrated. The second callback method tells us when the light level has changed, and we get a SensorEvent object to tell us the details of the new value or values from the sensor.

A SensorEvent object has several members, one of them being an array of float values. For a light sensor event, only the first float value has meaning, which is the SI lux value of the light that was detected by the sensor. For our sample app, we build up a message string by inserting the new messages on top of the older messages, and then display the batch of messages in a TextView. Our newest sensor values will always be displayed at the top of the screen.

When you run this application (on a real device, of course, since the emulator does not have a light sensor), you may notice that nothing is displayed at first. Just change the light that is shining on the upper-left corner of your device. This is most likely where your light sensor is. If you look very carefully, you might see the dot behind the screen that is the light sensor. If you cover this dot with your finger, the light level will probably change to a very small value (although it may not reach zero). The messages should display on the screen, telling you about the changing light levels.

> **NOTE:** You might also notice that, by covering up the light sensor, your buttons light up (if you have a device with lighted buttons). This is because Android has detected the darkness and lights up the buttons to make the device easier to use "in the dark."

Issues with Getting Sensor Data

The Android sensor framework has problems that you need to be aware of. This is the part that's not fun. In some cases, we have ways of working around the problem; in others we don't, or it's very difficult.

onAccuracyChanged() Always Says the Same Thing

Up until Android 2.2, the `onAccuracyChanged()` callback would get called every time there was a new sensor reading, and the accuracy parameter would always be `3` (for high). It's a good idea to accommodate changing accuracies of sensor data, but don't be surprised if this method gets called all the time even though the accuracy has not changed. This issue appears to be fixed since Android 3.0.

No Direct Access to Sensor Values

You may have noticed that there is no direct way to query the sensor's current value. The only way to get data from a sensor is through a listener. This means that, even once we've set up the listener, there are no guarantees that we'll get a new datum within a set period of time. At least the callback is asynchronous so we won't block the UI thread waiting for a piece of data from a sensor. However, your application has to accommodate the fact that sensor data may not be available at the exact moment that you want it.

It is possible to directly access sensors using native code and the JNI feature of Android. You'll need to know the low-level native API calls for the sensor driver you're interested in, plus be able to set up the interface back to Android. So it can be done, but it's not easy.

Sensor Values Not Sent Fast Enough

Even at `SENSOR_DELAY_FASTEST`, we might not get new values more often than every 20ms (it depends on the device). If you need more rapid sensor data than you can get with a rate setting of `SENSOR_DELAY_FASTEST`, it is possible to use native code and JNI to get to the sensor data faster, but similar to the previous situation, it is not easy.

Sensors Turn Off with the Screen

There have been problems in Android 2.x with sensor updates that get turned off when the screen is turned off. Apparently someone thought it was a good idea to not send sensor updates if the screen is off, even if your application (most likely using a service) has a wake lock. Basically, your listener gets unregistered when the screen turns off.

There are several workarounds to this problem. For more information on this issue and possible resolutions and workarounds, please refer to Android Issue 11028:

`http://code.google.com/p/android/issues/detail?id=11028`

Now that you know how to get data from sensors, what can you do with the data? As we said earlier, depending on which sensor you're getting data from, the values returned in the values array mean different things. The next section will explore each of the sensor types and what their values mean.

Interpreting Sensor Data

Now that we understand how to get data from a sensor, we must do something meaningful with the data. The data we get, however, will depend on which sensor we're getting the data from. Some sensors are simpler than others. In the sections that follow, we will describe the data that you'll get from the sensors we currently know about. As new devices come into being, new sensors will undoubtedly be introduced as well. The sensor framework is very likely to remain the same, so the techniques we show here should apply equally well to the new sensors.

Light Sensors

The light sensor is one of the simplest sensors on a device, and one we've used in our first sample applications of this chapter. The sensor gives a reading of the light level detected by the light sensor of the device. As the light level changes, the sensor readings change. The units of the data are in SI lux units. To learn more about what this means, please see the "References" section at the end of this chapter for links to more information.

For the values array in the `SensorEvent` object, a light sensor uses just the first element, `values[0]`. This value is a `float` and ranges technically from `0` to the maximum value for the particular sensor. We say *technically* because the sensor may only send very small values when there's no light, and never actually send a value of 0.

Remember also that the sensor can tell us the maximum value that it can return and that different sensors can have different maximums. For this reason, it may not be useful to consider the light-related constants in the SensorManager class. For example, SensorManager has a constant called LIGHT_SUNLIGHT_MAX, which is a float value of 120,000; however, when we queried our device earlier, the maximum value returned was 10,240, clearly much less than this constant value. There's another one called LIGHT_SHADE at 20,000, which is also above the maximum of the device we tested. So keep this in mind when writing code that uses light sensor data.

Proximity Sensors

The proximity sensor either measures the distance that some object is from the device (in centimeters) or represents a flag to say whether an object is close or far. Some proximity sensors will give a value ranging from 0.0 to the maximum in increments, while others return either 0.0 or the maximum value only. If the maximum range of the proximity sensor is equal to the sensor's resolution, then you know it's one of those that only returns 0.0, or the maximum. There are devices with a maximum of 1.0 and others where it's 6.0. Unfortunately, there's no way to tell before the application is installed and run which proximity sensor you're going to get. Even if you put a <uses-feature> tag in your AndroidManifest.xml file for the proximity sensor, you could get either kind. Unless you absolutely need to have the more granular proximity sensor, your application should accommodate both types gracefully.

Here's an interesting fact about proximity sensors: the proximity sensor is sometimes the same hardware as the light sensor. Android still treats them as logically separate sensors, though, so if you need data from both you will need to set up a listener for each one. Here's another interesting fact: the proximity sensor is often used in the phone application to detect the presence of a person's head next to the device. If the head is that close to the touchscreen, the touchscreen is disabled so no keys will be accidently pressed by the ear or cheek while the person is talking on the phone.

The source code projects for this chapter include a simple proximity sensor monitor application, which is basically the light sensor monitor application modified to use the proximity sensor instead of the light sensor. We won't include the code in this chapter, but feel free to experiment with it on your own.

Temperature Sensors

The old temperature sensor provided a temperature reading and also returned just a single value in values[0]. This sensor usually read an internal temperature, such as at the battery. There is a new temperature sensor called TYPE_AMBIENT_TEMPERATURE. The new value represents the temperature outside the device in degrees Celsius. You can get to Fahrenheit degrees from Celsius by multiplying by 9/5 and adding 32. For example, 0 degrees Celsius is 32 Fahrenheit (the temperature at which water freezes), and 100 degrees Celsius is 212 degrees Fahrenheit (the temperature at which water boils).

The placement of the temperature sensor is device-dependent, and it is possible that the temperature readings could be impacted by the heat generated by the device itself. The projects for this chapter include one for the temperature sensor called `TemperatureSensor`. It takes care of calling the correct temperature sensor based on which version of Android is running.

Pressure Sensors

This sensor measures barometric pressure, which could detect altitude for example or be used for weather predictions. This sensor should not be confused with the ability of a touchscreen to generate a `MotionEvent` with a pressure value (the pressure of the touch). We covered this touch type of pressure sensing in Chapter 27. Touchscreen pressure sensing doesn't use the Android sensor framework.

The unit of measurement for a pressure sensor is atmospheric pressure in hPa (millibar), and this measurement is delivered in `values[0]`.

Gyroscope Sensors

Gyroscopes are very cool components that can measure the twist of a device about a reference frame. Said another way, gyroscopes measure the rate of rotation about an axis. When the device is not rotating, the sensor values will be zeroes. When there is rotation in any direction, you'll get non-zero values from the gyroscope. By itself, a gyroscope can't tell you everything you need to know. And unfortunately, errors creep in over time with gyroscopes. But coupled with accelerometers, you can determine the path of movement of the device.

Kalman filters can be used to link data from the two sensors together. Accelerometers are not terribly accurate in the short term, and gyroscopes are not very accurate in the long term, so combined they can be reasonably accurate all the time. While Kalman filters are very complex, there is an alternative called *complementary filters* that are easier to implement in code and produce results that are pretty good. These concepts are beyond the scope of this book.

The gyroscope sensor returns three values in the values array for the x, y, and z axes. The units are radians per second, and the values represent the rate of rotation around each of those axes. One way to work with these values is to integrate them over time to calculate an angle change. This is a similar calculation to integrating linear speed over time to calculate distance.

Accelerometers

Accelerometers are probably the most interesting of the sensors on a device. Using these sensors, our application can determine the physical orientation of the device in space relative to gravity's pull straight down, plus be aware of forces pushing on the device. Providing this information allows an application to do all sorts of interesting things, from game play to augmented reality. And of course, the accelerometers tell Android when to switch the orientation of the user interface from portrait to landscape and back again.

The accelerometer coordinate system works like this: the accelerometer's x axis originates in the bottom-left corner of the device and goes across the bottom to the right. The y axis also originates in the bottom-left corner and goes up along the left of the display. The z axis originates in the bottom-left corner and goes up in space away from the device. Figure 29–2 shows what this means.

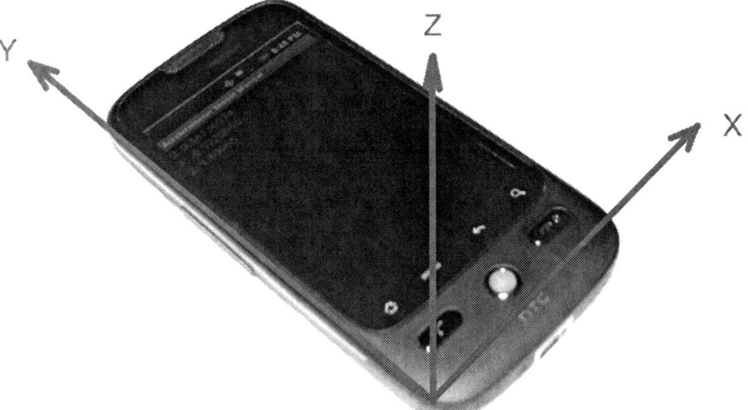

Figure 29–2. *Accelerometer coordinate system*

This coordinate system is different than the one used in layouts and 2D graphics. In that coordinate system, the origin (0, 0) is at the top-left corner, and y is positive in the direction down the screen from there. It is easy to get confused when dealing with coordinate systems in different frames of reference, so be careful.

We haven't yet said what the accelerometer values mean, so what *do* they mean? Acceleration is measured in meters per second squared (m/s^2). Normal Earth gravity is $9.81 m/s^2$, pulling down toward the center of the Earth. From the accelerometer's point of view, the measurement of gravity is -9.81. If your device is completely at rest (not moving) and is on a perfectly flat surface, the x and y readings will be 0 and the z reading will be +9.81. Actually, the values won't be exactly these because of the sensitivity and accuracy of the accelerometer, but they will be close. Gravity is the only force acting on the device when the device is at rest, and because gravity pulls straight down, if our device is perfectly flat, its effect on the x and y axes is zero. On the z axis, the accelerometer is measuring the force on the device minus gravity. Therefore, 0 minus -9.81 is +9.81, and that's what the z value will be (a.k.a. `values[2]` in the `SensorEvent` object).

The values sent to our application by the accelerometer always represent the sum of the forces on the device minus gravity. If we were to take our perfectly flat device and lift it straight up, the z value would increase at first, because we increased the force in the up

(z) direction. As soon as our lifting force stopped, the overall force would return to being just gravity. If the device were to be dropped (hypothetically—please don't do this), it would be accelerating toward the ground, which would zero out gravity so the accelerometer would read 0 force.

Let's take the device from Figure 29–2 and rotate it up so it is in portrait mode and vertical. The x axis is the same, pointing left to right. Our y axis is now straight up and down, and the z axis is pointing out of the screen straight at us. The y value will be +9.81, and both x and z will be 0.

What happens when we rotate the device to landscape mode and continue to hold it vertically, so the screen is right in front of our face? If you guessed that y and z are now 0 and x is +9.81, you'd be correct. Figure 29–3 shows what it might look like.

Figure 29–3. *Accelerometer values in landscape vertical*

When the device is not moving, or is moving with a constant velocity, the accelerometers are only measuring gravity. And in each axis, the value from the accelerometer is gravity's component in that axis. Therefore, using some trigonometry, you could figure out the angles and know how the device is oriented relative to gravity's pull. That is, you could tell if the device were in portrait mode or in landscape mode or in some tilted mode. In fact, this is exactly what Android does to figure out which display mode to use (portrait or landscape). Note, however, that the accelerometers do not say how the device is oriented with respect to magnetic north. That's where the magnetic field sensor will come in, which we will cover in the next section.

Accelerometers and Display Orientation

Accelerometers in a device are hardware, and they're firmly attached, and as such have a specific orientation relative to the device that does not change as the device is turned this way or that. The values that the accelerometers send into Android will change of course as a device is moved, but the coordinate system of the accelerometers will stay the same relative to the physical device. The coordinate system of the display, however, changes as the user goes from portrait to landscape and back again. In fact depending on which way the screen is turned, portrait could be right-side up, or 180 degrees upside-down. Similarly, landscape could be in one of two different rotations 180 degrees apart.

When our application is reading accelerometer data and wanting to affect the user interface correctly, our application must know how much rotation of the display has occurred to properly compensate. As our screen is reoriented from portrait to landscape, the screen's coordinate system has rotated with respect to the coordinate system of the accelerometers. To handle this, our application must use the method `Display.getRotation()`, which was introduced in Android 2.2. The return value is a simple integer but not the actual number of degrees of rotation. The value will be one of `Surface.ROTATION_0`, `Surface.ROTATION_90`, `Surface.ROTATION_180`, or `Surface.ROTATION_270`. These are constants with values of `0`, `1`, `2`, and `3`, respectively. This return value tells us how much the display has rotated from the "normal" orientation of the device. Because not all Android devices are normally in portrait mode, we cannot assume that portrait is at `ROTATION_0`.

Not all devices will give you all four return values. On the HTC Droid Eris running Android 2.1, `Display.getOrientation()` (the precursor to `Display.getRotation()` and now deprecated) will return `0` or `1`, and that's it. In normal portrait mode, the value returned is `0`. If you turn the device 90 degrees counterclockwise, the screen will rotate and `Display.getOrientation()` will return `1`. If you turn the device clockwise 90 degrees from portrait mode, the screen stays in portrait mode, and you still get a return value of `0` from `Display.getOrientation()`.

On the Motorola Droid running Android 2.2, `Display.getRotation()` returns `0`, `1`, or `3`. It does not return a `2` and will not show portrait upside down. Here is a disappointing result though: if you rotate the device 270 degrees in the counterclockwise direction from straight-up portrait, `Display.getRotation()` returns a `1` at 90 degrees and the display switches to landscape mode; at 180 degrees you still get a `1` and the display does not change; and at 270 degrees the display flips to the other landscape mode, but `Display.getRotation()` still returns `1`. If you rotate the device 90 degrees in the clockwise direction from normal portrait mode, then you'll get a 3 from `Display.getRotation()`. This last position looks exactly the same as 270 degrees counterclockwise, but you get a different return value from `Display.getRotation()` depending on how you got there.

Accelerometers and Gravity

So far, we've only briefly touched on what happens to the accelerometer values when the device is moved. Let's explore that further. All forces acting on the device will be detected by the accelerometers. If we lift the device, the initial lifting force is positive in the z direction, and we get a z value greater than +9.81. If we push the device on its left side, we'll get an initial negative reading in the x direction.

What we'd like to be able to do is separate out the force of gravity from the other forces acting on the device. There's a fairly easy way to do this, and it's called a *low-pass filter*. Forces other than gravity acting on the device will do so in a way that is typically not gradual. In other words, if the user is shaking the device, the shaking forces are reflected in the accelerometer values quickly. A low-pass filter will in effect strip out the shaking forces and leave only the steady force, which for us is gravity. Let's use a sample application to illustrate this concept. It's called GravityDemo. Listing 29–3 shows the layout XML and the Java code.

Listing 29–3. *Measuring Gravity from the Accelerometers*

```
// This file is MainActivity.java
public class MainActivity extends Activity implements SensorEventListener {
    private SensorManager mgr;
    private Sensor accelerometer;
    private TextView text;
    private float[] gravity = new float[3];
    private float[] motion = new float[3];
    private double ratio;
    private double mAngle;
    private int counter = 0;

    @Override
    public void onCreate(Bundle savedInstanceState) {
        super.onCreate(savedInstanceState);
        setContentView(R.layout.main);

        mgr = (SensorManager) this.getSystemService(SENSOR_SERVICE);

        accelerometer = mgr.getDefaultSensor(Sensor.TYPE_ACCELEROMETER);

        text = (TextView) findViewById(R.id.text);
    }

    @Override
    protected void onResume() {
        mgr.registerListener(this, accelerometer,
                    SensorManager.SENSOR_DELAY_UI);
        super.onResume();
    }

    @Override
    protected void onPause() {
        mgr.unregisterListener(this, accelerometer);
        super.onPause();
    }

    public void onAccuracyChanged(Sensor sensor, int accuracy) {
        // ignore
    }

    public void onSensorChanged(SensorEvent event) {
        // Use a low-pass filter to get gravity.
        // Motion is what's left over
        for(int i=0; i<3; i++) {
            gravity [i] = (float) (0.1 * event.values[i] +
                                   0.9 * gravity[i]);
            motion[i] = event.values[i] - gravity[i];
        }

        // ratio is gravity on the Y axis compared to full gravity
```

```
        // should be no more than 1, no less than -1
        ratio = gravity[1]/SensorManager.GRAVITY_EARTH;
        if(ratio > 1.0) ratio = 1.0;
        if(ratio < -1.0) ratio = -1.0;

        // convert radians to degrees, make negative if facing up
        mAngle = Math.toDegrees(Math.acos(ratio));
        if(gravity[2] < 0) {
            mAngle = -mAngle;
        }

        // Display every 10th value
        if(counter++ % 10 == 0) {
            String msg = String.format(
                "Raw values\nX: %8.4f\nY: %8.4f\nZ: %8.4f\n" +
                "Gravity\nX: %8.4f\nY: %8.4f\nZ: %8.4f\n" +
                "Motion\nX: %8.4f\nY: %8.4f\nZ: %8.4f\nAngle: %8.1f",
                event.values[0], event.values[1], event.values[2],
                gravity[0], gravity[1], gravity[2],
                motion[0], motion[1], motion[2],
                mAngle);
            text.setText(msg);
            text.invalidate();
            counter=1;
        }
    }
}
```

The result of running this application is a display that looks like Figure 29–4. This screenshot was taken as the device lay flat on a table.

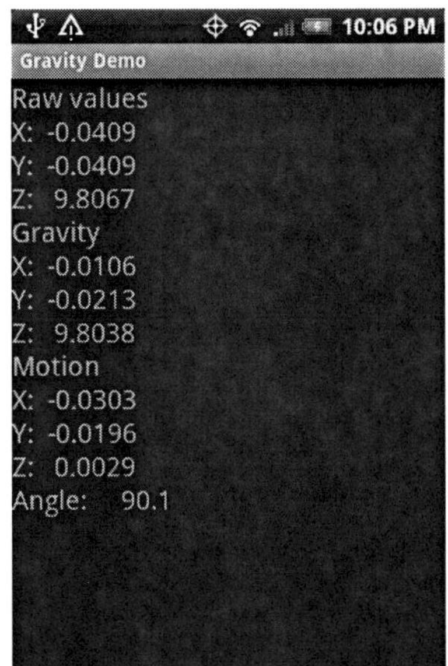

Figure 29–4. *Gravity, motion, and angle values*

Most of this sample application is the same as our Accel Sensor application from before. The differences are in the `onSensorChanged()` method. Instead of simply displaying the values from the event array, we attempt to keep track of gravity and motion. We get gravity by using only a small portion of the new value from the event array, and we use a large portion of the previous value of the gravity array. The two portions used must add up to 1.0. We used 0.9 and 0.1. You could try other values, too, such as 0.8 and 0.2. Our gravity array cannot possibly change as fast as the actual sensor values are changing. But this is closer to reality. And this is what a low-pass filter does. The event array values would only be changing if forces were causing the device to move, and we don't want to measure those forces as part of gravity. We only want to record into our gravity array the force of gravity itself. The math here does not mean we're magically recording only gravity, but the values we're calculating are going to be a lot closer than the raw values from the event array.

Notice also the motion array in the code. By tracking the difference between the raw event array values and the calculated gravity values, we are basically measuring the active, non-gravity, forces on the device in the motion array. If the values in the motion array are zero or very close to zero, it means the device is probably not moving. This is useful information. Technically, a device moving in a constant speed would also have values in the motion array close to zero, but the reality is that if a user is moving the device, the motion values will be somewhat larger than zero. Users can't possibly move a device at a perfect constant speed.

Using Accelerometers to Measure the Device's Angle

We wanted to show you one more thing about the accelerometers before we move on. If we go back to our trigonometry lessons, we remember that the cosine of an angle is the ratio of the near side and the hypotenuse. If we consider the angle between the y axis and gravity itself, we could measure the force of gravity on the y axis and take the arccosine to determine the angle. We've done that in this code as well, although here we have to deal yet again with some of the messiness of sensors in Android. There are constants in `SensorManager` for different gravity constants, including Earth's. But our actual measured values could possibly exceed the defined constants. We will explain what we mean by this next.

In theory, our device at rest would measure a value for gravity equal to the constant value, but this is rarely the case. At rest, the accelerometer sensor is very likely to give us a value for gravity that is larger or smaller than the constant. Therefore, our ratio could end up greater than one, or less than negative one. This would make the `acos()` method complain, so we fix the ratio value to be no more than 1 and no less than -1. The corresponding angles in degrees range from 0 to 180. That's fine except that we don't get negative angles from 0 to -180 this way. To get the negative angles, we use another value from our gravity array, which is the z value. If the z value of gravity is negative, it means the device's face is oriented downward. For all those values where the device face is pointed down, we make our angle negative as well, with the result being that our angle goes from -180 to +180, just as we would expect.

Go ahead and experiment with this sample application. Notice that the value of the angle is 90 when the device is laid flat, and it's zero (or close to it) when the device is held straight up and down in front of us. If we keep rotating down past flat, we will see the value of the angle exceed 90. If we tilt the device up more from the 0 position, the value of angle goes negative until we're holding the device above our heads and the value of the angle is -90. Finally, you may have noticed our counter that controls how often the display is updated. Because the sensor events can come rather frequently, we decided to only display every tenth time we get values.

Magnetic Field Sensors

The magnetic field sensor measures the ambient magnetic field in the x, y, and z axes. This coordinate system is aligned just like the accelerometers, so x, y, and z are as shown in Figure 29–2. The units of the magnetic field sensor are microteslas (uT). This sensor can detect the Earth's magnetic field and therefore tell us where north is. This sensor is also referred to as the *compass*, and in fact the `<uses-feature>` tag uses `android.hardware.sensor.compass` as the name of this sensor. Because this sensor is so tiny and sensitive, it can be affected by magnetic fields generated by things near the device, and even to some extent to components within the device. Therefore the accuracy of the magnetic field sensor may at times be suspect.

We've included a simple `CompassSensor` application in the download section of the web site, so feel free to import that and play with it. If you bring metal objects close to the device while this application is running, you might notice the values changing in response. Certainly if you bring a magnet close to the device you will see the values change, but we don't recommend that you mix Android devices and magnets.

You might be asking, can I use the compass sensor as a compass to detect where north is? And the answer is: not by itself. While the compass sensor can detect magnetic fields around the device, if the device is not being held perfectly flat in relation to the Earth's surface, you'd have no way of correctly interpreting the compass sensor values. But we have accelerometers that can tell us the orientation of the device relative to the Earth's surface! Therefore, we can create a compass from the compass sensor, but we need help from the accelerometers too. So let's see how to do that.

Using Accelerometers and Magnetic Field Sensors Together

The `SensorManager` provides some methods that allow us to combine the compass sensor and the accelerometers to figure out orientation. As we just discussed, you can't use just the compass sensor alone to do the job. So `SensorManager` provides a method called `getRotationMatrix()`, which takes the values from the accelerometers and from the compass and returns a matrix that can be used to determine orientation.

Another `SensorManager` method, `getOrientation()`, takes the rotation matrix from the previous step and gives an orientation matrix. The values from the orientation matrix tell us our device's rotation relative to the Earth's magnetic north, as well as the device's pitch and roll relative to the ground. This would be terrific if it did the job for us. Unfortunately, at least until Android 2.2, using this mechanism has some big challenges, not the least of which is the discontinuity when the device is in front of us and it goes from facing us, to where we've tilted it up a bit as if we're looking up at the screen. This discontinuity is basically saying that as soon as we tip up past the 0 degree mark (where it seems we're still facing forward), our orientation is now pointing behind us. This is not intuitive at all. Fortunately, Android 2.3 came along and provided additional methods to clear this all up for us (see "Rotation Vector Sensors"). But in the meantime, as long as you deploy applications to pre-Android 2.3 devices, you'll need to worry about what values to use with your sensors.

Orientation Sensors

We've avoided the orientation sensors until now, but it's time we introduced them. We've just explained how the magnetic field and the accelerometer sensors can be combined and made to work together to produce orientation values to tell you in which direction the phone is facing. There is another sensor that does the same thing: the orientation sensor. The orientation sensor is actually a combination of the magnetic field and accelerometer sensors at the driver level of Android. In other words, there is no extra hardware for the orientation sensor, but within the Android OS, there is code to expose these two sensors as if they were another sensor for orientation.

> **NOTE:** We avoided talking about orientation sensors until now because they were deprecated as of Android 2.2 and you're not supposed to use them anymore. However, this sensor is very easy to use, as you'll soon see. So we thought "what the heck, let's show it."

We just discussed how using the preferred method of calculating orientation is challenging. In our next sample application, we'll expose the orientation values from the preferred method as well as the orientation sensor so you can see for yourself the differences between them.

We're going to have a little fun with this application. While we can easily show the values returned from the sensors, we're also going to do something interesting with them. Imagine you're standing in a street in Jacksonville, FL. Our application is going to show you pictures from Streetview as if you were there, using the orientation of your phone to select which way you're facing. As you change the orientation of your phone, the view in Streetview will change accordingly. Listing 29–4 shows the Java code for our sample application, which we call `VirtualJax`.

Listing 29–4. *Getting Orientation from Sensors*

```
// This file is MainActivity.java
public class MainActivity extends Activity implements SensorEventListener {
    private static final String TAG = "VirtualJax";
    private SensorManager mgr;
    private Sensor accel;
```

```
private Sensor compass;
private Sensor orient;
private TextView preferred;
private TextView orientation;
private boolean ready = false;
private float[] accelValues = new float[3];
private float[] compassValues = new float[3];
private float[] inR = new float[9];
private float[] inclineMatrix = new float[9];
private float[] orientationValues = new float[3];
private float[] prefValues = new float[3];
private float mAzimuth;
private double mInclination;
private int counter;
private int mRotation;

@Override
public void onCreate(Bundle savedInstanceState) {
    super.onCreate(savedInstanceState);
    setContentView(R.layout.main);

    preferred = (TextView)findViewById(R.id.preferred);
    orientation = (TextView)findViewById(R.id.orientation);

    mgr = (SensorManager) this.getSystemService(SENSOR_SERVICE);

    accel = mgr.getDefaultSensor(Sensor.TYPE_ACCELEROMETER);
    compass = mgr.getDefaultSensor(Sensor.TYPE_MAGNETIC_FIELD);
    orient = mgr.getDefaultSensor(Sensor.TYPE_ORIENTATION);

    WindowManager window = (WindowManager)
            this.getSystemService(WINDOW_SERVICE);
    int apiLevel = Integer.parseInt(Build.VERSION.SDK);
    if(apiLevel < 8) {
        mRotation = window.getDefaultDisplay().getOrientation();
    }
    else {
        mRotation = window.getDefaultDisplay().getRotation();
    }
}

@Override
protected void onResume() {
    mgr.registerListener(this, accel,
            SensorManager.SENSOR_DELAY_GAME);
    mgr.registerListener(this, compass,
            SensorManager.SENSOR_DELAY_GAME);
    mgr.registerListener(this, orient,
            SensorManager.SENSOR_DELAY_GAME);
    super.onResume();
}

@Override
protected void onPause() {
    mgr.unregisterListener(this, accel);
    mgr.unregisterListener(this, compass);
    mgr.unregisterListener(this, orient);
```

```
        super.onPause();
    }

    public void onAccuracyChanged(Sensor sensor, int accuracy) {
        // ignore
    }

    public void onSensorChanged(SensorEvent event) {
        // Need to get both accelerometer and compass
        // before we can determine our orientation
        switch(event.sensor.getType()) {
        case Sensor.TYPE_ACCELEROMETER:
            for(int i=0; i<3; i++) {
                accelValues[i] = event.values[i];
            }
            if(compassValues[0] != 0)
                ready = true;
            break;
        case Sensor.TYPE_MAGNETIC_FIELD:
            for(int i=0; i<3; i++) {
                compassValues[i] = event.values[i];
            }
            if(accelValues[2] != 0)
                ready = true;
            break;
        case Sensor.TYPE_ORIENTATION:
            for(int i=0; i<3; i++) {
                orientationValues[i] = event.values[i];
            }
            break;
        }

        if(!ready)
            return;

        if(SensorManager.getRotationMatrix(
                inR, inclineMatrix, accelValues, compassValues)) {
            // got a good rotation matrix

            SensorManager.getOrientation(inR, prefValues);

            mInclination = SensorManager.getInclination(inclineMatrix);

            // Display every 10th value
            if(counter++ % 10 == 0) {
                doUpdate(null);
                counter = 1;
            }
        }
    }

    public void doUpdate(View view) {
        if(!ready)
            return;

        mAzimuth = (float) Math.toDegrees(prefValues[0]);
        if(mAzimuth < 0) {
```

```
            mAzimuth += 360.0f;
        }

        String msg = String.format(
    "Preferred:\nazimuth (Z): %7.3f \npitch (X): %7.3f\nroll (Y): %7.3f",
                mAzimuth, Math.toDegrees(prefValues[1]),
                Math.toDegrees(prefValues[2]));
        preferred.setText(msg);

        msg = String.format(
    "Orientation Sensor:\nazimuth (Z): %7.3f\npitch (X): %7.3f\nroll (Y): %7.3f",
            orientationValues[0],
            orientationValues[1],
            orientationValues[2]);
        orientation.setText(msg);

        preferred.invalidate();
        orientation.invalidate();
    }

    public void doShow(View view) {
        // google.streetview:cbll=30.32454,-81.6584&cbp=1,yaw,,pitch,1.0
        // yaw = degrees clockwise from North
        // For yaw we can use either mAzimuth or orientationValues[0].
        //
        // pitch = degrees up or down. -90 is looking straight up,
        // +90 is looking straight down
        // except that pitch doesn't work properly
        Intent intent=new Intent(Intent.ACTION_VIEW, Uri.parse(
            "google.streetview:cbll=30.32454,-81.6584&cbp=1," +
            Math.round(orientationValues[0]) + ",,0,1.0"
            ));
        startActivity(intent);
        return;
    }
}
```

The user interface is two buttons and a pair of sensor value listings, one for the preferred method and one for the orientation sensor output. When you run this, you should see something like Figure 29–5.

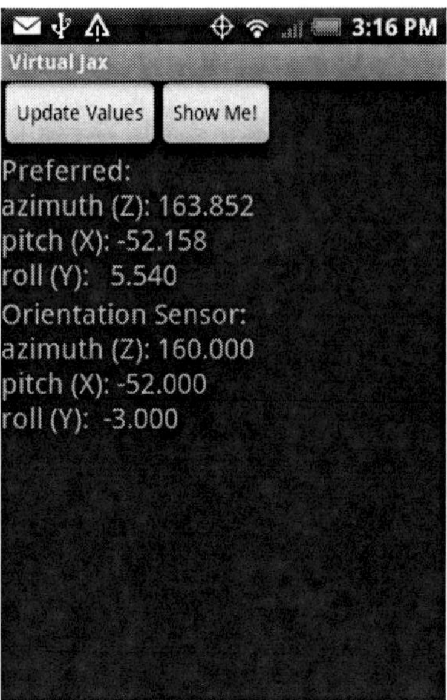

Figure 29–5. *Orientation done two ways*

Before we look at the results, let's explain what this application is doing. In the onCreate() method, we're doing the same sorts of things we did before: we're getting references to our text views, a SensorManager, and the three sensors we want to use here: accelerometers, compass, and the orientation sensor. We're also defining a variable to hold a rotation value. We'll get to that in a minute.

In onResume(), we activate the sensors; and in onPause(), we disable them. Notice how we can unregister our listener for specific sensors. We could have used the following to unregister our listener for all sensors all at once:

```
mgr.unregisterListener(this);
```

When we get a sensor value update, we switch on which type it is and record the values into local members: accelValues, compassValues, or orientationValues. Note that we could have cloned the event array to keep local copies of the values; however, that would mean instantiating objects constantly, which we don't really want to do. The cost of creating new objects and garbage cleaning up after them could really hurt performance, so we simply update our existing arrays.

Notice how we make sure we have values for both accelValues and compassValues, using the boolean ready, before we proceed into the next section of code. Now we see the getRotationMatrix() method call, followed by the getOrientation() method call. We also included the getInclination() method call. We're not going to use that here, but know that it represents the angle of the magnetic waves relative to the Earth's surface. The closer you are to the Earth's poles, the larger an angle this returns. Next we check a counter, as before, to only update the display every tenth update. Again, this is to prevent too much UI activity, which might cause our application to behave very poorly.

Within our doUpdate() method, which can also be called via the button in the UI, we're doing a few calculations and displaying the results. Using the preferred method, the first value, the azimuth, has a value in radians from negative pi to positive pi, representing −180 degrees to +180 degrees. The orientation sensor provides a value from 0 (north) to 360 degrees. To make these values comparable, we took the first value from the prefValues array, converted from radians to degrees, and added 360 if the value was negative. Now we're comparable to the orientation sensor. The rest of this method simply displays the sensor values in the UI.

Our last method in this sample application is doShow(). This is the fun one. We're going to set the yaw value for Streetview to indicate which way we want to be facing when displaying the image. Now we can show you how to pass in the yaw value as well as the pitch value.

For the latitude and longitude, we've preselected a location in Jacksonville, FL. You're free of course to substitute your own value. For yaw, we need to pass the number of degrees from north (0 - 360), so we use the value from either mAzimuth or orientationValues[0], converted to an integer. For pitch, in theory we could use the second value from either array, after adding 90 to it. However, the Streetview application doesn't seem to like pitch values other than 0, at least in this location. So we chose to set it to 0 for now. If you click the Show Me! button, you will get Streetview, and the image will be as if you were facing in the same direction as you are now, but in that location. If you click the Back button, rotate yourself, and click Show Me! again, you'll see the image from your new perspective. Now let's look more closely at the actual values from the sensors.

The values between the preferred method and the orientation sensor seem to be the same or very close to it. The values from the orientation sensor appear to be more stable. They also appear to be integer values. Looks pretty good right? But not so fast. When you start moving the device around, you'll find that if you tilt it such that you're looking up at it, the values get quite different. Now rotate the device so it's in landscape mode. You might see something that looks like Figure 29–6.

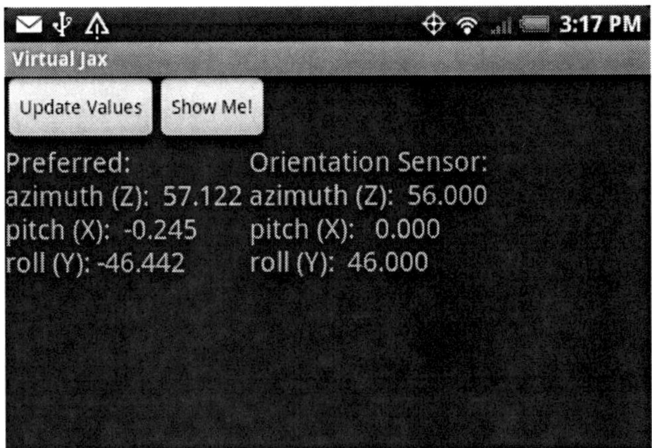

Figure 29–6. *Orientation done two ways in landscape mode*

What happened? Our roll value is opposite between the preferred method and the orientation sensor. What's going on is that the frames of reference are different between the two.

We haven't discussed yet what happens if we're not in portrait mode but rather in landscape mode. If the device is right in front of us in landscape mode, the accelerometers are still fixed in position, so instead of y going up, it's really x. We could do some math gymnastics to make everything work out for us, but fortunately, the `SensorManager` class has yet another method to help us out. This time the method is called `remapCoordinateSystem()`. It would be called in between getting the rotation matrix and calling `getOrientation()`. The basic function of `remapCoordinateSystem()` is to modify the rotation matrix by swapping axes around. The method signature looks like this:

```
public static boolean remapCoordinateSystem (float[] inR, int X, int Y, float[] outR)
```

We pass in our rotation matrix, plus values to indicate how to swap our x and y axes, and we get back a new rotation matrix (`outR`) plus a boolean return value that indicates if the remapping was successful. The values for x and y are constants from `SensorManager`, such as `AXIS_Z` and `AXIS_MINUS_Y`.

We've included a new sample application called `VirtualJaxWithRemap` with the downloads on the web site so you can see what this looks like.

Magnetic Declination and GeomagneticField

There's another topic we want to cover with regard to orientation and devices. The compass sensor will tell you where magnetic north is, but it won't tell you where true north is (a.k.a., geographic north). Imagine you are standing at the midpoint between the magnetic north pole and the geographic north pole. They'd be 180 degrees apart. The further away you get from the two north poles, the smaller this angle difference becomes. The angle difference between magnetic north and true north is called *magnetic declination*. And the value can only be computed relative to a point on the planet's surface. That is, you have to know where you're standing to know where geographic north is in relation to magnetic north. Fortunately, Android has a way to help us out, and it's the `GeomagneticField` class.

In order to instantiate an object of the `GeomagneticField` class, you need to pass in a latitude and longitude. Therefore, in order to get a magnetic declination angle, we need to know where the point of reference is. You also need to know the time at which you want the value. Magnetic north drifts over time. Once instantiated, you simply call this method to get the declination angle (in degrees):

```
float declinationAngle = geoMagField.getDeclination();
```

The value of `declinationAngle` will be positive if magnetic north is to the east of geographic north.

Gravity Sensors

Android 2.3 introduced the gravity sensor. This isn't really a separate piece of hardware. It's a virtual sensor based on the accelerometers. In fact, this sensor uses logic similar to what we described earlier for accelerometers to produce the gravity component of the forces acting on a device. We cannot access this logic, however, so whatever factors and logic are used inside the gravity sensor class are what we must accept. It's possible, though, that the virtual sensor will take advantage of other hardware such as a gyroscope to help it calculate gravity more accurately. The values array for this sensor reports gravity just like the accelerometer sensor reports its values.

Linear Acceleration Sensors

Similar to the gravity sensor, the linear acceleration sensor is a virtual sensor that represents the accelerometer forces minus gravity. Again, we did our own calculations earlier on the accelerometer sensor values to strip out gravity to get just these linear acceleration force values. This sensor makes that more convenient for us. And it could take advantage of other hardware, such as a gyroscope, to help it calculate linear acceleration more accurately. The values array reports linear acceleration just like the accelerometer sensor reports its values.

Rotation Vector Sensors

The rotation vector sensor is like the deprecated orientation sensor in that it represents the orientation of the device in space, with angles relative to the frame of reference of the hardware accelerometer (see Figure 29–2). However, unlike the orientation sensor, this sensor returns a set of values that represents the last three components of a unit quaternion. Quaternions are a subject that could fill a book, so we won't be going into them here.

Thankfully, Google has provided a few methods within `SensorManager` to help with this sensor. The `getQuaternionFromVector()` method converts a rotation vector sensor output to a normalized quaternion. The `getRotationMatrixFromVector()` method converts a rotation vector sensor output to a rotation matrix, and that can be used with `getOrientation()` to be used like the orientation sensor output we used earlier. When converting rotation vector sensor output to an orientation vector, though, you need to realize that it goes from -180 degrees to +180 degrees, just like the preferred values in the `VirtualJax` example.

The ZIP file of sample apps for this chapter includes a version of `VirtualJax` that shows the rotation vector in use.

Near Field Communication Sensors

With the introduction of Android 2.3, we got the ability to work with special tags using Near Field Communications (NFC). NFC tags are similar to Radio Frequency ID tags (RFID), except that the range for NFC is less than four inches. This means the sensor in the Android device must come very close to the tag to be scanned. NFC tags can be programmed to give out text information, URIs, and metadata, such as the language of the information. Certain NFC operations can also be secured.

There are actually three modes of NFC operation:

- *Reading and writing contactless tags:* These tags are generally very tiny and do not require any battery power. They can be embedded cheaply in all sorts of objects such as movie posters, products, stickers, and so on.
- *Card emulation mode:* Think smart credit cards. This allows an Android device to act like a smart card. The obvious benefit of this is that your device could act like one card, and then act like a different card at the touch of a button. This is one way an Android device could replace your wallet. Whatever credit card you own, or bus pass, or ticket, your Android device could impersonate (securely, of course) that item, so the reader on the other side of the transaction would think it was working with your credit card when in fact it was dealing with your Android device.
- *Peer-to-peer communication:* Each side recognizes that it is talking to another device and not just a tag. The protocol has been developed by Google and allows two devices to send messages back and forth.

Beyond using NFC to conduct financial transactions, NFC tags could be used in many other scenarios. For example, a museum could place an NFC tag next to items in its collection, allowing visitors to wave their phone close to the tag in order to access a web page that could provide multimedia information about that item. Bus stops could display an NFC tag allowing people to find out when the next bus is coming and where it is going. Businesses could display an NFC tag allowing easy check-in for location-aware services as a person walks in. Perhaps hotel room keys will be irrelevant when you can use your phone to unlock an NFC-equipped door. Even products on store shelves could come with NFC tags to allow shoppers to get more information on that product, such as nutritional information, or perhaps technical specifications and promotional videos.

Enabling the NFC Sensor

The support in Android for NFC is not like the other sensor types. Instead of working with `SensorManager`, you work with `NfcAdapter`. There is typically only one adapter on a device, and its job is to manage the reading and writing of tags, and the distribution of tags to activities on the device. The adapter can be either on or off, and there are controls under Settings to enable or disable it. The NFC adapter setting is with the Wireless settings.

If the adapter is on, and an NFC tag is detected, a somewhat complicated process is followed to determine which activity, if any, should receive an intent informing the activity about the detected NFC tag. Everything hinges on what sort of data is in the NFC tag and what intent filters exist for the installed applications on the device. And there is one other bit of information that is considered, and that is whether or not the activity currently in the foreground on the device has expressed a specific desire to receive NFC tags. We'll explore this more fully very soon.

To access the adapter, you first acquire an `NfcManager` instance using `getSystemService()`. Then you call the `getDefaultAdapter()` method on that, like so:

```
NfcManager manager = (NfcManager)
    context.getSystemService(Context.NFC_SERVICE);
NfcAdapter adapter = manager.getDefaultAdapter();
```

This returns the singleton object that is the `NfcAdapter`. To determine if the `NfcAdapter` is currently enabled, use the `isEnabled()` method, which returns a boolean answer telling you whether the NFC adapter is enabled in Settings.

There is no documented way to programmatically turn on (or off) the NFC adapter. If the NFC adapter is off and you want it turned on, you'll need to notify the user to ask them to enable the NFC adapter under Settings. To launch the appropriate Settings screen for the user from your application, you could use code like the following:

```
startActivityForResult(new Intent(
     android.provider.Settings.ACTION_WIRELESS_SETTINGS), 0);
```

When this runs, the appropriate Settings screen will be displayed, and the user can choose to enable NFC or not. Your activity's `onActivityResult()` callback will be called when the user is finished with the wireless settings screen. Keep in mind that the user may choose *not* to enable NFC even though you asked them to. Your application should take appropriate action if the NFC adapter stays disabled.

Routing NFC Tags

This seems like a good time to discuss the different types of NFC tags and technologies. NFC is not one single standard. In fact, there are several types of NFC tags that a user could come across. There is variation among the tag types, which means that Android must support them with different classes related to each tag type. If you look inside the `android.nfc.tech` package, you will find several different tag technology classes, from MiFare Classic to NfcV to ISO-DEP. The internal structures of each tag type can be different, and there are different methods for accessing and manipulating data in these tag types. Fortunately, Android provides a `Tag` class to help manage NFC communications, and each specific type of tag can be created from a `Tag` object. Once you have an instance of a specific NFC tag, you can perform operations on it that are specific to that tag type. This also means that to choose which activity to send a tag to, several factors must be considered. We'll first describe how an NFC tag intent is created, and then you can understand how to create an appropriate intent filter.

When an intent is being sent with tag data, a `Tag` object is always parceled into the intent's extras bundle, with a key of `EXTRA_TAG`. If the tag contains NDEF data, another extras value is set with a key of `EXTRA_NDEF_MESSAGES`. Last, the intent could have an extras value of the tag's ID with a key of `EXTRA_ID`. These last two extras values are optional and depend on the existence of the data on the tag. All NFC intents are sent using `startActivity()`. Note that you never need to actually access the NFC adapter to receive NFC messages. The intent messages will come into your application just like other intents that are sent from other sources, as long as they match your intent filter(s).

> **NOTE:** It is important to note that there is an NFC ecosystem in an Android device that supports NFC. The logic to create these NFC intents uses capabilities that are not exposed in the Android SDK. That means you cannot easily create a fake sender activity yourself. What we're about to explain is what happens in the NFC ecosystem, and it is not something you can write your own code for. This also means that if you really want to test an NFC application, you will need to use a real device with real NFC tags—unless Google someday provides some support in the emulator or in DDMS or both.

The action value of the tag intent depends on what information was discovered about the detected tag. There are three possible action values for the intent:

- `ACTION_NDEF_DISCOVERED` is the action if an NDEF payload is found in the tag. If this is the case, Android then looks for the existence of a `NdefRecord` in the first `NdefMessage`. If that `NdefRecord` is a URI or `SmartPoster` record, the intent will get the URI in its data field. If a MIME record is found, the intent's type field will be set to the MIME type of the tag. Android then looks for a suitable activity to start using this intent and the intent matching algorithm. If no activity can be found, this intent is abandoned, and Android tries to create the next type of NFC intent.
- `ACTION_TECH_DISCOVERED` is the action if NDEF is not detected or no NDEF activity could be found, but a tag technology exists. In this scenario, Android adds metadata to the intent indicating which tag technologies were detected. An NFC tag can implement more than one technology, especially since NDEF is more like a virtual technology. Android looks for an activity that will match this intent and, if found, sends it on. If not, Android throws this intent away and tries the third type of NFC intent.

- `ACTION_TAG_DISCOVERED` is the final action choice for an NFC tag. This is the action when all others failed to match an activity. This intent also does not carry data or a MIME type. If this intent does not match an activity on the device, then the NFC ecosystem gives up, and the tag information is thrown away.

Android Application Records

There's another way that NFC tags can be routed, starting with Android 4.0. An NFC tag can contain an Android Application Record (AAR) value anywhere within the NDEF message. The tag specifies an Android application package name. When a tag with an AAR is received on an Android device, Android will search locally for an application with that package name, and launch it if it is found. If it is not found, Android will direct the user to the Android Market to download the application so the tag can be processed. You would want to do this if you really needed to lock a user into your application to process your NFC tags. If the user chooses not to download your application, the tags become useless, because Android will not deliver them anywhere else. Actually, that last part is not entirely true: it is possible to override the AAR by using foreground dispatch.

As tempting as it might be to use AARs with your tags as the main mechanism to connect tags to your application, you should use MIME types and URIs in addition to support a broader range of devices. In this case, you'd probably be better off sticking with MIME types and URIs alone and not worrying about AARs.

Receiving NFC Tags

Whether you decide to create your intent filters in code or in the `AndroidManifest.xml` file, you will need to know what you are looking for and prepare your intent filters carefully. For example, if you specify too rigidly, you won't get notified for tags that you are interested in. If you specify too loosely, you'll get called for tags that you don't want to handle. And if your app is sent an NFC tag that you don't want to handle, that means another app might possibly exist on the device that could handle it, but didn't get it. This could happen if the intent-matching logic found more than one app and asked the user which one to run, and the user chose yours. That's yet another reason you want to be careful when defining your intent filters for NFC tags; if the user is prompted for which app to run, they very likely need to move the device away from the NFC tag to make the choice, and now the tag is out of range. If you have choice in what data the tags are going to have on them, you could make that data very specific to your needs, using a custom URI scheme or a custom MIME type for example.

Your choice of intent filter depends on which action was put into the NFC tag intent (see above). Listing 29–5 shows a sample intent filter for an NDEF tag that would go into your `AndroidManifest.xml` file.

Listing 29–5. *Intent Filter for an NDEF Tag with a MIME Type*

```
<intent-filter>
  <action android:name="android.nfc.action.NDEF_DISCOVERED"/>
  <data android:mimeType="type/subtype" />
</intent-filter>
```

Instead of "type/subtype" you would of course put the specific MIME type that you are looking for, or use wildcards if you will accept any type or subtype. For example you could set mimeType to "text/*" to match all text types. But you don't need to specify a MIME type for an NDEF tag. If the tag has a URI instead of a MIME type, you would want to use an intent filter like in Listing 29–6.

Listing 29–6. *Intent Filter for an NDEF Tag with a URI*

```
<intent-filter>
  <action android:name="android.nfc.action.NDEF_DISCOVERED"/>
  <data android:scheme="geo" />
</intent-filter>
```

In this example, we use the geo scheme so our activity would be launched if a tag with a geo: URI was detected. You could use any of the other attributes of the <data> tag to specify what NFC data your activity is looking for.

If your activity is looking for NFC tags that have a particular technology, you would use an intent filter as in Listing 29–7. It is also possible that a tag with NDEF was detected, but no activity could be found to process the NDEF_DISCOVERED intent. That could also result in your activity receiving the intent, as long as it matches your intent filter. In other words, if an NDEF_DISCOVERED tag intent could not be delivered to an activity looking for NDEF tags, an activity looking for a particular technology could end up receiving a technology intent for that tag.

Listing 29–7. *Intent Filter for an NFC Tag with Technology*

```
<intent-filter>
  <action android:name="android.nfc.action.TECH_DISCOVERED"/>
</intent-filter>
<meta-data android:name="android.nfc.action.TECH_DISCOVERED"
                android:resource="@xml/nfc_tech_filter" />
```

Notice that we have a different action now to match technology, and instead of a <data> tag, we have a <meta-data> tag, and it's outside of the <intent-filter> tag. The attributes of the <meta-data> tag are different too, and refer to another file that we must create under the /res/xml directory of our application's project. Listing 29–8 shows a sample nfc_tech_filter.xml file.

Listing 29–8. *A Sample NFC Tech Filter XML File*

```
<resources xmlns:xliff="urn:oasis:names:tc:xliff:document:1.2">
    <tech-list>
        <tech>android.nfc.tech.NfcA</tech>
        <tech>android.nfc.tech.MifareUltralight</tech>
    </tech-list>
</resources>

<resources xmlns:xliff="urn:oasis:names:tc:xliff:document:1.2">
    <tech-list>
```

```
        <tech>android.nfc.tech.NfcB</tech>
        <tech>android.nfc.tech.Ndef</tech>
    </tech-list>
</resources>
```

What this filter file does is specify two types of tags that our activity wants to see. An NFC tag usually has its list of technologies that it enumerates. If any one of the tech-lists in Listing 29–8 is a subset of our tag's tech-list, then this is a match and our activity will get that NFC tag intent.

In Listing 29–8, the first type of tag has NfcA and MifareUltralight technologies, and the second type of tag has NfcB and Ndef technologies. We could add additional <resources> to this file to specify additional tags that our activity could want to see. The list of available technologies to put into this file are the tag class names that are available in the android.nfc.tech package, but only put in what you want your activity to receive. The child tags of a <tech-list> specify all of the technologies that a tag must report for its intent to match our activity. All of the technologies in a specific tech-list must exist in the list of technologies enumerated by the tag. Therefore, the tech-list in the intent filter could have fewer technologies than the tag specifies but could not have more and still match. For the example in Listing 29–8, if a tag presented just the Ndef technology, it would not match either specification, and your activity would not receive the intent. None of the intent filter tech-lists is a subset of the tag's list. If a tag had NfcA, NfcB, and Ndef technologies, it would match the second specification, and your activity would receive the intent. The second tech-list is a subset of the tag's tech-list. We would match even though the tag enumerates one more technology than is in the intent filter's tech-list.

The final intent filter that you might use is shown in Listing 29–9, and it represents the catch-all intent filter. That is, if a tag was received and no NDEF or tech activity could be found to process the intent, or if the tag was an unknown type, an intent would be created with the ACTION_TAG_DISCOVERED action.

Listing 29–9. *Intent Filter for an Unknown or Unprocessed NFC Tag*

```
<intent-filter>
  <action android:name="android.nfc.action.TAG_DISCOVERED"/>
</intent-filter>
```

Notice that there is no <data> or <meta-data> tag for this intent filter, because there will not be any data in an intent that has an action of ACTION_TAG_DISCOVERED. This would normally mean that we must have a <category> tag. However, this is not the case with NFC tag intents. NFC tag intents are special, so no <category> tags are required in intent filters for NFC tag intent matching.

Getting back to our tag matching flow, when we're getting an ACTION_TAG_DISCOVERED intent, Android has almost given up trying to find an activity for the detected NFC tag. At this point, any activity that will take an ACTION_TAG_DISCOVERED action will receive these tag intents. In most normal operations, you won't ever see an ACTION_TAG_DISCOVERED tag intent, because almost all NFC tags that you'll come across will match on NDEF or on TECH.

There is another way that your activity could receive an NFC tag intent, and that is by using the foreground dispatch system. If your activity is in the foreground (which means onResume() is firing or has fired and the user can interact with your activity), you make a call like the following

```
mAdapter.enableForegroundDispatch(this, pendingIntent,
                       intentFiltersArray, techListsArray);
```

where mAdapter is the NFC adapter, and this is a reference to your activity. By making this call, you effectively insert your activity in front of all others, and if any of this activity's intent filters match a detected tag, your activity will get to process it. If your activity does not get the NFC tag intent because it doesn't match the setup of this call, the NFC tag intent will be tried with other activities using the previous logic. You must call this method from the UI thread, and the best place to do so is from the onResume() method of your activity. You would also need to call

```
mAdapter.disableForegroundDispatch(this);
```

from the onPause() callback of your activity, so that your activity won't get an intent it can't process. When your activity does get an intent in this way, the onNewIntent() callback will be used to receive it into your activity.

The pending intent is a standard one. The intentFiltersArray would be the collection of IntentFilter objects that you desire, each one specifying an appropriate action and any data or MIME types as needed. For example, Listing 29–10 shows some code to create an intent filter for Ndef and then add it to an array. This code would most likely be in your onCreate() method.

Listing 29–10. *Code for an Intent Filter for Ndef*

```
IntentFilter ndef = new IntentFilter(NfcAdapter.ACTION_NDEF_DISCOVERED);
try {
    ndef.addDataType("text/*");
}
catch (MalformedMimeTypeException e) {
    throw new RuntimeException("fail", e);
}
intentFiltersArray = new IntentFilter[] {
        ndef,
};
```

Keep in mind that the intent filter array can contain multiple instances of IntentFilter, each set with the same or different action, and with or without data and/or type field values.

The techListsArray is an array of arrays, where each inside array is the list of class names that a tag would enumerate, and you can have multiple lists of class names to match against. Listing 29–11 shows a sample of this, which is equivalent to the tech-list resource file shown in Listing 29–8. This code would also most likely be in your onCreate() method.

Listing 29–11. *Code for a* `tech-list` *Array*

```
techListsArray = new String[][] {
    new String[] { NfcA.class.getName(),
                   MifareUltralight.class.getName() },
    new String[] { NfcB.class.getName(),
                   Ndef.class.getName() }
    };
```

When all of this setup has been done, if this activity does receive an NFC tag intent, it will be the `onNewIntent()` callback that will be triggered to receive it. From there, you would access the extras bundle to read the tag, which we'll cover next. This is a lot of setup to do a dynamic claim for an NFC tag intent, but on the flip side, if you only want this activity to receive tags if it has already been started by the user, this is the way to do it. Note that it probably doesn't make sense to use this method and to also have intent filters in the manifest to receive NFC tag intents, but technically it is possible.

Reading NFC Tags

As alluded to earlier, the reading of NFC tags is somewhat complicated. Or rather, the process by which a tag gets delivered to your application can be complicated. At the most basic level, when an NFC tag is detected, the system will determine an activity to send the tag to, and then send it. Unlike with the sensors covered earlier in this chapter, the activity interested in NFC tags may not be running at the time of tag detection, and it certainly won't receive the tag information through a sensor listener. A notified activity will receive an intent, and this may mean launching the activity in order for it to process the NFC tag intent.

One of your first considerations when designing an application that receives and processes NFC tag information is that you are dealing with a physical tag in the environment of the device through a hardware interface. The NFC API has blocking calls, which means they might not return as quickly as you'd like, so you need to run the tag methods on a separate thread from the main UI thread.

The NFC tag data will be in the extras bundle of the received intent. Upon receiving the intent, you would access the NFC data using something like this:

```
Tag tag = intent.getParcelableExtra(NfcAdapter.EXTRA_TAG);
String[] techlists = tag.getTechLists();
```

If your intent filter was very precise, you already know what type of tag you have. But if there is a selection of tag technologies that could be present, you can now interrogate `techlists` to find out what technologies are in the tag. Each string is the class name of the tag technology that is enumerated by the detected tag.

If you find out that `android.nfc.tech.Ndef` is supported in this tag, you could do the following to get to the NDEF data more directly:

```
NdefMessage[] ndefMsgs = intent.getParcelableArrayExtra(NfcAdapter.EXTRA_NDEF_MESSAGES);
```

In theory, you could get a `null` value if no NDEF messages were in the intent. Otherwise, you should now be able to parse the NDEF messages sent to you. You could read the

NdefMessages from the intent, count them, and for each one, retrieve the NdefRecords contained within.

The NdefRecords are where things get interesting. You would be well-served to refer to the NFC specifications located here: www.nfc-forum.org/specs/. To access these specifications, you will need to accept a licensing agreement with the NFC Forum. It is free, but you will need to provide your name, address, telephone number, and e-mail address. Your other option is to look at the NfcDemo application that Google provides. That sample is included with the Android 2.3.3 SDK package under the samples folder. You can also view the source of that application here: http://developer.android.com/resources/samples/NFCDemo/index.html. This sample application receives NFC intents and displays the contents of the NdefRecords in a ListView. The reason this gets complicated is that there are several types of NdefRecords that you could receive in each NdefMessage. Each type serves a different purpose. For example, the Text type contains text in a specified language. The Uri type contains a URI. Of the known NDEF record types, the NfcDemo sample application uses just three: the two just described and SmartPoster, which we'll describe shortly.

The format of an NdefRecord includes a 3-bit Type Name Format (TNF) field, a variable-length type field, a variable-length ID field, and a variable-length payload field. Yes, there are two type fields. The TNF field is the top-level type of this record, and it tells you what the rest of the record is. For example, it could be an absolute URI record (TNF_ABSOLUTE_URI) or an official RTD record (TNF_WELL_KNOWN). The next type field gets more specific about what this record is, based on the value of TNF. If the TNF value is TNF_WELL_KNOWN, this next type field will be one of the RTD_* constants of the NdefRecord class, such as RTD_SMART_POSTER. If the TNF value is TNF_ABSOLUTE_URI, the next type field will follow the absolute-URI BNF construct defined by RFC 3986.

NOTE: The TNF_UNCHANGED record type is used when the message payload spans multiple NdefRecords because of its size. Google has taken care of handling chunked NdefRecords for you, so you should never see a type value of TNF_UNCHANGED. The android.nfc package combines the pieces of the payload into one big, single NdefRecord.

The next field in an NdefRecord is an identifier for this NdefRecord. The NdefRecord you're reading may or may not have an identifier.

Finally, there is the payload. This can be a rather large byte array, but it has some internal structure to it that you must be aware of, depending on which type of NdefRecord this is. For an RTD URI record type, the first byte of the payload byte array represents the beginning of the URI. For example, the byte value of 1 represents "http://www.", and this would precede the rest of the URI in the rest of the payload. For a Text record type, the first byte of the payload byte array represents the "status byte encodings" value, which identifies the text encoding value (UTF-8 or UTF-16), as well as the length of the language byte array which immediately follows this status field. After the language field is the text. For SmartPoster, things get more complicated, with the NdefRecord containing an NdefMessage which in turn contains more NdefRecords. The bottom NdefRecords can include Title records (just like a Text record), a URI record (just like before), a recommended action record, a size record, an icon record, and a type record. The recommended action value indicates what your application might want to do with the SmartPoster data. Note that these values are not provided as part of Android's NdefRecord class documentation. And they are as follows:

```
-1   UNKNOWN
0    DO_ACTION
1    SAVE_FOR_LATER
2    OPEN_FOR_EDITING
```

What you do with them is up to you, although obviously you probably want to attempt to perform the recommended action for the tag being read. For example, if TNF is TNF_WELL_KNOWN, the type is RTD_SMART_POSTER, and the recommended action is 0 (DO_ACTION) combined with a web page URL, you might want to launch the browser with that URL. The size record allows the tag to say how big the thing is at the other end of that URL. If the tag is referring to a downloadable executable, the size record could say how big the download file is. The icon record holds an icon image that can be used by a device to display an image along with the title and the URI.

The type record is yet another type value, different from the TNF and the NdefRecord's type. The type record is for SmartPoster tags, and in this case, the type represents the MIME type of the thing at the other end of the URI. A device could decide it can't support that object type, so avoid downloading it in the first place.

The only mandatory subrecord for a SmartPoster tag is the URI record, and there can be only one per SmartPoster. You can have multiple Title records, as long as each record is for a different language. You can also have multiple icon records as long as each has a different MIME type for its format.

For all types of NFC tags, including the NDEF tags, you can use something like the following to get an instance of that particular tag type:

```
NfcA nfca = NfcA.get(tag);
```

From this new object, we can access the specific methods that are appropriate for that tag type. For Ndef and NdefFormatable tags, the NdefMessage and NdefRecord classes are very helpful to deal with the tag data. The other tag classes have appropriate methods to help deal with those tags and their data. There are methods for reading and for writing data to a tag. Note that writing to a tag is not the same as the device doing card emulation. Writing a tag means that the device is close enough to some other tag to be able to write to it (with proper permissions, of course). Card emulation is different.

NFC Card Emulation

Card emulation means that the device will appear to another NFC reader as if the device is a smart card like a PayPass MasterCard credit card. This means our local device has a place in the hardware to store some data and programs, and if an NFC reader gets within range of our device and asks for the data, our device will talk to the reader.

Emulating credit cards demands an extraordinary amount of security. Typically, a piece of hardware called the Secure Element (SE) exists in NFC-equipped devices that do card emulation. Its purpose is to run secured programs that can interact with the NFC antenna on the device to participate in financial transactions. These are not programs that you write using the Android SDK, nor can you simply open up the SE and write programs to it. Only authorized third parties can do that, such as First Data for the Google Wallet program. In fact, too many attempts to unlock or tamper with the SE will cause it to self-destruct.

There is no published date when NFC card emulation might be supported in the SDK. Our recommendation is to not attempt to do this yourself. We've seen a few ideas on the Internet, but we won't be covering those in this book.

NFC Peer-to-Peer (P2P)

The Android SDK provides support for peer-to-peer (P2P) communication over NFC between two devices. There are some caveats to this feature, namely that P2P only works when your application is running and in the foreground, and also that your application must format with NDEF. Other tag technologies may be supported in P2P in the future, but for now it's just NDEF. This also means that your phone must be turned on and running your application for it to be able to talk NFC with another device.

To implement the P2P feature, you will use either the NfcAdapter method called `setNdefPushMessage()` or `setNdefPushMessageCallback()`. The first method takes an `NdefMessage` and at least one activity (can be multiple activities, one after another). This is the NDEF message that will be pushed across if any of the provided activities are in the foreground when an NFC reader requests a message. The second method is like the first, but instead of providing an NDEF message up front, the callback will be called to provide the message at that time. While it is technically possible to set up both of these methods on the same activity at the same time, the callback takes precedence and will be the only one called if both are active. To deactivate either of these, call them again with a null NDEF message or a null callback. Using the callback method is the better approach when the NDEF message to send is more dynamic, relying more on up-to-date data in your application.

Similar to the foreground dispatch system described earlier, these methods should be called in `onResume()` and disabled in `onPause()`. Your `NdefMessage` can be whatever you want it to be, but your activity should be in the foreground when the reader attempts to get your data.

There's one more new method in Android 4.0 that you should know about: the `setOnNdefPushCompleteCallback()` method. This is supposed to set up a callback for your activity (or activities) so you can know when an NDEF message has been pushed over P2P. However, it will be called on a binder thread, not a UI thread. This means it's sort of floating within your application. You will need to use a handler to actually communicate back to your activity, except that you don't get anything to work with as an argument to this callback. The parameter for the callback is an `NfcEvent` that only has the `nfcAdapter` field set. This isn't terribly helpful for figuring out which handler to use to communicate back to the appropriate activity. So your best bet is to use only one activity when setting up P2P within your application.

Earlier we covered the use of the uses-feature tag and sensors, so you can make sure that a device has the appropriate sensor in order to see your application. The NFC sensor is no exception. You should use the following in your AndroidManifest.xml to ensure that the device for your application has the necessary NFC hardware:

```
<uses-feature android:name="android.hardware.nfc" />
```

You should also ensure that your AndroidManifest.xml file contains an appropriate permission to allow your application to access the NFC hardware:

```
<uses-permission android:name="android.permission.NFC" />
```

Android Beam

One of the new features in Android 4.0 is something called Android Beam. This is more of a concept than an API, since it uses the APIs we've already discussed. It's basically the P2P mechanism of Android, with a little extra UI provided by Android to allow the user to control exactly when an NDEF message is sent between two NFC-capable devices.

Testing NFC with NFCDemo

We've covered much of the NFC API for Android, but the question now is, how do you test your application? For NFC tags, maybe you could find some objects that already have NFC tags in them. In countries that have been using NFC for a while, this might not be too difficult. In the United States, it's likely much harder. You could buy your own NFC tags; several vendors around the world sell tags as well as developer kits so you could write what you want onto your tags. Unfortunately, DDMS does not yet come with support for sending tag-discovery intents to the emulator. The NfcDemo sample application that is available in the Android SDK was first released with Android 2.3, back when there was only ACTION_TAG_DISCOVERED for the intents. Android advanced a lot with the release of 2.3.3, and unfortunately the NfcDemo couldn't keep up. There is some useful information in there about NFC tag layouts and what the bytes mean for NDEF tags. We hope this will get an update soon and will work with real tags and the new NFC ecosystem.

If you do decide to load the NfcDemo sample application, you may need to add an external library to your project. The download file for this library is located here: http://code.google.com/p/guava-libraries/. When you open the ZIP file, you will find JAR files. Save the Guava JAR file, the one without gwt, onto your workstation. You need to refer to the Guava JAR file from your Eclipse project by right-clicking the project, choosing Build Path, and then selecting Configure Build Path and the Libraries tab. Next, click Add External JARs, navigate to the Guava JAR file, select it, and click Open. Now rebuild the NfcDemo project by right-clicking the project and choosing Build Project.

References

Here are some helpful references to topics you may wish to explore further:

- www.androidbook.com/proandroid4/projects: A list of downloadable projects related to this book. For this chapter, look for a ZIP file called ProAndroid4_Ch29_Sensors.zip. This file contains all the projects from this chapter, listed in separate root directories. There is also a README.TXT file that describes exactly how to import projects into Eclipse from one of these ZIP files.
- http://en.wikipedia.org/wiki/Lux: The Wikipedia entry for lux, the unit of light measurement.
- www.ngdc.noaa.gov/geomag/faqgeom.shtml: Information about geomagnetism from NOAA.
- www.youtube.com/watch?v=C7JQ7Rpwn2k: A Google Tech Talk from David Sachs on accelerometers, gyroscopes, compasses, and Android development.
- http://stackoverflow.com/questions/1586658/combine-gyroscope-and-accelerometer-data: A nice posting on stackoverflow.com that talks about combining gyroscope and accelerometer sensor data for use in applications.
- http://en.wikipedia.org/wiki/Quaternions_and_spatial_rotation. The Wikipedia page on quaternions and how they can be used in representing spatial rotation, such as an Android device.
- www.nfc-forum.org/specs: The official site for the NFC specifications.
- www.slideshare.net/tdelazzari/architecture-and-development-of-nfc-applications: A very thorough SlideShare presentation about NFC by Thomas de Lazzari.
- www.youtube.com/watch?v=am8t6iZ7upO: The launch video for Google Wallet.

Summary

In this chapter, we covered the following topics:

- What sensors are in Android.
- Finding out what sensors are on a device.
- Specifying the sensors that are required for an application before it will be loadable onto an Android device.
- Determining the properties of a sensor on a device.
- How to get sensor events.
- The fact that events come whenever the sensor value changes, so it is important to understand there could be a lag before you get your first value.

- The different speeds of updates from a sensor and when to use each one.
- The details of a SensorEvent and how these can be used for the various sensor types.
- Virtual sensors, made up of data from other sensors. The ROTATION_VECTOR sensor is one of these.
- Determining the angle of the device using sensors, and telling which direction the device is facing.
- The NFC sensor and other components used with NFC.
- How to read and write tags.
- Card emulation, and how difficult it is to access as a developer due to the security involved.
- NFC P2P, and how to exchange messages with another NFC device that also has P2P capability.
- How NFC tag data is routed in Android, and how NFC tags are constructed and parsed.
- Android Beam

Interview Questions

The following questions are provided to help you solidify your understanding of this topic area:

1. True or false. There is one sensor framework for all sensors in Android, from light sensor to GPS sensor to NFC sensor.
2. Which sensors can be emulated in the Android emulator?
3. What mechanism should you use in AndroidManifest.xml to ensure that your application is only installed on devices that have the sensors you need?
4. Is it possible to directly read sensor values anytime you want? Why or why not?
5. Why is it a bad idea to clone the sensor event values array that's passed to the sensor callback method?
6. What's dangerous about using a constant like LIGHT_SUNLIGHT_MAX to tell when the device is outside?

7. Accelerometers are fixed to the device and don't compensate for device rotation even as the user interface adjusts itself to device rotation. What methods exist for doing your own compensation of accelerometer values as a device rotates?
8. What is a low-pass filter, and what does it do?
9. What sensors have already been deprecated?
10. For those sensors that have been deprecated, what took their place?
11. Why is it important to know where the device is when calculating the difference between magnetic north and true north?
12. What are the three modes of NFC?
13. What is the last possible way an NFC tag can match to an activity?
14. What is the one way to override AAR routing of NFC tags?
15. What is a secure element used for?

Chapter 30

Exploring the Contacts API

In Chapter 4, we covered content providers. In that chapter, we listed the benefits of exposing data through content provider abstraction. In a content provider abstraction, data is exposed as a series of URLs. These data URLs can be used to read, query, update, insert, and delete. These URLs and their corresponding cursors become the API for that content provider.

The Contacts API is one such content provider API for working with contact data. Contacts in Android are maintained in a database and exposed through a content provider whose authority is rooted at

```
content://com.android.contacts
```

The Android SDK documents the various URLs and the data they return using a set of Java interfaces and classes that are rooted at the Java package

```
android.provider.ContactsContract
```

You will see numerous classes whose parent context is `ContactsContract` that are useful in querying, reading, updating, and inserting contacts into and from the content database. The primary documentation for using the Contacts API is available on the Android site at

```
http://developer.android.com/resources/articles/contacts.html
```

The primary API entry point `ContactsContract` is appropriately named because this class defines the contract between the clients of the contacts and the provider and protector of the contacts database.

This chapter explores this contract in a fair amount of detail but does not cover every nuance. The Contacts API is large and its tentacles far-reaching. However, when you approach the Contacts API, it will take a few weeks of research to realize that it is simple in its underlying structure. This is where we would like to contribute the most and explain these basics in the time it takes to read this chapter.

Android 4.0 has extended the idea of contacts to include a user profile, similar to a user profile in a social network. A user profile is a dedicated contact that represents the owner of the device. Most of the general contact-based concepts remain the same. We will cover how the Contacts API is extended to support a user profile.

Understanding Accounts

All contacts in Android work in the context of an account. What is an account? Well, for example, if you have your e-mail through Google, you are said to have an account with Google. If you set up yourself as a user of Facebook, you are said to have an account with Facebook.

Even though you use only the e-mail service with Google, the same login and password could be used to access other Google services. Your Google e-mail account is not limited to just e-mail. However, some accounts are restricted to just one type of service such as a Post Office Protocol (POP) e-mail account. On your mobile device, you may be able to register with a variety of these account-based services.

You will be able to set up some of these accounts such as Google or a corporate Microsoft Exchange account through the "Accounts & sync" Settings option on the device. See the *Android User's Guide* to get more details around accounts and how to set them up. We have included a URL for the *Android User's Guide* in the "References" section at the end of this chapter.

A Quick Tour of Account Screens

To solidify the nature of accounts, let's look at a few account related screens from the emulator. To start off, Figure 30–1 shows the Account Settings options screen. Some of the screens you see here are from Android 2.3.x. We have included Android 4.0 screens only where they differ significantly.

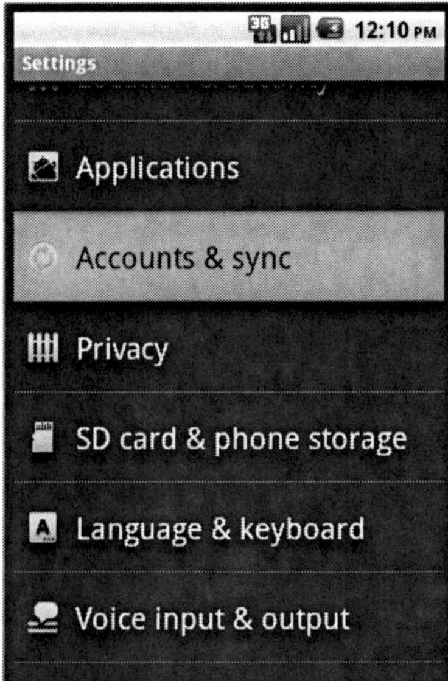

Figure 30–1. *Invoking "Accounts & sync" application settings*

When you choose the "Accounts & sync" menu item, you will see the "Accounts & sync settings" screen shown in Figure 30–2. This screen displays, along with some account-based options, a list of available accounts.

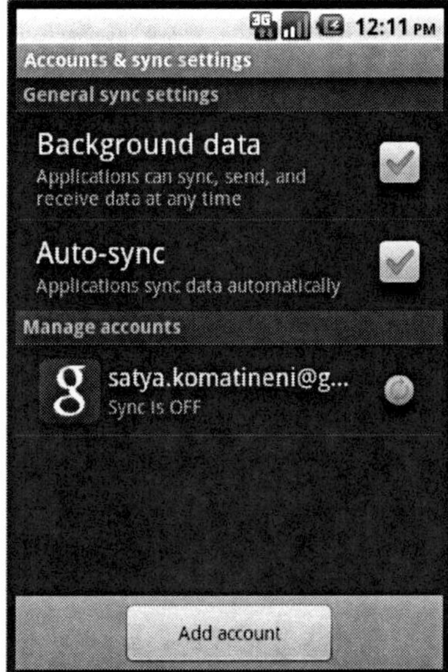

Figure 30–2. *"Accounts & sync settings" screen*

In Figure 30–2, we are mainly interested in the list of available accounts. Just to see what is involved in adding a new account, click the "Add account" button, and you will see the screen in Figure 30–3 with a list of possible accounts that can be set up or added.

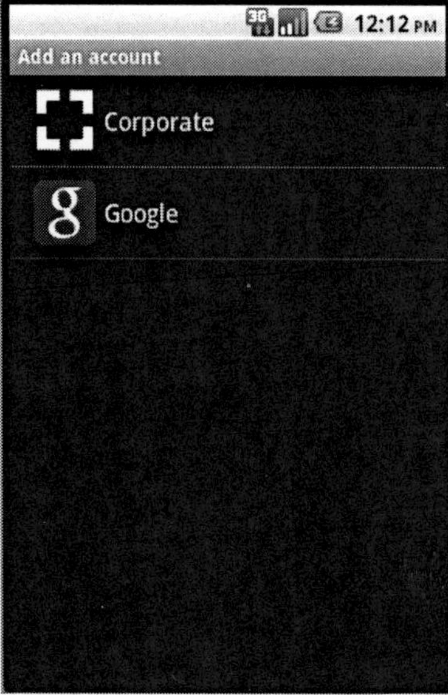

Figure 30–3. *List of accounts that can be set up*

This list of possible accounts to add will vary based on the type of device and what is available. The list in Figure 30–3 shows what is available in the Android 2.3 emulator when it is set up with Google API 9 as the target. If you have only downloaded the core SDK, you will not see the option to choose the Google API as the target for that emulator, so you won't see the option for setting up the Google account in Figure 30–3. This also means that this picture of available accounts could change with each Android release, device maker, and carrier or service provider. However, the concept of accounts remains largely the same, and these figures are included here only to aid your understanding of contacts. Refer to the user guide of each Android release to see the latest screens.

In addition, the fields needing to be set up for each account vary by account provider. For example, if you click to add a Google account in our emulator example, you will be presented with an option to create or sign in to a Google account (see Figure 30–4).

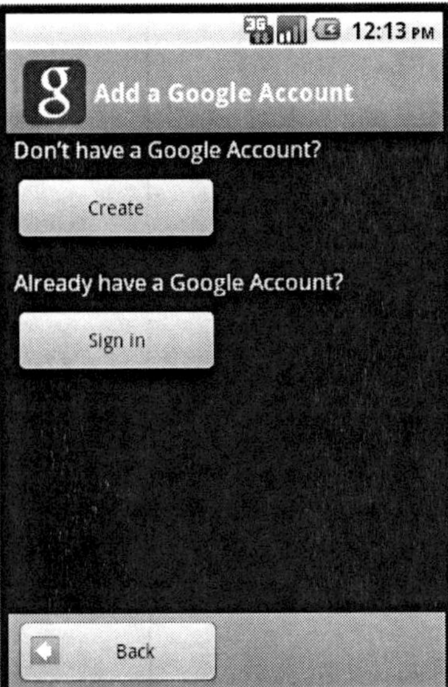

Figure 30–4. *Adding a Google account*

If you click the Create button, the fields to create a Google account appear, as shown in Figure 30–5.

Figure 30–5. *Creating a Google account*

Figure 30–5 illustrates the fields required to set up a Google account if you don't have one already. As stated, these fields could clearly vary from account type to account type. For example, we'll show you the account settings if you already have a Google account. In this case, the account setup merely involves signing into the account, as shown in Figure 30–6.

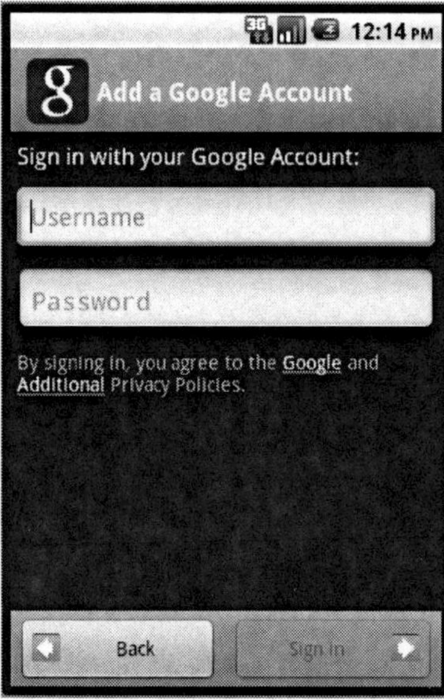

Figure 30–6. *Signing into an existing Google account*

Now that we have demonstrated the basics of an account and how it might end up on a device, the next section goes into how accounts become relevant to contacts.

Relevance of Accounts to Contacts

The contacts you manage are tied to a specific account. In other words, each account you have registered on the device can hold a number of contacts that are specific to that account. An account owns its set of contacts—or an account is said to be the *parent* of a contact. An account may have zero or more contacts.

An account is identified by two strings: the account name and the account type. In the case of Google, your account name is your e-mail user name at Gmail and your account type is `com.google`. Clearly, the account type must be unique across the device. Your account name is unique with in that account type. Together, an account type and an account name form an account, and only once the account is formed can a set of contacts be inserted for that account.

Enumerating Accounts

The Contacts API primarily deals with contacts that exist in various accounts. The mechanism of creating accounts is outside of the Contacts API, so explaining the ability to write your own account providers and how to sync the contacts within those accounts is outside the scope of this chapter. You can understand and benefit from this chapter without going the detail of how accounts get set up. However, when you want to add a contact or a list of contacts, you do need to know what accounts exist on the device. You can use the code in Listing 30–1 to enumerate the accounts and their necessary properties (the account name and type). The code in Listing 30–1 lists the account name and type given a context variable such as an activity.

Listing 30–1. *Code to Display a List of Accounts*

```
public void listAccounts(Context ctx)
{
    AccountManager am = AccountManager.get(ctx);
    Account[] accounts = am.getAccounts();
    for(Account ac: accounts)
    {
        String acname=ac.name;
        String actype = ac.type;
        Log.d("accountInfo", acname + ":" + actype);
    }
}
```

To run the code in Listing 30–1, the manifest file needs to ask for permission using the line in Listing 30–2.

Listing 30–2. *Permission to Read Accounts*

```
<uses-permission android:name="android.permission.GET_ACCOUNTS"/>
```

The code from Listing 30–1 will print something like the following:

```
Your-email-at-gmail:com.google
```

This assumes that you have only one account (Google) configured. If you have more than one account, all of those accounts will be listed in a similar manner.

Before diving more deeply into the contact details, let's consider how end users create contacts using the contacts application that comes with the Android platform.

Understanding the Contacts Application

The contacts application has changed a bit in 4.0. The primary change comes from the introduction of personal profiles. The contacts application has also been renamed People.

In the 4.0 emulator, the People icon is available right on the home page (in the bottom icon row), as shown in Figure 30–7.

Figure 30–7. *Accessing the People application in 4.0*

You can also access the People application from the roster of applications, as shown in Figure 30–8, where the icon is also titled People.

Figure 30–8. *People application icon*

Introducing the Personal Profile

When you invoke the People application for the first time on the emulator, it will look like Figure 30–9. (On a real device, the registration process most likely will add your personal profile as part of the device setup.)

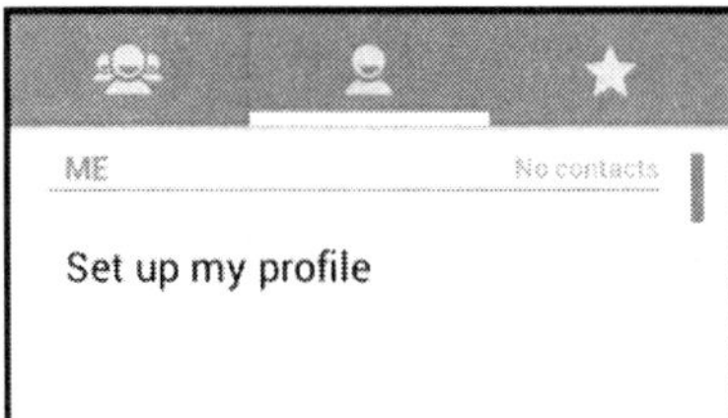

Figure 30–9. *Empty profile*

You can set up your profile by clicking "Set up my profile" in Figure 30–9, which will bring up the edit-profile screen in shown Figure 30–10.

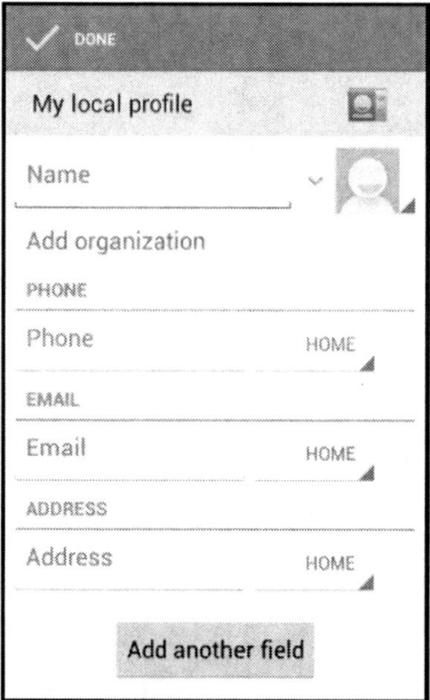

Figure 30–10. *Creating a profile*

This screen is similar to the one used to add any contact to the device. When you're finished, the home page of People will look like Figure 30–11.

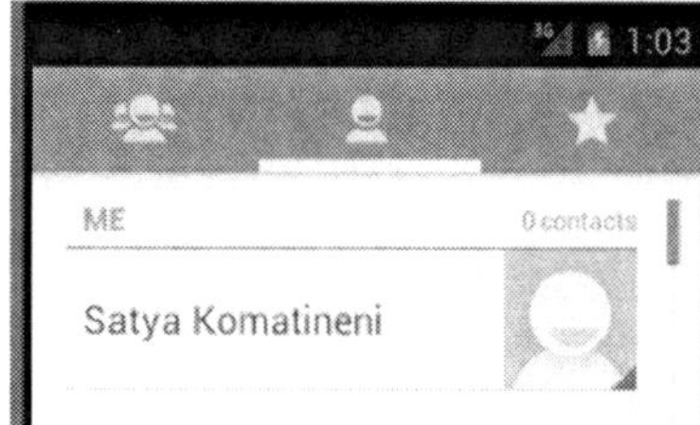

Figure 30–11. *Profile after profile creation*

Now if you go to your phone application and add a contact and return to this profile home page, you will see a screen like the one shown in Figure 30–12.

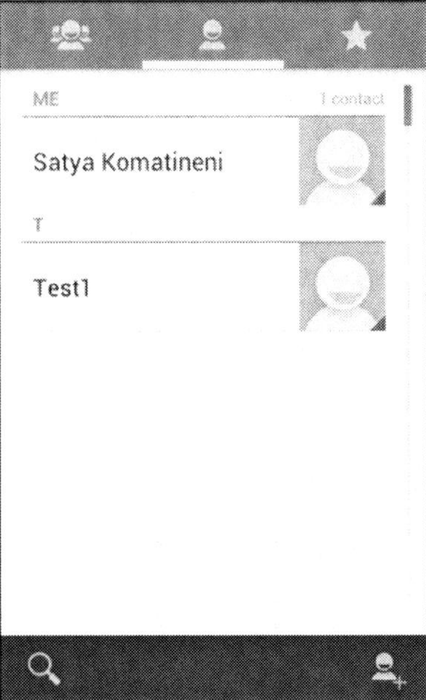

Figure 30–12. *List of contacts along with a personal profile*

The contact Test1 was added as a phone contact.

Showing Contacts

When you choose the People application, the first screen you see has a reference to your own contact (ME) and a list of other contacts (see Figure 30–12). If you have more than one account, the screen in Figure 30–12 will list all contacts from all accounts. By looking at this screen, you will not know what contact came from what account. Unless explicitly prevented, Android tries not to repeat contacts if they appear similar between two different accounts. We will cover this "appear similar" heuristic in the next main section.

Showing Contact Details

If you click one of the contacts in Figure 30–12, the contacts/people application will show the details of that contact, as shown in Figure 30–13. Here the contact being shown is called C1-First C1-Last.

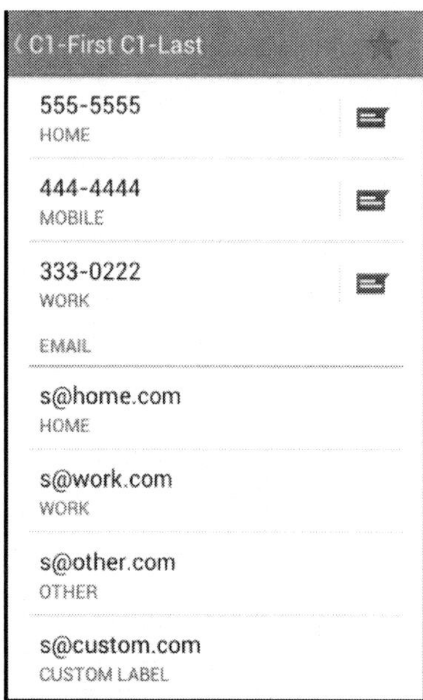

Figure 30–13. *A contact's details*

Figure 30–13 illustrates some of the sets of information a contact can carry. The contact details screen also shows that action icons can be provided that are specific to each contact type, such as what you can do for an e-mail contact versus what you can do for a telephone contact.

Editing Contact Details

Let's now look at how a contact, like the one in Figure 30–13, can be edited (or a new one created). You can do this by clicking the menu and choosing Edit or New contact. This will bring up the screen shown in Figure 30–14.

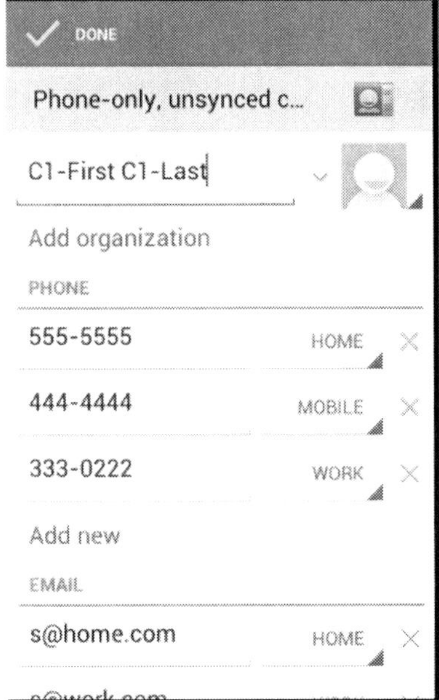

Figure 30–14. *Editing a contact*

In Figure 30–14, at the top of the edit-contact screen, you see the account under which this contact is being edited or created. For this contact, the account is shown as phone only. Usually the contacts for Gmail-like accounts are created on the server side and get synched to the phone based on the sync settings. It is quite possible to create a contacts client that can create contacts for other accounts as well on the device and sync them to the source account.

As the contact details screen in Figure 30–13 illustrates, it is possible to have different types of phone numbers and e-mail addresses. You may also be wondering if the contacts allow an arbitrary set of rows containing arbitrary data. (For example, in Figure 30–13, phone and e-mail are well-known predefined data types. What if you want store some data that is not anticipated? This is what we mean by *arbitrary*.) The Contacts API does allow arbitrary data, such as address information with a variety of address details for a contact.

Setting a Contact's Photo

You can also set up the photo for a contact. Figure 30–15 shows the photo setting screen that opens when you click the photo icon shown in Figure 30–13 (the first page of the contact details).

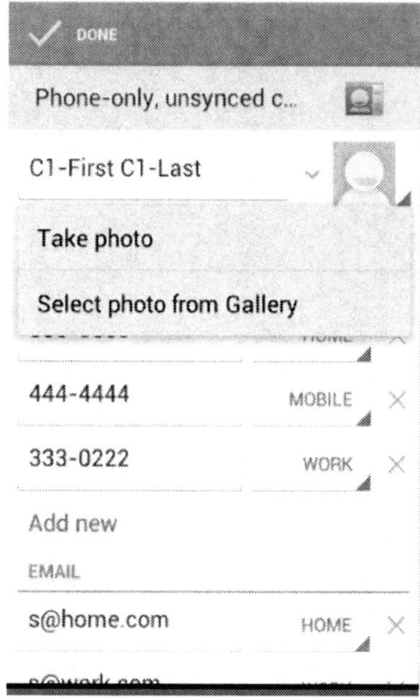

Figure 30–15. *Editing a contact's photo*

Exporting Contacts

Let's conclude this tour of the contacts/people application by showing how you can export contacts to external storage such as an SD card. Among other things, this SD card export facility allows you to see what kind of information is captured for a contact and how it is exposed as text. You can access the export-contacts functionality by using the menu options available on the home page of the personal profile, as shown in Figure 30–16.

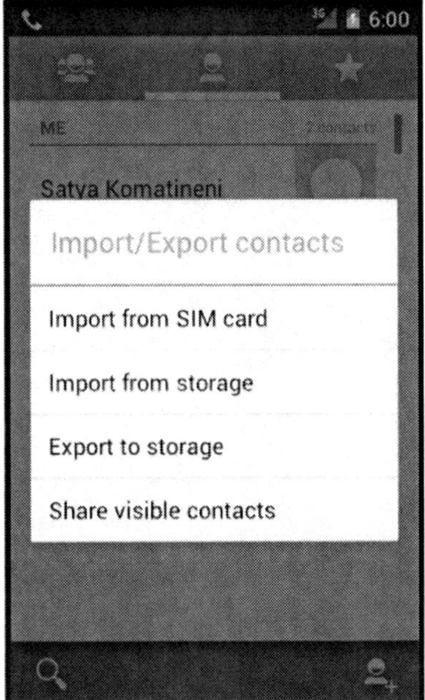

Figure 30–16. *Exporting contacts*

Once you export the contacts to an SD card, you can explore the SD card files using the Eclipse ADT. See Figure 30–17, where one of the exported .vcf files is visible in the Eclipse File Explorer.

Probl | @ Java | Decla | Cons | LogC | Searc | Prop | File E

Name	Size	Date	Time	Permissions	Info
data		2010-06-21	11:37	drwxrwx--x	
mnt		2010-11-17	09:02	drwxrwxr-x	
asec		2010-11-17	09:02	drwxr-xr-x	
sdcard		1969-12-31	19:00	d---rwxr-x	
00001.vcf	358	2010-10-28	13:13	----rwxr-x	
Android		2010-08-02	22:06	d---rwxr-x	
DCIM		2010-07-27	13:55	d---rwxr-x	
LOST.DIR		2010-06-21	11:38	d---rwxr-x	
secure		2010-11-17	09:02	drwx------	
system		2010-05-12	23:58	drwxr-xr-x	

Figure 30–17. *Contact information on an SD card*

You can copy the .vcf file in Figure 30–17 from the device to a local file using the icons in the top-right corner of the File Explorer tab. The sample contents of the .vcf file are displayed in Listing 30–3.

Listing 30–3. *Exported Contacts in VCF Format*

```
BEGIN:VCARD
VERSION:2.1
N:C1-Last;C1-First;;;
FN:C1-First C1-Last
TEL;TLX:55555
TEL;WORK:66666
EMAIL;HOME:test@home.com
EMAIL;WORK:test@work.com
ORG:WorkComp
TITLE:President
ORG:Work Other
TITLE:President
URL:www.com
NOTE:Note1
X-AIM:aim
X-MSN:wlive
END:VCARD

BEGIN:VCARD
VERSION:2.1
N:C2-Last;C2-first;;;
FN:C2-first C2-Last
END:VCARD
```

Various Contact Data Types

In the figures so far, you have seen how to add distinct sets of information for a contact. Listing 30–4 shows a list of these data types as defined in the API (this list could grow with new releases and is current as of version 4.0).

Listing 30–4. *Standard Contact Data Types*

```
email
event
groupmemebership
identity
im
nickname
note
organization
phone
photo
relation
SipAddress
structuredname
structuredpostal
website
```

Each data type, such as `email` or `structuredpostal` (indicating a postal address), has its own set of fields. So how do you know what these fields are? They are defined in the helper classes available in

```
android.provider.ContactsContract.CommonDataKinds
```

The URL for this class is

http://developer.android.com/reference/android/provider/ContactsContract.CommonDataKinds.html.

For example, the class `CommonDataKinds.Email` defines the fields shown in Listing 30–5.

Listing 30–5. *Specific Fields of an E-mail Contact*

```
Email address
Type of email: type_home, type_work, type_other, type_mobile
Label: to support type_other
```

Now that you have the background and tools necessary to work with accounts and contacts, let's get into the real details of the Contacts API.

Understanding Contacts

We indicated early in this chapter that your personal profile on the device is treated differently than any other regular contact. However, the personal profile contact and the rest of contacts share the same underlying architecture. When you understand contacts, it becomes easier to understand personal profiles as well. With that insight, we will cover the details of contacts first and follow up this discussion with the details of personal profiles in a subsequent section.

As declared, contacts are owned by an account. Each account has its own set of contacts. These are called *raw contacts*. Each raw contact then has its own set of data elements (for example, e-mail address, phone number, name, and postal address). Furthermore, Android presents an aggregated view of raw contacts by listing only once any raw contacts that seem to match. These aggregated contacts form the set of contacts you see on the first page of the contacts/people application (see Figure 30–12).

We will now examine how contacts and their related data are stored in various tables. Understanding these contact tables and their associated views is key to understanding the Contacts API.

Examining the Contacts SQLite Database

One way to understand and examine the contact database tables is to download the contact database from the device or the emulator and open it using one of the SQLite explorer tools.

To download the contacts database, use the File Explorer shown in Figure 30–17, and navigate to the following directory on your emulator:

```
/data/data/com.android.providers.contacts/databases
```

Depending on the release, the database file name may differ slightly, but it should be called `contacts.db`, `contacts2.db`, or something similar. In 4.0, the contacts provider uses a similarly structured but separate database file called `profile.db` to hold the contacts related to the personal profile.

In theory, all you have to do is open one of these databases, say `contacts2.db`, with a SQLite tool. However, we found a problem opening this database. Most tools we tried barfed (figuratively speaking). The problem has to do with the custom collation sequences defined by Android for such things as comparing phone numbers.

Apparently, for SQLite, the custom collation sequences are compiled as part of the SQLite distribution. If you don't have the DLL files that were compiled with the Android distribution, the general-purpose explorer tools won't be able to read the database accurately. Because the tools are using the Windows SQLite DLL files to open the database that was created with Linux distribution of Android, they are not successful. And the Windows distribution of SQLite does not have the collations sequences that are defined as required by the contacts database.

However, we are lucky enough that a program called SQLite Explorer has a glitch that allowed us to browse the tables even though it refused to publish the schema for the database. You may have better luck with other pricier tools. If you would like to explore further options, here is a link to see a list of available tools for SQLite:

`www.sqlite.org/cvstrac/wiki?p=ManagementTools`

Should you be really inquisitive, you can read more about the collation sequences from our research article "Exploring Contacts db" at

`www.androidbook.com/item/3582`

If you do have difficulties exploring the database, all is not lost, because we have listed all the important tables in this chapter. With that, we will start with exploring raw contacts first.

Raw Contacts

Again, the contacts we have seen when opening the contacts application are called aggregated contacts. Underneath each aggregated contact lies a set of contacts called raw contacts. An aggregated contact is merely a view on a set of similar raw contacts. To understand the aggregated contacts, you have to understand the raw contacts and the data that belongs to a raw contact. So, we will talk about the raw contacts first.

The set of contacts belonging to an account are called raw contacts. Each raw contact points to the details of one person that you know in the context of that account. This is in contrast to an aggregated contact, which crosses account boundaries and ends up belonging to the device as a whole.

This relationship between an account and its set of raw contacts is maintained in the raw contacts table. Listing 30–6 shows the structure of the raw contacts table in the contacts database.

Listing 30–6. *Raw Contact Table Definition*

```
CREATE TABLE raw_contacts
(_id INTEGER PRIMARY KEY AUTOINCREMENT,
is_restricted INTEGER DEFAULT 0,
account_name STRING DEFAULT NULL,
account_type STRING DEFAULT NULL,
sourceid TEXT,
version INTEGER NOT NULL DEFAULT 1,
dirty INTEGER NOT NULL DEFAULT 0,
deleted INTEGER NOT NULL DEFAULT 0,
contact_id INTEGER REFERENCES contacts(_id),
aggregation_mode INTEGER NOT NULL DEFAULT 0,
aggregation_needed INTEGER NOT NULL DEFAULT 1,
custom_ringtone TEXT
send_to_voicemail INTEGER NOT NULL DEFAULT 0,
times_contacted INTEGER NOT NULL DEFAULT 0,
last_time_contacted INTEGER,
starred INTEGER NOT NULL DEFAULT 0,
display_name TEXT,
display_name_alt TEXT,
display_name_source INTEGER NOT NULL DEFAULT 0,
phonetic_name TEXT,
phonetic_name_style TEXT,
sort_key TEXT COLLATE PHONEBOOK,
sort_key_alt TEXT COLLATE PHONEBOOK,
name_verified INTEGER NOT NULL DEFAULT 0,
contact_in_visible_group INTEGER NOT NULL DEFAULT 0,
sync1 TEXT, sync2 TEXT, sync3 TEXT, sync4 TEXT )
```

In Listing 30–6, important fields are bolded/highlighted. As with most Android tables, the raw contacts table has the `_ID` column that uniquely identifies a raw contact. Together, the fields `account_name` and `account_type` identify the account to which this contact (specifically, the raw contact) belongs. The `sourceid` field indicates how this raw contact is uniquely identified in the account identified by the account name and account type fields. For example, assume you need to know how a raw contact ID is identified in the Google e-mail account. Typically, in that case, this field would have carried the user's e-mail ID.

The field `contact_id` refers to the aggregated contact that this raw contact is one of. An aggregated contact points to one or more similar contacts that are essentially the same person set up among multiple accounts.

The field `display_name` points to the display name of the contact. This is primarily a read-only field. It is set by triggers based on the data rows added in the data table (which is covered in the next subsection) for this raw contact.

The `sync` fields are used by the account to sync contacts between the device and the server-side account such as Google mail.

Although we have used SQLite tools to explore these fields, there is more than one way to discover these fields. The recommended way is to follow the class definitions as declared in the `ContactsContract` API. To explore the columns belonging to a raw contact, you can look at the class documentation for `ContactsContract.RawContact`.

There are advantages and disadvantages to this approach. A significant advantage is that you get to know the fields published and acknowledged by the Android SDK. The database columns may get added or dropped without changing the public interface. So if you use the database columns directly, they may or may not be there. Instead, if you use the public definitions for these columns, you are safe between releases.

One disadvantage, however, is that the class documentation has many other constants interspersed with column names; we kind of got lost in figuring out what was what. These numerous class definitions give the impression that the API is complex when, in reality, 80 percent of the class documentation for the Contacts API is to define constants for these columns and the URIs to access these rows.

When we exercise the Contacts API in later sections, we will use the class-documentation-based constants instead of direct column names. However, we felt the direct exploration of the tables was the quickest way to help you understand the Contacts API.

Let's talk next about how the data relating to a contact (such as e-mail and phone number) is stored.

Data Table

As indicated from the raw contact table definition, the raw contact (in an anticlimactic sense) is just an ID indicating what account it belongs to. Most of the data pertaining to the contact is not in the raw contact table at all, but saved in the data table. Each data element, such as e-mail and phone number, is stored as separate rows in the data table. All of these related data rows are tied to a raw contact through the raw contact ID, which is one of the columns of the data table and also the primary ID of the raw contact table.

This data table contains 16 generic columns that can store any 16 different data points for any given data element, such as e-mail. Listing 30–7 describes how the data table is organized.

Listing 30–7. *Contact Data Table Definition*

```
CREATE TABLE data
(_id                INTEGER PRIMARY KEY AUTOINCREMENT,
package_id          INTEGER REFERENCES package(_id),
mimetype_id         INTEGER REFERENCES mimetype(_id) NOT NULL,
raw_contact_id      INTEGER REFERENCES raw_contacts(_id) NOT NULL,
is_primary          INTEGER NOT NULL DEFAULT 0,
is_super_primary    INTEGER NOT NULL DEFAULT 0,
data_version        INTEGER NOT NULL DEFAULT 0,
data1 TEXT,data2 TEXT,data3 TEXT,data4 TEXT,data5 TEXT,
data6 TEXT,data7 TEXT,data8 TEXT,data9 TEXT,data10 TEXT,
data11 TEXT,data12 TEXT,data13 TEXT,data14 TEXT,data15 TEXT,
data_sync1 TEXT, data_sync2 TEXT, data_sync3 TEXT, data_sync4 TEXT )
```

Critical columns in the data table shown in Listing 30–7 are bolded. As you might have anticipated, raw_contact_id points to the raw contact to which this data row belongs.

The mimetype_id points to the MIME type entry indicating one of the types identified in the contact data types in Listing 30–4. The columns data1 through data15 are generic string-based tables that can store anything that is necessary based on the MIME type. Again, the sync fields are there to support contact syncing. The table that resolves the MIME type IDs is in Listing 30–8.

Listing 30–8. *MIME Type Lookup Table Definition*

```
CREATE TABLE mimetypes
(_id INTEGER PRIMARY KEY AUTOINCREMENT,
mimetype TEXT NOT NULL)
```

As with the raw contacts table, you can discover the data table columns through the helper class documentation for ContactsContract.Data.

Although you can figure out the columns from this class definition, you will not know what is stored in each of the generic columns from data1 through data15. To know this, you will need to see the class definitions for a number of classes under the namespace ContactsContract.CommonDataKinds.

Some examples of these classes follow:

- ContactsContract.CommonDataKinds.Email
- ContactsContract.CommonDataKinds.Phone

In fact, you will see one class for each of the listed common data types in Listing 30–4. Ultimately, all the CommonDataKinds classes do is indicate which generic data fields (data1 through data15) are in use and what for.

Aggregated Contacts

Ultimately, a contact and its related data are unambiguously stored in the raw contacts table and the data table. An aggregated contact, on the other hand, is more of heuristic in nature and could be a bit ambiguous.

When there is a contact that is the same between multiple accounts, you may want to see one name instead of seeing the same or similar name repeated once for every account. Android addresses this by aggregating contacts into a read-only view. Android stores these aggregated contacts in a table called contacts. Android uses a number of triggers on the raw contact table and the data table to populate or change this aggregated contact table.

Before going into explaining the logic behind aggregation, let us show you the contact table definition (see Listing 30–9).

Listing 30–9. *Aggregated Contact Table Definition*

```
CREATE TABLE contacts
(_id                  INTEGER PRIMARY KEY AUTOINCREMENT,
name_raw_contact_id   INTEGER REFERENCES raw_contacts(_id),
photo_id              INTEGER REFERENCES data(_id),
custom_ringtone       TEXT,
send_to_voicemail     INTEGER NOT NULL DEFAULT 0,
```

```
times_contacted       INTEGER NOT NULL DEFAULT 0,
last_time_contacted   INTEGER,
starred               INTEGER NOT NULL DEFAULT 0,
in_visible_group      INTEGER NOT NULL DEFAULT 1,
has_phone_number      INTEGER NOT NULL DEFAULT 0,
lookup                TEXT,
status_update_id INTEGER REFERENCES data(_id),
single_is_restricted INTEGER NOT NULL DEFAULT 0)
```

In Listing 30–9, important columns are highlighted. No client directly updates this table. When a raw contact is added with its concomitant detail, Android searches other raw contacts to see if there are similar raw contacts. If there is one, it will use the aggregated contact ID of that raw contact as the aggregated contact ID of the new raw contact as well. No entry is made into the aggregated contact table. If none is found, it will create an aggregated contact and use that aggregated contact as the contact ID for that raw contact.

Android uses the following algorithm to determine which raw contacts are similar:

1. The two raw contacts have matching names, both first and last.
2. The words in the name are the same but vary in order: "first last" or "first, last" or "last, first."
3. The shorter versions of the names match, such as "Bob" for "Robert."
4. If one of the raw contacts has just a first or last name, this will trigger a search for other attributes, such as phone number or e-mail, and if the other attributes match, the contact will be aggregated.
5. If one of the raw contacts is missing the name altogether, this will also trigger a search for other attributes as in step 4.

Because these rules are heuristic, some contacts may be aggregated unintentionally. The client applications need to provide a mechanism to separate the contacts in such a case. If you refer to the *Android User's Guide*, you will see that the default contacts application allows you to separate contacts that are unintentionally merged.

You can also prevent the aggregation by setting the aggregation mode when you insert the raw contact. The available aggregation modes are shown in Listing 30–10.

Listing 30–10. *Aggregation Mode Constants*

```
AGGREGATION_MODE_DEFAULT
AGGREGATION_MODE_DISABLED
AGGREGATION_MODE_SUSPENDED
```

The first option is obvious; it is how aggregation works.

The second option (`disabled`) keeps this raw contact out of aggregation. Even if it is aggregated already, Android will pull it out of aggregation and allocate a new aggregated contact ID dedicated to this raw contact.

The third option (suspended) indicates that even though the properties of the contact may change, which will make it invalid for the aggregation into that batch of contacts, it should be kept tied to that aggregated contact.

The last point brings out the volatile dimension of the aggregated contact. Say you have a unique raw contact with a first name and a last name. Right now, it doesn't match any other raw contact, so this unique raw contact gets its own allocation of an aggregated contact. The aggregated contact ID will be stored in the raw contact table against that raw contact row.

However, you go and change the last name of this raw contact, which makes it a match to another set of contacts that are aggregated. In that case, Android will remove the raw contact from this aggregated contact and move it to the other one, abandoning this single aggregated contact by itself. In this case, the ID of the aggregated contact becomes entirely abandoned, as it will not match anything in the future because it is just an ID without an underlying raw contact.

So an aggregated contact is volatile. There is not a significant value to hold on to this aggregated contact ID over time.

Android offers some respite from this predicament by providing a field called lookup in the aggregated contacts tables. This lookup field is an aggregation (concatenation) of the account and the unique ID of this raw contact in that account for each raw contact. This information is further codified so that it can be passed as a URL parameter to retrieve the latest aggregated contact ID. Android looks at the lookup key and sees which underlying raw contact IDs are there for this lookup key. It then uses a best-fit algorithm to return a suitable (or perhaps new) aggregated contact ID.

While we are explicitly examining the contacts database, let's consider a couple of contact-related database views that are useful.

view_contacts

The first of these views is the view_contacts. Although there is a table that holds the aggregated contacts (contacts table), the API doesn't expose the contacts table directly. Instead, it uses view_contacts as the target for reading the aggregated contacts. When you query based on the URI ContactsContract.Contacts.CONTENT_URI, the columns returned are based on this view view_contacts. The definition of this view is shown in Listing 30–11.

Listing 30–11. *A View to Read Aggregated Contacts*

```
CREATE VIEW view_contacts AS

SELECT contacts._id AS _id,
contacts.custom_ringtone AS custom_ringtone,
name_raw_contact.display_name_source AS display_name_source,
name_raw_contact.display_name AS display_name,
name_raw_contact.display_name_alt AS display_name_alt,
name_raw_contact.phonetic_name AS phonetic_name,
name_raw_contact.phonetic_name_style AS phonetic_name_style,
```

```
name_raw_contact.sort_key AS sort_key,
name_raw_contact.sort_key_alt AS sort_key_alt,
name_raw_contact.contact_in_visible_group AS in_visible_group,
has_phone_number,
lookup,
photo_id,
contacts.last_time_contacted AS last_time_contacted,
contacts.send_to_voicemail AS send_to_voicemail,
contacts.starred AS starred,
contacts.times_contacted AS times_contacted, status_update_id

FROM contacts JOIN raw_contacts AS name_raw_contact
ON(name_raw_contact_id=name_raw_contact._id)
```

Notice that this view combines the contacts table with the raw contact table based on the aggregated contact ID.

contact_entities_view

Another useful view is the view that combines the raw contacts table with the data table. This view allows us to retrieve all the data elements of a given raw contact one time, or even the data elements of multiple raw contacts belonging to the same aggregated contact. Listing 30–12 presents the definition of the entities view.

Listing 30–12. *Contact Entities View*

```
CREATE VIEW contact_entities_view AS

SELECT raw_contacts.account_name AS account_name,
raw_contacts.account_type AS account_type,
raw_contacts.sourceid AS sourceid,
raw_contacts.version AS version,
raw_contacts.dirty AS dirty,
raw_contacts.deleted AS deleted,
raw_contacts.name_verified AS name_verified,
package AS res_package,
contact_id,
raw_contacts.sync1 AS sync1,
raw_contacts.sync2 AS sync2,
raw_contacts.sync3 AS sync3,
raw_contacts.sync4 AS sync4,
mimetype, data1, data2, data3, data4, data5, data6, data7, data8,
data9, data10, data11, data12, data13, data14, data15,
data_sync1, data_sync2, data_sync3, data_sync4,

raw_contacts._id AS _id,

is_primary, is_super_primary,
data_version,
data._id AS data_id,
raw_contacts.starred AS starred,
raw_contacts.is_restricted AS is_restricted,
groups.sourceid AS group_sourceid

FROM raw_contacts LEFT OUTER JOIN data
   ON (data.raw_contact_id=raw_contacts._id)
```

```
LEFT OUTER JOIN packages
  ON (data.package_id=packages._id)
LEFT OUTER JOIN mimetypes
  ON (data.mimetype_id=mimetypes._id)
LEFT OUTER JOIN groups
  ON (mimetypes.mimetype='vnd.android.cursor.item/group_membership'
    AND groups._id=data.data1)
```

The URIs needed to access this view are available in the class `ContactsContract.RawContacts.RawContactsEntity`.

Working with the Contacts API

So far, we have explored the basic idea behind the Contacts API by exploring its tables and views. We will now present a number of code snippets that can be used to explore contacts. These snippets are taken from the sample application that is developed to support this chapter. Although they are taken from the sample application, they are self sufficient to give a full picture. You can download the full sample program using the project download URL at the end of this chapter.

Exploring Accounts

We will start our exercise by writing a program that can print out the list of accounts. We have already given the code snippets necessary to get a list of accounts. Consider the class `AccountsFunctionTester` in Listing 30–13. You can see the corresponding Java file in the downloadable project as well. That file is reproduced here as is.

Listing 30–13. *AccountsFunctionTester*

```
public class AccountsFunctionTester extends BaseTester
{
    private static String tag = "tc>";
    public AccountsFunctionTester(Context ctx, IReportBack target)
    {
        // ctx saved in BaseTester as mContext
        // target saved in BaseTester as mReportTo
        super(ctx, target);
    }
    public void testAccounts()
    {
        AccountManager am = AccountManager.get(this.mContext);
        Account[] accounts = am.getAccounts();
        for(Account ac: accounts)
        {
            String acname=ac.name;
            String actype = ac.type;
            this.mReportTo.reportBack(tag,acname + ":" + actype);
        }
    }
}
```

NOTE: As we present and explore the Java code necessary to work with contacts, you will see three variables repeatedly used:

mContext: A variable pointing to an activity

mReportTo: A variable implementing a logging interface (IReportBack—you can see this Java file in the downloadable project) that can be used to log messages to the test activity that is used for this chapter

Utils: A static class that encapsulates very simple utility methods

We have chosen not to list these classes here because they will distract you from understanding the core functionality of the Contacts API. You are more than welcome to examine these classes in the downloadable project.

The key function to watch is testAccounts(). The rest of the code, such as BaseTester, IReportBack, and so on is a small scaffolding to keep the details out of the main code. You can see the details of these classes in the downloadable project. To understand what is presented in this section, you don't need to look at those classes. We will provide a brief explanation of these supporting classes now.

The BaseTester class keeps a reference to two objects. The first is the activity (passed in as ctx), and the second is a simple object that implements IReportBack. The IReportBack takes a string message and displays it either on an activity's screen or in the LogCat. So calling IReportBack is merely calling a log function. These details are secondary; you can largely treat the reportBack() calls as calls to log a message.

The code in Listing 30–13 is getting the name and type for each account and calling the report-back interface to log it. As long as there is a driver activity that can call the method testAccounts(), this code can report back the account name and type.

When you run the sample program that you can download for this chapter, you will see a main activity (Figure 30–18) that appears with a number of menu options. Each menu item will invoke a method from one of the function testers, like the one in Listing 30–13. More menu items that elicit other parts of this chapter are shown in Figure 30–19. The screen in Figure 30–19 appears when you click the More button from Figure 30–18. These two figures list all the menu items available in the sample application.

The menu item that invokes the method testAccounts() in the account function tester from Listing 30–13 is called Accounts in Figure 30–18.

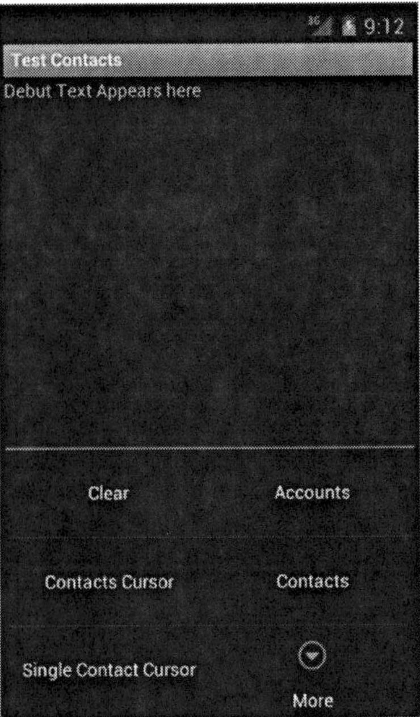

Figure 30–18. *Main driver activity with the menu*

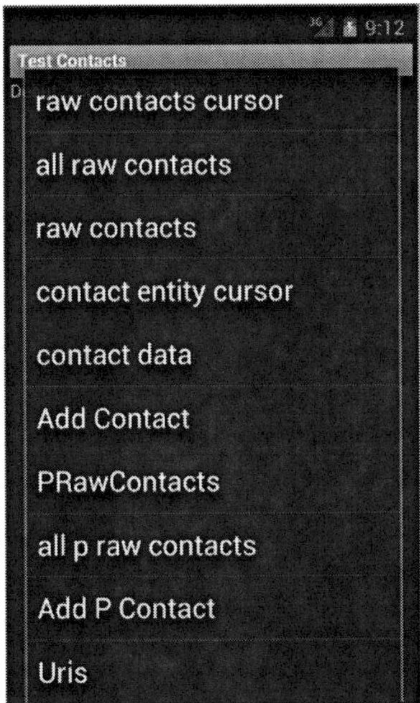

Figure 30–19. *Remaining menu items for the sample app for this chapter*

The function `testAccounts()` in Listing 30–13 will print the available accounts, as shown in Figure 30–20.

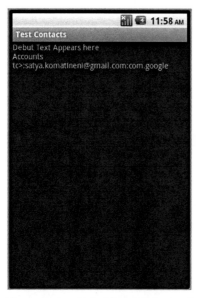

Figure 30–20. *Main driver activity showing a list of accounts*

The emulator we tested on has only one account set up, which is a Google account. So, Figure 30–20 shows that one account.

Exploring Aggregated Contacts

Let's see how we can explore aggregated contacts through code snippets. We will demonstrate three things related to aggregated contacts:

- Discover all the fields returned by firing off a URI that knows how to read aggregated contacts.
- List all aggregated contacts.
- Discover all the fields returned by a cursor based on a lookup URI.

To read contacts, you need to request the following permission in the manifest file:

```
android.permission.READ_CONTACTS
```

As the functionality we are testing deals with content providers, URIs, and cursors, let's look at some useful code snippets. (These code snippets are available either in `utils.java` or in some of the base classes derived from `BaseTester` for the chapter's downloadable project.)

The function `getACursor()` takes a URI string and a string-based `where` clause and returns a cursor (see Listing 30–14).

Listing 30–14. *Getting a Cursor Given a URI and a* `where` *Clause*

```
protected Cursor getACursor(String uri,String clause, Activity activity)
{
    // Run query

    return activity.managedQuery(Uri.parse(uri), null, clause, null, null);
}
```

The function in Listing 30–15, `getColumnValue()`, returns the value of a column, given its name, from the current row of the cursor. It returns the value of the column as a string, regardless of its fundamental type.

Listing 30–15. *Retrieving a Column from an Android Cursor*

```
public static String getColumnValue(Cursor cc, String cname)
{
    int i = cc.getColumnIndex(cname);
    return cc.getString(i);
}
```

The function in Listing 30–16, `getCursorColumnNames()`, takes any cursor and returns a separated list of all its available columns. It is useful when we explore new URIs to discover the type of fields those URIs return. Although we are supposed to document such columns in the Java code, this method of discovering them at runtime could come in handy.

Listing 30–16. *Examining an Android Cursor for Its Column Names*

```
protected static String getCursorColumnNames(Cursor c)
{
    int count = c.getColumnCount();
    StringBuffer cnamesBuffer = new StringBuffer();
    for (int i=0;i<count;i++)
    {
        String cname = c.getColumnName(i);
        cnamesBuffer.append(cname).append(';');
    }
    return cnamesBuffer.toString();
}
protected void printCursorColumnNames(Cursor c, IReportBack reportBackInterface)
{
    reportBackInterface.reportBack(tag,getCursorColumnNames(c));
}
```

In this section, we are primarily exploring the cursor returned by aggregated contact URIs. Each row returned by the resulting contact cursor will have a number of fields. For our example, we are not interested in all the fields but only a few. You can abstract this out into another class called an `AggregatedContact`. Listing 30–17 shows this class.

Listing 30–17. *An Object Definition for a Few Fields of an Aggregated Contact*

```
public class AggregatedContact
{
    public String id;
    public String lookupUri;
    public String lookupKey;
```

```
    public String displayName;

    public void fillinFrom(Cursor c)
    {
        //the following function from Utils.java returns a value for a column
        //given a cursor "c" and a column name like "_ID"
        id = Utils.getColumnValue(c,"_ID");
        lookupKey = Utils.getColumnValue(c,ContactsContract.Contacts.LOOKUP_KEY);
        lookupUri = ContactsContract.Contacts.CONTENT_LOOKUP_URI + "/" + lookupKey;
        displayName = Utils.getColumnValue(c,ContactsContract.Contacts.DISPLAY_NAME);
    }
}
```

Listing 30–17 is nothing too complex. In this code, we use the cursor to load up the fields that we are interested in.

Getting the Aggregated Contacts Cursor

The code snippet in Listing 30–18 shows how we can retrieve a cursor that is a collection of aggregated contacts.

Listing 30–18. *Getting a Cursor for All Aggregated Contacts*

```
/*
 * Get a cursor of all contacts.
 * Specify the where clause as null to indicate all rows.
 * Don't use it on a large set.
 */
private Cursor getContacts()
{
    // Run query
    Uri uri = ContactsContract.Contacts.CONTENT_URI;

    //Specify ascending or descending way to sort names
    String sortOrder = ContactsContract.Contacts.DISPLAY_NAME
                       + " COLLATE LOCALIZED ASC";

    //the variable mContext is an Activity held
    //as a local variable in the class
    Activity a = (Activity)this.mContext;
    return a.managedQuery(uri, null, null, null, sortOrder);
}
```

The URI used to read all the contacts is `ContactsContract.Contacts.CONTENT_URI`.

You can pass this URI to the `managedQuery()` function to retrieve a cursor. You can pass `null` as the column projection to receive all columns. Although this is not recommended in practice, in our case, it makes sense because we want to know about all the columns it returns. We have also used the display name of the contact as the sort order. Notice how, again, we have used `ContactContract.Contacts` to get the column name for the contact display name.

Listing the Fields of the Aggregated Contact Cursor

Listing 30–19 shows a code snippet that prints the names of the columns available in the cursor returned by the code in Listing 30–18.

Listing 30–19. *Exploring the Column Names of the Aggregated Contacts URI*

```
/*
 * Use the getContacts above
 * to list the set of columns in the cursor
 */
public void listContactCursorFields() {
    Cursor c = null;
    try {
        c = getContacts();
        int i = c.getColumnCount();
        //as indicated earlier use IReportBack to log
        this.mReportTo.reportBack(tag, "Number of columns:" + i);
        this.printCursorColumnNames(c);
    }
    finally
    {
        if (c!= null) c.close();
    }
}
```

In this method, we take the cursor from `getContacts()` and use the `getCursorColumnNames(cursor c)` method to print the names to the screen and to the LogCat. The function `printCursorColumnNames(cursor c)` is just a wrapper around `getCursorColumnNames()`.

> **NOTE:** In the sample program for this chapter, the method `listContactCursorField()` in Listing 30–19 is invoked by choosing the Contacts Cursor menu item.

Listing 30–20 contains the list of columns returned by the contacts content URI in Listing 30–19.

Listing 30–20. *Aggregated Contacts Content URI Cursor Columns*

```
times_contacted;
contact_status;
custom_ringtone;
has_phone_number;
phonetic_name;
phonetic_name_style;
contact_status_label;
lookup;
contact_status_icon;
last_time_contacted;
display_name;
sort_key_alt;
in_visible_group;
_id;
starred;
```

```
sort_key;
display_name_alt;
contact_presence;
display_name_source;
contact_status_res_package;
contact_status_ts;
photo_id;
send_to_voicemail;
```

The sample snippet in Listing 30–19 will print these columns both to the screen and to LogCat. We copied these fields from LogCat and formatted them as shown in Listing 30–20.

> **NOTE:** When working with content providers, the technique of going after the URIs and printing the columns they return is useful.

Reading Aggregated Contact Details

Now that we've explored the columns available with the contacts content URI, let's pick a few columns and see what contact rows are available. We are interested in the following columns from a contact cursor:

- `display name`
- `lookup key`
- `lookup uri`

We are considering these fields because we want to see what the lookup key and lookup key URI look like based on what is covered in the theory part of this chapter. Specifically, we are interested in firing off the lookup URI to see what type of a cursor it returns.

The function `listContacts()` in Listing 30–21 gets a contacts cursor and prints these three columns for each row of the cursor. Note that this listing is taken from a class that holds a local variable called `mContext` to indicate the activity and a local variable called `mReportTo` to be able to log any messages to the activity.

Listing 30–21. *Printing the Lookup Keys for an Aggregated Contact*

```
/*
 * Use the getContacts() function to get a cursor and print all
 * the contact names followed by lookup keys
 * uses the printLookupKeys() function
 */
public void listContacts()
{
    Cursor c = null;
    try
    {
        c = getContacts();
        int i = c.getColumnCount();
        //log the line
```

```
        this.mReportTo.reportBack(tag, "Number of columns:" + i);
        this.printLookupKeys(c);
    }
    finally
    {
        if (c!= null) c.close();
    }
}
/*
 * Given a cursor worth of contacts, print the contact names followed by
 * their lookup keys.
 */
private void printLookupKeys(Cursor c)
{
    for(c.moveToFirst();!c.isAfterLast();c.moveToNext())
    {
        String name=this.getContactName(c);
        String lookupKey = this.getLookupKey(c);
        String luri = this.getLookupUri(lookupKey);
        //log
        this.mReportTo.reportBack(tag, name + ":" + lookupKey);
        this.mReportTo.reportBack(tag, name + ":" + luri);
    }
}

private String getLookupKey(Cursor cc) {
    int lookupkeyIndex =
        cc.getColumnIndex(ContactsContract.Contacts.LOOKUP_KEY);
    return cc.getString(lookupkeyIndex);
}
private String getContactName(Cursor cc){
    //Utils is a class to encapsulate some reusable functions
    //The name of the function should tell you what this could be doing
    //See the downloadable project for details
    return Utils.getColumnValue(cc,ContactsContract.Contacts.DISPLAY_NAME);
}
private String getLookupUri(String lookupkey) {
    String luri = ContactsContract.Contacts.CONTENT_LOOKUP_URI + "/" + lookupkey;
    return luri;
}
```

NOTE: You can see the function `listContacts()` in Listing 30–21 in action by clicking the Contacts menu item.

Exploring the Lookup URI-Based Cursor

Now that we know how to extract lookup URIs for a given aggregated contact, let's see what we can do with a lookup URI.

The function `listLookupUriColumns()` in Listing 30–22 will take the first contact from the list of all contacts and then formulate a lookup URI for that contact and fire off the URI to see what kind of a cursor it returns by printing the column names from that cursor

Listing 30–22. *Exploring the Lookup URI Cursor*

```
/*
 * A function to see if the URI constructed by the lookup
 * uri returns a cursor that has a different set of columns.
 * It returns a similar cursor with similar columns
 * as one would expect.
 */
public void listLookupUriColumns()
{
    Cursor c = null;
    try
    {
        c = getContacts();
        String firstContactLookupUri = getFirstLookupUri(c);
        printLookupUriColumns(firstContactLookupUri);
    }
    finally
    {
        if (c!= null) c.close();
    }
}

/*
 * Take a list of contacts and  look up the first contact,
 * return null if there are no contacts.
 */
private String getFirstLookupUri(Cursor c)
{
    c.moveToFirst();
    if (c.isAfterLast())
    {
        Log.d(tag,"No rows to get the first contact");
        return null;
    }
    //There is a row
    String lookupKey = this.getLookupKey(c);
    String luri = this.getLookupUri(lookupKey);
    return luri;
}

public void printLookupUriColumns(String lookupuri)
{
    Cursor c = null;
    try
    {
        c = getASingleContact(lookupuri);
        int i = c.getColumnCount();
        this.mReportTo.reportBack(tag, "Number of columns:" + i);
        int j = c.getCount();
        this.mReportTo.reportBack(tag, "Number of rows:" + j);
        this.printCursorColumnNames(c);
    }
    finally
    {
        if (c!=null)c.close();
    }
}
```

```
/**
 * Use the lookup uri, retrieve a single aggregated contact
 */
private Cursor getASingleContact(String lookupUri)
{
    // Run query
    Activity a = (Activity)this.mContext;
    return a.managedQuery(Uri.parse(lookupUri), null, null, null, null);
}
```

NOTE: To see this code executed in this snippet, you can choose the menu item Single Contact Cursor.

As it turns out, it just returns a cursor that is identical in columns for that of the aggregated contact cursor as in Listing 30–20, except that it has only one row pointing to the contact for which this is the lookup key. Also notice that we have used the following lookup URI definition:

```
ContactsContract.Contacts.CONTENT_LOOKUP_URI
```

You know from the discussion of the contact lookup URIs that each lookup URI represents a collection of raw contact identities that have been concatenated. That being the case, you might have expected the lookup URI to return a series of matching raw contacts. However, the test in Listing 30–22 is showing that it is not returning a cursor of raw contacts but instead a cursor of contacts.

NOTE: A lookup based on the contact lookup URI returns an aggregated contact and not a raw contact.

Another tidbit is that the lookup process for the aggregated contact based on the lookup URI is not linear or exact. This means Android will not look for an exact match of the lookup key. Instead, Android parses the lookup key into its constituent raw contacts and then finds the aggregated contact ID that matches the most of the raw contact records and returns that aggregated contact record.

One consequence of this is that no public mechanism is available to go from the lookup key to its constituent raw contacts. Instead, you have to find the contact ID for that lookup key and then fire off a raw contact URI for that contact ID to retrieve the corresponding raw contacts.

Here is another code snippet that shows what is returned from a cursor as an object instead of as a set of columns. The code in Listing 30–23 returns the first aggregated contact as an object.

Listing 30–23. *Code Testing Aggregated Contacts*

```
/*
 * Take a list of contacts
 * look up the first contact and return it
 * as an object AggregatedContact.
 */
protected AggregatedContact getFirstContact()
{
    Cursor c=null;
    try
    {
        c = getContacts();
        c.moveToFirst();
        if (c.isAfterLast())
        {
            Log.d(tag,"No contacts");
            return null;
        }
        //contact is there
        AggregatedContact firstcontact = new AggregatedContact();
        firstcontact.fillinFrom(c);
        return firstcontact;
    }
    finally
    {
        if (c!=null) c.close();
    }
}
```

NOTE: The code for `AggregatedContact` was given in Listing 30–17 when we started talking about exploring aggregated contacts. Most of the code snippets in this section are in the file `AggregatedContactFunctionTester.java`.

Exploring Raw Contacts

In this section, let's explore raw contacts. All of the code presented in this section is available in the file `RawContactFunctionTester.java`, unless we explicitly name a different file.

We will demonstrate three things related to aggregated contacts in this section:

- Discover all the fields returned by firing off a URI that knows how to read raw contacts.
- Show all raw contacts.
- List all raw contacts for a set of aggregated contacts.

Listing 30–24, `RawContact.java`, captures a few important fields from the raw contacts table cursor. (This file, like all other code snippets in this chapter, is available in the downloadable project for this chapter.)

Listing 30–24. *RawContact.java*

```
//In the following code Utils is a utility class
//See the downloadable project to see its method
//Method names are self explanatory to what they do
public class RawContact
{
    public String rawContactId;
    public String aggregatedContactId;
    public String accountName;
    public String accountType;
    public String displayName;

    public void fillinFrom(Cursor c)
    {
        rawContactId = Utils.getColumnValue(c,"_ID");
        accountName = Utils.getColumnValue(c,ContactsContract.RawContacts.ACCOUNT_NAME);
        accountType = Utils.getColumnValue(c,ContactsContract.RawContacts.ACCOUNT_TYPE);
        aggregatedContactId = Utils.getColumnValue(c,
                                ContactsContract.RawContacts.CONTACT_ID);
        displayName = Utils.getColumnValue(c,"display_name");
    }
    public String toString()
    {
        return displayName
            + "/" + accountName + ":" + accountType
            + "/" + rawContactId
            + "/" + aggregatedContactId;
    }
}
```

Showing the Raw Contacts Cursor

As with the aggregated contact URIs, let's first examine the nature of the raw contact URI and what it returns. The signature for the raw contact URI is defined as follows:

```
ContactsContract.RawContacts.CONTENT_URI;
```

The function showRawContactsCursor() in Listing 30–25 prints the cursor columns for a raw contacts URI.

Listing 30–25. *Exploring the Raw Contacts Cursor*

```
public void showRawContactsCursor()
{
    Cursor c = null;
    try
    {
        c = this.getACursor(getRawContactsUri(),null);
        this.printCursorColumnNames(c);
    }
    finally
    {
        if (c!=null) c.close();
    }
}
private Uri getRawContactsUri()
```

```
{
    return ContactsContract.RawContacts.CONTENT_URI;
}
```

In the sample, you click the **Raw Contacts Cursor** menu item to invoke this functionality. This will show that the raw contact cursor has the fields shown in Listing 30–26.

Listing 30–26. *Raw Contacts Cursor Fields*

```
times_contacted;
phonetic_name;
phonetic_name_style;
contact_id;version;
last_time_contacted;
aggregation_mode;
_id;
name_verified;
display_name_source;
dirty;
send_to_voicemail;
account_type;
custom_ringtone;
sync4;sync3;sync2;sync1;
deleted;
account_name;
display_name;
sort_key_alt;
starred;
sort_key;
display_name_alt;
sourceid;
```

Once you know the columns of a raw contacts cursor, you may be curious to see the table's rows.

Seeing the Data Returned by a Raw Contacts Cursor

Listing 30–27 shows the method showAllRawContacts() that prints all the rows in the raw contacts cursor. We have used the data object RawContact (Listing 30–24) to select the columns we would like to be printed for each row.

This method walks the cursor with no WHERE clause (so that it can get all the rows), creates a RawContact object for each row, and prints it out. You can see these raw contacts both on the screen and in LogCat.

Listing 30–27. *Displaying Raw Contacts Fields*

```
public void showAllRawContacts()
{
    Cursor c = null;
    try
    {
        c = this.getACursor(getRawContactsUri(), null);
        this.printRawContacts(c);
    }
```

```
    finally
    {
        if (c!=null) c.close();
    }
}
private void printRawContacts(Cursor c)
{
    for(c.moveToFirst();!c.isAfterLast();c.moveToNext())
    {
        RawContact rc = new RawContact();
        rc.fillinFrom(c);
        //log it. Standard Android logging pattern
        //except we are using IReportBack interface that was talked about
        //many times earlier in the chapter.
        this.mReportTo.reportBack(tag, rc.toString());
    }
}
```

The menu item All Raw Contacts invokes the functionality demonstrated by the method `showAllRawContacts()`.

Constraining Raw Contacts with a Corresponding Set of Aggregated Contacts

Using the columns of the cursor in Listing 30–26, let's see if we can refine our query to retrieve the contacts for a given aggregated contact ID. The code in Listing 30–29 will look up the first aggregated contact and then issue a raw contact URI with a `where` clause specifying a value for the `contact_id` column. Note that we have already included the listings for the following three functions used in Listing 30–29. Listing 30–28 lists the functions and where they are referenced.

Listing 30–28. *Utility Functions Used to Display Raw Contacts*

```
getFirstContact() //listing 30-23
getACursor() //listing 30-14
printRawContacts() //listing 30-27
```

> **NOTE:** If you download and run the sample for this project, you can test Listing 30–29 by clicking the Raw Contacts menu item.

You will see a list of raw contacts belonging to the very first aggregated contact both in the UI and also in the LogCat. If you haven't created any raw contacts yet, wait to run this test until you add a few raw contacts using the code presented later in the chapter. Or you can create contacts using the emulator UI, and the code in Listing 30–29 will work to show those raw contacts if they belong to the first raw contact. You can see the results of this both in the UI and also in LogCat.

Listing 30–29. *Testing Raw Contacts*

```
public void showRawContactsForFirstAggregatedContact()
{
    AggregatedContact ac = getFirstContact();
```

```
    this.mReportTo.reportBack(tag, ac.displayName + ":" + ac.id);

    Cursor c = null;
    try {
        c = this.getACursor(getRawContactsUri(), getClause(ac.id));
        this.printRawContacts(c);
    }
    finally {
        if (c!=null) c.close();
    }
}
private String getClause(String contactId)
{
    return "contact_id = " + contactId;
}
```

Although we have explored aggregated contacts and raw contacts, we haven't retrieved the important parts of a contact, such as the e-mail address and phone number. You'll see how to do this in the next section.

Exploring Raw Contact Data

In this example, you'll see how we can explore the data values corresponding to raw contacts. We will do the following in this section:

- Discover all the fields returned by firing off a URI that knows how to read raw contact data.
- Retrieve the data elements for a set of aggregated contacts.

Because a data row belonging to a raw contact contains a number of fields, we have created a file called ContactData.java in Listing 30–30. This file captures a representative set of contact data, and not all fields.

Listing 30–30. *ContactData.java*

```
public class ContactData
{
    public String rawContactId;
    public String aggregatedContactId;
    public String dataId;
    public String accountName;
    public String accountType;
    public String mimetype;
    public String data1;

    public void fillinFrom(Cursor c)
    {
        rawContactId = Utils.getColumnValue(c,"_ID");
        accountName = Utils.getColumnValue(c,ContactsContract.RawContacts.ACCOUNT_NAME);
        accountType = Utils.getColumnValue(c,ContactsContract.RawContacts.ACCOUNT_TYPE);
        aggregatedContactId =
                Utils.getColumnValue(c,ContactsContract.RawContacts.CONTACT_ID);
        mimetype = Utils.getColumnValue(c,ContactsContract.RawContactsEntity.MIMETYPE);
        data1 = Utils.getColumnValue(c,ContactsContract.RawContactsEntity.DATA1);
```

```
        dataId = Utils.getColumnValue(c,ContactsContract.RawContactsEntity.DATA_ID);
    }
    public String toString()
    {
        return data1 + "/" + mimetype
            + "/" + accountName + ":" + accountType
            + "/" + dataId
            + "/" + rawContactId
            + "/" + aggregatedContactId;
    }
}
```

NOTE: The code to work with raw contact data elements is in the file `ContactFunctionTester.java` in the download for this chapter.

Android uses a special view called a `RawContactEntity` view to retrieve data from a raw contact table and the corresponding data tables as indicated in the section "Contact_entities_view" in this chapter. The URI to access this view is in Listing 30–31.

Listing 30–31. *Raw Entities Content URI*

```
ContactsContract.RawContactsEntity.CONTENT_URI
```

Let's see how this URI can be used to discover field names returned by this URI:

```
public void showRawContactsEntityCursor()
{
    Cursor c = null;
    try
    {
        Uri uri = ContactsContract.RawContactsEntity.CONTENT_URI;
        c = this.getACursor(uri,null);
        this.printCursorColumnNames(c);
    }
    finally
    {
        if (c!=null) c.close();
    }
}
```

The code in Listing 30–31 prints out the list of columns shown in Listing 30–32. So, these are the columns returned by the raw contacts entity cursor.

Listing 30–32. *Contact Entities Cursor Columns*

```
data_version;
contact_id;
version;
data12;data11;data10;
mimetype;
res_package;
_id;
data15;data14;data13;
name_verified;
is_restricted;
is_super_primary;
```

```
data_sync1;dirty;data_sync3;data_sync2;
data_sync4;account_type;data1;sync4;sync3;
data4;sync2;data5;sync1;
data2;data3;data8;data9;
deleted;
group_sourceid;
data6;data7;
account_name;
data_id;
starred;
sourceid;
is_primary;
```

> **NOTE:** You can see the set of fields in Listing 30–32 by clicking the Contact Entity Cursor menu item in the sample for this chapter.

Once you know this set of columns, you can narrow down the result set of this cursor by formulating a proper where clause. For example, in Listing 30–33 we retrieve the data elements pertaining to contact IDs 3, 4, and 5.

Listing 30–33. *Displaying Data elements from `RawContactsEntity`*

```
public void showRawContactsData()
{
    Cursor c = null;
    try
    {
        Uri uri = ContactsContract.RawContactsEntity.CONTENT_URI;
        c = this.getACursor(uri,"contact_id in (3,4,5)");
        this.printRawContactsData(c);
    }
    finally
    {
        if (c!=null) c.close();
    }
}
protected void printRawContactsData(Cursor c)
{
    for(c.moveToFirst();!c.isAfterLast();c.moveToNext())
    {
        ContactData dataRecord = new ContactData();
        dataRecord.fillinFrom(c);
        this.mReportTo.reportBack(tag, dataRecord.toString());
    }
}
```

Notice the where clause in Listing 30–33:

```
"contact_id in (3,4,5)"
```

> **NOTE:** You can run the function in Listing 30–33 by clicking the Contact Data menu item in the sample for this chapter.

The code in Listing 30–33 will print such things as name, e-mail address, and MIME type as defined the RawContactData object in Listing 30–24.

Adding a Contact and Its Details

So far, we have shown only the code snippets to retrieve the contacts. Let's look at a code snippet that shows what it takes to add a contact with name, e-mail, and phone number.

To write to contacts, you need the following permission in the manifest file:

```
android.permission.WRITE_CONTACTS
```

The code in Listing 30–34 adds a raw contact followed by adding two data rows (name and phone number).

Listing 30–34. *Adding a Contact*

```
public void addContact()
{
    long rawContactId = insertRawContact();
    this.mReportTo.reportBack(tag, "RawcontactId:" + rawContactId);
    insertName(rawContactId);
    insertPhoneNumber(rawContactId);
    showRawContactsDataForRawContact(rawContactId);
}
private long insertRawContact()
{
    ContentValues cv = new ContentValues();
    cv.put(RawContacts.ACCOUNT_TYPE, "com.google");
    cv.put(RawContacts.ACCOUNT_NAME, "satya.komatineni@gmail.com");
    Uri rawContactUri =
        this.mContext.getContentResolver()
            .insert(RawContacts.CONTENT_URI, cv);
    long rawContactId = ContentUris.parseId(rawContactUri);
    return rawContactId;
}
private void insertName(long rawContactId)
{
    ContentValues cv = new ContentValues();
    cv.put(Data.RAW_CONTACT_ID, rawContactId);
    cv.put(Data.MIMETYPE, StructuredName.CONTENT_ITEM_TYPE);
    cv.put(StructuredName.DISPLAY_NAME,"John Doe_" + rawContactId);
    this.mContext.getContentResolver().insert(Data.CONTENT_URI, cv);
}
private void insertPhoneNumber(long rawContactId)
{
    ContentValues cv = new ContentValues();
    cv.put(Data.RAW_CONTACT_ID, rawContactId);
    cv.put(Data.MIMETYPE, Phone.CONTENT_ITEM_TYPE);
    cv.put(Phone.NUMBER,"123 123 " + rawContactId);
    cv.put(Phone.TYPE,Phone.TYPE_HOME);
    this.mContext.getContentResolver().insert(Data.CONTENT_URI, cv);
}
private void showRawContactsDataForRawContact(long rawContactId)
{
```

```
    Cursor c = null;
    try
    {
        Uri uri = ContactsContract.RawContactsEntity.CONTENT_URI;
        c = this.getACursor(uri,"_id = " + rawContactId);
        this.printRawContactsData(c);
    }
    finally
    {
        if (c!=null) c.close();
    }
}
```

NOTE: The code in Listing 30–34 is in the file `AddContactFunctionTester.java`. You can add one of these contacts by invoking the **Add Contact** menu item in the sample.

The code in Listing 30–34 does the following:

1. Adds a new raw contact for a predefined account using its name and type, represented by the method `insertRawContact()`. Notice how it uses the URI `RawContact.CONTENT_URI`.

2. Takes the raw contact ID, and inserts a name record—the `insertName()` method—in the data table. Notice how it uses the URI `Data.CONTENT_URI`.

3. Takes the raw contact ID, and inserts a phone number record—the `insertPhone()` method—in the data table. Being a data row, it uses `Data.CONTENT_URI` as the URI.

Listing 30–34 also demonstrates the column aliases used in inserting records. These columns aliases are extracted and repeated in Listing 30–35 for quick review.

Listing 30–35. *Using Column Aliases for Standard Contact Data Structures*

```
cv.put(Data.RAW_CONTACT_ID, rawContactId);
cv.put(Data.MIMETYPE, StructuredName.CONTENT_ITEM_TYPE);
cv.put(StructuredName.DISPLAY_NAME,"John Doe_" + rawContactId);

cv.put(Data.RAW_CONTACT_ID, rawContactId);
cv.put(Data.MIMETYPE, Phone.CONTENT_ITEM_TYPE);
cv.put(Phone.NUMBER,"123 123 " + rawContactId);
cv.put(Phone.TYPE,Phone.TYPE_HOME);

cv.put(RawContacts.ACCOUNT_TYPE, "com.google");
cv.put(RawContacts.ACCOUNT_NAME, "satya.komatineni@gmail.com");
```

It's especially important to know that constants like `Phone.TYPE` and `Phone.NUMBER` actually point to the generic data table column names `data1` and `data2`. Finally, notice that the data fields are displayed by using the `ContactData` (Listing 30–30) data structure in the function `showRawContactsDataForRawContact()`.

Controlling Aggregation

It should be clear by now that clients that update or insert contacts do not explicitly change the contact table. The contact table is updated by triggers that look into the raw contact table and raw contact data table.

Raw contacts that get added or changed, in turn, affect the aggregated contacts in the contacts table. However, you may not want to allow two contacts to be aggregated.

You can control the aggregation behavior of a raw contact by setting the aggregation mode when that contract is created. As you can see from the raw contact table columns in Listing 30–26, the raw contact table contains a field called aggregation_mode. The values for these aggregation modes are shown in Listing 30–37 and explained in the section "Aggregated Contacts."

You can also keep two contacts always apart by inserting rows into a table called agg_exceptions. The URIs needed to insert into this table are defined in the Java class ContactsContract.AggregationExceptions. The table structure of agg_exceptions is shown in Listing 30–36.

Listing 30–36. *Aggregate Exceptions Table Definition*

```
CREATE TABLE agg_exceptions
(_id INTEGER PRIMARY KEY AUTOINCREMENT,
type INTEGER NOT NULL,
raw_contact_id1 INTEGER REFERENCES raw_contacts(_id),
raw_contact_id2 INTEGER REFERENCES raw_contacts(_id))
```

The type column can hold one of the constants in Listing 30–37.

Listing 30–37. *Aggregation Types in the Aggregation Exception Table*

```
TYPE_KEEP_TOGETHER
TYPE_KEEP_SEPARATE
TYPE_AUTOMATIC
```

The type definition and what they indicate are fairly clear. TYPE_KEEP_TOGETHER says the two raw contacts should never be broken apart. TYPE_KEEP_SEPARATE says that these raw contacts should never be joined. TYPE_AUTOMATIC says to use the default algorithm to aggregate contacts.

They URI you will use to insert, read, and update this table is defined as

```
ContactsContract.AggregationExceptions.CONTENT_URI
```

Constants for the field definitions to work with this table are also available in the Java class ContactsContract.AggregationExceptions.

Impacts of Syncing

So far, we have mainly talked about manipulating the contacts on the device. However, accounts and their contacts typically work hand in hand with syncing. For example, if you have created a Google account on your Android phone, the account will pull all your Gmail contacts and make them available to you on your device.

Every time you add a new contact on the device or a new server account, those contacts will be synced and reflected in both places.

However, we have not covered the syncing API and how it works in this edition of the book. Like contacts, it is a large topic. Knowing how contacts work significantly helps to understand the sync API. Please check our updates at `www.androidbook.com`.

The nature of a sync also has an impacts on deleting contacts on the device. When you delete a contact using the aggregated contact URI, it will delete all its corresponding raw contacts and the data elements of each of those raw contacts. However, Android will only mark them as deleted on the device and expects the background sync to actually sync with the server and then delete the contacts permanently from the device. This cascading of deletes also happens at the raw contact level where the corresponding data elements of that raw contact are deleted.

Understanding the Personal Profile

Now that we have presented all of the essentials of how contacts work, we can get to the business of how the personal profile is implemented in 4.0.

A personal profile works just like any other contact, except that there is only one personal profile contact. That is the singular you, on your device.

However, as an implementation detail, all the data pertaining to the singular personal profile contact is maintained in a separate database called `profile.db`. Our research shows that this database has a structure identical to `contacts2.db`. This means you already know what relevant tables are available and what the columns of each table are.

Being a single contact, the aggregation is lot more straightforward. Every raw contact that is added to the personal profile is expected to belong to the singular aggregated contact. If one doesn't exist, then a new aggregated contact is created and placed in the new raw contact. If one exists, that contact ID is used as the aggregated contact ID for the raw contact.

The Android SDK uses the same base class `ContactsContract` to define the necessary URIs to read/update/delete/add raw contacts to the personal profile. These URIs parallel their counterparts but with the string `"PROFILE"` somewhere in them. Listing 30–38 shows a few of these URIs.

Listing 30–38. *Profile-Based URIs Introduced in 4.0*

```
//Relates to profile aggregated contact
ContactsContract.Profile.CONTENT_URI

//Relates to profile based raw contact
ContactsContract.Profile.CONTENT_RAW_CONTACTS_URI

//Relates to profile based raw contact + profile based data table
ContactsContract.RawContactsEntity.PROFILE_CONTENT_URI
```

Clearly we have separate URIs when dealing with aggregated contact and a raw contact. However, there isn't a corresponding URI for the third spoke—the Data table. The same Data URI, Data.CONTENT_URI, is applicable to both regular contact data and also the profile contact data.

Also note that the same content provider serves the needs of both the personal profile and regular contacts. Internally, this content provider knows based on the raw contact ID if the data URI belongs to the profile data or the regular contact data.

Let's look next at some code snippets to read and add contact data to the personal profile. You will need the permissions from Listing 30–39 to read from and write to the profile data.

Listing 30–39. *Permissions Reading/Writing Profile Data*

```
<uses-permission android:name="android.permission.READ_PROFILE"/>
<uses-permission android:name="android.permission.WRITE_PROFILE"/>
```

Reading Profile Raw Contacts

Let's use the following URI to read the raw contacts that belong to the personal profile:

```
ContactsContract.Profile.CONTENT_RAW_CONTACTS_URI
```

Listing 30–40 is the code snippet to read raw contact entries.

Listing 30–40. *Showing All Profile Raw Contacts*

```
public void showAllRawProfileContacts()
{
    Cursor c = null;
    try {
        String whereClause = null;
        c = this.getACursor(
            ContactsContract.Profile.CONTENT_RAW_CONTACTS_URI,
            whereClause);
        this.printRawContacts(c);
    }
    finally {
        if (c!=null) c.close();
    }
}
private void printRawProfileContacts(Cursor c)
{
    for(c.moveToFirst();!c.isAfterLast();c.moveToNext())
    {
        RawContact rc = new RawContact();
        rc.fillinFrom(c);
        this.mReportTo.reportBack(tag, rc.toString());
    }
}
```

Notice that once we retrieve the cursor, the data it contains matches the RawContact that we defined earlier for a regular raw contact.

NOTE: The code in Listing 30–40 is in the file `ProfileRawContactFunctionTester.java`. You run the code in this listing by using the PRawContacts menu item.

Reading Profile Contact Data

Let's use the following URI to read the various data elements (such as e-mail, MIME type, and so on) of raw contacts that belong to the personal profile:

```
ContactsContract.RawContactsEntity.PROFILE_CONTENT_URI
```

Notice how we are using a similar view as in the case of regular contacts. The `RawContactEntity` is a join between raw contacts and the data rows belonging to that raw contact. We will see one row for each data element such as name, e-mail, MIME type, and so on.

Listing 30–41 shows the code snippet to read raw contact entries.

Listing 30–41. *Showing Data Elements for Profile Contacts*

```
public void showProfileRawContactsData()
{
    Cursor c = null;
    try {
        Uri uri = ContactsContract.RawContactsEntity.PROFILE_CONTENT_URI;
        String whereClause = null;
        c = this.getACursor(uri,whereClause);
        this.printProfileRawContactsData(c);
    }
    finally {
        if (c!=null) c.close();
    }
}
protected void printProfileRawContactsData(Cursor c)
{
    for(c.moveToFirst();!c.isAfterLast();c.moveToNext())
    {
        ContactData dataRecord = new ContactData();
        dataRecord.fillinFrom(c);
        this.mReportTo.reportBack(tag, dataRecord.toString());
    }
}
```

Notice that once we retrieve the cursor, the data it contains matches the `ContactData` object (Listing 30–30) that we defined earlier for a regular raw contact data element.

NOTE: The code in Listing 30–41 is in file `ProfileContactFunctionTester.java`. You run the code in this listing by using the menu item "all p raw contacts".

Adding Data to the Personal Profile

Let's use the following URI to add a raw contact to a personal profile:

```
ContactsContract.RawContactsEntity.PROFILE_CONTENT_URI
```

We will also add a few data elements such as a phone number and a nickname to that raw contact so they appear in the details for ME (Figure 30–12).

Listing 30–42 shows the code snippet.

Listing 30–42. *Adding a Profile Raw Contact*

```
public void addProfileContact()
{
    long rawContactId = insertProfileRawContact();
    this.mReportTo.reportBack(tag, "RawcontactId:" + rawContactId);
    insertProfileNickName(rawContactId);
    insertProfilePhoneNumber(rawContactId);
    showProfileRawContactsDataForRawContact(rawContactId);
}
private void insertProfileNickName(long rawContactId)
{
    ContentValues cv = new ContentValues();
    cv.put(Data.RAW_CONTACT_ID, rawContactId);
    //cv.put(Data.IS_USER_PROFILE, "1");
    cv.put(Data.MIMETYPE, CommonDataKinds.Nickname.CONTENT_ITEM_TYPE);
    cv.put(CommonDataKinds.Nickname.NAME,"PJohn Nickname_" + rawContactId);
    this.mContext.getContentResolver().insert(Data.CONTENT_URI, cv);
}
private void insertProfilePhoneNumber(long rawContactId)
{
    ContentValues cv = new ContentValues();
    cv.put(Data.RAW_CONTACT_ID, rawContactId);
    cv.put(Data.MIMETYPE, Phone.CONTENT_ITEM_TYPE);
    cv.put(Phone.NUMBER,"P123 123 " + rawContactId);
    cv.put(Phone.TYPE,Phone.TYPE_HOME);
    this.mContext.getContentResolver().insert(Data.CONTENT_URI, cv);
}
private long insertProfileRawContact()
{
    ContentValues cv = new ContentValues();
    cv.put(RawContacts.ACCOUNT_TYPE, "com.google");
    cv.put(RawContacts.ACCOUNT_NAME, "satya.komatineni@gmail.com");
    Uri rawContactUri =
        this.mContext.getContentResolver()
             .insert(ContactsContract.Profile.CONTENT_RAW_CONTACTS_URI, cv);
    long rawContactId = ContentUris.parseId(rawContactUri);
    return rawContactId;
}
private void showProfileRawContactsDataForRawContact(long rawContactId)
{
    Cursor c = null;
    try {
        Uri uri = ContactsContract.RawContactsEntity.PROFILE_CONTENT_URI;
        c = this.getACursor(uri,"_id = " + rawContactId);
        this.printRawContactsData(c);
    }
```

```
        finally {
            if (c!=null) c.close();
        }
}
```

The code in Listing 30–42 parallels the code we used to add a regular contact and its details (Listing 30–34). Although we have used a profile-specific URI to add a raw contact, we have used the same `Data.CONTENT_URI` to add the individual data elements.

Note the following commented-out code in Listing 30–42:

```
//cv.put(Data.IS_USER_PROFILE, "1");
```

Because `Data.CONTENT_URI` is not specific to the profile, how does the underlying content provider know whether to insert this data into a regular raw contact or a personal profile raw contact? We thought that specifying a column called `IS_USER_PROFILE` would help the content provider. Apparently not. This new column is available primarily for read purposes. Your inserts will fail if you specify this during inserts. The only conclusion then is that the content provider is relying on the raw contact ID to see whether that raw contact came from `profile.db` or `contacts2.db`.

Notice that once we retrieve the cursor, the data it contains matches the `ContactData` object (Listing 30–30) that we defined earlier for a regular raw contact data element. In fact, the function `printRawContactsData()` that is used in Listing 30–42 uses the same definition provided for it in Listing 30–27.

> **NOTE:** The code in Listing 30–42 is in the file `AddProfileContactFunctionTester.java`. You run the code in this listing by using the Add P Contact menu item.

Although we have shown you how to add a contact to the personal profile, this is usually done by your accounts such as Gmail, Google+, and so on. It is also unlikely that third-party programs will ask for permission to read your personal profile.

References

The following annotated references are useful for supporting and enhancing the material in this chapter. The last URL allows you to download projects developed for this chapter:

- `www.google.com/googlephone/AndroidUsersGuide.pdf`: The 2.2.1 version of the *Android User's Guide*. You can use this guide to read about the contacts application that lets you manage your contacts. Although we have covered the basic information here on how to use the contacts application, this user's guide is the authority, and you may pick up things that we have overlooked.
- `www.google.com/help/hc/pdfs/mobile/AndroidUsersGuide-30-100.pdf`: The Android 3.0 user's guide.

- www.google.com/support/ics/nexus/: The PDF user guide for 4.0-based phones is not out yet. You can use this link in the interim to learn about the same topics in HTML format.
- `http://developer.android.com/resources/articles/contacts.html`: The primary documentation for the Contacts API from Google.
- http://developer.android.com/sdk/android-4.0.html#Contacts: Documentation for the changes to the Contacts API in 4.0.
- http://developer.android.com/reference/android/provider/ContactsContract.Profile.html: A reference on how to use the new Profile URIs introduced in 4.0.
- `www.androidbook.com/item/3917`: The entry point for our research on the Contacts API. You will find here our research, a summary of the Contacts API, tables used in the contacts database, how to explore the contact databases, contacts application screen shots, how to explore sources for contact providers, and other useful links.
- `http://developer.android.com/reference/android/provider/ContactsContract.html`: The Java doc for the entry class for the published contacts contract. You will need this URL often as you code to the Contacts API.
- `www.androidbook.com/item/2865`: Because of the paucity of information on how contacts are treated, you may want to see the source code of the contacts content provider. This URL from our site will help you download the sources for the contact provider implementation.
- `www.androidbook.com/item/3537`: If you were to go through the source code of contact provider, you would see that Java generics will generously befuddle you. This URL contains a summarization of Java generics that could be of some help.
- `www.androidbook.com/proandroid4/projects`: You can use this URL to download the test project dedicated for this chapter. The name of the ZIP file is `ProAndroid4_ch30_TestContacts.zip`.

Summary

In this chapter we have covered the following:

- The nature of the Contacts API
- Exploring the contact database
- Exploring the Contacts API URIs and their cursors
- Reading and adding contacts
- Aggregating raw contacts
- The relationship between the personal profile and contacts
- Reading and adding contacts to a personal profile

Interview Questions

The following detailed set of questions should further consolidate what you learned in this chapter:

1. What are accounts, and what API do you use to list accounts on your device?
2. What is `ContactsContract`?
3. What is the name of the database where contacts are kept?
4. What is `profile.db`?
5. Where are `profile.db` and `contacts2.db` stored?
6. What is the best way to understand how the stock contacts/People application works?
7. How can you discover the fields for a particular contact data element such as an e-mail contact or a web site contact?
8. What is `CommonDataKinds`?
9. Draw out the relationship between contact, raw contact, and data entities.
10. How can you discover the fields returned by various Contacts API URI cursors?
11. List all the URIs relevant to the Contacts API.
12. Is the contacts table in the contacts database read-only for clients? Why?
13. When are two raw contacts likely to be merged?
14. Why is an aggregated contact ID volatile?
15. What is the need for a lookup key?
16. Does a lookup key retrieve a set of raw contacts?
17. Given a URI, how can you retrieve a cursor for it?
18. Given a URI, how can you get a cursor that is specific to a `where` clause?
19. What URI do you use to read such things as a contact's e-mail and phone number?
20. What is an aggregation exception table?

21. How is a personal profile similar to a contact?
22. Does a personal profile have the same table structure as a regular contact?
23. Why is there no profile specific URI to work with the `Data` table?
24. What permissions do you need to read and write profile contacts?
25. What permission do you need to read/write regular contacts?

Chapter 31

Deploying Your Application: Android Market and Beyond

Creating a great application that people will love is one thing, but you also need an easy way for people to find and download it. Google created Android Market for this purpose. From an icon right on the device, users can click straight into the Market to browse, search, review, and download applications. Users can also access Android Market over the Internet to do those same things, although the downloading is not to the computer but rather is sent to the user's device. Many applications are free; for those that are not, the Market provides payment mechanisms for easy purchasing.

The Market is even accessible from intents inside of applications, making it easy for applications to reach out to the Market to guide the user into getting what they need for your application to be successful. For example, when a new version of your application becomes available, you can make it easy for the user to go straight to that Market page to get or buy the new version. Android Market is not the only way to get applications to devices, however; other channels are all over the Internet.

The Android Market application is not available from within the emulator (although hacks exist to make it available). This makes things a little more difficult for a developer. Ideally you will have a device of your own that you can use with Android Market. In this chapter, we'll explore how to get you set up for publishing applications to the Market; how to prepare your application for sale through the Market; how you can protect yourself from piracy; how users will find, download, and use your applications; and finally, alternative ways to make your applications available.

Becoming a Publisher

Before you can upload an application to Android Market, you need to become a publisher. To do so, you must create a Developer Account. Once that's done, you will be able to upload your applications to the Market so they can be found and downloaded by users. Google has made the process to get a Developer Account relatively painless and reasonably priced.

To publish anything, you first need to have a Google account—for example, a `gmail.com` e-mail account. Next, you establish an identity with Android Market. You do this by going to `http://market.android.com/publish/signup`. You will need to provide a developer name, an e-mail address, a web site address, and a phone number where you can be contacted. You will be able to change these values later, once your account is set up. You will also need to pay the registration fee. This is done via Google Checkout. In order to continue with the transaction, you will be required to log in with a Google account.

One of the options presented to you during the payment process is "Keep my email address confidential." This refers to the current transaction between you and Google Android Market to "purchase" publisher access. If you choose yes, you'll keep your e-mail address secret from Google Android Market. This has nothing to do with keeping your e-mail address secret from buyers of your application. Buyers' ability to see your e-mail address has nothing to do with this option. More on that later.

Next up is the Android Market Developer Distribution Agreement (AMDDA). This is the legal contract between Google and you. It spells out the rules for distributing apps, collecting payments, granting refunds, feedback, ratings, user rights, developer rights, and so on. There's more on these in the "Following the Rules" section of this chapter.

Upon accepting the Agreement, you will be taken to a page commonly called the Developer Console at `http://market.android.com/publish/Home`.

Following the Rules

The AMDDA spells out a lot of rules. You might want legal counsel to review the contract before agreeing to it, depending on how seriously you plan to operate within Android Market. This section describes some highlights you might be interested in:

- You have to be a developer in good standing to use Android Market. This means you must go through the process as described to get registered, you must accept the Agreement, and you must abide by the rules in the Agreement. Breaking the rules could get you barred and your products removed from the Market.

- You can distribute products for free or for a price. The Agreement applies either way. If selling products, you must have a payment processor such as Google Checkout. When Android 2.0 was introduced, Google Checkout was the only way to collect money through Android Market. It has become possible for users to simply charge to their phone bill for downloading applications from Android Market, as first announced by T-Mobile in 2009 and AT&T in 2010. PayPal announced integration with Android Market in October of 2010, but over a year later it still isn't an option. This may change in a future release, however.
- Paid apps will incur a transaction fee, and possibly a fee from the device carrier, to be deducted from the sale price. As of October 2011, the transaction fee is 30 percent, so if the sale price is $10, Google collects $3 and you get $7 (assuming no carrier fees).
- It is your responsibility to remit appropriate taxes to your taxing authorities. When you set up your merchant account, you specify the appropriate tax rates to apply to purchases from people in other locations. Google Checkout will collect the appropriate taxes based on how you set up Google Checkout. This money will be provided to you, and you must remit it appropriately. For additional information on sales taxes in the U.S., try `http://biztaxlaw.about.com/od/businesstaxes/f/onlinesalestax.htm` and `www.thestc.com`.
- You are allowed to distribute a free demo version of your application, with an option to pay to unlock the application's full set of features; however, you must collect the payment via an authorized Android Market Payment Processor. You are not allowed to redirect users of your free application to some other payment processor to collect upgrade fees. You could think of it this way: if you're making money via Android Market, Google wants its share.
- In February 2011, Google announced in-app billing. This is an add-on SDK that allows an application to charge for digital goods or assets used within the application. A digital asset could be something like a virtual weapon or new levels for a game, or a music or graphics file. The checkout process is the same as for purchasing applications, which means users could pay from their phone bill for these digital assets.
- If your application requires a user to have a login on a web server somewhere, and that web server charges the user a subscription fee, that web server could collect the subscription fee any way it wants to. In this way, you have disconnected the subscription fee from the application, and it's OK by Google to make the application available in Android Market—as long as your free application is not directing users to the web site. But really, why not just distribute your free Android app from the same web server as the service?

- It seems that you can use alternate payment processors to accept donations from users of your free app, but you cannot create incentives within your app to encourage those donations.
- Refunds are a nasty subject with Android Market. Originally, users had 24 hours to request a refund of the purchase price. Then it was changed to 48 hours. In December 2010, it was changed to 15 minutes! And that's 15 minutes from when the purchase is made, not from when the download has successfully completed. There have been cases where a user hasn't even been able to finish downloading the application and the refund window has passed. Strangely, the AMDDA was not updated in December 2010 to reflect 15 minutes and still said 48 hours. Refunds are not given to users who can preview the product prior to download. This includes ringtones and wallpapers. Google Checkout, however, does allow the developer to issue a refund even if the refund window has passed, so users do have a way to get a refund no matter what. But developers don't want to be issuing refunds manually.
- You are required to provide adequate support for your product. If adequate support is not provided, users can request refunds, and these will be charged back to you, possibly including handling fees.
- Users get unlimited reinstalls of applications downloaded from Android Market. If a user does a factory reset of their device, this feature allows them to get all their apps back without having to repurchase.
- Developers agree to protect the privacy and legal rights of users. This includes protecting (securing) any data that might be collected in the process of using the application. It is possible to change the rules regarding users' data protection, but only by displaying and having the user accept a separate agreement between you and that user.
- Your application must not compete with Android Market. Google does not want an application from within Android Market to sell Android products from outside Android Market, thus bypassing its payment processor. This does not mean that you can't also sell your application through other channels, but your application on Android Market cannot itself be doing the selling of Android products outside of Android Market.
- Google will assign product ratings to your products. The ratings could be based on user feedback, install rates, uninstall rates, refund rates, and/or a Developer Composite Score. The Developer Composite Score may be calculated by Google using past history across applications, and this could influence the rating of new applications. For this reason, it is important to release good-quality applications associated with you, even the free ones. It's not clear that the Developer Composite Score even exists, but if it does there's no way to see yours.

- By selling your application through Android Market, you are granting the user a "non-exclusive, worldwide, perpetual license to perform, display and use the Product on the device." However, it is quite all right for you to write a separate End User License Agreement (EULA) that supersedes this statement. Make this EULA available on your web site, or provide another way for shoppers and users to be able to read it.
- Google requires that you abide by the branding rules for Android. These include restrictions on the use of the word *Android*, as well as use of the robot graphic, logo, and custom typeface. For more details, go to `www.android.com/branding.html`.

Developer Console

The Developer Console is your landing page for controlling your applications in Android Market. From the Developer Console, you can set up a merchant account in Google Checkout (so you can charge for your applications), upload applications, and get information about your uploaded applications. You can also edit your account details including developer name, e-mail address, web address, and phone number. Figure 31–1 shows the Developer Console.

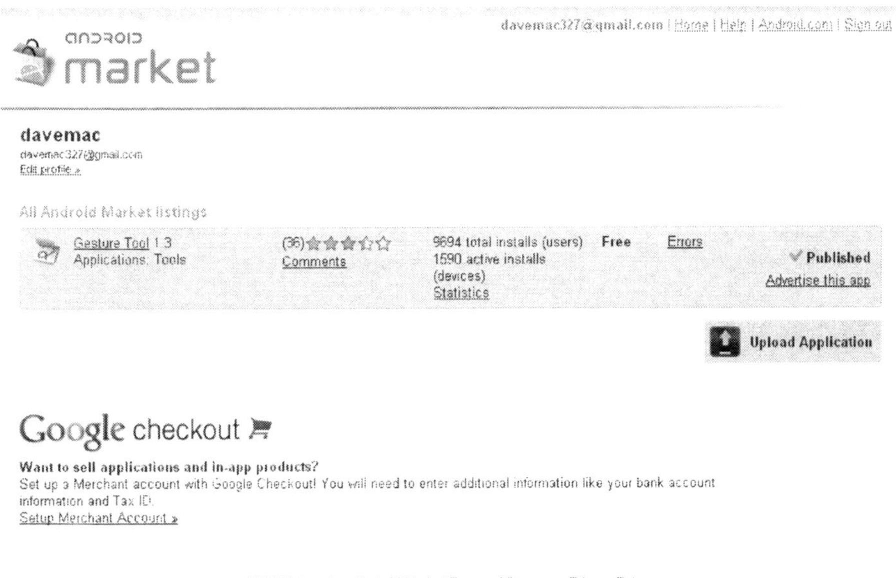

Figure 31–1. *The Android Market Developer Console*

If you do not set up a merchant account using Google Checkout, you will be unable to charge for your products in Android Market. Setting up a merchant account is not difficult. Click the link from the Developer Console, fill out the application, agree to the Terms of Service, and you're all set. You will need to provide a U.S. Federal tax ID (EIN), a credit card number plus a U.S. Social Security Number (SSN), or just a credit card number. The tax information is used to verify your credit status to ensure timely deposits. The credit card information is used to handle chargebacks due to buyer disputes when there are insufficient funds in your Google Checkout account. You can also supply bank account information to enable electronic funds transfers from the proceeds of your sales.

Note that Google Checkout is a service for more than just Android Market. Therefore, do not get confused by the transaction fee information for Google Checkout for non-Android Market sales. The 30 percent mentioned previously is the transaction fee rate for Android Market. There is also additional Google Checkout transaction fee information for non-Android Market sales, and those do not apply to Android Market.

Uploading and monitoring your applications are probably the main functions of the Developer Console that you will use. We'll discuss uploading applications later in this chapter. For monitoring, the Market provides tools to see how your application is doing in terms of total downloads and how many users still have it installed. You can see the overall rating of your apps in terms of 0 to 5 stars, and how many people have submitted a rating. In March 2011, Google added charts and graphs to the Developer Console so you can see how your application is doing in different versions of Android, on different devices, in different countries and in different languages.

Users can submit comments in addition to rating your application. It is in your best interest to read the comments in order to address any problems quickly. Included with a comment is the user's rating of your app, a name of the user as typed by them, and the date of the comment. Unfortunately, there is no way to reply to commenters directly, or even comment on the comment. In an extreme case, where a comment is particularly harmful or inappropriate, you can contact Google support by starting here: `http://market.android.com/support/`.

The Developer Console allows you to republish your application—for upgrades, for example—or to unpublish the application. Unpublishing does not remove it from devices, nor does it even necessarily remove the app from the Google servers, especially if it's a paid app. A user who has paid for your application and who has uninstalled it, but not requested a refund, is allowed to reinstall it later even if you've unpublished it. The only way it is truly unavailable to users is if Google pulls it due to violation of the rules.

You can also look at errors that were generated by your application and see application freezes and crashes. Figure 31–2 shows the Application Error Reports screen.

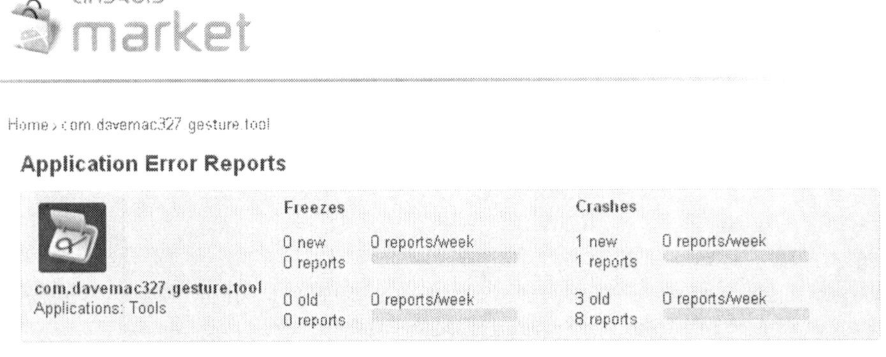

Figure 31–2. *The Application Error Reports screen*

Drilling into the details of a crash report, you can see the stack trace of the crash, as well as which type of device was running the application and the time of the crash. But as with user comments, you cannot communicate back to the user who experienced the problem to get additional details or to help them get the issue resolved. You have to hope that the affected users will get in touch with you through e-mail or your web site. Otherwise, you'll just have to figure out from the crash report what went wrong and try to fix it.

There's one more feature of the Developer Console you may need to use: the Help portion of the web site. The Help button is in the upper-right corner. Clicking it takes you to a Help web site that has a lot of decent documentation on how to use Android Market, and it also has a forum where you can search for questions and answers and post your own. For example, the forum is where you can read up on the latest refund policies, issues, and complaints. If the forum is not helpful, there is a Contacting Support link that will take you to a page where you can send a message specifically to Google for help.

We've now introduced you to some of the nice features of the Developer Console, but you probably want to get into the most useful part, which is getting your applications into Android Market so users can find them and download them. But before we do that, let's go over how to prepare your application for upload and sale.

Preparing Your Application for Sale

There are quite a few things to think about and do to take an application from code complete to Android Market. This section will help you through those items.

Testing for Different Devices

With more and more Android devices becoming available, and each one potentially having some new hardware configuration, it is very important that you test for those devices you want to support. The ideal case would be to get access to one of each type of device to test your application on. There are some online services that make real devices available over the Internet, but that's an expensive proposition. The next best choice is to configure Android Virtual Devices (AVDs) for each type of device, specify the appropriate hardware configuration, and then test with the emulator and each AVD. Some device manufacturers make Android emulator packages available that are specific to their devices, so check out their web sites for download options.

The Android SDK provides the `Instrumentation` class to assist with testing, as well as the UI/Application Exerciser Monkey. These tools will help you do automated testing so you don't spend forever testing your application. Before you begin testing, you probably want to remove any testing artifacts that you no longer need from your code and from `/res`. You want your application to be as small as possible and to run as quickly as possible with the least amount of memory. Finally, be sure to disable or remove any debugging features from your application that you don't want distributed to production.

Supporting Different Screen Sizes

When Android SDK 1.6 came out, developers had to contend with new screen sizes. In order to run on the new smaller size, you must set a specific `<supports-screens>` element as a child element of `<manifest>` within the `AndroidManifest.xml` file. Without this new tag specifying that your application supports the small screen size, your application will not be visible in the Market to devices that have a small screen.

To support different screen sizes, you may need to create alternate resource files under `/res`. For example, for files in `/res/layout`, you may need to create corresponding files in `/res/layout-small` to support small screens. This does not mean you must also create corresponding files in `/res/layout-large` and `/res/layout-normal`, since Android will look in `/res/layout` if it can't find what it needs in a more specific resource directory such as `/res/layout-large`. Remember, too, that you can have combinations of qualifiers for these resource files; for example, `/res/layout-small-land` would contain layouts for small screens in landscape mode. We talked about this in Chapter 6. Supporting small screens probably means creating alternate versions of drawables such as icons, too. For drawables, you may need to create alternate resource directories, taking into account screen resolution as well as screen size.

Tablets of course go in the opposite direction in terms of screen size, using the label `xlarge`. The same `<supports-screens>` tag as before is used to specify if your application will run on extra large screens, and the attribute to use inside of this tag is `android:xlargeScreens`. In some cases, you may have a tablet-only application, in which case you would specifically indicate that for the other sizes, the attribute value is `false`.

Preparing AndroidManifest.xml for Uploading

Your AndroidManifest.xml file may need to be tweaked a little bit before you can upload it to Android Market. ADT normally puts the android:icon attribute in the <application> tag, and not in <activity> tags. If you have more than one activity that can be launched, you'll want to specify separate icons for each activity so the user can more easily tell them apart. But you'll still need an icon specified in <application>, which also serves as the default activity icon for any activities that don't specify their own icon. Your application will work fine on devices and in the emulator with the android:icon only specified in the <activity> tags, but when Android Market inspects your application's .apk file when uploading, it looks for icon information in the <application> tag. Android Market also prevents uploading your application if the package name you've used starts with com.google, com.android, android, or com.example, but we hope you didn't use one of those in your application.

There are many other compatibilities to consider for your application. Some devices have cameras, some don't have physical keyboards, and some have trackballs instead of directional pads. Use <uses-configuration> and <uses-feature> tags in your AndroidManifest.xml file as needed to define what hardware/platform requirements your application has. Android Market will enforce this and not let your application be shown to a user on a device that won't support your application. Note that these tags are different and separate from the <uses-permission> tags of the AndroidManifest.xml file. While the user's device may come equipped with a camera, that doesn't mean the user wants to grant your application permission to use it. At the same time, declaring that your application needs permission to use the camera does not tell Android Market that your application requires a camera on the device. In most cases, you would end up with both tags in your AndroidManifest.xml file, for specifying that a camera is required, and for specifying that permission to use the camera is required. But not all features require permission, so it is in your best interest to specify the features you need.

There is another big difference between <uses-permissions> and <uses-feature>: the <uses-feature> tag can say that your application requires that feature or that your application can function without it. That is, there is an attribute called android:required that can be set to either true or false; by default it's true. For example, your application may take advantage of Bluetooth if it's available, but will work just fine if it is not. Therefore, in the manifest file, you'd have something like this:

```
<uses-feature android:name="android.hardware.bluetooth" android:required="false" />
```

Within your application's code, you should make a call to the PackageManager to find out if Bluetooth is available or not, which you could do with the following:

```
boolean hasBluetooth = getPackageManager().hasSystemFeature(
                PackageManager.FEATURE_BLUETOOTH);
```

Then take appropriate action in your application if Bluetooth is not there. The Android documentation can be confusing in this area. If you look at the Developer Guide page for <uses-feature>, you will not see as many features as are described on the

PackageManager reference page, which defines a FEATURE_* constant for each available feature.

The <uses-configuration> tag is a little different. It specifies what sort of keyboard, touchscreen, and/or navigational controls the device must have. But instead of being independent choices such as <uses-feature>, you would put the combinations of configuration choices together into what your application requires. For example, if your application requires a five-way navigation control (that is, a D-pad or a trackball) and a touchscreen (using either a stylus or a finger), you would specify two tags as follows:

```
<uses-configuration android:reqFiveWayNav="true" android:reqTouchScreen="stylus" />
<uses-configuration android:reqFiveWayNav="true" android:reqTouchScreen="finger" />
```

Localizing Your Application

If your application will be used in other countries, you might want to consider localizing it. This is relatively easy to do technically. Finding someone to do the localizing is another matter. From the technical point of view, you simply create another folder under /res—for example, /res/values-fr to hold a French version of strings.xml. Take your existing strings.xml file, translate the string values to the new language, and save the new translated file under the new resource folder using the same file name as the original file. At runtime, if the device's language is set to French, Android will look for strings that were placed under /res/values-fr. If it can't find strings from there, it will then look for strings from /res/values.

The same technique works for the other types of resource files—for example, drawables and menus. Images and colors may work better for your users if they are different for different countries or cultures. For this reason, it is a good idea to not use true color names for your resource names for colors. In the online documentation for colors, it is common to see something like this:

```
<color name="solid_red">#f00</color>
```

This means that in your code or other resource files, you're referring to the color by the actual name of the color, in this case, solid_red. In order to localize the color to something more appropriate for the other country or culture, it would be better to use a color name such as accent_color1 or alert_color. In English, red might be the appropriate color value to use, while in Spanish it might be better to use a shade of yellow. Because a color name like alert_color does not reveal the actual color that you're using, it is less confusing when you want to change the actual color value to something else. At the same time, you can design a pleasing color scheme, with base colors and accent colors, and be more confident that you're using the correct colors in the correct places.

Menu choices might need to be changed in different countries, using fewer or more menu items, or be organized differently, depending on where the application is being used. Menus are typically stored under /res/menu. If you are faced with this situation, you are probably better off putting all your string text into strings.xml, or other files located under the /res/values directory, and using string IDs in the appropriate resource files everywhere else. This makes it far less likely that you will miss translating a string value in some obscure resource file. Your language translation work is then limited to the files under /res/values.

Preparing Your Application Icon

Shoppers and your users will see your application's icon and label prominently in both Android Market and on their device once they've downloaded it. Please take special care to create good icons and good labels for your application and its activities. Localize them as necessary or desired. And remember that for different screen sizes, your icons may need to be tweaked to look good. Check out what other developers have done with their icons, especially those applications in the same category as your application. You want your application to get noticed, so it's better not to blend in with all the others. At the same time, you want your icon and label to work well on a device when surrounded by lots of other application icons that do other things. You don't want a user to be confused about what your application does, so make the icon representative of the functionality of your application.

When creating any image for your application, but especially your icon, you need to consider the screen density of the target device. *Density* means the number of pixels per inch. Don't think that a small screen is low density and a large screen is high density—you could see any combination of size and density. For a low-density screen, making an icon appear to be the right size means making the icon with fewer pixels, typically 36 × 36. For a high-density screen, you will probably choose an icon with 72 × 72 pixels. The medium-density icon will usually be of size 48 × 48 pixels. And for extra-high density, it's 96 × 96 pixels. In general, you'll find it easiest to only worry about density for images such as icons. You'll worry about screen size when defining layouts.

Considerations for Making Money from Apps

If you are selling your application for a price, you have some other considerations to think about. Do you offer separate free and paid applications, requiring you to build and manage two applications? Or do you keep one code base and use some sort of technique to tell if this application was paid for or not? No matter which approach you take, how do you protect your application from being copied and installed on other devices for other people? Due to security vulnerabilities in phones, and due to the ability of certain people to get inside devices, foolproof guarantees of copy protection are extremely difficult to manage.

One technique for maintaining a single code base, but allowing for separate free and paid modes, is to take advantage of the PackageManager:

```
this.getPackageManager().checkSignatures(mainAppPkg, keyPkg)
```

This method compares the signatures of the two named packages and returns PackageManager.SIGNATURE_MATCH if they both exist and are the same. The package names must be different for each app to coexist in Android Market, but that's fine. In your code, when you need to decide whether or not to allow functionality, you can call this method and provide the package name of your main application as well as the package name of your unlocking application. You then make the unlocking application a paid app in Android Market. If the user buys the unlocking application and downloads it to their device, the main application will then get a signature match and unlock the extra functionality.

Another way to deal with a single code base is to use source code versioning systems to configure appropriate sharing of common elements, and build scripts to handle creating the free and paid versions of your application.

Another way that you can make money from Android apps is with in-app advertising. There are many opportunities for embedding ads in your app. A couple of common examples are AdMob and AdSense. The process is basically to incorporate their SDK into your application, figure out where and when to display ads in your app, and add the INTERNET permission to your app (so the ad SDK can get to the ads to display); then you get paid as users click the ads. Your app can be free, so it's easier to get it into Android Market, plus you don't need to worry as much about piracy. Many developers report making some decent money from ads.

Another feature introduced in February 2011 is Buyer's Currency. Prior to this time, buyers had to pay in the currency of the seller, which could easily get confusing for buyers who had a hard time converting from the seller's currency amount to their own. It also meant that a seller could only really have one price for the world. Now that the seller can specify a price for a country, not only can the selling price be higher or lower in other countries, but the buyers' experience is much nicer and more convenient.

Directing Users Back to the Market

Android has introduced a new URI scheme to help facilitate finding applications in Android Market: market://. For example, if you want to direct your users to the Market to locate a needed component, or to upsell to an additional app that unlocks features in your application, you would do something as shown here, where MY_PACKAGE_NAME would be replaced by your real package name:

```
Intent intent = new Intent(Intent.ACTION_VIEW,
        Uri.parse("market://search?q=pname:MY_PACKAGE_NAME"));
startActivity(intent);
```

This will launch the Market app on the device and take the user to that package name. The user can then choose to download or buy the application. Note that this scheme does not work in a normal web browser. In addition to searching using package name (pname), you can search by developer name using market://search?q=pub:\"Fname Lname\" or against any of the public fields (application title, developer name, and application description) in Android Market using market://search?q=<querystring>.

If we combine what we've just learned with the technique in the previous section, our code could look for the unlocking package on the device. If we don't find it, we could prompt the user to see if they want to go get the unlocking app. If they answer yes, we invoke an intent, which opens the Market app and takes the user straight to our unlocking app to be purchased and downloaded.

The Android Licensing Service

The way that Android apps are constructed unfortunately makes them targets for piracy. It is possible to make copies of Android apps that can then be distributed to other devices. So how can you ensure that users who have not purchased your application cannot run it? The Android team has created something called the License Verification Library (LVL) to meet this need. Here's how it works.

If your application was downloaded via Android Market, then there must be a copy of the Android Market app on the device. In addition, the Android Market app has elevated permissions to be able to read values from the device such as the user's Google account name, the IMSI, and other information. The Android Market app has been modified, going back to Android version 1.5, to respond to a license verification request from an application. You make calls into the LVL from your application, LVL communicates with the Android Market app, the Android Market app communicates with Google servers, and your application gets an answer back indicating whether or not this user on this device is licensed to use your application. This means the app must have been purchased through Android Market; otherwise the Google servers won't know about it. There are settings under your control to decide what to do if the network is unavailable. A full description of the process of implementing LVL can be found at `http://developer.android.com/guide/publishing/licensing.html`.

One thing to be aware of, though, is that the LVL mechanism is subject to hacking. If someone can get to your application's `.apk` file, they can disassemble the app and then patch it if they know where to look for the return value from the LVL call. If you use the obvious pattern of a `switch` statement after getting the response from LVL, to branch to the appropriate logic based on the return code, a hacker can simply force a successful return code value, and they own your app. For this reason, the Android team highly recommends that you implement obfuscation of your app to hide the part of your application where you check the return code from LVL. This gets fairly complicated, as you can imagine.

Using ProGuard for Optimizing, Fighting Piracy

With Android 2.3, Google provided some support for obfuscation in the form of the ProGuard feature. ProGuard is not a Google product but has been integrated into ADT so it's easy to use. ProGuard does more than just provide obfuscation for fighting piracy; it also makes your application smaller and faster. It does all this by stripping out debugging information, cutting out code that will never run, and changing names (of classes, methods, and so on) to meaningless strings. Examples of code that will never run include library classes and methods that are never called, and logging that depends on a constant that you set to `false` (for production). It can also recognize optimizations such as binary-shifting a value left by one bit position instead of multiplying it by 2. By stripping out debugging information and changing the names, the resulting compiled `.apk` file won't reveal variable names, class names, methods, and so on, so it becomes extremely difficult to figure out what the code does and therefore how to steal it, modify it, and release it as something else.

When you set the target build of your application to 2.3 or later, your application should automatically get a proguard.cfg file. The default file will look something like Listing 31–1.

Listing 31–1. *Sample proguard.cfg File*

```
-optimizationpasses 5
-dontusemixedcaseclassnames
-dontskipnonpubliclibraryclasses
-dontpreverify
-verbose
-optimizations !code/simplification/arithmetic,!field/*,!class/merging/*

-keep public class * extends android.app.Activity
-keep public class * extends android.app.Application
-keep public class * extends android.app.Service
-keep public class * extends android.content.BroadcastReceiver
-keep public class * extends android.content.ContentProvider
-keep public class * extends android.app.backup.BackupAgentHelper
-keep public class * extends android.preference.Preference
-keep public class com.android.vending.licensing.ILicensingService

-keepclasseswithmembernames class * {
    native <methods>;
}

-keepclasseswithmembers class * {
    public <init>(android.content.Context, android.util.AttributeSet);
}

-keepclasseswithmembers class * {
    public <init>(android.content.Context, android.util.AttributeSet, int);
}

-keepclassmembers class * extends android.app.Activity {
   public void *(android.view.View);
}

-keepclassmembers enum * {
    public static **[] values();
    public static ** valueOf(java.lang.String);
}

-keep class * implements android.os.Parcelable {
  public static final android.os.Parcelable$Creator *;
}
```

You also need to set the proguard.config property in the application's default.properties file to the location of the proguard.cfg file. The line looks like this:

```
proguard.config=proguard.cfg
```

As mentioned, ProGuard does its work by stripping stuff out. Sometimes it strips out too much, and that is why you see the -keep options specified in the proguard.cfg file. When you produce an .apk file, you need to test it to make sure ProGuard didn't take out too much. If you find errors due to missing classes or methods, you can edit the `proguard.cfg` file to include another `-keep` option for the item you're missing. Rebuild your `.apk` file, and test again. We recommend using the Export Signed Application Package option under the Android Tools menu option in Eclipse, because it will take care of calling ProGuard for you as it builds the `.apk` file. Exporting is covered in the next section.

You can also configure Ant to obfuscate using ProGuard if you use Ant to do your builds.

When ProGuard does its thing, you'll get a file called `mapping.txt` along with your `.apk` file. Hang on to this file because you will need it to de-obfuscate a stack trace from your application. If you use Eclipse to export your `.apk` file, you will see a new `proguard` directory created within your Eclipse project. The `mapping.txt` file will be in there. The command to use is `retrace`, and it's located in the Android SDK directory under `tools/proguard/bin`. The arguments to retrace include the `mapping.txt` file and the stacktrace file, but be aware that you need to specify the full pathname to each. Also, you should keep track of which version of your application goes with which `mapping.txt` file.

Preparing Your .apk File for Uploading

To get your tested application ready for uploading—that is, to create the `.apk` file to upload—you need to do the following things (all covered in Chapter 14):

1. Create (if you haven't already) a production certificate with which to sign your application.
2. If you're using maps, replace the MAP API key (in your layout files that contain `MapViews`) with your production MAP API key. If you forget to do this, none of your users will be able to see maps.
3. Export your application by right-clicking your project in Eclipse, choosing Android Tools ➤ Export Unsigned Application Package, and choosing an appropriate file name. It is convenient to give this file a temporary name, because when you run `zipalign` in step 5, you need to provide an output file name and that should be your production `.apk` file name.
4. Run `jarsigner` on your new `.apk` file to sign it with the production certificate from step 1.
5. Run `zipalign` on your new `.apk` file to adjust any uncompressed data to the appropriate memory boundaries for better performance at runtime. This is where you will provide the final file name for your application's `.apk` file.

6. Android now provides an Export Signed Application Package option in Eclipse, which uses a wizard to do steps 3, 4, and 5.

Uploading Your Application

Uploading is easy to do but takes some preparation. Before you begin an upload, there are some things you will need to have ready and decisions you have to make. This section covers that preparation and those decisions. Then, when you've got everything you need, go to the Developer Console and choose Upload Application. You'll be prompted to supply lots of information about your application, the Market will run some processing of your application and the information, and then your application will be ready to publish to the Market.

The previous section covered preparing your application `.apk` file for uploading. Making your application attractive to shoppers requires some marketing on your part. You need good descriptions of what it is and does, and you need good images so shoppers understand what they might download.

Graphics

One of the first items you'll be asked for when uploading an application is screenshots. The easiest way to capture screenshots of your application is to use DDMS. Fire up Eclipse, launch your application in the emulator or on a real device, and then switch Eclipse perspectives to DDMS and the Device view. From within the Device view, select the device where your application is running, and then click the Screen Capture button (it looks like a little painting in the upper-right corner) or choose it from the View menu. If you have a choice when saving, choose 24-bit color. Android Market will convert your screenshots to compressed JPEG; starting with 24-bit will produce better results than starting with 8-bit color. Choose screenshots that will make your application stand out from the rest but that also show the important functionality. You must supply at least two screenshots, and you can provide up to eight.

Next up is a high-res application icon. This could be the exact same design as your application icon, but Android Market wants a 512 × 512 icon image. This is required.

You can provide a promotional graphic as well, but its size is smaller than a screenshot. Although this graphic is optional, it is a good idea to include it. You never know when the graphic could be displayed; without one, you don't know what will be displayed in its place, if anything. One place the Promo Graphic appears is at the top of your application's Details page in Android Market.

The feature graphic is another optional field and is a large 1024 × 500 in size. This graphic is used in the Featured section of Android Market so you want this to look really good.

The last bit of graphics related to your application is an optional video that you can put out on YouTube and link to from your Android Market page.

Listing Details

Android Market asks for textual information about your application to display to shoppers, including the title, descriptive text, and promotional text. Promotional text can only be provided if you already provided a promotional graphic. Text can be provided in multiple languages, since you can choose to distribute your application to countries all over the world. The graphics mentioned can only be supplied to Android Market once, so if your screenshots look different in different locales, you'll need to consider other ways to make those available to shoppers, perhaps on your own web site. This may change in the future.

If you have written a separate EULA for your users, provide a link to it in your descriptive text so shoppers can view it prior to downloading your application. Consider that shoppers will likely use search to locate applications, so be sure to put appropriate words into your text to maximize your hit rate on searches related to your application's functionality. Finally, it's worthwhile to put a short comment in the text that says to e-mail you if the user runs into problems. Without this simple prompt, people are more likely to leave a negative comment, and a negative comment really limits your ability to troubleshoot and solve the problem, as compared to an e-mail exchange with the affected user.

One drawback to the user comments mechanism described earlier is that it does not distinguish the version of your application. If negative reviews are received against version 1, and you release version 2 with everything fixed, the reviews from version 1 are still there, and shoppers may not realize that those comments don't apply to the new version. When releasing a new version of an application, the application rating (number of stars) does not get reset, either. Partly for this reason, Google started providing a Recent Changes text field where you can describe what's new in this release. This is where you could indicate that a certain problem has been fixed, or tell what the new features are.

There's also a separate Promo Text field that has only 80 characters. When your app is shown at the top of a list in Android Market, it's the Promo Graphic and the Promo Text that get displayed here. It's definitely a good idea to supply these.

One of your responsibilities when writing the text for your application is to disclose the permissions that are required. These are the same permissions as set in the <uses-permission> tags of your `AndroidManifest.xml` file within your application. When the user downloads your application to their device, Android will check the `AndroidManifest.xml` file and ask the user about all of the `uses-permission` requirements before completing the install. So you might as well disclose this up front. Otherwise, you risk negative reviews from users surprised that an application requires some permission that they are not prepared to grant. Not to mention the refunds, which also count against your Developer Composite Score. Similar to permissions, if your application requires a certain type of screen, a camera, or other device feature, this should be disclosed in your text descriptions of your application. As a best practice, you should disclose not only what permissions and features your application needs, but also what your application will do with them. You should answer the user's question in advance: why does this app require *X*?

When uploading your application, you will need to choose an application type and a category. As these values change with time, we won't list them here, but it's easy to go to the Upload Application screen to see what they are.

Publishing Options

Android Market provides an option to set copy protection on applications when you are uploading them. The Market takes care of applying this copy protection for you, but note that the copy protection will make your application use more device memory. It is also not foolproof, and there are no guarantees that your application cannot be copied off of a device. Since the copy protection method is being deprecated, you probably want to consider additional or alternative ways to prevent pirating of your application, such as the Android Licensing Service described earlier.

In late 2010, Google introduced an application rating scheme. The idea is to give consumers an idea of the appropriateness of an application for certain age groups. The ratings are High, Medium, and Low Maturity, and Everyone. Choosing the right level depends on the content in your application and how much of that content there is. Google has rules about location-awareness and posting or publishing locations. It's best to read the rules for yourself here: `www.google.com/support/androidmarket/bin/answer.py?hl=en&answer=188189`.

Next you set the price of your application. By default the price is Free, and you must have previously set up a Merchant Account in Google Checkout if you want to charge for your application. Setting the right price for an application is tricky, unless you've got some sophisticated market research capabilities, and even then it's still tricky. Prices set too high could turn people off, and you risk the effects of refunds if people don't feel the price was worth it. Prices set too low could also turn people off because they might think it's a cheap application.

One of the last decisions to make before uploading your application is to choose the locations and carriers for your application to be visible to. By choosing All, your application will be available everywhere. However, you may want to restrict distribution geographically or by carrier. Depending on what functionality is in your application, you may need to restrict by location in order to comply with United States export law. You may choose to restrict your application by carrier if your application has compatibility issues with certain carriers' devices or policies. To see carriers, click a country link, and the available carriers for that country will be displayed, allowing you to choose the ones you want. Choosing All also means that any new locations or carriers that Google adds will automatically see your application with no intervention from you.

In addition to country and carrier choices, Android Market also allows you to restrict your application to certain devices. By default, the devices list is filtered based on your manifest file, in which you've specified the features and so on that your application requires. This section of the Upload screen allows you to further restrict other devices.

You would probably only want to do this if there was a known issue with a particular device such that you were unable to get your application to work on that device even though it ought to.

Android also offers the option to upload multiple APKs for the same application. It enables you to have a single entry on Android Market but to have separate build for phones and tablets. See `http://android-developers.blogspot.com/2011/07/multiple-apk-support-in-android-market.html`.

Contact Information

Even though your developer profile contains your contact information, you can set different information when uploading each application. The Market asks for a web site, e-mail address, and phone number as contact information related to this application. You must supply at least one of these so buyers can get support, but you don't need to supply all three. It is a good idea to not use your personal e-mail address here, just as you probably wouldn't really want to give out your personal phone number. When you've made millions of dollars from selling your application, you'll want to let someone else receive and deal with the e-mails from users. By setting up an application-support type of e-mail address in advance, you can easily separate the support e-mails from your personal e-mails. Of course, you can always change these values later if you need/want to.

Consent

With all these decisions made, you must then attest that your application abides by Android's Content Guidelines (basically no nasty stuff) and make a second attestation that the software is OK for export from the United States. U.S. export laws apply because Google's servers are located inside the United States, even if you are outside of the United States, and even if both you and your customer are outside of the United States. Remember that you can always choose to distribute your application through other channels. When all your information is in and your graphics are uploaded, go ahead and click the Save button. This will prepare everything for your application to be ready to go live.

You can then publish your application by clicking the Publish button. Android Market will perform some checks on your application—for instance, checking your application's certificate for the expiration date. If all goes well, your application will now be available for download. Congratulations!

User Experience on Android Market

Android Market has been available on devices for some time now, and it is available over the Internet. Developers don't have any control over how Android Market works, other than to provide good text and graphics for their application's listing in the Market. Therefore, the user experience is pretty much up to Google. From a device, a user can search by keyword; look at top downloaded applications (both free and paid), featured applications, or new applications; or browse by categories. Once they find an application they want, they simply select it, which pops up an item details screen allowing them to install it or buy it. Buying will take the user to Google Checkout to conduct the financial part of the transaction. Once downloaded, the new application shows up with all the other applications.

From the Internet web site for Android Market (`http://market.android.com`), the user interface looks about the same, albeit much larger than most device screens. One difference is that the web-based Android Market expects the user to log in to their Google account to use the Market. This allows Google to connect your web experience on Android Market to your actual device. This means two things: when using the web site, Android Market knows what applications are already installed on your device; and when you make a purchase on the Android Market web site, the download can be sent to your device (or devices) and not to whatever computer you happen to be browsing on.

Android Market has an option to view downloaded applications in My Apps. This area contains all installed apps and any apps that you've purchased, even if you've removed them (perhaps you removed them just to make room for other applications). This means you could delete a paid app from your phone and then reinstall it later without having to repurchase it. Of course, if you opted for a refund, the app will not show up in My Apps. Also, free apps that you remove from your device will not show up in My Apps. The list of apps in My Apps is tied to your Google Account used for the device. This means you could switch to a new physical device and still have access to all the apps you've paid for. But beware. Since you might have multiple identities with Google, you must use the exact same identity as before to get your apps on a new device. When viewing apps in My Apps, any that have upgrades available will indicate this and allow you to get the upgrade.

Android Market filters applications available to users. It does this in a number of ways. Users in some countries can only see free applications because of the commerce legalities involved for Google in that country. Google is trying hard to overcome commerce hurdles so all paid apps will be available everywhere. Until that time comes, users in some countries will be unable to access paid apps. Users with devices running older versions of Android will not be able to see applications that require a newer version of the Android SDK. Users with device configurations that are not compatible with the requirements of the application (expressed via <uses-feature> tags in the `AndroidManifest.xml` file) will not be able to see those applications. For example, applications not specifically supporting small screens cannot be seen in Android Market by users on devices with small screens. This filtering is mostly intended to protect users from downloading applications that will not work on their device.

If you are purchasing apps in Android Market from other countries, your transaction may be subject to currency conversion, which can also carry an additional fee, unless the seller has specified pricing in your local currency. You're really purchasing using the Google Checkout from the seller's country. Android Market will display an approximate amount, but the actual charges could vary depending on when the transaction is placed and with which payment processor. Buyers may notice a pending transaction against their account for a small amount (for example, U.S. $1). This is done by Google to ensure that the payment information provided is correct, and this pending charge will not actually go through.

A few web sites are available that mirror Android Market. Shoppers can search, browse categories, and find out about Android Market applications over the Internet without having a device. This gets around the filtering that Android Market does based on your device configuration and location. However, this does not get apps onto your device. Examples of these mirror sites are `www.cyrket.com`, `www.androlib.com`, and `www.androidzoom.com`.

Beyond Android Market

Android Market is not the only game in town. You are not forced into using Android Market at all. You should consider utilizing other channels of distribution, not only to make your app available to more people in more countries, but also to take advantage of other payment processors and opportunities to make money.

There are Android app stores completely separate from Android Market, the biggest of which is probably Amazon. Other examples of Android app stores are `http://mall.soc.io/apps`, `http://slideme.org`, `www.getjar.com`, and `www.handango.com`. From these sites, you can search, browse, find out about apps, and also download apps, either from a device or via a web browser. These sites don't have to abide by Google's rules, including the transaction fees for paid apps and methods of payment. PayPal and other payment processors can be used to purchase apps on these separate sites. These sites also don't necessarily restrict by location or device configuration. Some of them provide an Android client that can be installed, or in some cases may come preinstalled on a device. Users can simply launch a browser on their device and find the app they want to download via the web site; when the file is saved to the device, Android knows what to do with it. That is to say, a downloaded `.apk` file is treated as an Android application. If you click it in the Download history of the browser (not to be confused with My Apps, covered earlier), you will be prompted to see if you want to install it or not. This freedom means you can set up your own methods of downloading Android applications to users, even from your own web site and with your own payment methods. You must still deal though with collecting any necessary sales tax and remitting it to the appropriate authorities.

While not restricted by Google's rules, these alternate methods of app distribution may not offer the same sort of buyer protections that are found in Android Market. It may be possible to purchase an application through an alternate market that will not work on the buyer's device. The buyer may also be responsible for creating backups, in case they lose the application from their device, or for transferring applications if they switch to a new device.

These other markets allow you to make money on the sale of each app. You've also got the ability within these other markets to implement alternate payment mechanisms. You can of course implement ads as we described earlier and make money that way. You can also embed other payment mechanisms right into your application. For example, PayPal introduced a payment library for Android apps (see `http://www.x.com`). With it you could allow users to purchase add-ons, content, or upgrades from right inside your app. They could make donations, too. You could implement a mobile store using PayPal for checkout.

Remember that Google does not restrict developers from selling their applications in multiple markets at the same time they sell through Android Market. So consider all your options to make the most of your efforts.

References

Here are some helpful references to topics you may wish to explore further:

- `http://developer.android.com/guide/topics/manifest/manifest-intro.html`: The Developer Guide page to the `AndroidManifest.xml` file, with descriptions of how to use the `supports-screens`, `uses-configuration`, and `uses-feature` tags.
- `http://developer.android.com/guide/practices/screens_support.html`: The Developer Guide page "Supporting Multiple Screens," which contains lots of good information on dealing with different screen sizes and densities.
- `http://developer.android.com/guide/practices/ui_guidelines/icon_design.html`: The Developer Guide page "Icon Design Guidelines," which contains lots of good information on designing effective icons for your application.
- `http://android-developers.blogspot.com/2010/09/securing-android-lvl-applications.html` and `http://android-developers.blogspot.com/2010/09/proguard-android-and-licensing-server.html`: Blog posts on how to use the License Verification Library (LVL) in ways that prevent piracy.
- `http://proguard.sourceforge.net/`: The main site for ProGuard, which includes documentation.
- `http://developer.android.com/guide/market/billing/index.html`: Documentation for the in-app billing module.

Summary

You are now equipped to take on the world with your Android applications! Here is a rundown of the topics we covered in this chapter:

- How to get established as an Android Market Publisher (that is, Developer) so you can publish to Android Market.
- The rules as laid out in the Android Market Developer Distribution Agreement.

- Giving Google its share of your revenue if you are selling through Android Market. We also discussed how Google does not want to see competition from within the Market.
- Your responsibility for paying taxes on revenues from your applications.
- The Android Market refund policy, both the published and the real one.
- How users can get copies of your application anytime in the future as long as they paid for it once.
- The Android branding rules. Make sure you don't violate any copyright associated with Android, images, or fonts.
- The Developer Console and its features. The Developer Console collects user feedback and error reports from users.
- Preparing your application for production, including testing, LVL and ProGuard to fight piracy, and using resource variations and tags in `AndroidManifest.xml` to filter which devices your application will be available to.
- Advice regarding localizing your application by language and/or culture.
- How to monetize your application while staying within the guidelines of Android Market.
- The Android Market user interface, both on device and on the Internet/Web.
- The fact that Android Market is not the only game in town, and that you can sell your application in other places on the Internet, all at the same time.

Interview Questions

These are questions that you may find helpful to go through to solidify your understanding of this chapter:

1. True or false. You can create your own End User License Agreement for your application.
2. What percentage of sales does Google take for applications in Android Market?
3. Who is responsible for sending tax money to your local taxing authorities?

4. True or false. If a user deletes your application from their device to make room for more applications, they have to buy it again if they want it back.
5. Is it possible for you to cause your application to be removed from certain devices?
6. What kinds of information and feedback can you get from Android Market about your applications?
7. If a user complains about your application through Android Market Comments, can you reply to them within the Market?
8. If you need support for Android Market, where do you go and how do you get there?
9. What's the difference between screen density and screen size?
10. Which is more important when it comes to icons?
11. Where do you specify things like the fact your application does not work on `xlarge` screen devices?
12. How would you convert an application from the English language to French?
13. Why is it a bad idea to use a color name that is a specific color when using color resources?
14. What does the URI look like that takes a user to Android Market for a specific developer name?
15. What does LVL stand for, and what does it do?
16. What else do you need to use with LVL to keep your application safe (or at least safer) from piracy?
17. How soon after an application is uploaded by you to Android Market is it available to be downloaded by users?
18. Does Google review all uploaded applications for content or for violating its rules?
19. Is it possible to sell your application in markets other than Android Market? What considerations should be made?

Index

Special Characters and Numerics

A

C

D

E

G

H

I

J

K

L

M

N

O

Q

R

S

T

U

X

Y

Z

CPSIA information can be obtained at www.ICGtesting.com
Printed in the USA
LVOW112351280612

288160LV00001B/3/P

9 781430 23930